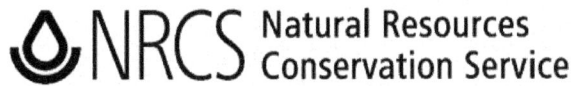

NRCS Natural Resources Conservation Service

SOIL SURVEY LABORATORY

METHODS MANUAL

Rebecca Burt, Editor

Soil Survey Investigations Report No. 42
Version 4.0
November 2004

Cover Photo:The cover illustrates the continuum of data collected or produced for the soil survey of a geographic area. The soil survey map, landscape, and pedon are from northern Iredell County, North Carolina and are representative of the Banister pedon. This series is classified as a fine, mixed, active, mesic Aquic Hapludalf. This soil is a very deep, moderately to somewhat poorly drained soil on stream terrace positions in the Piedmont physiographic province of Virginia and North Carolina. Laboratory data sheet illustrates particle-size data produced by the Soil Survey Laboratory. The photomicrograph of soil fabric features a slightly altered, exfoliated biotite grain under cross-polarized light. Bar scale in photo is 0.25 mm. *(Photo credit of Banister landscape and pedon: John Kelly and Michael A. Wilson, NRCS, Raleigh, NC and NRCS, Lincoln, NE.)*

FOREWORD

Laboratory data are critical to the understanding of the properties and genesis of a single pedon, as well as to the understanding of fundamental soil relationships based on many observations of a large number of soils. Development of both an analytical database and the soil relationships based on those data are the cumulative effort of generations of soil scientists at the Soil Survey Laboratory (SSL).

The purpose of the Soil Survey Investigations Report (SSIR) No., Soil Survey Laboratory Methods Manual is to document methodology and to serve as a reference for the laboratory analyst. It is expected that this document will continue to evolve and change over time as new methods are developed and old methods are modified or retired based on new knowledge or technologies. This manual is an historical document in that it describes both current and obsolete methods at the SSL. This manual also provides historical perspective, documenting the contributions of many soil scientists who have gone before us. Many of these scientists are noted in the section on contributors.

Dr. Rebecca Burt, author of the Soil Survey Information Manual (SSIR No. 45, 1995), served as technical editor for all versions of the SSIR No. 42 (1989, 1992, 1996, 2004). Her contribution is significant in scope; she wrote some of the methods; is responsible for the review process encompassing additions, corrections, and consistency of other methods; and provided leadership in the assemblage of this document.

PREFACE

The standard methods described in the Soil Survey Investigations Report (SSIR) No. 42, Soil Survey Laboratory Methods Manual are those used by the Soil Survey Laboratory (SSL), National Soil Survey Center (NSSC). Included in the SSIR No. 42 are descriptions of current as well as obsolete methods, both of which are documented by method codes and linked with analytical results that are stored in the SSL database. This linkage between laboratory method codes and the respective analytical results is reported on the SSL data sheets.

The methods in current use at this laboratory are described in enough detail that they can be performed in many laboratories without reference to other sources. An introduction to each group of related methods describes common characteristics. However, some repetition is included in order to make the method descriptions complete in themselves and to minimize reference to other parts of the manual.

Some analytical results in the SSL national database were obtained using procedures that are no longer used at the SSL. Descriptions for these procedures are located in a section following the current methods. Information is not available to describe these procedures in the same detail as used to describe the current methods in the laboratory.

Since the publication of the SSIR No. 42 (1996), there has been a significant increase in the number and kinds of methods performed at the SSL, resulting in a re-structuring of the method codes. As in past versions of the SSIR No. 42, the current method codes are hierarchical and alphanumerical. However, the obsolete method code structure had only a maximum of four characters, e.g., 6A2b, whereas the new structure has more characters, carrying more information about the method, e.g., particle-size and sample weight bases for reporting data. This version of the SSIR No. 42 carries not only the new method codes but also the older ones as well. The table for cross-referencing these older codes precede the method descriptions of the obsolete SSL methods. This linkage between the method code systems is important to maintain, as many older SSL data sheets and scientific publications report these older codes. The table of contents presents the ensemblage of this manual, beginning with the current methods and codes followed by obsolete methods with respective obsolete codes.

The SSL data have been provided in reports, e.g., Primary and Supplementary Characterization Data Sheets, and in electronic forms, including tapes, disks, and CD-ROMs. More recently, other reports have been developed, e.g., Soil Taxonomy Characterization Data Sheets, and data are available via the NRCS Soils web site http://soils.usda.gov/. Historically, the SSIR No. 42 has described and assigned method codes only to those data reported on the Primary Characterization Data Sheets. This tradition is followed in this version of the SSIR No. 42. With the exception of some SSL primary analytical data included for user convenience, the data on the Supplementary and Taxonomy Characterization Data Sheets are derived data, using analytical data as a basis for calculation and do not carry method codes. Data on the Supplementary and Taxonomy Characterization Data Sheets are not described in this manual. For more detailed information about the calculation and application of these derived values, refer to the other important source documents provided by the United States Department of Agriculture, Natural Resources Conservation Service (USDA-NRCS) referenced in this maual.

The methods described herein identify the specific type of analytical or calculated data. Most of these methods are analytical in nature, i.e., quantitative or semi-quantitative measurements, and include physical, chemical, mineralogical, and biological analyses. Sample collection and preparation in the field and the laboratory are also described. Historically, the SSIR No. 42 has described some derived values, e.g., coefficient of linear extensibility (COLE) and water retention difference (WRD) and reported these values along with the analytical data on the SSL Primary Characterization Data Sheets. This version of the SSIR No. 42 follows this tradition. Since the publication of the SSIR No. 42 (1996), several more derived values have been added to the Primary Characterization Data Sheets. While these data have been assigned method codes, detailed descriptions are not included in this version of the SSIR No. 42. For more detailed information about the calculation and application of these derived values, refer to other important source documents provided by USDA-NRCS referenced in this manual.

The purpose of this manual is to document methodology and to serve as a reference for the laboratory analyst. The standard methods described in this SSIR No. 42, Soil Survey Laboratory Methods Manual, Version 4.0 replaces as a methods reference all earlier versions of the SSIR No. 42 (1989, 1992, and 1996, respectively) and SSIR No. 1, Procedures for Collecting Soil Samples and Methods of Analysis for Soil Survey (1972, 1982, and 1984). All SSL methods are performed with methodologies appropriate for the specific purpose. The SSL SOP's are standard methods, peer-recognized methods, SSL-developed methods, and/or specified methods in soil taxonomy (Soil Survey Staff, 1999). An earlier version of this manual (1996) also served as the primary document from which a companion manual, Soil Survey Laboratory Information Manual (SSIR No. 45, 1995), was developed. The SSIR No. 45 describes in greater detail the application of SSL data.

Trade names are used in the manual solely for the purpose of providing specific information. Mention of a trade name does not constitute a guarantee of the product by USDA nor does it imply an endorsement by USDA.

Rebecca Burt, Editor
Reseach Soil Scientist
Soil Survey Investigations
USDA-NRCS
Lincoln, Nebraska

CONTRIBUTORS

Laboratory data are critical to the understanding of the properties and genesis of a single pedon, as well as to the understanding of fundamental soil relationships based on many observations of a large number of soils. The development of laboratory methods, the analytical database, and the soil relationships based on those data are the cumulative effort of several generations of scientists. These efforts may be defined as methods development, database design and development, and investigations of data relationships. Methods development in the Soil Survey Laboratory results from a broad knowledge of soils, encompassing topical areas of pedology, geomorphology, micromorphology, physics, chemistry, mineralogy, biology, and field and laboratory sample collection and preparation. The following lists many of these contributing scientists, some of whom have since retired from USDA-NRCS and/or are deceased. Some of these contributions are cited throughout this manual as scientific publications. The following list is not comprehensive for developmental work preceding the formation of the Soil Survey Laboratory in Lincoln, Nebraska. In addition to the scientists, the current Soil Survey Laboratory technicians are also listed. These analysts play a key role not only in performing the analyses but also in developing new or modifying old methods, and ensuring data quality. Scientists (past and current) and physical science technicians (current) are listed alphabetically as follows:

Scientists:
Lyle T. Alexander, Steven L. Baird, Otto W. Baumer, Ellis C. Benham, C. Reese Berdanier Jr., Wade A. Blankenship, Benny R. Brasher, Laurence E. Brown, Rebecca Burt, John G. Cady, Frank J. Carlisle, James H. DeMent, Joe D. Dixon, Moustafa A. Elrashidi, Richard R. Ferguson, Klaus W. Flach, Carol D. Franks, Donald P. Franzmeier, Erling E. Gamble, Don W. Goss, Robert B. Grossman, George C. Holmgren, C. Steven Holzhey, Robert H. Jordan, R. Leo Juve, Fredrick M. Kaisaki, John M. Kimble, Leo C. Klameth, Ellis G. Knox, Susan E. Samson-Liebig, Warren C. Lynn, Maurice J. Mausbach, M. Dewayne Mays, Dean C. McMurty, Milton M. Meyer, John L. Millet, Reuben E. Nelson, W. Dennis Nettleton, Carolyn G. Olson, Ronald F. Paetzold, Roger B. Parsons, Richard L. Pullman, Ivan W. Ratcliff Jr., Thomas G. Reinsch, Earnest D. Rivers, Sam J. Ross, Philip J. Schoeneberger, Leo G. Shields, Christopher W. Smith, Terrence M. Sobecki, John A. Thompson, Geroge W. Threlkeld, Michael A. Wilson, Douglas A. Wysocki, Ronald D. Yeck, and Keith K. Young.

Physical Science Technicians:
Larry V. Arnold, Michelle A. Etmund, Kristina Goings, Diane G. Hooper, Patricia E. Jones, Janis L. Lang, Christopher W. Lee, Valerie M. Murray, James F. Neal, Kathyrn L. Newman, Michael J. Pearson, Susan J. Reidel, Crystal A. Schaecher, Emily K. Rose Seifferlein, James E. Thomas, Pamela A. VanNeste, Lester E. Williams, and Thomas J. Zimmer.

ACKNOWLEDGEMENTS

We gratefully acknowledge those individuals who provided technical review of the Soil Survey Laboratory methods by topical area. Their thoughtful comments and suggestions are deeply appreciated. These reviewers of the Soil Survey Laboratory Methods Manual, Soil Survey Investigations Report No. 42, Version 4.0 are listed alphabetically as follows:

Reviewers

Jacob H. Dane, Department of Agronomy and Soils, 202 Funchess Hall, Auburn University, Auburn, Alabama 36849-5412

Richard P. Dick, Department of Crop and Soil Science, Agricultural and Life Sciences Building 3067, Oregon State University, Corvallis, Oregon 97331

Glendon W. Gee, 1637 Birch Avenue, Richland, Washington, 99352

Mark Flock, Vice-President, Soil and Plant Analysis Council, Brookside Laboratory, Inc., 308 S. Main Street, New Knoxsville, Ohio 45871

Willie G. Harris, Soil and Water Science Department University of Florida, 2169 McCarty Hall, Gainesville, Florida 32611-0000

Yash Kalra, Canadian Forest Service, 5320 – 122 Street, Edmonton, Alberta, Canada T6H 3S5

Robert Miller, Coordinator, North American Proficiency Testing Program, Soil and Crop Science Department, Colorado State University, Fort Collins, Colorado 80523

METHODS OF ANALYSIS FOR SOILS, WATER, AND BIOLOGICAL MATERIALS

CONTENTS

CONTENTS

CONTENTS

CONTENTS

CONTENTS

CONTENTS

CONTENTS

CONTENTS

CONTENTS

CONTENTS

CONTENTS

CONTENTS

CONTENTS

CONTENTS

CONTENTS

CONTENTS

SOIL SURVEY LABORATORY METHODS
(Obsolete Methods with Old Method Codes)

CONTENTS

CONTENTS

CONTENTS

CONTENTS

CONTENTS

CONTENTS

CONTENTS

CONTENTS

METHOD TITLE	METHOD CODE	PAGE

CONTENTS

CONTENTS

SAMPLE COLLECTION AND PREPARATION (1)

Field Sample Collection and Preparation (1A)
Site Selection (1A1)
Geomorphology (1A1a)
Pedon (1A1b)
Water (1A1c)
Biological (1A1d)

1. Application

General: The United States National Cooperative Soil Survey (NCSS) Program has prepared soil maps for much of the country. Both field and laboratory data are used to design map units and provide supporting information for scientific documentation and predictions of soil behavior. A soil map delineates areas occupied by different kinds of soil, each of which has a unique set of interrelated properties characteristic of the material from which it is formed, its environment, and its history (Soil Survey Division Staff, 1993). The soils mapped by the NCSS are identified by names that serve as references to a national system of soil taxonomy (Soil Survey Staff, 1999). Coordination of mapping, sampling site selection, and sample collection in this program contributes to the quality assurance process for laboratory characterization (Burt, 1996). Requisites to successful laboratory analysis of soils occur long before the sample is analyzed (Soil Conservation Service, 1984; Soil Survey Staff, 1996). In the field, these requisites include site selection, descriptions of site and soil pedon, and careful sample collection. A complete description of the sampling site not only provides a context for the various soil properties determined but it is also a useful tool in the evaluation and interpretation of the soil analytical results (Patterson, 1993). Landscape, landform, and pedon documentation of the sampling site serves as a link in a continuum of analytical data, sampled horizon, pedon, landscape, and overall soil survey area. This continuum is illustrated on the cover of this manual.

The objectives of a project or study form the basis for designing the sampling strategy. A carefully designed sampling plan is required to provide reliable samples for the purpose of sampling. The plan needs to address the site selection, depth of sampling, type and number of samples, details of collection, and sampling and sub-sampling procedures to be followed. The Soil Survey Laboratory (SSL) primarily serves the NCSS, which is conducted jointly by USDA Natural Resources Conservation Service (NRCS), the Bureau of Land Management (BLM), Forest Service, and representatives of U.S. Universities and Agricultural Experiment Stations. In this context, the primary objective of SSL sampling programs has been to select sites and pedons that are representative of a soil series and to collect samples that are representative of horizons within the pedon.

There are various kinds of sampling plans, e.g., intuitive and statistical, and many types of samples, e.g., representative, systematic, random, and composite. In the field, the SSL has more routinely used intuitive sampling plans to obtain representative samples. The intuitive sampling plan is one based on the judgment of the sampler, wherein general knowledge of similar materials, past experience, and present information about the universe of concern, ranging from knowledge to guesses, are used (Taylor, 1988). A representative sample is one that is considered to be typical of the universe of concern and whose composition can be used to characterize the universe with respect to the parameter measured (Taylor, 1988).

In the laboratory, the primary objectives of sample collection and preparation are to homogenize and obtain a representative soil sample to be used in chemical, physical, and mineralogical analyses (method 1B). The analyst and the reviewer of data assume that the sample is representative of the soil horizon being characterized. Concerted effort is made to keep analytical variability small. Precise laboratory work means that the principal variability in characterization data resides in sample variability, i.e., sampling is the precision-limiting

1

variable. As a result, site-selection and sample collection and preparation are critical to successful soil analysis.

Geomorphic Considerations[1]: Soils form a vital, complex continuum across the Earth's landscape. The prime goal of the Soil Survey is to segregate the soil continuum into individual areas that have similar properties, and therefore, similar use and management. Soils can not be fully understood or studied using a single observation scale. Instead, soil scientists use multiple scales to study and segregate soils and to transfer knowledge to soil users. To accomplish the task of Soil Survey at reasonable cost and time, soil scientists extend knowledge from point observations and descriptions to larger land areas.

Soil map-unit delineations are the individual landscape areas defined during and depicted in a soil survey. Soil observation, description, and classification occur at the pedon scale (1 to $\approx$ 7 m), and represent a small portion of any map unit (10's to 1000's hectares). Further, pedons selected, described, and sampled for laboratory analysis represent only a small subset of the observation points. Pedon descriptions and classifications along with measured lab data, however, accurately apply to a named soil map unit or landscape areas (soil component) within the map unit. Soil scientists can reliably project ("scale up") pedon information to soil map units based on experience and the strong linkages among soils, landforms, sediment bodies, and geomorphic processes.

Thus, soil geomorphology serves several key functions in Soil Survey, which can be summarized as:

1. Provides a scientific basis for quantitatively understanding soil landscape relationships, stratigraphy, parent materials, and site history.
2. Provides a geologic and geographic context or framework that explains regional soil patterns.
3. Provides a conceptual basis for understanding and reliably predicting soil occurrence at the landscape scale.
4. Communicates effectively and succinctly soil location within a landscape.

During a Soil Survey soil scientists achieve these functions both tacitly and by deliberate effort. Geomorphic functions are best explained by citing examples. The first function listed above involves planned, detailed soil landscape studies (e.g., Ruhe et al, 1967; Daniels et al, 1970; Gamble et al, 1970; Parsons et al., 1970; Gile et al., 1981; Lee et al., 2001; 2003a; 2003b), which are an important component of the Soil Survey. Such studies quantify and explain the links between soil patterns and stratigraphy, parent materials, landforms, surface age, landscape position, and hydrology. Studies of this nature provide the most rigorous, quantitative, and complete information about soil patterns and landscapes. The required time and effort are significant, but justified by the quantitative information and scientific understanding acquired. Soil survey updates by MLRA can and should involve similar studies.

The three remaining geomorphic functions are tacit and to a degree inherent in a soil survey. A number of earth science sources (Fenneman 1931, 1938, 1946; Hunt, 1967; Wahrheftig, 1965;) identify and name geomorphic regions, which are grouped by geologic and landform similarity. The value of relating soil patterns to these regions is self-evident. Terms such as Basin and Range, Piedmont, Columbia Plateau, and Atlantic Coastal Plain provide both a geologic and geographic context for communicating regional soil and landform knowledge.

Soils occurrence can be accurately predicted and mapped using observable landscape features (e.g., landforms, vegetation, slope inflections, parent material, bedrock outcrops,

---

[1] Douglas A. Wyosocki and Philip J. Schoeneberger, Research Soil Scientists, NSSC, NRCS, Lincoln, NE, wrote the geomorphic component of this procedure.

stratigraphy, drainage, photo tonal patterns). During a soil survey soil scientists develop a tacit knowledge of soil occurrence generally based on landscape relationships. Soil occurrence is consistently linked to a number of geomorphic attributes. Among these are landform type, landscape position, parent material distribution, slope shape and gradient, and drainage pattern. This tacit soil landscape knowledge model is partially encapsulated in block diagrams and map unit and pedon descriptions. In turn a clear, concise geomorphic description effectively conveys soil location within a landscape to other soil scientists and soil users. Recent publications (Soil Survey Staff, 1998; Schoeneberger et al., 2002; Wysocki et al., 2000) provide a comprehensive and consistent system for describing geomorphic and landscape attributes for soil survey. The Geomorphic Description Systems (GDS) is not discussed here. Refer to Soil Survey Staff, 1998; Wysocki et al., 2000; and Schoeneberger et al, 2002 for more detailed information.

Geomorphology is an integral part of all soil survey processes and stages. Preliminary or initial soil pattern knowledge is commonly based on landscape or geomorphic relationships. Observations during a soil survey refine existing landscape models, or sometimes compel and create new models. Map unit design includes landform recognition and naming, and observations on landscape position, parent materials, and landscape and soil hydrology. Soil scientists capture this observational and expert knowledge through soil map unit and pedon descriptions, which should convey soil property, soil horizon, landscape and geomorphic relationships, and parent material properties.

Any study plan, site selection, or pedon sampling must also consider and address the geomorphology. Study or sampling objectives can vary. Every sample pedon should include both a complete soil and geomorphic description. In a characterization project, the sample pedons should be representative of the landscape unit (e.g., stream terrace, backslope etc.) it occurs upon. Note that the landscape unit that is sampled can be multi-scale. The unit could be a landform (e.g., stream terrace, dune, or drumlin), a geomorphic component (e.g., nose slope), a hillslope position (e.g., footslope), or all of these.

Keep in mind that the sampled pedon represents both a taxonomic unit and landscape unit. Both the landscape and taxonomic unit should be considered in site selection. Note that a single landscape unit (e.g., backslope) may contain one or more taxonomic units. A landscape unit is a more easily field recognizable and mappable feature than a soil taxonomic unit. For characterization project select the dominant taxonomic unit within a given landscape unit. The existence of other soils or taxa can and should be included in the soil description and also the map unit description.

Soil patterns on landscapes follow catenary relationships. It is important to characterize both individual pedon properties and the soil relationships both above and below on the landscape. This goal requires that soils be sampled as a catenary sequence (i.e., multiple samples across the same hillslope). This appears intensive, but serves multiple purposes. A sample pedon or set of pedons provides vital characterization data and also can quantify the catenary pattern and processes. This is an efficient use of both sampling time and effort, as well as, laboratory resources. Moreover, it provides an understanding of the entire soil landscape.

Lastly, and perhaps most importantly, soil geomorphic relationships deserve and sometimes demand specific study during a soil survey. Crucial problems can be addressed by appropriately designed geomorphic, stratigraphic, or parent material study. For example, a silty or sandy mantle over adjacent soils and/or landforms may be of eolian origin. A well-designed geomorphic study can test this hypothesis. In another geomorphic setting soil distribution and hydrology may be controlled by stratigraphic relationships rather than by elevation or landscape patterns. A drill core or backhoe pit sequence can address this hypothesis. These studies need not be elaborate, but they require fore thought and planning. Such studies are applicable and necessary to the Major Land Resource Area (MLRA) soil survey approach.

Pedon, Water, and Soil Biology Sampling: The pedon is presented in soil taxonomy (Soil Survey Staff, 1999) as a unit of sampling within a soil, i.e., the smallest body of one kind of soil large enough to represent the nature and arrangement of horizons and variability in the other properties that are preserved in samples (Soil Survey Division Staff, 1993). In the U.S.

3

soil Survey Program, laboratory pedon data combined with field data (e.g., transects and pedon descriptions) are used to define map unit components, establish ranges of component properties, establish or modify property ranges for soil series, and answer taxonomic and interpretive questions (Wilson et al., 1994).

Biological samples are also collected for analysis at the SSL, either in conjunction with pedon sampling or independently. Measurable biological indices have been considered as a component to assess soil quality (Gregorich et al., 1997; Pankhurst et al., 1997). A large number of soil biological properties have been evaluated for their potential use as indicators of soil quality/health (Doran and Parkin, 1994; Pankhurst et al., 1995). The USDA NRCS has utilized soil biology and carbon data in macro-nutrient cycling, soil quality determinations, resource assessments, global climate change predictions, long-term soil fertility assessments, impact analysis for erosion effects, conservation management practices, and carbon sequestration (Franks et al., 2001).

Water samples are analyzed by the SSL on a limited basis in the support of specific research projects. These projects are typically in conjunction with soil investigations and have involved monitoring seasonal nutrient flux to evaluate movement of N and P via subsurface and overland flow from agricultural lands into waterways and wetlands.

2. Summary of Method

A site is selected that meets the objectives of the laboratory sampling. The site and soil pedon are described and georeferenced. These descriptions include observations of specific soil properties such as texture, color, slope, and depth. Descriptions may also include inferences of soil quality (soil erodibility and productivity) as well as soil-forming factors (climate, topography, vegetation, and geologic material).

A soil pit is often excavated with a back-hoe (Fig. 1). Depth and breadth of pit depend on the soil material and objectives of sampling. Soil horizons or zones of uniform morphological characteristics are identified for sampling (Fig. 2). Photographs are typically taken of soil profile after the layers have been identified (Fig. 3) but before the vertical section by the sampling process (Fig. 4).

The variable nature or special problems of the soil itself, e.g., Vertisols, Histosols, or permafrost-affected soils, may require the use of specific excavation and sampling techniques. For example, the shear failure that forms slickensides in Vertisols also disrupts the soil to the point that conventional soil horizons do not adequately describe the morphology.

Representative samples are collected and mixed for chemical, physical, and mineralogical analyses. A representative sample is collected using the boundaries of the horizon to define the vertical limits and the observed short-range variability to define the lateral limits. The tag on the sample bag is labeled to identify the site, pedon, and soil horizon for the sample.

In the field, the 20- to 75-mm fraction is generally sieved, weighed, and discarded. In the laboratory, the <20-mm fraction is sieved and weighed. The SSL estimates weight percentages of the >2-mm fractions from volume estimates of the >20-mm fractions and weight determinations of the <20-mm fractions by method 3A2.

Undisturbed clods are collected for bulk density and micromorphological analysis. Clods are obtained in the same part of the pit as the mixed, representative sample. Bulk density clods are used for water retention data; to convert from a weight to volume basis; to determine the coefficient of linear extensibility (COLE); to estimate saturated hydraulic conductivity; and to identify compacted horizons. Microscope slides of soil prepared for micromorphology are used to identify fabric types, skeleton grains, weathering intensity, illuviation of argillans, and to investigate genesis of soil or pedological features.

Biological samples are also be collected for analysis at the laboratory, either in conjunction with pedon sampling or independently. As with pedon sampling, sampling for root biomass includes selecting a representative site, sampling by horizon, and designating and sampling a sub-horizon if root mass and morphology change. The same bulk sample collected for soil mineralogical, physical, and chemical analyses during pedon sampling can also be used

Fig. 1. Excavated pit for pedon sampling.

Fig. 2. Soil horizons or zones of uniform morphological characteristics are identified for sampling.

Fig. 3. Photographs are typically taken of soil profile after the layers have been identified but before the vertical section by the sampling process.

Fig. 4. Pedon sampling activities.

for some soil biological analyses. Alternatively, a separate bio-bulk sample can be collected in the field. Surface litter and O horizons are sampled separately, as with pedon sampling. If certain biological analyses, e.g., microbial biomass, are requested, these samples require expedited transport under ice or gel packs and are refrigerated (4°C) immediately upon arrival at the laboratory to avoid changes in the microbial communities.

Water samples can also be collected for laboratory analyses at the same time as pedon sampling. Choice of water sampling site depends not only on the purpose of the investigation but also on local conditions, depth, and the frequency of sampling (Velthorst, 1996). Specific recommendations are not applicable, as the details of collection can vary with local conditions. Nevertheless, the primary objective of water sampling is the same as with soil and biological sampling, i.e., to obtain a representative sample in laboratory analyses. Water samples require expedited transport under ice or gel packs and are refrigerated (4°C) immediately upon arrival at the laboratory.

3. Interferences

In the process of sampling, a number of obstacles may arise from external sources, e.g., weather, accessibility, steep terrain, wet terrain, insects, and large rock fragments. Sometimes pits have to be hand-excavated. Common sense and the guidelines for obtaining representative samples are applied to the extent possible.

Preservation of sample integrity, avoiding changes or contamination during sampling and transport is important. Sampling for trace element analysis requires the use of clean, non-metallic equipment. Extreme care and precision are required for samples with low natural elemental concentrations.

Do not allow soils to dry, as some soils irreversibly harden upon drying, affecting some laboratory analyses such as particle size (Kubota, 1972; Espinoza et al., 1975; and Nanzyo, 1993). High temperatures can also alter microbial populations and activity (Wollum, 1994).

Avoid contamination of water samples by not touching the inner part of the sample container, screw cap, or sample water. Gloves (powderless) may be used. Water samples are affected by microbial activity, resulting in a change in the concentration of some elements (e.g., nitrate, phosphate, and ammonium); the reduction of sulfate to sulfide and chlorine to chloride; and the loss of iron through precipitation or oxidation (Velthorst, 1996). The addition of microbial inhibitors may be necessary.

In general, plastic bags will suffice for most biological samples, as they are generally permeable to CO_2 and O_2, preventing sample drying, i.e., aerobic samples will remain aerobic during transport to the laboratory (Wollum, 1994). The SSL recommends double-bagging zip locked plastic bags to prevent loss of sample water content.

The kind of water sample container (adsorption, desorption) as well as the bottle volume can affect the analytical results, e.g., polyethylene bottles increase the chlorine content with time or adsorb organic material; errors increase with the permeability of bottle wall; glass bottles release sodium and silicon with time; small sample volume has more contact with larger bottles compared to small bottles (Velthorst, 1996). Water sample containers should be acid washed and capped in the laboratory prior to collection in the field. The drying of these containers should also be considered with regards to interferences or contaminants. Ceramic cups for collection of soil water may require an acid pretreatment prior to installation in the field, as these cups have a small cation exchange capacity, sorbing dissolved organic carbon, and releasing aluminum and silica (Velhorst, 1996). Refer to the respective manufacturer's manual, e.g., Soil Moisture Corporation, for the appropriate treatment of these cups before use.

Avoid long periods between collection and laboratory analysis of water and some types of biological (e.g., microbial biomass) and soil samples (e.g., sulfidic materials). To prevent significant changes (e.g., degradation, volatilization, alteration in microbial community), these samples require expedited transport under ice or gel packs and are refrigerated (4°C) immediately upon arrival at the laboratory. Avoid freezing water samples, which can influence pH and the separation of dissolved organic matter from the water phase.

4. Safety

Sampling pits deeper than 125 cm (5 feet) need to be shored to meet U.S. Department of Labor Occupational Safety and Health Administration (OSHA) standards, or one side has to be opened and sloped upward to prevent entrapment. Acetone used in the saran mix is flammable and should be used down wind from a site to keep fumes from collecting in the bottom of the pit. Take precautions when operating or in the proximity of machinery, e.g., backhoe, drill rig, or hydraulic probe, and when lifting sample bags.

5. Equipment

5.1 Plastic bags, for mixed soil samples
5.2 Zip locked plastic freezer bags, for biological samples
5.3 Tags, for bagged samples
5.4 Plastic bags, for bulk density and thin section clods
5.5 Aluminum case, for shipping clod boxes
5.6 Shipping bags (canvas, leather, or burlap) for mixed samples
5.7 Clod boxes, cardboard with dividers
5.8 Core boxes, to transport cores from drill rig or hydraulic probe
5.9 Stapler, with staples
5.10 Hair nets
5.11 Rope
5.12 Clothes pins
5.13 Felt markers, permanent
5.14 Sampling pans
5.15 Sampling knives
5.16 Chisel
5.17 Rock hammer
5.18 Nails
5.19 Measuring tape
5.20 Photo tape
5.21 Sieves (3-inch and 20-mm)
5.22 Plastic sheets
5.23 Canvas tarp
5.24 Camera
5.25 Frame, 50 cm x 50 cm
5.26 Garden clippers
5.27 Pruning shears
5.28 Bucket
5.29 Scale, 100-lb capacity, for rock fragments
5.30 Analytical balance, ± 0.01 g sensitivity, for weighing roots and plant residue
5.31 Cooler, with ice or gel packs, for biological samples
5.32 Containers, with screw caps, acid-washed, for water samples
5.33 Gloves, plastic, powderless
5.34 Bulk density equipment, if natural clods are not appropriate technique, e.g., bulk density frame or ring excavations, compliant cavity (procedure 3B5, 3B4, 3B3, respectively)

6. Reagents

6.1 Acetone
6.2 Water, in spray bottle
6.3 Dow Saran F-310 Resin. Available from Dow Chemical Company.

7. Procedures

Project and Sampling Objectives

The number and types of samples collected from a site are governed in part by the objectives of the information needed. Example sampling schemes presented as general project categories based on project needs are as follows:

Reference Projects: These projects are designed to answer specific questions on mapping or soil classification, provide data for transect of a mapping unit, or for collection of calibration standards. Samples are typically collected from specific horizons in 3 to 5 locations that relate to the question or are representative of the map unit. A limited number of analyses, specific to the questions asked, are performed on these samples.

If a transect is used to test map unit composition, an appropriate sample from each transect point may be collected for analyses that are critical to distinguishing between map unit components. Also, samples may be collected as standards for the survey project for texture, organic carbon, or for calibration of field office analyses such as base saturation.

Characterization Projects: These projects are designed to obtain comprehensive soil characterization data for a representative pedon of a map unit or a pedon which is included in a research study. Samples collected from each horizon include bulk samples of approximately 3 kg, as well as clods of natural fabric for bulk density and micromorphology. A standard suite of laboratory analyses are performed on each horizon. In addition, specific analyses, such as mineralogy or andic properties, may be requested to provide more complete information on the specific pedon sampled.

Geomorphology and Stratigraphy Projects: These projects are designed to study relationships between soils, landforms, and/or the stratigraphy of their parent materials. For example, a specific project may be designed to study the relationships between a catena of soils, their morphological properties, e.g., redoximorphic features, and the hydrology of the area. Another study may be designed to determine the lateral extent of stratigraphic breaks. Site or pedon selection is governed by the objectives of the study but often is selected to represent typical segments of the landform. Sampling and analytical requests may be similar to the scheme used in a characterization or reference project. Often, core samples may be collected to several meters in depth through the use of a hydraulic probe.

Pedon Sampling Techniques

Excavated Pits: A pit may be excavated by hand or with a backhoe. Hand-digging may be necessary depending on site location, type of soil material, or availability of a backhoe. Pedons are generally excavated through the solum and into the parent material, or to a maximum depth of 2 meters. When using a backhoe, dig the pit in the form of an arc with a minimum working face deeper than about 150 cm (5 ft). Slope the pit upward toward the backhoe for an escape route. The pit can also be modified from the back side to form a T with the back of the trench opened and widened for an escape route. If this is not practical, shoring is required to meet OSHA standards if the pit is deeper than 125 cm (5 ft).

The sampling procedure is the same for hand-dug and backhoe pits. Mark horizons or zones to be sampled. Take a representative sample from boundary to boundary of a horizon and for a lateral extent to include the observed short-range variability. Unless the soil exhibits little short range variability, the best procedure is to place 4 to 5 kg soil on the plastic sheet or canvas tarp, mix thoroughly by rolling action, and place a representative sub-sample, minimum of 3 kg (3 qt), in a plastic sample bag. Label a tag with soil name, soil survey number, horizon (zone), and depth (as a minimum). Double fold the top of the plastic bag (forward and reverse) and staple the top of the tag under the folds. The sampling may be extended deeper by bucket auger or hydraulic probe as appropriate to meet the objectives of the project. If the soil has rock fragments in one or more horizons, the soil and coarse fragments need to be sieved and weighed as described below.

Collect 3 bulk density clods from each horizon. Two clods are used in the primary analysis. The third clod is reserved for a rerun, if needed. Clods should be roughly fist-sized and should fit into the cell (8 x 6 x 6 cm) of a clod box fairly snugly. Take the clods in the same vicinity of the pit as the mixed sample. Carve out a working section in the pit wall to remove an undisturbed block. Break the block into fist-sized pieces and pare into a ovoid (egg-shaped) clod. Place the clod in a hair net. If the clod is dry, mist the clod with water just until the surface glistens to inhibit saran penetration of the clod. Dip once, briefly, in saran mix to coat the clod, and hang from a rope with a clothes pin to dry. Clods can be dipped and then hung, or can be hung and then dipped by raising the container up to immerse the clod, briefly. Keep the saran container covered except when dipping clods to prevent acetone evaporation. Coat only once in the field. Additional coats are applied in the laboratory. When the clod is dry (bottom is not sticky to the touch), place the clod in a plastic bag, and put into a cell of a clod box. Label the appropriate cell on the inside of the lid of the box to identify the soil survey number and horizon (zone) for the clod. Clod boxes are designed to identify sequences of 3 clods per horizon.

Collect 2 clods from each horizon for preparation of thin sections and micromorphological examination. Place a staple in the top of each clod for orientation. Clods should be roughly fist size, but kept unmodified otherwise. If the soil fabric is fragile, the clod can be placed in a hairnet and dipped briefly in saran as described above. Place the clod in a plastic bag and put into a cell of a clod box. The sampler should make special note of any features to be studied by thin section. Label the appropriate cell on the inside of the lid of the box to identify the soil survey number and horizon (zone) for the clod.

If the material is too sandy and/or too dry to hold together in a clod, bulk density samples can be collected with an aluminum can or other small can of known volume. Sampling is easier if the can has a small hole in the bottom to allow air to escape as the can is inserted. Smooth a planar area in the pit face, or, if sampling from the top down, smooth a planar horizontal area. In either case, choose an area that appears representative of the horizon. With the palm of the hand, gently push the can into the smoothed area until the bottom of the can is flush with the wall or until resistance stops you. In this case, lay a board across the bottom of the can and tap lightly with a hammer or geology pick until the bottom of the can is flush with the pit wall. Then dig out the sampling can plus extra sample and, with a knife blade, smooth off the sample flush with the top of the can. Empty the contents of the can into a plastic bag, tie the top of the bag in a single knot and put into a cell in a clod box. Label the appropriate cell on the inside of the lid of the box to identify the soil survey number and horizon (zone) for the sample. Collect two samples per horizon. Indicate the volume of the sampling can in the sampling notes. It is assumed that there is no volume change with water content in sandy soils. Therefore, one density is representative for all water contents of coarse-textured soils.

Avoid leaving empty cells in a clod box. Fill empty cells with a wadded paper. This prevents clods from shifting in transit. Tape down the top of a filled clod box with nylon filament tape (one short piece on each end and two short pieces in front). Label the top of the box to identify type of sample (bulk density or thin section) and appropriate soil survey numbers and horizons (zones) for the samples. Place six clod boxes in an aluminum case for shipment. Single clod boxes also ship well.

Hand Probe: Remove surface if it is not suitable for coring. Remove core sections and lay in order on plastic sheet. Measure core length against depth in hole to determine if the core has been compressed. Mark horizon breaks on the plastic. Mix the horizon or zone to be sampled. Place sample in a plastic bag and label with soil survey number, horizon (zone), and depth for the core. Samples need to be a minimum of 500 g (1 pt), and are generally suitable for a limited number of analyses only.

Hydraulic probe: Remove surface if it is not suitable for coring. Remove core sections and lay in order on plastic sheet. With a sharp knife trim the exterior to remove any oil and contaminating soil material. Split one core open to mark horizons, describe and then sample. Measure core length against depth in hole to determine if the core has been compressed. Mark horizon breaks on the plastic. Mix the horizon or zone to be sampled. Place in a plastic bag and label with soil survey number, horizon (zone), and depth for the core. Obtain a minimum of 500 g (1 pt) for a reference sample or 3 kg (3 qt) for a characterization sample.

If the core has not been compressed, and the core diameter is 3 inches or more, samples for bulk density can be taken from a second core. Mark a segment 8-cm long on an undisturbed section and slice a cylindrical segment. Measurements of core diameter and length can be used to calculate volume and density at the field-state water content. Core segments can be placed in a hair net, dipped once briefly in saran mix to coat the clod, hung from a rope with a clothes pin to dry, placed in a plastic bag and then put into a cell of a clod box.

Rotary drill (hollow stem): Remove drill core sections and lay in order on plastic sheet. Measure core length against depth in hole to determine if the core has been compressed. Mark horizon breaks on the plastic. Mix the horizon or zone to be sampled. Place in a plastic bag and label with soil survey number, horizon (zone), and depth for the core. Obtain a minimum of 500 g (1 pt) for a reference sample or 3 kg (3 qt) for a characterization sample.

If the core has not been compressed, and the core diameter is 3 inches or more, samples for bulk density can be taken from the core. Mark a segment 8 cm long on an undisturbed section and slice a cylindrical segment. Note the core diameter and length in the soil description. Place the core segment in a plastic bag and place bag into a bulk density (clod) box for shipment. Measurements of core diameter and length can be used to calculate volume and density at the field-state water content. Core segments can be placed in a hair net, dipped once, briefly in saran mix to coat the clod, hung from a rope with a clothes pin to dry, placed in a plastic bag and then put into a cell of a clod box. Label the appropriate cell number on the inside of the box lid to identify the site, pedon, and horizon.

A core segment can be taken for thin section. Place a staple in the top of the core, place the core in a plastic bag and put the bag into a cell in a clod box. Label the appropriate cell number on the inside of the box lid to identify the site, pedon, and horizon.

Bucket Auger: Remove surface if it is not suitable for auguring. Remove auger loads and lay in order on plastic sheet. When horizon breaks are detected, measure depth in hole and mark it on the plastic. Mix the horizon or zone to be sampled. Place in a plastic bag and label with soil survey number, horizon (zone), and depth for the sample. Obtain a minimum of 500 g (1 pt) for a reference sample or 3 kg (3 qt) for a characterization sample. Sampling depth in a pit can be extended by the use of an auger in the pit bottom.

Specific Pedon Sampling Techniques

Soils with Rock Fragments: If coarse fragments up to 75 mm (3 in) in diameter are to be weighed in the field, place excavated sample in a bucket of known weight (tare) and weigh. Sieve the sample through both a 75-mm and 20-mm sieve (3/4 in) onto a canvas tarp which can be suspended from a scale. Estimate the coarse fragment volume percent of both the 75 to 250 mm (10 in) fraction and > 250-mm fraction, and record these values in the description or sampling notes. Weigh the 20 to 75 mm and the <20 mm fractions in pounds or kilograms, and record these weights. Weights are calculated to an oven-dry base in the laboratory. Place a minimum of 4 kg (1 gal) in a plastic bag, double fold the bag, and staple. The water content is determined on the sample in the laboratory. If the 20 to 75-mm fraction is not weighed in the field, estimate the volume percent and record in the sampling notes or description.

Organic Soils: If the soils are drained or the natural water table is below the surface, obtain samples of upper layers from a pit. If the hydraulic conductivity is slow enough, dig and remove samples below the water table as far as practical with due haste and place on a plastic sheet in an orderly fashion for describing and processing. If undisturbed blocks can be removed for bulk density, carve out cubes of known dimension (e.g., 5 cm on a side), place the block in a plastic bag and tie the top in a knot. Place in a second plastic bag if soil is saturated, tie the top in a knot. Put the double bagged sample in a clod box and label the appropriate cell on the inside of the lid to identify the soil survey number and horizon (zone) for the sample. Note the sample dimensions in the sampling notes.

Collect samples from below the water table with a Macaulay peat sampler. If the samples appear undisturbed, mark 10-cm segments, slice with a knife, and place a single segment in a plastic bag. Tie the top in a knot and place in a second plastic bag and tie the top of that bag in a knot. Put the double bagged sample in a clod box and label the appropriate cell on the inside of the lid to identify the soil survey number and horizon (zone) for the sample. Note the sampler diameter and length of core in sampling notes. The sample shape is a half-cylinder. As an alternative, carve a block to fit snugly in a tared water can. Place lid on can, put can in a plastic bag, tie the top, and put in a clod box. Identify can number, depth in sampling notes along with the tare weight. Take replicate samples for the mixed sample, as necessary.

Larger samples can be taken below the water table by removing the surface mat with a spade and sampling lower layers with a post-hole digger. Place samples of each layer on plastic for examination. Transfer samples to small plastic bags, knead to remove air. Put two small bags of sample into one large plastic bag, fold top, staple, and tag.

Sulfidic Soil Materials: These materials as defined by soil taxonomy (Soil Survey Staff, 1999) commonly occur in intra-tidal zones adjacent to oceans, and are saturated most or all of the time. Use containers with an airtight cover. Mason jars and plastic containers with a positive sealing mechanism work well. Glass containers must be adequately packed for shipment to prevent breakage. Fill the container nearly full of sample and add ambient soil water so that all air is eliminated when the lid is secured. Keep containers in the dark and cool. Sulfidic soil samples require expedited transport in a cooler and are refrigerated (4 °C) immediately upon arrival at the laboratory. Once in the lab, if it appears air remained in the container, nitrogen gas can be bubbled through the sample for a few minutes to displace air, then replace the lid. The intent is to keep the material at the field pH prior to running the oxidized pH test.

Permafrost Affected Soils: Soils with permafrost present two special sampling problems. The permafrost is very resistant to excavation and the cryoturbation disrupts horizon morphology. In many cases the surface layers are organic materials. The following sampling approach is suggested.

Test the depth to the frost table with a small (1 to 2 mm) diameter steel rod. Excavate a small pit (about 0.7 by 1.3 m) to leave about 10 cm of unfrozen material over the permafrost. If a cyclic pattern (up to a few meters) is evident in the surface topography, extend the pit through at least one cycle. The organic layers can be carved out with a sharp knife or shovel in many cases and removed. Save the large chunks, if possible.

The objective is to record the morphology of the unfrozen soil before the permafrost is disturbed. Examine the surface and designate horizons. If the soil is disrupted to the extent that lateral horizons do not represent the morphology, impose a grid over the pit face and sketch the morphology on graph paper. Describe the soil down to the frost table. When the description of the unfrozen material is complete, remove all unfrozen material to examine the conformation of the frost table. Note on graph paper if necessary and photograph.

Frozen earth can be removed in successive steps with a gasoline-powered jackhammer. Place pieces from each step on a separate plastic sheet. Examine pieces and describe the morphology as they are removed. Note thickness of segregated ice lenses and make a visual estimate of relative volume of segregated ice. Place representative pieces into a water-tight container so that the sample can be weighed, dried, and weighed again to calculate the amount of water and volume of ice. Excavate to a depth of 30 to 50 cm below the frost table, if practical. Clean off the pit face and be ready to photograph immediately. Sample each horizon or zone for mixed sample, bulk density, and thin section as is practical.

Vertisols: The shear failure that forms slickensides in Vertisols also disrupts the soil to the point that conventional horizons do not adequately describe the morphology. A gilgai surface topography is reflected in the subsurface by bowl-shaped lows and highs. One convention is to sample pedons out of the low and the high areas which represent extremes in the cyclic morphology.

In order to examine morphology and associated soil properties in more spatial detail, the following procedure is suggested: Dig a trench long enough to cover two or three cycles of morphological expression. From the bottom of the pit remove soil from the non-work face so it slopes up and away. Use nails and string to outline boundaries of morphological cells. Assign a number and a horizon designation to each cell.

Construct a level line about one meter below the highest point on the surface. Hammer a spike into the wall at one end of the pit. Tie a loop in string, place the loop over the spike, and run the string to the far end of the pit. Place a line level on the string, tie another loop in the string, place a second spike through the loop, pull the string taut, raise or lower the spike until the string is level, and hammer the spike into the pit face.

Place a marker at each meter along the string from one end to the other. Transfer the morphology outlined by the string to graph paper by measuring the x-coordinate along the string and the y-coordinate above or below the string, both in centimeters. Use a level or a plumb bob to make the y measurement vertical.

Sample each cell for characterization analysis as described above. The sampling scheme can include traditional pedon sequences by sampling vertical sequences of cells at low, high, and intermediate positions along the cycle.

Soil Biology Sampling

Biological samples are also be collected for laboratory analysis, either in conjunction with pedon sampling or independently. At the time of sampling for above-ground biomass, the plants should be identified either in the field or later using a plant identification key so as to determine which plants are associated with the soil microbial communities. Typically a 50 x 50-cm area is sampled. All vegetation is clipped to the soil surface, separated by genus or species and by live and dead fractions. Each plant fraction is weighed, dried, and re-weighed to determine above-ground biomass. As with pedon sampling, sampling for root biomass includes selecting a representative site, sampling by horizon, and designating and sampling a sub-horizon if root mass and morphology change. The sampling area is approximately 1 m². These samples are weighed, dried, and re-weighed to determine root biomass. Typically the roots are separated by hand sieving at the laboratory. The same bulk sample collected for soil mineralogical, physical, and chemical analyses during pedon sampling can also be used for some soil biological analyses, e.g., particulate organic matter (POM), total N, C, and S. Alternatively, a separate bio-bulk sample can be collected in the field. As with pedon sampling, surface litter and O horizons are sampled separately by cutting out a 50 x 50-cm area in a square to a measured depth for bulk density determinations. Include replicate samples in the sampling plan, the primary purpose of which is to identify and/or quantify the variability in all or part of the sampling and analysis system. Properly label samples to show important information, e.g., soil, depth, horizon, etc. If certain biological analyses, e.g., microbial biomass, are requested, these samples require expedited transport under ice or gel packs and are refrigerated (4°C) immediately upon arrival at the laboratory to avoid changes in the microbial communities. Other USDA-NRCS field procedures and sampling protocols for samples that do not require analysis at the SSL are not covered in this manual. Refer to http://soils.usda.gov for more detailed discussion of these topics.

Water Sampling

Water samples can also be collected for laboratory analyses, either in conjunction with pedon sampling or independently. The amount and composition of water samples vary strongly with small changes in location. Choice of water sampling site depends not only on the purpose of the investigation but also on local conditions, depth, and the frequency of sampling (Velthorst, 1996). Specific recommendations are not applicable, as the details of collection can vary with local conditions. Nevertheless, the primary objective of water sampling is the same as with soil sampling, i.e., to obtain a representative sample for use in laboratory analyses. The USDA-NRCS projects requiring collection of water samples have typically been in conjunction with special soil investigations. For more detailed discussion of sampling protocols and investigations of water quality, refer to the U.S. Geological Survey field manual, available online at http://pubs.water.usgs.gov/. Detailed information about the elements of a water quality monitoring and assessment program may be found at the U.S. Environmental Protection Agency's website http://www.epa.gov/.

Preserve samples in the field-state until analysis at the laboratory, without the introduction of change or contamination. Before water sampling in the field, rinse the containers several times with the sample water and completely fill the container and screw cap with the sample water. Avoid touching the sample water, inner part of the container or screw cap. Gloves (powderless) may be used. Include blank samples in the sampling plan, the primary purpose of which is to identify potential sources of sample contamination and assess the magnitude of contamination with respect to concentration of target analytes. There are many possible types of blanks (e.g., source-solution, equipment, trip, ambient, and field blanks). Include replicate samples in the sampling plan, the primary purpose of which is to identify and/or quantify the variability in all or part of the sampling and analysis system. Common types of replicate samples include concurrent, sequential, and split. Refer to Wilde et al. (1999) for more detailed descriptions of the purpose and processing procedures for blanks and replicate samples. Properly label sample containers to show important information, e.g., location, depth, time, etc. Water samples require expedited transport under ice or gel packs and are refrigerated (4 °C) immediately upon arrival at the laboratory.

Some water analyses, e.g., electrical conductivity, total and inorganic C, need to be performed promptly, as optimal preservation is not possible (Velthorst, 1996). Upon completion of these analyses, sample filtration (0.45-μm membrane) is used to separate dissolved from suspended material. The sample is then split into two sub-samples, with one acidified to pH 2 for cation analyses (e.g., Al, Fe, Mn) and the other for anion analyses. These other water analyses also need to be performed as promptly as possible.

8. References

Burt, R. 1996. Sample collection procedures for laboratory analysis in the United States Soil Survey Program. Commun. Soil. Sci. Plant Anal. 27:1293-1298.

Daniels, R.B., E.E. Gamble, and J.G. Cady. 1970. Some relationships among Coastal Plains soils and geomorphic surfaces in North Carolina. Soil. Sci. Soc. Am. Proc.34: 648-653.

Doran, J.W., and T.B. Parkin. 1994. Defining and assessing soil quality. p. 3-21. *In* J.W. Doran, D.C. Coleman, D.F. Bezdicek, and B.A. Stewart (eds.) Defining soil quality for a sustainable environment. SSSA, Madison, WI.

Espinoza, W., R.H. Rust, and R.S. Adams, Jr. 1975. Characterization of mineral forms in Andepts from Chile. Soil Sci. Soc. Am. Proc. 39:556-561.

Fenneman, N.M. 1931. Physiography of the western United States. McGraw-Hill Co., New York, NY. 534 pp.

Fenneman, N.M. 1938. Physiography of the eastern United States. McGraw-Hill Co., New York, NY. 714 pp.

Fenneman, N.M. 1946. (Reprinted 1957) Physical division of the Unites States. US Geological Survey, US GPO, Washington DC. (1 sheet), 1:7,000,000.

Franks, C.D., J.M. Kimble, S.E. Samson-Liebig, and T.M. Sobecki. 2001. Organic carbon methods, microbial biomass, root biomass, and sampling design under development by NRCS. p.105-113. *In* R. Lal, J.M. Kimble, R.F. Follett, and B.A. Stewart (eds.) Assessment methods for soil carbon. CRC Press LLC, Boca Raton, Fl.

Gamble, E. E., R.B. Daniels, and W.D. Nettleton. 1970. Geomorphic surfaces and soils in the Black Creek Valley, Johnston County North Carolina. Soil Sci. Soc. Am. Proc., 34: 276-281.

Gregorich, E.G., M.R. Carter, M.R. Doran, J.W. Pankhurst, C.E. and L.M. Dwyer. 1997. Biological attributes of soil quality. p. 81-113. *In* E.G. Gregorich and M.R. Carter (eds.) Soil quality for crop production and ecosystem health. Elsevier, NY.

Gile, L.H., J.W. Hawley, and R.B. Grossman, R.B., 1981. Soils and geomorphology in the basin and range area of southern New Mexico—Guidebook to the desert project . N.M. Bur. Mines and Miner. Resour. Mem. 39, 222 p.

Hunt, C.B. 1967. Physiography of the United States. W.H. Freeman & Co., London, England. 480 p.

Kubota, T. 1972. Aggregate formation of allophanic soils: Effects of drying on the dispersion of soils. Soil Sci. and Plant. Nutr. 18:79-87.

Lee, B.D., R.C. Graham, T.E. Laurent, C. Amrhein, and R.M. Creasy. 2001. Spatial distributions of soil chemical conditions in a serpentinitic wetland and surrounding landscape. Soil Sci. Soc. Am. J. 65:1183-1196.

Lee, B.D., R.C. Graham, T.E. Laurent, and C. Amrhein. 2003a. Pedogenesis in a wetland meadow and surrounding serpentinitic landslide terrain, northern California, USA. Geoderma. 118:303-320.

Lee, B.D., S.K. Sears, R.C. Graham, C. Amrhein, and H. Vali. 2003b. Secondary mineral genesis from chlorite and serpentine in an ultramafic soil toposequence. Soil Sci. Soc. Am. J. 67:1309-1317.

Nanzyo, M. S. Shoji, and R. Dahlgren. 1993. Physical characteristics of volcanic ash soils. *In* S. Shoji, M. Nanzyo, and R. Dahlgren (eds.) Volcanic ash soils, genesis, properties and utilization. Developments in soil science 21. Elsevier Sci. Publ., Amsterdam, The Netherlands. pp 189-207.

Pankhurst, C.E., B.M. Doube, and V.V.S.R. Gupta (eds.). 1997. Biological indicators of soil health. 437 p. CAB International Wallingford, UK.

Pankhurst, C.E., B.G. Hawke, H.J. McDonald, C.A. Kirkby, J.C. Buckerfield, P. Michelsen, K.A. O'Brien, V.V.S.R. Gupta, and B.M. Doube. 1995. Evaluation of soil biological properties as potential bioindicators of soil health. Aust. J. Exp. Agric. 35:1015-1028.

Parsons, R.B., C.A. Balster, and A.O. Ness. 1970. Soil development and geomorphic surfaces, Willamette Valley, Oregon. Soil Sci. Soc. Am. Proc. 34: 485-491.

Patterson, G.T. 1993. Collection and preparation of soil samples: Site description. p. 1-4. *In* Martin R. Carter (ed.) Soil sampling and methods of analysis. Can. Soc. Soil Sci. Lewis Publ., Boca Raton, FL.

Ruhe, R.V., R.B. Daniels, J.G. Cady. 1967. Landscape evolution and soil formation in southwestern Iowa. USDA Tech. Bull. 1349.

Schoeneberger, P.J., and D.A. Wysocki. 2002. Geomorphology. *In* P.J. Schoeneberger, D.A. Wysocki, E.C. Benham, and W.D. Broderson (eds.). Field book for describing and sampling soils, Version 2.0. USDA-NRCS, National Soil Survey Center, Lincoln, NE.

Soil Conservation Service. 1984. Procedures for collecting soil samples and methods of analysis for soil survey. USDA-SCS Soil Survey Investigations Report No. 1. U.S. Govt. Print. Office, Washington, DC.

Soil Survey Division Staff. 1993. Soil survey manual. USDA. Handb. No. 18. U.S. Govt. Print. Office, Washington, DC.

Soil Survey Staff. 1996. Soil survey laboratory methods manual. Version No. 3.0. USDA-NRCS. Soil Survey Investigations Report No. 42. U.S. Govt. Print. Office, Washington, DC.

Soil Survey Staff. 1998. Glossary of landforms and geologic materials. Part 629, National soil survey handbook USDA-NRCS, National Soil Survey Center, Lincoln, NE.

Soil Survey Staff. 1999. Soil taxonomy: A basic system of soil classification for making and interpreting soil surveys. 2nd ed. Agric. Handb. No. 436. USDA-NRCS. U.S. Govt. Print. Office, Washington, DC.

Taylor, J.K. 1988. Quality assurance of chemical measurements. Lewis Publ., Inc., Chelsea, MI.

Velthorst, E.J. 1996. Water analysis. p. 121-242. *In* P. Buurman, B. van Lagen, and E.J. Velthorst (eds.) Manual for soil and water analysis. Backhuys Publ. Leiden.

Wahrhaftig, C. 1965. Physiographic divisions of Alaska. US Geological Survey Professional paper 482, 52 p.

Wilde, F.D., D.B. Radtke, J. Gibs, and R.T. Iwatsubo. 1999. Collection of water samples: U.S. Geological survey techniques of water resources investigations, book 9, chap. A4, accessed May 2004 at http://pubs.water.usgs.gov/.

Wilson, M.A., S.P. Anderson, K.D. Arroues, S.B. Southard, R.L. D'Agostino, and S.L. Baird. November 15, 1994. No. 6 publication of laboratory data in soil surveys. National Soil Survey Center Soil Technical Note Handbook 430-VI Amendment 5.

Wollum, A.G., 1994. Soil sampling for microbiological analysis. *In* R.W. Weaver, J.S. Angle, and P.S. Bottomley (eds.) Methods of soil analysis. Part 2. Microbiological and biochemical properties. p. 1-14. ASA and SSSA, Madison, WI.

Wysocki, D.A., P.J. Schoeneberger, and H.E. LaGarry. 2000. Geomorphology of soil landscapes. *In* M.E. Sumner (ed.) Handbook of soil science. CRC Press LLC, Boca Raton, FL, ISBN: 0-8493-3136.

Laboratory Sample Collection and Preparation (1B)
Soil (1B1)
Samples (1B1a)

The purpose of any soil sample is to obtain information about a particular soil and its characteristics. Sampling provides a means to estimate the parameters of these soil characteristics with an acceptable accuracy at the lowest possible cost (Petersen and Calvin, 1986). Sub-sampling also may be used, as it permits the estimation of some characteristics of the larger sampling unit without the necessity of measurement of the entire unit. Sub-sampling reduces the cost of the investigation, but usually decreases the precision with which the soil characteristics are estimated. Efficient use of sub-sampling depends on a balance between cost and precision (Petersen and Calvin, 1986).

Soil variability and sample size are interferences to sample collection and preparation. The objective of laboratory preparation is to homogenize the soil samples used in chemical, physical, and mineralogical analyses. At each stage of sampling, an additional component of variability, the variability within the larger units, is added to the sampling error (Petersen and Calvin, 1986). Soil material needs to be adequate in amount and thoroughly mixed in order to obtain a representative sample.

The SSL receives bulk soil samples from across the U.S. and internationally for a wide variety of chemical, physical, and mineralogical analyses. The SSL also typically receives natural fabrics, clods, and cores. Undisturbed clods are used to investigate micromorphology and determine some physical properties, e.g., bulk density.

Bulk Samples (1B1a1)

Laboratory identification numbers and preparation codes are assigned to bulk soil samples. These identification numbers are unique client- and laboratory-assigned numbers that carry important information about the soil sample (e.g., pedon, soil horizon, location, and year sampled). Laboratory preparation codes depend on the properties of the sample and on the requested analyses. These codes carry generalized information about the characteristics of the analyzed fraction, i.e., the water content (e.g., air-dry, field-moist) and the original and final particle-size fraction (e.g., sieved <2-mm fraction processed to 75 μm) and by inference, the type of analyses performed. Identification numbers and preparation codes are reported on the SSL Primary Characterization Data Sheets. Refer to the Soil Survey Investigations Report (SSIR) No. 45 (1995) for a detailed explanation of sample identification numbers. Since the publication of the SSIR No. 42 (1996), there has been a significant revision of these preparation codes. This version of the SSIR No. 42 (2004) does not describe in detail the revised preparation codes. Detailed information on the current preparation codes as they appear on the Primary Characterization Data Sheets may be obtained from the SSL upon request.

All soils from quarantined areas are strictly controlled under APHIS quarantine regulations 7 CFR 330. For preparation procedures for soil bulk samples, proceed to 1B1b.

Natural Fabrics, Clods, and Cores (1B1a2-4)

Laboratory identification number and preparation codes are assigned to natural fabrics (NF), clods, and cores. These identification numbers typically relate to a corresponding bulk sample. Refer to 1B1a1 for information on these identification numbers, preparation codes, and soil quarantine regulations.

Soil Sample Preparation (1B1b)

1. Application

In the SSIR No. 42 (1996), laboratory preparation procedures were described as stand-alone methods based on various procedures summarized by specific preparation codes that are reported on the SSL Primary Characterization Data Sheets. In this version of the SSIR No. 42 (2004), these procedures are described more as a procedural process. A process approach is appropriate in that any one sample received from the field may result in a number of laboratory sub-samples being collected and prepared based on analytical requests and type of materials. This approach is the logic base whereby laboratory procedures are described in this version of the SSIR No. 42. The intent herein is not to detail all possibilities of the universe but to describe some of the master preparation procedures that are typically requested for laboratory analyses.

Laboratory analyses of soil samples are generally determined on the air-dry, fine-earth (<2-mm) fraction (method 1B1b2b1). Air-dry is generally the optimum water content to handle and to process soil. In addition, the weight of air-dry soil remains relatively constant, and biological activity is low during storage. For routine soil analyses, most U.S. and Canadian laboratories homogenize and process samples to pass a 2-mm sieve (Bates, 1993). For some standard air-dry analyses, the <2-mm fraction is further processed so as to be in accordance with a standard method, e.g., Atterberg Limits (1B1b2c1); to meet the sample preparation requirements of the analytical instrument, e.g., total C, N, and S (1B1b2d1); or to achieve greater homogeneity of sample material, e.g., total elemental analysis (1B1b2e1) and carbonates and/or gypsum (1B1b2d2). Additionally, some standard air-dry analyses by definition may require non-sieved material, e.g., whole-soil samples for aggregate stability (1B1b2a1)

A field-moist, <2-mm sample is prepared when the physical properties of a soil are irreversibly altered by air-drying, e.g., water retention (1B1b1b2), particle-size analysis, and plasticity index for Andisols and Spodosols (1B1b1c1), and/or when moist chemical analyses (1B1b1b1) are appropriate. Some biological analyses require field-moist samples (1B1b1a2), as air-drying may cause significant changes in the microbial community. The decomposition state of organic materials is used in soil taxonomy (Soil Survey Staff, 1999) to define sapric, hemic, and fibric organic materials, and as such the evaluation of these materials (Histosol analysis) require a field-moist, whole-soil sample (1B1b1a1).

Knowing the amount of rock fragments is necessary for several applications, e.g., available water capacity and linear extensibility. Generally, the >2-mm fractions are sieved, weighed, and discarded (1B1b2f1a) and are excluded from most chemical, physical, and mineralogical analysis. Some exceptions include but are not limited to samples containing coarse fragments with carbonate- or gypsum-indurated material or material from Cr and R soil horizons. In these cases, the coarse fragments may be crushed to <2 mm and analytical results reported on that fraction, e.g., 2 to 20 mm (1B1b2f1a2a), or the coarse fragments and fine-earth material are homogenized and crushed to < 2 mm with laboratory analyses made on the whole-soil (1B1b2f1b1a). Additionally, depending on the type of soil material, samples can be tested for the proportion and particle-size of air-dry rock fragments that resist abrupt immersion in tap water (1B1b2f1a3).

2. Summary of Method

Any one soil sample received from the field may result in a number of laboratory sub-samples being collected and prepared based on the properties of the sample and on the requested analyses. For most standard chemical, physical, and mineralogical analysis, the field sample is air-dried, crushed, and sieved to <2 mm. Field-moist, fine-earth fraction samples are processed by forcing the material through a 2-mm screen by hand or with a large, rubber stopper and placed in a refrigerator for future analysis. Depending on the nature of the soil material and requested analyses, air-dry and/or field-moist samples may also be prepared as whole-soil samples or processed further to finer fractions than <2 mm, e.g., <0.425 mm for Atterberg Limits and ≈ 180 µm for chemical analysis of organic materials. Air-dry, <75 µm sub-samples for major and trace elements are processed metal-free.

Generally, weight measurements are made and recorded on the 20- to 75-mm, 5- to 20-mm, and 2- to 5-mm fractions, with these fractions then discarded. In some cases, these weight measurements may not be recorded. Additionally, some or all of these >2-mm fractions may be processed to a finer fraction and saved for chemical, physical, and mineralogical analysis. For example, after the respective weights of the 5- to 20-mm and 2- to 5-mm fractions are recorded, these fractions may be recombined and crushed to <2 mm in a laboratory jaw crusher, with the recombined material saved for laboratory analysis. In other cases, the fine-earth fraction and the >2-mm fractions are homogenized and passed through a laboratory jaw crusher to reduce all material to pass a 2-mm sieve, with the processed material saved for laboratory analysis.

3. Interferences

Soil variability and sample size are interferences to sample collection and preparation. At each stage of sampling, an additional component of variability, the variability among smaller elements within the larger units, is added to the sampling error (Petersen and Calvin, 1986). Soil material needs to be in adequate amount and thoroughly mixed to obtain a representative sample.

Soil is mixed by moving it from the corners to the middle of processing area and then by redistributing the material. This process is repeated four times. Enough soil material needs to be sieved and weighed to obtain statistically accurate rock fragment content. In order to accurately measure rock fragments with a maximum particle diameter of 20 mm, the minimum specimen size ("dry" weight) that needs to be sieved and weighed is 1.0 kg. Refer to ASTM Standard Practice D 2488 (American Society for Testing and Materials, 2004). A homogenized soil sample is more readily obtained from air-dry material than from field-moist material. Whenever possible, "moist" samples or materials should have weights two to four times larger than for "dry" specimens (American Society for Testing and Materials, 2004). Refer to ASTM Standard Practice D 2488 (American Society for Testing and Materials, 2004).

4. Safety

Dust from sample processing is a nuisance. A mask should be worn in order to avoid breathing dust. Keep clothing and hands away from the crusher and pulverizer when these machines are in use. Use a face-shield and goggles when operating the jaw crusher. Use goggles when operating the air compressor. The HCl used to check carbonates can destroy clothing and irritate skin. Immediately rinse acid with water from clothing or skin and seek professional medical help, if needed.

5. Equipment
5.1   Electronic Balance, ±1-g sensitivity and 15-kg capacity
5.2   Cardboard trays for sample storage
5.3   Trays, plastic, tared
5.4   Sieves, square-hole, stainless steel
5.4.1 80 mesh, 180 µm

5.4.2 10 mesh, 2 mm

5.4.3 4 mesh, 4.75 mm

5.4.4 19 mm, 3/4 in

5.4.5 76 mm, 3 in

5.4.6 200-mesh, 75 μm, nylon cloth sieve

5.4.7 40-mesh, 0.425 mm

5.5 Pulverizer

5.6 Wooden rolling pin

5.7 Rubber roller

5.8 Laboratory jaw crusher, Retsch, Model BB2/A, Brinkmann Instruments Inc., Des Plaines, IL.

5.9 Metal plate, 76 x 76 x 0.5 cm

5.10 Containers, paper, 12-oz, with lids

5.11 Containers, plastic, 1-pint, 1, 4, and/or 8 oz with tops

5.12 Scintillation glass vials, 20-mL

5.13 Metal weighing cans, 2-oz

5.14 Brown Kraft paper

5.15 Air compressor, Cast-iron Series, SpeedAire, Campbell Hausfeld Mfg. Co., Harrison, OH.

5.16 Planetary ball mill, Fritsch, Model P-5, Gilson, Lewis Center, OH

5.17 Syalon balls, 12- to 15-mm, and bowls, 80-mL

5.18 Metal weighing cans, 2-oz

5.19 Cross beater mill, Retsch, Brinkmann Instruments Inc., Des Plaines, IL.

6. Reagents

6.1 Reverse osmosis (RO) water

6.2 1 N HCl

6.3 Sodium hexametaphosphate solution. Dissolve 35.7 g of sodium hexametaphosphate $(NaPO_3)_6$ and 7.94 g of sodium carbonate (Na_2CO_3) in 1 L of RO water.

7. Procedures

Soil Bulk Sample Preparation (1B1b)

7.1 Weigh soil sample in sample bag to nearest g when logged-in and record weight.

7.2 Weigh sample in sample bag to nearest g before air-drying and record weight.

7.3 Remove soil sample from sample bag and distribute on a plastic tray. Thoroughly mix soil material.

7.4 If field-moist sub-samples are requested, proceed to procedure 1B1b1.

7.5 If only air-dry sub-samples are requested, proceed to procedure 1B1b2.

Soil Bulk Sample Preparation (1B1b)
Field-Moist Preparation (1B1b1)

7.6 If field-moist, whole-soil sub-samples are requested, proceed to procedure 1B1b1a.

7.7 If only field-moist, <2-mm sub-samples are requested, proceed to procedure 1B1b1b.

Soil Bulk Sample Preparation (1B1b)
Field-Moist Preparation (1B1b1)
Whole-soil (1B1b1a)

7.8 Procedures for field-moist, whole-soil sub-samples include but are not limited to 1B1b1a1 and 1B1b1a2 as follows:

Soil Bulk Sample Preparation (1B1b)
Field-Moist Preparation (1B1b1)
Whole-soil (1B1b1a)
Histosol Analysis (1B1b1a1)

7.9 For Histosol analysis, select material for representative sub-samples from at least five different areas on the plastic tray. Prepare a sub-sample of the field-moist, whole-soil prepared in an 8-oz container. Store in the refrigerator for future analysis.

Soil Bulk Sample Preparation (1B1b)
Field-Moist Preparation (1B1b1)
Whole-soil (1B1b1a)
Biological Analysis (1B1b1a2)

7.10 For biological analyses, select material for representative sub-samples from at least five different areas on the plastic tray. Prepare a sub-sample of field-moist, whole-soil prepared in a plastic container. This sub-sample is termed a biology bulk sample, which may be obtained from the bulk soil sample upon arrival at the laboratory. Store the biology bulk sample in the refrigerator for future analysis. For additional information on biology collection and preparation procedures, proceed to 1B3.

Soil Bulk Sample Preparation (1B1b)
Field-Moist Preparation (1B1b1)
<2-mm Fraction (1B1b1b)

7.11 Procedures for field-moist, <2-mm sub-samples include but are not limited to 1B1b1b1, 1B1b1b2, 1B1b1b3, and 1B1b1c1 as follows:

Soil Bulk Sample Preparation (1B1b)
Field-Moist Preparation (1B1b1)
<2-mm Fraction (1B1b1b)
Chemical and Selected Physical Analysis (1B1b1b1)

7.12 For moist chemical analysis, select material for representative sub-samples from at least five different areas on the plastic tray. Process a sub-sample of field-moist material by forcing the material through a 2-mm screen by hand or with a large, rubber stopper and place in a 1-pint plastic container. Store in the refrigerator for future analysis.

Soil Bulk Sample Preparation (1B1b)
Field-Moist Preparation (1B1b1)
<2-mm Fraction (1B1b1b)
1500-kPa Water Content (1B1b1b2)

7.13 For moist 1500-kPa water content of mineral and organic material, select material for representative sub-samples from at least five different areas on the plastic tray. Process a sub-sample of field-moist material by forcing the material through a 2-mm screen by hand or with a large, rubber stopper and place in a 4-oz plastic container. Store in the refrigerator for future analysis.

Soil Bulk Sample Preparation (1B1b)
Field-Moist Preparation (1B1b1)
<2-mm Fraction (1B1b1b)
Field-Moist/Oven-Dry Ratio (1B1b1b3)

7.14 For field-moist/oven-dry (FMOD) ratio (required if any moist analyses are determined), select material for representative sub-samples from at least five different areas on the plastic tray. Process a sub-sample of field-moist material by forcing the material through a 2-mm screen by hand or with a large, rubber stopper and place in a 2-oz metal weighing can.

Soil Bulk Sample Preparation (1B1b)
Field-Moist Preparation (1B1b1)
<2-mm Fraction Sieved to <0.425 mm (1B1b1c)
Atterberg Limits (1B1b1c1)

7.15 For moist Atterberg Limits, select material for representative sub-samples from at least five different areas on the plastic tray. Process a sub-sample of field-moist material by forcing the material through a 2-mm screen by hand or with a large, rubber stopper and then sieved to 40-mesh (0.425 mm). Place sub-sample in a 12-oz plastic container and store in the refrigerator for future analysis.

Soil Bulk Sample Preparation (1B1b)
Air-Dry Preparation (1B1b2)

7.16 Before air-drying, weigh sample on a tared tray (tray weight) to nearest g and record weight.

7.17 Air-dry the sample in an oven at 30 to 35° C for 3 to 7 days.

7.18 Weigh sample to nearest g after air-drying and record weight.

7.19 If air-dry, whole-soil sub-samples are requested, proceed to procedure 1B1b2a.

7.20 If only air-dry, <2-mm sub-samples are requested, proceed to procedure 1B1b2b.

Soil Bulk Sample Preparation (1B1b)
Air-Dry Preparation (1B1b2)
Whole-Soil (1B1b2a)

7.21 Procedures for air-dry, whole-soil sub-samples include but are not limited to 1B1b2a1 and 1B1b2a2 as follows:

Soil Bulk Sample Preparation (1B1b)
Air-Dry Preparation (1B1b2)
Whole-Soil (1B1b2a)
Aggregate Stability Analysis (1B1b2a1)

7.22 For aggregate stability analysis, select material for representative sub-samples from at least five different areas on the plastic tray. Prepare an air-dry, whole-soil sample in a 12-oz paper container.

Soil Bulk Sample Preparation (1B1b)
Air-Dry Preparation (1B1b2)
Whole-Soil (1B1b2a)
Repository Samples (1B1b2a2)

7.23 For comprehensive sampling projects, prepare a 12-oz paper container of fine-earth material. Also prepare a 12-oz container for natural fabric (NF) sample. Store soil samples in repository. If a NF sample was not selected in the field, select one from air-dried bulk sample. Generally, do not select NF samples for reference projects.

Soil Bulk Sample Preparation (1B1b)
Air-Dry Preparation (1B1b2)
<2-mm Fraction (1B1b2b)

7.24 Weigh sample to nearest g and record weight. This weight includes the >2-mm fractions.

7.25 Roll soil material on a flat, metal plate that is covered with brown Kraft paper with wooden rolling pin and/or rubber roller to crush clods to pass a 2-mm sieve. For samples with easily crushed coarse fragments, substitute a rubber roller for a wooden rolling pin. Roll and sieve until only the coarse fragments that do not slake in sodium hexametaphosphate solution remain on sieve. Crush clayey soils that contain no coarse fragments in a laboratory jaw crusher.

7.26 Process air-dry soil by sieving to <2-mm. Thoroughly mix material by moving the soil from the corners to the middle of the processing area and then by redistributing the material. Repeat process four times. For preparation of the >2-mm fractions, proceed to procedure 1B1b2f .

7.27 Procedures for air-dry, <2-mm fractions include but are not limited to 1B1b2b1, 1B2b2b2, 1B2b2b3, 1B2b2b4, 1B2b2b5, 1b2b2c1, 1B2b2d1, 1b2b2d2, 1B2b2d3, and 1B2b2e1 as follows:

Soil Bulk Sample Preparation (1B1b)
Air-Dry Preparation (1B1b2)
<2-mm Fraction (1B1b2b)
Standard Chemical, Physical and Mineralogical Analysis (1B1b2b1)

7.28 For standard chemical, physical, and mineralogical analysis, select material for representative sub-samples from at least five different areas on the plastic tray. Prepare one sub-sample of the air-dry, sieved <2-mm fraction in a 12-oz paper container.

Soil Bulk Sample Preparation (1B1b)
Air-Dry Preparation (1B1b2)
<2-mm Fraction (1B1b2b)
Salt Analysis (1B1b2b2)

7.29 For a saturation paste when salt analyses are requested, select material for representative sub-samples from at least five different areas on the plastic tray. Prepare one sub-sample of the air-dry, sieved <2-mm fraction in a 1-pint plastic container.

Soil Bulk Sample Preparation (1B1b)
Air-Dry Preparation (1B1b2)
<2-mm Fraction (1B1b2b)
1500-kPa Water Content (1B1b2b3)

7.30 For air-dry 1500-kPa water content of mineral and organic materials, select material for representative sub-samples from at least five different areas on the plastic tray. Prepare a sub-sample of air-dry, sieved <2-mm fraction in a 1-oz plastic cup.

Soil Bulk Sample Preparation (1B1b)
Air-Dry Preparation (1B1b2)
<2-mm Fraction (1B1b2b)
Air-Dry/Oven-Dry Ratio (1B1b2b4)

7.31 For air-dry/oven-dry (ADOD) ratio (required if any air-dry analysis are determined), select material for representative sub-samples from at least five different areas on the plastic tray. Prepare a sub-sample of the air-dry, sieved <2-mm fraction in a 2-oz metal weighing can.

Soil Bulk Sample Preparation (1B1b)
Air-Dry Preparation (1B1b2)
<2-mm Fraction (1B1b2b)
Presence of Carbonates (1B1b2b5)

7.32 Use a sub-sample of the ADOD sample (procedure 1B1b2b4) and check for the presence of carbonates. Reference samples (knowns) are available for comparisons. Place 1 g of the air-dry fine-earth fraction in porcelain spot plate, add reverse osmosis water, and stir to remove entrapped air. Add 1 *N* HCl to soil, observe amount of effervescence, and record as follows:

None – No visual efferevescence.

Very Slight - Bubbles rise at a few points in the sample and consistently appear at the same point in either a steady stream of tiny bubbles or in a slower stream of larger bubbles. Do not mistake trapped air bubbles for a positive test. Generally, these air bubbles appear immediately after the addition of 1 *N* HCl.

Slight - More small bubbles, and possibly a few larger bubbles, appear throughout the sample than with a *very slight* reaction.

Strong - More large bubbles are evident than with a *slight* reaction. Often the reaction is violent at first and then quickly decreases to a reaction that produces many small bubbles.

Violent - The sample effervesces violently. Many large bubbles appear to burst from the spot plate.

Soil Bulk Sample Preparation (1B1b)
Air-Dry Preparation (1B1b2)
<2-mm Fraction Sieved to 0.425 mm (1B1b2c)
Atterberg Limits (1B1b2c1)

7.33 For Atterberg Limits, select material for representative sub-samples from at least five different areas on the plastic tray. Prepare one sub-sample of the air-dry, sieved <2-mm material sieved to 40-mesh (0.425 mm) in a 12-oz paper container.

Soil Bulk Sample Preparation (1B1b)
Air-Dry Preparation (1B1b2)
<2-mm Fraction Processed to ≈ 180 μm (1B1b2d)
Total Carbon, Nitrogen, and Sulfur Analysis (1B1b2d1)

7.34 For total C, N, and S analyses, select material for representative sub-samples from at least five different areas on the plastic tray. Prepare one sub-sample of the air-dry, sieved <2-mm fraction processed in a cross beater mill to (≈ 80 mesh, 180 μm) in a 20-mL scintillation glass vial.

Soil Bulk Sample Preparation (1B1b)
Air-Dry Preparation (1B1b2)
<2-mm Fraction Processed to ≈ 180 μm (1B1b2b)
Calcium Carbonate and Gypsum (1B1b2d2)

7.35 Use the prepared sub-sample (procedure 1B1b2d1) for the determination of the amount of carbonates and/or gypsum.

Soil Bulk Sample Preparation (1B1b)
Air-Dry Preparation (1B1b2)
<2-mm Fraction Processed to ≈ 180 μm (1B1b2b)
Chemical Analysis of Organic Materials (1B1b2d3)

7.36 For chemical analysis of organic materials, select material for representative sub-samples from at least five different areas on the plastic tray. Prepare a sub-sample of the air-dry, sieved <2-mm fraction processed in a cross beater to a fine grind (≈ 80 mesh, 180 μm) in a 12-oz paper container.

Soil Bulk Sample Preparation (1B1b)
Air-Dry Preparation (1B1b2)
<2-mm Fraction Sieved to 75 μm (1B1b2e)
Total Major and/or Trace Elements (1B1b2e1)

7.37 For total major and/or trace element analyses, select material for representative sub-samples from at least five different areas on the plastic tray. Prepare one sub-sample of the air-dry, metal free, sieved <2-mm fraction processed in a planetary ball mill for 2 min and sieved to <75 μm (200 mesh) in a 20-mL scintillation glass vial.

Soil Bulk Sample Preparation (1B1b)
Air-Dry Preparation (1B1b2)
>2-mm Fractions (1B1b2f)

7.38 The following procedures are used for samples with >2 mm fractions. These fractions include mineral coarse fragments as well wood fragments that are >20 mm in cross section and cannot be crushed and shredded with the fingers. If the >2-mm fractions are to be weighed, recorded, and discarded, with no further laboratory analysis, proceed to procedure 1B1b2f1a. When the data for >2-mm fractions are not recorded, proceed to procedure 1B1b2f1b. If the >2-mm fractions contain carbonate- or gypsum-indurated material, and laboratory analysis is requested, proceed to procedure 1B1b2f1a1a. If the >2-mm fractions are Cr or R material (Soil Survey Staff, 1999), and laboratory analysis is requested, proceed to procedure 1B1b2f1b1a. If testing is requested for the proportion and particle-size of the air-dry, >2-mm fraction that resist abrupt immersion in tap water, proceed to procedure 1B1b2f1a3.

Soil Bulk Sample Preparation (1B1b)
 Air-Dry Preparation (1B1b2)
>2-mm Fractions (1B1b2f)
Particle-Size Analysis (1B1b2f1)
Particle-Size Analysis, Recorded (1B1b2f1a)

7.39 Process the air-dry soil by sieving to <2 mm as described in procedure 1B1b2b. In this procedure (1B1b2f1a), weight measurements are made on the 20- to 75-mm, 5- to 20-mm, and 2- to 5-mm fractions. Weigh soil material with diameters of 2 to 5 mm. If difficult to separate the <2-mm fraction from fragments, soak (100 g of 2- to 5-mm fraction) in sodium hexametaphosphate solution for 12 h. Air-dry, weigh the material that does not slake, record weight, and discard. Weigh, record weight, and discard particles with diameters of 20 to 75 mm and 5 to 20 mm. The <2-mm fraction is saved for chemical, physical, and mineralogical analysis (1B1b2b)

Soil Bulk Sample Preparation (1B1b)
Air-Dry Preparation (1B1b2)
>2-mm Fractions (1B1b2f)
Particle-Size Analysis (1B1b2f1)
Particle-Size Analysis, Not Recorded (1B1b2f1b)

7.40 This procedure (1B1b2f1b) is the same as described in procedure 1B1b2f1a except the weight of the >2-mm fractions are not recorded and all analytical results are reported on a <2-mm basis.

Soil Bulk Sample Preparation (1B1b)
Air-Dry Preparation (1B1b2)
>2-mm Fractions (1B1b2f)
Particle-Size Analysis (1B1b2f1)
Particle-Size Analysis, Recorded (1B1b2f1a)
2- to 20-mm Fraction Processed to <2-mm (1B1b2f1a1)
Chemical, Physical, and Mineralogical Analysis (1B1b2f1a1a)

7.41 This procedure (1B1b2f1a1a) is commonly used for samples with carbonate- or gypsum-indurated material. Process the air-dry soil by sieving to <2 mm as described in procedure 1B1b2b. Weigh soil material with diameters of 20- to 75-mm, 5 to 20-mm, and 2 to 5-mm and record weights as described in procedure 1B1b2f1a. The 5- to 20-mm and 2- to 5-mm fractions are then recombined after their respective weights are recorded. The recombined, 2- to 20-mm, material is crushed to <2 mm in a laboratory jaw crusher. This material is saved for laboratory analysis and analytical results reported on the 2- to 20-mm basis. The <2-mm material is also saved for chemical, physical, and mineralogical analysis as described in procedure 1B1b2b. If carbonate or gypsum accumulations are soft and easily pass a 2-mm sieve, the standard procedure (1B1b2b) is usually requested.

Soil Bulk Sample Preparation (1B1b)
Air-Dry Preparation (1B1b2)
>2-mm Fractions (1B1b2f)
Particle-Size Analysis (1B1b2f1)
Particle-Size Analysis, Recorded (1B1b2f1a)
2- to 20-mm Fraction Processed to <2-mm and Recombined with <2-mm
Fraction (1B1b2f1a2)
Chemical, Physical, Mineralogical Analysis (1B1b2f1a2a)

7.42 This procedure (1B1b2f1a2a) is rarely requested. Process the air-dry soil by sieving to <2 mm as described in procedure 1B1b2b. Weigh soil material with diameters of 20- to 75-mm, 5 to 20-mm, and 2 to 5-mm and record weights as described in procedure 1B1b2f1a. The 5- to 20-mm and 2- to 5-mm fractions are then recombined after their respective weights are recorded. The recombined, 2- to 20-mm, material is crushed to <2 mm in a laboratory jaw crusher. This material is then recombined with the <2-mm fraction. These analytical results are reported on the <20-mm basis or are calculated to the <2-mm basis.

Soil Bulk Sample Preparation (1B1b)
Air-Dry Preparation (1B1b2)
>2-mm Fractions (1B1b2f)
Particle-Size Analysis (1B1b2f1)
Particle-Size Analysis, Not Recorded (1B1b2f1b)
Whole-Soil Processed to <2-mm (1B1b2f1b1)
Chemical, Physical, and Mineralogical Analysis (1B1b2f1b1a)

7.43 This procedure (1B1b2f1b1a) is mainly used to prepare samples from Cr or R soil horizons (Soil Survey Staff, 1999). Homogenize particles with diameters >2 mm and the fine-earth material (<2-mm) and crushed to <2-mm in a laboratory jaw crusher. This material is saved for laboratory analysis and analytical results reported on the whole-soil basis.

8. Calculations
    Calculations for coarse fragments are reported in 3A2.

9. Report
    Reported data include but are not limited to the following:

9.1 Weight (g) of field-moist soil sample
9.2 Weight (g) of air-dry soil sample
9.3 Weights (g) of processed air-dry soil
9.4 Weight (g) of 20- to 75-mm fraction
9.5 Weight (g) of 5- to 20-mm fraction
9.6 Weight (g) of 2- to 5-mm fraction
9.7 Weight (g) of sub-sample of 2- to 5-mm fraction before slaking
9.8 Weight (g) of sub-sample of 2- to 5-mm fraction after slaking
9.9 Effervescence with HCl (None, Very Slight, Slight, Strong, Violent)

10. Precision and Accuracy
    Precision and accuracy data are available from the SSL upon request.

11. References
    American Society for Testing and Materials. 2004. Standard practice for description and
        identification of soils (visual-manual procedure). D 2488.. Annual book of ASTM
        standards. Construction. Section 4. Soil and rock; dimension stone; geosynthesis. Vol.
        04.08. ASTM, Philadelphia, PA.

Bates, T.E. 1993. Soil handling and preparation. p. 19-24. *In* M.R. Carter (ed.) Soil sampling and methods of analysis. Can. Soc. Soil Sci. Lewis Publ., CRC Press, Boca Raton, FL.

Petersen, R.G., and L.D. Calvin. 1986. Sampling. p. 33-51. *In* A. Klute (ed.) Methods of soil analysis. Part 1. Physical and mineralogical methods. 2nd ed. Agron. Monogr 9. ASA and SSSA, Madison, WI.

Soil Survey Staff. 1995. Soil survey laboratory information manual. Version No. 1.0. USDA-NRCS. Soil Survey Investigations Report No. 42. U.S. Govt. Print. Office, Washington, DC.

Soil Survey Staff. 1996. Soil survey laboratory methods manual. Version No. 3.0. USDA-NRCS. Soil Survey Investigations Report No. 42. U.S. Govt. Print. Office, Washington, DC.

Soil Survey Staff. 1999. Soil taxonomy: A basic system of soil classification for making and interpreting soil surveys. 2nd ed. Agric. Handb. No. 436. USDA-NRCS. U.S. Govt. Print. Office, Washington, DC.

Soil Bulk Sample Preparation (1B1b)
Air-Dry Preparation (1B1b2)
>2-mm Fractions (1B1b2f)
Particle-Size Analysis (1B1b2f1)
Particle-Size Analysis, Recorded (1B1b2f1a)
Proportion and Particle-Size of Air-Dry Rock Fragments that Resist Abrupt Immersion in Tap Water (1B1b2f1a3)

1. Application

It is widely accepted that the mechanical preparation of soil for laboratory analysis is difficult to standardize for samples that contain >2-mm particles with rupture resistance intermediate between earthy bodies and fragments of highly resistant rock (Grossman, 2004). There is a need for a procedure that is both reproducible and subjects the >2-mm fraction to an intermediate stress less than standard mechanical preparation. Abrupt immersion in tap water of the initially air-dry >2-mm fraction has been chosen as the manner in which stress is applied. For some situations, comparison of the results to use a dispersing agent may be valuable; for this another procedure would need establishment.

This method (1B1b2f1a3) provides a comparison to mechanical preparation methods and should aid in the development of more precise mechanical preparation dependent on sample properties (Grossman, 2004). Certain analyses can be run on the >2-mm fraction after separation by the water-immersion method. The resulting values can be subtracted from analyses on the whole ground sample to obtain estimates for the <2-mm fraction as removed by the water immersion treatment.

2. Summary of Method

A representative sub-sample of the air-dry field sample is weighed and passed through a No.10 sieve. The >2-mm fraction is placed abruptly in tap water, left overnight, gently agitated, passed through a No. 10 sieve, and the wet >2-mm passed through a nest of No. 10, No. 4, 9.5-mm (nominally 10 mm), and 20-mm sieves.

3. Interferences

There are no known interferences.

4. Safety

Dust from sample processing is a nuisance. A mask should be worn in order to avoid breathing dust.

5. Equipment
5.1 Buckets, plastic, 10-L with 1-L marks
5.2 Bucket, plastic, in which 20-cm diameter sieve fits snugly
5.3 Bucket, 19-L (5-gal), straight-sided, with 30-cm diameter
5.4 Drying trays, fiberglass, 35 x 48 cm
5.5 Cake pans, aluminum foil, 20 x 20 x 4cm
5.6 Sieves: 30-cm diameter No. 10, 20-mm diameter No. 4, 9.5- and 20-mm plus bottom pan
5.7 Top loading balance, 1-g sensitivity and >10,000-g capacity, with pan large enough to mount the trays listed under Section 5.4
5.8 Pan, plastic, 60 x 40 x 15 cm
5.9 Plastic sheet, 8-mm, large enough to line plastic container
5.10 Plastic beads, 9 x 6 mm, segregated by different colors
5.11 Teaspoon, tablespoon

6. Reagents
6.1 $CaCl_2 \cdot 2H_2O$
6.2 Tap water of acceptable dispersability (taken as Zone A in Flanagan and Holmgren, 1977).

7. Procedure

7.1 Remove roughly a one-fourth representative sub-sample inclusive of the >20-mm fraction from the field sample and weigh.

7.2 Pass through a No. 10 sieve. Weigh the >2-mm particles and discard the<2-mm fraction. Immerse the >2-mm fraction abruptly in tap water in a 10-L bucket in which there is approximately 1-L of water for roughly every 500 g of >2-mm particles. Leave the sample overnight.

7.3 Insert the hand to the bottom of the bucket and rotate the mass at one rotation per second 20 times. In the sink, place the 30-cm diameter No. 10 sieve over the top of a 19-L diameter bucket. Quantitatively transfer the >2-mm particles into the No.10 sieve and wash with a water stream to remove the material not resistant to the immersion treatment. Use a minimum amount of water.

7.4 After the 19-L bucket is full, add about 30 g of $CaCl_2$. Allow to stand until settled out. Pour off into the sink as much of the water as possible without transfer of solids to the sink. Dry the sediment in the bucket by evaporation, followed by disposal as a solid.

7.5 Follows 7.3 directly if <u>no</u> strength measurements. Quantitatively transfer from the 30-cm diameter sieve to a tray. Air-dry the sample on the tray at 30 to 35°C. Pass through a nest of 20-mm, No. 4 and No. 10 sieves.

7.6 Weigh each separate to the nearest g and report. Subtract the >20-mm weight from the initial >2-mm weight. Obtain the 2 to 5, 5 to 20, and 2- to 20-mm particles as a percent of the <20-mm fraction.

8. Calculations

8.1 $A = [B/(C - D)] \times 100$

8.2 $E = [F/(C - D)] \times 100$

8.3 $G = A + E$

where:

A = Weight percentage 2- to 5-mm fraction on <20-mm basis (%)
B = Air-dry weight of 2- to 5-mm fraction after water immersion (g)
C = Air-dry total weight of whole soil sub-sample (g)
D = Air-dry weight of >20-mm fraction after water immersion (g)
E = Weight percentage of 5- to 20-mm fraction on <20-mm basis (%)
F = Air-dry weight of 5- to 20-mm fraction after water immersion (g)
G = Weight percentage 2- to 20-mm fraction on <20-mm basis (%)

9. Report

Report the 2 to 5, 5 to 20, and 2- to 20-mm fractions as weight percentages of the <20-mm fraction.

10. Precision and Accuracy

Precision and accuracy data are available from the SSL upon request.

11. References

Flanagan, C.P., and G.G.S. Holmgren. 1977. p. 121-134. Field methods for determination of soluble salts and percent sodium from extract for identifying dispersive clay soils. *In* J.L. Sherard and R.S. Decker (eds.) Dispersive clays, related piping, and erosion in geotechnical projects, ASTM STP 623. American Society of Testing Materials, West Conshohocken, PA.

Grossman, R.B. 2004. Proportion and particle-size of air-dry rock fragments that resist abrupt immersion in tap water. USDA-NRCS, Lincoln, NE.

Laboratory Sample Collection and Preparation (1B)
Water (1B2)
Samples (1B2a)

As with soil samples, laboratory identification numbers and preparation codes are assigned to water samples. Refer to 1B1a1 for information on these identification numbers and preparation codes.

Water (1B2)
Water Sample Preparation (1B2b)

Avoid long periods between collection and laboratory analysis of water. To prevent significant changes (e.g., degradation, volatilization), water samples require expedited transport under ice or gel packs and are refrigerated (4°C) immediately upon arrival at the laboratory. Avoid freezing water samples, which can influence pH and the separation of dissolved organic matter from the water phase.

Some water analyses, e.g., electrical conductivity, total and inorganic C, need to be performed promptly, as optimal preservation is not possible (Velthorst, 1996). Upon completion of these analyses, sample filtration (0.45-μm membrane) is used to separate dissolved from suspended material. The sample is then split into two sub-samples, with one acidified to pH 2 for cation analyses (e.g., Al, Fe, Mn) and the other for anion analyses. These other water analyses also need to be performed as promptly as possible.

Biological Materials (1B3)
Samples (1B3a)
Biology Bulk Sample (1B3a1)
Microbial Biomass Sample (1B3a2)

As with soil and water samples, laboratory identification numbers and preparation codes are assigned to biology samples. Refer to 1B1a1 for information on these identification numbers and preparation codes. Some biology samples arrive at the laboratory as part of the soil bulk sample. If this is the case, biological sub-samples are collected and prepared as described in 1B1b1a2. In other cases, biology bulk samples may be split in the field (1B3a1) and are separate sampling units from the soil bulk sample (1B1a1). Additionally, some biological samples, e.g., microbial biomass (1B3a2), are separate units from the soil bulk or other biology samples, requiring expedited transport under ice or gel packs and are refrigerated (4°C) immediately upon arrival at the laboratory.

Biological Materials (1B3)
Biological Material Preparation (1B3b)

Avoid long periods between collection and laboratory analysis of biological samples to prevent significant changes (e.g., microbial community). Store biology samples in the refrigerator (4°C) for future analysis. Biological preparation includes but is not limited to the following:

Biological Materials (1B3)
Biological Material Preparation (1B3b)
Field-Moist Preparation (1B3b1)
<2 mm (1B3b1a)
Microbial Biomass (1B3b1a1)

Refer to the section on soil biological and plant analyses (6) for additional information on the further processing and preparation of these biological samples for laboratory analysis.

Biological Materials (1B3)
Biological Material Preparation (1B3b)
Air-Dry Preparation (1B3b2)
<2-mm Fraction Sieved to <53 μm, with ≥ 53 μm (Particulate Organic Matter) and <53 μm Processed to ≈ 180 μm (1B3b2a)
Total Carbon, Nitrogen, and Sulfur (1B3b2a1)

Refer to the section on soil biological and plant analyses (6) for additional information on the further processing and preparation of these biological samples for laboratory analysis.

Biological Materials (1B3)
Biological Material Preparation (1B3b)
Air-Dry Preparation (1B3b2)
<2-mm Fraction (1B3b2b)
Other Biological Analyses (1B3b2b1)

Refer to the section on soil biological and plant analyses (6) for additional information on the further processing and preparation of biological samples for laboratory analysis.

Biological Materials (1B3)
Biological Material Preparation (1B3b)
Dry (50°C) Preparation (1B3b3)
Roots (1B3b3a)
Root Biomass (1B3b3a1)
Roots Processed to ≈ 180 μm (1B3b3a1a)
Total Carbon, Nitrogen, and Sulfur (1B3b3a1a1)
Plants (1B3b3b)
Plant Biomass (1B3b3b1)
Plants Processed to ≈ 180 μm (1B3b3b1a)
Total Carbon, Nitrogen, and Sulfur (1B3b3b1a1)

Refer to the section on soil biological and plant analyses (6) for additional information on the further processing and preparation of these biological samples for laboratory analysis.

References
Velthorst, E.J. 1996. Water analysis. p. 121-242. *In* P. Buurman, B. van Lagen, and E.J. Velthorst (eds.) Manual for soil and water analysis. Backhuys Publ. Leiden.

CONVENTIONS (2)

Methods and Codes (2A)

The SSL ensures continuity in its analytical measurement process with the use of standard operating procedures (SOP's). A standard method is defined herein as a method or procedure developed by an organization, based on consensus opinion or other criteria and often evaluated for its reliability by a collaborative testing procedure (Taylor, 1988). A SOP is a procedure written in a standard format and adopted for repetitive use when performing a specific measurement or sampling operation, i.e., a SOP may be a standard or one developed by a user (Taylor, 1988).

The use of SOP's provides consistency and reproducibility in soil preparations and analyses and helps to ensure that these preparations and analyses provide results of known quality. The standard methods described in the SSIR No. 42, Soil Survey Laboratory Methods Manual, Version 4.0 replaces as a methods reference all earlier versions of the SSIR No. 42 (1989, 1992, and 1996, respectively) and SSIR No. 1, Procedures for Collecting Soil Samples and Methods of Analysis for Soil Survey (1972, 1982, and 1984). All SSL methods are performed with methodologies appropriate for the specific purpose. The SSL SOP's are standard methods, peer-recognized methods, SSL-developed methods, and/or specified methods in soil taxonomy (Soil Survey Staff, 1999). This manual also serves as the primary document from which a companion manual, Soil Survey Laboratory Information Manual (SSIR No. 45, 1995), has been developed. The SSIR No. 45 describes in greater detail the application of SSL data.

Included in the SSIR No. 42 are descriptions of current as well as obsolete methods, both of which are documented by method codes and linked with analytical results that are stored in the SSL database. This linkage between laboratory method codes and the respective analytical results is reported on the SSL data sheets. Reporting the method by which the analytical result is determined helps to ensure user understanding of SSL data. In addition, this linkage provides a means of technical criticism and traceability if data are questioned in the future.

The methods in current use at the SSL are described in the SSIR No. 42 in enough detail that they can be performed in many laboratories without reference to other sources. Descriptions of the obsolete methods are located at the back of this methods manual.

Information is not available to describe some of these obsolete procedures in the same detail as used to describe the current methods in the laboratory.

Since the publication of the SSIR No. 42 (1996), there has been a significant increase in the number and kind of methods performed at the SSL, resulting in a re-structuring of the method codes. As in past versions of the SSIR No. 42, the current method codes are hierarchical and alphanumerical. However, the older method code structure had only a maximum of four characters, e.g., 6A1b, whereas the new structure has more characters, carrying more information about the method, e.g., particle-size and sample weight bases for reporting data. This version of the SSIR No. 42 carries not only the new method codes but also the older ones as well. These older codes are cross-referenced in a table preceding the method descriptions of the obsolete SSL methods This linkage between the two method code systems is important to maintain, as many older SSL data sheets and scientific publications report these older codes.

The SSL data have been provided in reports, e.g., Primary and Supplementary Characterization Data Sheets, and in electronic forms, including tapes, disks, and CD-ROMs. More recently, other reports have been developed, e.g., Soil Taxonomy Characterization Data Sheets, and data are available via the NRCS Soils web site http://soils.usda.gov/. Historically, the SSIR No. 42 has described and assigned method codes to only those data reported on the Primary Characterization Data Sheets. This tradition is followed in this version of the SSIR No. 42. With the exception of some SSL primary analytical data included for user convenience, the data on the Supplementary and Taxonomy Characterization Data Sheets are derived data, using analytical data as a basis for calculation and do not carry method codes. Data on the Supplementary and Taxonomy Characterization Data Sheets are not described in this manual. Additionally, there are other calculated data on the Primary Characterization Data Sheets that appear as "Pedon Calculations", e.g., Weighted Average Clay. These calculated data are not assigned method codes nor described in this manual. For more detailed information about the calculation and application of these derived values, refer to the SSIR No. 45, Soil Survey Information Manual (1995) and Soil Taxonomy (Soil Survey Staff, 1999). Additional information may also be obtained from the SSL upon request.

Data Types (2B)

The methods described herein identify the specific type of analytical or calculated data. Most of these methods are analytical in nature, i.e., quantitative or semi-quantitative measurements, and include physical, chemical, mineralogical, and biological analyses. Sample collection and preparation in the field and the laboratory are also described. Historically, the SSIR No. 42 has described some derived values, e.g., coefficient of linear extensibility (COLE) and water retention difference (WRD) and reported these values along with the analytical data on the SSL Primary Characterization Data Sheets. This version of the SSIR No. 42 follows this tradition. Since the publication of the SSIR No. 42 (1996), there have been a few more derived
values added to the Primary Characterization Data Sheets. While these data have been assigned method codes, detailed descriptions are not included in this version of the SSIR No. 42. For more detailed information about the calculation and application of these derived values, refer to the SSIR No. 45 (1995) and Soil Taxonomy (Soil Survey Staff, 1999).

Size-Fraction Base for Reporting Data (2C)
Particles <2 mm (2C1)
Particles <Specified Size>2mm (2C2)

The reporting conventions for particle-size fractions for the < 2 mm and > 2mm fractions are herein designated as 2C1 and 2C2, respectively. Unless otherwise specified, all data are reported on the basis of the <2-mm material. Other size fractions reported on the Primary Characterization Data Sheets include but are not limited to the <0.4, <20, <75 mm, and the whole-soil base. The maximum coarse-fragment size for the >2-mm base varies. The base usually includes those fragments as large as 75 mm (3 in), if present in the soil. The maximum size for fragments >75 mm, commonly termed whole soil, includes boulders with maximum horizontal dimensions less than those of the pedon. The maximum particle-size set is recorded in the parentheses in the column heading. The base with which to calculate the reported >2-mm percentages includes all material in the sample smaller than the particle size recorded in the column heading.

Sample Weight Base for Reporting Data (2D)
Air-Dry/Oven-Dry (2D1)
Field-Moist/Oven-Dry (2D2)
Correction for Crystal Water (2D3)

Unless otherwise specified, all data are reported on an oven-dry weight or volume basis for the designated particle-size faction. The calculation of the air-dry/oven-dry (AD/OD) ratio is used to adjust AD results to an OD weight basis and, if required in a procedure, to calculate the sample weight that is equivalent to the required OD soil weight. The AD/OD ratio is calculated by procedure 3D1. The AD/OD ratio is converted to a crystal water basis by procedure 3D3 for gypsiferous soils (Nelson et al., 1978). The calculation of the field-moist/oven-dry (FM/OD) ratio is used to adjust FM results to an OD weight basis, and if required in a procedure, to calculate the sample weight that is equivalent to the required OD soil weight. The FM/OD ratio is calculated by procedure 3D2.

AD and OD weights are defined herein as constant sample weights obtained after drying at 30±5°C (≈ 3 to 7 days) and at 110±5°C (≈ 12 to 16 h), respectively. As a rule of thumb, air-dry soils contain about 1 to 2 percent water and are drier than soils at 1500-kPa water content. FM weight is defined herein as the sample weight obtained without drying prior to laboratory analysis. In general, these weights are reflective of the water content at the time of sample collection.

Significant Figures and Rounding (2E)

Unless otherwise specified, the SSL uses the procedure of significant figures to report analytical data. Historically, significant figures are said to be all digits that are certain plus one, which contains some uncertainty. If a value is reported as 19.4 units, the 0.4 is not certain, i.e., repeated analyses of the same sample would vary more than one-tenth but generally less than a whole unit.

Data Sheet Symbols (2F)

The analytical result of "zero" is not reported by the SSL. The following symbols are used or have been used for trace or zero quantities and for samples not tested.

tr, Tr, TR Trace, either is not measurable by quantitative procedure used or is less than reported amount.

tr(s) Trace, detected only by qualitative procedure more sensitive than quantitative procedure used.

- Analysis run but none detected.

-- Analysis run but none detected.

-(s) None detected by sensitive qualitative test.

blank Analysis not run.

nd Not determined, analysis not run.

< Either none is present or amount is less than reported amount, e.g., <0.1 is in fact <0.05 since 0.05 to 0.1 is reported as 0.1.

References

Soil Conservation Service. 1972. Soil survey laboratory methods and procedures for collecting soil samples. USDA-SCS. Soil Survey Investigations Report No. 1. U.S. Govt. Print. Office, Washington, DC.

Soil Conservation Service. 1982. Soil survey laboratory methods and procedures for collecting soil samples. USDA-SCS. Soil Survey Investigations Report No. 1. U.S. Govt. Print. Office, Washington, DC.

Soil Conservation Service. 1984. Soil survey laboratory methods and procedures for collecting soil samples. USDA-SCS. Soil Survey Investigations Report No. 1. U.S. Govt. Print. Office, Washington, DC.

Soil Survey Staff, 1989. Soil survey laboratory methods manual. Version No. 1.0. USDA-NRCS. Soil Survey Investigations Report No. 42. U.S. Govt. Print. Office, Washington, DC.

Soil Survey Staff, 1992. Soil survey laboratory methods manual. Version No. 2.0. USDA-NRCS. Soil Survey Investigations Report No. 42. U.S. Govt. Print. Office, Washington, DC.

Soil Survey Staff. 1995. Soil survey laboratory information manual. Version No. 1.0. USDA-NRCS. Soil Survey Investigations Report No. 42. U.S. Govt. Print. Office, Washington, DC.

Soil Survey Staff. 1996. Soil survey laboratory methods manual. Version No. 3.0. USDA-NRCS. Soil Survey Investigations Report No. 42. U.S. Govt. Print. Office, Washington, DC.

Soil Survey Staff. 1999. Soil taxonomy: A basic system of soil classification for making and interpreting soil surveys. 2nd ed. Agric. Handb. No. 436. USDA-NRCS. U.S. Govt. Print. Office, Washington, DC.

SOIL PHYSICAL AND FABRIC-RELATED ANALYSES (3)

Particle-Size Distribution Analysis (3A)
Particles <2 mm (3A1)
Pipet Analysis (3A1a)

1. Application

General: One of the most requested SSL characterization analysis is particle-size distribution analysis (PSDA). The behavior of most soil physical and many chemical properties are sharply influenced by the particle-size distribution classes present and their relative abundance. Precise meaning is given to the term soil texture only through the concept of particle-size distribution (Skopp, 1992).

Particle-size distribution analysis is a measurement of the size distribution of individual particles in a soil sample. These data may be presented on a cumulative PSDA curve. These distribution curves are used in many kinds of investigations and evaluations, e.g., geologic, hydrologic, geomorphic, engineering, and soil science (Gee and Bauder, 1986). In soil science, particle-size is used as a tool to explain soil genesis, quantify soil classification, and define soil texture.

In the USDA classification system (Soil Survey Staff, 1953, 1993), soil texture refers to the relative proportions of clay, silt, and sand on a <2-mm basis. It also recognizes proportions of five subclasses of sand. In addition to the USDA soil classification scheme, there are other classification systems, e.g., the particle-size classes for differentiation of families in soil taxonomy (Soil Survey Staff, 1999); International Union of Soil Science (IUSS); the Canadian Soil Survey Committee (CSSC); and American Society for Testing and Materials (ASTM). In reporting and interpreting data, it is important to recognize that these other classification systems are frequently cited in the literature, especially engineering systems, e.g., American Association of State Highway and Transportation Officials (AASHTO) and the ASTM Unified Soil Classification System (USCS) (Gee and Bauder, 1986).

Standard SSL PSDA (3A1a1): The procedure described herein covers the destruction or dispersion of <2-mm diameter soil aggregates into discrete units by chemical, mechanical, or ultrasonic means, followed by the separation or fractionation of these particles according to size limits through sieving and sedimentation (Gee and Bauder, 1986). Upon isolation of these particle-sizes or size increments, the amount of each size-fraction is then gravimetrically measured as a percent of the total sample weight on an oven-dry basis. The Kilmer and Alexander (1949) pipet method was chosen by the USDA Soil Conservation Service because it is reproducible in a wide range of soils.

The SSL routinely determines the soil separates of total sand (0.05 to 2.0 mm), silt (0.002 to 0.05 mm), and clay (< 2 μm), with five subclasses of sand (very coarse, coarse, medium, fine, and very fine) and two subclasses of silt (coarse and fine). The coarse silt is a soil separate with 0.02 to 0.05-mm particle diameter. The 0.02-mm (20 μm) is the break between sand and silt in the International Classification system. The particle-size separation at 20 μm also has significance in optical microscopy, as this class limit represents the optical limits of the polarizing light microscope. Particle-size classes are a compromise between engineering and pedologic classes (Soil Survey Staff, 1999). In engineering classifications, the limit between sand and silt is a 0.074-mm diameter. The break between sand and silt is 0.05 and 0.02 mm in the USDA and International classification systems, respectively. In engineering classes, the very fine sand (VFS) separate is split. In particle-size classes of soil taxonomy (Soil Survey Staff, 1999), the VFS is allowed to *float*, i.e., the VFS is treated as sand if the texture is fine sand, loamy fine sand, or a coarser class and is treated as silt if the texture is very fine sand, loamy very fine sand, sandy loam, silt loam, or a finer class (Soil Survey Staff, 1999). Refer to the Soil Survey Staff (1999) for additional discussion on particle-size classes.

Fine-Clay and Carbonate Clay: In addition to the routine soil separates of sand, silt, and clay, the SSL determines the fine-clay and/or carbonate clay fractions, depending on analytical requests and properties of the sample. The fine-clay fraction consists of mineral soil particles with an effective diameter of <0.2 μm. The percentage fine clay is determined for soils that are suspected of having illuviated clay. Fine-clay data can be used to determine the presence of argillic horizons or as a tool to help explain soil genesis. The carbonate-clay fraction is considered important in PSDA because clay-size carbonate particles have properties that are different from non-carbonate clay. The cation exchange capacity of carbonate clay is very low compared to non-carbonate clay. Water holding capacity of carbonate clay is ≈ two thirds that of non-carbonate clay. Since carbonate clay is a diluent, it is often subtracted from the total clay in order to make inferences about soil genesis and clay activities.

Pretreatment and Dispersion Techniques: The phenomena of flocculation and dispersion (deflocculation) are very important in determining the physical behavior of the colloidal fraction of soils and thus indirectly, have a major bearing on the physical properties which soils exhibit (Sumner, 1992). In the standard SSL procedure (3A1a1), soils are pretreated to remove organic matter and soluble salts. Samples are chemically treated with hydrogen peroxide and sodium hexametaphosphate to effect dispersion. The primary objectives of dispersion are the removal of cementing agents; re-hydration of clays; and the physical separation of individual soil particles (Skopp, 1992). The hydrogen peroxide oxidizes the organic matter. The sodium hexametaphosphate complexes any calcium in solution and replaces it with sodium on the exchange complex. Upon completion of the chemical treatments, mechanical agitation through shaking is used to enhance separation of particles and facilitates fractionation.

Complete dispersion by procedure 3A1a1 may be prevented in the presence of cementing agents such calcium carbonate, Fe, and Si. In these instances, special pretreatment and dispersion procedures (3A1a2-5) may be performed upon request by the project coordinator on the sample as follows:

Carbonate Removal (3A1a2): Soils high in carbonate content do not readily disperse. Pretreatment of these soils with acid removes the carbonates (Grossman and Millet, 1961; Jackson, 1969; Gee and Bauder, 1986; Gee and Or, 2002). The determination of particle-size distribution after the removal of carbonates is used primarily for studies of soil genesis and parent material.

Iron Removal (3A1a3): Iron and other oxides coat and bind particles of sand, silt, and clay form aggregates. Soils with iron cementation do not readily disperse. The iron oxides are removed using bicarbonate-buffered, sodium dithionite-citrate solution (Mehra and Jackson, 1960; Gee and Bauder, 1986; Gee and Or, 2002).

Silica Removal (3A1a4): Soils that are cemented by Si do not completely disperse with hydrogen peroxide pretreatment and sodium hexametaphosphate. A pretreatment with a weak base dissolves the Si bridges and coats and increases the soil dispersion. The determination is used for soil parent material and genesis studies.

Ultrasonic Dispersion (3A1a5): Soils that do not completely disperse with standard PSDA can be dispersed using ultrasonic dispersion (Gee and Bauder, 1986; Gee and Or, 2002). Pretreatments coupled with ultrasonic dispersion yield maximum clay concentrations (Mikhail and Briner, 1978). This is a developmental procedure as no standard method has been adopted using ultrasonic dispersion.

Water Dispersible PSDA (3A1a6): This method provides a means of evaluating the susceptibility of a soil to water erosion. The degree to which a soil disperses without the oxidation of organic matter or the removal of soluble salts, or the addition of a chemical dispersant may be compared with results from chemical dispersion (Bouyoucos, 1929).

Field-Moist PSDA (3A1a1b): The standard SSL procedure for particles with <2-mm diameter is the air-dry method (3A1a1a). While a homogenized sample is more easily obtained from air-dry material than from moist material, some soils irreversibly harden when dried and as such moist PSDA (3A1a1b) may be used. The phenomenon of aggregation through oven or air-drying is an important example of irreversibility of colloidal behavior in the soil-water system (Kubota, 1972; Espinoza et al. 1975). Drying such soils decreases the measured clay content. This can be attributed to the cementation upon drying (Maeda et al., 1977). The magnitude of the effect varies with the particular soil (Maeda et al., 1977).

SSL PSDA Process: In the SSIR No. 42 (1996), stand-alone methods were described for the non-routine pretreatment and dispersion techniques as well as for the analysis of particles not routinely reported, e.g., fine and/or carbonate-clay fractions. In this version of the SSIR No. 42 (2004), these procedures are described more as a procedural process. This approach is appropriate in that certain procedural steps may be modified, omitted, or enhanced by the investigator, depending on the properties of the sample and on the requested analyses. The process by which specific procedural steps are selected for sample analysis is based upon knowledge or intuition of certain soil properties or related to specific questions, e.g., special studies of soil genesis and parent material.

2. Summary of Method

Standard SSL PSDA: The standard SSL procedure for analysis of particles with <2-mm diameters is the air-dry method (procedure 3A1a1a), whereby a 10-g sample is pretreated to remove organic matter and soluble salts. The sample is dried in the oven to obtain the initial weight, dispersed with a sodium hexametaphosphate solution, and mechanically shaken. The sand fraction is removed from the suspension by wet sieving and then fractionated by dry sieving. The clay and fine silt fractions are determined using the suspension remaining from the wet sieving process. This suspension is diluted to 1 L in a sedimentation cylinder, stirred, and 25-mL aliquots removed with a pipet at calculated, predetermined intervals based on Stokes' law (Kilmer and Alexander, 1949). The aliquots are dried at 110°C and weighed. Coarse silt is the difference between 100% and the sum of the sand, clay, and fine silt percentages.

Fine-Clay Determination: The soil suspension from procedure 3A1a1a is used to determine the fine-clay fraction. This suspension is stirred, poured into a centrifuge bottle, and centrifuged at 1500 rpm. A 25-mL aliquot is withdrawn with a pipet. The aliquot is dried in the oven, weighed, and the percentage of fine clay is calculated based on the total sample weight. The time of centrifugation is determined from the following equation modified from Stokes' law (Jackson, 1969).

$$t_m = (63.0 \times 10^8 \eta \, \log \, (rs^{-1})) \, (N_m^2 \, D\mu^2 \, \Delta\rho)^{-1}$$

where:

t_m = Time in minutes
η = Viscosity in poises
r = Radius in cm from center of rotation to sampling depth (3 cm + s)
s = Radius in cm from center of rotation to surface of suspension
N_m = rpm (1500)
$D\mu$ = Particle diameter in microns (0.2 μm)
$\Delta\rho$ = Difference in specific gravity between solvated particles and suspension liquid
63.0×10^8 = Combination of conversion factors for convenient units of time in minutes, t_{min}, N_m
          as rpm, and particle diameter in microns, $D\mu$.

Carbonate-Clay Determination: The residue from procedure 3A1a1a is used to determine the carbonate-clay fraction. This residue is treated with acid in a closed system. The pressure of the evolved gas is measured. The pressure is related linearly to the CO_2 content in the carbonates. A manometer is used to measure the pressure.

Pretreatment and Dispersion Techniques: In the standard PSDA, an air-dry soil sample is pretreated to remove organic matter and soluble salts (3A1a1a). There are additional non-routine chemical pretreatments for the removal of cementing agents that often prevent complete dispersion. These pretreatments may be requested by the project coordinator as follows:

Carbonate Removal (3A1a2a): Carbonates are destroyed with a 1 N NaOAc solution buffered to pH 5. The NaOAc solution is added to sample until carbonate bubbles no longer evolve. The NaOAc solution is then washed from the sample. After destruction of carbonates, the remainder of procedure 3A1a1 is followed.

Iron Removal (3A1a3a): Soil samples are pretreated with H_2O_2 to remove organic matter. Iron oxides are removed with bicarbonate-buffered, sodium dithionite-citrate solution and heated until the sample color changes to a grayish color. The suspension is flocculated with saturated NaCl solution and filtered to remove soluble salts. After removal of iron oxides, the remainder of procedure 3A1a1 is followed.

Silica Removal (3A1a4a): Soils are pretreated with H_2O_2 to remove organic matter. Soils with Si cementation or coatings are pretreated with a weak NaOH solution overnight. After removal of siliceous cementing agents, the remainder of procedure 3A1a1 is followed.

Ultrasonic Dispersion (3A1a5a): A soil sample is pretreated to remove organic matter and soluble salts. The sample is dried in the oven and weighed to obtain the initial weight. Sodium hexametaphosphate solution is added to the sample and then made to 100-mL volume with RO water. The sample is subjected to ultrasonic vibration for 5 min. After dispersion with the ultrasonic probe, procedure 3A1a1 is followed.

Water Dispersible PSDA (3A1a6a): Water dispersible particle-size distribution analysis may also be determined from a soil suspension without the removal of organic matter or soluble salts, or without the use of a chemical dispersant. Upon omitting these procedural steps, the remainder of procedure 3A1a1 is followed.

SSL Field-moist PSDA: For soils that irreversibly harden when dried, moist particle-size analysis (procedure 3A1a1b) may be requested by project coordinator. This procedure requires two 10-g samples of <2-mm, moist soil to be pretreated to remove organic matter and soluble salts. One sample is dried in the oven to obtain the oven-dry weight, and the other sample is further processed as outlined in procedure 3A1a1a. Similar to an air-dry sample, a field-moist soil may also be subjected to special pretreatment and dispersion techniques

(3A1a2-6b). Both the air-dry and moist PSDA data are determined as percent of the <2-mm fraction on an oven-dry basis.

3. Interferences

Standard SSL PSDA: The sedimentation equation that is used to measure the settling rates of particles of different sizes is as follows:

$$v = 2r^2g(\rho_s-\rho_l)/(9\eta)$$

where:
v = Velocity of fall
r = Particle radius
g = Acceleration due to gravity
ρ_s = Particle density
ρ_l = Liquid density
η = Fluid viscosity

This formula results from an application of Stokes' law and is referred to as Stokes' law. Assumptions used in applying Stokes' law to soil sedimentation measurements are as follows:

1. Terminal velocity is attained as soon as settling begins.

2. Settling and resistance are entirely due to the viscosity of the fluid.

3. Particles are smooth and spherical.

4. There is no interaction between individual particles in the solution (Gee and Bauder, 1986; Gee and Or, 2002).

Since soil particles are not smooth and spherical, the radius of the particle is considered an equivalent rather than an actual radius. In this method, particle density is assumed to be 2.65 g cc[-1].

Gypsum interferes with PSDA by causing flocculation of particles. Gypsum is removed by stirring and washing the soil with reverse osmosis water. This procedure is effective if the soil contains <25% gypsum.

Partial flocculation may occur in some soils if excess H_2O_2 is not removed from the soil after its use in organic matter oxidation.

Treatment of micaceous soils with H_2O_2 causes exfoliation of the mica plates and a matting of particles when dried in the oven. Since exfoliation occurs in these soils, a true measurement of fractions is uncertain (Drosdoff and Miles, 1938).

Fine-Clay Determination: In the fine-clay determination, the distance from the center of rotation to the surface of the suspension must be constant for each centrifuge bottle. The particle density (ρ_p) of the fine clay is assumed to be 2.5 g cc[-1] (Jackson, 1969). The suspension temperature must be used to enter the correct liquid viscosity in the equation. Position the bottle under pipet without sudden movement of the centrifuge rotor, which causes disturbance of solution. The withdrawal rate with pipet should be constant.

Carbonate-Clay Determination: The carbonate-clay analysis is semi-quantitative. It is assumed that all of the carbonates are converted to CO_2. This method measures all forms of carbonates. In addition to Ca, the carbonates of Mg, Na, and K also react with the acid. Analytical interferences may be caused by temperature changes within the reaction vessel. The analyst should not touch the glass of the vessel when reading the pressure. When sealing the vessel, the analyst should not hold onto the vessel any longer than necessary to tighten the cap. The internal pressure must be equalized with the atmosphere. Approximately 3 to 5 s are required to equalize the internal pressure of the bottle when piercing the septa with a needle. The analyst should replace septa and O rings at regular intervals, as they develop leaks after extensive use.

Pretreatment and Dispersion Techniques: The PSDA results are dependent on the pretreatments used to disperse the soil. The presence of cementing agents such as carbonates, Fe, and Si often prevent complete dispersion. In these cases, special pretreatment and dispersion procedures may be performed upon request on either an air-dry or field-moist sample. However, these special techniques in themselves may interfere with PSDA as follows:

Carbonate Removal (3A1a2): The removal of carbonates with 1 N NaOAc (pH 5) results in sample acidification. This pretreatment can destroy the primary mineral structure of clay (Gee and Bauder, 1986).

Iron Removal (3A1a3): If the temperature of the water bath exceeds 80° C during Fe removal, elemental S can precipitate (Mehra and Jackson, 1960). This pretreatment can destroy primary mineral grains in the clay fraction (El-Swaify, 1980).

Silica Removal (3A1a4): The effects of Si removal with 0.1 N NaOH on the clay fraction and particle-size distribution are unknown.

Ultrasonic Dispersion (3A1a5): Ultrasonic dispersion has been reported to destroy primary soil particles. Watson (1971) summarized studies that reported the destruction of biotite and breakdown of microaggregates by ultrasonic dispersion. However, Saly (1967) reported that ultrasonic vibration did not cause the destruction of the clay crystalline lattice or the breakdown of primary grains. The samples ranged from sandy to clayey soils. The cementing agents represented humus, carbonates, and hydroxides of Fe and Al. No standard procedures have been adopted using ultrasonic dispersion.

Field-Moist PSDA: Soils that irreversibly harden when dried are difficult to disperse. The PSDA for these soils can be determined on moist samples (procedure 3A1a1b) upon the request of the project coordinator.

4. Safety

Wear protective clothing (coats, aprons, sleeve guards, and gloves) and eye protection (face shields, goggles, or safety glasses) when handling acid and H_2O_2. Mix acids in ventilated fume hoods. Use sodium bicarbonate and water to neutralize and dilute spilled acids. Heat samples in ventilated fume hoods for removal of organic matter or cementing agents in ventilation hoods. Handle heated samples with leather gloves. Perform the transfer of acid to gelatin capsules near a sink in case of leakage or spills.

Users should be familiar with centrifuge operation. Opposite centrifuge bottles need to be balanced. Centrifuge should not be opened until centrifuge rotor has completely stopped

5. Equipment
5.1 Fleakers, 300 mL, tared to 1 mg
5.2 Ceramic filter candles, .3 μm absolute retention (source currently unavailable)
5.3 Rack to hold ceramic filter candle and sample container.
5.4 Mechanical shaker, horizontal, 120 oscillations min^{-1}, 1 ½ in strokes, Eberbach 6000, Eberbach Corp., Ann Arbor, MI
5.5 Cylinders, 1 L, white line fused onto glass at 1-L mark
5.6 Oven, 110°C
5.7 Hot plate, 100°C
5.8 Vacuum, 0.8 bars (80kPa)
5.9 Thermometer, 0 to 150°C
5.10 Desiccator
5.11 Motor driven stirrer, (Kilmer and Mullins, 1954)
5.12 Hand stirrer, perforated disk fastened to a rod
5.13 Adjustable pipet rack (Shaw, 1932; Fig.1-3)
5.14 Lowy pipets, 25 mL, with overflow bulb
5.15 Polyurethane foam, pipe insulation that fits snugly around cylinder.

5.16 Sieve shaker with 12.7-mm (1/2 in) vertical and lateral movement at 500 oscillations min⁻¹. Accommodates a nest of 76-mm (3 in) sieves.

5.17 Weighing bottles, 90 mL, with screw caps, tared to 1 mg

5.18 Weighing bottles, 90 mL, tared to 1 mg

5.19 Weighing bottles, 90 mL, tared to 0.1 mg

5.20 Drying dishes, aluminum

5.21 Timer or clock with second hand

5.22 Electronic balance, ±0.10-mg sensitivity

5.23 Electronic balance, ±1.0-mg sensitivity

5.24 Watch glass, 50- and 65-mm diameters

5.25 Evaporating dish, porcelain, 160-mm diameter, 31-mm height, with lip

5.26 Set of 76-mm (3 in) sieves, square weave phosphor bronze wire cloth except 300 mesh which is twilled weave. U.S. series and Tyler Screen Scale equivalent designations are as follows:

Sand Size	Opening (mm)	U.S. No.	Tyler Mesh Size
VCS	1.0	18	16
CS	0.5	35	32
MS	0.25	60	60
FS	0.105	140	150
VFS	0.047	300	300

5.27 Centrifuge, International No. 11, with No. 949 rotor head, International Equip. Co., Boston, MA

5.28 Centrifuge bottle, 500 mL

5.29 Torsion balance

5.30 Manometer, hand-held gauge and differential pressure, PCL-200A/C Series, Omega Engineering, Stamford, CT.

5.31 Gelatin capsules, 5 mL

5.32 Threaded weighing bottles, 90 mL

5.33 Machined PVC caps for threaded 90-mL weighing bottles, 3.2-cm (1 1/4 in) diameter with 1.1-cm (7/16 in) diameter hole drilled in center, O-ring seal

5.34 O-rings, 3.2 x 38.1 mm (1/8 x 1 1/2 in)

5.35 Septa, rubber, 7.9-mm (5/16 in) diameter. Place in machined cap.

5.36 Hypodermic needle, 25.4 mm (1 in), 23 gauge

5.37 Ultrasonic probe, 19-mm (3/4 in) horn, 20 kHz, 300 watts

6. Reagents

6.1 Reverse osmosis (RO) water, ASTM Type III grade of reagent water

6.2 Hydrogen peroxide (H_2O_2), 30 to 35%

6.3 Sodium hexametaphosphate (($NaPO_3$)$_6$), reagent grade

6.4 Sodium carbonate (Na_2CO_3), reagent grade

6.5 Sodium hexametaphosphate solution. Dissolve 35.7 g of ($NaPO_3$)$_6$ and 7.94 g of Na_2CO_3 in 1 L of RO water. See Section 7.12 for standardization of sodium hexametaphosphate solution.

6.6 Ethyl alcohol

6.7 Calcium sulfate (anhydrous) or equivalent desiccant

6.8 Hydrochloric acid (HCl), 6 N, technical grade. Dilute 1 L of concentrated HCl with 1 L of RO water

6.9 Sodium carbonate (Na_2CO_3) reagent. Dissolve 10.6 g Na_2CO_3 in RO water and make to 1 L (10 mg $CaCO_3$

6.10 1 N sodium acetate (NaOAc) solution, buffered to pH 5. Dissolve 680 g of NaOAc in 4 L of RO water. Add $\approx$ 250 mL of acetic acid. Make to 5-L volume with RO water.

6.11 Sodium citrate solution, 0.3 M $Na_3C_6H_5O_7 \cdot 2H_2O$ (88.4 g L^{-1})

6.12 Sodium bicarbonate buffer solution, 1 M $NaHCO_3$ (84 g L^{-1})

6.13 Sodium dithionite ($Na_2S_2O_4$ - hydrosulphite)

6.14 Saturated NaCl solution (solubility at 20°C; 360 g L^{-1})

6.15 Sodium hydroxide solution (NaOH), 0.1 N. Dissolve 4 g NaOH pellets in 1 L of RO water.

7. Procedures

Particle-Size Distribution Analysis (3A)
Particles <2 mm (3A1)
Pipet Analysis (3A1a)

7.1 If the standard air-dry PSDA procedure for the <2-mm fraction is requested, proceed to procedure 3A1a1a.

7.2 If PSDA is requested on a field-moist sample, proceed to procedure 3A1a1b.

7.3 If special pretreatment and dispersion techniques (removal of carbonate, Fe, or Si; ultrasonic dispersion; or water dispersible) are requested on air-dry soil samples, proceed to procedures 3A1a2-6a, respectively.

7.4 If special pretreatment and dispersion techniques (removal of carbonate, Fe, or Si; ultrasonic dispersion; or water dispersible) are requested on field-moist soil samples, proceed to procedures 3A1a2-6b, respectively.

Particle-Size Distribution Analysis (3A)
Particles <2 mm (3A1)
Pipet Analysis (3A1a)
Standard Pretreatments and Dispersion (3A1a1)
Air-Dry (3A1a1a)

7.5 Weigh 10 g of <2-mm, air-dry soil to nearest mg on an electronic balance and place into a numbered, tared 300-mL, fleaker. Wash and tare these fleakers once every two months. A quality control sample is included in each batch (≤24 samples).

7.6 Add $\approx$ 50 mL of RO water and 5 to 7.5 mL of H_2O_2 to the soil sample at ambient temperature. Cover the soil sample with a 50-mm watch glass. Allow initial oxidation of organic matter to complete and then place sample on hot plate. If the froth from the reaction exceeds the capacity of the fleaker, transfer the sample to a larger beaker.

7.7 Place the sample on a hot plate and heat to 90°C. Add four 5 to 7.5-mL increments of H_2O_2 at 30-min intervals. If oxidation is incomplete, add additional H_2O_2 until organic matter oxidation is complete. Heat the sample for an additional 45 min to decompose excess H_2O_2. If the reaction is violent, do one or any combination of the following: (a) add small increments of ethyl alcohol to the sample; (b) remove the sample from the hot plate to slow the reaction; (c) transfer sample to a 1000-mL beaker; or (d) reduce the amount of H_2O_2 to sample. Record any unusual sample reactions.

7.8 Place the sample on the filter rack. Add 150 mL of RO water. Insert a filter candle, connect to the vacuum trap assembly with tubing, and turn on the vacuum. Aspirate until liquid is removed and only slightly dampened sample remains. Wash the sample four additional times with ≈ 150 mL of RO water. Stir the sample with filter candle to ensure all soil particles will be rinsed. During aspiration, it may be necessary to occasionally apply back-pressure to filter candle and remove build-up of soil which inhibits aspiration. If the sample contains gypsum and flocculates, then the following additional washings may be used. If the sample contains 1 to 5% gypsum, stir the sample with a magnetic stirrer for 5 min and wash 5 times with ≈ 250 mL of RO water each time. If the sample contains >5% gypsum, place the sample in a 1000-mL beaker and stir the sample with a magnetic stirrer for 5 min then wash 5 times with ≈ 750 mL of RO water each time to remove soluble gypsum.

7.9 Place sample in oven. Dry the sample overnight at 110°C. Remove the sample from the oven, place in a desiccator, and cool to ambient temperature.

7.10 Record the total weight (TW) of the sample to the nearest mg.

7.11 Add the exact volume of sodium hexametaphosphate solution (≈ 10 mL), equivalent to 0.4408 g of sodium hexametaphosphate, to each sample. Subtract the weight of the sodium hexametaphosphate (DW) that is contained in the extracted aliquot from the silt and clay weights to calculate silt and clay percentages. To determine the exact volume of sodium hexametaphosphate to add to each sample, refer to Section 7.12. Let stand until sample is completely moistened by sodium hexametaphosphate. Add ≈ 175 mL of RO water.

7.12 A sodium hexametaphosphate standardization is performed with each new batch of solution. Use only designated weighing bottles for standardization. Wash and tare these bottles after each standardization. Add duplicate aliquots (8.5, 9.0, 9.3, 9.6, 10.0, 10.3, 10.6, 11.0 mL) of sodium hexametaphosphate solution to numbered tared, 90-mL weighing bottles. Oven-dry aliquots overnight and record dry residue weight of sodium hexametaphosphate. Determine the exact volume of sodium hexametaphosphate to add to each sample by regressing the volume of sodium hexametaphosphate against the dry residue weight of sodium hexametaphosphate and then by predicting the volume needed to dispense 0.4408 g of sodium hexametaphosphate into each sample.

7.13 Place the sample in a horizontal shaker set at 120 oscillations min^{-1} and shake for 15 h (overnight).

7.14 Remove the sample from the shaker and pour through a 300-mesh (0.047-mm) sieve mounted on a ring stand. Finger-rubbing of sample may be required during transfer to speed washing of sample. Place a funnel below the sieve and a 1-L cylinder below the funnel. Collect the silt and clay in the 1-L cylinder. Avoid using jets of water in washing the sample. Wash and rub all particles from the fleaker into the sieve. Continue to wash until the suspension volume in the cylinder is ≈ 800 mL. Sand and some of the coarse silt remain on the sieve. Rinse all <20-μm particles into the cylinder. Fill the cylinder to 1 L and cover with a 65-mm watch glass. Place pipe insulation around sample and blank cylinders to prevent rapid changes in temperatures. Prepare a RO water blank to measure temperature fluctuations. Allow the cylinder to stand overnight to equilibrate the suspension with the room temperature. Wash the sand into an evaporation dish and dry the sand at 110°C overnight.

7.15 Transfer the dried sand to a nest of sieves that has a top-to-bottom order of 1.0, 0.5, 0.25, 0.1, and 0.047 mm. Shake the sand for 3 min on a shaker that has 1.3-cm vertical and lateral movements and oscillates at 500 strokes min^{-1}. Record the weight of each separate sand fraction (SW$_i$) to the nearest mg. If optical analysis is requested, place the very fine sand and fine sand fractions in gelatin capsules and the remaining sand fractions in a labeled vial. Store capsules in the labeled vial. Wash sand dishes after every use.

7.16 Determine the percentage of fine silt and clay gravimetrically by removing an aliquot from the suspension in the 1-L cylinder with a Lowy, 25-mL pipet. Periodically, gravimetrically calibrate the delivery volume of the pipet by weighing the amount of RO water dispensed from the pipet. Record the delivery volume (DV) and use the value to calculate the results. Regulate the vacuum such that the pipet fills in ≈ 12 s. Record temperature (T_1) of blank. Mount the pipet on an adjustable pipet rack (Shaw, 1932). Stir the silt and clay suspension with mechanical stirrer for at least 5 min. Place the cylinder on a stable, vibrationless table and stir with a hand stirrer in an up-and-down motion for 30 s. Timing is started upon completion of the stirring. Record the time that stirring is stopped. For the <20-µm fraction, slowly lower the closed pipet to a 10-cm depth in the suspension, turn on the vacuum, and withdraw an aliquot at the calculated time (Table 1). Dispense the aliquot into a tared and numbered, 90-mL weighing bottle. Rinse the pipet twice with RO water and dispense into the tared, weighing bottle with the aliquot. For the <2-µm fraction, pipet after a time of 4.5, 5, 5.5, or 6.5 h. Record temperature (T_2) of blank. Use the average of T_1 and T_2 and adjust the pipet depth in the suspension as indicated in Table 2. Repeat the procedure described for the <20-µm fraction. If determination of carbonate is required, use weighing bottle with screw threads. Dry the aliquots at 110°C overnight and cool in a desiccator that contains calcium sulfate or an equivalent desiccant. Record the weight of the residue (RW) to the nearest 0.1mg.

7.17 Use the 90-mL, round-bottomed, weighing bottles for the <20-µm aliquots. Wash and tare after every fourth use. Use the 90-mL, square-bottomed, weighing bottles for the <2-µm aliquots. Wash and tare after every use.

7.18 If optical mineralogy, fine-clay and/or carbonate clay determinations are not requested, the procedural aspects of the standard air-dry PSDA for the <2-mm fraction are complete. If optical mineralogy is requested, save the sediment and proceed on with the section on optical mineralogy. If fine-clay and/or carbonate clay are requested, proceed on with the sections on these analyses. .

Optical Mineralogy
7.19 If optical mineralogy is requested, decant the suspension and transfer the sediment to a 400-mL beaker. Fill the beaker to a 5.5-cm height. Stir the sediment and allow to settle for 5 min. Discard the supernatant. Refill the beaker to 5.5-cm height. Stir again, allow to settle for 3 min, and then decant. Repeat the filling and the stirring; allow to settle for 2 min; and then decant until top half of suspension is clear. Transfer the sediment, which is dominantly 20 to 50µm, to a labeled drying dish. Wash with ethanol, air-dry, and save in the drying dish for optical mineralogy.

Fine Clay Determination (<0.2 µm)
7.20 Stir the silt and clay suspension with mechanical stirrer for 5 min. Remove sample from mechanical stirrer and place on table. Stir with the hand stirrer in an up-and-down motion for 30 s and allow the suspension to settle for 15 min.

7.21 Pour the suspension into a centrifuge bottle and fill to the line marked on the bottle. The marked line on each bottle is 13 cm which is the distance from the center of rotation to the surface of the suspension. Stopper and shake well to mix the suspension.

7.22 Balance opposite centrifuge loads, which consist of centrifuge bottle, trunnion carrier and bucket. Place loads on a torsion balance and add water to the lighter bucket until both loads weigh the same.

7.23 Read the temperature of the suspension.

7.24 Centrifuge at 1500 rpm.

7.24.1 For procedures 3A1a1-5a and 3A1a1-5b, vary the centrifuge time according to the temperature as follows:

Temp (°C)	Viscosity η	Delta-Density $\Delta\rho$	Time Min
18	0.01055	1.501	39.0
19	0.01029	1.501	38.0
20	0.01004	1.502	37.1
21	0.00980	1.502	36.2
22	0.00957	1.502	35.3
23	0.00934	1.502	34.5
24	0.00913	1.502	33.7
25	0.00892	1.503	32.9
26	0.00872	1.503	32.2
27	0.00853	1.503	31.4
28	0.00834	1.503	30.8
29	0.00816	1.504	30.1
30	0.00799	1.504	29.4

7.24.2 For procedures, 3A1a6a and 3A1a6b, vary the centrifuge time according to the temperature as follows:

Temp (°C)	Viscosity η	Delta-Density $\Delta\rho$	Time Min
18	0.01053	1.501	38.9
19	0.01028	1.502	37.9
20	0.01002	1.502	37.0
21	0.00978	1.502	36.1
22	0.00955	1.502	35.2
23	0.00933	1.502	34.4
24	0.00911	1.503	33.6
25	0.00890	1.503	32.8
26	0.00871	1.503	32.1
27	0.00851	1.503	31.4
28	0.00833	1.504	30.7
29	0.00815	1.504	30.0
30	0.00798	1.504	29.4

where:
s = 15 cm
r = 18 cm
N_m = 1500 rpm
ρ_p = 2.5 g cc^{-1}

7.25 After centrifuging, lower the pipet to a 3-cm depth in the suspension. Withdraw a 25-mL aliquot at a rate of $\approx$ 12 s. Avoid turbulence. Transfer the aliquot to a weighing bottle

7.26 Place weighing bottle with aliquot in oven. Dry overnight at 110°C. Remove sample from oven, place in desiccator with calcium sulfate or equivalent desiccant, and cool to ambient temperature.

7.27 Weigh residue weight (RW) to nearest 0.1 mg.

7.28 Use the 90-mL, round-bottomed, weighing bottles for the <0.2μm aliquots. Wash and tare after every fourth use.

Carbonate Clay (<2 μm): Manometer Calibration

7.29 Calibrate the manometer quarterly or whenever equipment changes. Calibrate by placing replicated aliquots of 0.0 to 20.0 mL of the Na_2CO_3 reagent into numbered, tared, 90-mL weighing bottles. Dry the standard samples in the oven overnight at 110°C. Remove samples from oven, place in desiccator and cool to ambient temperature. Record the weight of the standard samples to nearest 0.1mg.

7.30 Lubricate the lip of the 90-mL, weighing bottle that contains the Na_2CO_3 with a thin film of glycerine. Dispense 3 mL of 6 N HCl into a gelatin capsule and place the top on the capsule. If HCl leaks from the capsule, discard the capsule. Place the capsule into the glass bottle and immediately cap the bottle. Release pressure in the bottle by piercing the septa with a hypodermic needle which is not connected to the manometer. Allow 3 to 5 s for internal pressure in bottle to equalize.

7.31 After the gelatin capsule has dissolved (several minutes), slowly tip the bottle and rotate it to saturate the standard sample adhering to the sides of the bottle. Avoid changing the temperature of the container by only handling the cap. Allow sample to stand for at least 30 min.

7.32 Adjust the manometer to zero before taking measurements. Insert the hypodermic needle in the septa stopper which is connected to the transducer. Measure the pressure inside the weighing bottle. Record the manometer readings (mm Hg) to the nearest whole number.

7.33 Calculate the linear regression equation, i.e., the dependent variable is the Na_2CO_3 weights (regressed or predicted values) and the independent variable is the corresponding manometer readings.

Carbonate Clay: Analysis

7.34 Determine the presence of carbonates in <2-mm soil by placing soil on a spot plate and adding two or three drops of 1 N HCl. The rate of CO_2 evolution indicates the relative amount of carbonates (1B1b2b5).

7.35 If the soil contains more than a "slight" amount of carbonates, determine the amount of carbonate clay in the <2-μm dry residue. Use the 90-mL, square-bottomed, weighing bottles for the <2-μm aliquots and carbonate determination. Wash and tare after every use.

7.36 With a thin film of glycerine, lubricate the lip of the 90-mL, weighing bottle that contains the <2-μm residue. In each analysis batch, include an empty weighing bottle as a blank. Dispense 3 mL of 6 N HCl into a gelatin capsule and place the top on the capsule. If HCl leaks from the capsule, discard the capsule. Place the capsule into the glass bottle and immediately cap the bottle. Release any pressure in the bottle by piercing the septa with a hypodermic needle that is not connected to the manometer. Approximately 3 to 5 s are required to equalize the internal pressure of the bottle.

7.37 After the gelatin capsule has dissolved (several minutes), slowly tip the bottle and rotate it to saturate the clay adhering to the sides of the bottle. Handle only the cap to avoid changing the temperature of the container. Allow sample to stand for at least 30 min.

7.38 Adjust the manometer to zero before taking measurements. Insert the hypodermic needle in septa stopper which is connected to the transducer. Measure the pressure inside the weighing bottle and record the manometer readings (MR) to the nearest whole number (mm Hg). Begin readings with the blank (BR).

7.39 Compare the sample readings with those of a standard curve prepared by measuring CO_2 evolved from a series of Na_2CO_3 aliquots with a range 0 to 200 mg.

Particle-Size Distribution Analysis (3A)
Particles <2 mm (3A1)
Pipet Analysis (3A1a)
Carbonate Removal (3A1a2)
Air-Dry (3A1a2a)

7.40 Weigh sufficient sample to yield 10 g of <2-mm, air-dry carbonate-free soil sample, e.g. if the sample contains 50% carbonates, weigh 20 g of soil. Place the <2-mm, air-dry sample into a 300-mL, tared, fleaker.

7.41 Add ≈ 200 mL of the 1 N NaOAc solution to the sample, mix with a stirring rod, and cover with a watch glass. Allow the sample to stand overnight.

7.42 Place the sample on the hot plate and heat to ≈ 90°C until bubbles are no longer evident. Do not boil. Heating accelerates reaction. Decant the solution and add more 1 N NaOAc solution. If a reaction occurs, repeat the heating procedure. Continue to decant, add NaOAc solution, and heat until all the carbonates are removed. The speed of dissolution can be increased by lowering the pH of the 1 N NaOAc solution (Rabenhorst and Wilding, 1984).

7.43 When no more carbonate bubbles are observed, insert the ceramic filter candle into the solution. Apply vacuum and candle the sample to dryness. Rinse once with 200 mL of RO water. Proceed to procedure 3A1a1a for remaining PSDA procedural steps.

Particle-Size Distribution Analysis (3A)
Particles <2 mm (3A1)
Pipet Analysis (3A1a)
Iron Removal (3A1a3)
Air-Dry (3A1a3a)

7.44 A maximum of ≈ 0.5 g of Fe_2O_3 can be dissolved in 40 mL of the citrate solution. Adjust the weight of the <2-mm, air-dry soil sample so that any one fleaker does not contain more than 0.5 g of Fe_2O_3. Split the sample into different fleakers if necessary. Total sample weight after dissolution should be ≈ 10 g. Place the sample into a tared, labeled, 300-mL fleaker.

7.45 Add ≈ 50 mL of RO water and 5 mL of H_2O_2 to the soil sample at ambient temperature. Cover the soil sample with a 50-mm watch glass. Allow initial oxidation of organic matter to complete, and then place sample on hot plate. If the froth from the reaction exceeds the capacity of the fleaker, transfer the sample to a larger beaker.

7.46 Place the sample on a hot plate and heat to 90°C. Add 5-mL increments of H_2O_2 at 45-min intervals until oxidation has completed or until 30 mL of H_2O_2 have been added. Heat the sample for an additional 45 min to decompose excess H_2O_2. If the reaction is violent, add small increments of ethyl alcohol to the sample or remove the sample from the hot plate to slow the reaction.

7.47 Add 40 mL of the citrate solution and 5 mL of the sodium bicarbonate. Heat to 80°C in a water bath, but do not exceed 80°C. Add 1 g of sodium dithionite powder with a calibrated scoop. Stir constantly with a glass rod for 1 min and then occasionally for 15 min. Add 10 mL of saturated NaCl solution and mix.

7.48 Centrifuge or candle the sample to remove the dissolved Fe_2O_3. Combine the split samples into fewer fleakers.

7.49 If the sample contains less than 0.5 g of Fe_2O_3, repeat the dissolution treatment. For samples with more than 0.5 g of Fe_2O_3, repeat the dissolution treatment two more times. Upon completion of Fe removal, proceed to procedure 3A1a1a for remaining PSDA procedural steps.

Particle-Size Distribution Analysis (3A)
Particles <2 mm (3A1)
Pipet Analysis (3A1a)
Silica Removal (3A1a4)
Air-Dry (3A1a4a)

7.50 Weigh 10 g of <2-mm, air-dry soil to nearest mg on an electronic balance and place in a 300-mL, tared fleaker.

7.51 Add ≈ 50 mL of RO water and 5 mL of H_2O_2 to soil sample at ambient temperature. Cover the soil sample with 50-mm watch glass. Allow initial oxidation of organic matter to complete, and then place sample on hot plate. If the froth from the reaction exceeds the capacity of the fleaker, transfer the sample to a larger beaker.

7.52 Place the sample on a hot plate and heat to 90°C. Add 5-mL increments of H_2O_2 at 45-min intervals until oxidation has completed or until 30 mL of H_2O_2 have been added. Heat the sample for an additional 45 min to decompose excess H_2O_2. If the reaction is violent, add small increments of ethyl alcohol to the sample or remove the sample from the hot plate to slow the reaction.

7.53 Soak the sample overnight in 100 mL of 0.1 N NaOH. Upon removal of siliceous cementing agents, proceed to procedure 3A1a1a for remaining PSDA procedural steps.

Particle-Size Distribution Analysis (3A)
Particles <2 mm (3A1)
Pipet Analysis (3A1a)
Ultrasonic Dispersion (3A1a5)
Air-Dry (3A1a5a)

7.54 Weigh 10 g of <2-mm, air-dry soil to nearest mg on an electronic balance and place into a numbered, tared, 300-mL, fleaker. Wash and tare these fleakers once every two months. A quality control sample is included in each batch (≤24 samples).

7.55 Add ≈ 50 mL of RO water and 7.5 mL of H_2O_2 to the soil sample at ambient temperature. Cover the soil sample with a 50-mm watch glass. Allow initial oxidation of organic matter to complete and then place sample on hot plate. If the froth from the reaction exceeds the capacity of the fleaker, transfer the sample to a larger beaker.

7.56 Place the sample on a hot plate and heat to 90°C. Add four 5 to 7.5-mL increments of H_2O_2 at 30-min intervals. If oxidation is incomplete, add additional H_2O_2 until organic matter oxidation is complete. Heat the sample for an additional 45 min to decompose excess H_2O_2. If the reaction is violent, do one or any combination of the following: (a) add small increments of ethyl alcohol to the sample; (b) remove the sample from the hot plate to slow the reaction; (c) transfer sample to a 1000-mL beaker; or (d) reduce the amount of H_2O_2 to sample. Record any unusual sample reactions. Record any unusual sample reactions.

7.57 Place the sample on the filter rack. Add 150 mL of RO water. Insert a filter candle, connect to the vacuum trap assembly with tubing, and turn on the vacuum. Aspirate until liquid is removed and only slightly dampened sample remains. Wash the sample four additional times with ≈ 150 mL of RO water. Stir the sample with filter candle to ensure all soil particles will be rinsed. During aspiration, it may be necessary to occasionally apply back-pressure to filter candle and remove build-up of soil which inhibits aspiration. If the sample contains gypsum and flocculates, then the following additional washings may be used. If the sample contains 1 to 5% gypsum, stir the sample with a magnetic stirrer for 5 min and wash 5 times with ≈ 250 mL of RO water each time. If the sample contains >5% gypsum, place the sample in a 1000-mL beaker and stir the sample with a magnetic stirrer for 5 min then wash 5 times with ≈ 750 mL of RO water each time to remove soluble gypsum.

7.58 Place sample in oven. Dry the sample overnight at 110°C. Remove the sample from the oven, place in a desiccator, and cool to ambient temperature.

7.59 Record the total weight (TW) of the sample to the nearest mg.

7.60 Add ≈ 100 mL of RO water and the exact volume of sodium hexametaphosphate solution (≈ 10 mL), equivalent to 0.4408 g of sodium hexametaphosphate, to sample. Subtract the weight of the sodium hexametaphosphate (DW) that is contained in the extracted aliquot from the silt and clay weights to calculate silt and clay percentages. To determine the exact volume of sodium hexametaphosphate to add to each sample, refer to Section 7.61.

7.61 A sodium hexametaphosphate standardization is performed with each new batch of solution. Use only designated weighing bottles for standardization. Wash and tare these bottles after each standardization. Add duplicate aliquots (8.5, 9.0, 9.3, 9.6, 10.0, 10.3, 10.6, 11.0 mL) of sodium hexametaphosphate solution to numbered tared, 90-mL weighing bottles. Oven-dry aliquots overnight and record dry residue weight of sodium hexametaphosphate. Determine the exact volume of sodium hexametaphosphate to add to each sample by regressing the volume of sodium hexametaphosphate against the dry residue weight of sodium hexametaphosphate and then by predicting the volume needed to dispense 0.4408 g of sodium hexametaphosphate into each sample.

7.62 Disperse the suspension with ultrasonic vibrations. Ensure the power supply is properly tuned. Consult the instruction manual. Immerse the probe in the suspension to a 3-cm depth. Set timer to 5 min. Press start button. Adjust output control as required. Between samples, clean the probe by placing it in water or alcohol and energizing it for a few seconds.

7.63 After ultrasonic dispersion, add ≈ 75 mL RO water and pour the suspension through a 300-mesh (0.047-mm) sieve mounted on a ring stand. Finger-rubbing of sample may be required during transfer to speed washing of sample. Place a funnel below the sieve and a 1-L cylinder below the funnel. Collect the silt and clay in the 1-L cylinder. Avoid using jets of water in washing the sample. Wash and rub all particles from the fleaker into the sieve. Continue to wash until the suspension volume in the cylinder is ≈ 800 mL. Sand and some of the coarse silt remain on the sieve. Rinse all <20-μm particles into the cylinder. Fill the cylinder to 1 L and cover with a 65-mm watch glass. Place pipe insulation around sample and blank cylinders to prevent rapid changes in temperatures. Prepare a RO water blank to measure temperature fluctuations. Allow the cylinder to stand overnight to equilibrate the suspension with the room temperature. Wash the sand into an evaporation dish and dry the sand at 110°C overnight. Proceed to procedure 3A1a1a for remaining PSDA procedural steps.

Particle-Size Distribution Analysis (3A)
Particles <2 mm (3A1)
Pipet Analysis (3A1a)
Water Dispersion (3A1a6)
Air-Dry (3A1a6a)

7.64 Weigh 10 g of <2-mm, air-dry soil to nearest mg on an electronic balance and place into a numbered, tared 300-mL, fleaker. Wash and tare these fleakers once every two months. A quality control sample is included in each batch ($\leq$24 samples).

7.65 Dry the sample in an oven at 110°C overnight. Remove the sample from the oven, place in a desiccator, and cool to ambient temperature.

7.66 Record the total weight (TW) of the sample to the nearest mg.

7.67 Add $\approx$ 175 mL of RO water to sample. Place the sample in a horizontal shaker set at 120 oscillations min^{-1} and shake for 15 h (overnight).

7.68 Remove the sample from the shaker and pour through a 300-mesh (0.047-mm) sieve mounted on a ring stand. Finger-rubbing of sample may be required during transfer to speed washing of sample. Place a funnel below the sieve and a 1-L cylinder below the funnel. Collect the silt and clay in the 1-L cylinder. Avoid using jets of water in washing the sample. Wash and rub all particles from the fleaker into the sieve. Continue to wash until the suspension volume in the cylinder is $\approx$ 800 mL. Sand and some of the coarse silt remain on the sieve. Rinse all <20-μm particles into the cylinder. Fill the cylinder to 1 L and cover with a 65-mm watch glass. Place pipe insulation around sample and blank cylinders to prevent rapid changes in temperatures. Prepare a RO water blank to measure temperature fluctuations. Allow the cylinder to stand overnight to equilibrate the suspension with the room temperature. Wash the sand into an evaporation dish and dry the sand at 110°C overnight.

7.69 Transfer the dried sand to a nest of sieves that has a top-to-bottom order of 1.0, 0.5, 0.25, 0.1, and 0.047 mm. Shake the sand for 3 min on a shaker that has 1.3-cm vertical and lateral movements and oscillates at 500 strokes min^{-1}. Record the weight of each separate sand fraction (SW$_i$) to the nearest mg. If optical analysis is requested, place the very fine sand and fine sand fractions in gelatin capsules and the remaining sand fractions in a labeled vial. Store capsules in the labeled vial. Wash sand dishes after every use.

7.70 Determine the percentage of fine silt and clay gravimetrically by removing an aliquot from the suspension in the 1-L cylinder with a Lowy, 25-mL pipet. Periodically, gravimetrically calibrate the delivery volume of the pipet by weighing the amount of RO water dispensed from the pipet. Record the delivery volume (DV) and use the value to calculate the results. Regulate the vacuum such that the pipet fills in $\approx$ 12 s. Record temperature (T$_1$) of blank. Mount the pipet on an adjustable pipet rack (Shaw, 1932). Stir the silt and clay suspension with mechanical stirrer for at least 5 min. Place the cylinder on a stable, vibrationless table and stir with a hand stirrer in an up-and-down motion for 30 s. Timing is started upon completion of the stirring. Record the time that stirring is stopped. For the <20-μm fraction, slowly lower the closed pipet to a 10-cm depth in the suspension, open the pipet, turn on the vacuum, and withdraw an aliquot at the calculated time (Table 3). Dispense the aliquot into a tared and numbered, 90-mL weighing bottle. Rinse the pipet twice with RO water and dispense into the tared, weighing bottle with aliquot. For the <2-μm fraction, pipet after a time of 4.5, 5, 5.5, or 6.5 h. Record temperature (T_2) of blank. Use the average of T_1 and T_2 and adjust the pipet depth in the suspension as indicated in Table 4. Repeat the procedure described for the <20-μm fraction. Dry the aliquots at 110°C overnight and cool in a desiccator that contains calcium sulfate or an equivalent desiccant. Record the residue weight (RW) to the nearest 0.1 mg. Proceed to procedure 3A1a1a for remaining PSDA procedural steps.

Particle-Size Distribution Analysis (3A)
Particles <2 mm (3A1)
Pipet Analysis (3A1a)
Standard Pretreatments and Dispersion (3A1a1)
Field-Moist (3A1a1b)

7.71 Weigh enough <2mm, moist soil to achieve two ≈ 10-g samples of air-dry soil. Weigh to nearest mg on an electronic balance and place into numbered, tared, 300-mL, tared fleakers. Wash and tare these fleakers once every two months. A quality control sample is included in each batch (≤24 samples).

7.72 Add ≈ 50 mL of RO water and 7.5 mL of H_2O_2 to both soil sub-samples at ambient temperature. Cover the soil samples with a 50-mm watch glass. Allow initial oxidation of organic matter to complete and then place sample on hot plate. If the froth from the reaction exceeds the capacity of the fleaker, transfer the sample to a larger beaker.

7.73 Place the samples on a hot plate and heat to 90°C. Add four 5 to 7.5-mL increments of H_2O_2 at 30-min intervals. If oxidation is incomplete, add additional H_2O_2 until organic matter oxidation is complete. Heat the sample for an additional 45 min to decompose excess H_2O_2. If the reaction is violent, do one or any combination of the following: (a) add small increments of ethyl alcohol to the sample; (b) remove the sample from the hot plate to slow the reaction; (c) transfer sample to a 1000-mL beaker; or (d) reduce the amount of H_2O_2 to sample. Record any unusual sample reactions.

7.74 Place the sample on the filter rack. Add 150 mL of RO water. Insert a filter candle, connect to the vacuum trap assembly with tubing, and turn on the vacuum. Aspirate until liquid is removed and only slightly dampened sample remains. Wash the sample four additional times with ≈ 150 mL of RO water. Stir the sample with filter candle to ensure all soil particles will be rinsed. During aspiration, it may be necessary to occasionally apply back-pressure to filter candle and remove build-up of soil which inhibits aspiration. If the sample contains gypsum and flocculates, then the following additional washings may be used. If the sample contains 1 to 5% gypsum, stir the sample with a magnetic stirrer for 5 min and wash 5 times with ≈ 250 mL of RO water each time. If the sample contains >5% gypsum, place the sample in a 1000-mL beaker and stir the sample with a magnetic stirrer for 5 min then wash 5 times with ≈ 750 mL of RO water each time to remove soluble gypsum.

7.75 Place one of the samples in the oven and dry overnight at 110°C. Remove the sample from the oven, place in a desiccator, and cool to ambient temperature.

7.76 Record the total weight (TW) of the H_2O_2-treated, oven-dry sample to the nearest mg. The H_2O_2-treated, oven-dry sample is not used in the remaining PSDA procedural steps. Proceed to procedure 3A1a1a and use the H_2O_2-treated sample that was not dried in oven for remaining PSDA procedural steps.

Particle-Size Distribution Analysis (3A)
Particles <2 mm (3A1)
Pipet Analysis (3A1a)
Carbonate Removal (3A1a2)
Field-Moist (3A1a2b)

7.77 Weigh two 10-g samples of <2-mm, moist soil to nearest mg on an electronic balance and place into 300-mL, tared fleakers. Weigh sufficient samples to yield 10 g of <2-mm, moist, carbonate-free soil sample, e.g., if the sample contains 50% carbonates, weigh 20 g of soil.

7.78 Add $\approx$ 200 mL of the 1 N NaOAc solution to both samples, mix with a stirring rod, and cover with a watch glass. Allow the samples to stand overnight.

7.79 Place the samples on the hot plate and heat to $\approx$ 90°C until bubbles are no longer evident. Do not boil. Heating accelerates reaction. Decant the solution and add more 1 N NaOAc solution. If a reaction occurs, repeat the heating procedure. Continue to decant, add NaOAc solution, and heat until all the carbonates are removed. The speed of dissolution can be increased by lowering the pH of the 1 N NaOAc solution (Rabenhorst and Wilding, 1984).

7.80 When no more carbonate bubbles are observed, insert the ceramic filter candle into the solution. Apply vacuum and candle the samples to dryness. Rinse once with 200 mL of RO water.

7.81 Add $\approx$ 50 mL of RO water and 5 mL of H_2O_2 to both soil samples at ambient temperature. Cover the soil samples with 50-mm watch glass. Allow initial oxidation of organic matter to complete and then place samples on hot plate. If the froth from the reaction exceeds the capacity of the fleakers, transfer the samples to larger beakers.

7.82 Place the samples on the hot plate and heat to $\approx$ 90°C. Add 5-mL increments of H_2O_2 at 45-min intervals until oxidation has completed or until 30 mL of H_2O_2 have been added. Heat the sampled for an additional 45 min to decompose excess H_2O_2. If the reaction is violent, add small increments of ethyl alcohol to the samples or remove the samples from the hot plate to slow the reaction.

7.83 Place the samples on the filter rack. Add 150 mL of RO water. Insert filter candle, connect to the vacuum trap assembly with tubing, and turn on vacuum. Wash the samples four additional times with $\approx$ 150 mL of RO water. If the samples contain gypsum and flocculates, then the following additional washings may be used. If the samples contain 1 to 5% gypsum, stir the samples with a magnetic stirrer for 5 min and wash 5 times with $\approx$ 250 mL of RO water each time. If the samples contain >5% gypsum, stir the samples with a magnetic stirrer for 5 min then wash 5 times with $\approx$ 750 mL of RO water each time to remove soluble gypsum.

7.84 Place one of the samples in the oven and dry overnight at 110°C. Remove the sample from the oven, place in a desiccator, and cool to ambient temperature.

7.85 Record the total weight (TW) of the H_2O_2-treated, oven-dry sample to the nearest mg. The H_2O_2-treated, oven-dry sample is not used in the remaining PSDA procedural steps. Proceed to procedure 3A1a1a and use the H_2O_2-treated sample that was not dried in oven for remaining PSDA procedural steps.

Particle-Size Distribution Analysis (3A)
Particles <2 mm (3A1)
Pipet Analysis (3A1a)
Iron Removal (3A1a3)
Field-Moist (3A1a3b)

7.86 Weigh two 10-g samples of <2-mm, moist soil to nearest mg on an electronic balance and place into 300-mL, tared fleakers. As a maximum of ≈ 0.5 g of Fe_2O_3 can be dissolved in 40 mL of the citrate solution, adjust the weight of the <2-mm, moist soil sample so that any one fleaker does not contain more than 0.5 g of Fe_2O_3. Split the sample into different fleakers if necessary. Total sample weight after dissolution should be ≈ 10 g.

7.87 Add ≈ 50 mL of RO water and 5 mL of H_2O_2 to both soil samples at ambient temperature. Cover the soil samples with 50-mm watch glass. Allow initial oxidation of organic matter to complete, and then place samples on hot plate. If the froth from the reaction exceeds the capacity of the fleaker, transfer the samples to larger beakers.

7.88 Place the samples on a hot plate and heat to 90°C. Add 5-mL increments of H_2O_2 at 45-min intervals until oxidation has complete or until 30 mL of H_2O_2 have been added. Heat the samples for an additional 45 min to decompose excess H_2O_2. If the reaction is violent, add small increments of ethyl alcohol to the samples or remove the samples from the hot plate to slow the reaction.

7.89 Add 40 mL of the citrate solution and 5 mL of the sodium bicarbonate. Heat to 80°C in a water bath, but do not exceed 80°C. Add 1 g of sodium dithionite powder with a calibrated scoop. Stir constantly with a glass rod for 1 min and then occasionally for 15 min. Add 10 mL of saturated NaCl solution and mix.

7.90 Centrifuge or candle the samples to remove the dissolved Fe_2O_3. Combine the split samples into fewer fleakers.

7.91 If the samples contain less than 0.5 g of Fe_2O_3, repeat the dissolution treatment. For samples with more than 0.5 g of Fe_2O_3, repeat the dissolution treatment two more times.

7.92 Place the samples on the filter rack. Add 150 mL of RO water. Insert a filter candle, connect to the vacuum trap assembly with tubing, and turn on the vacuum. Wash the samples four additional times with ≈ 150 mL of RO water. If the samples contain gypsum and flocculates, then the following additional washings may be used. If the sample contains 1 to 5% gypsum, stir the sample with a magnetic stirrer for 5 min and wash 5 times with ≈ 250 mL of RO water each time. If the sample contains >5% gypsum, stir the sample with a magnetic stirrer for 5 min then wash 5 times with ≈ 750 mL of RO water each time to remove soluble gypsum.

7.93 Place one sample in the oven and dry overnight at 110°C. Remove the sample from the oven, place in a desiccator, and cool to ambient temperature.

7.94 Record the total weight (TW) of the sample to the nearest mg. The H_2O_2 oven-dry sample is not used in the remaining PSA procedural steps. Proceed to procedure 3A1a1a and use the H_2O_2-treated sample that was not dried in oven for remaining PSDA procedural steps.

Particle-Size Distribution Analysis (3A)
Particles <2 mm (3A1)
Pipet Analysis (3A1a)
Silica Removal (3A1a4)
Field-Moist (3A1a4b)

7.95 Weigh two 10-g of <2-mm, moist soil to nearest mg on an electronic balance and place in 300-mL, tared fleakers.

7.96 Add ≈ 50 mL of RO water and 5 mL of H_2O_2 to both soil samples at ambient temperature. Cover the soil samples with 50-mm watch glass. Allow initial oxidation of organic matter to complete, and then place samples on hot plate. If the froth from the reaction exceeds the capacity of the fleaker, transfer the sample to larger beakers.

7.97 Place the samples on a hot plate and heat to 90°C. Add 5-mL increments of H_2O_2 at 45-min intervals until oxidation has completed or until 30 mL of H_2O_2 have been added. Heat the samples for an additional 45 min to decompose excess H_2O_2. If the reaction is violent, add small increments of ethyl alcohol to the samples or remove the samples from the hot plate to slow the reaction.

7.98 Soak the samples overnight in 100 mL of 0.1 N NaOH.

7.99 Place the samples on the filter rack. Add 150 mL of RO water. Insert filter candle, connect to the vacuum trap assembly with tubing, and turn on the vacuum. Wash the samples four additional times with ≈ 150 mL of RO water. If the sample contains gypsum and flocculates, then the following additional washings may be used. If the samples contain 1 to 5% gypsum, stir the samples with a magnetic stirrer for 5 min and wash 5 times with ≈ 250 mL of RO water each time. If the samples contain >5% gypsum, stir the sample with a magnetic stirrer for 5 min then wash 5 times with ≈ 750 mL of RO water each time to remove soluble gypsum.

7.100 Place one sample in the oven and dry overnight at 110°C. Remove the sample from the oven, place in a desiccator, and cool to ambient temperature.

7.101 Record the total weight (TW) of the sample to the nearest mg. The H_2O_2-treated, oven-dry sample is only used for calculation of results and is not used in the remaining PSDA procedural steps. Proceed to procedure 3A1a1a and use the H_2O_2-treated sample that was not dried in oven for all of the remaining PSDA procedural steps.

Particle-Size Distribution Analysis (3A)
Particles <2 mm (3A1)
Pipet Analysis (3A1a)
Ultrasonic Dispersion (3A1a5)
Field-Moist (3A1a5b)

7.102 Weigh enough <2mm, moist soil to achieve two ≈ 10-g samples of air-dry soil. Weigh to nearest mg on an electronic balance and place into numbered, tared, 300-mL, tared fleakers. Wash and tare these fleakers once every two months. A quality control sample is included in each batch (≤24 samples).

7.103 Add ≈ 50 mL of RO water and 7.5 mL of H_2O_2 to the soil samples at ambient temperature. Cover the soil samples with 50-mm watch glass. Allow initial oxidation of organic matter to complete and then place samples on hot plate. If the froth from the reaction exceeds the capacity of the fleaker, transfer the samples to a larger beaker.

7.104 Place the samples on a hot plate and heat to 90°C. Add four 5 to 7.5-mL increments of H_2O_2 at 30-min intervals. If oxidation is incomplete, add additional H_2O_2 until organic matter oxidation is complete. Heat the sample for an additional 45 min to decompose excess H_2O_2. If the reaction is violent, do one or any combination of the following: (a) add small increments of ethyl alcohol to the sample; (b) remove the sample from the hot plate to slow the reaction; (c) transfer sample to a 1000-mL beaker; or (d) reduce the amount of H_2O_2 to sample. Record any unusual sample reactions.

7.105 Place the sample on the filter rack. Add 150 mL of RO water. Insert a filter candle, connect to the vacuum trap assembly with tubing, and turn on the vacuum. Aspirate until liquid is removed and only slightly dampened sample remains. Wash the sample four additional times with ≈ 150 mL of RO water. Stir the sample with filter candle to ensure all soil particles will be rinsed. During aspiration, it may be necessary to occasionally apply back-pressure to filter candle and remove build-up of soil which inhibits aspiration. If the sample contains gypsum and flocculates, then the following additional washings may be used. If the sample contains 1 to 5% gypsum, stir the sample with a magnetic stirrer for 5 min and wash 5 times with ≈ 250 mL of RO water each time. If the sample contains >5% gypsum, place the sample in a 1000-mL beaker and stir the sample with a magnetic stirrer for 5 min then wash 5 times with ≈ 750 mL of RO water each time to remove soluble gypsum.

7.106 Place one sample in oven and dry overnight at 110°C. Remove sample from oven, place in a desiccator, and cool to ambient temperature.

7.107 Record the total weight (TW) of the H_2O_2-treated, oven-dry sample to the nearest mg. The H_2O_2-treated, oven-dry sample is only used for the calculations of results and is not used in the remaining PSDA procedural steps. Use the H_2O_2-treated sample that was not dried in oven for the following PSDA procedural steps.

7.108 Add ≈ 100 mL of RO water and the exact volume of sodium hexametaphosphate solution (≈ 10 mL), equivalent to 0.4408 g of sodium hexametaphosphate, to sample. Subtract the weight of the sodium hexametaphosphate (DW) that is contained in the extracted aliquot from the silt and clay weights to calculate silt and clay percentages. To determine the exact volume of sodium hexametaphosphate to add to each sample refer to Section 7.109.

7.109 A sodium hexametaphosphate standardization is performed with each new batch of solution. Use only designated weighing bottles for standardization. Wash and tare these bottles after each standardization. Add duplicate aliquots (8.5, 9.0, 9.3, 9.6, 10.0, 10.3, 10.6, 11.0 mL) of sodium hexametaphosphate solution to numbered tared, 90-mL weighing bottles. Oven-dry aliquots overnight and record dry residue weight of sodium hexametaphosphate. Determine the exact volume of sodium hexametaphosphate to add to each sample by regressing the volume of sodium hexametaphosphate against the dry residue weight of sodium hexametaphosphate and then by predicting the volume needed to dispense 0.4408 g of sodium hexametaphosphate into each sample.

7.110 Disperse the suspension with ultrasonic vibrations. Ensure the power supply is properly tuned. Consult the instruction manual. Immerse the probe in the suspension to a 3-cm depth. Set timer to 5 min. Press start button. Adjust output control as required. Between samples, clean the probe by placing it in water or alcohol and energizing it for a few seconds.

7.111 After ultrasonic dispersion, add ≈ 75 mL RO water and pour the suspension through a 300-mesh (0.047-mm) sieve mounted on a ring stand. Finger-rubbing of sample may be required during transfer to speed washing of sample. Place a funnel below the sieve and a 1-L cylinder below the funnel. Collect the silt and clay in the 1-L cylinder. Avoid using jets of water in washing the sample. Wash and rub all particles from the fleaker into the sieve. Continue to wash until the suspension volume in the cylinder is ≈ 800 mL. Sand and some of the coarse silt remain on the sieve. Rinse all <20-μm particles into the cylinder. Fill the cylinder to 1 L and cover with a 65-mm watch glass. Place pipe insulation around sample and blank cylinders to prevent rapid changes in temperatures. Prepare a RO water blank to measure temperature fluctuations. Allow the cylinder to stand overnight to equilibrate the suspension with the room temperature. Wash the sand into an evaporation dish and dry the sand at 110°C overnight. Proceed to procedure 3A1a1a for remaining PSDA procedural steps.

Particle-Size Distribution Analysis (3A)
Particles <2 mm (3A1)
Pipet Analysis (3A1a)
Water Dispersion (3A1a6)
Field-Moist (3A1a6b)

7.112 Weigh enough <2mm, moist soil to achieve two ≈ 10-g samples of air-dry soil. Weigh to nearest mg on an electronic balance and place into numbered, tared, 300-mL, tared fleakers. Wash and tare these fleakers once every two months. A quality control sample is included in each batch (≤24 samples).

7.113 Dry one sample in an oven at 110°C overnight. Remove the sample from the oven, place in a desiccator, and cool to ambient temperature.

7.114 Record the total weight (TW) of the sample to the nearest mg. The oven-dry sample is not used in the remaining PSA procedural steps.

7.115 Add ≈ 175 mL of RO water to sample that was not dried in the oven. Place sample in a horizontal shaker set at 120 oscillations min^{-1} and shake for 15 h (overnight).

7.116 Remove the sample from the shaker and pour through a 300-mesh (0.047-mm) sieve mounted on a ring stand. Finger-rubbing of sample may be required during transfer to speed washing of sample. Place a funnel below the sieve and a 1-L cylinder below the funnel. Collect the silt and clay in the 1-L cylinder. Avoid using jets of water in washing the sample. Wash and rub all particles from the fleaker into the sieve. Continue to wash until the suspension volume in the cylinder is ≈ 800 mL. Sand and some of the coarse silt remain on the sieve. Rinse all <20-μm particles into the cylinder. Fill the cylinder to 1 L and cover with a 65-mm watch glass. Place pipe insulation around sample and blank cylinders to prevent rapid changes in temperatures. Prepare a RO water blank to measure temperature fluctuations. Allow the cylinder to stand overnight to equilibrate the suspension with the room temperature. Wash the sand into an evaporation dish and dry the sand at 110°C overnight.

7.117 Transfer the dried sand to a nest of sieves that has a top-to-bottom order of 1.0, 0.5, 0.25, 0.1, and 0.047 mm. Shake the sand for 3 min on a shaker that has 1.3-cm vertical and lateral movements and oscillates at 500 strokes min^{-1}. Record the weight of each separate sand fraction (SW$_i$) to the nearest mg. If optical analysis is requested, place the very fine sand and fine sand fractions in gelatin capsules and the remaining sand fractions in a labeled vial. Store capsules in the labeled vial. Wash sand dishes after every use.

7.118 Determine the percentage of fine silt and clay gravimetrically by removing an aliquot from the suspension in the 1-L cylinder with a Lowy, 25-mL pipet. Periodically, gravimetrically calibrate the delivery volume of the pipet by weighing the amount of RO water dispensed from the pipet. Record the delivery volume (DV) and use the value to calculate the results. Regulate the vacuum such that the pipet fills in ≈ 12 s. Record temperature (T$_1$) of blank. Mount the pipet on an adjustable pipet rack (Shaw, 1932). Stir the silt and clay suspension with mechanical stirrer for at least 5 min. Place the cylinder on a stable, vibrationless table and stir with a hand stirrer in an up-and-down motion for 30 s. Timing is started upon completion of the stirring. Record the time that stirring is stopped. For the <20-μm fraction, slowly lower the closed pipet to a 10-cm depth in the suspension, turn on the vacuum, open the pipet, and withdraw an aliquot at the calculated time (Table 1). Place the aliquot into a tared and numbered, 90-mL weighing bottle. Rinse the pipet with RO water into the tared, weighing bottle. For the <2-μm fraction, pipet after a time of 4.5, 5, 5.5, or 6.5 h. Record the temperature (T_2) of blank. Use the average of T_1 and T_2 and adjust the pipet depth in the suspension as indicated in Table 2. Repeat the procedure described for the <20-μm fraction. Dry the aliquots at 110°C overnight and cool in a desiccator that contains calcium sulfate or an equivalent desiccant. Record the residue weight (RW) to the nearest 0.1 mg. Proceed to procedure 3A1a1a for remaining PSDA procedural steps. Also proceed to procedure 3A1a1a for determination of fine-clay and/or carbonate clay.

8. Calculations

Use calculations in sections 8.1 – 8.6 for all PSDA procedures (3A1a1-5a and 3A1a1-5b) except water dispersible PSDA. Use calculations in sections 8.7 – 8.10 for water dispersible PSDA (3A1a6a and 3A1a6b).

Calculations 8.2 – 8.6 for procedures 3A1a1-5a and 3A1a1-5b as follows:

8.1 Clay % = 100 x ((RW_2 - DW) x (CF/TW))

where:
RW_2 = Residue weight (g), <2-μm fraction
DW   = Dispersing agent weight (g) = (0.4408/CF)
CF   = 1000 mL/DV
DV   = Dispensed pipet volume
TW   = Total weight (g), H_2O_2-treated, oven-dry sample

8.2 Fine Silt % = 100 x ((RW_{20}-DW) x (CF/TW)) - Clay %

where:
RW_{20} = Residue weight (g) of <20-μm fraction

8.3 Sand % = $\sum$ (SW_i /TW) x 100

where:
SW_i = Weight of sand fractions (1.0, 0.5, 0.25, 0.1, and 0.047 mm)

8.4 Coarse silt % = 100 - (Clay % + Fine Silt % + Sand %)

8.5 Fine Clay (%) = 100 x ((RW-DW) x (CF/TW))

where:
RW = Residue weight (g) of <0.2-μm fraction
DW = Dispersing agent weight (g) = (0.4364/CF)
CF = 1000 mL/DV
DV = Dispensed pipet volume
TW = Total weight of H_2O_2-treated, oven-dry sample

8.6 Calculate carbonate clay percentage as follows:

8.6.1 Correct the manometer reading as follows:

CR = (MR - BR)

where:
CR = Corrected reading
MR = Manometer reading
BR = Blank reading

Three blanks are run with each batch ($\leq$24 samples). The average of three blanks is used as BR.

8.6.2 Calculate two regression equations, i.e., one for corrected manometer readings <100 and another for corrected readings $\geq$100. Use the Na_2CO_3 weights as the dependent variable (regressed or predicted values) and the corresponding manometer readings as the independent variable.

8.6.3 Use the corrected (CR) linear regression equations to estimate the g of $CaCO_3$ in the sample.

8.6.4 Carbonate Clay Equivalent (<2 μm) (%) = ((g $CaCO_3$) x 100 x CF)/TW

where:
CF = 1000 mL/dispensed pipet volume (mL)
TW = Total weight of H_2O_2-treated oven-dry sample

8.6.5 Noncarbonate Clay (<2 μm) (%) = Total Clay (%) - Carbonate Clay Equivalent (%)

Calculations 8.7 – 8.10 for procedures 3A1a6a and 3A1a6b as follows:

8.7 Clay % = 100 x ((RW_2 x CF)/TW)

where:
RW_2 = Residue weight (g), <2-μm fraction
CF   = 1000 mL/DV
DV   = Dispensed pipet volume
TW   = Total weight (g), oven-dry sample

8.8 Fine Silt % = (100 x (RW_{20} x CF)/TW)) - Clay %

where:
RW_{20} = Residue weight (g) of <20-μm fraction

8.9 Sand % = $\sum$ (SW_i/TW) x 100

where:
SW_i = Weight of sand fractions (1.0, 0.5, 0.25, 0.1, and 0.047 mm)

8.10 Coarse silt % = 100 - (Clay % + Fine Silt % + Sand %)

9. Report
        Report each particle-size fraction to the nearest 0.1 percent.

10. Precision and Accuracy

Precision and accuracy data are available from the SSL upon request.

11. References

Bouyoucos, G.B. 1929. The ultimate natural structure of soil. Soil Sci. 28:27-37.

Drosdoff, M., and E.F. Miles. 1938. Action of hydrogen peroxide on weathered mica. Soil Sci. 46:391-395.

El Swaify, S.A. 1980. Physical and mechanical properties of Oxisols. p. 303-324. *In* B.K.G. Theng (ed.) Soils with variable charge. N.Z. Soc. Soil Sci., Lower Hutt, N.Z.

Espinoza, W., R.H. Rust, and R.S. Adams Jr. Characterization of mineral forms in Andepts from Chile. Soil Sci. Soc. Am. Proc. 39:556-561.

Gee, G.W., and J.W. Bauder. 1986. Particle-size analysis. p. 383-411. *In* A. Klute (ed.) Method of soil analysis. Part 1. Physical and mineralogical methods. 2nd ed. Agron. Monogr. 9. ASA and SSSA, Madison, WI.

Gee, G.W., and D. Or. 2002. Particle-size analysis. p. 255-293. *In* J.H. Dane and G.C. Topp (eds.) Methods of soil analysis. Part 4. Physical methods. Soil Sci. Am. Book Series No. 5. ASA and SSSA, Madison, WI.

Grossman, R.B., and J.L. Millet. 1961. Carbonate removal from soils by a modification of the acetate buffer method. Soil Sci. Soc. Am. Proc. 25:325-326.

Jackson, M.L. 1969. Soil chemical analysis—advanced course. 2nd ed. Univ. Wisconsin, Madison.

Kilmer, V.J., and L.T. Alexander. 1949. Methods of making mechanical analyses of soils. Soil Sci. 68:15-24.

Kilmer, V.J., and J.K. Mullins. 1954. Improved stirring and pipeting apparatus for mechanical analysis of soils. Soil Sci. 77:437-441.

Kubota, T. 1972. Aggregate formation of allophonic soils; Effects of drying on the dispersion of soils. Soil Sci. and Plant Nutr. 18:79-87.

Maeda, T., H. Takenaka, and B.P. Warkentin. 1977. p. 229-263. Physical properties of allophone soils. In N.C. Brady (ed.) Adv. Agron. Acad. Press, Inc., New York, NY.

Mehra, O.P., and M.L. Jackson. 1960. Iron oxide removal from soils and clays by a dithionite-citrate system buffered with sodium bicarbonate. p. 237-317. *In* Clays and clay minerals. Proc. 7th Conf. Natl. Acad. Sci. Natl. Res. Counc. Pub., Washington, DC.

Mikhail, E.H., and G.P. Briner. 1978. Routine particle size analysis of soils using sodium hypochlorite and ultrasonic dispersion. Aust. J. Soil Res. 16:241-244.

Rabenhorst, M.C., and L.P. Wilding. 1984. Rapid method to obtain carbonate-free residues from limestone and petrocalcic materials. Soil Sci. Soc. Am. J. 48:216-219.

Saly, R. 1967. Use of ultrasonic vibration for dispersing of soil samples. Sov. Soil Sci. 11:1547-1559.

Shaw, T.M. 1932. New aliquot and filter devices for analytical laboratories. Ind. Eng. Chem. Anal. Ed. 4:409.

Shields, L.G., and M.W. Meyer. 1964. Carbonate clay: Measurement and relationship to clay distribution and cation exchange capacity. Soil Sci. Soc. Am. Proc. 28:416-419.

Skopp, J. 1992. Concepts of soil physics. University of Nebraska, Lincoln, NE.

Soil Survey Division Staff. 1993. Soil survey manual. USDA. Handb. No. 18. U.S. Govt. Print. Office, Washington, DC.

Soil Survey Staff. 1996. Soil survey laboratory methods manual. Version No. 3.0. USDA-NRCS. Soil Survey Investigations Report No. 42. U.S. Govt. Print. Office, Washington, DC.

Soil Survey Staff. 1999. Soil taxonomy: A basic system of soil classification for making and interpreting soil surveys. 2nd ed. Agric. Handb. No. 436. USDA-NRCS. U.S. Govt. Print. Office, Washington, DC.

Sumner, M.E., 1992. The electrical double layer and clay dispersion. p. 1-32. *In* M.E. Sumner and B.A. Stewart (eds.) Soil crusting, chemical, and physical processes. CRC Press Inc., Boca Raton, FL.

Watson, J.R. 1971. Ultrasonic vibration as a method of soil dispersion. Soil Fertil. 34:127-134.

Williams, D.E. 1948. A rapid manometric method for the determination of carbonate in soils. Soil Sci. Soc. Am. Proc. 13:127-129.

Socket Set Screws

①

⑮

Top View

⑪ diameter - 1.27 cm

⑥ Precision Knurled Head Thumb Screw No. 10-32 1/2 Long

⑫ Parallel Rails diameter - 63 mm

⑬ diameter - 63 mm (scale optional)

④

⑥ Precision Knurled Head Thumb Screw No. 10-32 1/2 Long (place on opposite side)

⑭

⑲

Thumb Screw No. 1/4-20 1/2 Long

Fil. Hd. Mach. Scr. No. 6-32 N.P. 3/8 Long (4 required)

Precision Knurled Head Thumb Screw No. 10-32 1/2 Long

⑥

⑯

Nos. 1, 2, 14, 15, and 16 are 12.5 mm thick.

No. 19 is 3 cm square.

③

④

5 cm

1   3   5 cm
Scale

②

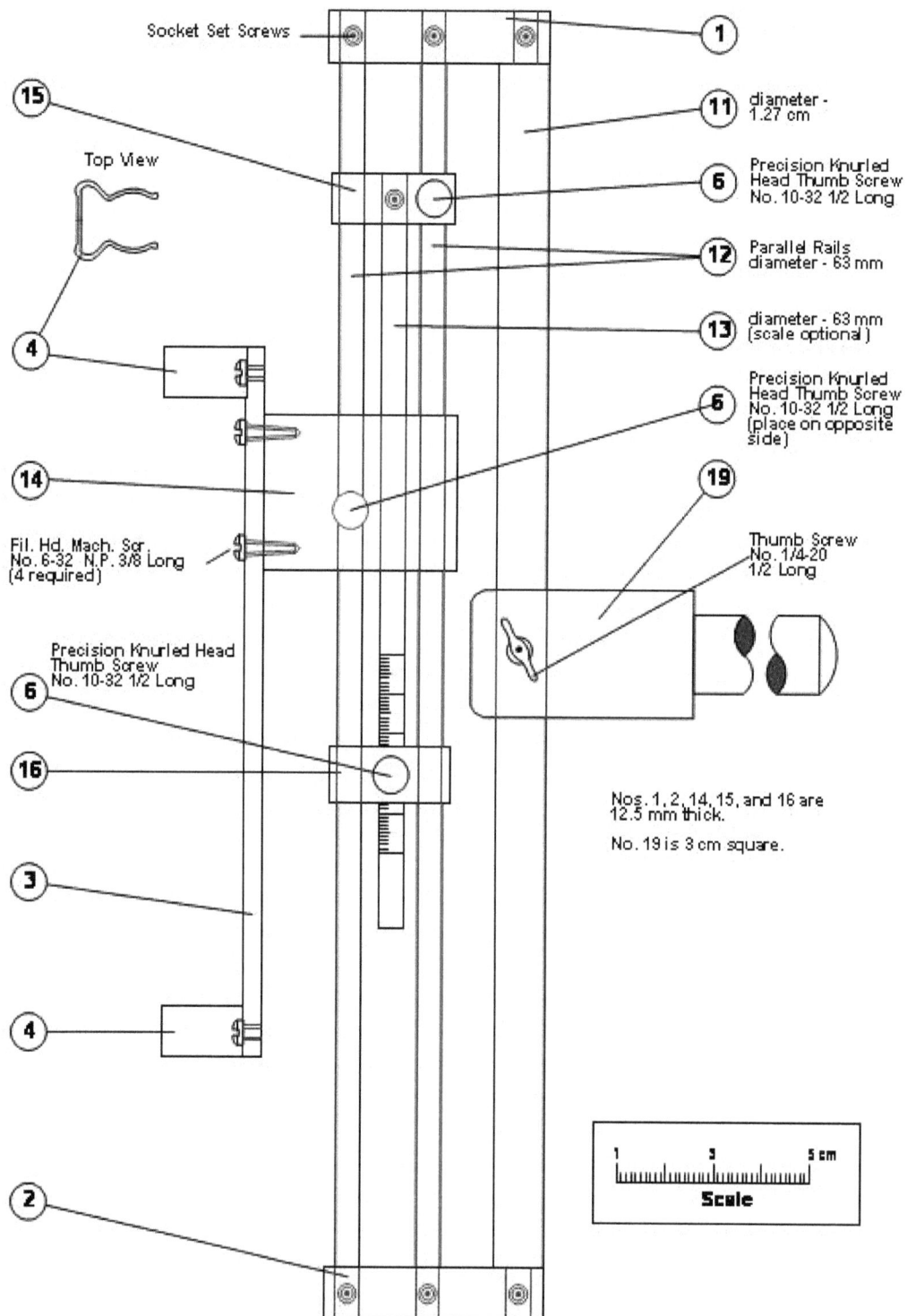

Fig 1. Shaw Pipet (Shaw, 1932).

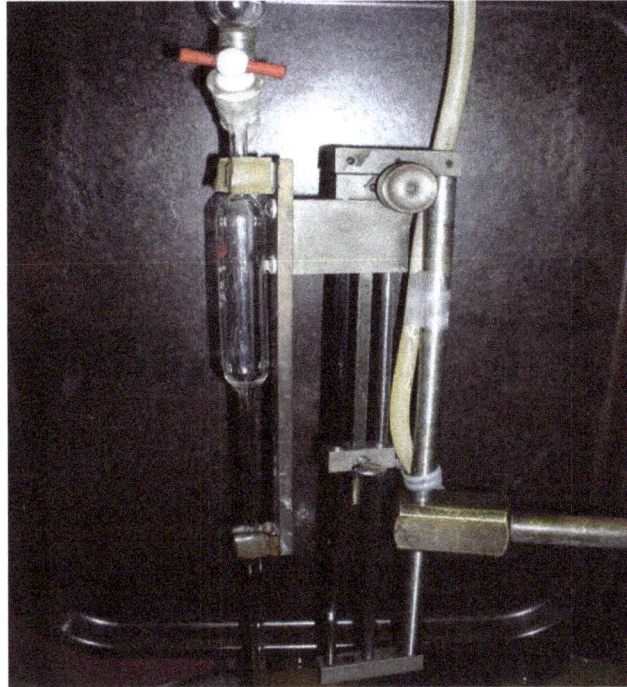

Fig. 2. Close-up of the Shaw Pipet apparatus.

Fig. 3. Shaw Pipet used for particle-size analysis of the <2-mm fraction at the USDA-NRCS Soil Survey Laboratory.

Table 1. Sampling times at 10-cm depth, 0.4408 g L^{-1} NaHMP solution, and 2.65 g cc^{-1} particle density.[1]

Temp (°C)	20 μm		5 μm	
	min	s	min	s
15.00	5	17	84	36
16.00	5	9	82	22
17.00	5	1	80	16
18.00	4	53	78	12
18.50	4	50	77	13
19.00	4	46	76	17
19.50	4	42	75	18
20.00	4	39	74	23
20.25	4	37	73	56
20.50	4	36	73	29
20.75	4	34	73	2
21.00	4	32	72	35
21.25	4	31	72	9
21.50	4	29	71	43
21.75	4	27	71	18
22.00	4	26	70	52
22.25	4	24	70	27
22.50	4	23	70	2
22.75	4	21	69	37
23.00	4	20	69	13
23.25	4	18	68	48
23.50	4	17	68	24
23.75	4	15	68	0
24.00	4	14	67	37
24.25	4	12	67	13
24.50	4	11	66	50
24.75	4	9	66	27
25.00	4	8	66	4
25.25	4	6	65	42
25.50	4	5	65	19
25.75	4	4	64	57
26.00	4	2	64	35
26.25	4	1	64	13
26.50	3	59	63	52
26.75	3	58	63	30
27.00	3	57	63	9
27.25	3	56	62	48
27.50	3	54	62	27
27.75	3	53	62	7
28.00	3	52	61	46
28.25	3	50	61	26
28.50	3	49	61	6
28.75	3	48	60	46
29.00	3	47	60	26
29.25	3	45	60	6
29.50	3	44	59	47
29.75	3	43	59	28
30.00	3	42	59	9

[1]Use this table with procedures 3A1a1-5a and 3A1a1-5b.

Table 2. Sampling depths (cm) for 2μm clay, 0.4408 g L^{-1} NaHMP solution, and 2.65 g cc^{-1} particle density.[1]

Temp (°C)	Time (h)				
	4.5	**5.0**	**5.5**	**6.0**	**6.5**
15.00	5.11	5.67	6.24	6.81	7.38
16.00	5.25	5.83	6.41	6.99	7.58
17.00	5.38	5.98	6.58	7.18	7.77
18.00	5.52	6.14	6.75	7.37	7.98
18.50	5.59	6.22	6.84	7.46	8.08
19.00	5.66	6.29	6.92	7.55	8.18
19.50	5.74	6.37	7.01	7.65	8.29
20.00	5.81	6.45	7.10	7.74	8.39
20.25	5.84	6.49	7.14	7.79	8.44
20.50	5.88	6.53	7.19	7.84	8.49
20.75	5.92	6.57	7.23	7.89	8.54
21.00	5.95	6.61	7.27	7.93	8.60
21.25	5.99	6.65	7.32	7.98	8.65
21.50	6.02	6.69	7.36	8.03	8.70
21.75	6.06	6.73	7.41	8.08	8.75
22.00	6.10	6.77	7.45	8.13	8.81
22.25	6.13	6.81	7.49	8.18	8.86
22.50	6.17	6.85	7.54	8.22	8.91
22.75	6.21	6.89	7.58	8.27	8.96
23.00	6.24	6.94	7.63	8.32	9.02
23.25	6.28	6.98	7.67	8.37	9.07
23.50	6.32	7.02	7.72	8.42	9.12
23.75	6.35	7.06	7.76	8.47	9.18
24.00	6.39	7.10	7.81	8.52	9.23
24.25	6.43	7.14	7.85	8.57	9.28
24.50	6.46	7.18	7.90	8.62	9.34
24.75	6.50	7.22	7.95	8.67	9.39
25.00	6.54	7.26	7.99	8.72	9.44
25.25	6.58	7.31	8.04	8.77	9.50
25.50	6.61	7.35	8.08	8.82	9.55
25.75	6.65	7.39	8.13	8.87	9.61
26.00	6.69	7.43	8.18	8.92	9.66
26.25	6.73	7.47	8.22	8.97	9.72
26.50	6.76	7.52	8.27	9.02	9.77
26.75	6.80	7.56	8.31	9.07	9.83
27.00	6.84	7.60	8.36	9.12	9.88
27.25	6.88	7.64	8.41	9.17	9.94
27.50	6.92	7.69	8.45	9.22	9.99
27.75	6.96	7.73	8.50	9.27	10.05
28.00	6.99	7.77	8.55	9.32	10.10
28.25	7.03	7.81	8.59	9.38	10.16
28.50	7.07	7.86	8.64	9.43	10.21
28.75	7.11	7.90	8.69	9.48	10.27
29.00	7.15	7.94	8.74	9.53	10.33
29.25	7.19	7.99	8.78	9.58	10.38
29.50	7.23	8.03	8.83	9.63	10.44
29.75	7.27	8.07	8.88	9.69	10.49
30.00	7.30	8.12	8.93	9.74	10.55

[1]Use this table with procedures 3A1a1-5a and 3A1a1-5b.

Table 3. Sampling times at 10-cm depth and 2.65 g cc⁻¹ particle density.[1]

Temp (°C)	20 μm		5 μm	
	min	s	min	s
15.00	5	17	84	26
16.00	5	8	82	12
17.00	5	0	80	7
18.00	4	53	78	3
18.50	4	49	77	4
19.00	4	45	76	8
19.50	4	42	75	9
20.00	4	38	74	14
20.25	4	37	73	47
20.50	4	35	73	20
20.75	4	33	72	53
21.00	4	32	72	27
21.25	4	30	72	0
21.50	4	28	71	35
21.75	4	27	71	9
22.00	4	25	70	43
22.25	4	24	70	18
22.50	4	22	69	53
22.75	4	21	69	29
23.00	4	19	69	4
23.25	4	17	68	40
23.50	4	16	68	16
23.75	4	15	67	52
24.00	4	13	67	28
24.25	4	12	67	5
24.50	4	10	66	42
24.75	4	9	66	19
25.00	4	7	65	56
25.25	4	6	65	34
25.50	4	4	65	11
25.75	4	3	64	49
26.00	4	2	64	27
26.25	4	0	64	6
26.50	3	59	63	44
26.75	3	58	63	23
27.00	3	56	63	1
27.25	3	55	62	40
27.50	3	54	62	20
27.75	3	52	61	59
28.00	3	51	61	39
28.25	3	50	61	18
28.50	3	39	60	58
28.75	3	47	60	38
29.00	3	46	60	19
29.25	3	45	59	59
29.50	3	44	59	40
29.75	3	43	59	21
30.00	3	41	59	1

[1]Use this table with procedures 3A1a6a and 3A1a6b.

Table 4. Sampling depths (cm) for 2-μm clay and 2.65 g cc^{-1} particle density.[1]

Temp (°C)	Time (h)				
	4.5	**5.0**	**5.5**	**6.0**	**6.5**
15.00	5.12	5.69	6.25	6.82	7.39
16.00	5.26	5.84	6.42	7.01	7.59
17.00	5.39	5.99	6.59	7.19	7.79
18.00	5.54	6.15	6.77	7.38	8.00
18.50	5.61	6.23	6.85	7.47	8.10
19.00	5.67	6.31	6.94	7.57	8.20
19.50	5.75	6.39	7.03	7.66	8.30
20.00	5.82	6.47	7.11	7.76	8.41
20.25	5.86	6.51	7.16	7.81	8.46
20.50	5.89	6.55	7.20	7.86	8.51
20.75	5.93	6.59	7.24	7.90	8.56
21.00	5.96	6.63	7.29	7.95	8.61
21.25	6.00	6.67	7.33	8.00	8.67
21.50	6.04	6.71	7.38	8.05	8.72
21.75	6.07	6.75	7.42	8.10	8.77
22.00	6.11	6.79	7.47	8.14	8.82
22.25	6.14	6.83	7.51	8.19	8.88
22.50	6.18	6.87	7.55	8.24	8.93
22.75	6.22	6.91	7.60	8.29	8.98
23.00	6.25	6.95	7.64	8.34	9.03
23.25	6.29	6.99	7.69	8.39	9.09
23.50	6.33	7.03	7.73	8.44	9.14
23.75	6.37	7.07	7.78	8.49	9.19
24.00	6.40	7.11	7.83	8.54	9.25
24.25	6.44	7.16	7.87	8.59	9.30
24.50	6.48	7.20	7.92	8.64	9.36
24.75	6.51	7.24	7.96	8.69	9.41
25.00	6.55	7.28	8.01	8.74	9.46
25.25	6.59	7.32	8.05	8.79	9.52
25.50	6.63	7.36	8.10	8.84	9.57
25.75	6.66	7.41	8.15	8.89	9.63
26.00	6.70	7.45	8.19	8.94	9.68
26.25	6.74	7.49	8.24	8.99	9.74
26.50	6.78	7.53	8.28	9.04	9.79
26.75	6.82	7.57	8.33	9.09	9.85
27.00	6.85	7.62	8.38	9.14	9.90
27.25	6.89	7.66	8.42	9.19	9.96
27.50	6.93	7.70	8.47	9.24	10.01
27.75	6.97	7.74	8.52	9.29	10.07
28.00	7.01	7.79	8.57	9.34	10.12
28.25	7.05	7.83	8.61	9.40	10.18
28.50	7.09	7.87	8.66	9.45	10.23
28.75	7.12	7.92	8.71	9.50	10.29
29.00	7.16	7.96	8.75	9.55	10.35
29.25	7.20	8.00	8.80	9.60	10.40
29.50	7.24	8.05	8.85	9.65	10.46
29.75	7.28	8.09	8.90	9.71	10.52
30.00	7.32	8.13	8.95	9.76	10.57

[1]Use this table with procedures 3A1a6a and 3A1a6b.

Particle-Size Distribution Analysis (3A)
Particles >2 mm (3A2)

Rock and pararock fragments are defined as particles >2mm in diameter and include all particles with horizontal dimensions less than the size of a pedon (Soil Survey Division Staff, 1993). Rock fragments are further defined as strongly cemented or more resistant to rupture, whereas pararock fragments are less cemented than the strongly cemented class, with most of these fragments broken into particles 2 mm or less in diameter during the preparation of samples for particle-size analysis in the laboratory. Rock fragments are generally sieved and excluded from most chemical, physical, and mineralogical analyses. Exceptions are described in method 1B1b2f. It is necessary to know the amount of rock fragments for several applications, e.g., available water capacity and linear extensibility (Grossman, 2002).

The SSL determines weight percentages of the >2-mm fractions by field and laboratory weighings by procedure 3A2a1. In the field or in the laboratory, the sieving and weighing of the >2-mm fraction are limited to the <75-mm fractions. In the field, fraction weights are usually recorded in pounds, whereas in the laboratory, fraction weights are recorded in grams. The 20- to 75-mm fraction is generally sieved, weighed, and discarded in the field. This is the preferred and usually the most accurate method. Less accurately, the 20- to 75-mm fraction is estimated in the field as a volume percentage of the whole soil. If it is sieved and weighed in the laboratory, the results are usually not reliable because of small sample size.

The SSL estimates weight percentages of the >2-mm fractions from volume estimates of the >20-mm fractions and weight determinations of the <20-mm fractions by procedure 3A2a2. The volume estimates are visual field estimates. Weight percentages of the >20-mm fractions are calculated from field volume estimates of the 20- to 75-mm, 75- to 250-mm, and >250-mm fractions. The >250-mm fraction includes stones and boulders that have horizontal dimensions that are smaller than the size of the pedon. Weight measurements for the 2- to 20-mm fraction are laboratory measurements. Weight measurements of the 20- to 75-mm fractions in the field are more accurate than visual volume estimates. Weight measurements of this fraction in the laboratory are not reliable. The volume estimates that are determined in the field are converted to dry weight percentages. For any >2-mm fractions estimated by volume in the field, the SSL calculates weight percentages by procedure 3A2b. The visual volume estimates of the >20-mm fraction are subjective. The conversion of a volume estimate to a weight estimate assumes a particle density of 2.65 g cc-1 and a bulk density for the fine-earth fraction of 1.45 g cc-1. Measured values can be substituted in this volume to weight conversion, if required.

Soil variability and sample size are interferences to weight determinations of the >2-mm particles. Enough soil material needs to be sieved and weighed to obtain statistically accurate rock fragment content. In order to accurately measure rock fragments with maximum particle diameters of 20 and 75 mm, the minimum dry specimen sizes that need to be sieved and weighed are 1.0 and 60.0 kg, respectively. Refer to ASTM method D 2488 (American Society for Testing and Materials, 2004). Whenever possible, the field samples or "moist" material should have weights two to four times larger (American Society for Testing and Materials, 2004). Therefore, sieving and weighing the 20- to 75-mm fraction should be done in the field. The <20-mm fractions are sieved and weighed in the laboratory.

Procedures for reporting data for a size fraction base are outlined in Section 2C. Unless otherwise specified, the particle-size fractions 2 to 5, 5 to 20, 20 to 75, and 0.1 to 75 mm are reported on a <75-mm oven-dry weight percentage basis. The total >2-mm fraction is reported on a whole soil oven-dry weight percentage base.

Particle-Size Distribution Analysis (3A)
Particles >2 mm (3A2)
Weight Estimates (3A2a)
By Field and Laboratory Weighing (3A2a1)

1. Application

Procedure 3A2a1 is used to determine weight percentages of the >2 mm fractions by field and laboratory weighings. The 20- to 75-mm fraction is generally sieved, weighed, and discarded in the field or is obtained from a field volume percentage estimate. However, the 20- to 75-mm fraction can be sieved and weighed in the laboratory. The <20-mm fractions are sieved and weighed in the laboratory.

2. Summary of Method

Field weights are determined for the 20- to 75-mm fraction. This is the preferred method. When field determinations are not possible, weight measurements for the 20- to 75-mm fraction can be determined in the laboratory. The <20-mm fractions are sieved and weighed in the laboratory. The percentage of any 2- to 75-mm fraction on a <75-mm oven-dry weight basis is calculated.

3. Interferences

Soil variability and sample size are interferences to weight determinations of the >2-mm particles. Enough soil material needs to be sieved and weighed to obtain statistically accurate rock fragment content. In order to measure accurately rock fragments with maximum particle diameters of 20 and 75 mm, the minimum specimen sizes ("dry" weights) that need to be sieved and weighed are 1.0 and 60.0 kg, respectively. Refer to ASTM method D 2488 (American Society for Testing and Materials, 2004). Samples received in the laboratory generally have a maximum weight of 4 kg. Therefore, sieving and weighing the 20- to 75-mm fraction should be done in the field. The <20-mm fractions are sieved and weighed in the laboratory.

4. Safety

The main hazards are in the field during sample collection. Some hazards are sharp-edged excavation tools, snake bites, and falls.

5. Equipment
5.1 Electronic balance, ±1-g sensitivity and 15-kg capacity
5.2 Trays, plastic, tared
5.3 Sieves, square-hole
5.3.1 9 mesh, 2 mm
5.3.2 4 mesh, 4.76 mm
5.3.3 20 mm, 3/4 in
5.3.4 76 mm, 3 in
5.4 Mechanical shaker with 9-mesh and 4-mesh sieves
5.5 Rubber roller
5.6 Metal plate, 76 x 76 x 0.5 cm
5.7 Scale, 100-lb (45-kg) capacity
5.8 Brown Kraft paper

6. Reagents
6.1 Reverse osmosis (RO) water
6.2 1 N HCl. Refer to procedure 1B.
6.3 Sodium hexametaphosphate solution. Dissolve 35.7 g of sodium hexametaphosphate $(NaPO_3)_6$ and 7.94 g of sodium carbonate (Na_2CO_3) in L of distilled water.

7. Procedure

Field

7.1 Sieve a representative horizon sample with a 76-mm sieve. Sieve about 60 kg of material to accurately measure rock fragments that have a maximum particle diameter of 75 mm. As a 60-kg sample may not be feasible because of limitations of time and/or soil material, actual sample size may be 30 or 40 kg. Discard the >75-mm material. Weigh and record weight (lbs) of <75-mm fraction. Sieve this material with a 20-mm sieve. Discard the 20- to 75-mm fraction. Weigh and record weight (lbs) of <20-mm fraction. Place a sub-sample of the <20-mm material in a plastic bag. Label and send to laboratory for analyses.

Laboratory

7.2 Distribute the field sample on a plastic tray, weigh, and record moist weight. Air-dry, weigh, and record weight.

7.3 Process air-dry material on a flat, metal plate that is covered with brown Kraft paper. Thoroughly mix material by moving the soil from the corners to the middle of the processing area and then by redistributing the material. Repeat process four times. Roll material with wooden rolling pin to crush clods to pass a 2-mm sieve. For samples with easily crushed coarse fragments, substitute rubber roller for wooden rolling pin. Sieve clayey soils that contain many coarse fragments in the mechanical shaker. Roll and sieve until only the coarse fragments that do not slake in sodium hexametaphosphate solution on the sieve.

7.4 If more sample is received than is needed for processing, select a sub-sample for preparation. Weigh sub-sample and record weight.

7.5 Weigh soil material with diameters of 2 to 5 mm. Soak in sodium hexametaphosphate solution for 12 h. Air-dry, weigh the material that does not slake, and discard. Weigh, record weight, and discard coarse fragments with diameters of 20 to 75 mm and 5 to 20 mm. Most laboratory samples do not contain 20- to 75-mm fragments, as this fraction is generally sieved, weighed, and discarded in the field.

8. Calculations

8.1 If field weight measurements are determined for the <75-mm and the 20- to 75-mm fraction, convert these weights in pounds to grams. If laboratory measurements are determined for the <75 mm and the 20- to 75-mm fractions, these weights are already in grams.

8.2 Determine field-moist weight of the sub-sample as received in the laboratory. Determine air-dry weight of sub-sample. Air-dry weight is defined as a constant sample weights obtained after drying at $30\pm5°C$ ($\approx$ 3 to 7 days).

8.3 Determine ratio of slaked, air-dried weight (g) to unslaked, air-dried weight (g) for the 2- to 5-mm fraction. Using this ratio, adjust weight of coarse fragments with <5-mm diameters.

8.4 Base coarse fragment calculation on oven-dry weight-basis. Use the AD/OD (air-dry/oven-dry ratio) (procedure 3D1) to calculate the oven-dry weight of <2-mm fraction. Use the following equation to determine the percentage of any 2- to 75-mm fraction on a <75-mm oven-dry weight-basis.

Percentage >2 mm fraction(<75-mm basis) = (A/B) x 100

where:

A = Weight of 2- to 75- mm fraction (g)
B = Weight of <75-mm fraction (g)

8.5 Determine oven-dry weight by weighing the sample after oven-drying at 110°c for 24 h or by calculating as follows:

Oven-dry weight (g) = [Air-dry weight (g)]/ADOD

where:
ADOD = Air-dry/oven-dry weight

8.6 Similarly, determine oven-dry weight from the field-moist weight of a sample by calculating as follows:

Oven-dry weight (g) = [Field-moist weight (g)]/[Field-moist weight (g)/Oven-dry weight (g)]

8.7 In calculations of the oven-dry weight percentages of the >2-mm fraction, make corrections for the field-water content of the <75-mm sample at sampling and for the water content of the air-dry bulk laboratory sample. Base the corrections for the field-water content on the difference between the field-moist weight and air-dry weight of the bulk sample.

9. Report

Field
9.1 Weight (lbs) of field-moist, <75-mm fraction
9.2 Weight (lbs) of field-moist, 20- to 75-mm fraction

Laboratory
9.3 Weight (g) of field-moist soil sample
9.4 Weight (g) of air-dry soil sample
9.5 Weight (g) of air-dry processed soil sample
9.6 Weight (g) 20-to 75-mm fraction
9.7 Weight (g) 5- to 20-mm fraction
9.8 Weight (g) 2- to 5-mm fraction
9.9 Weight (g) of sub-sample 2- to 5-mm fraction before slaking
9.10 Weight (g) of sub-sample 2- to 5-mm fraction after slaking

10. Precision and Accuracy
Precision and accuracy data are available from the SSL upon request.

11. References
American Society for Testing and Materials. 2004. Standard Practice for Description and Identification of Soils (Visual-Manual Procedure). D 2488. Annual book of ASTM standards. Construction. Section 4. Soil and rock; dimension stone; geosynthesis. Vol. 04.08. ASTM, Philadelphia, PA.
Grossman, R.B., and T.G. Reinsch. 2002. Bulk density and linear extensibility. p. 201-228. *In* J.H. Dane and G.C. Topp (eds.) Methods of soil analysis. Part 4. Physical methods. Soil Sci. Am. Book Series No. 5. ASA and SSSA, Madison, WI.
Soil Survey Division Staff. 1993. Soil survey manual. USDA. Handb. No. 18. U.S. Govt. Print. Office, Washington, DC.

Particle-Size Distribution Analysis (3A)
Particles >2 mm (3A2)
Weight Estimates (3A2a)
From Volume and Weight Estimates (3A2a2)
Volume Estimates (3A2b)

1. Application

Procedure 3A2a2 is used to determine weight percentages of the >2 mm fractions from volume estimates and weight determinations. The volume estimates are visual field estimates for any fractions that are >20 mm. The weight estimates are laboratory measurements for the 2-to 20-mm or 2- to 75-mm fractions. The volume estimates for any fractions that are >20 mm are converted to weight percentages. The total >2-mm fraction is reported on an oven-dry weight basis for whole soil. Method 3A2b is the calculations used to derive weight percentages from volume percentages of all the >2-mm material.

2. Summary of Method

Visual field volume estimates are determined for any fractions that are >20 mm. These volume estimates include, if applicable, the 20- to 75-mm, 75- to 250-mm, and the >250-mm fractions. The >250-mm fraction includes stones and boulders that have horizontal dimensions that are less than those of the pedon. Instead of visual field volume estimates, field weights for the 20- to 75-mm fraction may be determined. This is the preferred method. If these measurements are unavailable, visual field volume estimates of the 20- to 75-mm fraction are used rather than laboratory weights of this fraction. The <20-mm fractions are sieved and weighed in the laboratory.

3. Interferences

Soil variability and sample size are interferences to weight determinations of the >2-mm particles. Enough soil material needs to be sieved and weighed to obtain statistically accurate rock fragment content. In order to accurately measure rock fragments with maximum particle diameters of 20 and 75 mm, the minimum specimen sizes ("dry" weights) that need to be sieved and weighed are 1.0 and 60.0 kg, respectively. Refer to ASTM Standard Practice D 2488 (American Society for Testing and Materials, 2004). Samples received in the laboratory generally have a maximum weight of 4 kg. Therefore, sieving and weighing the 20- to 75-mm fraction should be done in the field.

The visual volume estimates of the >75-mm fractions are subjective. The conversion of a volume estimate to a weight estimate assumes a particle density of 2.65 g cc^{-1} and a bulk density for the fine-earth fraction of 1.45 g cc^{-1}. If particle density and bulk density measurements are available, they are used in the calculations.

4. Safety

The main hazards are in the field during sample collection. Some hazards are sharp-edged excavation tools, snake bites, and falls.

5. Equipment

5.1 Electronic balance, ±1-g sensitivity and 15-kg capacity
5.2 Trays, plastic, tared
5.3 Sieves, square-hole
5.3.1 9 mesh, 2 mm
5.3.2 4 mesh, 4.76 mm
5.3.3 20 mm, 3/4 in
5.3.4 76 mm, 3 in
5.4 Mechanical shaker with 9-mesh and 4-mesh sieves
5.5 Rubber roller

5.6 Metal plate, 76 x 76 x 0.5 cm
5.7 Scale, 45-kg (100-lb) capacity
5.8 Brown Kraft paper

6. Reagents
6.1 Reverse osmosis (RO) water
6.2 1 N HCl. Refer to procedure 1B.
6.3 Sodium hexametaphosphate solution. Dissolve 35.7 g of sodium hexametaphosphate $(NaPO_3)_6$ and 7.94 g of sodium carbonate (Na_2CO_3) in L of distilled water.

7. Procedure
Field
7.1 Determine volume estimates as percentages of soil mass for the 75- to 250-mm and for the >250-mm fractions. The >250-mm fraction includes stones and boulders with horizontal dimensions less than those of the pedon.

7.2 Determine either weight measurements in pounds or visual field volume estimates in percentages for the 20- to 75-mm fragments. Weight measurements for the 20- to 75-mm fraction are the preferred method. However, volume estimates are more accurate than laboratory weights using small samples.

7.3 If field weight measurements are determined for the 20- to 75-mm fraction, sieve an entire horizon sample with a 76-mm sieve. Sieve ≈ 60 kg of material to accurately measure rock fragments that have a maximum particle diameter of 75 mm. A 60-kg sample may not be possible because of limitations of time and/or soil material. Actual sample size may be 30 or 40 kg. Discard the >75-mm material. Weigh and record weight of <75-mm fraction. Sieve this material with a 20-mm sieve. Discard the 20- to 75-mm fraction. Weigh and record weight of <20-mm fraction. Place a sub-sample of the <20-mm material in an 8-mL, plastic bag. Label and send to laboratory for analyses.

Laboratory
7.4 Distribute the field sample on a plastic tray, weigh, and record moist weight. Air-dry, weigh, and record weight.

7.5 Process air-dry material on a flat, metal plate that is covered with brown Kraft paper. Thoroughly mix material by moving the soil from the corners to the middle of the processing area and then by redistributing the material. Repeat process four times. Roll material with wooden rolling pin to crush clods to pass a 2-mm sieve. For samples with easily crushed coarse fragments, substitute rubber roller for wooden rolling pin. Sieve clayey soils that contain many coarse fragments in the mechanical shaker. Roll and sieve until only the coarse fragments that do not slake in sodium hexametaphosphate solution remain on the sieve.

7.6 If more sample is received than is needed for processing, select sub-sample for preparation. Weigh sub-sample and record weight.

7.7 Weigh soil material with diameters of 2 to 5 mm. Soak in sodium hexametaphosphate solution for 12 h. Air-dry, weigh the material that does not slake, and discard. Weigh, record weight, and discard coarse fragments with diameters of 20 to 75 mm and 5 to 20 mm. Most laboratory samples do not contain 20- to 75-mm fragments as this fraction is generally weighed, sieved, and discarded in the field.

8. Calculations

From Volume and Weight Estimates (3A2a2)

8.1 Calculate weight percentages from volume percentages using measured bulk density (Db_m) and particle density (Dp). If measurements are unavailable, assume a Db_m of 1.45 g cc^{-1} and a D_p of 2.65 g cc^{-1}.

8.2 Use the following equation to convert all volume estimates to weight percentages for specified fractions.

$$\text{Percentage} >2 \text{ mm (wt basis)} = [100 \, Dp \, (x)]/[Dp \, (x) + Db_m \, (1-x)]$$

where:
D_p = Particle density (2.65 g cc^{-1}, unless measured)
Db_m = Bulk density (1.45 g cc^{-1} for <2-mm fraction, unless measured)

x = [volume fragments > i mm]/[volume whole soil]

where:
i = size fraction above which volume estimates are made and below which weight percentages are determined, usually 20 or 75 mm in diameter

8.3 Use the preceding equation to calculate any individual fraction >j mm (j = any size fraction) by substituting an appropriate value of Db_m representing the fabric <j mm.

Volume Estimates (3A2b)

8.4 Use the following equation to determine the volume of the <2-mm fraction per unit volume of whole soil.

$$Cm = [\text{Volume}_{\text{moist <2-mm fabric}}]/[\text{Volume}_{\text{moist whole-soil}}] = [Dp \, (1-y) \, (1-x)]/[Dp \, (1-y) + Db_m \, (y)]$$

where:
Cm = Rock fragment conversion factor
Volume moist whole soil = Volume of fine earth + rock fragments on moist whole-soil basis

y = [weight material between 2 mm and i mm]/[weight material < i mm]

8.5 Use the following formula to convert laboratory data on a <2-mm weight basis to moist whole soil volume basis.

$Cm \times Db_m \times \text{lab datum}$

8.6 Use the following formula to determine the volume percentage of <2-mm fabric in whole soil.

$Cm \times 100$

8.7 Use the following formula to determine the volume percentage of >2-mm fabric in whole soil.

$100 \, (1-Cm)$

8.8 Use the following formula to report weight of <2-mm fabric per unit volume of whole soil for some soils.

$$(C_m \times Db_m)$$

9. Report

Field
9.1 Volume (%) >250-mm fraction (includes stones and boulders with horizontal dimensions smaller than size of a pedon)
9.2 Volume (%) 75- to 250-mm fraction
9.3 Volume (%) 20- to 75-mm fraction (not needed if weighed in field)
9.4 Weight (lbs) <75-mm fraction
9.5 Weight (lbs) 20-to 75-mm fraction

Laboratory
9.6 Weight (g) of field moist soil sample
9.7 Weight (g) of air-dry soil sample
9.8 Weight (g) of air-dry processed soil sample
9.9 Weight (g) 20-to 75-mm fraction
9.10 Weight (g) 5- to 20-mm fraction
9.11 Weight (g) 2- to 5-mm fraction
9.12 Weight (g) of sub-sample 2- to 5-mm fraction before slaking
9.13 Weight (g) of sub-sample 2- to 5-mm fraction after slaking

10. Precision and Accuracy
Precision and accuracy data are available from the SSL upon request.

11. References
American Society for Testing and Materials. 2004. Standard Practice for Description and Identification of Soils (Visual-Manual Procedure). D 2488. Annual book of ASTM standards. Construction. Section 4. Soil and rock; dimension stone; geosynthesis. Vol. 04.08. ASTM, Philadelphia, PA.

SOIL PHYSICAL AND FABRIC RELATED ANALYSES (3)

Bulk Density (3B)

Density is defined as mass per unit volume. Soil bulk density of a sample is the ratio of the mass of solids to the total or bulk volume. This total volume includes the volume of both solids and pore space. Bulk density is distinguished from particle density which is mass per unit volume of only the solid phase. Particle density excludes pore spaces between particles. As bulk density (Db) is usually reported for the <2-mm soil fabric, the mass and volume of rock fragments are subtracted from the total mass and volume.

Bulk density maybe highly dependent on soil conditions at the time of sampling. Changes in soil volume due to changes in water content will alter bulk density. Soil mass remains fixed, but the volume of soil may change as water content changes (Blake and Hartge, 1986). Bulk density, as a soil characteristic, is actually a function rather than a single value. Therefore, subscripts are added to the bulk density notation, Db, to designate the water state of the sample when the volume was measured. The SSL uses the bulk density notations of Db_f, Db_{33}, Db_{od}, and Db_r for field-state, 33-kPa equilibration, oven-dry, and rewet, respectively.

Field-state (Db_f) is the bulk density of a soil sample at field-soil water content at time of sampling. The 33-kPa equilibration (Db_{33}) is the bulk density of a soil sample that has been desorbed to 33kPa (1/3 bar). The oven-dry (Db_d) is the bulk density of a soil sample that has been dried in an oven at 110°C. The rewet (Db_r) is the bulk density of soil sample that has been equilibrated, air-dried, and re-equilibrated. The Db_r is used to determine the irreversible shrinkage of soils and subsidence of organic soils. The determinations of these bulk density values, Db_f, Db_{33}, Db_{od}, and Db_r, are described in procedures 3B1a, 3B1b, 3B1c, and 3B1d, respectively. Bulk density also may be determined for field-moist soil cores of known volume (procedure 3B6a). The bulk density of a weak or loose soil material for which the clod or core method is unsuitable may be determined by the compliant cavity method (procedure 3B3a).

In general, there are two broad groupings of bulk density methods as follows: (1) one for soil materials coherent enough that a field-sample can be removed; and (2) the other for soils too fragile to remove a sample and an excavation operation must be performed. Under the former, there are clod methods in which the sample has an undefined volume, is coated, and the volume is determined by submergence. Also under the former there are various methods in which a cylinder of known volume is obtained of soil sufficiently coherent that it remains in the cylinder. The complete cylinder may be inserted (3B6a) or only part of the cylinder is inserted and the empty volume is subtracted from the total volume of the core (e.g., variable height method, Grossman and Reinsch, 2002). Three excavation procedures have been used to determine Db_f as follows: (1) compliant cavity (3B3a); (2) ring excavation (3B4a); and (3) frame excavation (3B5a) (Grossman and Reinsch, 2002). The frame-excavation provides for a larger sample area and is advantageous where there is large, very local variability as found in the O horizons (Soil Survey Staff, 1999) of woodlands.

The complication concerning the difference between bulk density of the soil and that of the sample is particularly important for the clod method as presented here which permits determination of the volume at different water contents and hence volumes. If the water content is at or near field capacity, desiccation cracks are closed and the bulk density (Db_{33} or Db_f if field-water is near field capacity) of the soil and of the sample are considered the same. However, if the sample is at a water content below field-capacity through drying after sampling or because the sample was taken below field-capacity, then desiccation cracks that occur in place are excluded from the soil and the sample bulk density exceeds that of the soil. If the sample is large and inclusive of the desiccation cracks, as in some excavation procedures, then again the sample and soil bulk density are the same. The difference between sample and soil bulk density is particularly large for oven-dry clods (D_{od}) of soils with high extensibility. And may also be hard for soils subject to a large increase if taken through a rewet cycle. Grossman and Reinsch (2002) discuss the manipulation of clod bulk densities (the sample) at water contents below field capacity to obtain an estimate of the soil bulk density at such water contents. Similarly, estimates of soil bulk density at intermediate field-water contents between field capacity and oven-dryness inclusive of desiccation crack space are discussed by Grossman et al. (1990).

References

Blake, G.R., and K.H. Hartge. 1986. Bulk density. p. 363-382. *In* A. Klute (ed.) Methods of soil analysis. Part 1. Physical and mineralogical methods. 2nd ed. Agron. Monogr. 9. ASA and SSSA, Madison, WI.

Grossman, R.B., F.B. Pringle, R.J. Bigler, W.D. Broderson, and T.M. Sobecki. 1990. Systematics and field morphology for a use-dependent, temporal soil data base for edaphological and environmental applications. p. 452-460. *In* Int. Symp. On Advanced Technology in Natural Resource Management, 2nd. ed. Georgetown Univ., Washington, DC.

Grossman, R.B., and T.G. Reinsch. 2002. Bulk density and linear extensibility. p. 201-228. *In* J.H. Dane and G.C. Topp (eds.) Methods of soil analysis. Part 4. Physical methods. Soil Sci. Am. Book Series No. 5. ASA and SSSA, Madison, WI.

Soil Survey Staff. 1999. Soil taxonomy: A basic system of soil classification for making and interpreting soil surveys. 2nd ed. Agric. Handb. No. 436. USDA-NRCS. U.S. Govt. Print. Office, Washington, DC.

Bulk Density (3B)
Saran-Coated Clods (3B1)
Field-State (Db_f) (3B1a)

1. Application

Bulk density is used to convert data from a weight to a volume basis; to estimate saturated hydraulic conductivity; and to identify compacted horizons. The procedure 3B1a determines the bulk density value Db_f of a soil sample at field-soil water content at time of sampling. Field-bulk density (Db_f) offers the opportunity to obtain relatively cheaply bulk density information without the expense incurred to obtain water retention. Db_f is particularly useful if the soil layers are at or above field capacity and/or the soils have low extensibility and do not exhibit desiccation cracks even if below field capacity.

2. Summary of Method

Field-occurring fabric (clods) is collected from the face of an excavation. One coat of plastic lacquer is applied in the field. Additional coats of plastic lacquer are applied in the laboratory. In its field-water state or after equilibration, the clod is weighed in air to measure its mass and in water to measure its volume. After the clod is dried in oven at 110°C, its mass and volume are determined again. A correction is made for the mass and volume of rock fragments and plastic coatings (Brasher et al., 1966; Blake and Hartge, 1986; Grossman and Reinsch, 2002).

3. Interferences

Errors are caused by non-representative samples. Only field-occurring fabric (clods) should be sampled. The whole soil bulk density may be overestimated because sampled clods frequently exclude the crack space between clods (Grossman and Reinsch, 2002).

The penetration of plastic lacquer into the voids of sandy or organic soils interferes with the corrections for mass and volume of the plastic coat and with the accuracy of water content determinations. Penetration can be reduced by spraying water on the clod followed by immediately dipping the clod in the plastic lacquer.

Loss of soil during the procedure will void the analyses because all calculations are based on the oven-dry soil mass. Holes in the plastic coating, which are detected by escaping air bubbles from submerged clod, introduce errors in volume measurement. An inadequate evaporation of the plastic solvent results in overestimation of the soil mass. A drying time of 1 h is usually sufficient time for evaporation of solvent. However, clods with high organic matter content may need to dry longer.

Bulk density is reported for <2-mm soil fabric. Correction for rock fragments >2-mm requires either knowledge or assumption of the rock fragment density. Estimate or measurement errors of rock fragment density will affect the accuracy of the value for soil bulk density. Rock fragments may contain water which complicates the application to actual water-holding capacity.

4. Safety

Methyl ethyl ketone (MEK) is extremely flammable. A type B fire extinguisher should be in close proximity in the laboratory. No open flames or nearby operation of electrical equipment are permitted while using MEK. The MEK vapor is classified as a sensory irritant. The 8-h time-weighted average (TWA) exposure limit is 200 ppm, and the short-term exposure limit (STEL) is 300 ppm (Occupational Safety and Health Administration, 1989). Avoid physical contact. Use with adequate ventilation. In closed buildings, use a fume hood. Keep in tightly closed containers. Use safety glasses, proper gloves, and a lab coat. Wash hands immediately after handling MEK. Additional information on the safe handling of MEK is available in Chemical Safety Data Sheet SD-83, Manufacturing Chemists' Association, Inc., 1825 Connecticut Ave. NW, Washington, DC.

Saran F-310 resin will decompose rapidly at temperatures >200°C releasing hydrogen chloride gas. Avoid contact with Fe, Zn, Cu, and Al in solution. Avoid all contact with strong bases.

5. Equipment
5.1 Electronic balance, ±0.01-g sensitivity
5.2 Rigid shipping containers. The SSL uses a corrugated box with compartments.
5.3 Plastic bags, 1 mL, 127 x 89 x 330 mm
5.4 Wire. The SSL uses a 28-awg coated copper wire.
5.5 Hairnets
5.6 Stock tags, 25.4-mm (1-in) diameter paper tag, with metal rim
5.7 Hook assembly for weighing below balance
5.8 Plexiglass water tank mounted on a fulcrum and lever to elevate tank
5.9 Oven, 110°C
5.10 Sieve, No. 10 (2-mm openings)
5.11 Rope, 3 m
5.12 Clothespins
5.13 Silt loam soil
5.14 Hot plate
5.15 Spray bottle
5.16 Liquid vapor trap. The SSL constructs a tin enclosure over hot plate with a chimney and duct to transfer vapor to water stream.

6. Reagents
6.1 Methyl ethyl ketone (MEK), practical (2-butanone)
6.2 Water
6.3 Alcohol
6.4 Liquid detergent. The SSL uses Liqui-Nox.
6.5 Dow Saran F-310 Resin, available from Dow Chemical Company.
6.6 Plastic lacquer. Prepare plastic lacquer with resin to solvent ratios of 1:4 and 1:7 on a weight basis. Fill a 3.8-L (1-gal) metal paint can with 2700 ± 200 mL of solvent. Fill to the bottom of handle rivet. Add 540 or 305 g of resin to make 1:4 or 1:7 plastic lacquer, respectively. For the initial field and laboratory coatings, use the 1:4 plastic lacquer. Use 1:7 plastic lacquer for the last two laboratory coats. The 1:7 plastic lacquer is used to conserve the resin and to reduce cost. In the field, mix solvent with a wooden stick. In the laboratory, stir solvent with a non-sparking, high speed stirrer while slowly adding resin. Stir plastic lacquer for 15 min at 25°C. Store plastic lacquer in covered plastic or steel containers. Acetone may be substituted for MEK.

7. Procedure

Field
7.1 Collect field-occurring clods, ≈ 100 to 200 cm³ in volume (fist-sized), from the face of the excavation. Three clods per horizon are recommended. It is important that these clods be as representative of the bulk sample volume as possible. Remove a piece of soil larger than clod from the face of sampling pit. From this piece, prepare a clod by gently cutting or breaking protruded peaks and compacted material from clod. If roots are present, trim roots with shears. No procedure for sampling clods is applicable to all soils. Adjust field-sampling techniques to meet field-conditions at time of sampling.

7.2 Make a clothesline by stretching a rope between two fixed points. Tie clod with fine copper wire or place clod in a hairnet. If clod is dry, moisten surface with a fine mist of water. Quickly dip entire clod into plastic lacquer (Fig. 1). Suspend clod from clothesline to dry (Fig. 2). Dry clod for 30 min or until odor of solvent dissipates. If the value of Db_f is required, store clods in waterproof plastic bags as soon as coating dries because coating is permeable to water vapor.

7.3 Pack clods in rigid containers to protect them during transport.

Laboratory

7.4 Prepare a round stock tag with sample identification number. Cut the copper wire and loop around the clod. Record weight (TAG) of tag and wire. Loop fine copper wire around clod, leaving a tail to which round stock tag is attached. Record weight of clod (CC1).

7.5 Dip clod in 1:4 plastic lacquer. Wait 7 min and then dip clod in 1:7 plastic lacquer. Wait 12 min and then dip clod in 1:7 plastic lacquer. Wait 55 min and then reweigh clod. If clod has adsorbed >3% plastic by clod weight or smells excessively of solvent, allow longer drying time, then reweigh clod and record weight (CC2).

7.6 The clod should be waterproof and ready for volume measurement by water displacement. Suspend the clod below the balance, submerge in water, and record weight (WMCW).

7.7 Dry clod in an oven at 110°C until weight is constant. Weigh oven-dry clod in air (WODC) and in water (WODCW) and record weights.

7.8 If clod contains >5% rock fragments by weight, remove them from clod. Place clod in a beaker and place on hot plate. Cover hot plate with a liquid vapor trap. Use a fume hood. Heat clod on hot plate in excess of 200°C for 3 to 4 h. The plastic coating disintegrates at temperatures above 200°C. After heating, clod should appear black and charred. Remove clod from hot plate, lightly coat with liquid detergent, and add hot water.

7.9 Wet sieve the cool soil through a 2-mm, square-hole sieve. Dry and record weight (RF) of rock fragments that are retained on the sieve. Determine rock fragment density by weighing them in air to obtain their mass and in water to obtain their volume. If rock fragments are porous and have a density similar to soil sample, do not correct clod mass and volume measurement for rock fragments. Correct for rock fragments if these fragments can withstand breakdown when dry soil is placed abruptly in water or calgon.

7.10 Correct bulk density for weight and volume of plastic coating. The coating has an air-dry density of $\approx$ 1.3 g cm^{-3}. The coating loses 10 to 20% of its air-dry weight when dried in oven at 110°C.

8. Calculations

8.1 Db$_f$ = [WODC - RF - ODPC – TAG]/{[(CC2 - WMCW)/WD] - (RF/PD) - (MPC/1.3)}

where:
Db$_f$ = Bulk density in g cc^{-1} of <2-mm fabric at field-sampled water state
WODC = Weight of oven-dry coated clod
RF   = Weight of rock fragments
ODPC = MPC x 0.85, weight of oven-dry plastic coat
TAG   = Weight of tag and wire
CC2   = Weight of tag and wire
WMCW = Weight of coated clod in water before oven drying
WD   = Water density
PD   = Density of rock fragments

8.2 MPC = [(CC2 - CC1) + FCE] x RV

where:
MPC = Weight of plastic coat before oven-drying
CC1 = Weight of clod before three laboratory plastic coats
RV = Percent estimate of remaining clod volume after cutting to obtain flat surface ($\approx 80\%$)

8.3 $FCE = 1.5 \times [(CC2 - CC1)/3]$

where:
FCE = Estimate of field-applied plastic coat

8.4 $Db_{od} = [WODC - RF - ODPC - TAG]/\{[(WODC - WODCW)/WD] - (RF/PD) - (MPC/1.3)\}$

where:
Db_{od} = Bulk density in g cm^{-3} <2-mm fabric at oven dryness
WODCW = Weight of oven-dry coated clod in water

8.5 $W_f = \{[(CC2 - MPC) - (WODC - ODPC)]/[WODC - RF - ODPC - TAG]\} \times 100$

where:
W_f = Percent water weight in sampled clod

9. Report
Bulk density is reported to the nearest 0.01g cm^{-3}.

10. Precision and Accuracy
Precision and accuracy data are available from the SSL upon request.

11. References
Blake, G.R. and K.H. Hartge. 1986. Bulk density. p. 363-382. *In* A. Klute (ed.) Methods of soil analysis. Part 1. Physical and mineralogical methods. 2nd ed. Agron. Monogr. 9. ASA and SSSA, Madison, WI.

Brasher, B.R., D.P. Franzmeier, V.T. Volassis, and S.E. Davidson. 1966. Use of saran resin to coat natural soil clods for bulk density and water retention measurements. Soil Sci. 101:108.

Grossman, R.B., and T.G. Reinsch. 2002. Bulk density and linear extensibility. p. 202-228. *In* J.H. Dane and G.C. Topp (eds.) Methods of soil analysis, Part 4. Physical methods. Soil Sci. Am. Book Series No. 5. ASA and SSSA, Madison, WI.

Occupational Safety and Health Administration. 1989. Air contaminants; final rule. Federal Register. 29 CFR Part 1910. Vol. 54. No. 12. p. 2452-2453. U.S. Govt. Print. Office, Washington, DC.

Fig. 1. Dipping clods with hairnet in plastic-lacquer.

Fig. 2. After dipping, clods are tied to clothesline to dry.

Bulk Density (3B)
Saran-Coated Clods (3B1)
33-kPA Desorption (Db_{33}) (3B1b)

1. Application

Bulk density is used to convert data from a weight to a volume basis; to determine the coefficient of linear extensibility; to estimate saturated hydraulic conductivity; and to identify compacted horizons. The procedure 3B1b determines the bulk density value ($Db_{1/3}$) of a soil sample equilibrated at 33 kPa.

2. Summary of Method

Field-occurring fabric (clods) are collected from the face of an excavation. One coat of plastic lacquer is applied in the field. Additional coats of plastic lacquer are applied in the laboratory. The clod is desorbed to 33 kPa. After equilibration, the clod is weighed in air to measure its mass and in water to measure its volume. After the clod is dried in the oven at 110°C, its mass and volume are determined again. A correction is made for the mass and volume of rock fragments and for plastic coatings (Brasher et al., 1966; Blake and Hartge, 1986).

3. Interferences

Errors are caused by non-representative samples. Only field-occurring fabric (clods) should be sampled. The whole soil bulk density may be overestimated because sampled clods frequently exclude the crack space between clods (Grossman and Reinsch, 2002).

The penetration of plastic lacquer into the voids of sandy and organic soils interferes with the corrections for mass and volume of the plastic coat and with the accuracy of water content determinations. Penetration can be reduced by spraying water on the clod followed by immediately dipping the clod in the plastic lacquer. Dipping should be done as quickly as possible to reduce penetration of plastic.

Loss of soil during the procedure will void the analyses because all calculations are based on the oven-dry soil mass. Holes in the plastic coating, which are detected by escaping air bubbles from submerged clod, introduce errors in volume measurement. An inadequate evaporation of plastic solvent results in overestimation of the soil mass. A drying time of 1 h is usually sufficient time for evaporation of solvent. However, clods with high organic matter content may need to dry longer.

Clods placed in an unsealed plastic bag can lose moisture during storage prior to analysis. If clods irreversibly dry below 33-kPa-water content, then $Db_{1/3}$ values for 33 kPa will be erroneous. Completely seal the plastic storage bag to prevent drying.

Bulk density is reported for <2-mm soil fabric. Correction for rock fragments >2-mm requires either knowledge or assumption of the rock fragment density. Estimate or measurement errors of rock fragment density will affect the accuracy of the value for soil bulk density. Rock fragments may contain water, which complicates the application to actual water-holding capacity.

4. Safety

Methyl ethyl ketone (MEK) is extremely flammable. A type B fire extinguisher should be in close proximity in the laboratory. No open flames or nearby operation of electrical equipment are permitted while using MEK. The MEK vapor is classified as a sensory irritant. The 8-h time weighted average (TWA) exposure limit is 200 ppm, and the short-term exposure limit (STEL) is 300 ppm (Occupational Safety and Health Administration, 1989). Avoid physical contact. Use with adequate ventilation. In closed buildings, use a fume hood. Keep in tightly closed containers. Use safety glasses, proper gloves, and a lab coat. Wash hands immediately after handling MEK. Additional information on the safe handling of MEK is available in Chemical Safety Data Sheet SD-83, Manufacturing Chemists' Association, Inc., 1825 Connecticut Ave. NW, Washington, DC.

Saran F-310 resin will decompose rapidly at temperatures >200°C releasing hydrogen chloride gas. Avoid contact with Fe, Zn, Cu, and Al in solution. Avoid all contact with strong bases.

5. Equipment
5.1 Electronic balance, ±0.01-g sensitivity
5.2 Pressure plate extractor with porous ceramic plate.
5.3 Air pressure, 33-kPa
5.4 Rigid shipping containers. The SSL uses a corrugated box with compartments.
5.5 Plastic bags, 1 mil, 127 x 89 x 330 mm
5.6 Wire. The SSL uses a 28-awg coated copper wire.
5.7 Hairnets
5.8 Stock tags, 25.4-mm (1-in) diameter paper tag, with metal rim
5.9 Hook assembly for weighing below balance
5.10 Plexiglass water tank mounted on a fulcrum and lever to elevate tank
5.11 Oven, 110°C
5.12 Sieve, No. 10 (2-mm openings)
5.13 Rope, 3 m
5.14 Clothespins
5.15 Knife
5.16 Tile cut-off saw with diamond blade
5.17 Silt loam soil
5.18 Hot plate
5.19 Desiccator with ceramic plate
5.20 Vacuum, 80 kPa (0.8 bar)
5.21 Metal probe
5.22 Spray bottle
5.23 Liquid vapor trap. The SSL constructs a tin enclosure over hot plate with a chimney and duct to transfer vapor to water stream.
5.24 Reinforced paper towels or cheesecloth
5.25 Tension table. The SSL constructs a tension table by placing porous firebricks, covered with reinforced paper towels, in a tub of water.

6. Reagents
6.1 Methyl ethyl ketone (MEK), practical (2-butanone)
6.2 Water
6.3 Alcohol
6.4 Liquid detergent. The SSL uses Liqui-Nox.
6.5 Dow Saran F-310 Resin, available from Dow Chemical Company.
6.6 Plastic lacquer. Prepare plastic lacquer with resin to solvent ratios of 1:4 and 1:7 on a weight basis. Fill a 3.8-L (1-gal) metal paint can with 2700 ±200 mL of solvent. Fill to the bottom of handle rivet. Add 540 or 305 g of resin to make 1:4 or 1:7 plastic lacquer, respectively. For the initial field and laboratory coatings, use the 1:4 plastic lacquer. Use 1:7 plastic lacquer for the last two laboratory coats. The 1:7 plastic lacquer is used to conserve the resin and to reduce cost. In the field, mix solvent with a wooden stick. In the laboratory, stir solvent with a non-sparking, high-speed stirrer while slowly adding resin. Stir plastic lacquer for 15 min at 25°C. Store plastic lacquer in covered plastic or steel containers. Acetone may be substituted for MEK.

7. Procedure

Field

7.1 Collect field-occurring clods, ≈ 100 to 200 cm^3 in volume (fist-sized), from the face of the excavation. Three clods per horizon are recommended. It is important that these clods be as representative of the bulk sample volume as possible. Remove a piece of soil larger than clod from the face of sampling pit. From this piece, prepare a clod by gently cutting or breaking protruded peaks and compacted material from clod. If roots are present, trim roots with shears. No procedure for sampling clods is applicable to all soils. Adjust field-sampling techniques to meet field-conditions at time of sampling.

7.2 Make a clothesline by stretching a rope between two fixed points. Tie clod with fine copper wire or place clod in a hairnet. If the clod is dry, moisten surface with a fine mist of water. Quickly dip entire clod into plastic lacquer (See Fig. 1, procedure 3B1a). Suspend clod from clothesline to dry (See Fig. 2, procedure 3B1a). Dry clod for 30 min or until odor of solvent dissipates. If the value of Db$_f$ is required, store clods in waterproof plastic bags as soon as coating dries because coating is permeable to water vapor.

7.3 Pack clods in rigid containers to protect them during transport.

Laboratory

7.4 Prepare a round stock tag with sample identification number. Cut the copper wire and loop around the clod (Fig. 1). Record weight (TAG) of tag and wire. Loop fine copper wire around clod, leaving a tail to which round stock tag is attached. Record weight of clod (CC1).

7.5 Dip clod in 1:4 plastic lacquer. Wait 7 min and then dip clod in 1:7 plastic lacquer. Wait 12 min and then dip clod in 1:7 plastic lacquer. Wait 55 min and then reweigh clod. If clod has adsorbed >3% plastic by clod weight or smells excessively of solvent, allow longer drying time, then reweigh clod and record weight (CC2).

7.6 With a diamond saw, cut a flat surface on the clod. Place cut clod surface on a tension table, maintained at 5-cm tension (Fig.2). Periodically check clod to determine if it has reached equilibrium by inserting metal probe, touching, or by weight comparison. When clod has reached equilibrium, remove clod and record weight (WSC).

7.7 If cut clod does not adsorb water, place clod in a desiccator on a water-covered plate with a 0-cm tension. Submerge only the surface of clod in the water. Add a few mL of alcohol. Use in-house vacuum and apply suction until clod has equilibrated at saturation. Remove clod and record weight (WSC).

7.8 Place clod in a pressure plate extractor. To provide good contact between clod and ceramic plate, cover ceramic plate with a 5-mm layer of silt loam soil and saturate with water. Place a sheet of reinforced paper towel or cheesecloth over the silt loam soil. Place surface of cut clod on paper towel. Close container and secure lid. Apply gauged air pressure of 33 kPa. When water ceases to discharge from outflow tube, clod is at equilibrium. Extraction usually takes 3 to 4 wk. Remove clod and record weight (WMC). Compare WMC to WSC. If WMC ≥ WSC, equilibrate clod on tension table and repeat desorption process.

7.9 Dip clod in the 1:4 plastic lacquer. Wait 7 min and then dip clod in 1:7 plastic lacquer. Wait 12 min and then dip clod in 1:7 plastic lacquer. Wait 12 min and dip clod in 1:7 plastic lacquer. After 55 min, reweigh clod and record weight (CC3). If clod has adsorbed >3% plastic by weight or smells excessively of solvent, allow longer drying time, then reweigh clod.

7.10 The clod should be waterproof and ready for volume measurement by water displacement. Suspend clod below the balance, submerge in water, and record weight (WMCW).

7.11 Dry clod in an oven at 110°C until weight is constant. Weigh oven-dry clod in air (WODC) and in water (WODCW) and record weights.

7.12 If clod contains >5% rock fragments by weight, remove them from clod. Place clod in a beaker and place on hot plate. Cover hot plate with a liquid vapor trap. Use a fume hood. Heat clod on hot plate in excess of 200°C for 3 to 4 h. The plastic coating disintegrates at temperatures above 200°C. After heating, clod should appear black and charred. Remove clod from hot plate, lightly coat with liquid detergent, and add hot water.

7.13 Wet sieve the cool soil through a 2-mm, square-hole sieve. Dry and record weight (RF) of rock fragments that are retained on the sieve. Determine rock fragment density by weighing them in air to obtain their mass and in water to obtain their volume. If rock fragments are porous and have a density similar to soil sample, do not correct clod mass and volume measurement for rock fragments. Correct for rock fragments if these fragments can withstand breakdown when dry soil is placed abruptly in water or calgon.

7.14 Correct bulk density for weight and volume of plastic coating. The coating has an air-dry density of ≈ 1.3g cm^{-3}. The coating loses 10 to 20% of its air-dry weight when dried in oven at 110°C.

8. Calculations

8.1 Db_{33} = [WODC - RF - ODPC – TAG]/{[(CC3 - WMCW)/WD] - (RF/PD) - (MPC1/1.3)}

where:
Db_{33}	= Bulk density in g cc^{-1} of <2-mm fabric at 33-kPa tension
WODC	= Weight of oven-dry coated clod
RF	= Weight of rock fragments
TAG	= Weight of tag and wire
ODPC	= MPC1 x 0.85, weight of oven-dry plastic coat
CC3	= Weight of equilibrated clod after four additional plastic coats
WD	= Water density
PD	= Density of rock fragments
MPC1	= Weight of plastic coat before oven-drying
WMCW	= Weight in water of coated clod equilibrated at 33-kPa tension

8.2 MPC1 = {[(CC2 - CC1) + FCE] x RV} + (CC3-WMC)

where:
MPC1	= Weight of plastic coat before oven-drying
CC2	= Weight of clod after three laboratory plastic coats
CC1	= Weight of clod before three laboratory plastic coats
WMC	= Weight of coated clod equilibrated at 33-kPa tension
RV	= Percent estimate of remaining clod volume after cutting to obtain flat surface ($\approx 80\%$)

8.3 FCE = 1.5 x [(CC2 -CC1)/3]

where:
FCE = Estimate of field-applied plastic coat

8.4 $Db_{od} = [WODC - RF - ODPC - TAG]/ \{[(WODC - WODCW)/WD] - (RF/PD) - (MPC1/1.3)\}$

where:
Db_{od} = Bulk density in g cc^{-1} <2-mm fabric, oven-dry fabric
WODCW = Weight of oven-dry clod coated in water

8.5 $W_{33} = \{[(CC3 - MPC1) - (WODC - ODPC)]/[(WODC - RF - ODPC - TAG)]\} \times 100$

where:
W_{33} = Percent water weight retained at 33-kPa tension

9. Report
Bulk density is reported to the nearest 0.01g cm^{-3}.

10. Precision and Accuracy
Precision and accuracy data are available from the SSL upon request.

11. References

Blake, G.R., and K.H. Hartge. 1986. Bulk density. p. 363-382. *In* A. Klute (ed.) Methods of soil analysis. Part 1. Physical and mineralogical methods. 2nd ed. Agron. Monogr. 9. ASA and SSSA, Madison, WI.

Brasher, B.R., D.P. Franzmeier, V.T. Volassis, and S.E. Davidson. 1966. Use of saran resin to coat natural soil clods for bulk density and water retention measurements. Soil Sci. 101:108.

Grossman, R.B. and T.G. Reinsch. 2002. Bulk density and linear extensibility. p. 201-228. *In* J.H. Dane and G.C. Topp (eds.) Methods of soil analysis, Part 4. Physical methods. Soil Sci. Am. Book Series No. 5. ASA and SSSA, Madison, WI.

Occupational Safety and Health Administration. 1989. Air contaminants; final rule. Federal Register. 29 CFR Part 1910. Vol. 54. No. 12. p. 2452-2453. U.S. Govt. Print. Office, Washington, DC.

Fig. 1. A round stock tag with sample identification number is prepared. The cut copper wire is looped around the clod.

Fig. 2. After a flat surface on the clod is cut with a diamond saw, the clod is placed on a tension table, maintained at 5-cm tension.

Bulk Density (3B)
Saran-Coated Clods (3B1)
Oven-Dry (Db$_{od}$) (3B1c)

1. Application

Bulk density is used to convert data from a weight to a volume basis; to determine the coefficient of linear extensibility; to estimate saturated hydraulic conductivity; and to identify compacted horizons. The procedure 3B1c determines the bulk density value (Db$_{od}$) of an oven-dry soil sample.

2. Summary of Method

Field-occurring fabric (clods) is collected from the face of an excavation. One coat of plastic lacquer is applied in the field. Additional coats of plastic lacquer are applied in the laboratory. The clod is dried in an oven at 110°C and then weighed in air to measure its mass and in water to measure its volume. A correction is made for the mass and volume of rock fragments and for plastic coatings (Brasher et al., 1966; Blake and Hartge, 1986).

3. Interferences

Errors are caused by non-representative samples. Only field-occurring fabric (clods) should be sampled. The whole soil bulk density may be overestimated because sampled clods frequently exclude the crack space between clods (Grossman and Reinsch, 2002).

The penetration of plastic lacquer into the voids of sandy or organic soils interferes with the corrections for mass and volume of the plastic coat and with the accuracy of water content determinations. Penetration can be reduced by spraying water on the clod followed by immediately dipping the clod in the plastic lacquer. Dipping should be done as quickly as possible to reduce penetration of plastic.

Loss of soil during the procedure will void the analyses because all calculations are based on the oven-dry soil mass. Holes in the plastic coating, which are detected by escaping air bubbles from submerged clod, introduce errors in volume measurement. An inadequate evaporation of plastic solvent results in overestimation of the soil mass. A drying time of 1 h is usually sufficient time for evaporation of solvent. However, clods with high organic matter content may need to dry longer.

Bulk density is reported for <2-mm soil fabric. Correction for rock fragments >2-mm requires either knowledge or assumption of the rock fragment density. Estimate or measurement errors of rock fragment density will affect the accuracy of the value for soil bulk density. Rock fragments may contain water, which complicates the application to actual water-holding capacity.

4. Safety

Methyl ethyl ketone (MEK) is extremely flammable. A type B fire extinguisher should be in close proximity in the laboratory. No open flames or nearby operation of electrical equipment are permitted while using MEK. The MEK vapor is classified as a sensory irritant. The 8-h time weighted average (TWA) exposure limit is 200 ppm, and the short-term exposure limit (STEL) is 300 ppm (Occupational Safety and Health Administration, 1989). Avoid physical contact. Use with adequate ventilation. In closed buildings, use a fume hood. Keep in tightly closed containers. Avoid physical contact. Use safety glasses, proper gloves, and a lab coat. Wash hands immediately after handling MEK. Additional information on the safe handling of MEK is available in Chemical Safety Data Sheet SD-83, Manufacturing Chemists' Association, Inc., 1825 Connecticut Ave. NW, Washington, DC.

Saran F-310 resin will decompose rapidly at temperatures >200°C releasing hydrogen chloride gas. Avoid contact with Fe, Zn, Cu, and Al in solution. Avoid all contact with strong bases.

5. Equipment

5.1 Electronic balance, ±0.01-g sensitivity

5.2 Rigid shipping containers. The SSL uses a corrugated box with compartments.

5.3 Plastic bags, 1 mil, 127 x 89 x 330 mm

5.4 Wire. The SSL uses a 28-awg coated copper wire.

5.5 Hairnets

5.6 Stock tags, 25.4-mm (1-in) diameter paper tag with metal rim

5.7 Hook assembly for weighing below balance

5.8 Plexiglass water tank mounted on a fulcrum and lever

5.9 Oven, 110°C

5.10 Sieve, no. 10 (2-mm openings)

5.11 Rope, 3 m

5.12 Clothespins

5.13 Hot plate

5.14 Spray bottle

5.15 Liquid vapor trap. The SSL constructs a tin enclosure over hot plate with a chimney and duct to transfer vapor to water stream.

5.16 Tension table. The SSL constructs a tension table by placing porous firebricks, covered with reinforced paper towels, in a tub of water.

6. Reagents

6.1 Methyl ethyl ketone (MEK), practical (2-butanone).

6.2 Water.

6.3 Liquid detergent. The SSL uses Liqui-Nox.

6.4 Dow Saran F-310 Resin, available from Dow Chemical Company.

6.5 Plastic lacquer. Prepare plastic lacquer with resin to solvent ratios of 1:4 and 1:7 on a weight basis. Fill a 3.8-L (1-gal) metal paint can with 2700 ±200 mL of solvent. Fill to the bottom of handle rivet. Add 540 or 305 g of resin to make 1:4 or 1:7 plastic lacquer, respectively. For the initial field and laboratory coatings, use the 1:4 plastic lacquer. Use 1:7 plastic lacquer for the last two laboratory coats. The 1:7 plastic lacquer is used to conserve the resin and to reduce cost. In the field, mix solvent with a wooden stick. In the laboratory, stir solvent with a non-sparking, high-speed stirrer while slowly adding resin. Stir plastic lacquer for 15 min at 25°C. Store plastic lacquer in covered plastic or steel containers. Acetone may be substituted for MEK.

7. Procedure

Field

7.1 Collect field-occurring clods, ≈ 100 to 200 cm^3 in volume (fist-sized), from the face of the excavation. Three clods per horizon are recommended. It is important that these clods be as representative of the bulk sample volume as possible. Remove a piece of soil larger than clod from the face of sampling pit. From this piece, prepare a clod by gently cutting or breaking protruded peaks and compacted material from clod. If roots are present, trim roots with shears. No procedure for sampling clods is applicable to all soils. Adjust field-sampling techniques to meet field conditions at time of sampling.

7.2 Make a clothesline by stretching a rope between two fixed points. Tie clod with fine copper wire or place clod in a hairnet. If the clod is dry, moisten surface with a fine mist of water. Quickly dip entire clod into plastic lacquer (See Fig. 1, procedure 3B1a). Suspend clod from clothesline to dry (See Fig. 2, procedure 3B1a). Dry clod for 30 min or until odor of solvent dissipates. If the value of Db_f is required, store clods in waterproof plastic bags as soon as coating dries because coating is permeable to water vapor.

7.3 Pack clods in rigid containers to protect them during transport.

Laboratory

7.4 Prepare a round stock tag with sample identification number. Cut the copper wire and loop around the clod. Record weight (TAG) of tag and wire. Loop fine copper wire around clod, leaving a tail to which round stock tag is attached. Record weight of clod (CC1).

7.5 Dip clod in 1:4 plastic lacquer. Wait 7 min and then dip clod in 1:7 plastic lacquer. Wait 12 min and then dip clod in 1:7 plastic lacquer. Wait 55 min and then reweigh clod. If clod has adsorbed >3% plastic by clod weight or smells excessively of solvent, allow longer drying time, then reweigh clod and record weight (CC2).

7.6 Dry clod in an oven at 110°C until weight is constant. Weigh oven-dry clod in air (WODC) and in water (WODCW) and record weights.

7.7 If clod contains >5% rock fragments by weight, remove them from clod. Place clod in a beaker and place on hot plate. Cover hot plate with a liquid vapor trap. Use a fume hood. Heat clod on hot plate in excess of 200°C for 3 to 4 h. The plastic coating disintegrates at temperatures above 200°C. After heating, clod should appear black and charred. Remove clod from hot plate, lightly coat with liquid detergent, and add hot water.

7.8 Wet sieve the cool soil through a 2-mm, square- hole sieve. Dry and record weight (RF) of rock fragments that are retained on the sieve. Determine rock fragment density by weighing them in air to obtain their mass and in water to obtain their volume. If rock fragments are porous and have a density similar to soil sample, do not correct clod mass and volume measurement for rock fragments. If rock fragments are porous and have a density similar to soil sample, do not correct clod mass and volume measurement for rock fragments. Correct for rock fragments if these fragments can withstand breakdown when dry soil is placed abruptly in water or calgon.

7.9 Correct bulk density for weight and volume of plastic coating. The coating has an air-dry density of ≈ 1.3 g cm^{-3}. The coating loses 10 to 20% of its air-dry weight when dried in oven at 110°C.

8. Calculations

8.1 Db_{od} = [WODC- RF-ODPC – TAG]/{[(WODC – WODCW)/WD] - (RF/PD) - (MPC/1.3)}

where:
Db_{od} = Bulk density in g cm^{-3} of <2-mm, oven-dry fabric
WODC = Weight of oven-dry coated clod
RF = Weight of rock fragments
TAG = Weight of tag and wire
ODPC = MPC1 x 0.85, weight of oven-dry plastic coat
WD = Water density
PD = Density of rock fragments
WODCW = Weight of oven-dry coated clod in water

8.2 MPC = [(CC2 - CC1) + FCE] x RV

where:
MPC = Weight of plastic coat before oven-drying
CC2 = Weight of clod after three laboratory plastic coats
CC1 = Weight of clod before three laboratory plastic coats
RV = Percent estimate of remaining clod volume after cutting to obtain flat surface (≈ 80%)

8.3 FCE = 1.5 x [(CC2 -CC1)/3]

where:
FCE = Estimate of field-applied plastic coat

9. Report
Bulk density is reported to the nearest 0.01g cm^{-3}.

10. Precision and Accuracy
Precision and accuracy data are available from the SSL upon request.

11. References

Blake, G.R., and K.H. Hartge. 1986. Bulk density. p. 363-382. *In* A. Klute (ed.) Methods of soil analysis. Part 1. Physical and mineralogical methods. 2nd ed. Agron. Monogr. 9. ASA and SSSA, Madison, WI.

Brasher, B.R., D.P. Franzmeier, V.T. Volassis, and S.E. Davidson. 1966. Use of saran resin to coat natural soil clods for bulk density and water retention measurements. Soil Sci. 101:108.

Grossman, R.B., and T.G. Reinsch. 2002. Bulk density and linear extensibility. p. 201-228. *In* J.H. Dane and G.C. Topp (eds.) Methods of soil analysis, Part 4. Physical methods. Soil Sci. Am. Book Series No. 5. ASA and SSSA, Madison, WI.

Occupational Safety and Health Administration. 1989. Air contaminants; final rule. Federal Register. 29 CFR Part 1910. Vol. 54. No. 12. p. 2452-2453. U.S. Govt. Print. Office, Washington, DC.

Bulk Density (3B)
Saran-Coated Clods (3B1)
Rewet (Db$_r$) (3B1d)

1. Application
Bulk density is used to convert data from a weight to a volume basis; to determine the coefficient of linear extensibility; to estimate saturated hydraulic conductivity; and to identify compacted horizons. The rewet bulk density (Db$_r$) is used to determine irreversible shrinkage of soils and subsidence of organic soils. The procedure 3B1d determines the bulk density value (Db$_r$) of a re-wetted soil sample.

2. Summary of Method
Field-occurring fabric (clods) is collected from the face of an excavation. One coat of plastic lacquer is applied in the field. Additional coats of plastic lacquer are applied at the laboratory. After equilibration, the clod is weighed in air to measure its mass and in water to measure its volume. The clod is air-dried, re-equilibrated, and its mass and volume re-measured. After the clod is dried in the oven at 110°C, its mass and volume are determined again. A correction is made for the mass and volume of rock fragments and for plastic coatings (Brasher et al., 1966; Blake and Hartge, 1986; Grossman and Reinsch, 2002).

3. Interferences
Errors are caused by non-representative samples. Only field-occurring fabric (clods) should be sampled. The whole soil bulk density may be overestimated because sampled clods frequently exclude the crack space between clods (Grossman and Reinsch, 2002).

The penetration of plastic lacquer into the voids of sandy or organic soils interferes with the corrections for mass and volume of the plastic coat and with the accuracy of water content determinations. Penetration can be reduced by spraying water on the clod followed by immediately dipping the clod in the plastic lacquer. Dipping should be done as quickly as possible to reduce penetration of plastic.

Loss of soil during the procedure will void the analyses because all calculations are based on the oven-dry soil mass. Holes in the plastic coating, which are detected by escaping air bubbles from submerged clod, introduce errors in volume measurement. Inadequate drying results in overestimation of the soil mass. An inadequate evaporation of plastic solvent results in overestimation of the soil mass. A drying time of 1 h is usually sufficient time for evaporation of solvent. However, clods with high organic matter content may need to dry longer.

Bulk density is reported for <2-mm soil fabric. Correction for rock fragments >2-mm requires either knowledge or assumption of the rock fragment density. Estimate or measurement errors of rock fragment density will affect the accuracy of the value for soil bulk density. Rock fragments may contain water, which complicates the application to actual water-holding capacity.

4. Safety

Methyl ethyl ketone (MEK) is extremely flammable. A type B fire extinguisher should be in close proximity in the laboratory. No open flames or nearly operation of electrical equipment are permitted while using MEK. The MEK vapor is classified as a sensory irritant. The 8 h time weighted average (TWA) exposure limit is 200 ppm, and the short term exposure limit (STEL) is 300 ppm (Occupational Safety and Health Administration, 1989). Avoid physical contact. Use with adequate ventilation. In closed buildings, use a fume hood. Keep in tightly closed containers. Avoid physical contact. Use safety glasses, proper gloves, and a lab coat. Wash hands immediately after handling MEK. Additional information on the safe handling of MEK is available in Chemical Safety Data Sheet SD-83, Manufacturing Chemists' Association, Inc., 1825 Connecticut Ave. NW, Washington, DC.

Saran F-310 resin will decompose rapidly at temperatures >200°C releasing hydrogen chloride gas. Avoid contact with Fe, Zn, Cu, and Al in solution. Avoid all contact with strong bases.

5. Equipment
5.1 Electronic balance, ±0.01-g sensitivity
5.2 Pressure plate extractor with porous ceramic plate.
5.3 Air pressure, 33-kPa
5.4 Rigid shipping containers. The SSL uses a corrugated box with compartments.
5.5 Plastic bags, 1 mil, 127 x 89 x 330 mm
5.6 Wire. The SSL uses a 28-awg coated copper wire.
5.7 Hairnets
5.8 Stock tags, 25.4-mm (1-in) diameter paper tag, with metal rim
5.9 Hook assembly for weighing below balance
5.10 Plexiglass water tank mounted on a fulcrum and lever to elevate tank
5.11 Oven, 110°C
5.12 Sieve, No. 10 sieve (2-mm openings)
5.13 Rope, 3 m
5.14 Clothespins
5.15 Knife
5.16 Tile cut-off saw with diamond blade
5.17 Silt loam soil
5.18 Hot plate
5.19 Desiccator with ceramic plate
5.20 Vacuum, 80 kPa, (0.8 bar)
5.21 Metal probe
5.22 Spray bottle

5.23 Liquid vapor trap. The SSL constructs a tin enclosure over hot plate with a chimney and duct to transfer vapor to water stream.

5.24 Reinforced paper towels or cheesecloth

5.25 Tension table. The SSL constructs a tension table by placing porous firebricks, covered with reinforced paper towels, in a tub of water.

6. Reagents

6.1 Methyl ethyl ketone (MEK), practical (2-butanone).

6.2 Water.

6.3 Alcohol.

6.4 Liquid detergent. The SSL uses Liqui-Nox.

6.5 Dow Saran F-310 Resin, available from Dow Chemical Company.

6.6 Plastic lacquer. Prepare plastic lacquer with resin to solvent ratios of 1:4 and 1:7 on a weight basis. Fill a 3.8-L (1-gal) metal paint can with 2700 ±200 mL of solvent. Fill to the bottom of handle rivet. Add 540 or 305 g of resin to make 1:4 or 1:7 plastic lacquer, respectively. For the initial field and laboratory coatings, use the 1:4 plastic lacquer. Use 1:7 plastic lacquer for the last two laboratory coats. The 1:7 plastic lacquer is used to conserve the resin and to reduce cost. In the field, mix solvent with a wooden stick. In the laboratory, stir solvent with a non-sparking, high-speed stirrer while slowly adding resin. Stir plastic lacquer for 15 min at 25°C. Store plastic lacquer in covered plastic or steel containers. Acetone may be substituted for MEK.

7. Procedure

Field

7.1 Collect field-occurring clods, $\approx$ 100 to 200 cm^3 in volume (fist-sized), from the face of the excavation. Three clods per horizon are recommended. It is important that these clods be as representative of the bulk sample volume as possible. Remove a piece of soil larger than clod from the face of sampling pit. From this piece, prepare a clod by gently cutting or breaking protruded peaks and compacted material from clod. If roots are present, trim roots with shears. No procedure for sampling clods is applicable to all soils. Adjust field sampling techniques to meet field conditions at time of sampling.

7.2 Make a clothesline by stretching a rope between two fixed points. Tie clod with fine copper wire or place clod in a hairnet. If the clod is dry, moisten surface with a fine mist of water. Quickly dip entire clod into plastic lacquer (See Fig. 1, procedure 3B1a). Suspend clod from clothesline to dry (See Fig. 2, procedure 3B1a). Dry clod for 30 min or until odor of solvent dissipates. If the value of Db$_f$ is required, store clods in waterproof plastic bags as soon as coating dries because coating is permeable to water vapor.

7.3 Pack clods in rigid containers to protect them during transport.

Laboratory

7.4 Prepare a round stock tag with sample identification number. Cut the copper wire and loop around the clod. Record weight (TAG) of tag and wire. Loop fine copper wire around clod, leaving a tail to which round stock tag is attached. Record weight of clod (CC1).

7.5 Dip clod in 1:4 plastic lacquer. Wait 7 min and then dip clod in 1:7 plastic lacquer. Wait 12 min and then dip clod in 1:7 plastic lacquer. Wait 55 min and then reweigh clod. If clod has adsorbed >3% plastic by clod weight or smells excessively of solvent, allow longer drying time, then reweigh clod and record weight (CC2).

7.6 With a diamond saw, cut a flat surface on the clod.

7.7 Place cut clod surface on a tension table, maintained at 5-cm tension. Periodically check clod to determine if it has reached equilibrium by inserting metal probe, touching, or by weight comparison. When clod has reached equilibrium, remove clod and record weight (WSC).

7.8 If cut clod does not adsorb water, place clod in a desiccator on a water-covered plate with a 0-cm tension. Submerge only the surface of clod in the water. Add a few mL of alcohol. Use in-house vacuum and apply suction until clod has equilibrated at saturation. Remove clod and record weight (WSC).

7.9 Place clod in a pressure plate extractor. To provide good contact between clod and ceramic plate, cover ceramic plate with a 5-mm layer of silt loam soil and saturate with water. Place a sheet of reinforced paper towel or cheesecloth over the silt loam soil. Place surface of cut clod on paper towel. Close container and secure lid. Apply gauged air pressure of 33-kPa. When water ceases to discharge from outflow tube, clod is at equilibrium. Extraction usually takes 3 to 4 wk. Remove clod and record weight (WMC). Compare WMC to WSC. If WMC $\geq$ WSC, equilibrate clod on tension table and repeat desorption process.

7.10 Dip clod in the 1:4 plastic lacquer. Wait 7 min and then dip clod in 1:7 plastic lacquer. Wait 12 min and then dip clod in 1:7 plastic lacquer. Wait 12 min and dip clod in 1:7 plastic lacquer. After 55 min, reweigh clod and record weight (CC3). If clod has adsorbed >3% plastic by weight or smells excessively of solvent, allow longer drying time, then reweigh clod.

7.11 The clod should be waterproof and ready for volume measurement by water displacement. Suspend clod below the balance, submerge in water, and record weight (WMCW).

7.12 Remove layer of plastic from flat surface of clod. Air-dry clod at room temperature ($\sim$ 20 to 25°C) for 4 to 6 days. Dry clod at 40 to 50°C for 2 to 3 days or until weight is constant.

7.13 Repeat steps 7.7, 7.8, and 7.9. After equilibrium is obtained, remove clod and record weight (WAR).

7.14 Dip clod in the 1:4 plastic lacquer. Wait 7 min and then dip clod in 1:7 plastic lacquer. Wait 12 min and then dip clod in 1:7 plastic lacquer. Wait 12 min and dip clod in 1:7 plastic lacquer. After 55 min, reweigh clod and record weight (CC4). If clod has adsorbed >3% plastic by weight or smells excessively of solvent, allow longer drying time, then reweigh clod.

7.15 After coating, record weight of clod suspended in air (CC4) and in water (WARW).

7.16 Dry clod in an oven at 110°C until weight is constant. Weigh oven-dry clod in air (WODC) and in water (WODCW) and record weights.

7.17 If clod contains >5% rock fragments by weight, remove them from clod. Place clod in a beaker and place on hot plate. Cover hot plate with a liquid vapor trap. Use a fume hood. Heat clod on hot plate in excess of 200°C for 3 to 4 h. The plastic coating disintegrates at temperatures above 200°C. After heating, clod should appear black and charred. Remove clod from hot plate, lightly coat with liquid detergent, and add hot water.

7.18 Wet sieve the cool soil through a 2-mm, square- hole sieve. Dry and record weight (RF) of rock fragments that are retained on the sieve. Determine rock fragment density by weighing them in air to obtain their mass and in water to obtain their volume. If rock fragments are porous and have a density similar to soil sample, do not correct clod mass and volume measurement for rock fragments. If rock fragments are porous and have a density similar to soil sample, do not correct clod mass and volume measurement for rock fragments. Correct for rock fragments if these fragments can withstand breakdown when dry soil is placed abruptly in water or calgon.

7.19 Correct bulk density for weight and volume of plastic coating. The coating has an air-dry density of ≈ 1.3 g cm^{-3}. The coating loses 10 to 20% of its air-dry weight when dried in oven at 110°C.

8. Calculations

8.1 $Db_{33} = [WODC - RF - ODPC - TAG]/\{[(CC3 - WMCW)/WD] - (RF/PD) - (MPC1/1.3)\}$

where:
Db_{33} = Bulk density in g cm^{-3} of <2-mm fabric at 33-kPa tension
WODC = Weight of oven-dry coated clod
RF = Weight of rock fragments
ODPC = MPC1 x 0.85, weight of oven-dry plastic coat
TAG = Weight of tag and wire
CC3 = Weight of equilibrated clod after four additional plastic coats
WD = Water density
PD = Density of rock fragments
WMCW = Weight in water of coated clod equilibrated at 33-kPa tension

8.2 $MPC1 = \{[(CC2 - CC1) + FCE] \times RV\} + (CC3 - WMC)$

where:
MPC1 = Weight of plastic coat before air-drying and rewet
CC2 = Weight of clod after three laboratory plastic coats
CC1 = Weight of clod before three laboratory plastic coats
RV = Percent estimate of remaining clod volume after cutting to obtain flat surface ($\approx 80\%$)
WMC = Weight of coated clod equilibrated at 33-kPa tension

8.3 $FCE = 1.5 \times [(CC2 - CC1)/3]$

where:
FCE = Estimate of field-applied plastic coat.

8.4 $Db_r = [WODC - RF - ODPC - TAG]/\{[(CC4 - WARW)/WD] - (RF/PD) - (MPC2/1.3)\}$

where:
Db_r = Bulk density in g cm^{-3} <2-mm fabric at 33-kPa tension after rewetting
CC4 = Weight of clod after twelve plastic coats
WARW = Weight in water of coated clod equilibrated at 33-kPa tension after rewet

8.5 $MPC2 = \{[(CC2 - CC1) + FCE] \times RV2\} + (CC3 - WMC) + (CC4 - WAR)$

where:
MPC2 = Weight of plastic coat after rewet and before oven drying
WAR = Weight of clod after rewet equilibration
RV2 = Percent estimate of remaining clod volume after remaining layer of plastic (≈ 0.95)

8.6 $Db_{od} = [WODC - RF - ODPC - TAG]/\{[(WODC - WODCW)/WD] - (RF/PD) - (MPC2/1.3)\}$

where:
Db_{od} = Bulk density in g cm^{-3} of <2-mm fabric at oven dryness

WODCW = Weight in water of oven-dry coated clod

8.7 $W_{1/3}$ = {[(CC3 - MPC1) – (WODC - ODPC)]/(WODC - RF - ODPC – TAG)} x 100

where:
W_{33} = Percent water weight retained at 33-kPa tension

8.8 W_r ={[(CC4 - MPC2) – (WODC - ODPC)]/ [WODC - RF - ODPC – TAG]} x 100

where:
W_r = Percent water weight retained at 33-kPa tension after rewet

9. Report

Bulk density is reported to the nearest 0.01g cm^{-3}.

10. Precision and Accuracy

Precision and accuracy data are available from the SSL upon request.

11. References

Blake, G.R., and K.H. Hartge. 1986. Bulk density. p. 383-382. *In* A. Klute (ed.) Methods of soil analysis. Part 1. Physical and mineralogical methods. 2nd ed. Agron. Monogr. 9. ASA and SSSA, Madison, WI.

Brasher, B.R., D.P. Franzmeier, V.T. Volassis, and S.E. Davidson. 1966. Use of saran resin to coat natural soil clods for bulk density and water retention measurements. Soil Sci. 101:108.

Grossman, R.B., and T.G. Reinsch. 2002. Bulk density and linear extensibility. p. 201-228. *In* J.H. Dane and G.C. Topp (eds.). Methods of soil analysis, Part 4. Physical methods. Soil Sci. Am. Book Series No. 5. ASA and SSSA, Madison, WI.

Occupational Safety and Health Administration. 1989. Air contaminants; final rule. Federal Register. 29 CFR Part 1910. Vol. 54. No. 12. p. 2452-2453. U.S. Govt. Print. Office, Washington, DC.

Bulk Density (3B)
Reconstituted (3B2)
33 kPa Desorption (Db$_{33}$) (3B2a)
Oven-Dry (Db$_{od}$) (3B2b)

1. Application

Bulk density is used to convert data from weight to volume basis, to determine the coefficient of linear extensibility, to estimate saturated hydraulic conductivity, and to identify compacted horizons. Some models and programs require one bulk density to represent a given horizon. The reconstituted bulk density provides a single, reproducible value for horizons that are subject to tillage or other mechanical disturbances followed by an extreme water-state cycle (Reinsch and Grossman, 1995).

2. Summary of Method

The <2-mm sample is formed into a clod by wetting and dessiccation cycles that simulate reconsolidating by water in a field setting. Plastic lacquer is applied in the laboratory to form an impermeable coat on the clod. The clod is desorbed to 33-kPa. After equilibration, the clod is weighed in air to measure the mass and in water to measure the volume. After the clod is oven dried at 110°C, its mass and volume are determined again (Brasher et al., 1966; Blake and Hartge, 1986; Grossman and Reinsch, 2002).

3. Interferences
Some samples disintegrate when they are removed from the cells.

4. Safety
Methyl ethyl ketone (MEK) is extremely flammable. A type B fire extinguisher should be in close proximity in the laboratory. No open flames or nearby operation of electrical equipment are permitted while using MEK. The MEK vapor is classified as a sensory irritant. The 8-h time-weighted average (TWA) exposure limit is 200 ppm, and the short-term exposure limit (STEL) is 300 ppm (Occupational Safety and Health Administration, 1989). Avoid physical contact. Use with adequate ventilation. In closed buildings, use a fume hood. Keep in tightly closed containers. Use safety glasses, proper gloves, and a lab coat. Wash hands immediately after handling MEK. Additional information on the safe handling of MEK is available in Chemical Safety Data Sheet SD-83, Manufacturing Chemists' Association, Inc., 1825 Connecticut Ave. NW, Washington, DC.

5. Equipment
5.1 Electronic balance, ±0.01-g sensitivity
5.2 Pressure plate extractor with porous ceramic plate
5.3 Air pressure, 33-kPa
5.4 Clod forming cylinder. We construct the cell by attaching a brass ring or schedule 20 or 40 PVC pipe, 5.4-cm diameter and 6 to 7 cm high, to a 100-kPa ceramic plate with waterproof glue and caulk
5.5 Anti-sorting device. We construct a wire screen with 0.5-cm openings and 5.2 cm diameter with a perpendicular wire attached to the center to reduce natural sorting caused by placing the sample in the cell
5.6 Wire. SSL uses 28-awg-coated copper wire
5.7 Hairnets
5.8 Stock tags, 1-inch diameter paper tag with metal rim
5.9 Hook assembly for weighing below balance
5.10 Plexiglas water tank mounted on a fulcrum and lever to elevate tank
5.11 Oven, 110°C
5.12 Plastic tub at least 10 cm deep
5.13 Paper discs cut from water insoluble, permeable paper
5.14 Tile cut-off saw with diamond blade
5.15 Silt soil
5.16 Desiccator with ceramic plate
5.17 Vacuum, 80 kPa (0.8 bar)
5.18 Metal probe
5.19 Spray bottle
5.20 Reinforced paper towels or cheesecloth
5.21 Tension table. We construct a tension table by placing porous firebricks, covered with reinforced paper towels, in a tub of water

6. Reagents
6.1 Methyl Ethyl Ketone, practical (2-butanone).
6.2 Water.
6.3 Alcohol.
6.4 Dow Saran F-310 Resin, available from Dow Chemical Company.
6.5 Plastic lacquer. Prepared the plastic lacquer with resin to solvent ratios of 1:4 and 1:7 on a weight basis. We fill a metal gallon paint can with 2700 ± 200 ml of solvent (fill to the bottom of the handle rivet) and add 540 g or 305 g of resin to make 1:4 or 1:7 plastic lacquer, respectively. Use the 1:4 plastic lacquer for the first field and lab coats and the 1:7 plastic lacquer for the last two coats. The 1:7 plastic lacquer is used to conserve the resin and reduce cost. Stir the solvent with a non-sparking, high-speed stirrer while slowly

adding the resin. Stir the plastic lacquer for 15 min at 25°C. In the field, mix with a wooden stick. Store the plastic lacquer in covered plastic or steel containers. Acetone may be substituted for MEK.

7. Procedure

Reconstituted Clod Construction

7.1 Drape a hairnet in the cell. Place a paper disc in the bottom of the cell. Place the anti-sorting screen into the cell. Add <2-mm, prepared sample to within a few mm of the top of the cell. Lift the anti-sorting screen from the cell.

7.2 Place the cell on a tension table with the height 5-cm below the top of the table. After equilibration, place the cell into a tub and add water to a level higher than the surface of the soil in the cell but below the top lip of the cell. This allows the soil to become inundated from beneath. Allow the sample to equilibrate.

7.3 Remove the cell from the tub and allow to dry at room temperature. After drying remove the clod by lifting on the hairnet or inverting the cell and lightly tamping the base of the cell. The reconstituted clod is used to measure bulk density and water retention.

Bulk Density Measurement

7.4 Prepare a round stock tag with sample identification number. Cut the copper wire to loop around the clod. Record the weight of the tag and wire (TAG). Loop fine copper wire around the clod leaving a tail to which the round stock tag is attached. Record the weight of the clod (CC1).

7.5 Mist the clod with water to create a film of water on the surface of the clod. Dip the clod in the 1:4 plastic lacquer. After 7 min, dip the clod in the 1:4 plastic lacquer. After 7 min, dip the clod in the 1:7 plastic lacquer. After 12 min, dip the clod in the 1:7 plastic lacquer. After 55 min, reweigh the clod. If the clod has adsorbed more than 3 percent plastic by clod weight or smells excessively of solvent, allow longer drying time, then reweigh the clod and record the weight (CC2).

7.6 Cut a flat surface on the clod with a diamond saw. Place the cut clod surface on a tension table maintained at 5-cm tension. Check the clod periodically by inserting a metal probe, touching, or weight comparison to determine if it has reached equilibrium. When the clod has reached equilibrium, remove the clod and record the weight (WSC).

7.7 If the clod does not adsorb water, place the clod in a desiccator that has water covering the desiccator plate. Add a few ml of alcohol. Apply suction using in-house vacuum for 24 hours. Remove the clod and record the weight (WSC).

7.8 Place the clod in a pressure plate extractor. To provide good contact between the clod and ceramic plate, cover the ceramic plate with a 5-mm layer of silt and saturate with water. Place a sheet of reinforced paper towel or cheesecloth over the silt . Place the cut surface of the clod on the paper towel. Close the container and secure the lid. Apply gauged air pressure of 33-kPa. When water stops discharging from the outflow tube (usually after 3 or 4 wk in the extractor), the clod is at equilibrium. Remove the clod and record the weight (WMC). Compare WMC to WSC. If WMC is greater than or equal to WSC, equilibrate the clod on the tension table and repeat the desorption process.

7.9 Dip the clod in the 1:4 plastic lacquer. After 7 min, dip the clod in the 1:7 plastic lacquer. After 12 min, dip the clod in the 1:7 plastic lacquer. After 12 min, dip the clod in the 1:7 plastic lacquer. After 55 min, reweigh the clod and record the weight (CC3). If the clod has adsorbed more than 3 percent plastic by clod weight or smells excessively of solvent, allow longer drying time, then reweigh the clod.

7.10 The clod should now be waterproof and ready for the volume measurement by water displacement. Suspend the clod below the balance, submerge in water, and record the weight (WMCW).

7.11 Dry the clod in an oven at 110°C overnight. Weigh the oven dry clod in air (WODC) and in water (WODCW) and record the weights.

7.12 It is necessary to correct bulk density for weight and volume of the plastic coating. The coating has an air-dry density of about 1.3 g/cm^3. The coating loses 10 to 20 percent of its air-dry weight on oven drying at 110°C.

8. Calculations

8.1 Db_{33} = [WODC - RF - ODPC – TAG]/{[(CC2 - WMCW)/WD] - (RF/PD) - (MPC1/1.3)}

where:
Db_{33} = Bulk density in grams per cubic centimeter of < 2-mm fabric at 33-kPa tension
WODC = Weight of oven dry coated clod
ODPC = MPC x 0.85, weight of oven dry plastic coat
TAG = Weight of tag and wire
CC3 = Weight of equilibrated clod after four additional plastic coats
WMCW = Weight of coated clod equilibrated at 33-kPa tension in water
MPC1 = Weight of plastic coat before oven-drying
WD = Water density

8.2 MPC1 = {[(CC2-CC1)] x RV} + (CC3-WMC)

where:
MPC1 = Weight of plastic coat before oven-drying
CC2 = Weight of clod after four laboratory plastic coats
CC1 = Weight of clod before four laboratory plastic coats
RV = Percent estimate of remaining clod volume after cutting to obtain flat surface (≈95%)
WMC = Weight of coated clod equilibrated at 33-kPa tension

8.3 Db_{od} = [WODC – ODPC - TAG]/{[(WODC – WODCW)/WD] - (MPC/1.3)}

where:
Db_{od} = Bulk density in grams per cubic centimeter of < 2-mm fabric at oven-dryness
WODCW = Weight of oven dry coated clod in water

8.4 W_{33} = {[(CC3 – MPC1) - (WODC - ODPC)]/[WODC - ODPC – TAG)]} x 100

where:
W_{33} = Weight percentage of water retained at 33-kPa tension

9. Report
Bulk density is reported to the nearest 0.01 g cm^{-3}.

9. Precision and Accuracy
Precision and accuracy data are available from the SSL upon request.

11. References

Blake, G.R. and K.H. Hartge. 1986. Bulk density. p. 363-382. *In* A. Klute (ed.) Methods of soil analysis, Part 1. Physical and mineralogical methods. 2nd ed. Agron. Monogr. 9. ASA and SSSA, Madison, WI.

Brasher, B.R., D.P. Franzmeier, V.T. Volassis, and S.E. Davidson. 1966. Use of saran resin to coat natural soil clods for bulk density and water retention measurements. Soil Sci. 101:108.

Grossman, R.B., and T.G. Reinsch. 2002. Bulk density and linear extensibility. p. 201-228. *In* J.H. Dane and G.C. Topp (eds.) Methods of soil analysis, Part 4. Physical methods. Soil Sci. Am. Book Series No. 5. ASA and SSSA, Madison, WI.

Occupational Safety and Health Administration. 1989. Federal Register. Air contaminants; final rule. Federal Register. 29 CFR Part 1910. Vol. 54. No. 12. p. 2452-2453. U.S. Govt. Print. Office, Washington, DC.

Reinsch, T.G., and R.B. Grossman. 1995. A method to predict bulk density of tilled Ap horizons. Soil and Tillage Res. 34: 95-104.

Bulk Density (3B)
Compliant Cavity (3B3)
Field-State (3B3a)

1. Application

Bulk density is used to convert data from a weight to a volume basis; to estimate saturated hydraulic conductivity; and to identify compacted horizons. Excavation procedures (e.g., compliant cavity, ring, and frame) have applicability to layers that can be described as cohesionless, high in rock fragments > 5mm, or thin (<5 cm thick), and for which the clod method is unsuitable (Grossman and Reinsch, 2002). The compliant cavity method was designed for fragile cultivated near-surface layers and O horizons of forestland soils. This method has the important advantage that it is not necessary to flatten the ground surface on steep slopes or remove irregularities, i.e., the surficial zone is usually not altered (Grossman and Reinsch, 2002).

2. Summary of Method

By this procedure (3B3a), the cavity volume on the zone surface is lined with thin plastic and water is added to a datum level. Soil is quantitatively excavated in a cylindrical form to the required depth. The difference between the initial volume and that after excavation is the sample volume. The excavated soil is dried in an oven and then weighed. A correction is made for the weight and volume of rock fragments.

3. Interferences

Bulk density by compliant cavity can be made on soils with rock fragments but is more complex (Grossman and Reinsch, 2002).

4. Safety

Follow standard field and laboratory safety precautions.

5. Equipment

5.1 Fabricated Plexiglass rings, 9-mm thick, 130-mm inside diameter, and ≥200-mm outside diameter. Make three 16-mm diameter holes that are 10 mm from the outer edge of ring. Position holes equidistant apart. Use three, 25 x 50 mm, Plexiglass pieces as guides. Attach two pieces on one side to form an "L". Allow a 15-mm gap to permit removal of soil material. On the other side, position the single piece in line with the longer leg of the "L" so that an adjacent, parallel line forms a diameter.

5.2 Make 50-mm thick foam rings from flexible polyurethane with an "Initial Load Displacement" of 15 to 18 kg. The foam rings have the same inside diameter as the Plexiglass rings.

5.3 Fabricate a 240-mm crossbar from 5 x 18 mm metal stock to which legs (25-mm high and 180 x 180 mm in cross section) are welded. Drill a hole 100 mm from one end of the crossbar and 7 mm from the edge and through which a No. 6 machine bolt is placed.

5.4 Mount hook gauge on crossbar. Make hook gauge from No. 6, round-headed, 100-mm long machine bolts and from hexagonal nuts. Obtain the machine bolts from toggle bolt assemblies. Sharpen the machine bolt to a sharp point. Drill a hole in the center of the crossbar. Insert the machine bolt in the hole. Place nuts above and below the crossbar. The two nuts adjust the hook length below the crossbar and provide rigidity. Hold the machine bolt by the tightened nuts and heat the bolt. After softening, sharply bend the bolt upward to form a U-shape.

5.5 Use wing nuts and three, 250- to 400-mm long, 10- to 13-mm diameter, threaded rods to mount and position the compliant cavity. Sharpen the rods. Place two regular nuts at the end of threaded rod to increase the area of surface struck.

5.6 Syringe, 60 mL

5.7 Plastic film, 1/2 mil, 380-mm wide or wider; 460-mm wide for larger ring.

5.8 Plastic bags, 110°C-capability, with ties

5.9 Sharpie pen

5.10 Graduate cylinders, plastic, 250 to 2000 mL

5.11 Level, small

5.12 Kitchen knife, small

5.13 Scissors, small, to cut fine roots

5.14 Hack saw blade to cut large roots

5.15 Weights for plastic film

5.16 Clothespins. If wind, use clothespins for corners of plastic film.

5.17 Hard rubber or plastic mallet

5.18 Sieve, square-hole, 10 mesh, 2 mm

6. Reagents
6.1 Water

7. Procedure

7.1 Place a ring of plastic foam on ground and cover with rigid ring (130-mm inside diameter). Mount the assembly on the soil surface by securely driving threaded rods into the ground through holes in ring and by tightening ring with wing nuts.

7.2 Line cavity with 1/2-mL plastic. Fill cavity to tip of hook gauge with a known quantity of water from graduate cylinder.

7.3 Remove plastic film and water. Measure the volume of water to tip of hook gauge. This volume (Vd) is the measurement of cavity volume prior to excavation (dead space).

7.4 Excavate soil quantitatively and in a cylindrical form to the required depth. Fill the excavation cavity to tip of hook gauge with water from graduated cylinder. Measure the volume of water. This volume (Vf) is the measurement of excavated soil and dead space. The difference between the two water volumes (Vf - Vd) is the volume of excavated soil (Ve).

7.5 The excavated soil is dried in oven and weighed. If necessary, make a correction for weight and volume of >2-mm material (Vg) in sample and compute bulk density. The weight of macroscopic vegetal material (g cm^{-3}) also may be reported.

8. Calculations

8.1 $Ve = Vf - Vd - Vg$

where:
Ve = Excavation volume of <2-mm fraction (cc)
Vf = Water volume measurement of excavated soil and dead space (cc)
Vd = Water volume measurement of dead space (cc)
Vg = Gravel volume (>2mm- fraction) (cc). Calculate Vg by dividing the weight of >2-mm fraction by particle density of the >2-mm fraction. Default value of 2.65 g cc^{-1}.

8.2 $Wf = Wo - Wc$

where:
Wf = Oven-dry weight of <2-mm soil (g)
Wo = Oven-dry weight of excavated soil (g)
Wc = Oven-dry weight of rock fragments (g)

8.3 $Db = Wf/Ve$

where:
Db = Bulk density (g cc^{-1})
Wf = Oven-dry weight of <2-mm soil (g)
Ve = Excavation volume of <2-mm material (cc)

9. Report
Bulk density is reported to the nearest 0.01 g cm^{-3}.

10. Precision and Accuracy
Precision and accuracy data are available from the SSL upon request.

11. References
Grossman, R.B., and T.G. Reinsch. 2002. Bulk density and linear extensibility. p. 201-228. *In* J.H. Dane and G.C. Topp (eds.) Methods of soil analysis, Part 4. Physical methods. Soil Sci. Am. Book Series No. 5. ASA and SSSA, Madison, WI.

Bulk Density (3B)
Ring Excavation (3B4)
Field-State (3B4a)

1. Application
Bulk density is used to convert data from a weight to a volume basis; to estimate saturated hydraulic conductivity; and to identify compacted horizons. Excavation procedures (e.g., compliant cavity, ring, and frame) have applicability to layers that can be described as coheshionless, high in rock fragments > 5mm, or thin (<5 cm thick), and for which the clod method is unsuitable (Grossman and Reinsch, 2002). The ring excavation is a robust, simple, and rapid method. This method is good for O horizons in the woods where local variability is large. The diameter can range down to 15 cm and upwards to 30 cm or more. It is not necessary to excavate from the whole area within the ring. A limit of 2 cm on the minimum thickness of the sample should be considered.

2. Summary of Method
A 20-cm ring is inserted into the ground. A piece of shelf standard is placed across the ring near to a diameter. The distance to the ground surface is measured at eight points equally

spaced along the diameter using the depth-measurement tool to measure the distance. The piece of shelf is rotated 90° and eight more measurements are made. The 16 measurements are then averaged. The soil is excavated to the desired depth, and the distance measurements repeated. The change in distance is calculated on the removal of the soil. This change in distance is then multiplied by the inside cross-sectional area of the ring to obtain the volume of soil. The excavated soil is oven-dried and weighed. If rock fragments are present, the weight and volume of >2-mm material in sample are corrected and bulk density computed. Bulk density of soil is reported in g cm^{-3} by procedure 3B4a.

3. Interferences
Rock fragments may make insertion of ring into the ground impossible.

4. Safety
Follow standard field and laboratory safety precautions.

5. Equipment
5.1 Metallic cylinder, 20-cm diameter, 10- to 20-cm high, and about 1-mm depth
5.2 Shelf standard (slotted rod), 1.5-cm wide, 1-cm high, and 25-cm long
5.3 Piece of retractable ruler, 30-cm long with 0.1-mm divisions
5.4 Piece of wood, 10 x 10 x 30 cm
5.5 Hand digging equipment
5.6 Depth-measurement tool (Grossman and Reinsch, 2002)

6. Reagents
None.

7. Procedure

7.1 Insert a 20-cm diameter ring below the depth of excavation.

7.2 Place a piece of shelf standard across the ring near to or along a diameter. Measure the distance to the ground surface at eight points equally spaced along the diameter using the depth-measurement tool to measure the distance.

7.3 Rotate the piece of shelf standard 90° and make eight more measurements. Average the 16 measurements.

7.4 Excavate the soil to the desired depth. Repeat the distance measurements.

7.5 Calculate the change in distance on removal of the soil. Multiply the change in distance by the inside cross-sectional area of the ring to obtain the volume of the soil (Ve).

7.6 The excavated soil is dried in oven and weighed. If necessary, make a correction for weight and volume of >2-mm material in sample and compute bulk density. The weight of macroscopic vegetal material (g cm^{-3}) also may be reported.

8. Calculations

8.1 Wf = Wo – We

where:
Wf = Oven-dry weight of <2-mm soil (g)
Wo = Oven-dry weight of excavated soil (g)
Wc = Oven-dry weight of rock fragments (g)

8.2 $Db = Wf/Ve$

Db = Bulk density (g cm^{-3})
Wf = Oven-dry weight of <2-mm soil (g)
Ve = Excavation volume of <2-mm material (cm^{-3})

9. Report
Bulk density is reported to the nearest 0.01 g cm^{-3}.

9. Precision and Accuracy
Precision and accuracy data are available from the SSL upon request.

11. References
Grossman, R.B., and T.G. Reinsch. 2002. Bulk density and linear extensibility. p. 201-228. In J.H. Dane and G.C. Topp (eds.) Methods of soil analysis, Part 4. Physical methods. Soil Sci. Am. Book Series No. 5. ASA and SSSA, Madison, WI.

Bulk Density (3B)
Frame Excavation (3B5)
Field-State (3B5a)

1. Application
Bulk density is used to convert data from a weight to a volume basis; to estimate saturated hydraulic conductivity; and to identify compacted horizons. Excavation procedures (e.g., compliant cavity, ring, and frame) have applicability to layers that can be described as coheshionless, high in rock fragments > 5mm, or thin (<5 cm thick), and for which the clod method is unsuitable (Grossman and Reinsch, 2002). This method is good for O horizons in the woods where local variability is large and commonly rock fragments are present. The size of the 0.1 m^2 is sufficient to encompass considerable local variability.

2. Summary of Method
The assembled frame is placed on the ground surface. The four threaded rods are pushed through the holes in the corners of the frame deep enough to hold. The frame is then secured onto the soil surface by screwing down wing nuts and plastic placed over the frame and secured. The depth-measurement tool is placed on top of a slot to measure the distance to the soil surface. The slots are traversed, and measurements of the distance to the ground surface are made at about 40 regularly spaced intervals. The plate is then removed and soil is excavated and retained. Measurements of the distance to the ground surface are repeated. The volume of soil is determined by taking the difference in height and multiplying by 1000 cm2. The rock fragments up to 20 mm are included in the sample. Excavated soil is oven-dried and weighed. Bulk density of soil is reported in g cm^{-3} by procedure 3B5a.

3. Interferences
None.

4. Safety
Follow standard field and laboratory safety precautions..

5. Equipment
5.1 Lumber for square wooden frame with 0.1 m2 inside area. Frame is made from 8 pieces of wood: 2 pieces, 2 x 4 x 46 cm; 2 pieces, 2 x 4 x 53 cm; and 4 blocks, 4 x 5 x 9 cm
5.2 Square Plexiglass, 35 cm on edge x 0.6 cm thick, with 5 parallel equally spaced slots, 1.5 cm across x 28 cm long
5.3 Four threaded rods, 50 cm long x 0.6-cm diameter with wing nuts

5.4 Depth-measurement tool (Grossman and Reinsch, 2002, p. 209)

5.5 Hand digging equipment

6. Reagents
None.

7. Procedure

7.1 Assemble the square wooden frame by attaching the 4 x 5 x 9 x cm blocks to the 9 cm of each end of both 53-cm long pieces. Two-centimeter-wide cuts are made half-way across each of the 46- and 53-cm-long pieces to provide half-lap joints. The cuts are 5 cm in for the 46-cm-long pieces. Holes 1.0 to 1.5 cm in diameter are drilled in the center of the attached blocks. The four pieces are joined by the vertical half-lap joints to form a square frame.

7.2 Place the frame on the ground surface. Push the four threaded rods through the holes in the corners of the frame sufficiently deeply to hold. Secure onto the soil surface by screwing down wing nuts.

7.3 Place the plastic plate over the frame and secure.

7.4 Place the depth-measurement tool on top of a slot and measure the distance to the soil surface.

7.5 Traverse the slots, making measurements of the distance to the ground surface at about 40 regularly spaced intervals. Remove the plate.

7.6 Excavate and retain the soil. The walls of the cavity should be vertical and coincident with the edge of the frame.

7.7 Repeat the measurements of the distance to the ground surface. Determine the difference in height and multiply by 1000 cm^2 to obtain the volume of soil excavated. Usually rock fragments up to 20 mm are included in the sample.

7.8 The excavated soil is dried in oven and weighed. If necessary, make a correction for weight and volume of >2-mm material in sample and bulk density computed. The weight of macroscopic vegetal material (g cm^{-3}) also may be reported.

8. Calculations

8.1 Wf = Wo – We

where:
Wf = Oven-dry weight of <2-mm soil (g)
Wo = Oven-dry weight of excavated soil (g)
We = Oven-dry weight of rock fragments (g)

8.2 Db = Wf/Ve

Db = Bulk density (g cm^{-3})
Wf = Oven-dry weight of <2-mm soil (g)
Ve = Excavation volume of <2-mm material (cm^{-3})

9. Report
Bulk density is reported to the nearest 0.01 g cm^{-3}.

10. Precision and Accuracy

Precision and accuracy data are available from the SSL upon request.

11. References

Grossman, R.B., and T.G. Reinsch. 2002. Bulk density and linear extensibility. p. 201-228. In J.H. Dane and G.C. Topp (eds.) Methods of soil analysis, Part 4. Physical methods. Soil Sci. Am. Book Series No. 5. ASA and SSSA, Madison, WI.

Bulk Density (3B)
Soil Cores (3B6)
Field-State (3B6a)

1. Application

Bulk density is used to convert data from a weight to a volume basis; to determine the coefficient of linear extensibility; to estimate saturated hydraulic conductivity; and to identify compacted horizons. Procedure 3B6a determines the bulk density value of a moist soil core of known volume. Field bulk density (Db_f) offers the opportunity to obtain relatively cheaply bulk density information without the expense incurred to obtain water retention. Db_f is particularly useful if the soil layers are at or above field capacity and/or the soils have low extensibility and do not exhibit desiccation cracks even if below field capacity.

2. Summary of Method

A metal cylinder is pressed or driven into the soil. The cylinder is removed extracting a sample of known volume. The moist sample weight is recorded. The sample is then dried in a oven and weighed.

3. Interferences

During coring process, compaction of the sample is a common problem. Compression can be observed by comparing the soil elevation inside the cylinder with the original soil surface outside the cylinder. If compression is excessive, soil core may not be a valid sample for analysis. Rock fragments in the soil interfere with core collection. Dry or hard soils often shatter when hammering the cylinder into the soil. Pressing the cylinder into the soil reduces the risk of shattering the sample.

If soil cracks are present, select the sampling area so that crack space is representative of sample, if possible. If this is not possible, make measurements between the cracks and determine the aerial percentage of total cracks or of cracks in specimen.

4. Safety

No known hazard exists with this procedure.

5. Equipment

5.1 Containers, air-tight, tared, with lids
5.2 Electronic balance, ±0.01-g sensitivity
5.3 Oven 110°C
5.4 Sieve, No. 10 (2 mm-openings)
5.5 Coring equipment. Sources described in Grossman and Reinsch (2002).

6. Reagents

None

7. Procedure

7.1 Record the empty core weights (CW).

7.2 Prepare a flat surface, either horizontal or vertical, at the required depth in sampling pit.

7.3 Press or drive core sampler into soil. Use caution to prevent compaction. Remove core from the inner liner, trim protruding soil flush with ends of cylinder, and place in air-tight container for transport to laboratory. If soil is too loose to remain in the liner, use core sampler without the inner liner and deposit only the soil sample in air-tight container. Moisture cans can also be pushed directly into a prepared face. For fibrous organic materials, trim sample to fit snugly into a moisture can.

7.4 Dry core in an oven at 110°C until weight is constant. Record oven-dry weight (ODW).

7.5 Measure and record cylinder volume (CV).

7.6 If sample contains rock fragments, wet sieve sample through a 2-mm sieve. Dry and weigh the rock fragments that are retained on sieve. Record weight of rock fragments (RF). Determine density of rock fragments (PD).

8. Calculations

$$Db = (ODW - RF - CW)/[CV - (RF/PD)]$$

where:
Db = Bulk density of < 2-mm fabric at sampled, field water state (g cm^{-3})
ODW = Oven-dry weight
RF = Weight of rock fragments
CW = Empty core weight
CV = Core volume
PD = Density of rock fragments

9. Report
Bulk density is reported as g cc^{-1} to the nearest 0.01 g cm^{-3}.

10. Precision and Accuracy
Precision and accuracy data are available from the SSL upon request.

11. References
Grossman, R.B. and T.G. Reinsch. 2002. Bulk density and linear extensibility. p. 201-228. *In* J.H. Dane and G.C. Topp (eds.) Methods of soil analysis, Part 4. Physical methods. Soil Sci. Am. Book Series No. 5. ASA and SSSA, Madison, WI.

PHYSICAL AND FABRIC-RELATED ANALYSES (3)

Water Retention (3C)

Water retention is defined as the soil water content at a given soil water suction. By varying the soil suction and recording the changes in soil water content, a water retention function or curve is determined. This relationship is dependent on particle-size distribution, clay mineralogy, organic matter, and structure or physical arrangement of the particles as well as hysteresis, i.e., whether the water is absorbing into or desorbing from the soil. The data collected in these procedures are from water desorption. Water retention or desorption curves are useful directly and indirectly as indicators of other soil behavior traits such as drainage, aeration, infiltration, plant available water, and rooting patterns (Topp et al., 1993).

Two desorption procedures are commonly used to measure water retention, a suction method and a pressure method. The SSL uses the pressure method (U.S. Salinity Laboratory, 1954) with either a pressure-plate or pressure-membrane extractor. Procedures 3C1a-e1 (pressure-plate extraction) are used to determine water retention at 6, 10, 33, 100, or 200 kPa, respectively (0.06, 0.1, 1/3, 1, or 2 bar, respectively) for sieved, <2-mm, air-dry soil samples of non-swelling soils, loamy sand or coarser soil and for some sandy loams. Procedures 3C1a-d2 and 3C1a-d3 (pressure-plate extractions) are used to measure water retention of natural clods or cores that have been equilibrated at 6, 10, 33, or 100 kPa. Procedures 3C1a-d2 and 3C1a-d3 are usually used in conjunction with the bulk density procedure 3B1b.

Procedure 3C1c4 (pressure-plate extraction) is used to determine the water retention of a clod equilibrated at 33-kPa, air-dried, and re-equilibrated. The resulting data are called rewet water-retention data and are usually used in conjunction with the rewet bulk density data in procedure 3B1d to estimate changes in physical properties of a soil as it undergoes wetting and drying cycles. Procedure 3C2a1a (pressure-membrane extraction) is used to determine water retention at 1500 kPa (15 bar) for <2-mm (sieved), air-dry soil samples. Procedure 3C2a1b is used to measure water retention at 1500 kPa for <2-mm (sieved), field moist soil samples. Procedure 3C3 is used to determine field water content at the time of sampling for cores, clods, or bulk samples.

References

Topp, G.C., Y.T. Galganov, B.C. Ball, and M.R. Carter. 1993. Soil water desorption curves. p. 569-579. *In* M.R. Carter (ed.) Soil sampling and methods of analysis. Can. Soc. Soil Sci., CRC Press, Boca Raton, FL.

U.S. Salinity Laboratory Staff. 1954. L.A. Richards (ed.) Diagnosis and improvement of saline and alkali soils. USDA Handb. 60. U.S. Govt. Print. Office, Washington, DC.

Water Retention (3C)
Pressure-Plate Extraction (3C1)
6, 10, 33, 100, or 200 kPa (3C1a-e)
<2-mm (sieved), Air-Dry Samples (3C1a-e1a)

1. Application

The data collected are used for the water retention function, water-holding capacity, pore-size distribution, porosity, and saturated conductivity of a soil sample at specific water contents. The data are also used to calculate unsaturated hydraulic conductivity. Procedures 3C1a-e1a (pressure-plate extraction) are used to determine water retention at 6, 10, 33, 100, or 200 kPa, respectively, for <2-mm (sieved), air-dry soil samples of non-swelling soils, loamy sand or coarser soil and for some sandy loams.

2. Summary of Method

The pressure desorption method (U.S. Salinity Laboratory Staff, 1954) is used. A sample of <2-mm (sieved), air-dry soil is placed in a retainer ring sitting on a porous ceramic plate in a pressure-plate extractor. The plate is covered with water to wet the samples by capillarity. The sample is equilibrated at the specified pressures. The pressure is kept constant until equilibrium is obtained (Klute, 1986). The gravimetric water content is determined.

3. Interferences

A leaking pressure extractor prevents equilibration of samples. Check for outflow air to verify that the pressure-plate extractor is functioning properly and does not leak. The pressure should be monitored for stability. Equilibration must be done at constant temperature and humidity.

With extended use, the porous ceramic plate becomes clogged and water outflow is restricted. The plate is cleaned by flushing sequentially with 500 mL of 10% H_2O_2, 1000 mL of 1 *N* HCl, and 500 mL of reverse osmosis (RO) water. The solutions are pulled through the plate with a vacuum, and the waste is captured in a trap.

The rubber membrane on the bottom of the plate is checked for leaks. The membrane is inflated and then submerged in water. If air bubbles escape from the membrane, the plate is removed from service.

Laboratory-determined, water retention data are usually higher than field-determined water retention data, because the confining soil pressure is not present in the laboratory (Bruce and Luxmoore, 1986). Water retention data for soils with expansive clay is overestimated when sieved samples are used in place of natural soil fabric for tensions of 6, 10, and 33 kPa, respectively, (Young and Dixon, 1966).

Aerated 0.005 *M* CaSO$_4$ has also been recommended (Dane and Hopmans, 2002), especially for fine-textured soils that contain significant amounts of swelling clays. Distilled or deionized water can possibly promote dispersion of clays in samples and freshly drawn tap water is often supersaturated with air, affecting the water content at a given pressure head (Dane and Hopmans, 2002).

4. Safety

High pressure plumbing must be maintained in good working order. Ensure that the pressure is zero before removing bolts from the pressure extractor lid. Ensure that the bolts are tightened before applying pressure. Do not drop the heavy lid.

5. Equipment
5.1 Pressure plate extractor
5.2 Electronic balance, ±0.01-g sensitivity
5.3 Oven, 110°C
5.4 Pressure source, regulator, and gauge.
5.5 Retainer rings. Use 10-mm high and 50-mm diameter rings for organic soils and 10-mm high and 40-mm diameter rings for all other soils.
5.6 Metal weighing cans with lids

6. Reagents
6.1 Reverse osmosis (RO) water
6.2 Hydrogen peroxide (H$_2$O$_2$), 10% solution. Dilute 333 mL of 30% H$_2$O$_2$, technical grade, in 1 L of RO water.
6.3 Hydrochloric acid (HCl), 1 *N*. Dilute 83.3 mL of concentrated HCl in 1 L of RO water.
6.4 Ethyl alcohol, 95%, technical grade

7. Procedure

7.1 Saturate the ceramic plate by applying RO water through the adapter and apply enough pressure so that the rubber membrane is bulging a few centimeters. Care should be taken to remove all air.

7.2 Place the saturated ceramic plate in a pressure plate extractor. Place retainer rings on the ceramic plate. Use 50-mm diameter rings for soils that have >12% in organic matter. Use 40-mm diameter rings for all other soils.

7.3 Fill retaining ring with 10 to 15 g of <2-mm or fine-grind, air-dry soil. Include a quality control (QC) sample in each pressure-plate extractor.

7.4 Add water to cover the ceramic plate but not to cover the rings. Continue to add water until all samples have moistened by capillarity. If samples do not moisten, apply ethyl alcohol to the surface of the samples. Close the apparatus and let stand overnight.

7.5 Apply the specified pressure. Monitor the outflow tube for water discharge. Periodically submerge the outflow tube in water to monitor for air bubbles that indicate ceramic plate failure. Samples are equilibrated when water ceases to emit from the outflow tube. The outflow tube can be submerged under water in a burette to measure when water ceases to emit from the outflow tube.

7.6 When samples have equilibrated, quickly transfer the samples to tared water cans (M_c), cover with lids, and record the weights (M_{s+w}).

7.7 Remove lids, place samples in oven, and dry at 110°C overnight. Record weights (M_s).

8. Calculations

$$H_2O \% = 100 \times [(M_{s+w} - M_s)/(M_s - M_c)]$$

where:
$H_2O \%$ = Percent gravimetric water content
M_{s+w} = Weight of solids + H_2O + container
M_s = Weight of solids + container
M_c = Weight of container

9. Report
Report water content to the nearest 0.1 percent.

10. Precision and Accuracy
Precision and accuracy data are available from the SSL upon request.

11. References

Bruce, R.R., and R.J. Luxmoore. 1986. Water retention: field methods. p. 663-686. *In* A. Klute (ed.) Methods of soil analysis. Part 1. Physical and mineralogical methods. 2nd ed. Agron. Monogr. 9. ASA and SSSA, Madison, WI.

Dane, J.H., and J. W. Hopmans. 2002. Water retention and storage: Laboratory. p. 675-720. *In* J.H. Dane and G.C. Topp (eds.) Methods of soil analysis. Part 4. Physical methods. Soil Sci. Am. Book Series No. 5. ASA and SSSA, Madison, WI.

Klute, A. 1986. Water retention: Laboratory methods. p. 635-662. *In* Klute, A. (ed.) Methods of soil analysis. Part 1. Physical and mineralogical methods. Agron. Monogr. 9. ASA and SSSA, Madison, WI.

U.S. Salinity Laboratory Staff. 1954. L.A. Richards (ed.) Diagnosis and improvement of saline and alkali soils. USDA Handb. 60. U.S. Govt. Print. Office, Washington, DC.

Young, K.K., and J.D. Dixon. 1966. Overestimation of water content at field capacity from sieved-sample data. Soil Sci. 101:104-107.

Water Retention (3C)
Pressure-Plate Extraction (3C1)
6, 10, 33, or 100 kPa (3C1a-d)
Natural Clods (3C1a-d2)

1. Application
The data collected are used for the water retention function, water-holding capacity, pore-size distribution, porosity, and saturated conductivity of a soil sample at specific water contents. The data are also used to calculate unsaturated hydraulic conductivity. Procedures 3C1a-d2 are used to determine the water retention of natural clods at 6, 10, 33, or 100 kPa, respectively.

2. Summary of Method

The pressure desorption method (U.S. Salinity Laboratory Staff, 1954) is used. Natural clods are placed on a tension table and equilibrated at a 5-cm tension at the base of the sample. The clods are then transferred to a porous ceramic plate which is placed in a pressure-plate extractor. The sample is equilibrated at the specified pressures. The pressure is kept constant until equilibrium is obtained (Klute, 1986). The gravimetric water content is determined.

3. Interferences

A leaking pressure extractor prevents equilibration of samples. Check outflow air to verify that each pressure-plate extractor is functioning properly and does not leak. The pressure should be monitored for stability. Equilibration must be done at constant temperature and humidity.

With extended use, the porous ceramic plate becomes clogged and water outflow is restricted. The plate is cleaned by flushing sequentially with 500 mL of 10% H_2O_2, 1000 mL of 1 N HCl, and 500 mL of RO water. The solutions are pulled through the plate with a vacuum, and the waste is captured in a trap.

The rubber membrane on the bottom of the plate is checked for leaks in the gasket. The membrane is inflated and then submerged in water. If air bubbles escape from the membrane, the plate is removed from service.

Laboratory-determined, water retention data are usually higher than field-determined, water retention data because the confining soil pressure is not present in the laboratory (Bruce and Luxmoore, 1986).

Aerated 0.005 M $CaSO_4$ has also been recommended (Dane and Hopmans, 2002), especially for fine-textured soils that contain significant amounts of swelling clays. Distilled or deionized water can possibly promote dispersion of clays in samples and freshly drawn tap water is often supersaturated with air, affecting the water content at a given pressure head (Dane and Hopmans, 2002).

4. Safety

High pressure plumbing must be maintained in good working order. Ensure that the pressure is zero before removing bolts from the pressure-apparatus lid. Ensure that the bolts are tightened before applying pressure. Do not drop the heavy lid.

5. Equipment

5.1 Electronic balance, ±0.01-g sensitivity
5.2 Pressure plate extractor with porous ceramic plate (Fig.1 and 2)
5.3 Pressure source, regulator, and gauge.
5.4 Oven, 110°C
5.5 Retainer rings, 10-mm height and 40-mm diameter
5.6 Metal weighing cans with lids
5.7 Clothespins
5.8 Knife
5.9 Tile cut-off saw with diamond blade
5.10 Silt loam soil
5.11 Desiccator with ceramic plate
5.12 Vacuum, 80 kPa (0.8 bar)
5.13 Needle probe
5.14 Sieve, No. 10 (2-mm openings)
5.15 Hot plate
5.16 Fume hood
5.17 Reinforced paper towels with nylon fibers, GSA
5.18 Liquid vapor trap. The SSL constructs a tin enclosure over hot plate with a chimney and duct to transfer vapor to water stream.

5.19 Tension table. The SSL constructs a tension table by placing porous firebricks, covered with reinforced paper towels, in a tub of water.

5.20 Stock tags, 25.4-mm (1-in) diameter paper tag, with metal rim

5.21 Wire. The SSL uses a 28-awg coated copper wire.

5.22 Retainer rings. Use 10-mm high and 50-mm diameter rings for organic soils and 10-mm high and 40-mm diameter rings for all other soils.

6. Reagents

6.1 Reverse osmosis (RO) water

6.2 Hydrogen peroxide (H_2O_2), 10% solution. Dilute 333 mL of 30% H_2O_2, technical grade, in 1 L of RO water

6.3 Hydrochloric acid (HCl), 1 N. Dilute 83.3 mL of concentrated HCl in 1 L of RO water.

6.4 Ethyl alcohol, 95%, technical grade

6.5 Liquid detergent. The SSL uses Liqui-Nox

6.6 Methyl ethyl ketone (MEK), practical (2-butanone)

6.7 Dow Saran F-310 Resin, available from Dow Chemical Company

6.8 Plastic lacquer. Prepare plastic lacquer with resin to solvent ratios of 1:4 and 1:7 on a weight basis. Fill a 3.8-L (1-gal) metal paint can with 2700 ± 200 mL of solvent. Fill to the bottom of handle rivet. Add 540 g or 305 g of resin to make 1:4 or 1:7 plastic lacquer, respectively. For the initial field and laboratory coatings, use the 1:4 plastic lacquer. Use 1:7 plastic lacquer for the last two laboratory coats. The 1:7 plastic lacquer is used to conserve the resin and to reduce cost. In the field, mix solvent with a wooden stick. In the laboratory, stir solvent with a non-sparking, high speed stirrer while slowly adding resin. Stir plastic lacquer for 30 min at 25°C. Store plastic lacquer in covered plastic or steel containers. Acetone may be substituted for MEK.

6.9 Microbiocide, for tension table, Fisher Scientific

7. Procedure

7.1 This procedure is usually combined with the bulk density procedure (3B1b).

7.2 Prepare a round stock tag with sample identification number. Cut the copper wire and loop around the clod. Record the weight of the tag and wire (TAG). Loop fine copper wire around the clod, leaving a tail to which the round stock tag is attached. Record the weight of the clod (CC1).

7.3 Dip the clod in the 1:4 plastic lacquer. Wait 7 min and then dip the clod in the 1:7 plastic lacquer. Wait 12 min and then dip the clod in the 1:7 plastic lacquer. Wait 55 min and then reweigh the clod. If the clod has absorbed >3% in plastic by clod weight or smells excessively of solvent, allow longer drying time, then reweigh the clod and record the weight (CC2).

7.4 Cut a flat surface on the clod with a tile saw. It is necessary to wet the clod above the initial desorption point. This is accomplished by placing the cut clod surface on a tension table that is maintained at 5-cm tension. Periodically check the clod by inserting a needle probe, touching, or by weight comparison to determine if it has reached equilibrium (clod has wetted up). When the clod has reached equilibrium, remove the clod and record the weight (WSC).

7.5 If cut clod does not absorb water, place clod in a desiccator on a water-covered plate with a 0-cm tension. Submerge only the surface of clod in the water. Add a few mL of alcohol. Use in-house vacuum and apply suction until clod has equilibrated at saturation. Remove the clod and record the weight (WSC).

7.6 Saturate the ceramic plate by applying RO water through the adapter and apply enough pressure so that the rubber membrane is bulging a few centimeters. Care should be taken to remove all air.

7.7 Place the saturated ceramic plate in a pressure plate extractor. To provide good contact between the clod and ceramic plate, cover the ceramic plate with a 5-mm layer of silt loam soil and saturate with water. Place a sheet of reinforced paper towel over the silt loam soil. Place cut clod surface on the paper towel. Prepare a saturated, sieved soil as a quality control (QC) sample. Place several retaining rings in the extractor. Fill the retaining rings with the soil standard. Close the container and secure lid.

7.8 Apply gauged air pressure of 6, 10, 33, or 100 kPa. If more than one water retention point is requested, begin with the lowest pressure. Periodically submerge the outflow tube in water to monitor for air bubbles that indicate ceramic plate failure. Samples are equilibrated when water ceases to emit from the outflow tube. The outflow tube can be submerged under water in a burette to measure when water ceases to emit from the outflow tube. When water stops discharging from the outflow tube, the clod is at equilibrium. Determine the gravimetric water content of the QC. If the water content of the QC is higher than twice the standard deviation, apply pressure for additional time. Recheck the QC. If the water content of the QC is lower than twice the standard deviation, rewet the clods and desorb again. If the water content of the QC is within acceptable limits, then the apparatus has functioned properly.

7.9 Remove the clod and record the weight (WMC). Compare WMC to WSC. If WMC $\geq$ WSC, re-equilibrate the clod on the tension table and repeat the desorption process. If additional water retention points are requested, then repeat the desorption process at the next higher pressure. When the clod is equilibrated at 33 kPa and bulk density is to be measured, continue with procedure 3B1b.

7.10 Dry the clod in an oven at 110°C overnight and record oven-dry weight (WODC).

7.11 If the clods contains >5% in rock fragments by weight, remove the rock fragments from the clod. This determination of the percent rock fragments is based on particle-size data of >2-mm fraction.. Submerge the remaining soil material in a beaker of water and place on a hot plate. Use a fume hood. Boil $\approx$ 1 h. The plastic coating loosens from soil material upon heating. Remove beaker from the hot plate. Allow to cool. Discard plastic coating.

7.12 Wet sieve the cool soil through a 2-mm sieve. Dry and record the weight (RF) of the rock fragments that are retained on the sieve. If the rock fragments have similar properties to the soil sample, do not correct the clod mass for the rock fragments.

8. Calculations

8.1 H$_2$O % =[(WMC - MPC) - (WODC - ODPC) x 100]/(WODC - RF - ODPC – TAG)

where:
H$_2$O % = Percent gravimetric water content
WMC = Weight of equilibrated, coated clod
WODC= Weight of oven-dry coated clod
RF = Weight of rock fragments
ODPC = MPC x 0.85, weight of oven-dry plastic coat
TAG = Weight of tag and wire

8.2 MPC = {[(CC2 - CC1) + FCE] x RV}

where:
MPC = Weight of plastic coat before oven-drying
CC2 = Weight of clod after three laboratory plastic coats
CC1 = Weight of clod before three laboratory plastic coats
RV = Percent estimate of remaining clod volume after cutting to obtain flat surface ($\approx 80\%$)

8.3 FCE = 1.5 x [(CC2 - CC1)/3]

where:
FCE = Estimate of field-applied plastic coat, if applied

9. Report

Report water content to the nearest 0.1 percent.

10. Precision and Accuracy

Precision and accuracy data are available from the SSL upon request.

11. References

Bruce, R.R., and R.J. Luxmoore. 1986. Water retention: field methods. p. 663-686. *In* A. Klute (ed.) Methods of soil analysis. Part 1. Physical and mineralogical methods. 2nd ed. Agron. Monogr. 9. ASA and SSSA, Madison, WI.

Dane, J.H., and J. W. Hopmans. 2002. Water retention and storage: Laboratory. p. 675-720. *In* J.H. Dane and G.C. Topp (eds.) Methods of soil analysis. Part 4. Physical methods. Soil Sci. Am. Book Series No. 5. ASA and SSSA, Madison, WI.

Klute, A. 1986. Water retention: Laboratory methods. p. 635-662. *In* A. Klute (ed.) Methods of soil analysis. Part 1. Physical and mineralogical methods. 2nd ed. Agron. Monogr. 9. ASA and SSSA, Madison, WI.

U.S. Salinity Laboratory Staff. 1954. L.A. Richards (ed.) Diagnosis and improvement of saline and alkali soils. USDA Handb. 60. U.S. Govt. Print. Office, Washington, DC.

Fig 1. Clods placed in pressure-plate extractor following saturation.

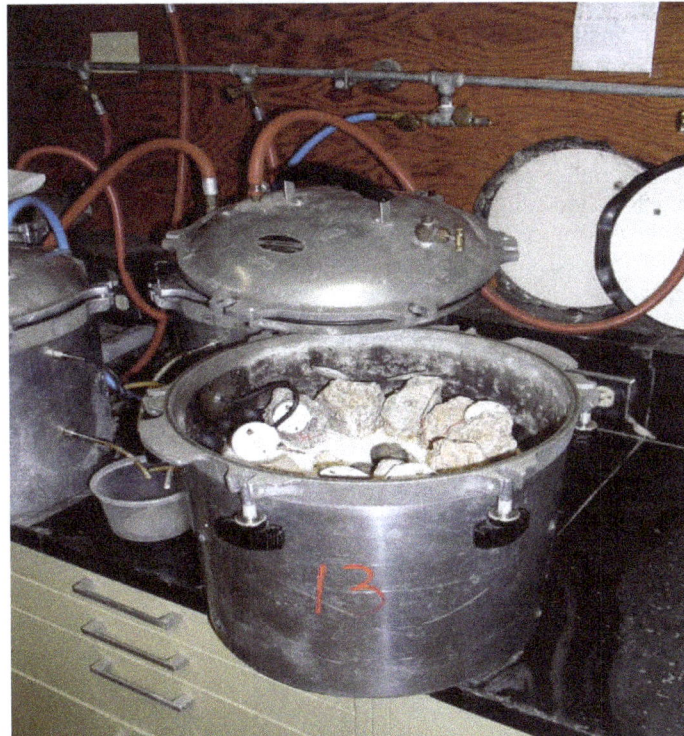

Fig. 2. Pressure-plate extraction at 33-kPa for clods.

Water Retention (3C)
Pressure-Plate Extraction (3C1)
6, 10, 33, or 100 kPa (3C1a-d)
Soil Cores (3C1a-d3)

1. Application

The data collected are used for the water retention function, water-holding capacity, pore-size distribution, porosity, and saturated conductivity of a soil sample at specific water contents. The data are also used to calculate unsaturated hydraulic conductivity. Procedures 3C1a-d3 are used to determine the water retention of soil cores at 6, 10, 33, or 100 kPa, respectively.

2. Summary of Method

The pressure desorption procedure (U.S. Salinity Laboratory Staff, 1954) is used. A metal cylinder is pressed or driven into the soil. Upon removal from the soil, the cylinder extracts a sample of known volume. The sample weight is recorded. The sample is dried in the oven and then weighed. Soil core is placed on a tension table and equilibrated at a 5-cm tension at the base of the sample. The cores are then transferred to a porous ceramic plate which is placed in a pressure-plate extractor. The sample is equilibrated at the specified pressures. The pressure is kept constant until equilibrium is obtained (Klute, 1986). The gravimetric water content is determined.

3. Interferences

A leaking pressure extractor prevents equilibration of samples. Check for outflow air to verify that the extractor is functioning properly and does not leak. The pressure should be monitored for stability. Equilibration must be done at constant temperature and humidity.

With extended use, the porous ceramic plate becomes clogged and water outflow is restricted. The plate is cleaned by flushing sequentially with 500 mL of 10% H_2O_2, 1000 mL of 1 N HCl, and 500 mL of RO water. The solutions are pulled through the plate with a vacuum, and the waste is captured in a trap.

The rubber membrane on the bottom of the plate is checked for leaks in the gasket. The membrane is inflated and then submerged in water. If air bubbles escape from the membrane, the plate is removed from service.

Laboratory-determined, water retention data are usually higher than field-determined, water retention data because the confining soil pressure is not present in the laboratory (Bruce and Luxmoore, 1986).

Compaction of the sample during the sampling process is a common problem. Compression can be observed by comparing the soil elevation inside the cylinder with the original soil surface outside the cylinder. If compression is excessive, soil core may not be a valid sample for analysis. Rock fragments in the soil interfere with core collection. Dry or hard soils often shatter when hammering the cylinder into the soil. Pressing the cylinder into the soil reduces the risk of shattering the sample.

Aerated 0.005 M CaSO$_4$, has also been recommended (Dane and Hopmans, 2002), especially for fine-textured soils that contain significant amounts of swelling clays. Distilled or deionized water can possibly promote dispersion of clays in samples and freshly drawn tap water is often supersaturated with air, affecting the water content at a given pressure head (Dane and Hopmans, 2002).

4. Safety

High pressure plumbing must be maintained in good working order. Ensure that the pressure is zero before removing bolts from the pressure-apparatus lid. Ensure that the bolts are tightened before applying pressure. Do not drop the heavy lid.

5. Equipment

5.1 Electronic balance, ±0.01-g sensitivity
5.2 Pressure plate extractor with porous ceramic plate
5.3 Pressure source, regulator, and gauge
5.4 Oven, 110°C
5.5 Retainer rings, 10-mm height and 40-mm diameter
5.6 Metal weighing cans with lids
5.7 Silt loam soil
5.8 Desiccator with ceramic plate
5.9 Vacuum, 80 kPa (0.8 bar)
5.10 Needle probe
5.11 Sieve, No. 10 (2-mm openings)
5.12 Fume hood
5.13 Coring equipment. Sources described in Blake and Hartge (1986).
5.14 Tension table. The SSL constructs a tension table by placing porous firebricks, covered with reinforced paper towels, in a tub of water.
5.15 Reinforced paper towels with nylon fibers, GSA
5.16 Retainer rings. Use 10-mm high and 50-mm diameter rings for organic soils and 10-mm high and 40-mm diameter rings for all other soils.

6. Reagents

6.1 Reverse osmosis (RO) water
6.2 Hydrogen peroxide (H_2O_2), 10% solution. Dilute 333 mL of 30% H_2O_2, technical grade, in 1 L of RO water.
6.3 Hydrochloric acid (HCl), 1 N. Dilute 83.3 mL of concentrated HCl in 1 L of RO water.
6.4 Alcohol
6.5 Liquid detergent. The SSL uses Liqui-Nox.
6.6 Microbiocide, for tension table, Fisher Scientific

7. Procedure

7.1 This procedure can be combined with the bulk density procedure (3B1b).

7.2 Record the weight (CW) of the sampling cylinders.

7.3 Prepare a flat surface in the sampling pit, either horizontal or vertical, at the required depth. Press or drive the core sampler into the soil. Use caution to prevent compaction. Remove the core from the sample holder, trim the protruding soil flush with the cylinder ends, and place core in an air-tight container for transport to the laboratory.

7.4 It is necessary to wet the core above the initial desorption point. This is accomplished by placing the flat core surface on a tension table maintained at 5-cm tension. Periodically, check the core by inserting a needle probe, touching, or by weight comparison to determine if core has reached equilibrium (core has wetted up). When the core has reached equilibrium, remove the core and record the weight (WSC).

7.5 If core does not absorb water, place core in a desiccator on a water-covered plate with a 0-cm tension. Submerge only the surface of core in the water. Add a few mL of alcohol. Use in-house vacuum and apply suction until core has equilibrated at saturation. Remove the core and record the weight (WSC).

7.6 Saturate the ceramic plate by applying RO water through the adapter and apply enough pressure so that the rubber membrane is bulging a few centimeters. Care should be taken to remove all air.

7.7 Place the saturated ceramic plate in a pressure plate extractor. To provide good contact between the core and ceramic plate, cover the ceramic plate with a 5-mm layer of silt loam soil and saturate with water. Place a sheet of reinforced paper towel over the silt loam soil. Place the flat core surface on the paper towel. Prepare a saturated, sieved soil as a quality control (QC) sample. Place several retaining rings in the extractor. Fill the retaining rings with the soil standard. Close the container and secure lid.

7.8 Apply gauged air pressure of 6, 10, 33, or 100 kPa. If more than one water retention point is requested, begin with the lowest pressure. When water stops discharging from the outflow tube, the core is at equilibrium. Determine the gravimetric water content of the standard. If the water content of the QC is higher than twice the standard deviation, apply pressure for additional time. Recheck the QC. If the water content of the QC is lower than twice the standard deviation, rewet the cores and desorb again. If the water content of the QC is within acceptable limits, then the apparatus has functioned properly.

7.9 Remove core and record the weight (WMC). Compare WMC to WSC. If WMC $\geq$ WSC, re-equilibrate core on the tension table and repeat the desorption process. If additional water retention points are requested, then repeat the desorption process at the next higher pressure. When the core is equilibrated at 33 kPa and bulk density is to be measured, continue with method 3B1b.

7.10 Dry core in an oven at 110°C overnight and record oven-dry weight (WODC).

7.11 If sample contains rock fragments, wet sieve the sample through a 2-mm sieve. Dry and weigh the rock fragments that are retained on the sieve. If the rock fragments have similar properties to the soil sample, do not correct the clod mass for the rock fragments. Record the weight of the rock fragments (RF).

8. Calculations

$$H_2O \% = 100 \times [(WMC - WODC)/(WODC - CW - RF)]$$

where:
$H_2O \%$ = Percent gravimetric water content
WMC = Weight of solids + H_2O + container
CW = Weight of solids + container
WODC = Weight of container
RF = Weight of rock fragments

9. Report
Report water content to the nearest 0.1 percent.

10. Precision and Accuracy
Precision and accuracy data are available from the SSL upon request.

11. References
Bruce, R.R., and R.J. Luxmoore. 1986. Water retention: field methods. p. 663-686. *In* A. Klute (ed.) Methods of soil analysis. Part 1. Physical and mineralogical methods. 2nd ed. Agron. Monogr. 9. ASA and SSSA, Madison, WI. .

Dane, J.H., and J. W. Hopmans. 2002. Water retention and storage: Laboratory. p. 675-720. *In* J.H. Dane and G.C. Topp (eds.) Methods of soil analysis. Part 4. Physical methods. Soil Sci. Am. Book Series No. 5. ASA and SSSA, Madison, WI.

Klute, A. 1986. Water retention: Laboratory methods. p. 635-662. *In* A. Klute (ed.) Methods of soil analysis. Part 1. Physical and mineralogical methods. 2nd ed. Agron. Monogr. 9. ASA and SSSA, Madison. WI.

U.S. Salinity Laboratory Staff. 1954. L.A. Richards (ed.) Diagnosis and improvement of saline and alkali soils. USDA Handb. 60. U.S. Govt. Print. Office, Washington, DC.

Water Retention (3C)
Pressure-Plate Extraction (3C1)
33 kPa (3C1c)
Rewet (3C1c4)

1. Application

The data collected are used for the water retention function, water-holding capacity, pore-size distribution, porosity, and saturated conductivity of a soil sample at specific water contents. The data are also used to calculate unsaturated hydraulic conductivity. The rewet water retention data (3C1c4) are used in conjunction with the rewet bulk density (3B1d) to estimate the change in physical properties of a soil as it undergoes wetting and drying cycles.

2. Summary of Method

The pressure desorption method (U.S. Salinity Laboratory Staff, 1954) is used. Natural clods are placed on a tension table and equilibrated at a 5-cm tension at the base of the sample. The clods are then transferred to a porous ceramic plate which is placed in a pressure-plate extractor. The sample is equilibrated at 33kPa. The pressure is kept constant until equilibrium is obtained (Klute, 1986). The equilibrated clod weight is recorded. The clod is air dried and then placed on a tension table and desorbed again. After the second equilibration, the gravimetric water content is determined.

3. Interferences

A leaking pressure extractor prevents equilibration of samples. Check outflow air to verify each pressure-plate extractor is functioning properly and does not leak. The pressure should be monitored for stability. Equilibration must be done at constant temperature and humidity.

With extended use, the porous ceramic plate becomes clogged and water outflow is restricted. The plate is cleaned by flushing sequentially with 500 mL of 10% H_2O_2, 1000 mL of 1 N HCl, and 500 mL of RO water. The solutions are pulled through the plate with a vacuum, and the waste is captured in a trap.

The rubber membrane on the bottom of the plate is checked for leaks in the gasket. The membrane is inflated and then submerged in water. If air bubbles escape from the membrane, the plate is removed from service.

Laboratory-determined, water retention data are usually higher than field-determined, water retention data because the confining soil pressure is not present in the laboratory (Bruce and Luxmoore, 1986).

Aerated 0.005 M $CaSO_4$, has also been recommended (Dane and Hopmans, 2002), especially for fine-textured soils that contain significant amounts of swelling clays. Distilled or deionized water can possibly promote dispersion of clays in samples and freshly drawn tap water is often supersaturated with air, affecting the water content at a given pressure head (Dane and Hopmans, 2002).

4. Safety

High pressure plumbing must be maintained in good working order. Ensure that the pressure is zero before removing bolts from the pressure-plate lid. Ensure that the bolts are tightened before applying pressure. Do not drop the heavy lid.

5. Equipment

5.1 Electronic balance, ±0.01-g sensitivity
5.2 Pressure plate extractor with porous ceramic plate
5.3 Pressure source, regulator, and gauge
5.4 Oven, 110°C-capability
5.5 Retainer rings, 10-mm height and 40-mm diameter
5.6 Metal weighing cans with lids
5.7 Clothespins
5.8 Knife
5.9 Tile cut-off saw with diamond blade
5.10 Silt loam soil
5.11 Desiccator with ceramic plate
5.12 Vacuum, 80 kPa (0.8 bar)
5.13 Needle probe
5.14 Sieve, No. 10 (2-mm openings)
5.15 Hot plate
5.16 Fume hood
5.17 Reinforced paper towels with nylon fibers, GSA
5.18 Liquid vapor trap. The SSL constructs a tin enclosure over hot plate with a chimney and duct to transfer vapor to water stream.
5.19 Tension table. The SSL constructs a tension table by placing porous firebricks, covered with reinforced paper towels, in a tub of water.
5.20 Stock tags, 25.4-mm (1-in) diameter paper tag, with metal rim
5.21 Wire. The SSL uses a 28-awg coated copper wire.
5.22 Retainer rings. Use 10-mm high and 50-mm diameter rings for organic soils and 10-mm high and 40-mm diameter rings for all other soils.

6. Reagents

6.1 Reverse osmosis (RO) water
6.2 Hydrogen peroxide (H_2O_2), 10% solution. Dilute 333 mL of 30% H_2O_2, technical grade, in 1 L of RO water.
6.3 Hydrochloric acid (HCl), 0.05 N. Dilute 8 mL of concentrated HCl in 1 L of RO water.
6.4 Ethyl alcohol, 95%, technical grade
6.5 Liquid detergent. The SSL uses Liqui-Nox.
6.6 Methyl ethyl ketone (MEK), practical (2-butanone)
6.7 Dow Saran F-310 Resin. Available from Dow Chemical Company.
6.8 Plastic lacquer. Prepare plastic lacquer with resin to solvent ratios of 1:4 and 1:7 on a weight basis. Fill a 3.8-L (1-gal) metal paint can with 2700 ±200 mL of solvent. Fill to the bottom of handle rivet. Add 540 g or 305 g of resin to make 1:4 or 1:7 plastic lacquer, respectively. For the initial field and laboratory coatings, use the 1:4 plastic lacquer. Use 1:7 plastic lacquer for the last two laboratory coats. The 1:7 plastic lacquer is used to conserve the resin and to reduce cost. In the field, mix solvent with a wooden stick. In the laboratory, stir solvent with a non-sparking, high speed stirrer while slowly adding resin. Stir plastic lacquer for 15 min at 25°C. Store plastic lacquer in covered plastic or steel containers. Acetone may be substituted for MEK.
6.9 Microbiocide, for tension table, Fisher Scientific

7. Procedure

7.1 This procedure is usually used in conjunction with the bulk density procedure 4A1i.

7.2 Prepare a round stock tag with sample identification number. Cut the copper wire. Record the weight of the tag and wire (TAG). Loop fine copper wire around the clod, leaving a tail to which the round stock tag is attached. Record the weight of the clod (CC1).

7.3 Dip the clod in the 1:4 plastic lacquer. Wait 7 min and then dip the clod in the 1:7 plastic lacquer. Wait 12 min and then dip the clod in the 1:7 plastic lacquer. Wait 55 min and then reweigh the clod. If the clod has absorbed >3% in plastic by clod weight or smells excessively of solvent, allow longer drying time, then reweigh the clod and record the weight (CC2).

7.4 Cut a flat surface on the clod with a tile saw.

7.5 It is necessary to wet the clod above the initial desorption point. This is accomplished by placing the cut clod surface on a tension table that is maintained at 5-cm tension. Periodically check the clod by inserting a needle probe, touching, or by weight comparison to determine if it has reached equilibrium (clod has wetted up). When the clod has reached equilibrium, remove the clod and record the weight (WSC).

7.6 If cut clod does not absorb water, place clod in a desiccator on a water-covered plate with a 0-cm tension. Submerge only the surface of clod in the water. Add a few mL of alcohol. Use in-house vacuum and apply suction until clod has equilibrated at saturation. Remove the clod and record the weight (WSC).

7.7 Saturate the ceramic plate by applying RO water through the adapter and apply enough pressure so that the rubber membrane is bulging a few centimeters. Care should be taken to remove all air.

7.8 To provide good contact between the clod and ceramic plate, cover the ceramic plate with a 5-mm layer of silt loam soil and saturate with water. Place a sheet of reinforced paper towel over the silt loam soil. Place cut clod surface on the paper towel. Prepare a saturated, sieved soil as a quality control (QC) sample. Place several retaining rings in the extractor. Fill the retaining rings with the soil standard. Close the container and secure lid.

7.9 Apply gauged air pressure of 6, 10, 33, or 100 kPa. If more than one water retention point is requested, begin with the lowest pressure. When water stops discharging from the outflow tube, the clod is at equilibrium. Determine the gravimetric water content of the QC. If the water content of the QC is higher than twice the standard deviation, apply pressure for additional time. Recheck the QC. If the water content of the QC is lower than twice the standard deviation, rewet the clods and desorb again. If the water content of the QC is within acceptable limits, then the apparatus has functioned properly.

7.10 Remove the clod and record the weight (WMC). Compare WMC to WSC. If WMC $\geq$ WSC, re-equilibrate the clod on the tension table and repeat the desorption process. If additional water retention points are requested, then repeat the desorption process at the next higher pressure. When the clod is equilibrated at 33 kPa and bulk density is to be measured, continue with procedure 3B1d

7.11 Air-dry the clod at room temperature ($\approx$ 20 to 25°C) for 4 to 6 days. Dry the clods at 40 to 50°C for 2 to 3 days or until weights are constant.

7.12 Repeat steps 7.5, 7.6, 7.7, and 7.8. Record clod weight after equilibration (WMC2). Determine bulk density as described in method 3B1d.

7.13 Dry the clod in oven at 110°C overnight and record oven-dry weight (WODC).

7.14 If the clod contains >5% in rock fragments by weight, remove the rock fragments from the clod. This determination of the percent rock fragments is based on particle-size data of >2-mm fraction. Submerge the remaining soil material in a beaker of water and place on a hot plate.

Use a fume hood. Boil ≈ 1 h. The plastic coating loosens from soil material upon heating. Remove beaker from the hot plate. Allow to cool. Discard plastic coating.

7.15 Wet sieve the cool soil through a 2-mm sieve. Dry and record the weight (RF) of the rock fragments that are retained on the sieve. If the rock fragments have similar properties to the soil sample, do not correct the clod mass for the rock fragments.

8. Calculations

8.1 H_2O % =[(WMC - MPC) - (WODC - ODPC) x 100]/(WODC - RF – ODPC – TAG)

where:
H_2O % = Percent gravimetric water content
WMC = Weight of equilibrated, coated clod
WODC= Weight of oven-dry coated clod
RF   = Weight of rock fragments
ODPC = MPC x 0.85, weight of oven-dry plastic coat
TAG   = Weight of tag and wire

8.2 MPC = {[(CC2 – CC1) + FCE] x RV}

where:
MPC   = Weight of plastic coat before oven-drying
CC2   = Weight of clod after three laboratory plastic coats
CC1   = Weight of clod before three laboratory plastic coats
RV   = Percent estimate of remaining clod volume after cutting to obtain flat surface (≈ 80%)

8.3 FCE = 1.5 x [(CC2 – CC1)/3]

where:
FCE   = Estimate of field applied plastic coat, if applied

8.4 H_2O_r % = {[(WMC2 - MPC2) - (WODC - ODPC2)] x 100}/(WODC - RF – ODPC – TAG)

where:
H_2O_r % = Percent water weight retained at 33-kPa tension after rewetting
WMC2   = Weight of equilibrated, coated clod after rewetting
MPC2   = Weight of moist plastic coat after rewetting. Same as MPC unless additional plastic coats were added.
OPC2   = MPC2 x 0.85, weight of oven-dry plastic coat.

9. Report
Report water content to the nearest 0.1 percent.

10. Precision and Accuracy
Precision and accuracy data are available from the SSL upon request.

11. References
Bruce, R.R., and R.J. Luxmoore. 1986. Water retention: Field methods. p. 663-686. *In* A. Klute (ed.) Methods of soil analysis. Part 1. Physical and mineralogical methods. 2nd ed. Agron. Monogr. 9. ASA and SSSA, Madison, WI.

Dane, J.H., and J. W. Hopmans. 2002. Water retention and storage: laboratory. p. 675-720. *In* J.H. Dane and G.C. Topp (ed.) Methods of soil analysis. Part 4. Physical methods. Soil Sci. Am. Book Series No. 5. ASAS and SSSA, Madison, WI.

Klute, A. 1986. Water retention: Laboratory methods. p. 635-662. *In* A. Klute (ed.), Methods of soil analysis. Part 1. Physical and mineralogical methods. 2nd ed. Agron. Monogr. 9. ASA and SSSA, Madison, WI.

U.S. Salinity Laboratory Staff. 1954. L.A. Richards (ed.) Diagnosis and improvement of saline and alkali soils. U.S. Dept. of Agric. Handb. 60. U.S. Govt. Print. Office, Washington, DC.

Water Retention (3C)
Pressure-Plate Extraction (3C1)
33 kPa (3C1c)
Reconstituted (3C1c5)

1. Application

The data collected are used for the water retention function, water-holding capacity, pore-size distribution, porosity, and saturated conductivity of a soil sample at specific water contents. The data are also used to calculate unsaturated hydraulic conductivity. Procedure 3C1c5 is used to determine the water retention of a reconstituted clod at 33 kPa.

2. Summary of Method

The pressure desorption method (U.S. Salinity Laboratory Staff, 1954) is used. Natural clods are placed on a tension table and equilibrated at a 5-cm tension at the base of the sample. The clods are then transferred to a porous ceramic plate which is placed in a pressure-plate extractor. The sample is equilibrated at the specified pressures. The pressure is kept constant until equilibrium is obtained (Klute, 1986). The gravimetric water content is determined.

3. Interferences

A leaking pressure extractor prevents equilibration of samples. Check outflow air to verify that each pressure-plate extractor is functioning properly and does not leak. The pressure should be monitored for stability. Equilibration must be done at constant temperature and humidity.

With extended use, the porous ceramic plate becomes clogged and water outflow is restricted. The plate is cleaned by flushing sequentially with 500 mL of 10% H_2O_2, 1000 mL of 1 N HCl, and 500 mL of RO water. The solutions are pulled through the plate with a vacuum, and the waste is captured in a trap.

The rubber membrane on the bottom of the plate is checked for leaks in the gasket. The membrane is inflated and then submerged in water. If air bubbles escape from the membrane, the plate is removed from service.

Laboratory-determined, water retention data are usually higher than field-determined, water retention data because the confining soil pressure is not present in the laboratory (Bruce and Luxmoore, 1986).

Aerated 0.005 M $CaSO_4$ has also been recommended (Dane and Hopmans, 2002), especially for fine-textured soils that contain significant amounts of swelling clays. Distilled or deionized water can possibly promote dispersion of clays in samples and freshly drawn tap water is often supersaturated with air, affecting the water content at a given pressure head (Dane and Hopmans, 2002).

4. Safety

High pressure plumbing must be maintained in good working order. Ensure that the pressure is zero before removing bolts from the pressure-apparatus lid. Ensure that the bolts are tightened before applying pressure. Do not drop the heavy lid.

5. Equipment

5.1 Electronic balance, ±0.01-g sensitivity

5.2 Pressure plate extractor with porous ceramic plate

5.3 Pressure source, regulator, and gauge

5.4 Oven, 110°C

5.5 Retainer rings, 10-mm height and 40-mm diameter

5.6 Metal weighing cans with lids

5.7 Clothespins

5.8 Knife

5.9 Tile cut-off saw with diamond blade

5.10 Silt loam soil

5.11 Desiccator with ceramic plate

5.12 Vacuum, 80 kPa (0.8 bar)

5.13 Needle probe

5.14 Sieve, No. 10 (2-mm openings)

5.15 Hot plate

5.16 Fume hood

5.17 Reinforced paper towels with nylon fibers, GSA

5.18 Liquid vapor trap. The SSL constructs a tin enclosure over hot plate with a chimney and duct to transfer vapor to water stream.

5.19 Tension table. The SSL constructs a tension table by placing porous firebricks, covered with reinforced paper towels, in a tub of water.

5.20 Stock tags, 25.4-mm (1-in) diameter paper tag, with metal rim

5.21 Wire. The SSL uses a 28-awg coated copper wire.

5.22 Retainer rings. Use 10-mm high and 50-mm diameter rings for organic soils and 10-mm high and 40-mm diameter rings for all other soils.

6. Reagents

6.1 Reverse osmosis (RO) water

6.2 Hydrogen peroxide (H_2O_2), 10% solution. Dilute 333 mL of 30% H_2O_2, technical grade, in 1 L of RO water.

6.3 Hydrochloric acid (HCl), 1 N. Dilute 83.3 mL of concentrated HCl in 1 L of RO water.

6.4 Ethyl alcohol, 95%, technical grade

6.5 Liquid detergent. The SSL uses Liqui-Nox.

6.6 Methyl ethyl ketone (MEK), practical (2-butanone)

6.7 Dow Saran F-310 Resin. Available from Dow Chemical Company.

6.8 Plastic lacquer. Prepare plastic lacquer with resin to solvent ratios of 1:4 and 1:7 on a weight basis. Fill a 3.8-L (1-gal) metal paint can with 2700 ± 200 mL of solvent. Fill to the bottom of handle rivet. Add 540 g or 305 g of resin to make 1:4 or 1:7 plastic lacquer, respectively. For the initial field and laboratory coatings, use the 1:4 plastic lacquer. Use 1:7 plastic lacquer for the last two laboratory coats. The 1:7 plastic lacquer is used to conserve the resin and to reduce cost. In the field, mix solvent with a wooden stick. In the laboratory, stir solvent with a non-sparking, high speed stirrer while slowly adding resin. Stir plastic lacquer for 30 min at 25°C. Store plastic lacquer in covered plastic or steel containers. Acetone may be substituted for MEK.

6.9 Microbiocide, for tension table, Fisher Scientific.

7. Procedure

7.1 This procedure is usually combined with the bulk density procedure (3B2b).

7.2 Prepare a round stock tag with sample identification number. Cut the copper wire and loop around the clod. Record the weight of the tag and wire (TAG). Loop fine copper wire around the clod, leaving a tail to which the round stock tag is attached. Record the weight of the clod (CC1).

7.3 Dip the clod in the 1:4 plastic lacquer. Wait 7 min and then dip the clod in the 1:7 plastic lacquer. Wait 12 min and then dip the clod in the 1:7 plastic lacquer. Wait 55 min and then reweigh the clod. If the clod has absorbed >3% in plastic by clod weight or smells excessively of solvent, allow longer drying time, then reweigh the clod and record the weight (CC2).

7.4 Cut a flat surface on the clod with a tile saw. It is necessary to wet the clod above the initial desorption point. This is accomplished by placing the cut clod surface on a tension table that is maintained at 5-cm tension. Periodically check the clod by inserting a needle probe, touching, or by weight comparison to determine if it has reached equilibrium (clod has wetted up). When the clod has reached equilibrium, remove the clod and record the weight (WSC).

7.5 If cut clod does not absorb water, place clod in a desiccator on a water-covered plate with a 0-cm tension. Submerge only the surface of clod in the water. Add a few mL of alcohol. Use in-house vacuum and apply suction until clod has equilibrated at saturation. Remove the clod and record the weight (WSC).

7.6 Saturate the ceramic plate by applying RO water through the adapter and apply enough pressure so that the rubber membrane is bulging a few centimeters. Care should be taken to remove all air.

7.7 Place the saturated ceramic plate in a pressure plate extractor. To provide good contact between the clod and ceramic plate, cover the ceramic plate with a 5-mm layer of silt loam soil and saturate with water. Place a sheet of reinforced paper towel over the silt loam soil. Place cut clod surface on the paper towel. Prepare a saturated, sieved soil as a quality control (QC) sample. Place several retaining rings in the extractor. Fill the retaining rings with the soil standard. Close the container and secure lid.

7.8 Apply gauged air pressure of 33 kPa. If more than one water retention point is requested, begin with the lowest pressure. Periodically submerge the outflow tube in water to monitor for air bubbles that indicate ceramic plate failure. Samples are equilibrated when water ceases to emit from the outflow tube. The outflow tube can be submerged under water in a burette to measure when water ceases to emit from the outflow tube. When water stops discharging from the outflow tube, the clod is at equilibrium. Determine the gravimetric water content of the QC. If the water content of the QC is higher than twice the standard deviation, apply pressure for additional time. Recheck the QC. If the water content of the QC is lower than twice the standard deviation, rewet the clods and desorb again. If the water content of the QC is within acceptable limits, then the apparatus has functioned properly.

7.9 Remove the clod and record the weight (WMC). Compare WMC to WSC. If WMC $\geq$ WSC, re-equilibrate the clod on the tension table and repeat the desorption process. If additional water retention points are requested, then repeat the desorption process at the next higher pressure. When the clod is equilibrated at 33 kPa and bulk density is to be measured, continue with procedure 3B1b.

7.10 Dry the clod in an oven at 110°C overnight and record oven-dry weight (WODC).

7.11 If the clods contains >5% in rock fragments by weight, remove the rock fragments from the clod. This determination of the percent rock fragments is based on particle-size data of >2-mm fraction.. Submerge the remaining soil material in a beaker of water and place on a hot plate. Use a fume hood. Boil $\approx$ 1 h. The plastic coating loosens from soil material upon heating. Remove beaker from the hot plate. Allow to cool. Discard plastic coating.

7.12 Wet sieve the cool soil through a 2-mm sieve. Dry and record the weight (RF) of the rock fragments that are retained on the sieve. If the rock fragments have similar properties to the soil sample, do not correct the clod mass for the rock fragments.

8. Calculations

8.1 H_2O % =[(WMC - MPC) - (WODC - ODPC) x 100]/(WODC - RF - ODPC – TAG)

where:
H_2O % = Percent gravimetric water content
WMC = Weight of equilibrated, coated clod
WODC= Weight of oven-dry coated clod
RF = Weight of rock fragments
ODPC = MPC x 0.85, weight of oven-dry plastic coat
TAG = Weight of tag and wire

8.2 MPC = {[(CC2 - CC1) + FCE] x RV}

where:
MPC = Weight of plastic coat before oven-drying
CC2 = Weight of clod after three laboratory plastic coats
CC1 = Weight of clod before three laboratory plastic coats
RV = Percent estimate of remaining clod volume after cutting to obtain flat surface ($\approx$ 80%)

8.3 FCE = 1.5 x [(CC2 - CC1)/3]

where:
FCE = Estimate of field-applied plastic coat, if applied

9. Report
Report water content to the nearest 0.1 percent.

10. Precision and Accuracy
Precision and accuracy data are available from the SSL upon request.

12. References
Bruce, R.R., and R.J. Luxmoore. 1986. Water retention: field methods. p. 663-686. *In* A. Klute (ed.) Methods of soil analysis. Part 1. Physical and mineralogical methods. 2nd ed. Agron. Monogr. 9. ASA and SSSA, Madison, WI.

Dane, J.H., and J. W. Hopmans. 2002. Water retention and storage: Laboratory. p. 675-720. *In* J.H. Dane and G.C. Topp (eds.) Methods of soil analysis. Part 4. Physical methods. Soil Sci. Am. Book Series No. 5. ASA and SSSA, Madison, WI.

Klute, A. 1986. Water retention: Laboratory methods. p. 635-662. *In* A. Klute (ed.) Methods of soil analysis. Part 1. Physical and mineralogical methods. 2nd ed. Agron. Monogr. 9. ASA and SSSA, Madison, WI.

U.S. Salinity Laboratory Staff. 1954. L.A. Richards (ed.) Diagnosis and improvement of saline and alkali soils. USDA Handb. 60. U.S. Govt. Print. Office, Washington, DC.

Water Retention (3C)
Pressure-Membrane Extraction (3C2)
1500-kPa (3C2a)
<2-mm (sieved), Air-Dry Samples (3C2a1a)

1. Application

The data collected are used for the water retention function, water-holding capacity, pore-size distribution, porosity, and saturated conductivity of a soil sample at specific water contents. The data are also used to calculate unsaturated hydraulic conductivity. Procedure 3C2a1a is used to determine the water retention at 1500 kPa for <2-mm (sieved), air-dry soil samples.

2. Summary of Method

The pressure desorption procedure (U.S. Salinity Laboratory Staff, 1954) is used. A sample of <2-mm (sieved), air-dry soil is placed in a retainer ring sitting on a cellulose membrane in a pressure-membrane extractor. The membrane is covered with water to wet the samples by capillarity. The sample is equilibrated at 1500 kPa. The pressure is kept constant until equilibrium is obtained (Klute, 1986). The gravimetric water content is determined.

3. Interferences

A leaking pressure extractor prevents equilibration of samples. Check outflow air to verify that the pressure membrane extractor is functioning properly and does not leak. The pressure should be monitored for stability. Equilibration must be done at constant temperature and humidity. Samples that do not wet by capillarity are moistened with ethyl alcohol.

Laboratory-determined, water retention data are usually higher than field-determined, water retention data because the confining soil pressure is not present in the laboratory (Bruce and Luxmoore, 1986).

Aerated 0.005 M $CaSO_4$ has also been recommended (Dane and Hopmans, 2002), especially for fine-textured soils that contain significant amounts of swelling clays. Distilled or deionized water can possibly promote dispersion of clays in samples and freshly drawn tap water is often supersaturated with air, affecting the water content at a given pressure head (Dane and Hopmans, 2002).

4. Safety

High-pressure plumbing must be maintained in good working order. Ensure that the pressure is zero before removing bolts from the pressure-apparatus lid. Ensure that the bolts are tightened before applying pressure. Do not drop the heavy lid.

5. Equipment

5.1 Pressure membrane extractor (Fig.1 and 2)
5.2 Cellulose membrane
5.3 Retainer rings. Use 10-mm height and 50-mm diameter rings for organic soils and 10-mm height and 40-mm diameter rings for all other soils.
5.4 Electronic balance, ±0.01-g sensitivity
5.5 Oven, 110°C
5.6 Pressure source, regulator, and gauge
5.7 Metal weighing cans, tared, with lids
5.8 Vacuum trap assembly
5.9 Vacuum, 80 kPa (0.8 bar)

6. Reagents

6.1 Ethyl alcohol, 95%, technical grade
6.2 Reverse osmosis (RO) water

7. Procedure

7.1 Submerge a cellulose membrane in RO water for 12 h or more before use. Install the wet cellulose membrane in the pressure extractor.

7.2 Add water and retaining rings. Add enough water to keep membrane moist. Water level should be less than height of retaining rings. Use 5-cm diameter rings for soils that are >12% in organic matter. Use 4-cm diameter rings for all other soils.

7.3 Fill retaining rings with 10 to 15 g of <2-mm or fine-grind, air-dry soil sample. Include a quality control (QC) sample with each plate. Continue to add water until all samples have moistened by capillarity. If samples do not moisten, apply ethyl alcohol to the surface of the sample. Allow ethyl alcohol to evaporate. Cover samples with a sheet of plastic to reduce evaporation, close the extractor, and let stand overnight.

7.4 Remove excess water on the plate with a vacuum and trap assembly.

7.5 Assemble the extractor and uniformly tighten the bolts. Torque the bolts on both sides of the hinge to 138.0 kPa (200 psi). Torque the remaining bolts to 103.5 kPa (150 psi).

7.6 Increase air pressure ≈ 150 kPa every 15 min until 1500 kPa is reached. After 4 h, apply the pressure differential by closing the valve that joins the mercury circuit and by opening the pressure release valve until air is forced through the mercury. Quickly close the pressure release valve. This forces the rubber diaphragm against the top of the samples. The samples are equilibrated when water ceases to emit from the outflow tube.

7.7 At equilibrium, open the extractor and quickly transfer the samples to water cans, cover with lids, and record the weights (M_{s+w}).

7.8 Remove the lids, place samples in the oven, and dry at 110°C overnight. Remove samples from the oven, replace the lids, allow cans to cool to ambient temperature, and record the weights (M_s).

7.9 Record the weights of the empty cans (M_c).

8. Calculations

$$H_2O \% = 100 \times [(M_{s+w} - M_s)/ (M_s - M_c)]$$

where:
$H_2O \%$ = Percent gravimetric water content
M_{s+w} = Weight of solids + H_2O + container
M_s = Weight of solids + can
M_c = Weight of container

Gypsiferous soils are a special case because gypsum ($CaSO_4 \cdot 2H_2O$) loses most of its two water molecules at 105°C. Properties of gypsiferous soils such as 1500-kPa water content that are reported on an oven-dry weight basis are converted to include the weight of crystal water in gypsum. Refer to procedure 3D3 for these conversion calculations. The 1500-kPa water content is corrected when the gypsum content of the soil is >1%. Gypsum content of the soil is determined in procedure 4E2a1a1.

9. Report
Report water content to the nearest 0.1 percent.

10. Precision and Accuracy

Precision and accuracy data are available from the SSL upon request.

11. References

Bruce, R.R., and R.J. Luxmoore. 1986. Water retention: Field methods. p. 663-686. *In* A. Klute (ed.) Methods of soil analysis. Part 1. Physical and mineralogical methods. 2nd ed. Agron. Monogr. 9. ASA and SSSA, Madison, WI.

Dane, J.H., and J. W. Hopmans. 2002. Water retention and storage: Laboratory. p. 675-720. *In* J.H. Dane and G.C. Topp (eds.) Methods of soil analysis. Part 4. Physical methods. Soil Sci. Am. Book Series No. 5. ASA and SSSA, Madison, WI.

Klute, A. 1986. Water retention: Laboratory methods. p. 635-662. *In* A. Klute (ed.) Methods of soil analysis. Part 1. Physical and mineralogical methods. 2nd ed. Agron. Monogr. 9. ASA and SSSA, Madison, WI.

Nelson, R.E., L.C. Klameth, and W.D. Nettleton. 1978. Determining soil content and expressing properties of gypsiferous soils. Soil Sci. Soc. Am. J. 42:659-661.

U.S. Salinity Laboratory Staff. 1954. L.A. Richards (ed.) Diagnosis and improvement of saline and alkali soils. USDA Agric. Handb. 60. U.S. Govt. Print. Office, Washington, DC.

Figure 1. Sieved (<2-mm) soil placed in pressure-membrane extractor.

Figure 2. Pressure-membrane extraction at 1500-kPa for <2-mm samples.

Water Retention (3C)
Pressure-Membrane Extraction (3C2)
1500-kPa (3C2a)
<2-mm (sieved), Field-Moist Samples (3C2a1b)

1. Application

The data collected are used for the water retention function, water-holding capacity, pore-size distribution, porosity, and saturated conductivity of a soil sample at specific water contents. The data are also used to calculate unsaturated hydraulic conductivity. Procedure 3C2a1b is used to determine the water retention at 1500 kPa for <2-mm (sieved), field-moist soil samples.

2. Summary of Method

The pressure desorption method (U.S. Salinity Laboratory Staff, 1954) is used. A sample of <2-mm (sieved) moist soil is placed in a retainer ring sitting on a cellulose membrane in a pressure-membrane extractor. The membrane is covered with water to wet the samples by capillarity. The sample is equilibrated at 1500 kPa. The pressure is kept constant until equilibrium is obtained (Klute, 1986). The gravimetric water content is determined.

3. Interferences

A leaking pressure extractor prevents equilibration of samples. Check outflow air to verify that the pressure membrane extractor is functioning properly and does not leak. The pressure should be monitored for stability. Equilibration must be done at constant temperature and humidity. Samples that do not wet by capillarity are moistened with ethyl alcohol.

Laboratory-determined, water retention data are usually higher than field-determined, water retention data because the confining soil pressure is not present in the laboratory (Bruce and Luxmoore, 1986).

Aerated 0.005 M $CaSO_4$ has also been recommended (Dane and Hopmans, 2002), especially for fine-textured soils that contain significant amounts of swelling clays. Distilled or deionized water can possibly promote dispersion of clays in samples and freshly drawn tap water is often supersaturated with air, affecting the water content at a given pressure head (Dane and Hopmans, 2002).

4. Safety

High-pressure plumbing must be maintained in good working order. Ensure that the pressure is zero before removing bolts from the pressure-plate apparatus lid. Ensure that the bolts are tightened before applying pressure. Do not drop the heavy lid.

5. Equipment

5.1 Pressure membrane extractor
5.2 Cellulose membrane
5.3 Retainer rings. Use 10-mm height and 50-mm diameter rings for organic soils and 10-mm height and 40-mm diameter rings for all other soils.
5.4 Electronic balance, ±0.01-g sensitivity
5.5 Oven, 110°C
5.6 Pressure source, regulator, and gauge.
5.7 Metal weighing cans, tared, with lids
5.8 Vacuum trap assembly
5.9 Vacuum, 80 kPa (0.8 bar)

6. Reagents

6.1 Ethyl alcohol, 95%, technical grade
6.2 Reverse osmosis (RO) water

7. Procedure

7.1 Submerge a cellulose membrane in RO water for 12 h or more before use. Install the wet cellulose membrane in the pressure apparatus.

7.2 Add water and retaining rings. Add enough water to keep membrane moist. Water level should be less than height of retaining rings. Use 5-cm diameter rings for soils that are >12% in organic matter. Use 4-cm diameter rings for all other soils.

7.3 Fill retaining rings with 10 to 15 g of <2-mm or fine-grind, field moist soil sample. Include a quality control (QC) sample with each plate. Continue to add water until all samples have moistened by capillarity. If samples do not moisten, apply ethyl alcohol to the surface of the sample. Cover samples with a sheet of plastic to reduce evaporation, close the extractor, and let stand overnight.

7.4 Remove excess water on the plate with a vacuum and trap assembly.

7.5 Assemble the extractor and uniformly tighten the bolts. Torque the bolts on both sides of the hinge to 138.0 kPa (200 psi). Torque the remaining bolts to 103.5 kPa (150 psi).

7.6 Increase air pressure $\approx$ 150 kPa every 15 min until 1500 kPa is reached. After 4 h, apply the pressure differential by closing the valve that joins the mercury circuit and by opening the pressure release valve until air is forced through the mercury. Quickly close the pressure release valve. This forces the rubber diaphragm against the top of the samples. The samples are equilibrated when water ceases to emit from the outflow tube.

7.7 At equilibrium, open the apparatus and quickly transfer the samples to water cans, cover with lids, and record the weights (M_{s+w}).

7.8 Remove the lids, place samples in the oven, and dry at 110°C until weights are constant. Remove samples from the oven, replace the lids, allow cans to cool to ambient temperature, and record the weights (M_s).

7.9 Record the weights of the empty cans (M_c).

8. Calculations

$$H_2O \% = 100 \times [(M_{s+w} - M_s)/(M_s - M_c)]$$

where:
$H_2O \%$ = Percent gravimetric water content
M_{s+w} = Weight of solids + H_2O + container
M_s = Weight of solids + container
M_c = Weight of container

9. Report
Report water content to the nearest 0.1 percent.

10. Precision and Accuracy
Precision and accuracy data are available from the SSL upon request.

11. References

Bruce, R.R., and R.J. Luxmoore. 1986. Water retention: Field methods. p. 663-686. *In* A. Klute (ed.) Methods of soil analysis. Part 1. Physical and mineralogical methods. 2nd ed. Agron. Monogr. 9. ASA and SSSA, Madison, WI.

Dane, J.H., and J. W. Hopmans. 2002. Water retention and storage: Laboratory. p. 675-720. *In* J.H. Dane and G.C. Topp (eds.) Methods of soil analysis. Part 4. Physical methods. Soil Sci. Am. Book Series No. 5. ASA and SSSA, Madison, WI.

Klute, A. 1986. Water retention: Laboratory methods. p. 635-662. *In* A. Klute (ed.) Methods of soil analysis. Part 1. Physical and mineralogical methods. 2nd ed. Agron. Monogr. 9. ASA and SSA, Madison, WI.

U.S. Salinity Laboratory Staff. 1954. L.A. Richards (ed.) Diagnosis and improvement of saline and alkali soils. USDA Agric. Handb. 60. U.S. Govt. Print. Office, Washington, DC.

Water Retention (3C)
Field-State (3C3)

1. Application

Field-water content can be determined by weighing, drying, and re-weighing a soil sample (3C3). The resulting data are used to estimate the water content at the time of sampling.

2. Summary of Method

Soil samples are collected in the field. The samples are stored in plastic or metal containers to prevent drying and then transported to the laboratory. Gravimetric water content is determined (Gardner, 1986).

3. Interferences

Leaks in the plastic or metal storage containers cause the samples to dry, resulting in an underestimation of the field water content.

4. Safety

Use insulated gloves to remove samples from the oven.

5. Equipment

5.1 Electronic balance, ±0.01-g sensitivity

5.2 Oven, 110°C

5.3 Moisture cans, tared

6. Reagents

None

7. Procedure

7.1 Collect soil samples in the field. Place samples in airtight, metal or plastic containers.

7.2 Record sample weight (M_{s+w}).

7.3 Dry sample in an oven at 110°C overnight. Record oven-dry weight (M_s).

7.4 Record weight of container (M_c).

8. Calculations

$$H_2O \% = 100 \times [(M_{s+w} - M_s)/(M_s - M_c)]$$

where:

H₂O % = Percent gravimetric water content

$H_2O\%$ = Percent gravimetric water content
M_{s+w} = Weight of solids + H_2O + container
M_s = Weight of solids + container
M_c = Weight of container

9. Report

Report water content to the nearest 0.1 percent.

10. Precision and Accuracy

Precision and accuracy data are available from the SSL upon request.

11. References

Gardner, W.H. 1986. Water content. p. 493-544. *In* A. Klute (ed.) Methods of soil analysis. Part 1. Physical and mineralogical methods. 2nd ed. Agron. Monogr. 9. ASA and SSA, Madison, WI.

PHYSICAL AND FABRIC-RELATED ANALYSES (3)

Ratios and Estimates Related to Particle-Size Analysis, Bulk Density, and Water Retention (3D)
Air-Dry/Oven-Dry Ratio (AD/OD) (3D1)
Field-Moist/Oven-Dry Ratio (FM/OD)(3D2)
Correction for Crystal Water (3D3)

1. Application

Soil properties generally are expressed on an oven-dry weight basis. The calculation of the air-dry/oven-dry (AD/OD) ratio (procedure 3D1) *or* field-moist/oven-dry (FM/OD) ratio (procedure 3D2) is used to adjust all results to an oven-dry basis and, if required in a procedure, to calculate the sample weight that is equivalent to the required oven-dry soil weight.

Gypsiferous soils are a special case because gypsum ($CaSO_4 \cdot 2H_2O$) loses most of its two water molecules at 105°C. Properties of gypsiferous soils that are reported on an oven-dry weight basis should be converted to include the weight of crystal water in gypsum. In procedure 3D1, the AD/OD ratio is calculated. This ratio is used to convert soil properties to an oven-dry basis. In procedure 3D3, the AD/OD ratio is converted to a crystal water basis (Nelson et al., 1978). The inclusion of weight of crystal water in gypsum allows the properties of gypsiferous soils to be compared with those properties of nongypsiferous soils. This conversion also avoids the possible calculation error of obtaining >100% gypsum when the data are expressed on an oven-dry basis (Nelson, 1982).

2. Summary of Method

A sample is weighed, dried to a constant weight in an oven, and reweighed. The moisture content is expressed as a ratio of the air-dry to the oven-dry weight (AD/OD) or as a ratio of field-moist to the oven-dry weight (FM/OD). Soil properties of gypsiferous soils that are reported on an oven-dry weight basis are converted to include the weight of the crystal water. When reporting the water content of gypsiferous soils, the crystal water content must be subtracted from the total oven-dry water content. The AD/OD ratio is corrected to a crystal water basis when the gypsum content of the soil is ≥1%. Gypsum content of the soil is determined in procedure 4E2a1a1.

3. Interferences

Traditionally, the most frequently used definition for a dry soil is the soil mass after it has come to a constant weight at a temperature of 100 to 110°C (American Society for Testing and Materials, 2004). Many laboratory ovens are not capable of maintaining this prescribed temperature range. Temperatures that are >50°C may promote oxidation or decomposition of some forms of organic matter.

Samples may not reach a constant weight with overnight drying. Do not add moist samples to an oven with drying samples unless the drying samples have been in the oven for at least 12 to 16 h. Soil samples may adsorb significant amounts of moisture from the atmosphere after cooling. Prompt weighing, i.e., <30 min after samples have cooled, helps to eliminate this problem. During the weighing or drying processes, the non-uniform weight of weighing vessels, sample contamination, or sample loss may lead to erroneous results.

The removal of structural water, most commonly in gypsum, can produce a positive error. When reporting the water content of gypsiferous soils, the crystal water content must be subtracted from the total oven-dry water content. Gypsum, hydrous oxides, and amorphous material may be affected.

4. Safety

Use heat resistant gloves to remove weighing containers from a hot oven. No other significant hazard is associated with this procedure. Follow standard laboratory procedures.

5. Equipment
5.1 Electronic balance, ± 1-mg sensitivity
5.2 Oven, thermostatically controlled, 110 ± 5°C
5.3 Thermometer, 0 to 200°C
5.4 Tin dishes, 4.5-cm diameter x 3-cm height, with covers

6. Reagents
No reagents are required for this determination.

7. Procedure

7.1 Tare the moisture dishes. Record each sample number and associated dish number.

7.2 Add 10 to 20 g <2mm or fine-grind, air-dry soil to each moisture dish for AD/OD determination. For FM/OD determination, add enough <2mm or fine-grind, moist soil to achieve ≈ 10 to 20 g sample of air-dry soil. Weigh the dish plus the sample and record the weight to the nearest 1mg. Place the sample dish in a drying oven set at 110 ± 5°C. Allow the sample to remain in the oven overnight (12 to 16 h).

7.3 Remove the sample dish and allow it to cool before re-weighing. Record the oven-dry weight to the nearest 1mg.

7.4 Do not allow the sample dish to remain at room temperature for >30 min before re-weighing.

7.5 Discard the sample.

7.6 Refer to the calculations for the correction for crystal water of gypsum in gypsiferous soils.

8. Calculations

Calculations 8.1 – 8.2 for AD/OD ratio (procedure 3D1) are as follows:

133

8.1 AD/OD ratio = AD/OD

where:
AD = (Air-dry weight) – (Tin tare weight)
OD = (Oven-dry weight) – (Tin tare weight)

8.2 $H_2O = [(AD - OD) \times 100]/OD$

where:
H_2O = % Water content
AD = (Air-dry weight) – (Tin tare weight)
OD = (Oven-dry weight) – (Tin tare weight)

Calculations 8.3 – 8.4 for FM/OD ratio (procedure 3D2) are as follows:

8.3 FM/OD ratio = FD/OD

where:
FM = (Field-moist weight) – (Tin tare weight)
OD = (Oven-dry weight) – (Tin tare weight)

8.4 $H_2O = [(FM - OD) \times 100]/OD$

where:
FM = (Field-moist weight) – (Tin tare weight)
OD = (Oven-dry weight) – (Tin tare weight)

Calculations 8.5 – 8.6 for gypsum H_2O correction (procedure 3D3) are as follows:

8.5 $(AD/OD)_c = (AD/OD)_{uc} /[1 + (Gypsum \times 0.001942)]$

where:
AD/OD_c = Air-dry/oven-dry ratio, corrected basis, gypsiferous soils
AD/OD_{uc} = Air-dry/oven-dry ratio, uncorrected basis
Gypsum = % Gypsum uncorrected (procedure 4E2a1a1)

8.6 $H_2O_c = [H_2O_{uc} - (Gypsum \times 0.1942)]/ [1 + (Gypsum \times 0.001942)]$

where:
H_2O_c = % Water content, corrected basis, gypsiferous soils
H_2O_{uc} = % Water content, uncorrected basis (calculation 8.2)
Gypsum= % Gypsum uncorrected (procedure 4E2a1a1)

AD/OD Data Use

The following equation is used to calculate the weight of air-dry soil needed to provide a given weight of oven-dry soil for other analytical procedures.

8.7 $AD = (OD_r)/[1-(H_2O/100)]$

where:
AD   = Required weight of air-dry soil
OD_r = Desired weight of oven-dry soil
H_2O = Percent water determined from AD/OD (calculation 8.2)

134

9. Report

Report the AD/OD and/or FM/OD ratio as a dimensionless value to the nearest 0.01 unit.

10. Precision and Accuracy

Precision and accuracy data are available from the SSL upon request.

11. References

Nelson, R.E. 1982. Carbonate and gypsum. p. 159-165. *In* A.L. Page, R.H. Miller, and D.R. Keeney (eds.) Methods of soil analysis. Part 2. Chemical and microbiological properties. 2nd ed. Agron. Monogr. 9. ASA and SSSA, Madison, WI.

Nelson, R.E., L.C. Klameth, and W.D. Nettleton. 1978. Determining soil content and expressing properties of gypsiferous soils. Soil Sci. Soc. Am. J. 42:659-661.

Ratios and Estimates Related to Particle-Size Analysis, Bulk Density, and Water Retention (3D)
Coefficient of Linear Extensibility (COLE) (3D4)

Coefficient of linear extensibility (COLE) is a derived value that denotes the fractional change in the clod dimension from a moist to a dry state (Franzmeier and Ross, 1968; Grossman et al., 1968; Holmgren, 1968). COLE may be used to make inferences about shrink-swell capacity and clay mineralogy. The COLE concept does not include irreversible shrinkage such as that occurring in organic and some andic soils. Certain soils with relatively high contents of smectite clay have the capacity to swell significantly when moist and to shrink and crack when dry. This shrink-swell potential is important for soil physical qualities (large, deep cracks in dry seasons) as well as for genetic processes and soil classification (Buol et al., 1980).

COLE can also be expressed as percent, i.e., linear extensibility percent (LEP). LEP = COLE x 100. The LEP is not the same as LE. In soil taxonomy (Soil Survey Staff, 1999), linear extensibility (LE) of a soil layer is the product of the thickness, in centimeters, multiplied by the COLE of the layer in question. The LE of a soil is defined as the sum of these products for all soil horizons (Soil Survey Staff, 1999). Refer to Soil Survey Staff (1999) for additional discussion of LE.

Ratios and Estimates Related to Particle-Size Analysis, Bulk Density, and Water Retention (3D)
Coefficient of Linear Extensibility (COLE) (3D4)
Air-Dry or Oven-Dry to 33-kPa Tension (3D4a)

The SSL calculates the COLE for the whole soil (air-dry or oven-dry to 33-kPa suction) by procedure 3D4a. The COLE value is reported in cm cm^{-1}. Calculate COLE when coarse fragments are present as follows:

8.1 $COLE_{ws} = \{1/[Cm \times (Db_{33<2mm}/Db_{d<2mm}) + (1 - Cm)]\}^{1/3} - 1$

where:
$COLE_{ws}$ = Coefficient of linear extensibility on a whole-soil base.
$Db_{/33<2mm}$ = Bulk density at 33-kPa water content on a <2-mm base (g cc^{-1}).
$Db_{d<2mm}$ = Bulk Density, oven-dry or air-dry, on a <2-mm base (g cc^{-1}).

Cm   = Coarse fragment (moist) conversion factor.

8.2 If no coarse fragments, Cm = 1. If coarse fragments are present, calculate Cm as follows:
$Cm = Vol_{<2mm} / Vol_{whole}$

where:
$Vol_{<2mm}$ = Volume moist <2-mm fabric (cm^3)
Vol_{whole} = Volume moist whole soil (cm^3)

OR (alternatively)

8.3 $Cm = (100 - Vol_{>2mm})/100$

where:
$Vol_{>2mm}$ = Volume percentage of the >2-mm fraction

8.4 If no coarse fragments, Cm = 1, the previous equation reduces as follows:

$$COLE_{ws} = (Db_{d<2mm}/Db_{33<2mm})^{1/3} - 1$$

where:
$COLE_{ws}$ = Coefficient of linear extensibility on a whole-soil base
$Db_{d<2mm}$ = Bulk Density, oven-dry or air-dry, on a <2-mm base (g cc^{-1})
$Db_{33<2mm}$ = Bulk Density at 33-kPa water content on a <2-mm base (g cc^{-1})

References
Buol, S.W., F.D. Hole, and R.J. McCracken. 1980. Soil genesis and classification. 2nd ed. Iowa
    State Univ. Press, Ames IA.
Franzmeier, D.P., and S.J. Ross, Jr. 1968. Soil swelling: Laboratory measurement and relation
    to other soil properties. Soil Sci. Soc. Amer. Proc. 32:573-577.
Grossman, R.B., B.R. Brasher, D.P. Franzmeier, and J.L. Walker. 1968. Linear extensibility as
    calculated from natural-clod bulk density measurements. Soil Sci. Soc. Am. Proc. 32:570-
    573.
Holmgren, George G.S. 1968. Nomographic calculation of linear extensibility in soils
    containing coarse fragments. Soil Sci. Soc. Amer. Proc. 32:568-570.
Soil Survey Staff. 1999. Soil taxonomy: A basic system of soil classification for making and
    interpreting soil surveys. Agric. Handb. No. 436. 2nd ed USDA-NRCS. Govt. Print. Office,
    Washington, DC.

Ratios and Estimates Related to Particle-Size Analysis, Bulk Density, and Water Retention (3D)
Water Retention Difference (WRD) (3D5)

The calculation of the water retention difference (WRD) is considered the initial step in the approximation of the available water capacity (AWC). WRD does not allow for restriction of roots from the soil layer or osmotic pressure. Usually the volume of rock fragments is considered a diluent containing no water between the suctions that define WRD. WRD is a calculated value that denotes the volume fraction for water in the whole soil that is retained between 1500-kPa suction and an upper limit of usually 33 or 10-kPa suction. The upper limit (lower suction) is selected so that the volume of water retained approximates the volume of water held at field capacity. The 10-, 33- and 1500-kPa gravimetric water contents are then converted to a whole soil volume basis by multiplying by the bulk density (Db_{33}) and adjusting downward for the volume fraction of rock fragments, if present in the soil. The lower suctions, e.g., 10 or 5-kPa, are used for coarse materials. Refer to Soil Survey Staff Division Staff

(1993) and Grossman et al. (1994) for additional discussion on coarse materials and the significance of soil water content at lower suctions, e.g., 5 and 10 kPa, as well as suggestions for the selection of these lower suctions for the determination of water retention difference (WRD).

Ratios and Estimates Related to Particle-Size Analysis, Bulk Density, and Water Retention (3D)
Water Retention Difference (WRD) (3D5)
Between 33-kPa and 1500-kPa Tension (3D5a)

The SSL calculates the WRD between 33 and 1500-kPa suctions in the whole soil by procedure 3D5a. The WRD is reported as centimeters of water per centimeter of depth of soil ($cm\ cm^{-1}$), but the numbers do not change when other units, e.g., in in^{-1} or ft ft^{-1} are needed. The WRD with W_{33} as the upper limit is reported as $cm\ cm^{-1}$. This WRD is calculated on a whole-soil base as follows:

$$WRD_{ws} = [(W_{33<2mm} - W_{1500<2mm}) \times (Db_{33<2mm}) \times Cm]/(P_w \times 100)$$

where:

WRD_{ws} = Volume fraction ($cm^3\ cm^{-3}$) of water retained in the whole soil between 33-kPa and 1500-kPa suction reported in $cm\ cm^{-1}$.

$W_{33<2mm}$ = Weight percentage of water retained at 33-kPa suction on a <2-mm soil basis.

$W_{1500<2mm}$ = Weight percentage of water retained at 1500-kPa suction on a <2-mm soil basis. If available, moist 1500-kPa (procedure 3C2a1b) is the first option in the WRD calculation; otherwise, dry 1500-kPa (procedure 3c2a1a) is used.

$Db_{33<2mm}$ = Bulk density at 33-kPa water content on a <2-mm base ($g\ cm^{-3}$).

P_w = Density of water (1 $g\ cm^{-3}$).

Cm = Coarse fragment material conversion factor. If no coarse fragments, $Cm = 1$. If coarse fragments are present, calculate Cm as follows:

$$Cm = Vol_{<2mm}/Vol_{whole}$$

where:
$Vol_{<2mm}$ = Volume moist <2mm fabric (cm^3)
Vol_{whole} = Volume moist whole soil (cm^3)

OR (alternatively)

$$Cm = (100 - Vol_{>2mm})/100$$

where:
$Vol_{>2mm}$ = Volume percentage of the >2-mm fraction.

Ratios and Estimates Related to Particle-Size Analysis, Bulk Density, and Water Retention (3D)
Water Retention Difference (WRD) (3D5)
Between 10-kPa and 1500-kPa Tension (3D5b)

The SSL also calculates the WRD between 10-kPa (W_{10}) and 1500-kPa suctions (W_{1500}) by procedure 3D5b. This WRD value can be calculated by substituting the W_{10} in place of W_{33} in the equation for procedure 3D5a. The W_{10} may be used as the upper limit of plant available water for coarse soil materials.

Ratios and Estimates Related to Particle-Size Analysis, Bulk Density, and Water Retention (3D)
Water Retention Difference (WRD) (3D5)
Between 33-kPa Rewet and 1500-kPa (Air-Dry) Tension (3D5c)

The SSL also calculates the WRD between 33-kPa rewet (W_r) and W_{15} by procedure 3D5c. This WRD value can be calculated by substituting the W_r in place of W_{33} in the equation for 3D5a. The W_r is used for organic materials.

References

Grossman, R.B., T. Sobecki, and P. Schoeneberger. 1994. Soil interpretations for resource soil scientists. Soil Water. Tech. Workbook. USDA-SCS, Lincoln, NE.
Soil Survey Division Staff. 1993. Soil survey manual. USDA. Handb. No. 18. U.S. Govt. Print. Office, Washington, DC.

Ratios and Estimates Related to Particle-Size Analysis, Bulk Density, and Water Retention (3D)
1500-kPa Water Content/Total Clay (3D6)

Divide the 1500-kPa water retention (procedure 3C2a) by the total clay percentage (procedure 3A1a). This ratio is reported as a dimensionless value. In the past, the ratios of 1500-kPa water:clay have been reported as g g^{-1}. For more detailed information on the application of this ratio, refer to Soil Survey Staff (1995, 1999).

Ratios and Estimates Related to Particle-Size Analysis, Bulk Density, and Water Retention (3D)
Total Silt Fraction (3D7)

Total silt is a soil separate with 0.002- to 0.05-mm particle diameter. The SSL determines the fine silt separate by pipet analysis and the coarse silt separate by difference (3A1a). Total silt is reported as a weight percentage on a <2-mm basis (3D7). For more information on these data, refer to Soil Survey Staff (1995).

Ratios and Estimates Related to Particle-Size Analysis, Bulk Density, and Water Retention (3D)
Total Sand Fraction (3D8)

Total sand is a soil separate with 0.05- to 2.0-mm particle diameter. The SSL determines the sand fractions by sieve analysis (3A1a). Total sand is the sum of the very fine sand (VFS), fine sand (FS), medium sand (MS), coarse sand (CS), and very coarse sand fractions VCS). The rationale for five subclasses of sand and the expansion of the texture classes of sand, e.g., sandy loam and loamy sand, is that the sand separates are the most visible to the naked eye and the most detectable by "feel" by the field soil scientist. Total sand is reported as a weight percentage on a <2-mm basis (3D8). For more information on the application of these data, refer to Soil Survey Staff (1995).

Ratios and Estimates Related to Particle-Size Analysis, Bulk Density, and Water Retention (3D)
2- to 5- mm Fraction (3D9)

The SSL determined coarse fraction with 2- to 5-mm particle diameter by procedures outlined in 3A2. The 2- to 5-mm divisions correspond to the size of opening of the No. 10 and No. 4 screen (4.76 mm), respectively, used in engineering. Coarse fractions with 2- to 5- mm particle-diameter correspond to the rock-fragment division, fine pebbles (Soil Survey Staff,

1993). Coarse frations with 2- to 5- mm particle diameter are reported as a weight percentage on a <75-mm basis (3D9). For more information on coarse fraction with >2-mm particle diameters. For more information on these data, refer to Soil Survey Staff (1995).

Ratios and Estimates Related to Particle-Size Analysis, Bulk Density, and Water Retention (3D)
5- to 20-mm Fraction (3D10)

The SSL determined coarse fraction with 5 to 20-mm particle diameter by procedures outlined in 3A2. The 5- to 20-mm divisions correspond to the size of opening of the No. 4 screen (4.76 mm) and the ¾-in screen (19.05 mm), respectively, used in engineering. Coarse fractions with 5- to 20- mm particle-diameter correspond to the rock fragment division, medium pebbles (Soil Survey Staff, 1993). Coarse frations with 5- to 20- mm particle-diameter are reported as a weight percentage on a <75-mm basis (3D10). For more information on these data, refer to Soil Survey Staff (1995).

Ratios and Estimates Related to Particle-Size Analysis, Bulk Density, and Water Retention (3D)
20- to 75-mm Fration (3D11)

The SSL determined coarse fraction with 20 to 75-mm particle diameter by procedures outlined in 3A2. The 20- to 75-mm divisions correspond to the size of opening of the ¾-in screen (19.05 mm) and the 3-in screen (76.1 mm), respectively, used in engineering. Coarse fractions with 20- to 75- mm particle-diameter correspond to the rock fragment division, coarse pebbles (Soil Survey Staff, 1993). Coarse frations with 20- to 75- mm particle diameter are reported as a weight percentage on a <75-mm basis (3D11). For more information on these data, refer to Soil Survey Staff (1995).

Ratios and Estimates Related to Particle-Size Analysis, Bulk Density, and Water Retention (3D)
0.1 to 75-mm Fraction (3D12)

The SSL determines coarse fractions with 0.1 to 75-mm particle diameter by procedures outlined in 3A1a and 3A2. The 75-mm division corresponds to the size of opening in the 3-in screen (76.1 mm) used in engineering. These data are listed for taxonomic placement for particle-size class, i.e., to distinguish loamy and silty family particle-size classes. Refer to Soil Survey Staff (1995, 1999) for additional discussion on particle-size classes. Coarse fractions with 0.1 to 75-mm particle diameter are reported as a weight percentage on a <75-mm basis (3D12).

Ratios and Estimates Related to Particle-Size Analysis, Bulk Density, and Water Retention (3D)
>2-mm Fraction (3D13)

The SSL determine coarse fractions with >2-mm particle diameter by procedure outlined in 3A2. Coarse fractions with >2-mm particle diameter are reported as a weight percent on a whole-soil basis (3D13). For more information on these data, refer to Soil Survey Staff (1993, 1995, 1999).

References
Soil Survey Division Staff. 1993. Soil survey manual. USDA. Handb. No. 18. U.S. Govt. Print. Office, Washington, DC.
Soil Survey Staff. 1995. Soil survey laboratory information manual. Version No. 1.0. USDA-NRCS. Soil Survey Investigations Report No. 42. U.S. Govt. Print. Office, Washington, DC

Soil Survey Staff. 1999. Soil taxonomy: A basic system of soil classification for making and interpreting soil surveys. 2nd ed. Agric. Handb. No. 436. USDA-NRCS. U.S. Govt. Print. Office, Washington, DC.

Micromorphology (3E)
Thin Sections (3E1)
Preparation (3E1a)

1. Application

Micromorphology is used to identify fabric types, skeleton grains, weathering intensity, illuviation of argillans, and to investigate genesis of soil or pedological features.

2. Summary of Method

In this procedure (3E1a), a soil clod is impregnated with a polymer resin (Innes and Pluth, 1970). A flat surface of the soil sample is glued to a glass slide. The soil sample is cut and ground to a thickness of ≈ 30 μm. The thin section is examined with a petrographic microscope (Anon. 1987; Cady, et al., 1986).

3. Interferences

Impregnation of the soil sample must be complete, or the sample will disintegrate during processing. Air bubbles interfere with petrographic examination. Bubbles are avoided by using proper temperature, pressure, and technique. The final, 30-μm thickness is estimated by examining the slide under polarized light. If the quartz interference colors are of first order, i.e., white, gray, and pale yellow, the sample is ≈ 30 μm (Anon. 1987).

4. Safety

Use adequate ventilation when mixing, heating, and applying the resins. Use tongs or heat resistant gloves when handling hot slides or resins.

5. Equipment

5.1 Petro-thin, thin sectioning system, Buehler, Lake Bluff, IL
5.2 Metallographic polisher with cast-iron laps
5.3 Diamond saw
5.4 Electric oven
5.5 Hot plate with temperature control or a petrographic slide warmer
5.6 Polarizing microscope
5.7 Vacuum, 0.8 bars (80 kPa)
5.8 Desiccator
5.9 Porcelain crucibles
5.10 Standard petrographic slides
5.11 Cover glass
5.12 Silicon carbide abrasives
5.13 Squares of thick, rough-textured plate glass, 305 mm
5.14 Metal probes or dissecting needles
5.15 Small forceps
5.16 Art brush
5.17 Razor blade
5.18 Small chisel, probe (ice pick), ordinary hacksaw, or jeweler's hacksaw

6. Reagent

6.1 Scotchcast resin, Industrial Electric Products Division, 3M Company, 3M Center, St. Paul, MN 55101.
6.2 Epoxide. The SSL uses EPO-MIX from Buehler.
6.3 Ethylene glycol, automotive coolant/anti-freeze

7. Procedure

Sample Collection

7.1 Collect samples by any procedure that does not disturb the natural structure. Core samplers are commonly used. A satisfactory procedure for some soils is to use a knife or trowel to carve a clod that fits in bulk density box cells, a tin box, matchbox, or a round ice-cream container.

7.2 Place clods in an upright position in the container. Mark the top of clod with a thumb tack, staple, or pin to ensure proper orientation.

7.3 Select clods from bulk samples if orientation is of no interest. Avoid coated clods with Saran or other plastic because coatings interfere with the grinding after the clod is sectioned.

7.4 Place clods in small plastic bags to avoid contamination from other samples during transit. Pack irregular clods in box cells or containers with light weight material to avoid breakage. Unless fragility is affected, the prevention of moisture loss is usually unnecessary as most samples are usually dried before impregnation.

7.5 The whole core or clod can be impregnated, but better results are generally obtained with specimens that are 5 cm^3 or smaller. Remove specimens from the sample with a small chisel, a probe (ice pick), an ordinary hacksaw, or a jeweler's hacksaw.

Sample Preparation

7.6 Place the soil sample in a disposable heat- and chemical-resistant beaker and place in a glass desiccator. Evacuate the air from the desiccator and dry the sample overnight at 80°C. The natural structure of the soil samples is better preserved by the technique of freeze-drying. Impregnation of the freeze-dried samples may be an improvement over oven-dried samples. In preparation for the proceeding step, bring the freeze-dried or oven-dried sample to ≈ 80°C.

Mixing plastic solution

7.7 Many good resins are on the market. The NSSL uses Scotchcast. Add two parts of part A by weight to three parts of part B. For best results, raise part A and part B to ≈ 80°C before mixing. Weigh parts to an accuracy of 2% before mixing. Mix until a uniform color is obtained.

Impregnation

7.8 Open the desiccator which contains the dry samples, add the heated plastic solution, and evacuate all the air from the sample. Release the vacuum and again evacuate the air. Do not mistake boiling solution under evacuated conditions for escaping air bubbles from the sample material.

7.9 Cure the impregnated soil overnight in the oven at 110°C. After curing, the block is ready for sectioning. Disposable containers can be cut with the cooled samples during sectioning.

Cutting and Rough Grinding

7.10 With a diamond saw blade, cut the sample block into 13-mm thick chips that are small enough to fit on a regular petrographic slide.

7.11 With a slurry of successively finer abrasives, grind one surface smooth on the revolving lap until the surface is highly polished. Experience is required to determine the mixture of abrasive and water that gives the best results for each grade of abrasive. If the sample surface tends to pull apart or to react with water, dry and re-impregnate the small chip or polish it by hand on a glass plate. Alternatively, grind the block with an abrasive and lubricate with ethylene glycol.

7.12 Clean the sample free of all abrasive material. An ultrasonic bath is recommended. Dry thoroughly.

Mounting
7.13 Burnish petrographic slides to a uniform thickness.

7.14 Firmly attach the chip to the burnished slide with a strong, transparent bonding agent. Use a thermoplastic cement or epoxide. Mix the resin and hardener according to the product instructions. Allow entrapped air to rise to the top. Apply a thin layer of epoxide to the slide and to the chip. Place the chip obliquely on the slide and lower slowly. Move the chip back and forth with moderate pressure to remove entrapped air bubbles and excess epoxide. Clamp with a spring clamp and cure at ~ 50°C for 1 h.

Final Grinding
7.15 Consult the Petro-thin operation and maintenance instructions for proper operation.

7.16 Clean the glass of excess cement.

7.17 Use the Petro-thin to cut off the excess sample and grind the sample to ≈ 30 μm with the diamond lap. Examine the section frequently under a polarizing microscope during the final stages of grinding. If quartz is present in the sample, use it to judge thickness. If the sample is ≈ 0.030 mm thick, the quartz interference colors are of the first order, i.e., white, gray, and pale yellow.

7.18 If a Petro-thin is not available, trim the mounted sample with a diamond saw blade to a thickness of 50 to 100 μm. Begin with the coarse abrasive and lap by hand until the sample is relatively thin. Use successively finer abrasives.

7.19 Care and considerable practice are needed to develop the dexterity required to handle an almost finished section without over-grinding. Use the finest abrasive to finish grinding on ground-glass plates. Wash the section free of abrasive and dry thoroughly.

Seating Cover Glass
7.20 Heat the finished section and the cover slip to ≈ 40°C. Spread a small quantity of epoxide over the surface of the thin section and the cover slip. Wait a few seconds for the air bubbles to escape. Place the cover glass obliquely on one end of the section and lower very slowly. If any air bubbles remain, remove them by pressing lightly on the cover glass with a soft eraser. As the section cools, but before the plastic hardens, remove excess epoxy with a razor blade. After the epoxy hardens, remove the final thin film with a razor blade. A thick film may cause the slide to break when the epoxy is removed.

7.21 Very dense soils and soils with clay fractions that have 30% or more montmorillonite require special handling. Use either a dry-grinding technique or a more penetrating impregnation procedure. Without using water, cut the sample to the appropriate size. Sprinkle a coarse abrasive (American Optical No. 190) on the ground-glass plate and commence to dry-grind one face of the sample by hand. Use a figure "8" or a counterclockwise motion for best results. Use successively finer abrasives and continue to grind until the surface is highly polished. Proceed with the standard mounting technique.

7.22 Aroclor 5460, a thermoplastic chlorinated diphenyl resin (Monsanto), seems to give better impregnation of dense soils. Place pieces of air-dry soil material in xylene and evacuate.

Submerge the xylene-saturated soil material in molten Aroclor 5460. Hold the sample in the Aroclor at $\approx 200°C$ for 1 to 2 days. Remove impregnated soil material, allow to cool, and prepare thin sections by dry-grinding.

8. Calculations
None

9. Report
Describe the thin section as outlined in procedure 3E1b.

10. Precision and Accuracy
Precision and accuracy data are available from the SSL upon request.

11. References
Anon. 1987. Petrographic sample preparation for micro-structural analysis. Buehler Digest, Vol. 24, No.1.

Cady, J.G., L.P. Wilding, and L.R. Drees. 1986. Petrographic microscope techniques. p. 198-204. *In* A. Klute (ed.) Methods of soil analysis. Part 1. Physical and mineralogical methods. 2nd ed. Agron. Monogr. 9. ASA and SSSA, Madison, WI.

Innes, R.P., and D.J. Pluth. 1970. Thin section preparation using an epoxy impregnation for petrographic and electron microprobe analysis. Soil Sci. Soc. Am. Proc. 34:483-485.

Micromorphology[1] (3E)
Thin Sections (3E1)
Interpretation (3E1b)

Application
Background: Micromorphology may be defined as the study of soils or regolith samples in their natural undisturbed arrangement using microscopic techniques (Cady, 1986; Stoops, 2003). This technique is also termed microfabric analysis and entails descriptive terminology that has been developed over the past 50 years. The science and terminology of microfabric analysis was initially documented by Kubiena (1938) and important publications documenting terminology since have included Brewer (1964); Fitzpatrick (1984, 1993); Bullock et al. (1985); and Stoops (2003). Methodological descriptions for producing thin sections can be found in Cady et al. (1986); Fitzpatrick (1984); Murphy (1986); Fox et al.(1993); and Fox and Parent (1993). An excellent book on examination of mineral weathering in thin sections is Delvigne (1998) and the Soil Science Society of America has a collection of images on a CD that illustrates many features of microfabrics (SSSA, 1993).

Examination of thin sections with a polarizing light microscope can be considered an extension of field morphological studies. The level of resolution increases from field examination to optical microscopic examination and finally to submicroscopic techniques (electron microscopy), but this sequence of techniques increasingly sacrifices field of view (Cady et al., 1986). Thus, the results of micromorphological studies are most useful when they are combined with other field (landscape description, pedon morphological description) and laboratory data (Cady, 1965). Micromorphology is used to identify types and sequences of active processes occurring in soils via identification of argillans, fabric types, skeleton grains, weathering intensity. It is an ideal tool to investigate genesis of soil or pedological features.

---

[1] W. Dennis Nettleton, Research Soil Scientist (retired) and Michael A.Wilson, Research Soil Scientist, NSSC, NRCS, Lincoln, NE, wrote the procedure for description and interpretation of soil micromorphology as seen in thin sections.

Initially, the investigator should scan the overall features of a thin section and determine those features that require emphasis. This initial scanning may include all the thin sections from a soil profile or all those related to a particular problem. Different kinds of illumination should be used with each magnification. Strong convergent light with crossed polarizers elucidates structures in dense or weakly birefringent material that may appear opaque or isotropic. Structures in translucent specimens become more clearly visible if plain light is used, and the condensers are stopped down. Everything should be viewed in several positions of the stage or during slow rotation with crossed polarized light.

A thin section is a two-dimensional slice through a three-dimensional body. The shapes of mineral grains and structural features are viewed in one plane, and the true shapes must be inferred. A grain that appears needle-shaped may be a needle or the edge of a flat plate. An elliptical pore may be an angular slice through a tube. A circular unit is probably part of a sphere. With a three-dimensional perspective in mind as well as an awareness of section thickness, repeated viewing of similar features that appear to be cut at different angles is the best way to accustom oneself to a volume rather than a planar interpretation of shape. A well-prepared section is 20- to 30-μm thick. Grains smaller in thickness are stacked and cannot be viewed as individual grains. Similarly, pores smaller than 20 to 30 μm cannot be seen clearly. A pore size of 20-μm diameter equates to a soil moisture tension of 15 kPa (0.15 bar) (Rode, 1969) so that visible pores in thin section are mostly drained at water contents below field capacity.

Sand and silt grains in thin sections are identified by standard methods presented in petrography texts. The general analytical approach is the same for grain studies (procedure 7B1) as it is for thin sections. However, in grain studies, the refractive index is used only as a relative indicator, and other optical and morphological properties are more important. Furthermore, in thin sections, a concern with minerals that occur in small quantities or an attempt to quantify mineralogical analysis is seldom necessary. The separate particle-size fractions should be used for the identification and mineralogical analyses that are important to a study, whereas the thin sections should be used mainly for information about component arrangement. Recognition of aggregates, concretions, secondary pseudomorphs, and weathered grains is more important in thin section studies than in sand and silt petrography. Recognition of these components in thin section are easier because interior structures are exposed. Although grain studies are important in soil genesis studies, the arrangement of components is destroyed or eliminated by sample preparation procedures that separate the sand, silt, and clay.

In the United States, emphasis in micromorphology has been on clay arrangement. Clay occurs not only in the form of aggregates but also in massive interstitial fillings, coatings, bridges, and general groundmass. Even though the clay particles are submicroscopic, they can be described, characterized and sometimes identified, e.g., the 1:1 and 2:1 lattice clays can be distinguished. Completely dispersed, randomly arranged clay of less than 1 μm exhibits no birefringence and appears isotropic in crossed polarized light. Clay in a soil is seldom all random and isotropic. Clay develops in oriented bodies, either during formation or as a result of pressure or translocation. If enough plate-shaped particles are oriented together in a body that is large enough to see, birefringence can be observed.

With the exception of halloysite, the silicate clay minerals in soils are platy. The a and b crystallographic axes are within the plane of the plate, and the c axis is almost perpendicular to this plane. Even though the crystals are monoclinic, the minerals are pseudohexagonal, as the distribution of stems along the a and b axes is so nearly the same, and the c axis is so nearly perpendicular to the other axes. The optical properties, crystal structure, and general habit of clay are analogous to those of the micas, which can be used as models to analyze and describe clay properties.

The speed of light that travels in the direction of the c axis and vibrates parallel to the a axis is almost the same as that light that vibrates parallel to the b axis. Therefore, the refractive indices are very close, and the interference effects in crossed polarized light are small when observed along the c axis. Light that vibrates parallel to the c axis travels faster than in other directions. Hence, the refractive index is lower. If the edge of the crystal or aggregate of crystals is viewed along the a-b plane between crossed polarizers, two straight extinction positions are viewed, and interference colors are manifested in other positions. If a clay concentration is organized so that most of the plates are parallel, the optical effects can be observed. The degree and quality of optical effects depend on the purity, continuity, and the orientation process of the clay body.

Kaolinite has low birefringence and has refractive indices slightly higher than quartz. In the average thin section, interference colors for kaolinite are gray to pale yellow. In residual soils that are derived from coarse-grained igneous rocks, kaolinite occurs as book-like and accordion-like aggregates of silt and sand size.

Even though halloysite can form oriented aggregates, it should not show birefringence because of its tubular habit. Halloysite may show very faint, patternless birefringence, which is caused by impurities or by refraction of light at the interfaces between particles.

The 2:1 lattice minerals (Fig. 1) have high birefringence and show bright, intermediate-order, interference colors if the edges of aggregates are viewed. In the clay-size range, distinctions among smectite, mica, vermiculite, and chlorite in thin section are seldom possible. These clay minerals are usually mixed in the soil and seldom occur pure. In many soils, these clay minerals are stained and mixed with iron oxide and organic matter.

Residual clay has been in place since its formation by weathering. Although it may have been transported within fragments of weathered material, it remains in place relative to the fabric of these fragments. This clay may be random, have no orientation, and thus be isotropic; however, more often, it shows some birefringence. In transported materials, silt-size flakes and other small aggregates are common. In many residual materials, clay is arranged either in forms that are pseudomorphs of rock minerals or in definite bodies of crystal aggregates, e.g., vermicular or accordion-like kaolin books. The regular, intact arrangement of these materials is usually diagnostic of residual material.

Clay rearrangement may result from differentially applied stress that produces shear (Fig. 2). Platy particles become oriented by slippage along a plane, e.g., slickenside faces in a Vertisol or in clayey layers. Platy particles also are oriented inside the blocks. Root pressure, mass movement, slump, and creep can produce stress orientation. If the faces on structural units are smooth and do not have separate coatings, stress orientation can be inferred. Otherwise, in plain light, stress orientation cannot be observed. In plain light, clay in the thin section may be homogeneous and featureless. In crossed polarized light, the orientation pattern is reticulate, consisting of bright lines showing aggregate birefringence, often intersecting at regular angles. The effect is that of a network in a plaid pattern. There may be numerous sets of these slippage planes, which appear in different positions as the stage is turned. Stress-oriented clay may be near rigid bodies, e.g., quartz grains, or along root channels. Stress-oriented clay is often strongly developed on ped faces. Stress can also orient mica flakes and any other small platy grains.

Location features that distinguish translocated clay from residual clay are its occurrence in separate bodies, usually with distinct boundaries, and its location on present or former pore walls, channel linings, or ped faces. Translocated clay may have a different composition than matrix clay, especially if its origin is another horizon. This clay is more homogeneous and is usually finer than the matrix clay. Translocated clay displays lamination, indicating deposition in successive increments, and manifests birefringence and extinction, indicating that these translocated clay bodies are oriented aggregates. If these bodies are straight, they have parallel extinction. If these bodies are curved, a dark band is present wherever the composite c axis and the composite a and b axes are parallel to the vibration planes of the polarizers. When the stage is rotated, these dark bands sweep through the clay aggregate.

Other substances such as goethite, gibbsite, carbonate minerals (Fig. 3), and gypsum may form pore linings and ped coatings. These substances can be identified by their mineralogical properties.

Amorphous coatings of organic matter, with or without admixed Fe and Al, are common, especially in spodic horizons. This material is dark brown to black, isotropic or faintly birefringent, and often flecked with minute opaque grains. Amorphous coatings of organic matter occur as the bridging and coating material in B horizons of sandy Spodosols and as thin coatings or stains on pore and ped faces in other soils.

2. Procedure

Description of Microfabrics

Terms have been defined for distribution patterns of the components of soil thin sections (Brewer, 1964 and 1976; Stoops and Jongerius, 1975; Brewer et al., 1983; Bullock et al., 1985; and Stoops, 2003). As these terms have become more widely adopted in the literature, the SSL increasingly uses them in Soil Survey Investigations Reports (SSIR's) and in soil project correspondence. Micromorphological descriptions often contain terminology from different sources to describe properties of the fabric.

Related Distribution Patterns: The five "coarse-fine related distribution patterns" of Stoops and Jongerius (1975) are in common usage. The nomenclature of these distribution patterns, as described by Stoops and Jongerius (1975), are intended to be broadly defined. There are no restrictions on material type, absolute size, orientation, granulation, or origin. The system may be used to describe the distribution of primary particles, e.g., quartz grains, as well as compound units, e.g., humic micro-aggregates. The coarser particles may be silt, sand, or gravel, whereas the finer material may be clay, silt, or sand. Figure 4 shows the average textures, linear extensibilities (LE), and drained pore to filled pore (DP/FP) ratios of some related distribution patterns of a number of U.S. soils.

The *monic type* (granic type of Brewer et al., 1983) consists of fabric units of only one size group, e.g., pebbles, sand, lithic fragments (coarse monic) or clays (fine monic). In the *gefuric type*, the coarser units are linked by bridges of finer material but are not surrounded by this material. In the *chitonic type* (chlamydic type of Brewer et al., 1983), the coarser units are surrounded by coatings of finer material. In the *enaulic type*, the larger units support one another, and the interstitial spaces are partially filled with finer material. The enaulic fabric consists of material finer than is found in either the gefuric or chitonic type but is not so fine as is found in the porphyric type. In the end member of the sequence, the *porphyric type*, the large fabric units occur in a dense groundmass of smaller units, and there is an absence of interstitial pores. This type is equivalent to the earlier porphyroskelic class of Brewer (1964) or to the current porphyric class (Brewer et al., 1983). The class may be divided into types based on the spacing of the coarser units.

Plasma Fabrics: Brewer (1976) divided soil materials into three groups for descriptive purposes: peds, pedological features, and s-matrices. *Peds* are the basic units in soils that contain organized structural units and are composed of skeletal grains, plasma, and pedological features. The *s-matrix* is the material within which pedological features occur, having no definite boundary, size, shape, or orientation (Brewer, 1976). *Skeleton grains* of a soil material are individual grains larger than colloidal size. The *soil plasma* includes all the colloidal size material as well as relatively soluble material not bound in skeleton grains.

The description of plasmic fabrics is based on the interpretations of optical properties under crossed-polarized light, especially extinction phenomena. Plasma concentrated or crystallized into pedological features is not included in the description of plasmic fabrics. In general, the descriptive terms for the s-matrix are those as defined by Brewer (1976). The s-matrix plasma fabrics are divided into two groups, the asepic and sepic types. *Asepic fabrics* are those with anisotropic plasma in which the *domains*, i.e., the plasma separations, are not oriented relative to each other. *Sepic fabrics* are those with anisotropic domains with various orientation patterns visible under cross-polarized light. Figure 5 shows some plasma fabrics and their clay, silt + clay, and linear extensibility averages of a number of U.S. soils.

Eswaran (1983) characterized the <25-μm^2 size domains of monomineralic soils using a scanning electron microscope (SEM). These features are smaller than some domains described by Brewer (1976). However, these small features provide the detail expected of the interparticle relationships present in the larger separations. The domains in allophanic soils are composed of globular aggregates. The halloysitic soils differ in that the halloysite tubes generally may be seen as protrusions from globular forms. The domains in micaceous soils retain the face-to-face packing that is common in micas and may retain some of the book-like forms as well. The domains in montmorillonitic soils are bent to conform to the shape of skeletan grains. However, the packing is essentially face-to-face and, upon drying, the fabric is very dense and compact. In kaolinitic soils, the domains frequently are present as booklets that are packed face-to-face, unless iron hydrous oxide has disrupted the platelets, in which case, the platelets may still be packed face-to-face in subparallel stacks.

Asepic plasmic fabrics are subdivided into two groups, argillasepic and silasepic types. *Argillasepic fabrics* are dominated by anisotropic clay minerals and have a random orientation pattern of clay-size domains. Overall, asepic fabrics have flecked extension patterns. *Silasepic fabrics* have a wider range of particle sizes than argillasepic types. However, a careful observer may view silt-size domains or plasma bodies that give the matrix an overall flecked extinction pattern (Fig. 6).

The *sepic plasmic fabrics* have recognizable domains with various patterns of orientation. Internally, the *domains*, i.e., plasma separations, have striated extinction patterns. Brewer (1964) recognizes seven kinds, most of which are widely adopted. *Insepic fabrics* consist of isolated, striated plasma domains within a flecked plasma matrix (Fig. 7). *Mosepic fabrics* consist of plasma domains with striated orientation that may adjoin each other or be separated by small plasma areas with flecked orientation that are not oriented relative to each other (Fig. 8). The fabric is *vosepic* when the plasma separations with striated orientation are associated with channel or pore (void) walls. The fabric is *skelsepic* when the plasma separations occur at the skeleton grain-matrix contact (Fig. 9).

The remaining three sepic plasmic fabrics are most common in fine-textured soils. In *masepic fabrics*, the plasma separations occur as elongated zones within the s-matrix and apparently are not associated with void walls or skeleton grains (Fig. 10). The striations have parallel orientations to zone length. *Lattisepic fabrics* are similar to masepic fabrics except that the acicular and prolate domains occur in lattice-like patterns. In *omnisepic fabrics*, all of the plasma has a complex striated orientation pattern.

Three other kinds of plasmic fabrics are characteristic of particular minerals or kinds of soils. *Undulic plasmic fabrics* have practically isotropic extinction patterns at low magnification, and the domains are indistinct even at high magnification. *Isotic plasmic fabrics* have isotropic plasma, even at highest magnifications with high light intensity. The *crystic plasmic fabrics* have anisotropic plasma with recognizable crystals, usually of soluble materials.

Pedological Features, Cutans: The term, cutan, and definitions of its respective types (Brewer, 1964) have been widely adopted by soil scientists. *Cutan* is defined by Brewer as a modification of the texture, structure, or fabric at natural surfaces in soil materials due to the concentration of particular soil constituents or as in-place modification of the plasma (Fig. 11). Generally, the cutans are subdivided on the basis of their location, composition, and internal fabric. Cutan locations are surfaces of grains, peds, channels, or voids. The mineralogical nature of cutans is characterized, e.g., argillans, ferri-argillans, or organo-argillans. *Argillans* are composed dominantly of clay minerals, *ferri-argillans* have iron oxides as a significant part of their composition, and *organo-argillans* have significant color addition by inclusion of organic matter.

Sesquan is a general term used for a cutan of sesquioxides or hydroxides. Sesquans that are specific for goethite, hematite, or gibbsite are called *goethans, hematans,* or *gibbsans,* respectively. Similarly, cutans of gypsum, carbonate, calcite, halite, quartz, silica, and chalcedony are called *gypsans, calcans, calcitans, halans, quartzans, silans, and chalcedans,* respectively. Skeleton grains that adhere to the cutanic surface are called *skeletans*.

Pedological Features, Glaebules: *Glaebules* (Brewer, 1964) are three dimensional pedological units (e.g., Fe oxide or carbonate nodules) within the s-matrix whose morphology is incompatible with the composition of the present matrix material. (The name is derived from the Latin term *glaebula,* meaning a small lump or aggregate of earth.) They are usually prolate to equant. A glaebule is recognized as a unit either because of a greater concentration of a constituent, or difference from the s-matrix fabric, or because of the presence of distinct boundaries of a constituent within the enclosing s-matrix. Glaebules include papules, nodules, concretions, and pedodes. *Papules* are pedogenic features composed of clay minerals with continuous and/or laminar fabric, sharp external boundaries, and commonly prolate to equant, somewhat rounded shapes. *Nodules* (Fig. 12) are pedological features with undifferentiated internal fabric. *Concretions* are pedological features with concentrically laminated structures about a center. *Pedodes* are pedological features with hollow interiors, often lined with crystals.

Pedological Features, Voids: *Voids* are the empty spaces within the s-fabric. Those voids with diameters of 20 μm to > 2 mm can be studied and measured in thin section. Brewer (1976) classifies these voids as follows: (1) *simple packing voids* (empty spaces due to random packing of single skeleton grains); (2) *compound packing voids* (Fig. 13) (empty spaces between peds or other compound individuals); (3) *vughs* (Fig. 14) (relatively large spaces that are not formed by packing of skeleton grains; (4) *vesicles* (Fig. 15) (relatively large empty spaces with smooth, regular outlines); (5) *chambers* (empty spaces with smooth, regular outlines that connect to other voids); (6) *joint planes* (plane-shaped, empty spaces that traverse the s-matrix in a regular pattern); (7) *skew planes* (plane-shaped, empty spaces that traverse the s-matrix in an irregular pattern); (8) *craze planes* (plane-shaped, empty spaces that traverse the s-matrix in a highly irregular pattern of short flat or curved planes); and (9) *channels* (mostly cylindrical-shaped, empty spaces that are larger than packing voids).

Interpretations

Related Distribution Patterns: Usually, the basic descriptive terms for soil fabrics do not imply any specific genesis of the feature. However, modifiers commonly are added when fabric descriptions are complete enough to understand the means of formation, i.e., *stress cutan*, or in-place plasma modification, is the result of differential forces, e.g., shearing, whereas an *illuviation cutan* is formed by movement of material in solution or suspension and later deposited (Brewer, 1964).

The average properties of some related distributions we have described are given in Figure 4. In an experimental study of soil microfabrics by anisotropic stresses of confined swelling and shrinking, Jim (1986) showed that with an increase in the activity and proportion of the clay fraction, the related distribution patterns alter from dominantly *matrigranic* (*monic*, with the units being aggregates) to *matrigranodic* (*enaulic*) to *porphyric*. Similarly our data for some U.S. soils show that the relative pore volumes at 30 kPa for some soil coarse-fine distributions increase from *enaulic* through *open porphyric* (Fig. 16).

Some *monic* fabrics are inherited, and include soil fabrics formed in sand dunes, sandy sediments deposited by streams and rivers, beach deposits, and gruss. Fauna can produce monic fabrics that are mostly fecal pellets. Monic fabrics also can form by fracturing and flaking of organic coatings in the upper B horizons of the Spodosols (Flach, 1960) and by freezing and thawing (Brewer and Pawluk, 1975).

Several kinds of finer material (plasma) can bridge the coarser particles (skeleton grains) to form *gefuric* related distribution patterns. Gefuric patterns are common in weakly developed argillic and spodic horizons and in duripans. Silicate clays can bridge skeleton grains in some argillic horizons; the organic matter, iron, and aluminum complexes in some kinds of spodic horizons; and the amorphous silica in some kinds of duripans.

In soils that are slightly more developed than those with gefuric patterns, *chitonic* related distribution patterns form. These are common in argillic and spodic horizons and in duripans. Bridges as well as complete coatings of skeleton grains are present. Usually, the cement or plasma is material that adheres to skeleton grains. These cements have covalent bonds and commonly include silica (Fig. 17) , iron, aluminum, and organic matter (Chadwick and Nettleton, 1990).

The *enaulic* related distribution patterns are more common in soil material in which the cement bonds to itself more strongly than to skeleton grains. In sandy soils, ionic-bonded calcite and gypsum tend to bond to themselves more strongly than to skeleton grains (Fig. 3), thereby producing *open porphyric* related distribution patterns (Chadwick and Nettleton, 1990). Even though organic matter has covalent bonds and usually surrounds grains, organic material forms pellets in void spaces between skeleton grains in some spodic horizons.

Porphyric related patterns form from the normal packing of grains in materials with a high proportion of fine material. These patterns can be the end member of several kinds of sequences (Brewer et al., 1983). In porphyric related patterns, there may or may not be skeleton grains of primary minerals, pedorelicts, organics, lithic fragments of shale, sandstone, or other rocks. In the porphyric related patterns, the material consists of silt and clay, and the interstices tend to be filled with minimal formation of coatings. In precursors of the porphyric related distribution patterns, the silt to clay ratio is used to identify the kind of sequences by which the porphyric pattern forms (Brewer et al., 1983). The porphyric patterns are common in loessial soils, especially in argillic and petrocalcic horizons, duripans, and ortstein.

Plasmic Fabrics: The *asepic plasmic fabrics* differ in composition mainly in silt to clay ratios. *Argillasepic fabrics* have the higher clay contents, usually <30 percent but may have as much as 70 percent (Brewer et al., 1983). Organic matter or iron stains, resulting in a flecked distribution pattern mask the birefringence of the plasma. Argillasepic fabrics are important fabrics in many fine-textured B horizons. *Silasepic plasmic fabrics* have low clay contents and have more silt than clay. The silasepic fabrics are common in porphyric related distribution patterns in A and B horizons of Solonetz, Solodized Solonetz and Solodic Soils, Soloths, Red Podzolic Soils, Lateritic Podzolic Soils, and are also associated with some sedimentary deposits (Brewer et al., 1983). Silasepic plasma fabrics are common in A and B horizons of loessial soils in association other kinds of plasma separations. Even if there is high clay content, the horizons with asepic plasmic fabrics have low effective linear extensibilities (LE) either because the clays are low-swelling types or because the soils do not dry enough to undergo the full range of laboratory-measured LE.

In soils that form in the same climate, the kind of *sepic plasmic fabrics* form a sequence relative to increasing linear extensibility (Nettleton et al., 1969; Holzhey et al., 1974). In increasing order of shrink-swell stress, the plasmic fabric sequence is insepic, mosepic, lattisepic, omnisepic, and masepic. Using X-ray diffraction (Clark, 1970) and scanning electron microscopy (Edil and Krizek, 1976), observations of deformation experiments indicate that the degree of clay orientation increases with an increase in applied stress. In an experimental study of soil microfabrics by anisotropic stresses of confined swelling and shrinking, Jim (1986) shows that with an increase in the activity and content of the clay fraction, there is an increase in the long and narrow plasma separations, i.e., a progression from *insepic* to *mosepic* to *masepic* plasmic fabrics.

Insepic plasmic fabrics are very common in finer-grained porphyric B horizons of a wide range of soil groups (Brewer et al., 1983). Soil horizons with insepic fabrics generally have a LE of <4 percent. In some insepic plasmic fabrics, the plasma islands or papules are pseudomorphs of some weatherable mineral, whereas in other insepic fabrics, the papules are clay skin fragments or are eolian sand-size clay aggregates (Butler, 1974). In some samples, the pseudomorphs do not disperse well in particle-size distribution analysis (PSDA).

Mosepic plasmic fabrics commonly have more clay than insepic fabrics do because they contain more islands of plasma. However, in mosepic plasmic fabrics, LE also remains low. Shrink-swell forces have not been sufficient or have not operated long enough to have homogenized the islands of plasma into the soil matrix.

Vosepic plasmic fabrics occur in soil horizons that have undergone stress either due to shrink-swell forces or to tillage. Even though root growth is adequate to increase the percentage of oriented clay near the root-soil interface (Blevins et al., 1970), root growth does not appear adequate to form vosepic or other highly stressed plasmic fabrics. Usually, vosepic fabrics are present in soil horizons in which the main fabric type is masepic or skelsepic. The vosepic plasmic fabric rarely occurs as the only fabric in a soil horizon.

There are at least two types of origins for orientation of plasma on sands. One is a result of clay illuviation. By definition, this type would not be included with skelsepic fabric. The related distribution patterns associated with this fabric commonly are *monic, gefuric,* or

149

enaulic. The other origin is commonly the porphyric related distribution patterns with LE's that are >4 percent for dryland soils, i.e., soils in aridic, xeric, or ustic soil moisture regimes. These are the true *skelsepic* fabrics. Shrink-swell forces have been involved in their formation as shown by relatively few papules or clay skins remaining, and there are vosepic areas.

Masepic, *lattisepic*, and *omnisepic* plasmic fabrics are evidence of stress >4 percent in dryland soils. Clay contents are usually >35 percent, but the threshold amount is dependent on clay mineral type and on degree of dryness common to the environment. In masepic, lattisepic, and omnisepic plasmic fabrics, papules and clay skins rarely are found, but areas of *skelsepic* and *vosepic* areas commonly are present.

Undulic plasmic fabrics seem to be associated with basic parent materials, especially basalt, and with moderate to strong weathering (Brewer et al., 1983). The fabric commonly is stained deeply by iron minerals, and kaolinite and halloysite are the important clay minerals. Clays in these horizons do not disperse well in PSDA, but high 1500-kPa (15 bar) water contents suggest that the horizons belong in clayey families. Some papules and clay skins commonly are present, but these plasma separations also are stained deeply by iron.

Isotic plasmic fabrics are common in spodic horizons and in Andisols. The clays in these horizons are amorphous and disperse poorly in PSDA. The water-holding capacities of these soil horizons are relatively high. Some unweathered volcanic ash may be present.

Crystic plasmic fabrics are common in B horizons of soils formed in dryland areas. In soil horizons with large areas of interlocking crystals, there is restricted soil permeability, increased unconfined compressive strength, and limited particle dispersion, depending on the degree of cementation.

Cutans and Pedogenic features: Most *argillans* (Fig. 11) are formed, at least in part, by illuviation. The content of strongly oriented clay (usually argillans plus *papules*), in texture-contrast soils (soils with argillic horizons) is usually <5 percent of the soil volume (Brewer et al., 1983). In some sandy soils that are low in silt, the argillans and papules are as much as 30 percent of the soil material (Brewer et al., 1983). The measured illuviated clay rarely accounts for the difference in clay content between the A and B horizons. Some of the clay may originate from weathering in place and some from a destruction of argillans and papules.

If argillans and papules are present in argillic horizons in dryland soils, the soil LE is usually <4 percent (Nettleton et al., 1969). In some humid environments, argillans and papules may be present even where the LE is >4 percent. As soils in humid environments do not dry to the same degree as those in the desert, the clay skins may survive because only part of the linear extensibility is effective.

Papules may originate by the weathering of primary minerals, the isolation of clay skins by the channel and void migration within the soil matrix (Nettleton et al., 1968; Nettleton et al., 1990) or by the introduction of eolian sands and silts that are composed of clays (Butler, 1974; Brewer and Blackmore, 1976). The comparison of size and shape of papules and minerals, as well as of parent material, may help to determine if the papules are pseudomorphs of one of the primary minerals. Internal fabric resemblances and residual parts of the primary mineral within the papules help to determine if a papule is a pseudomorph.

The determination of whether or not a papule is an illuvial feature is important for classification purposes. Arcuate forms and laminar internal fabrics are evidence that the feature is illuvial. If the feature partially surrounds an oval body of silt, illuvial origin of the feature is relatively certain (Nettleton et al., 1968).

The origin of the papule as eolian may be determined by studying its size and shape, its internal fabric, and the number and degree of its alterations relative to other particles. Microlaminae may suggest an origin as sediment. Unlike soil pedorelicts or rock fabrics (lithorelicts), nodules, or glaebules rich in soluble plasma, probably form by accretion (Brewer, 1976). Most concretions, as well as pedodes, are accretionary and usually form in place.

A study of soil *voids* may be useful in predicting the clay activity and shrink-swell behavior of soils. In an experimental study of soil microfabrics by anisotropic stresses of confined swelling and shrinking, Jim (1986) shows that with an increase in the activity and content of the clay fraction, there is a drastic decrease in void volume, especially the >30μm. Furthermore, the void shapes change from compound packing voids to planar voids and vughs. With an increase in stress from shrink-swell forces, aggregates become flattened at contacts, resulting in more angular and eventually fused compound units.

A possible objective of micromorphological studies may be the measurement of porosity and the prediction not only of soil water content at various suctions but also of hydraulic conductivity. In thin section studies of voids in sands and sandy soils, there is a close correlation between microscopic and suction methods (Swanson and Peterson, 1942). However, in those soils whose volumes change with changes in water content, pore size distribution is undefined, and no constant void size distribution exists (Brewer, 1976). Furthermore, there are several invalidated assumptions that commonly are made in relating porosity to permeability (Nielsen et al., 1972 p. 11). The assumptions that especially relate to soil fabric are that no pores are sealed off, pores are distributed at random, and pores are generally uniform in size. A more serious difficulty may be that a thin section, even if reduced to a 20-μm thickness, may make the examination of the <20-μm diameter pores impossible, if these pores pass through the section at an angle of ≤45°. This means that many voids that are involved in unsaturated water flow in soils will not be visible in thin section (Baver, 1956 p. 271).

The size, shape, and arrangement of skeleton grains determine the nature of simple packing voids, but the origin of compound packing voids is not so straightforward. The unaccommodated peds of the compound packing voids may be formed by faunal excreta, shrink-swell action, man's activities, or by other unknown causes.

Vughs usually occur in soil materials with a wide range in size of particles, including silicate clays. Some vughs form by the weathering and removal of carbonate, and others form by faunal activity or the normal packing of plasma and skeleton grains. The very regular outline of *vesicles* is of interest (Nettleton and Peterson, 1983). Lapham (1932) states that in Sierozems (Aridisols), the vesicles that are near the surface are the result of air entrapment by rainfall following dry dusty periods. Laboratory studies verify this phenomenon (Springer, 1958). If high silt soils are allowed to dry before each irrigation, the vesicle size increases with the number of irrigations (Miller, 1971). As a result of studies of infiltration rates and sediment production in rangeland in central and eastern Nevada, Blackburn and Skau (1974) and Rostagno (1989) conclude that the infiltration rates are the lowest and the sediment yields are the highest on sites that have vesicular surface horizons. The failure of most vesicles to connect to other voids and the low strength of the crust in which vesicles occur help to explain the low infiltration rates and the high sediment yields that commonly are found on these soils.

Joint planes (Fig. 18) are produced in relatively uniform fine-textured soils by a relatively regular system of cracking upon drying (Brewer, 1976). Once formed, these joint planes tend to open in the same place during successive drying cycles. *Skew planes* are produced in more heterogeneous materials or by irregular drying (Brewer, 1976). *Craze planes* often occur in Chernozems (Mollisols), possibly as a result of the high humic acid content (Brewer, 1976). Because of their size, cross-sectional shape, and kind of branching pattern, channels probably form by faunal activity, plant root systems, or by certain geological processes (Brewer and Sleeman, 1963).

3. References

Baver, L.D. 1956. Soil physics. 3rd ed. John Wiley & Sons, Inc., NY.

Blackburn, W.H., and C.M. Skau. 1974. Infiltration rates and sediment production of selected plant communities in Nevada. J. Range Manage. 27:476-480.

Blevins, R.L., N. Holowaychuk, and L.P. Wilding. 1970. Micromorphology of soil fabric at tree root-soil interface. Soil Sci. Soc. Am. Proc. 34:460-465Brewer, R. 1964. Fabric and mineral analysis of soils. John Wiley & Sons Inc., NY.

Brewer, R. 1964. Fabric and mineral analysis of soils. John Wiley & Sons Inc., NY.

Brewer, R. 1976. Fabric and mineral analysis of soils. Reprint of 1964 ed., with suppl. material. Robert E. Cringer Publ. Co., Huntington, NY.

Brewer, R., and A. V. Blackmore. 1976. Relationships between subplasticity rating, optically oriented clay, cementation and aggregate stability. Aust. J. Soil Res. 14:237-248.

Brewer, R., and S. Pawlak. 1975. Investigations of some soils developed in hummocks of the Canadian sub-artic and southern-arctic regions. I. Morphology and micromorphology. Can. J. Soil Sci. 55:301-319.

Brewer, R., and J.R. Sleeman. 1963. Pedotubules: their definition, classification and interpretation. J. Soil Sci. 15:66-78.

Brewer, R., J.R. Sleeman, and R.C. Foster. 1983. The fabric of Australian soils. p. 439-476. *In* Division of soils, CSIRO (ed.) Soils: an Australian viewpoint. CSIRO: Melbourne/Academic Press, London.

Bullock, P., N. Fedoroff, and A. Jongerius, G. Stoops, T. Tursina, and others. 1985. Handb. for soil thin section description. Waine Res. Publ., Walverhampton, England.

Butler, B.E. 1974. A contribution towards the better specification of parna and some other aeolian clays in Australia. Z. Geomorph. N.F. Suppl. Bd. 20:106-116. Berlin, Germany.

Cady, J.G. 1965. Petrographic microscope techniques. *In* C.A. Black, D.D. Evans, J.L. White, L.E. Ensminger, and F.E. Clark (eds.) Methods of soil analysis. Part 1. Physical and mineralogical properties, including statistics of measurement and sampling. 1st ed. Agron. 9:604-631.

Cady, J.G., L.P. Wilding, and L.R. Drees. 1986. Petrograhic microscope techniques. *In* A. Klute (ed.) Methods of soil analysis. Part 1. Physical and mineralogical properties. 2nd ed. Agron. 9:185-218.

Chadwick, O.A., and W.D. Nettleton. 1990. Micromorphologic evidence of adhesive and cohesive forces in soil cementation. p. 207-212. *In* L.A. Douglas (ed.) Developments in Soil Sci. 19. Soil micromorphology: a basic and applied science. Elsevier Sci. Publ., NY.

Clark, B.R. 1970. Mechanical formation of preferred orientation in clays. Am. J. Sci. 269:250-266.

Delvigne, J.E. 1998. Atlas of micromorphology of mineral alteration and weathering. Canadian Miner Special Pub. 3. Miner. Assoc. of Canada. Ontario.

Edil, T.B., and R.J. Krizek. 1976. Influence of fabric and soil suction on the mechanical behavior of a kaolinitic clay. Geoderma. 15:323-341.

Eswaran, H. 1983. Characterization of domains with the scanning electron microscope. Pedology 33:41-54.

FitzPatrick, E.A. 1984. Micromorphology of soils. Chapman and Hall, Ltd., London.

FitzPatrick, E.A. 1993. Soil microscopy and micromorphology. John Wiley and Sons, New York.

Flach, K.W. 1960. Sols bruns acides in the northeastern United States. Genesis, morphology, and relationships to associated soils. Unpubl. Ph.D. Thesis, Cornell, NY.

Fox, C.A., R.K. Guertin, E. Dickson, S. Sweeney, R. Protz, and A.R. Mermut. 1993. Micromorphology methodology for inorganic soils. p. 683-709. *In* M.R. Carter (ed.) Soil sampling and methods of Analysis. Lewis Publishers., Boca Raton.

Fox, C.A., and L.E. Parent. 1993. Micromorphology methodology for organic soils. p. 473-485. *In* M.R. Carter (ed.) Soil sampling and methods of Analysis. Lewis Publishers., Boca Raton.

Holzhey, C.S., W.D. Nettleton, and R.D. Yeck. 1974. Microfabric of some argillic horizons in udic, xeric, and torric soil environments of the United States. p. 747-760. *In* G.K. Rutherford (ed.) Soil Microscopy. Proc. of 4th Int. Working-meeting on Soil Micromorphology. Limestone Press, Kingston, Ontario, Canada.

Kubiena, W.L. 1938. Micropedology. Collegiate Press, Ames, IA.

Jim, C.Y. 1986. Experimental study of soil microfabrics induced by anistropic stresses of confined swelling and shrinking. Geoderma. 37:91-112.

Lapham, M.H. 1932. Genesis and morphology of desert soils. Am. Soil Surv. Assoc. Bull. 13:34-52.

Miller, D.E. 1971. Formation of vesicular structure in soil. Soil Sci. Soc. Am. J. 35:635-637.

Murphy, C.P. 1986. Thin section preparation of soils and sediments. Academic Publishers, Great Britain.

Nettleton, W.D., K.W. Flach, and B.R. Brasher. 1969. Argillic horizons without clay skins. Soil Sci. Soc. Am. Proc. 33:121-125.

Nettleton, W.D., R.B. Grossman, and B.R. Brasher. 1990. Concept of argillic horizons in Aridisols-taxonomic implications. p. 167-176. *In* J.M. Kimble and W.D. Nettleton (eds.) Proc. of 4th Int. Soil Correlation Meeting (ISCOM). Characterization, classification, and utilization of Aridisols in Texas, New Mexico, Arizona, and California, October 3-17, 1987. Part A: Papers.

Nettleton, W.D., R.J. McCracken, and R.B. Daniels. 1968. Two North Carolina Coastal Plain catenas. II. Micromorphology, composition, and fragipan genesis. Soil Sci. Soc. Am. Proc. 32:582-587.

Nettleton, W.D., and F.F. Peterson. 1983. Aridisols. p. 165-215. *In* L.P. Wilding, N.E. Smeck, and G.F. Hall (eds.). Pedogenesis and soil taxonomy. II. The soil orders. Elsevier, NY.

Nielsen, D.R., R.D. Jackson, J.W. Cary, and D.D. Evans (eds). 1972. Soil water. Am. Soc. Agron., Madison, WI.

Rode, A.A. 1969. Theory of soil moisture. Vol. 1. Moisture properties of soils and movement of soil moisture. (In English, translated from Russian.) Israel Prog. for Scientific Translations, Jerusalem. p. 38.

Rostagno, C.M. 1989. Infiltration and sediment production as affected by soil surface conditions in a shrubland of Patagonia, Argentina. J. Range Manage. 42:383-385.

Soil Science Society of America. 1993. A reference collection for soil micromorphology. CD Rom. ASA and SSSA, Madison, WI.

Springer, M.E. 1958. Desert pavement and vesicular layer of some soils of the desert of the Lahontan Basin, Nevada. Soil Sci. Soc. Am. Proc. 22:63-66.

Stoops, G., and A. Jongerius. 1975. Proposal for micromorphological classification in soil materials. I. A classification of the related distribution of coarse and fine particles. Geoderma. 13:189-200.

Swanson, C.L.W., and J.B. Peterson. 1942. The use of micrometric and other methods for the evaluation of soil structure. Soil Sci. 53:173-183.

153

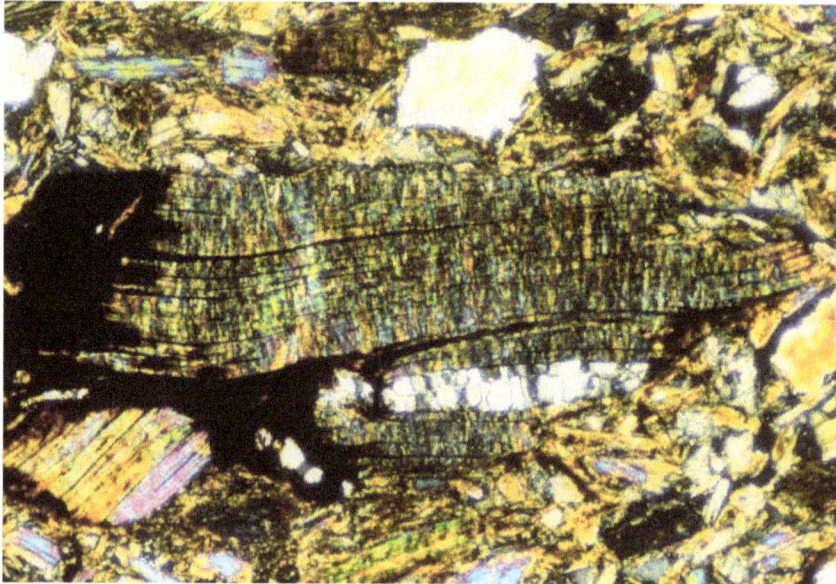

Figure 1. Large biotite grain undergoing expansion from weathering. Note the high birefringence due to the orientation of the grain in thin section. Frame width = 1.0 mm. (Series name not designated; Fremont Co., WY, Pedon 98P0456, Bt2 horizon under crossed polarized light).

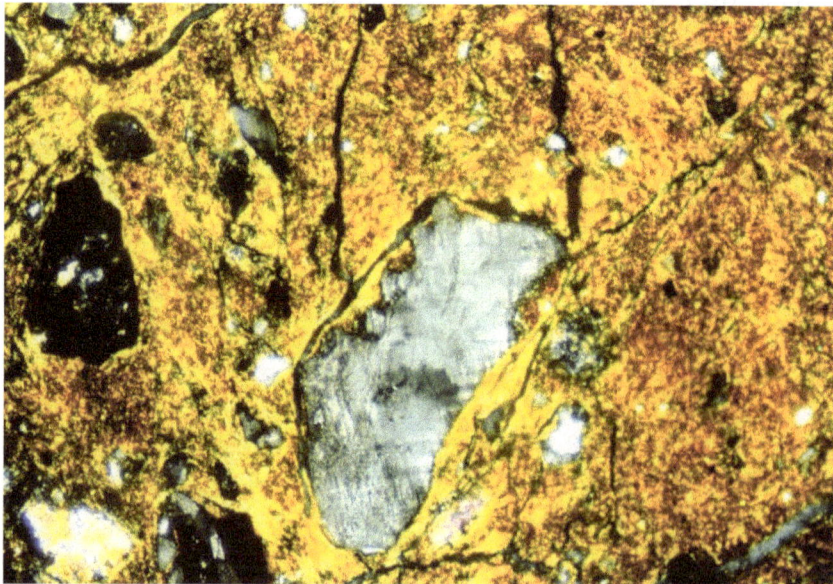

Figure 2. Horizons with high percentage of clay of expandable aluminosilicate clay-sized minerals become aligned through shrink-swell processes. This alignment results in preferred orientation of clay particles, making the plasma anisotropic (visible under crossed polarized light. This process results in a loss of argillans along ped faces in many soils. Frame width = 0.9 mm. (White House pedon, Cochise Co., AZ, Pedon 40A001, BCtk horizon under crossed polarized light).

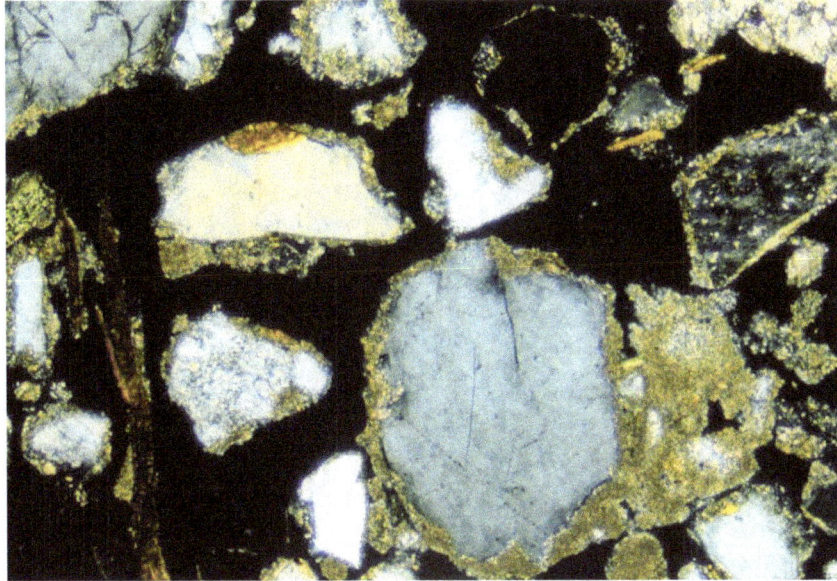

Figure 3. Calcium carbonate around skeletal (sand-sized) grains in a coarse-textured matrix. These carbonate coatings are referred to as calcitans (Brewer, 1976). Their formation illustrates attraction to and deposition of carbonates on mineral surfaces (e.g., quartz or feldspar grains) accessible to percolating water. Frame width = 0.9 mm. (Cax pedon, San Bernardino, Co., CA, Pedon 97P0420, Bkg2 horizon under crossed polarized light).

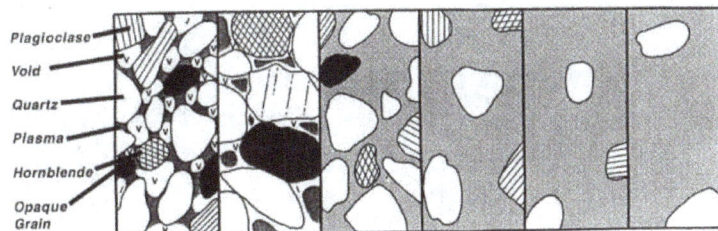

Variable[†]						
C/f Patterns	Chitonic	Enaulic	Close Porphyric	Single-Space Porphyric	Double-Space Porphyric	Open Porphyric
Clay, %	20C[‡]	23BC	20C	25BC	34AB	41A
Silt & Clay, %	37D	50BC	40CD	54B	74A	84A
LE, %	4.4AB	2.2B	2.9AB	2.7AB	4.0AB	6.6A
DP/FP	1.1BC	1.3B	0.7CD	0.6CD	0.5D	0.3D
Nos.	8	26	23	112	53	94

† C/F Patterns, are related distribution patterns of coarse and fine constituents, LE, linear extensibility; DP/FP, ratio of drained to filled pores at 33 kPa suction.
‡ Means with the same letter are not significantly different at the 95% confidence level (SAS Institute, 1988).

Figure 4. Kinds of related distribution patterns and a listing of their physical properties. Frame width of each idealized kind of fabric is 0.5mm. The lower size limit of coarse material in the C/Fpatterns was set at about 50 µm for most of the slides.

155

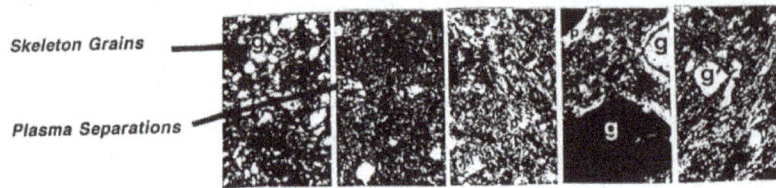

Variable[†]	Silasepic	Insepic	Mosepic	Skelsepic	Masepic
Clay, %	21C[‡]	23C	32B	33B	52A
Silt + Clay, %	58B	59B	65B	61B	84A
LE, %	2.2CD	2.2CD	3.5BCD	5.8B	9.2A
DP/FP	0.8A	0.8A	0.5AB	0.7AB	0.2B
LE + Clay x 100	1.0A	1.1A	1.1A	1.6A	1.8A
Number Observed	97	49	46	67	37

† L.E., linear extensibility; DP/FP, ratio of drained to filled pores.
‡ Means with the same letter are not significantly different at the 95% confidence level (SAS Institute, 1988).

Figure 5. Kinds of plasma fabrics and a listing of their physical properties. Frame width of each idealized kind of fabric is 0.5 mm.

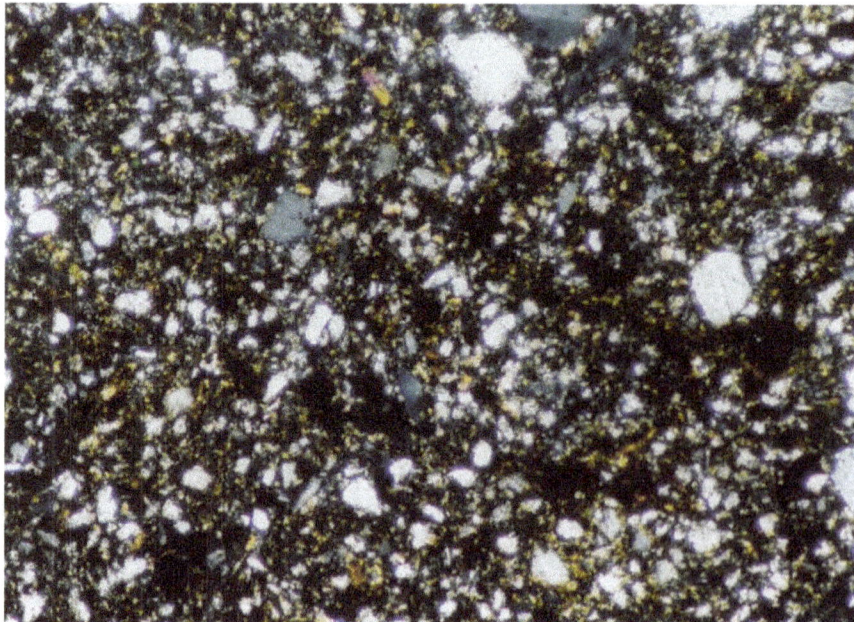

Figure 6. Silasepic plasma fabric. Frame width = 1.3 mm. (Southridge pedon, Allamakee Co., IA; 87P0075 Ap horizon under crossed polarized light).

Figure 7. Insepic plasma fabric. Frame width = 1.3 mm. (Mexico pedon, Macon Co., MO, 87P0771, BE horizon under crossed polarized light).

Figure 8. Mosepic plasma fabric. Frame width = 1.3 mm. (Leonard pedon, Macon Co., MO, 87P0770, 2Btg3 horizon under crossed polarized light).

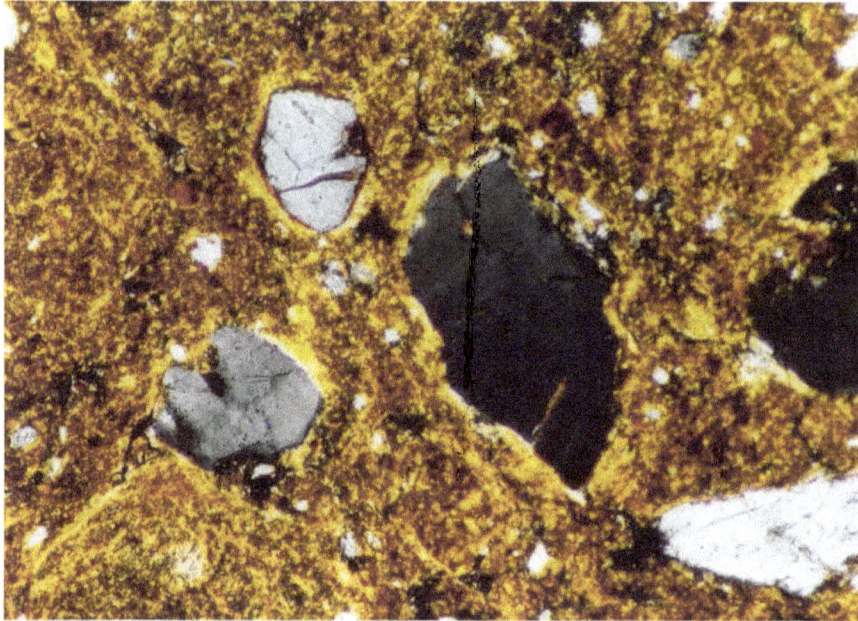

Figure 9. Skelsepic plasma fabric. Frame width = 1.3 mm. (Redding pedon, San Diego Co., CA, 40A2847, Bt horizon under crossed polarized light).

Figure 10. Masepic plasma fabric. Frame width = 1.3 mm. (Gloria pedon, Monterey Co., CA, , 40A2845, Bt horizon under crossed polarized light).

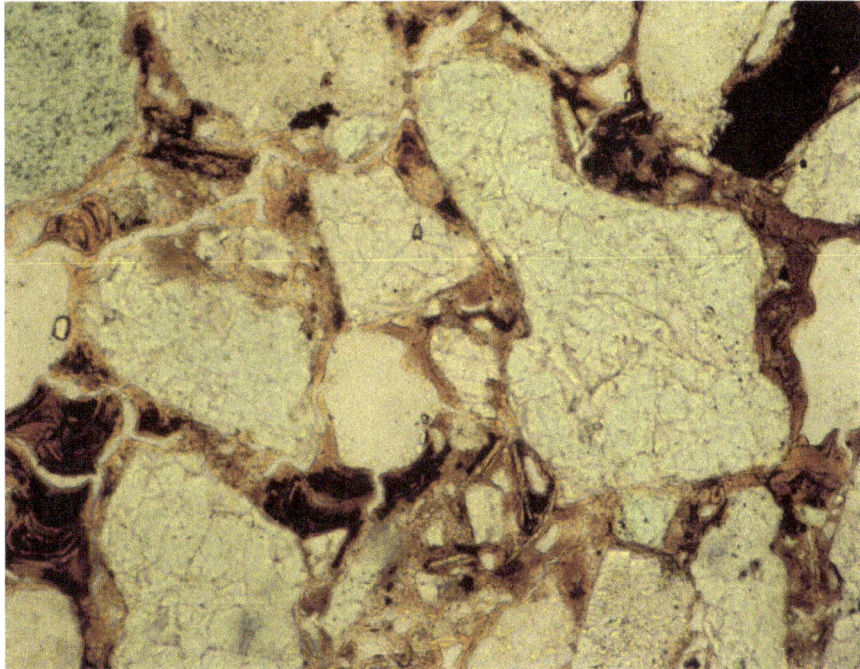

Figure 11. Oriented illuvial clay (argillans) surrounding skeletal grains. Frame width = 1.1 mm (Paxon pedon, New York Co., NY, 00P0001, 2Cd1 horizon under plane polarized light)

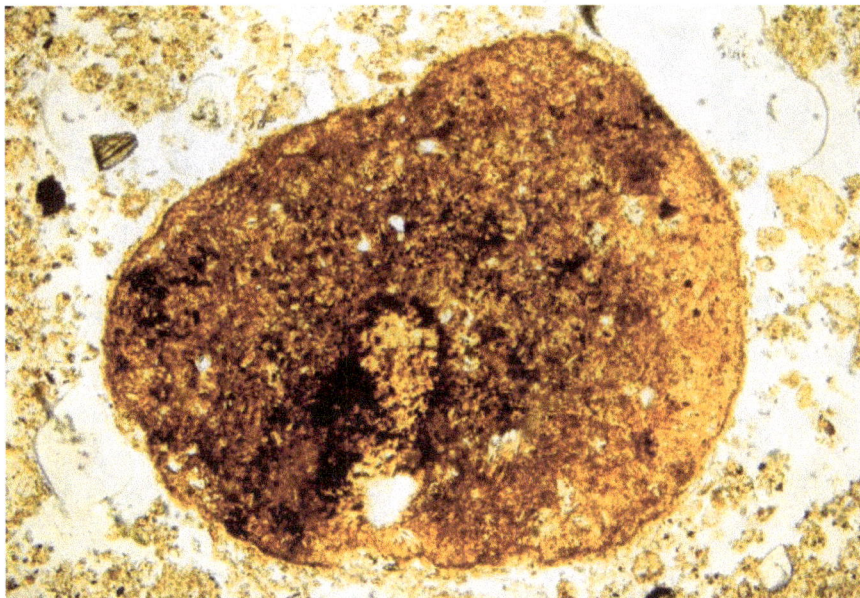

Figure 12. Fe oxide nodule from an Andisol in Blue Mountains of eastern Oregon. Frame width = 2.5 mm. (Tower pedon, Umatilla Co., OR, 97P0547, Bw1 horizon under plane polarized light).

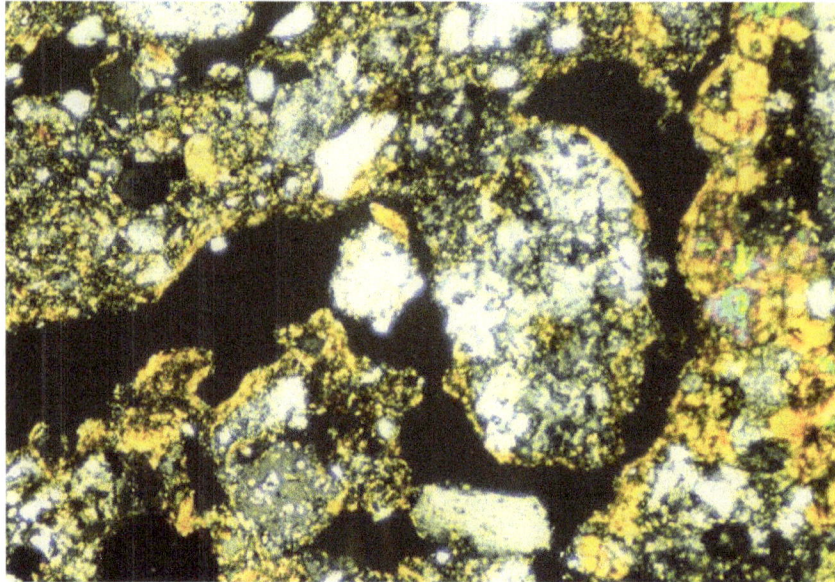

Figure 13. Compound packing voids, surrounded by illuvial clay (argillans). Clay lining channels is anisotropic due to orientation during deposition, while clay (plasma) in the s-matrix is partially anisotropic due to stress orientation (shrink-swell processes). Frame width = 1.0 mm. (Endlich pedon, 99P0001, Bt1 horizon under crossed polarized light).

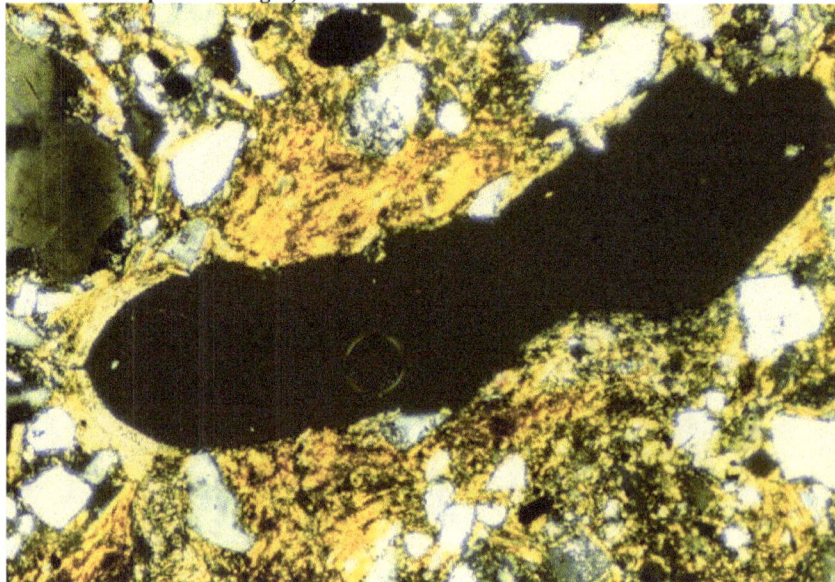

Figure 14. Void that has smooth edges and is elongated. This void type is described by Brewer (1976) as a vugh. Frame width = 1.0 mm. (Troutville pedon, Gunnison Co., CO, 99P0002, E&Bt horizon under cross polarized light).

Figure 15. The walls consisting of "smooth, simple curves" indicates that this void is a vesicle. These vesicles were formed in the thin, surface crust of a Typic Haplargid. Frame width = 3.2 mm (Dera pedon, Juab Co., UT, 81P0610, A1 horizon under crossed polarized light).

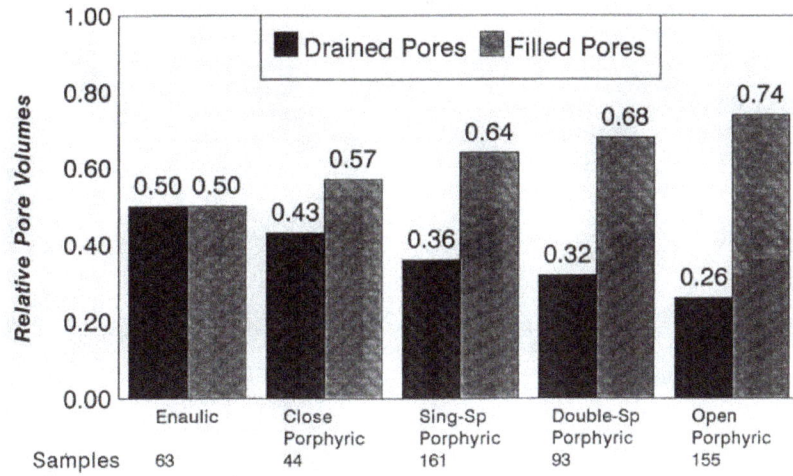

Figure 16. Relative pore volumes at 30 kPa for soil fabric coarse-fine distributions for some U.S. soils.

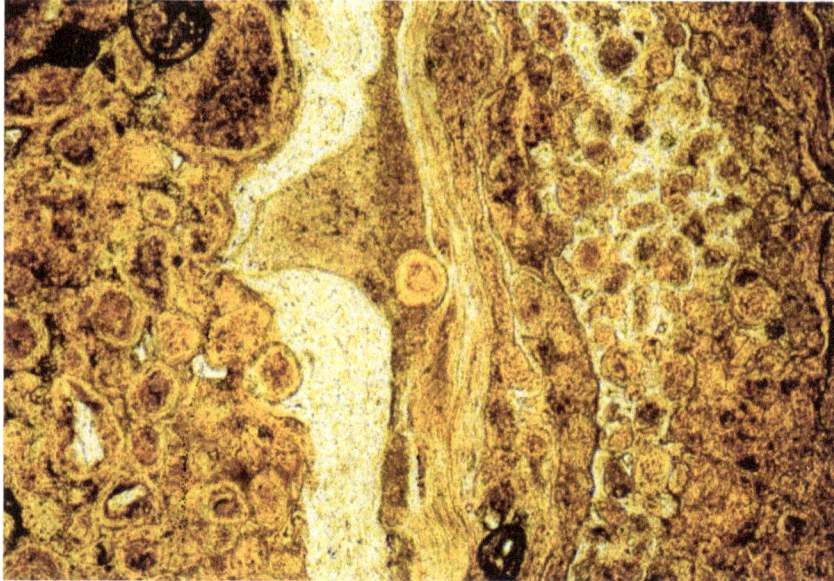

Figure 17. Horizon with duripan exhibiting silica cementation. Fabric has an opal and chalcedony laminar cap. The matrix above and below is composed of durinodes (non-crystalline silica) surrounded by moderately-oriented silicate clays. Clay can provide the initial absorption surface for silica in soil solution. The absorption of silica onto established silica phases leads to formation of nodules. Frame width = 1.0 mm. (Series name not designated, Jefferson Co., OR, Pedon 87P0513, 2Bkqm horizon under plane polarized light).

Figure 18. Joint planes (platy structure) formed in the surface horizon of a soil. Frame width = 2.5 mm. (Frisite pedon, Fremont County, WY, 98P0453, A horizon under cross polarized light).

Wet Aggregate Stability (3F)
Wet-Sieving (3F1)
Air-dry, 2 to 1 mm, 2- to 0.5-mm Aggregates Retained (3F1a1a)

1. Application

An aggregate is a group of primary particles that cohere to each other more strongly than to other surrounding soil particles (Soil Science Society of America, 1997). Disaggregration of soil mass into aggregates requires the application of a disrupting force. Aggregate stability is a function of whether the cohesive forces between particles can withstand the applied disruptive force. The analysis of soil aggregation can be used to evaluate or predict the effects of various agricultural techniques, e.g. tillage and organic-matter additions, and erosion by wind and water (Nimmo and Perkins, 2002). The measurement can serve as a predictor of infiltration and soil erosion potential. This method provides a measure of aggregate stability following a disruption of initially air-dry aggregates by abrupt submergence followed by wet sieving. This procedure was developed for use by the Natural Resources Conservation Service, Soil Survey field offices.

2. Summary of Method

This method (3F1a1a) measures the retention of air-dry aggregates (2 to 1 mm) on a 0.5-mm sieve after sample has been submerged in RO water overnight followed by agitation of sample.

3. Interferences

Air bubbles in the sieve can create tension in the water, thereby reducing the percentage of aggregates that are retained on the 0.5-mm sieve. Variation in the moisture content of air-dry soils can affect results. A correction should be made for the sand > 0.5 mm resistant to dispersion in sodium hexametaphosphate.

4. Safety

If ovens are used, hot surfaces can be a hazard. Follow standard laboratory safety precautions.

5. Equipment

5.1 Bowls, Rubbermaid or equivalent, 1800 mL
5.2 Electronic balance, ±0.01-g sensitivity and 500-g capacity
5.3 Sieves, square-hole
5.3.1 0.5 mm, stainless steel, no.35, 125-mm diameter, 50-mm height
5.3.2 1 mm, brass, 203-mm diameter, 50-mm height
5.3.3 2 mm, brass, 203-mm diameter, 50-mm height
5.4 Oven, 110°C
5.5 Camping plate, Coleman, stainless steel, 152-mm diameter, Peak 1, Model 8553-462.
5.6 Aluminum foil dish, 57-mm diameter x 15-mm deep, with lifting tab

6. Reagents

6.1 Reverse osmosis (RO) water
6.2 Sodium hexametaphosphate solution. Dissolve 35.7 g of sodium hexametaphosphate ($Na_4P_2O_7$) and 7.94 g of sodium carbonate (Na_2CO_3) in 1 L of RO water. Alternatively, use Calgon, water softener.

7. Procedure

7.1 Use natural fabric (NF) samples in pint containers. Assemble a 2-mm sieve on top of a 1-mm sieve. Crush the NF sample by hand or with mortar and pestle. Crush sample so as to pass the 2-mm sieve with a minimum reduction in size. Sieve entire NF sample.

7.2 Place the material that is retained on 1-mm sieve in pint container and discard the remaining material.

7.3 Sieve the material again with 1-mm sieve to remove dust and other small particles. Weigh a 3.00 ± 0.05-g sample of the 2- to 1-mm material in aluminum foil dishes.

7.4 Place 0.5-mm sieve in plastic bowl and fill bowl so that the water level is at a 20-mm height above the base of screen. Remove air bubbles with a syringe.

7.5 Distribute the 3.00-g sample (2 to 1 mm) on the 0.5-mm sieve. Aggregates should not touch. Allow sample to sit overnight.

7.6 Agitate the sample by raising and lowering the sieve in the water bowl 20 times in 40 s. On the upward strokes, drain sieve but do not raise so high that air enters to beneath the sieve.

7.7 Remove sieve from water bowl, place on Coleman plate, and dry in oven for 2 to 2.5 h at 110°C. During drying process, the plate retains the soil that drops through the sieve.

7.8 Remove the sample from the oven. Weigh sieve, plate, and sample. Sample is those aggregates retained on 0.5-mm sieve. Record weight. If no sand (>0.5 mm) is present, discard sample from sieve and plate by brushing. Weigh sieve and plate. Record weight.

7.9 Calculate the S_w from the particle-size data (procedure 3A1). If there is sand (>0.5 mm) and no particle-size data, discard sample on plate and disperse that retained on the sieve with sodium hexametaphosphate solution. Alternatively, place 3 g of Calgon in plastic bowl and stir until dissolved. Place the 0.5-mm sieve with sample in sodium hexametaphosphate (or Calgon) solution so that the solution line is at a 35-mm height above the base of screen.. Gently triturate the dispersing solution with the fingers to remove soft <0.5 mm material adhering to the $\geq$ 0.5 mm. Remove sieve from sodium hexametaphosphate (or Calgon) solution and rinse with RO water until all sodium hexametaphosphate (or Calgon) solution has passed through sieve, and only the sand (>0.5 mm) is left on sieve. Place sieve on Coleman plate, place in oven, and dry for 2 to 2.5 h at 110°C.

7.10 Remove sample from oven. Weigh the sieve, plate, and sample. Record weight. Discard sample and brush sieve and plate. Weigh sieve and plate. Record weight. Alternatively, calculate the S_w from the particle-size data (procedure 3A1).

7.11 Thoroughly wash sieve and plate with RO water, especially those sieves with sodium hexametaphosphate solution.

8. Calculations

Aggregates (%) = $[(W_R - S_W)/(3.00 - S_W)]$ x 100

where:
W_R = Total weight of aggregates retained on 0.5-mm sieve
S_W = Weight of 2- to 0.5-mm sand

9. Report

Report aggregate stability as a percentage of aggregates (2- to 0.5-mm) retained after wet sieving. Do not report determinations if the 2- to 0.5-mm fraction is $\geq$50% of the 2- to 1-mm sample.

10. Precision and Accuracy
Precision and accuracy data are available from the SSL upon request.

11. References
Nimmo, J.R., and K.S. Perkins. 2002. Aggegate stabililty and size distribution. p. 317-328. *In* J.H. Dane and G.C. Topp (eds.). Part 4. Physical methods. Soil Sci. Am. Book Series No. 5. ASA and SSSA, Madison, WI.

Soil Science Society of America. 1997. Glossary of soil science terms. Soil Sci. Soc. Am., Madison, WI.

Particle Density (3G)
Pycnometer Gas Displacement (3G1)
Oven-dry, <2 mm (3G1a1)
Oven-dry, >2 mm (3G1a2)

1. Application
Density is defined as mass per unit volume. Particle density refers to the density of the solid particles collectively (Flint and Flint, 2002). Particle density is required for sedimentation analysis; calculating soil volume or mass; and for mathematically correcting bulk soil samples containing significant amounts of rock fragments so as to determine fine-soil density, water content, or other soil properties affected by volume displacement of rock fragments (Flint and Childs, 1984).

2. Summary of Method
This method (3G1) determines particle density by the pycnometer gas displacement procedure. This is accomplished by employing Archimedes' principle of fluid displacement to determine the volume. The displaced fluid is a gas that can penetrate the finest pores, thereby assuring maximum accuracy (Quantachrome Instruments, 2003). Helium gas is the most commonly recommended gas since its small atomic dimensions assures penetration into crevices and pores approaching one Angstrom (10^{-10}m) in dimension, and its behavior as an ideal gas is also desirable.

3. Interferences
Sample should be dry. Displacement gas will evaporate water molecules and create additional partial pressure (Flint and Flint, 2002). Temperature should be relatively constant since the method uses the ideal gas equation of state. Instrument should be calibrated when environmental conditions change.

4. Safety
No significant hazards are associated with this procedure. Follow standard laboratory safety practices.

5. Equipment
5.1 Electronic balance, ±0.01-g sensitivity
5.2 Oven, 110°C
5.3 Gas pycnometer, Penta-pycnometer, Quantachrome Instruments, Boynton Beach, FL

6. Reagents
6.1 Helium gas

7. Procedure
7.1 Oven-dry the soil sample at 110°C overnight.

7.2 Allow the instrument to warm up for least 30 min prior to use.

7.3 Set the regulator pressure to slightly over 20 PSIG.

7.4 Validate the calibration by determining the volume of the calibration sphere.

7.5 If volumes are outside specification range, recalibrate the instrument per instruction manual.

7.6 Select the largest sample cell size of 135, 50, or 10 cc. The sample should fill at least half of the sample cell volume.

7.7 Weigh the sample cell to the nearest 0.01g. Place sample in the sample cell and reweigh to the nearest 0.01 g. The difference is the sample weight. Record the weight.

7.8 Place the sample cells into the sample cell holder. Properly seal the cell holder with the cover.

7.9 Define each cell with sample cell size, weight, and sample identification number.

7.10 Set the instrument to multi run and enter the number of runs between 3 and 5.

7.11 Set the purge mode to 3 pulse cycles.

7.12 Start the sample run.

7.13 Record the average volume for each sample.

7.14 Calculate the particle density.

8. Calculations

Particle density (g cm^{-3}) = Sample weight (g)/Sample volume

9. Report

Report particle density (g cm^{-3}) to the nearest 0.01 unit on either the <2-mm or >2-mm particle-size fraction.

10. Precision and Accuracy

Precision and accuracy data are available from the SSL upon request.

11. References

Flint, A.L., and S. Childs. 1984. Development and calibration of an irregular hole bulk density sampler. Soil Sci. Soc. Am. J. 48:374-378.

Flint, A.L., and L.E. Flint. 2002. Particle density. p. 229-240. *In* J.H. Dane and G.C. Topp (eds.) Methods of soil analysis, Part 4. Physical methods. Soil Sci. Am. Book Series No. 5. ASA and SSSA, Madison, WI.

Quantachrome Corporation. 2003. Penta-pycnometer instruction manual. Quantachrome Corporation, Boynton Beach, FL

Atterberg Limits (3H)
Liquid Limit (LL) (3H1)
Air-Dry, <0.4 mm (3H1a1)
Field-Moist, <0.4 mm (3H1b1)

Liquid Limit (LL) is the percent water content of a soil at the arbitrarily defined boundary between the liquid and plastic states. This water content is defined as the water content at which a pat of soil placed in a standard cup and cut by a groove of standard dimensions will flow together at the base of the groove for a distance of 13 mm (1/2 in) when subjected to 25 shocks from the cup being dropped 10 mm in a standard LL apparatus operated at a rate of 2 shocks s^{-1}. Refer to ASTM method D 4318 (American Society for Testing and Materials, 2004). The LL is reported as percent water on a <0.4-mm basis (40-mesh) (procedure 3H1).

Atterberg Limits (3H)
Plasticity Index (3H2)
Air-Dry, <0.4 mm (3H2a1)
Field-Moist, <0.4 mm (3H2b1)

The PI is the range of water content over which a soil behaves plastically. Numerically, the PI is the difference in the water content between the LL and the plastic limit (PL). Refer to procedure 3H1 for the definition of LL. The PL is the percent water content of a soil at the boundary between the plastic and brittle states. The boundary is the water content at which a soil can no longer be deformed by rolling into 3.2-mm (1/8-in) threads without crumbling. Refer to ASTM method D 4318 (American Society for Testing and Materials, 2004). The PL is reported as percent water on a <0.4-mm basis (procedure 3H2).

References
American Society for Testing and Materials. 2004. Standard tests for liquid limit, plastic limit, and plasticity index for soils. D 4318. Annual book of ASTM standards. Construction. Section 4. Soil and rock; dimension stone; geosynthesis. Vol. 04.08. ASTM, Philadelphia, PA.

SOIL AND WATER CHEMICAL EXTRACTIONS AND ANALYSES (4)

Acid Standardization (4A)

1. Application
The process by which the concentration of a solution is accurately ascertained is known as standardization (Dean, 1995). Commonly a solution is standardized by a titration in which it reacts with a weighed portion of a primary standard. The reaction between the titrant and the substance selected as a primary standard should fulfill the requirements for titrimetric analysis. In addition, a primary standard should be in a state of known high purity (typically <0.1 to 0.2% impurities), stable (easy to dry, not very hydroscopic, or lose weight upon exposure to air), and have a reasonably high equivalent weight in order to minimize the consequences of errors in weighing (Day and Underwood, 1980). For acid-base titrations, it is customary to prepare solutions of an acid and base of approximately the desired concentration and then to standardize one of the solutions against a primary standard. The solution thus standardized can be used a secondary standard to obtain the normality of the other solution. Some widely used primary standards are as follows:

Benzoic acid - ($C_6H_5CO_2H$). Molecular weight = 122.123. Dissolve about 0.5 g in 50% ethanol and titrate to phenolphthalein end point.

Borax - ($Na_2B_4O_7 \cdot 10H_2O$). Molecular weight = 381.360. Methyl red indicator. Dissolve in water. Weak acid.

Mercuric oxide - (HgO) Molecular weight = 216.599. Bromthymol blue indicator. Dissolve 0.5 g with 15 g of KBr in 25 mL of reverse osmosis deionized (RODI) water, excluding CO_2.

Potassium bicarbonate - ($KHCO_3$). Molecular weight = 100.116. Bromcresol green indicator, first tint of green end point.

Potassium biodate - ($KH(IO_3)_2$). Molecular weight = 389.912. Bromthymol blue end biodate point. Strong acid. Low solubility.

Potassium biphthalate - ($KHC_8H_4O_4$). Molecular weight = 204.224. Phenolphthalein indicator. Weak acid.

Potassium bitatrate - ($KHC_4H_4O_6$). Molecular weight = 188.178. Phenolphthalein bitatrate indicator. Molding problem.

Sodium carbonate - (Na_2CO_3). Molecular weight = 105.988 Bromocresol green indicator, first green end point.

Refer to Table 1 for a list of some common acids and bases.

2. Summary of Method

Dissolve a known amount of the primary standard (e.g., Na_2CO_3) in reverse osmosis deionized (RODI) water. Prepare 10 of these Na_2CO_3 solutions plus 8 blanks and titrate with the acid to be standardized (e.g., HCl, H_2SO_4). Calculate the normality of the acid from the mean blank and titers. Report the normality and standard deviation for the acid standardization.

3. Interferences

Clean the glass electrode by rinsing with distilled water. Wiping the electrode dry with a cloth, laboratory tissue, or similar material may cause electrode polarization.

Slow electrode response may cause the end point to be overshot. Cleaning the electrode with detergent solution may decrease the response time. If all else fails, change the electrode.

Dry primary standards and store in a dessicator to prevent hydration. Contamination of the primary standard may occur when drying, storing, or weighing the reagent. Use gloves to weigh the primary standard in weighing vessel.

4. Safety

Wear protective clothing and eye protection when preparing reagents. Dispense concentrated acids in a fume hood. Thoroughly wash hands after handling reagents. Be prepared to use safety showers and eyewash stations if necessary. Use sodium bicarbonate and water to neutralize and dilute spilled acids. Follow the manufacturer's safety precautions when using the automatic titrator. Review Material Safety Data Sheets (MSDS) for reagents.

5. Equipment
5.1 Electronic balance, ±0.10-mg sensitivity
5.2 Oven, 110°C
5.3 Weighing vessel, 40 x 50mm
5.4 Desiccator

5.5 Automatic titrator, with control unit, sample changer, and dispenser, Metrohm Ltd., Brinkmann Instruments, Inc.

5.6 Combination pH-reference electrode, Metrohm Ltd., Brinkmann Instruments, Inc.

5.7 Computer, with Titrino Workcell software, Metrohm Ltd., Brinkman Instruments, Inc., and printer

5.8 Titration beakers, 250 mL, borosilicate, Metrohm Ltd., Brinkman Instruments Inc.

6. Reagents
6.1 Reverse osmosis deionized (RODI) water, ASTM Type I grade of reagent water
6.2 pH buffers (4.00, 7.00, 9.18)
6.3 Sodium carbonate, anhydrous, 99.8% pure
6.4 Calcium sulfate (anhydrous) or equivalent desiccant

7. Procedure

7.1 Place a weighing vessel containing $\approx$ 10 g of the Na_2CO_3 in an oven at $110^{\circ}C$ overnight. Let cool in a desiccator to room temperature ($\approx$ 30 min).

7.2 Refer to Acid Standardization Form. Weigh the weighing vessel plus the Na_2CO_3 designated in Section 7.1 to the nearest 0.1mg (VW_{1a}).

7.3 Tare a 250-mL titration beaker on the electronic balance. Add 0.15 - 0.55 g Na_2CO_3 to the beaker. Record the weight of the Na_2CO_3 to the nearest 0.1mg (SW_1).

7.4 Verify the weighed SW_1 as follows:

7.4.1 Re-weigh the weighing vessel plus Na_2CO_3 designated in Section 7.1 to the nearest 0.1mg (VW_{1b}).

7.4.2 Calculate SWC_1 as follows: $VW_{1a} - VW_{1b} = SWC_1$.

7.4.3 If SWC_1 calculated in Section 7.4b is within 0.5 mg of the weighed SW_1 determined in Section 7.3, proceed to Section 7.5, but if not, return to Section 7.2 and begin again.

7.5 Prepare nine more Na_2CO_3 solutions ($SW_2 ... SW_{10}$) by this same procedure (Sections 7.2 - 7.4).

7.6 Add 100 mL of RODI water to each of the $SW_1 ...... SW_{10}$ and to eight empty beakers for blanks ($B_{1....8}$) and gently swirl to facilitate dissolution of the salt.

7.7 Prepare the titration system according to the manufacture's instructions. Calibrate the pH electrode with 4.00, 7.00, and 9.18 pH buffers. Recalibrate electrode if R^2 <0.950.

7.8 Set-up the automatic titrator to set end point titration mode. The "Set" pH parameters are listed as follows:

Parameter	Value
Measured value	pH
Titration rate	normal
Stop volume	100.0 mL
Stop End Point	9
Stop Potential	4.2 pH
Sensitivity	High

Fixed End Point   4.6 pH
Signal Drift   25
Equilibration time   5 s

7.9 Titrate each of the SW_1......SW_{10} and $B_{1.....8}$ and record the titres.

8. Calculations

8.1 $B_{mean} = B_{sum}/n$

where:
B   = Blank mean (mL)
B_{sum} = Sum of blank titres (mL)
n   = Number of blanks (n = 8)

8.2 $TCS_{1.....10} = TS_{1......10} - B_m$

where:
$TCS_{1.....10}$ = Corrected sample titre (mL)
$TS_{1......10}$ = Sample titre (mL)
B_m = Blank mean (mL)

8.3 $N_{1......10} = [(SW_{1.....10} \times 0.0188698 \text{ mole HCl/g Na}_2\text{CO}_3) / (TCS_{1.....10} \times 10^{-3} \text{ L ml}^{-1})] = (SW_{1....10} \times 18.8698)/ TCS_{1.....10}$

where:
$N_{1......10}$ = Normality calculated from titre (number of equivalents of solute per liter)
$SW_{1.....10}$ = Na_2CO_3 weight (g)

8.4 $N_{mean} = N_{sum}/n$

where:
N_{mean} = Normality mean
N_{sum} = Sum of normalities
n   = Number of samples (n = 10)

8.5 Calculate standard deviation (Std) for N_{mean}. If the N_{Std} is ≤ 0.0005, record N_{mean} to four decimals to left of decimal. If $N_{Std} > 0.0005$, discard no more than two suspicious values and return again to Sections 7.2 to 8.5 until 10 (n = 10) are achieved.

9. Report

Report the mean normality of acid (N_{mean}), standard deviation, and date of standardization.

10. Precision and Accuracy

Precision and accuracy data are available from the SSL upon request.

10. References

Day, R.A., Jr. and, A.L. Underwood. 1980. Quantitative analysis. Prentice-Hall, Inc., Englewood Cliffs, NJ.

Dean, J.A. 1995. Analytical chemistry handbook. McGraw Hill, Inc., New York, NY.

STANDARDIZATION OF ACIDS

Table 1. Common commercial strengths of acids and bases.

Name	Molecular Weight	Moles per Liter	Grams per Liter	Percent by Weight	Specific Gravity
Acetic acid	60.05	17.4	1045	99.5	1.05
Glacial acetic acid	60.05	6.27	376	36	1.045
Butyric acid	88.1	10.3	912	95	0.96
Formic acid	46.02	23.4	1080	90	1.20
		5.75	264	25	1.06
Hydriodic acid	127.9	7.57	969	57	1.70
		5.51	705	47	1.50
		0.86	110	10	1.1
Hydrobromic acid	80.92	8.89	720	48	1.50
		6.82	552	40	1.38
Hydrochloric acid	36.5	11.6	424	36	1.18
		2.9	105	10	1.05
Hydrocyanic acid	27.03	25	676	97	0.697
		0.74	19.9	2	0.996
Hydrofluoric acid	20.01	32.1	642	55	1.167
		28.8	578	50	1.155
Hydrofluosilic acid	144.1	2.65	382	30	1.27
Hypophosphorus acid	66.0	9.47	625	50	1.25
		5.14	339	30	1.13
		1.57	104	10	1.04
Lactic acid	90.1	11.3	1020	85	1.2
Nitric acid	63.02	15.99	1008	71	1.42
		14.9	938	67	1.40
			837	61	1.37
Perchloric acid	100.5	11.65	1172	70	1.67
		9.2	923	60	1.54
Phosphoric acid	98	14.7	1445	85	1.70
Sulfuric acid	98.1	18.0	1766	96	1.84
Sulfurous acid	82.1	0.74	61.2	6	1.02
Ammonia water	17.0	14.8	252	28	0.898
Potassium hydroxide	56.1	13.5	757	50	1.52
		1.94	109	10	1.09
Sodium carbonate	106.0	1.04	110	10	1.10
Sodium hydroxide	40	19.1	763	50	1.53
		2.75	111	10	1.11

Acid Standardization

Acid:_____		Date:_____			Technician:

		Vessel Weight	Sample Weight	Sample Titer	Blanks	CorrectedSample Titer
#1 Beginning Weight		VW_{1a}	SW_1	TS_1	B_1	TCS_1
Ending Weight		VW_{1b}			B_2	
#2 Beginning Weight		VW_{2a}	SW_2	TS_2	B_3	TCS_2
Ending Weight		VW_{2b}			B_4	
#3 Beginning Weight		VW_{3a}	SW_3	TS_3	B_5	TCS_3
Ending Weight		VW_{3b}			B_6	
#4 Beginning Weight		VW_{4a}	SW_4	TS_4	B_7	TCS_4
Ending Weight		VW_{4b}			B_8	
#5 Beginning Weight		VW_{5a}	SW_5	TS_5	B_{sum}	TCS_5
Ending Weight		VW_{5b}			B_{mean}	
#6 Beginning Weight		VW_{6a}	SW_6	TS_6		TCS_6
Ending Weight		VW_{6b}				
#7 Beginning Weight		VW_{7a}	SW_7	TS_7		TCS_7
Ending Weight		VW_{7b}				
#8 Beginning Weight		VW_{8a}	SW_8	TS_8		TCS_8
Ending Weight		VW_{8b}				
#9 Beginning Weight		VW_{9a}	SW_9	TS_9		TCS_9
Ending Weight		VW_{9b}				
#10 Beginning Weight		VW_{10a}	SW_{10}	TS_{10}		TCS_{10}
Ending Weight		VW_{10b}				
					$N_{1 \ .10}$	
					N_{Std}	

SOIL AND WATER CHEMICAL EXTRACTIONS AND ANALYSES (4)

Ion Exchange and Extractable Cations (4B)

Ion exchange is a reversible process by which one cation or anion held on the solid phase is exchanged with another cation or anion in the liquid phase, and if two solid phases are in contact, ion exchange may also take place between two surfaces (Tisdale et al., 1985). In most agricultural soils, the cation exchange capacity (CEC) is generally considered to be more important than anion exchange (AEC), with the anion molecular retention capacity of these soils usually much smaller than the CEC (Tisdale et al., 1985). Some soils with abundant goethite and gibbsite, as do some oxic horizons or subsoils of Oxisols (Soil Survey Staff, 1999), may have a CEC to AEC ratio approaching 1.0 (net charge of zero) or a small positive charge (Foth and Ellis, 1988).

Soil mineral and organic colloidal particles have negative valence charges that hold dissociable cations, and thus are "colloidal electrolytes" (Jackson, 1958). The CEC is a measure of the quantity of readily exchangeable cations that neutralize negative charges in the soil (Rhoades, 1982). CEC is a reversible reaction in soil solution, dependent upon negative charges of soil components arising from permanently charged or pH-dependent sites on organic matter and mineral colloid surfaces. The mechanisms for these negative charges are isomorphic substitution within layered silicate minerals; broken bonds at mineral edges and external surfaces; dissociation of acidic functional groups in organic compounds; and preferential adsorption of certain ions on particle surfaces (Rhoades, 1982). Isomorphic substitution produces permanent charge. The other charge mechanisms produce variable charge which is dependent on the soil solution phase as affected by soil pH, electrolyte level, valence of counter-ions, dielectric constant, and nature of anions (Rhoades, 1982). As a result of the variable charge in soils, the CEC is a property dependent on the method and conditions of determination. The method of determination is routinely reported with CEC data.

CEC is a measure of the total quantity of negative charges per unit weight of the material and is commonly expressed in units of milliequivalents per 100 g of soil (meq 100 g^{-1}) or centimoles per kg of soil (cmol(+) kg^{-1}). The SSL reports cmol(+) kg^{-1} on <2-mm base. The CEC can range from less than 1.0 to greater than 100 cmol(+) kg^{-1} soil. The term *equivalent* is defined as "1 gram atomic weight of hydrogen or the amount of any other ion that will combine with or displace this amount of hydrogen". The milliequivalent weight of a substance is one thousandth of its atomic weight. Since the equivalent weight of hydrogen is about 1 gram, the term *milliequivalent* may be defined as "1 milligram of hydrogen or the amount of any other ion that will combine with or displace it" (Tisdale et al., 1985).

Common CEC values for some soil components (NSSL, 1975) are as follows:

Soil Component	cmol(+) kg^{-1}
Organic Matter	200 to 400
"Amorphous" Clay	160 (at pH 8.2)
Vermiculite	100 to 150
Montmorillonite	60 to 100
Halloysite $4H_2O$	40 to 50
Illlite	20 to 40
Chlorite	10 to 40
Kaolinite	2 to 16
Halloysite $2H_2O$	5 to 10
Sesquioxides	0

These very broad CEC ranges are intended only as general guidelines. More narrow groupings of CEC values are possible as data are continually collected and correlated, e.g., the CEC of organic matter in Mollisols in the western United States ranges from 100 to 300 cmol (+) kg^{-1} (average 200), and the CEC of organic matter in Histosols ranges from 125 to 185 cmol (+) kg^{-1} and increases with decomposition of the organic matter (NSSL Staff, 1975).

Many procedures have been developed to determine CEC. These CEC measurements vary according to the nature of the cation employed, concentration of salt, and the equilibrium pH. The CEC measurement should not be thought of as highly exact but rather as an equilibrium measurement under the conditions selected (Jackson, 1958). Knowledge of the operational definition (procedure, pH, cation, and concentration) is necessary before evaluating the CEC measurement (Sumner and Miller, 1996). The more widely adopted methods of CEC determination are classified (Rhoades, 1982) as follows:

(1) cation summation
(2) direct displacement
(3) displacement after washing
(4) radioactive tracer

The SSL performs a number of CEC methods, using several different reagents and pH levels. The CEC's most commonly reported by the SSL are CEC-7 (4B1a1a1a1), CEC-8.2 (4B4b1), and effective cation exchange capacity (ECEC) (4B4b2). As a general rule, the CEC-8.2 > CEC-7 > ECEC.

Cation Exchange Capacity: NH$_4$OAc, pH 7.0 (CEC-7)

The CEC-7 is a commonly used method (4B1a1a1a1) and has become a standard reference to which other methods are compared (Peech et al., 1947). Displacement after washing is the basis for this procedure. The CEC is determined by saturating the exchange sites with an index cation (NH$_4^+$) using a mechanical vacuum extractor (Holmgren et al., 1977); washing the soil free of excess saturated salt; displacing the index cation (NH$_4^+$) adsorbed by the soil; and measuring the amount of the index cation (NH$_4^+$). An advantage of using this method is that the extractant is highly buffered so that the extraction is performed at a constant and known pH (pH 7.0). In addition, the NH$_4^+$ on the exchange complex is easily determined. CEC-7 is an analytically determined value and is usually used in calculating the CEC-7/clay ratios, although many SSL Primary Characterization Data Sheets predating 1975 show CEC-8.2/clay.

Cation Exchange Capacity: Sum of Cations (CEC-8.2)

CEC-8.2 is calculated (4B4b1) by summing the NH$_4$OAc extractable bases (4B1a1b1-4) plus the BaCl$_2$-TEA extractable acidity (4B2a1a1 or 4B2b1a1). Cation summation is the basis for this procedure. The CEC-8.2 minus the CEC-7 is considered the pH dependent charge from pH 7.0 to pH 8.2. The CEC-8.2 is not reported if significant quantities of soluble salts or carbonates are present in the soil. CEC-8.2 is calculated as follows:

CEC-8.2 = NH$_4$OAc extractable bases + Extractable acidity

Effective Cation Exchange Capacity: NH$_4$OAc Extractable Bases + Aluminum:

CEC can be measured by extraction with an unbuffered salt, which measures the effective cation exchange capacity (ECEC), i.e., CEC at the normal soil pH (Coleman et al., 1958). Since the unbuffered salt solution, e.g., 1 N KCl, only affects the soil pH one unit or less, the extraction is determined at or near the soil pH and extracts only the cations held at active exchange sites at the particular pH of the soil. Neutral NH$_4$OAc extracts the same amounts of Ca^{2+}, Mg^{2+}, Na$^+$, K$^+$ as KCl and therefore extractable bases by NH$_4$OAc is used at the SSL in place of KCl-extractable bases.

ECEC may be determined by extracting one soil sample with neutral normal NH$_4$OAc to determine the exchangeable basic cations (Ca^{2+}, Mg^{2+}, Na$^+$, and K$^+$) and by extracting another sample of the same soil with 1.0 N KCl to determine the exchangeable Al. The 1 N

KCl-extractable Al approximates exchangeable Al and is a measure of "active" acidity present in soils with a 1:1 pH <5.5. Aluminum is non-exchangeable at pH >5.5 due to hydrolysis, polymerization, and precipitation. For soils with pH <7.0, the ECEC should be less than the CEC measured with a buffered solution at pH 7.0. The ECEC is not reported for soils with soluble salts. ECEC (4B4b2a) is calculated by summing the NH_4OAc bases (4B1a1b1-4) plus the KCl extractable Al (4B3a1a1) as follows:

ECEC = NH_4OAc extractable bases + KCl-extractable Al

Effective Cation Exchange Capacity: NH₄Cl (ECEC)

The CEC using a neutral unbuffered salt (NH_4Cl) is also an analytically determined value (4B1b1a1a1). The CEC by NH_4Cl provides an estimate of the ECEC of the soil (Peech et al., 1947). For a soil with a pH of <7.0, the ECEC value should be < CEC measured with a buffered solution at pH 7.0. The NH_4Cl CEC is ≈ equal to the NH_4OAc extractable bases plus the KCl extractable Al for noncalcareous soils. This ECEC method is less commonly used at the SSL.

References

Coleman, T.R., E.J. Kamprath, and S.B. Weed. 1958. Liming. Adv. Agron. 10:475.

Foth, H.D., and B.G. Ellis. 1988. Soil fertility. John Wiley & Sons, Inc., NY.

Holmgren, G.S., R.L. Juve, and R.C. Geschwender. 1977. A mechanically controlled variable rate leaching device. Soil Sci. Am. J. 41:1207-1208.

Jackson, M.L. 1958. Soil chemical analysis. Prentice-Hall, Inc., Englewood Cliffs, NJ.

National Soil Survey Laboratory Staff. 1975. Proposed tables for soil survey reports. RSSIU, USDA-SCS, Lincoln, NE

Peech, M., L.T. Alexander, L.A. Dean, and J.F. Reed. 1947. Methods of soils analysis for soil fertility investigations. USDA Circ. 757, 25 pp.

Rhoades, J.D. Cation exchange capacity. 1982. p. 149-157. *In* A.L. Page, R.H. Miller, and D.R. Keeney (eds.) Methods of soil analysis. Part 2. Chemical and microbiological properties. 2nd ed. Agron. Monogr. 9. ASA and SSSA, Madison, WI.

Sumner, M.E., and W.P. Miller. 1996. Cation exchange capacity and exchange coefficients. p. 1201-1229. *In* D.L. Sparks (ed.) Methods of soil analysis. Part 3. Chemical methods. No. 5. ASA and SSSA, Madison, WI.

Tisdale, S.L., W.L. Nelson, and J.D. Beaton. 1985. Soil fertility and fertilizers. 4th ed. Macmillan Publ. Co., NY.

Ion Exchange and Extractable Cations (4B)
Displacement after Washing, NH₄OAc, pH 7 (4B1a)
Automatic Extractor, 2 M KCl Rinse (4B1a1a)
Steam Distillation, HCl Titration (4B1a1a1a)
Cation Exchange Capacity (CEC-7) (4B1a1a1a1)
Air-dry or Field-Moist, <2 mm (4B1a1a1a1a-b1)

1. Application

The cation exchange capacity (CEC) determined with 1 N NH_4OAc buffered at pH 7.0 (CEC-7), is a commonly used method and has become a standard reference to which other methods are compared (Peech et al., 1947). The advantages of using this method are that the extractant is highly buffered so that the extraction is performed at a constant, known pH (7.0) and that the NH_4^+ on the exchange complex is easily determined.

2. Summary of Method

Displacement after washing is the basis for this procedure. The CEC is determined by saturating the exchange sites with an index cation (NH_4^+); washing the soil free of excess saturated salt; displacing the index cation (NH_4^+) adsorbed by the soil; and measuring the amount of the index cation (NH_4^+). A sample is leached using 1 N NH_4OAc and a mechanical vacuum extractor (Holmgren et al., 1977). The extract is weighed and saved for analyses of the cations. The NH_4^+ saturated soil is rinsed with ethanol to remove the NH_4^+ that was not adsorbed. The soil is then rinsed with 2 M KCl. This leachate is then analyzed by steam distillation and titration to determine the NH_4^+ adsorbed on the soil exchange complex. The CEC by NH_4OAc, pH 7 is reported in meq 100 g^{-1} or (cmol (+) kg^{-1}) soil in procedure 4B1a1a1a1.

3. Interferences

Incomplete saturation of the soil with NH_4^+ and insufficient removal of NH_4^+ are the greatest interferences to this method. Ethanol removes some adsorbed NH_4^+ from the exchange sites of some soils. Isopropanol rinses have been used for some soils in which ethanol removes adsorbed NH_4^+. Soils that contain large amounts of vermiculite can irreversibly "fix" NH_4^+. Soils that contain large amounts of soluble carbonates can change the extractant pH and/or can contribute to erroneously high cation levels in the extract. This method overestimates the "field" CEC of soils with pH <7 (Summer and Miller, 1996)

4. Safety

Wear protective clothing (coats, aprons, sleeve guards, and gloves) and eye protection (face shields, goggles, or safety glasses) when preparing reagents, especially concentrated acids and bases. Dispense concentrated acids and bases in a fume hood. Thoroughly wash hands after handling reagents. Use the safety showers and eyewash stations to dilute spilled acids and bases. Use sodium bicarbonate and water to neutralize and dilute spilled acids. Nessler's reagent contains mercury which is toxic. Proper disposal of the Nessler's reagent and clean-up of equipment in contact with the reagent is necessary.

Ethanol is flammable. Avoid open flames and sparks. Standard laboratory equipment includes fire blankets and extinguishers for use when necessary. Follow the manufacturer's safety precautions when using the vacuum extractor and the Kjeltec Auto Analyzers.

5. Equipment

5.1 Electronic balance, ±1.0-mg sensitivity
5.2 Mechanical vacuum extractor, 24-place, SAMPLETEX, MAVCO Industries, Lincoln, NE (Fig. 1 and 2)
5.3 Tubes, 60-mL, polypropylene, for extraction (0.45-μm filter), reservoir, and tared extraction tubes
5.4 Rubber tubing, 3.2 ID x 1.6 OD x 6.4 mm (1/8 ID x 1/16 OD x 1 in) for connecting syringe barrels.
5.5 Kjeltec Auto 2300 Sampler System, Tecator, Perstorp Analytical
5.6 Digestion tubes, straight neck, 250 mL
5.7 Syringe filters, 0.45 μm, Whatman
5.8 Wash bottles
5.9 Vials, plastic
5.10 Centrifuge, Centra, GP-8, Thermo IEC, Needham Heights, MA

6. Reagents

6.1 Reverse osmosis deionized (RODI) water, ASTM Type I grade of reagent water
6.2 Ammonium acetate solution (NH_4OAc), 1 N, pH 7.0. Add 1026 mL of glacial acetic acid (CH_3COOH) to 15 L RODI water. Add 1224 mL of concentrated ammonium hydroxide (NH_4OH). Cool. Allow to stand one day to equilibrate to room temperature. Mix and adjust to pH 7.0 with CH_3COOH (typically, ≈ 40 mL) or NH_4OH and dilute with RODI water to 18 L.
6.3 Ethanol (CH_3CH_2OH), 95%, U.S.P.
6.4 Nessler's reagent. Add 4.56 g of potassium iodide (KI) to 30 mL RODI water. Add 5.68 g of mercuric iodide (HgI_2). Stir until dissolved. Dissolve 10 g of sodium hydroxide (NaOH) in 200 mL of RODI water. Transfer NaOH solution to a 250-mL volumetric flask

and slowly add K-Hg-I solution. Dilute to volume with RODI water and thoroughly mix. Solution should not contain a precipitate. Solution can be used immediately. Store in brown bottle to protect from light.

6.5 Potassium chloride solution, 2 M. Add 1341.9 g of KCl reagent in 8 L RODI water. Allow solution to equilibrate to room temperature. Dilute to 9 L with RODI water.

6.6 Boric acid, 4% (w:v), with bromcresol green-methyl red indicator (0.075 % bromcresol green and 0.05% methyl red), Chempure

6.7 Hydrochloric acid (HCl), 0.1 N, standardized. Dilute 167 mL of concentrated HCl in 20 L of RODI water. Refer to procedure for Standardization of Acids.

6.8 NaOH, 1 M. Add 500 mL of 50% NaOH solution to 8 L of RODI water. Dilute to 9 L with RODI water.

7. Procedure

Extraction of Bases

7.1 Weigh 2.5 g of <2-mm, air-dry soil to the nearest mg and place in a labeled extraction tube (ET). If sample is fine-grind, weigh 1 g to the nearest mg. If sample is moist, weigh enough soil to achieve ≈ 2.5 or 1g, respectively, of air-dry soil. Prepare one quality control check sample per 24 samples.

7.2 Place labeled ET on extractor and connect to corresponding tared extraction tube (TET$_{NH4OAc}$) with rubber tubing.

7.3 Use wash bottle to rinse inside of ET with NH$_4$OAc. All soil should be wetted and no air bubbles. Shaking, swirling, or stirring may be required to wet organic samples. Fill ET to the 20-mL mark with NH$_4$OAc solution (≈ 10 mL).

7.4 Secure reservoir tube (RT) to top of ET tube and let stand for 30 min. Extract at 30-min rate the NH$_4$OAc solution until 2 mL of this solution remains above soil level. Turn off extractor. Do not let soil dry.

7.5 Add 40 mL of NH$_4$OAc solution to the RT. Set extractor for an overnight (12h) extraction. Extractor will turn off automatically.

7.6 Next day, remove RT from top of extractor and place in a clean container. Carefully remove TET$_{NH4OAc}$. Leave the rubber tubing on the ET. Weigh each TET$_{NH4OAc}$ containing the NH$_4$OAc extract to the nearest mg.

7.7 Mix the extract in each TET$_{NH4OAc}$ by manually shaking. Fill a labeled plastic vial with extract solution and cap. Discard the excess properly. The solution in the vial is reserved for analyses of extracted cations (procedure 4B1a1b1-4) on the atomic absorption spectrophotometer (AAS). Some samples may be cloudy and need to be filtered prior to analysis on the AA. If extracts are not to be determined immediately after collection, then store samples at 4°C in plastic tubes.

Removal of Excess Ammonium Acetate

7.8 Re-connect the TET$_{NH4OAc}$ with paired ET. Use a wash bottle to rinse the sides of the ET with ethanol to remove any remaining NH$_4$OAc or soil particles adhering to the ET. All soil should be wetted and no air bubbles. Fill ET to the 20-mL mark with ethanol. Secure RT to top of ET tube and let stand for 30 min.

7.9 Extract at the 30-min extraction rate the ethanol solution until 2 mL of this solution remains above the soil level. Turn off the extractor. Do not let soil dry.

7.10 Add 45 mL of ethanol to the RT. Extract ($\approx$ 45 min) the ethanol until 2 mL of this solution remains above the soil level. Turn off the extractor. Do not let soil dry. Disconnect the TET$_{NH4OAc}$ from the ET and discard the ethanol properly.

7.11 Re-connect the TET$_{NH4OAc}$ to the ET and add 55 mL of ethanol to the RT. Set the extractor for 45 min. Turn off the extractor. Remove the TET$_{NH4OAc}$, leaving the tubing connected to the ET. Discard the ethanol properly.

7.14 After the final ethanol wash, collect a few drops of ethanol extract from the ET on a spot plate. Test for NH$_4^+$ by using Nessler's reagent. A yellow, red to reddish brown precipitate is a positive test. If the test is positive, repeat the ethanol wash and retest with Nessler's reagent. Repeat until a negative test is obtained.

2 M KCl Rinse

7.15 Connect a new labeled extraction tube (ET$_{KCl}$) with rubber tubing to ET on extractor.

7.16 Use a wash bottle to rinse inside of ET with 2 M KCl to remove any remaining ethanol or soil particles adhering to the ET. All soil should be wetted and no air bubbles. Fill ET to the 20-mL mark with KCl solution and let stand for 30.

7.17 Extract at the 30-min rate the KCl solution until 2 mL of this solution remains above soil level. Turn off extractor. Do not let soil dry.

7.18 Secure RT to top of ET tube. Add 40 mL KCl solution to RT and set the extract for 45 min. Remove the ET and ET$_{KCl}$ from the extractor.

Steam Distillation: Setup, Operation, and Analysis

7.19 Transfer the contents of the ET$_{KCl}$ to a 250-mL digestion tube. If extracts are not to be determined immediately after collection, then store samples at 4°C.

7.20 Refer to the manufacturer's manual for operation of the distillation unit. The following are only very general guidelines for instrument conditions.

Program: Kjeldahl 1

Receiving solution (boric acid): 30 mL
Water: 0 mL
Alkali (NaOH): 20 mL
Mode: Delay
Time: 1 s
Distillation: Volume
Tube Drain: Yes

7.21 When using new reagents, e.g., boric acid, reagent blanks are distilled in 2 sets of 6, one set per Kjeltec machine. Each set of 6 is averaged and recorded on bench worksheet and manually set on each machine. During the steam distillation, the mean reagent blank titer is automatically subtracted from the sample titer.

7.22 Record the normality of standardized acid.

7.23 Connect the tube to the distillation unit. Close the safety door. Distillation and titration are performed automatically. Record the titer in mL of titrant.

8. Calculations

CEC = [Titer x N x 100 x R]/[Sample Weight (g)]

where:
CEC = Cation Exchange Capacity (meq 100 g^{-1})
Titer = Titer of sample (mL)
N = Normality of HCl titrant
100   = Conversion factor to 100-g basis
R   = Air-dry/oven-dry ratio (procedure 3D1) or field-moist/oven-dry ratio (procedure 3D2)

9. Report
Report CEC-7 to the nearest 0.1 meq 100 g^{-1} (cmol (+) kg^{-1}).

10. Precision and Accuracy
Precision and accuracy data are available from the SSL upon request.

11. References
Holmgren, G.G.S., R.L. Juve, and R.C. Geschwender. 1977. A mechanically controlled variable rate leaching device. Soil Sci. Am. J. 41:1207-1208.

Peech, M., L.T. Alexander, L.A. Dean, and J.F. Reed. 1947. Methods of soil analysis for soil fertility investigations. USDA Circ. 757, 25 pp.

Sumner, M.E., and W.P. Miller. 1996. Cation exchange capacity and exchange coefficients. p. 1201-1229. *In* D.L. Sparks (ed.) Methods of soil analysis. Part 3. Chemical methods. No. 5. ASA and SSSA, Madison, WI.

Fig.1. Mechanical vacuum extractor,
SAMPLETEX, MAVCO Industries, Lincoln, NE.

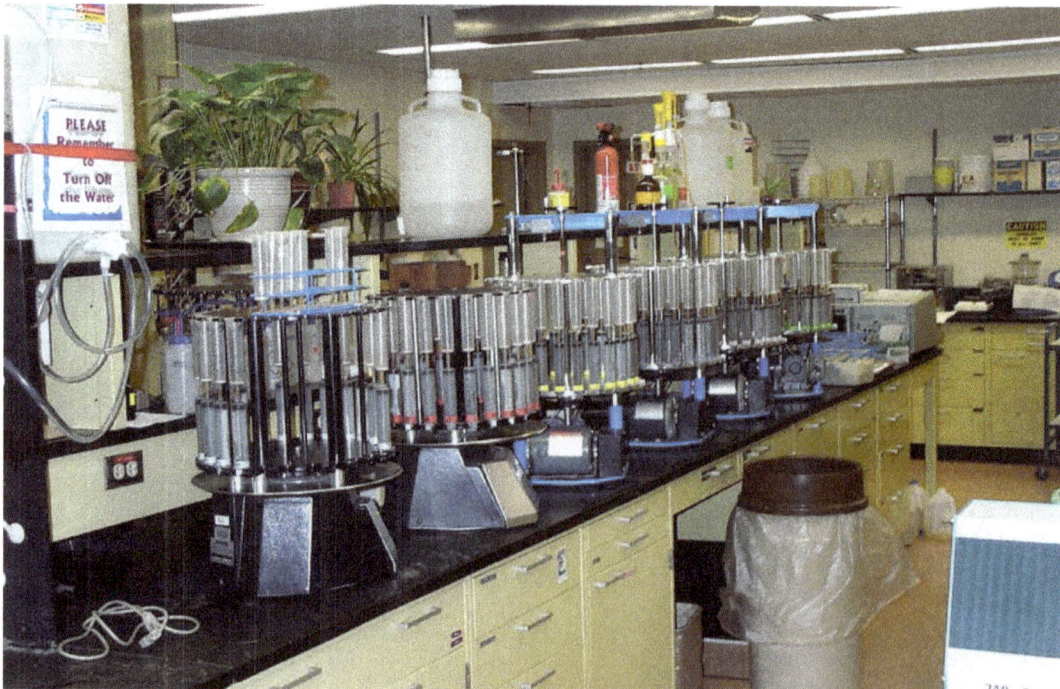

Fig. 2. Mechanical vacuum extractor used at the USDA-NRCS Soil Survey Laboratory
for chemical extractions and analyses, e.g., cation exchange capacity, NH_4OAc, pH 7.

Ion Exchange and Extractable Cations (4B)
Displacement after Washing, NH₄Cl (4B1b)
Automatic Extractor, 2 M KCl Rinse (4B1b1a)
Steam Distillation, HCl Titration (4B1b1a1a)
Cation Exchange Capacity (4B1b1a1a1)
Air-dry or Field-Moist, <2 mm (4B1b1a1a1a-b1)

1. Application
The cation exchange capacity (CEC) determined with a neutral cation unbuffered salt, e.g., 1 N NH₄Cl, is an estimate of the "effective" CEC (ECEC) of the soil (Peech et al., 1947). For a soil with a pH of <7.0, the ECEC values should be <CEC measured with a buffered solution at pH 7.0. The NH₄Cl CEC is approximately equal to the NH₄OAc extractable bases plus the KCl extractable Al for noncalcareous soils.

2. Summary of Method
Displacement after washing is the basis for this procedure. The CEC is determined by saturating the exchange sites with an index cation (NH₄⁺); washing the soil free of excess saturated salt; displacing the index cation (NH₄⁺) adsorbed by the soil; and measuring the amount of the index cation (NH₄⁺). A sample is leached using 1 N NH₄Cl and a mechanical vacuum extractor (Holmgren et al., 1977). The extract is weighed and saved for analyses of the cations. The NH₄⁺ saturated soil is rinsed with ethanol to remove the NH₄⁺ that was not adsorbed. The soil is then rinsed with 2 M KCl. This leachate is then analyzed by steam distillation and titration to determine the NH₄⁺ adsorbed on the soil exchange complex. The CEC by NH₄Cl is reported as meq 100 g⁻¹ or (cmol (+) kg⁻¹) soil in procedure 4B1b1a1a1.

3. Interferences
Incomplete saturation of the soil with NH₄⁺ and insufficient removal of NH₄⁺ are the greatest interferences to this method. Ethanol removes some adsorbed NH₄⁺ from the exchange sites of some soils. Isopropanol rinses have been used for some soils in which ethanol removes adsorbed NH₄⁺. Soils that contain large amounts of vermiculite can irreversibly "fix" NH₄⁺. Soils that contain large amounts of soluble carbonates can change the extractant pH and/or can contribute to erroneously high cation levels in the extract.

4. Safety
Wear protective clothing (coats, aprons, sleeve guards, and gloves) and eye protection (face shields, goggles, or safety glasses) when preparing reagents, especially concentrated acids and bases. Dispense concentrated acids and bases in a fume hood. Thoroughly wash hands after handling reagents. Use the safety showers and eyewash stations to dilute spilled acids and bases. Use sodium bicarbonate and water to neutralize and dilute spilled acids. Nessler's reagent contains mercury which is toxic. Proper disposal of the Nessler's reagent and clean-up of equipment in contact with the reagent is necessary.

Ethanol is flammable. Avoid open flames and sparks. Standard laboratory equipment includes fire blankets and extinguishers for use when necessary. Follow the manufacturer's safety precautions when using the vacuum extractor and the Kjeltec Auto Analyzers.

5. Equipment
5.1 Electronic balance, ±1.0-mg sensitivity
5.2 Mechanical Vacuum Extractor, 24-place, SAMPLETEX, MAVCO Industries, Lincoln, NE (See Fig. 1 and 2, procedure 4B1a1a1a1)
5.3 Tubes, 60-mL, polypropylene, for extraction (0.45-μm filter), reservoir, and tared extraction tubes
5.4 Rubber tubing, 3.2 ID x 1.6 OD x 6.4 mm (1/8 ID x 1/16 OD x 1 in) for connecting syringe barrels.
5.5 Kjeltec Auto 2300 Sampler System, Tecator, Perstorp Analytical

5.6 Digestion tubes, straight neck, 250 mL

5.7 Syringe filters, 0.45 μm, Whatman

5.8 Wash bottles

5.9 Vials, plastic

5.10 Centrifuge, Centra, GP-8, Thermo IEC, Needham Heights, MA

6. Reagents

6.1 Reverse osmosis deionized (RODI) water, ASTM Type I grade of reagent water

6.2 Ammonium chloride solution (NH_4Cl), 1 N. Dissolve 535 g of NH_4Cl reagent in RODI water and dilute to 10 L.

6.3 Ethanol (CH_3CH_2OH), 95%, U.S.P.

6.4 Nessler's reagent. Add 4.56 g of potassium iodide (KI) to 30 mL RODI water. Add 5.68 g of mercuric iodide (HgI_2). Stir until dissolved. Dissolve 10 g of sodium hydroxide (NaOH) in 200 mL of RODI water. Transfer NaOH solution to a 250-mL volumetric flask and slowly add K-Hg-I solution. Dilute to volume with RODI water and thoroughly mix. Solution should not contain a precipitate. Solution can be used immediately. Store in brown bottle to protect from light.

6.5 Potassium chloride solution, 2 N. Add 1341.9 g of KCl reagent in 8 L RODI water. Allow solution to equilibrate to room temperature. Dilute to 9 L with RODI water.

6.6 Boric acid, 4% (w:v), with bromcresol green-methyl red indicator (0.075 % bromcresol green and 0.05% methyl red), Chempure.

6.7 Hydrochloric acid (HCl), 0.1 N, standardized. Dilute 167 mL of concentrated HCl in 20 L of RODI water. Refer to procedure for Standardization of Acids.

6.8 NaOH, 1 M. Add 500 mL of 50% NaOH solution to 8 L of RODI water. Dilute to 9 L with RODI water.

7. Procedure

Extraction of Bases

7.1 Weigh 2.5 g of <2-mm, air-dry soil to the nearest mg and place in a labeled extraction tube (ET). If sample is fine-grind, weigh 1 g to the nearest mg. If sample is moist, weigh enough soil to achieve ≈ 2.5 or 1g, respectively, of air-dry soil. Prepare one quality control check sample per 24 samples.

7.2 Place labeled ET on extractor and connect to corresponding tared extraction tube (TET$_{NH4Cl}$) with rubber tubing.

7.3 Use wash bottle to rinse inside of ET with NH_4Cl. All soil should be wetted and no air bubbles. Shaking, swirling, or stirring may be required to wet organic samples. Fill ET to the 20-mL mark with NH_4Cl solution (≈ 10 mL).

7.4 Secure reservoir tube (RT) to top of ET tube and let stand for 30 min. Extract at 30-min rate the NH_4Cl solution until 2 mL of this solution remains above soil level. Turn off extractor. Do not let soil dry.

7.5 Add 40 mL of NH_4Cl solution to the RT. Set extractor for an overnight (12h) extraction. Extractor will turn off automatically.

7.6 Next day, remove RT from top of extractor and place in a clean container. Carefully remove TET$_{NH4Cl}$. Leave the rubber tubing on the ET. Weigh each TET$_{NH4Cl}$ containing the NH_4Cl extract to the nearest mg.

7.7 Mix the extract in each TET$_{NH4Cl}$ by manually shaking. Fill a labeled plastic vial with extract solution and cap. Discard the excess properly. The solution in the vial is reserved for analyses of extracted cations (procedure 4B1b1b1-4) on the atomic absorption (AA) spectrophotometer. Some samples may be cloudy and need to be filtered prior to analysis on the AA. If extracts are not to be determined immediately after collection, then store samples at 4°C in plastic tubes.

Removal of Excess Ammonium Chloride

7.8 Re-connect the TET_{NH4Cl} with paired ET. Use a wash bottle to rinse the sides of the ET with ethanol to remove any remaining NH_4Cl or soil particles adhering to the ET. All soil should be wetted and no air bubbles. Fill ET to the 20-mL mark with ethanol. Secure RT to top of ET and let stand for 30 min.

7.9 Extract at 30-min rate the ethanol solution until 2 mL of this solution remains above the soil level. Turn off the extractor. Do not let soil dry.

7.10 Add 45 mL of ethanol to the RT. Extract ($\approx$ 45 min) the ethanol until 2 mL of this solution remains above the soil level. Turn off the extractor. Do not let soil dry. Disconnect the TET_{NH4Cl} from the ET and discard the ethanol properly.

7.11 Re-connect the TET_{NH4Cl} to the ET and add 55 mL of ethanol to the RT. Set the extractor for 45 min. Turn off the extractor. Remove the TET_{NH4Cl}, leaving the tubing connected to the ET. Discard the ethanol properly.

7.12 After the final ethanol wash, collect a few drops of ethanol extract from the ET on a spot plate. Test for NH_4^+ by using Nessler's reagent. A yellow, red to reddish brown precipitate is a positive test. If the test is positive, repeat the ethanol wash and retest with Nessler's reagent. Repeat until a negative test is obtained.

2 *M* KCl Rinse

7.13 Connect a new labeled extraction tube (ET_{KCl}) with rubber tubing to ET on extractor.

7.14 Use a wash bottle to rinse inside of ET with 2 *M* KCl to remove any remaining ethanol or soil particles adhering to the ET. All soil should be wetted and no air bubbles. Fill ET to the 20-mL mark with KCl solution and let stand for 30 min.

7.15 Extract at the 30-min extraction rate the KCl solution until 2 mL of this solution remains above soil level. Turn off extractor. Do not let soil dry.

7.16 Secure RT to top of ET tube. Add 40 mL KCl solution to RT and set the extract for 45 min. Remove the ET and ET_{KCl} from the extractor.

Steam Distillation: Setup, Operation, and Analysis

7.17 Transfer the contents of the ET_{KCl} to a 250-mL digestion tube. If extracts are not to be determined immediately after collection, then store samples at 4°C.

7.18 Refer to the manufacturer's manual for operation of the distillation unit. The following are only very general guidelines for instrument conditions.

Program: Kjeldahl 1

Receiving Solution (boric acid): 30 mL
Water: 0 mL
Alkali (NaOH):20 mL
Mode: Delay
Time: 1 s
Distillation: Volume
Tube Drain: Yes

7.19 When using new reagents, e.g., boric acid, reagent blanks are distilled in 2 sets of 6, one set per Kjeltec machine. Each set of 6 is averaged and recorded on bench worksheet and manually set on each machine. During the steam distillation, the mean reagent blank titer is automatically subtracted from the sample titer.

7.20 Record the normality of standardized acid.

7.21 Connect the tube to the distillation unit. Close the safety door. Distillation and titration are performed automatically. Record the titer in mL of titrant.

8. Calculations

CEC = [Titer x N x 100 x R]/[Sample Weight (g)]

where:
CEC = Cation Exchange Capacity (meq 100 g^{-1})
Titer = Titer of sample (mL)
N = Normality of HCl titrant
100   = Conversion factor to 100-g basis
R   = Air-dry/oven-dry ratio (procedure 3D1) or field-moist/oven-dry ratio (procedure 3D2)

9. Report
Report neutral salts CEC to the nearest 0.1 meq 100 g^{-1} (cmol (+) kg^{-1}).

10. Precision and Accuracy
Precision and accuracy data are available from the SSL upon request.

11. References
Holmgren, G.G.S., R.L. Juve, and R.C. Geschwender. 1977. A mechanically controlled variable rate leaching device. Soil Sci. Am. J. 41:1207-1208.
Peech, M., L.T. Alexander, L.A. Dean, and J.F. Reed. 1947. Methods of soil analysis for soil fertility investigations. USDA. Agric. Circ. 757, 25 pp.

Ion Exchange and Extractable Cations (4B)
Displacement after Washing, NH₄OAc, pH 7 (4B1a)
Automatic Extractor (4B1a1)
Atomic Absorption Spectrophotometer (4B1a1b)
Calcium, Magnesium, Potassium, and Sodium (4B1a1b1-4)
Air-dry or Field-Moist, <2-mm (4B1a1b1-4a-b1)

1. Application
The extractable bases (Ca^{2+}, Mg^{2+}, K^+, and Na^+) from the NH_4OAc extraction (procedure 4B1a1) are generally assumed to be those exchangeable bases on the cation exchange sites of the soil. The term extractable rather than exchangeable bases is used because

any additional source of soluble bases influences the results. The abundance of these cations usually occurs in the sequence of $Ca^{2+} > Mg^{2+} > K^+ > Na^+$. Deviation from this usual order signals that some factor or factors, e.g., free $CaCO_3$ or gypsum, serpentine (high Mg^{2+}), or natric material (high Na^+), have altered the soil chemistry. The most doubtful cation extractions with this method are Ca^{2+} in the presence of free $CaCO_3$ or gypsum and K^+ in soils that are dominated by mica or vermiculite (Thomas, 1982).

2. Summary of Method

The NH_4OAc extract from procedure 4B1a1 is diluted with an ionization suppressant (La_2O_3). The analytes are measured by an atomic absorption spectrophotometer (AAS). The analyte is measured by absorption of the light from a hollow cathode lamp. An automatic sample changer is used to aspirate a series of samples. The AAS converts absorption to analyte concentration. Data are automatically recorded by a microcomputer and printer. The NH_4OAc extracted cations, Ca^{2+}, Mg^{2+}, K^+, and Na^+, are reported in meq 100 g^{-1} soil or (cmol (+) kg^{-1}) in procedures 4B1a1b1-4, respectively.

3. Interferences

There are four types of interferences (matrix, spectral, chemical, and ionization) in the analyses of these cations. These interferences vary in importance, depending upon the particular analyte selected. Do not use borosilicate tubes because of potential leaching of analytes.

4. Safety

Wear protective clothing and safety glasses. When preparing reagents, exercise special care. Restrict the use of concentrated HCl to a fume hood. Many metal salts are extremely toxic and may be fatal if ingested. Thoroughly wash hands after handling these metal salts.

Follow standard laboratory procedures when handling compressed gases. Gas cylinders should be chained or bolted in an upright position. Acetylene gas is highly flammable. Avoid open flames and sparks. Standard laboratory equipment includes fire blankets and extinguishers for use when necessary. Follow the manufacturer's safety precautions when using the AAS.

5. Equipment

5.1   Electronic balance, ±1.0-mg sensitivity
5.2   Atomic absorption spectrophotometer (AAS), double-beam optical system, AAnalyst, 300, Perkin-Elmer Corp., Norwalk, CT
5.3   Autosampler, AS-90, Perkin-Elmer Corp., Norwalk, CT
5.4   Computer, with AA WinLab software, Perkin-Elmer Corp., Norwalk, CT, and printer
5.5   Single-stage regulator, acetylene
5.6   Digital diluter/dispenser, with syringes 10000 and 1000 µL, gas tight, MicroLab 500, Hamilton Co., Reno, NV
5.7   Plastic test tubes, 15-mL, 16 mm x 100, for sample dilution and sample changer
5.8   Containers, polyethylene
5.9   Peristaltic pump

6. Reagents

6.1   Reverse osmosis deionized (RODI) water, ASTM Type I grade of reagent water
6.2   Hydrochloric acid (HCl), concentrated 12 N
6.3   HCl, 1:1 HCl:RODI, 6 N. Carefully mix 1 part of concentrated HCl to 1 part RODI water.
6.4 NH_4OH, reagent-grade, specific gravity 0.90
6.5   Glacial acetic acid, 99.5%
6.6   Ammonium acetate solution (NH_4OAc), 1 N, pH 7.0. Add 1026 mL of glacial acetic acid (CH_3COOH) to 15 L RODI water. Add 1224 mL of concentrated ammonium

hydroxide (NH_4OH). Cool. Allow to stand one day to equilibrate to room temperature. Mix and adjust to pH 7.0 with CH_3COOH (typically, ≈ 40 mL) or NH_4OH and dilute with RODI water to 18 L. The NH_4OAc solution is used for extraction of cations (procedure 4B1a1).

6.7 NH_4OAc solution, 2.0 N, pH 7.0. Mix 228 mL of glacial acetic acid in 1200 mL of RODI water. While stirring, carefully add 272 mL of concentrated NH_4OH. Cool. Allow to stand one day to equilibrate to room temperature. Mix and adjust pH 7.0 using CH_3COOH or NH_4OH. Dilute to 2 L with RODI water.

6.8 Stock lanthanum ionization suppressant solution (SLISS), 65,000 mg L^{-1}. Wet 152.4 g lanthanum oxide (La_2O_3) with 100 mL RODI water. Slowly and cautiously add 500 mL of 6 N HCl to dissolve the La_2O_3. Cooling the solution is necessary. Dilute to 2 L with RODI water. Filter solution. Store in polyethylene container.

6.9 Working lanthanum ionization suppressant solution (WLISS), 2000 mg L^{-1}. Dilute 61.5 mL of SLISS with 1800 mL of RODI water (1:10). Dilute to final volume of 2-L with RODI water. Store in polyethylene container.

6.10 Primary stock standards solutions (PSSS), high purity, 1000 mg L^{-1}: Ca, Mg, K, and Na.

6.11 Working stock mixed standards solution (WSMSS), High, Medium, Low, Very Low, and Blank. In five 500-mL volumetric flasks, add 250 mL of 2 N NH_4 OAc and the following designated amounts of Ca PSSS, Mg PSSS, K PSSS, and Na PSSS. Dilute to volume with RODI. Invert to thoroughly mix. Store in polyethylene containers. Prepare fresh weekly. Store in the refrigerator. Allow to equilibrate to room temperature before use. Prepare WSMSS as follows:

6.11.1 High Standard WSMSS: 90 mL Ca PSSS, 7.5 mL Mg PSSS, 20.0 mL K PSSS, and 100.0 mL Na PSSS = 180 mg L^{-1} Ca, 15 mg L^{-1} Mg, 40 mg L^{-1} K, and 200 mg L^{-1} Na

6.11.2 Medium Standard WSMSS: 60 mL Ca PSSS, 5.0 mL Mg PSSS, 10.0 mL K PSSS, and 50.0 mL Na PSSS = 120 mg L^{-1} Ca, 10 mg L^{-1} Mg, 20 mg L^{-1} K, and 100 mg L^{-1} Na

6.11.3 Low Standard WSMSS: 30 mL Ca PSSS, 2.5 mL Mg PSSS, 5.0 mL K PSSS, and 10.0 mL Na PSSS = 60 mg L^{-1} Ca, 5 mg L^{-1} Mg, 10 mg L^{-1} K, and 20 mg L^{-1} Na

6.11.4 Low/Low Standard WSMSS: 12.5 mL Ca PSSS, 0.25 mL Mg PSSS, 0.125 mL K PSSS, and 5.0 mL Na PSSS = 25 mg L^{-1} Ca, 0.5 mg L^{-1} Mg, 0.25 mg L^{-1} K, and 10 mg L^{-1} Na

6.11.5 Blank WSMSS = 0 mL of Ca, Mg, K, and Na PSSS.

6.12 Mixed calibration standard solutions (MCSS), High, Medium, Low, Very Low, and Blank. Dilute 1 part WSMSS with 19 parts of WLISS (1:20) dilution with resulting concentrations for MCSS as follows:

6.12.1 MCSS High Standard: 9.0 mg L^{-1} Ca, 0.75 mg L^{-1} Mg, 2.0 mg L^{-1} K, and 10.0 mg L^{-1} Na

6.12.2 MCSS Medium Standard: 6.0 mg L^{-1} Ca, 0.5 mg L^{-1} Mg, 1.0 mg L^{-1} K, and 5.0 mg L^{-1} Na

6.12.3 MCSS Low Standard: 3.0 mg L^{-1} Ca, 0.25 mg L^{-1} Mg, 0.5 mg L^{-1} K, and 1.0 mg L^{-1} Na

6.12.4 MCSS Very Low Standard: 1.25 mg L^{-1} Ca, 0.025 mg L^{-1} Mg, 0.125 K, and 0.5 mg L^{-1} Na

6.12.5 Blank = 0 mg L^{-1} Ca, Mg, K, and Na

6.13 Compressed air with water and oil traps.
6.14 Acetylene gas, purity 99.6%.

7. Procedure

Dilution of Calibration Standards and Sample Extracts
7.1 The 10-mL syringe is for diluent (WLISS). The 1-mL syringe is for the MCSS and NH4OAc extracts (procedure 4B1a1). Set the digital diluter at a 1:20 dilution. See reagents for preparation of the MCSS (High, Medium, Low, Very Low, Blank). Dilute 1 part NH4OAc sample extract with 19 parts of WLISS (1:20 dilution).

7.2 Dispense the diluted sample solutions into test tubes which have been placed in the sample holders of the sample changer.

AAS Set-up and Operation
7.3 Refer to the manufacturer's manual for operation of the AAS. The following are only very general guidelines for instrument conditions for the various analytes.

Analyte	Conc. (mg L^{-1})	Burner & angle	Wavelength (nm)	Slit (mm)	Fuel/Oxidant (C$_2$H$_2$/Air)
Ca	9.0	10 cm @ 0°	422.7	0.7	1.5/10.0
Mg	0.75	10 cm @ 0°	285.2	0.7	1.5/10.0
K	2.0	10 cm @ 0°	766.5	0.7	1.5/10.0
Na	10.0	10 cm @ 30°	589.0	0.2	1.5/10.0

7.4 Use the computer and printer to set instrument parameters and to collect and record instrument readings.

AAS Calibration and Analysis
7.5 Calibrate the instrument by using the MCSS (High, Medium, Low, Very Low, Blank). The data system will then associate the concentrations with the instrument responses for each MCSS. Rejection criteria for MCSS, if R^2 <0.99.

7.6 If sample exceeds calibration standard, the sample is diluted 1:5, 1:20, 1:100, etc., with 1 N NH4OAc followed by 1:20 dilution with WLISS.

7.7 Perform one quality control (QC) (Low Standard MCSS) every 12 samples. If reading is not within 10%, the instrument is re-calibrated and QC re-analyzed.

7.8 Record analyte readings to 0.01 unit.

8. Calculations

The instrument readings for analyte concentration are in mg L^{-1}. These analyte concentrations are converted to meq 100 g^{-1} as follows:

Soil Analyte Concentration (meq 100 g^{-1}) = [A x [(B$_1$ - B$_2$)/B$_3$] x C x R x 100]/[1000 x E x F]

where:
A   = Analyte (Ca, Mg, K, Na) concentration in extract (mg L^{-1})
B$_1$ = Weight of extraction syringe and extract (g)
B$_2$ = Weight of tared extraction syringe (g)
B$_3$ = Density of 1 N NH4OAc at 20°C (1.0124 g cm^{-3})

C = Dilution, if performed
100 = Conversion factor (100-g basis)
R = Air-dry/oven-dry ratio (procedure 3D1) or field-moist/oven-dry ratio (procedure 3D2)
1000 = mL L^{-1}
E = Soil sample weight (g)
F = Equivalent weight (mg meq^{-1})

where:
Ca^{+2} = 20.04 mg meq^{-1}
Mg^{+2} = 12.15 mg meq^{-1}
Na^{+1} = 22.99 mg meq^{-1}
K^{+1} = 39.10 mg meq^{-1}

9. Report

Report the extractable Ca^{2+}, Mg^{2+}, Na^+, and K^+ to the nearest 0.1 meq 100 g^{-1} (cmol (+) kg^{-1}).

10. Precision and Accuracy

Precision and accuracy data are available from the SSL upon request.

11. References

Thomas, G.W. 1982. Exchangeable cations. p. 159-165. *In* A.L. Page, R.H. Miller, and D.R. Keeney (eds.) Methods of soil analysis. Part 2. Chemical and microbiological properties. 2nd ed. Agron. Monogr. 9. ASA and SSSA, Madison, WI.

Ion Exchange and Extractable Cations (4B)
Displacement after Washing, NH₄Cl (4B1b)
Automatic Extractor (4B1ba)
Atomic Absorption Spectrophotometer (4B1b1b)
Calcium, Magnesium, Potassium, and Sodium (4B1b1b1-4)
Air-dry or Field-Moist, <2-mm (4B1b1b1-4a-b1)

1. Application

The extractable bases (Ca^{2+}, Mg^{2+}, Na^+, and K^+) from the NH₄Cl extraction (procedure 4B1b1) are generally assumed to be those exchangeable bases on the cation exchange sites of the soil. The abundance of these cations usually occurs in the sequence of $Ca^{2+} > Mg^{2+} > K^+ > Na^+$. Deviation from this usual order signals that some factor or factors, e.g., free $CaCO_3$ or gypsum, serpentine (high Mg^{2+}), or natric material (high Na^+), have altered the soil chemistry. The most doubtful cation extractions with this method are Ca^{2+} in the presence of free $CaCO_3$ or gypsum and K^+ in soils that are dominated by mica or vermiculite (Thomas, 1982).

2. Summary of Method

The NH₄Cl extract from procedure 4B1b1 is diluted with an ionization suppressant (La_2O_3). The analytes are measured by an atomic absorption spectrophotometer (AAS). The analyte is measured by absorption of the light from a hollow cathode lamp. An automatic sample changer is used to aspirate a series of samples. The AAS converts absorption to analyte concentration. Data are automatically recorded by a microcomputer and printer. The NH₄Cl extracted cations, Ca^{2+}, Mg^{2+}, K^+, and Na^+, are reported in meq 100 g^{-1} soil or (cmol (+) kg^{-1}) in procedures 4B1b1b1-4, respectively.

3. Interferences

There are four types of interferences (matrix, spectral, chemical, and ionization) in the analyses of these cations. These interferences vary in importance, depending upon the particular analyte selected. Do not use borosilicate tubes because of potential leaching of analytes.

4. Safety

Wear protective clothing and safety glasses. When preparing reagents, exercise special care. Restrict the use of concentrated HCl to a fume hood. Many metal salts are extremely toxic and may be fatal if ingested. Thoroughly wash hands after handling these metal salts.

Follow standard laboratory procedures when handling compressed gases. Gas cylinders should be chained or bolted in an upright position. Acetylene gas is highly flammable. Avoid open flames and sparks. Standard laboratory equipment includes fire blankets and extinguishers for use when necessary. Follow the manufacturer's safety precautions when using the AAS.

5. Equipment

5.1 Electronic balance, ±1.0-mg sensitivity
5.2 Atomic absorption spectrophotometer (AAS), double-beam optical system, AAnalyst, 300, Perkin-Elmer Corp., Norwalk, CT
5.3 Autosampler, AS-90, Perkin-Elmer Corp., Norwalk, CT
5.4 Computer, with AA WinLab software, Perkin-Elmer Corp., Norwalk, CT, and printer
5.5 Single-stage regulator, acetylene
5.6 Digital diluter/dispenser, with syringes 10000 and 1000 µL, gas tight, MicroLab 500, Hamilton Co., Reno, NV
5.7 Plastic test tubes, 15-mL, 16 mm x 100, for sample dilution and sample changer
5.8 Containers, polyethylene
5.9 Peristaltic pump

6. Reagents

6.1 Reverse osmosis deionized (RODI) water, ASTM Type I grade of reagent water
6.2 Hydrochloric acid (HCl), concentrated 12 N
6.3 HCl, 1:1 HCl:RODI, 6 N. Carefully mix 1 part of concentrated HCl to 1 part RODI water.
6.4 Ammonium chloride solution (NH_4Cl), 1 N. Dissolve 535 g of NH_4Cl reagent in RODI water and dilute to 10 L.
6.5 Stock lanthanum ionization suppressant solution (SLISS), 65,000 mg L^{-1}. Wet 152.4 g lanthanum oxide (La_2O_3) with 100 mL RODI water. Slowly and cautiously add 500 mL of 6 N HCl to dissolve the La_2O_3. Cooling the solution is necessary. Dilute to 2 L with RODI water. Filter solution. Store in polyethylene container.
6.6 Working lanthanum ionization suppressant solution (WLISS), 2000 mg L^{-1}. Dilute 61.5 mL of SLISS with 1800 mL of RODI water (1:10). Dilute to final volume of 2-L with RODI water. Store in polyethylene container.
6.7 Primary stock standards solutions (PSSS), high purity, 1000 mg L^{-1}: Ca, Mg, K, and Na.
6.8 Working stock mixed standards solution (WSMSS), High, Medium, Low, Very Low, and Blank. In five 500-mL volumetric flasks, add 250 mL of 2 N NH_4Cl and the following designated amounts of Ca PSSS, Mg PSSS, K PSSS, and Na PSSS. Dilute to volume with RODI. Invert to thoroughly mix. Store in polyethylene containers. Prepare fresh weekly. Store in the refrigerator. Allow to equilibrate to room temperature before use. Prepare WSMSS as follows:

6.8.1 High Standard WSMSS: 90 mL Ca PSSS, 7.5 mL Mg PSSS, 20.0 mL K PSSS, and 100.0 mL Na PSSS = 180 mg L^{-1} Ca, 15 mg L^{-1} Mg, 40 mg L^{-1} K, and 200 mg L^{-1} Na

6.8.2 Medium Standard WSMSS: 60 mL Ca PSSS, 5.0 mL Mg PSSS, 10.0 mL K PSSS, and 50.0 mL Na PSSS = 120 mg L^{-1} Ca, 10 mg L^{-1} Mg, 20 mg L^{-1} K, and 100 mg L^{-1} Na

6.8.3 Low Standard WSMSS: 30 mL Ca PSSS, 2.5 mL Mg PSSS, 5.0 mL K PSSS, and 10.0 mL Na PSSS = 60 mg L^{-1} Ca, 5 mg L^{-1} Mg, 10 mg L^{-1} K, and 20 mg L^{-1} Na

6.8.4 Low/Low Standard WSMSS: 12.5 mL Ca PSSS, 0.25 mL Mg PSSS, 0.125 mL K PSSS, and 5.0 mL Na PSSS = 25 mg L^{-1} Ca, 0.5 mg L^{-1} Mg, 0.25 mg L^{-1} K, and 10 mg L^{-1} Na

6.8.5 Blank WSMSS = 0 mL of Ca, Mg, K, and Na PSSS

6.9 Mixed calibration standard solutions (MCSS), High, Medium, Low, Very Low, and Blank. Dilute 1 part WSMSS with 19 parts of WLISS (1:20) dilution with resulting concentrations for MCSS as follows:

6.9.1 MCSS High Standard: 9.0 mg L^{-1} Ca, 0.75 mg L^{-1} Mg, 2.0 mg L^{-1} K, and 10.0 mg L^{-1} Na

6.9.2 MCSS Medium Standard: 6.0 mg L^{-1} Ca, 0.5 mg L^{-1} Mg, 1.0 mg L^{-1} K, and 5.0 mg L^{-1} Na

6.9.3 MCSS Low Standard: 3.0 mg L^{-1} Ca, 0.25 mg L^{-1} Mg, 0.5 mg L^{-1} K, and 1.0 mg L^{-1} Na

6.9.4 MCSS Very Low Standard: 1.25 mg L^{-1} Ca, 0.025 mg L^{-1} Mg, 0.125 K, and 0.5 mg L^{-1} Na

6.9.5 Blank = 0 mg L^{-1} Ca, Mg, K, and Na

6.10 Compressed air with water and oil traps.
6.11 Acetylene gas, purity 99.6%.

7. Procedure

Dilution of Calibration Standards and Sample Extracts

7.1 The 10-mL syringe is for diluent (WLISS). The 1-mL syringe is for the MCSS and NH_4Cl extracts (procedure 4B1b). Set the digital diluter at a 1:20 dilution. See reagents for preparation of the MCSS (High, Medium, Low, Very Low, Blank). Dilute 1 part NH_4Cl sample extract with 19 parts of WLISS (1:20 dilution).

7.2 Dispense the diluted sample solutions into test tubes which have been placed in the sample holders of the sample changer.

AAS Set-up and Operation

7.3 Refer to the manufacturer's manual for operation of the AAS. The following are only very general guidelines for instrument conditions for the various analytes.

Analyte	Conc. (mg L^{-1})	Burner & angle	Wavelength (nm)	Slit (mm)	Fuel/Oxidant (C_2H_2/Air)
Ca	9.0	10 cm @ 0°	422.7	0.7	1.5/10.0
Mg	0.75	10 cm @ 0°	285.2	0.7	1.5/10.0
K	2.0	10 cm @ 0°	766.5	0.7	1.5/10.0
Na	10.0	10 cm @ 30°	589.0	0.2	1.5/10.0

7.4 Use the computer and printer to set instrument parameters and to collect and record instrument readings.

AAS Calibration and Analysis

7.5 Calibrate the instrument by using the MCSS (High, Medium, Low, Very Low, Blank). The data system will then associate the concentrations with the instrument responses for each MCSS. Rejection criteria for MCSS, if R^2 <0.99.

7.6 If sample exceeds calibration standard, the sample is diluted 1:5, 1:20, 1:100, etc., with 1 N NH_4Cl followed by 1:20 dilution with WLISS.

7.7 Perform one quality control (QC) (Low Standard MCSS) every 12 samples. If reading is not within 10%, the instrument is re-calibrated and QC re-analyzed.

7.8 Record analyte readings to 0.01 unit.

8. Calculations

The instrument readings for analyte concentration are in mg L^{-1}. These analyte concentrations are converted to meq 100 g^{-1} as follows:

Soil Analyte Concentration (meq 100 g^{-1}) = [A x [(B_1 - B_2)/B_3] x C x R x 100]/[1000 x E x F]

where:

A = Analyte (Ca, Mg, K, Na) concentration in extract (mg L^{-1})
B_1 = Weight of extraction syringe and extract (g)
B_2 = Weight of tared extraction syringe (g)
B_3 = Density of 1 N NH_4Cl at 20°C (1.0166 g cm^{-3})
C = Dilution, if performed
100 = Conversion factor (100-g basis)
R = Air-dry/oven-dry ratio (procedure 3D1) or field-moist/oven-dry ratio (procedure 3D2)
1000 = mL L^{-1}
E = Soil sample weight (g)
F = Equivalent weight (mg meq^{-1})

where:
Ca^{+2} = 20.04 mg meq^{-1}
Mg^{+2} = 12.15 mg meq^{-1}
Na^{+1} = 22.99 mg meq^{-1}
K^{+1} = 39.10 mg meq^{-1}

9. Report

Report the extractable Ca^{2+}, Mg^{2+}, K^+, and Na^+ to the nearest 0.1 meq 100 g^{-1} (cmol (+) kg^{-1}).

10. Precision and Accuracy

Precision and accuracy data are available from the SSL upon request.

11. References

Thomas, G.W. 1982. Exchangeable cations. p. 159-165. *In* A.L. Page, R.H. Miller, and D.R. Keeney (eds.) Methods of soil analysis. Part 2. Chemical and microbiological properties. 2nd ed. Agron. Monogr. 9. ASA and SSSA, Madison, WI.

BaCl$_2$-Triethanolamine, pH 8.2 Extraction (4B2)
Automatic Extractor (4B2a)
Automatic Titrator (4B2a1)
Back Titration with HCl (4B2a1a)
Extractable Acidity (4B2a1a1)
Air-Dry or Field-Moist, <2 mm (4B2a1a1a-b1)

1. Application

The extractable acidity is the acidity released from the soil by a barium chloride-triethanolamine (BaCl$_2$-TEA) solution buffered at pH 8.2 and includes all the acidity generated by replacement of the H and Al from permanent and pH dependent exchange sites. Extractable acidity may be measured at any pH, and a variety of methods have been used to measure it. The Soil Conservation Service adopted a pH of 8.2 because it approximates the calculated pH of a soil containing free CaCO$_3$ in equilibrium with the normal CO$_2$ content (0.03%) of the atmosphere. A pH of 8.2 also closely corresponds to the pH of complete neutralization of soil hydroxy-Al compounds. Although other pH values are valid for some types of soils, and the BaCl$_2$-TEA, pH 8.2 method (4B2a1a1) may not always accurately reflect the nature of soils as they occur in the environment, this method has become a standard reference to which other methods are compared.

2. Summary of Method

A soil sample is leached with a BaCl$_2$-TEA solution buffered at pH 8.2. Sample is allowed to stand overnight and extracted using a mechanical vacuum extractor (Holmgren et al., 1977). The extract is back-titrated with HCl. The difference between a blank and the extract is the extractable acidity. Extractable acidity is reported in meq 100 g^{-1} soil or (cmol (+) kg^{-1}).

3. Interferences

No significant interferences are known to exist with this method. However, for some very acid soils, the buffer capacity of the BaCl$_2$-TEA solution may be exceeded.

4. Safety

Wear protective clothing (coats, aprons, sleeve guards, and gloves) and eye protection (face shields, goggles, or safety glasses) when preparing reagents, especially concentrated acids and bases. Dispense concentrated acids in a fume hood. Thoroughly wash hands after handling reagents. Use the safety showers and eyewash stations to dilute spilled acids and bases. Use sodium bicarbonate and water to neutralize and dilute spilled acids.

5. Equipment

5.1 Electronic balance, ±1.0-mg sensitivity
5.2 Mechanical vacuum extractor, 24 place, SAMPLETEX, MAVCO Industries, Lincoln, NE
5.3 Pipettes or dispenser, adjustable volume to 20 mL
5.4 Titration beakers, 250-mL, plastic, Metrohm Ltd., Brinkmann Instruments Inc.
5.5 Automatic titrator, with control unit, sample changer, and dispenser, Metrohm Ltd., Brinkmann Instruments, Inc.
5.6 Combination pH-reference electrode, Metrohm Ltd., Brinkmann Instruments, Inc.
5.7 Computer, with Titrino Workcell software, Metrohom Ltd., Brinkmann Instruments, Inc., and printer
5.8 Titration beakers, 250-mL, plastic, Metrohm Ltd., Brinkmann Instruments Inc.
5.9 Tubes, 60-mL, polypropylene, for extraction, with 0.45-μm filter
5.10 Rubber tubing, 3.2 ID x 6.4 OD x 25.4 mm, (1/8 ID x 1/16 OD x 1 in) for connecting syringe barrels

6. Reagents

6.1 Reverse osmosis deionized (RODI) water

6.2 Reverse osmosis (RO) water

6.3 Hydrochloric acid (HCl), concentrated, 12 N

6.4 HCl, 0.13 N, standardized. Dilute 193 mL of concentrated HCl to 16-L volume with RODI water.

6.5 Buffer solution [0.5 N BaCl$_2$, 0.2 N Triethanolamine (TEA), pH 8.2]. Dissolve 977 g of BaCl$_2$·2H$_2$O in 8 L of RODI water. Dissolve 477 g of TEA in 4 L of RODI water. Mix two solutions and bring to nearly 16-L volume with RODI water. Adjust to pH 8.2 with ≈ 33 mL of concentrated HCl or barium hydroxide. Bring to 16-L volume with RODI water.

6.6 Replacement solution. Dissolve 977 g of BaCl$_2$·2H$_2$O in 8 L of RODI water. Add 80 mL of buffer solution and dilute 16-L volume with RODI water.

7. Procedure

Extraction of Acidity

7.1 Weigh 2.5 g of <2-mm or fine-grind, air-dry soil to the nearest mg and place in a labeled extraction (ET$_1$). If sample is moist, weigh enough soil to achieve ≈ 2.5 g of air-dry soil. Prepare at least one reagent blank (no sample in syringe) and one quality control check sample per 24 samples.

7.2 Place labeled ET on extractor and connect to corresponding extraction tube (ET$_{Acidity}$) with rubber tubing.

7.3 Use a dispenser to add 20.00 mL of BaCl$_2$-TEA solution to the ET$_1$. During the addition, wash the sides of the tube and wet the sample. For organic soils, shaking, swirling, or stirring may be required to wet the sample.

7.4 Let ET$_1$ tube stand overnight.

7.5 Set the extractor for a 30-min extraction rate. Extract solution to a 0.5- to 1.0-cm height above the sample. Turn off the extractor. Do not allow the sample to become dry.

7.6 Use a dispenser to add 20.00 mL of replacement solution to ET$_1$. Extract the sample at 30-min rate, pulling the solution almost completely through the sample.

7.7 Add a second 20.00-mL aliquot of replacement solution to ET$_1$. Extract at 30-min rate until all the solution has been drawn through the sample.

7.8 Carefully remove ET$_{Acidity}$. Leave rubber tubing on the ET$_1$.

Titration of BaCl$_2$-TEA Extract

7.9 Transfer the BaCl$_2$-TEA extract from the ET$_{Acidity}$ to a 250-mL polyethylene titration beaker.

7.10 Add 100 mL of RO water to the beaker. The solution is ready to be titrated.

7.11 Refer to manufacturer's manual for operation of the automatic titrator.

7.12 Calibrate automatic titrator with with 9.18, 7.00 and 4.00 pH buffers. Set-up the automatic titrator to set end point titration mode. The "Set" pH parameters are listed as follows:

Parameter	Value
Ep_1	pH 4.60
Dyn change pH	1.5 units
Drift	0.4 mV s^{-1}
Time delay	10 s
Drift	0.4 mV s^{-1}
Temp	25°C
Stop Volume	75 mL

7.13 If pre-titration pH is 0.3 units lower than the average pH of the blanks, re-run using a 0.25 g sample

7.14 Record the titer to the nearest 0.01 mL. Record the normality of the HCl solution. Average the titer of the reagent blanks and record.

8. Calculations

Extractable acidity (meq 100 g^{-1}) = $\{[(B - T) \times N \times R]/C\} \times 100$

where:
B = Average reagent blank titer (mL)
T = Sample titer (mL)
N = Normality of HCl
C = Sample Weight (g)
100 = Conversion factor (100-g basis)
R = Air-dry/oven-dry ratio (procedure 3D1) or field-moist/oven-dry ratio (procedure 3D2)

9. Report
Report extractable acidity to the nearest 0.1 meq 100 g^{-1} (cmol (+) kg^{-1}).

10. Precision and Accuracy
Precision and accuracy data are available from the SSL upon request.

11. References
Holmgren, G.G.S., R.L. Juve, and R.C. Geschwender. 1977. A mechanically controlled variable rate leaching device. Soil Sci. Am. J. 41:1207-1208.

$BaCl_2$-Triethanolamine, pH 8.2 Extraction (4B2)
Centrifuge (4B2b)
Automatic Titrator (4B2b1)
Back Titration with HCl (4B2b1a)
Extractable Acidity (4B2b1a1)
Air-Dry or Field-Moist, <2 mm (4B2b1a1a-b1)

1. Application
The extractable acidity is the acidity released from the soil by a barium chloride-triethanolamine ($BaCl_2$-TEA) solution buffered at pH 8.2 and includes all the acidity generated by replacement of the H and Al from permanent and pH dependent exchange sites. Extractable acidity may be measured at any pH, and a variety of methods have been used to measure it. The Soil Conservation Service adopted a pH of 8.2 because it approximates the calculated pH of a soil containing free $CaCO_3$ in equilibrium with the normal CO_2 content (0.03%) of the atmosphere. A pH of 8.2 also closely corresponds to the pH of complete neutralization of soil hydroxy-Al compounds. Although other pH values are valid for some types of soils, and the

BaCl$_2$-TEA, pH 8.2 method (4B2b1a1) may not always accurately reflect the nature of soils as they occur in the environment, this method has become a standard reference to which other methods are compared.

2. Summary of Method

A soil sample is leached with a BaCl$_2$-TEA solution buffered at pH 8.2. Sample is allowed to stand overnight, shaken, and centrifuged. The extract is back-titrated with HCl. The difference between a blank and the extract is the extractable acidity. Extractable acidity is reported in meq 100 g^{-1} soil or (cmol (+) kg^{-1}).

3. Interferences

No significant interferences are known to exist with this method. However, for some very acid soils, the buffer capacity of the BaCl$_2$-TEA solution may be exceeded.

4. Safety

Wear protective clothing (coats, aprons, sleeve guards, and gloves) and eye protection (face shields, goggles, or safety glasses) when preparing reagents, especially concentrated acids and bases. Dispense concentrated acids in a fume hood. Thoroughly wash hands after handling reagents. Use the safety showers and eyewash stations to dilute spilled acids and bases. Use sodium bicarbonate and water to neutralize and dilute spilled acids.

5. Equipment

5.1 Electronic balance, ±1.0-mg sensitivity
5.2 Pipettes or dispenser, adjustable volume to 40 mL
5.3 Vortexer, mini, Analog, VRW Scientific Products
5.4 Centrifuge tubes, 50-mL, polyethylene
5.5 Centrifuge, Centra, GP-8, Thermo IEC, Needham Heights, MA
5.6 Titration beakers, 250-mL, plastic, Metrohm Ltd., Brinkmann Instruments Inc.
5.7 Automatic titrator, with control unit, sample changer, and dispenser, Metrohm Ltd., Brinkmann Instruments, Inc.
5.8 Combination pH-reference electrode, Metrohm Ltd., Brinkmann Instruments, Inc.
5.9 Computer, with Titrino Workcell software, Metrohom Ltd., Brinkmann Instruments, Inc., and printer

6. Reagents

6.1 Reverse osmosis deionized (RODI) water
6.2 Hydrochloric acid (HCl), concentrated, 12 N
6.3 HCl, 0.13 N, standardized. Dilute 193 mL of concentrated HCl to 16-L volume with RODI water.
6.4 Buffer solution (0.5 N BaCl$_2$, 0.2 N Triethanolamine (TEA), pH 8.2). Dissolve 977 g of BaCl$_2$·2H$_2$O in 8 L of RODI water. Dissolve 477 g of TEA in 4 L of RODI water. Mix two solutions and bring to nearly 16-L volume with RODI water. Adjust to pH 8.2 with ≈ 33 mL of concentrated HCl or barium hydroxide. Bring to 16-L volume with RODI water.
6.5 Replacement solution. Dissolve 977 g of BaCl$_2$·2H$_2$O in 8 L of RODI water. Add 80 mL of buffer solution and dilute to 16-L volume with RODI water.

7. Procedure

Extraction of Acidity

7.1 Weigh 5 g of <2-mm or fine-grind, air-dry soil to the nearest mg and place in a centrifuge tube. If sample is moist, weigh enough soil to achieve ≈ 5 g of air-dry soil. Prepare at least two reagent blanks (no sample in tube) and one quality control check sample per 21 samples.

7.2 Add 40.00 mL of BaCl$_2$-TEA solution to sample. Cap the tube and shake to ensure all soil is wetted. Place tube in a rack.

7.3 Place tube rack on its side and gently shake to stratify the mixture lengthwise along the tube. Allow to stand overnight on its side.

7.4 Centrifuge sample at 2000 rpm for 5 min.

7.5 Decant extract into numbered titration beakers.

7.6 Add 40 mL of replacement solution to sample.

7.7 Cap tube and use a vortexer to loosen soil. Manually shake.

7.8 Repeat Sections 7.4 – 7.7.

7.9 Repeat Sections 7.4 – 7.5. Total volume in titration beaker should be ≈ 120 mL.

Titration of BaCl$_2$-TEA Extract

7.10 Place titration beakers on automatic sample changer.

7.11 Refer to the manufacturer's manual for operation of the automatic titrator.

7.12 Calibrate the titrator meter with 9.18, 7.00 and 4.00 pH buffers. Set-up the automatic titrator to sent end point mode. The "Set" pH parameters are listed as follows:

Parameter	Value
Ep$_1$	pH 4.60
Dyn change pH	1.5 units
Drift	0.4 mV s^{-1}
Time delay	10 s
Drift	0.4 mV s^{-1}
Temp	25°C
Stop Volume	75 mL

7.13 If pre-titration pH is 0.3 units lower than the average pH of the blanks, re-run using a 0.5-g sample.

7.14 Record the titer to the nearest 0.01 mL. Record the normality of the HCl solution. Average the titer of the reagent blanks and record.

8. Calculations

Extractable acidity (meq 100 g^{-1}) = {[(B – T) x N x R]/C} x 100

where:
B = Average reagent blank titer (mL)
T = Sample titer (mL)
N = Normality of HCl
C = Sample Weight (g)
100 = Conversion factor (100-g basis)
R = Air-dry/oven-dry ratio (procedure 3D1) or field-moist/oven-dry ratio (procedure 3D2)

9. Report

Report extractable acidity to the nearest 0.1 meq 100 g^{-1} (cmol (+) kg^{-1}).

10. Precision and Accuracy

Precision and accuracy data are available from the SSL upon request.

11. References

Holmgren, G.G.S., R.L. Juve, and R.C. Geschwender. 1977. A mechanically controlled variable rate leaching device. Soil Sci. Am. J. 41:1207-1208.

1 N KCl Extraction (4B3)
Automatic Extractor (4B3a)
Inductively Coupled Plasma Atomic Emission Spectrophotometer (4B3a1)
Radial Mode (4B3a1a)
Al and Mn (4B3a1a1-2)
Air-Dry or Field-Moist, <2 mm (4B3a1a1-2a-b1)

1. Application

The Al extracted by 1 N KCl approximates exchangeable Al and is a measure of the "active" acidity present in soils with a 1:1 water pH <5.5. Above pH 5.5, precipitation of Al occurs during analysis. This method does not measure the acidity component of hydronium ions (H_3O^+). If Al is present in measurable amounts, the hydronium is a minor component of the active acidity. Because the 1 N KCl extractant is an unbuffered salt and usually affects the soil pH one unit or less, the extraction is determined at or near the soil pH. The KCl extractable Al is related to the immediate lime requirement and existing CEC of the soil. The "potential" acidity is better measured by the BaCl$_2$-TEA method (procedure 4B2a1a1) (Thomas, 1982). The use of NH$_4$Cl in place of KCl is useful where a single exttractant for exchangeable bases and Al is preferred since NH$_4^+$ is as effective as K at displacing Al (Lee et al., 1985: Bertsch and Bloom, 1996). The Mn extracted by 1 N KCl approximates exchangeable Mn. Mn is an essential trace metal for plant nutrition. Soil analysis for Mn is of interest from both deficiency and toxicity perspectives (Gambrell, 1996).

2. Summary of Method

In this procedure (4B3a1a1-2), a soil sample is leached with 1 N KCl using the mechanical vacuum extractor (Holmgren et al., 1977). The extract is weighed. The KCl extracted solution is diluted with 0.5 N HCl. The analytes are measured by inductively coupled plasma atomic emission spectrophotometer (ICP-AES). The Mn and Al are reported in mg kg^{-1} and cmol(+) kg^{-1}, respectively, in the soil.

3. Interferences

There are four types of interferences (matrix, spectral, chemical, and ionization) in the ICP-AES analyses of these cations. These interferences vary in importance, depending upon the particular analyte selected.

The soil:extractant ratio must remain constant. A soil:extractant ratio of 1:10 (w:v) for batch procedures is most commonly used. Using a leaching technique, a 1:20 (w:v) ratio gives comparable results. If the sample size is changed, the amount of extractable Al is changed. No other significant interferences have been identified for this procedure.

4. Safety

Wear protective clothing and eye protection. When preparing reagents, exercise special care. Follow standard laboratory practices when handling compressed gases. Gas cylinders should be chained or bolted in an upright position. Follow the manufacturer's safety precautions when using the ICP-AES.

5. Equipment
5.1 Electronic balance, ±1.0-mg sensitivity
5.2 Mechanical vacuum extractor, 24-place, SAMPLETEX, MAVCO Industries, Lincoln, NE (See Fig.1 - 2, procedure 4B1a1a1a1)
5.3 Tubes, 60-mL, polypropylene, for extraction (0.45-μm filter), reservoir, and tared extraction tubes
5.4 Rubber tubing, 3.2 ID x 1.6 OD x 6.4 mm (1/8 ID x 1/16 OD x 1 in) for connecting syringe barrels.
5.5 Dispenser, 10 mL
5.6 Inductively coupled plasma atomic emission spectrophotometer (ICP-AES), Perkin-Elmer Optima 4300 Dual View (DV), Perkin-Elmer Corp., Norwalk, CT.
5.7 Scott spray chamber with end cap and gem cone x-flow nebulizer
5.8 Torch coupler at -3 position
5.9 RF generator, floor mounted power unit, 45 MHz free running, Perkin-Elmer Corp.,Norwalk, CT.
5.10 Computer, with WinLab software ver. 4.1, Perkin-Elmer Corp., Norwalk, CT, and printer
5.11 Recirculating chiller, Neslab, CFT Series
5.12 Autosampler, AS-93, Perkin-Elmer Corp., Norwalk, CT.
5.13 Single-stage regulator, high-purity, high-flow, argon
5.14 Pipettes, electronic digital, 10000 and 1000 μL, with tips 10000 and 1000 μL
5.15 Test tubes, 15-mL, 16 mm x 100, for sample dilution and sample changer
5.16 Containers, polyethylene
5.17 Vortexer, mini, MV1, VWR Scientific Products
5.18 Disposable tubes, glass, 10 mL

6. Reagents
6.1 Reverse osmosis deionized (RODI) water, ASTM Type I grade of reagent water
6.2 Potassium chloride solution (KCl), 1.0 N. Dissolve 1341.9 g of KCl reagent in 16 L RODI water. Allow solution to equilibrate to room temperature. Dilute to 18 L with RODI water. Use 1.0 N KCl for Al and Mn extraction.
6.3 Potassium chloride solution (KCl), 2.0 N. Dissolve 298.2 g of KCl reagent in 1.5 L RODI water. Allow solution to equilibrate to room temperature. Dilute to 2 L with RODI water. Use 2.0 N KCl for standards.
6.4 HCl, 0.5 N. In a 1-L volumetric, add 41.67 mL concentrated HCl 12 N to RODI water and dilute to volume with RODI water.
6.5 Primary Stock Standard Solution (PSSS), high purity, 1000 mg L^{-1}: Al and Mn.
6.6 Mixed calibration standards solution (MCSS) for Al and Mn as follows:

6.6.1 MCSS High: In 1-L volumetric flask, mix 40 mL Al PSSS, 5 mL Mn PSSS, 14.91 g KCl, and 33.3 mL concentrated HCl and dilute to volume with RODI water. Invert to thoroughly mix. Final concentration = 40 and 5 mg L^{-1} Al and Mn, respectively. Store in polyethylene containers. Prepare fresh weekly. Store in the refrigerator. Allow to equilibrate to room temperature before use.

6.6.2 MCSS Medium: In 1-L volumetric flask, mix 20 mL Al PSSS, 2 mL Mn PSSS, 14.91 g KCl, and 33.3 mL concentrated HCl and dilute to volume with RODI water. Invert to thoroughly mix. Final concentration = 20 and 2 mg L^{-1} Al and Mn, respectively. Store in polyethylene containers. Prepare fresh weekly. Store in the refrigerator. Allow to equilibrate to room temperature before use.

6.6.3 MCSS Low: In 1-L volumetric flask, mix 10 mL Al PSSS, 1mL Mn PSSS, 14.91 g KCl and 33.3 mL concentrated HCl and dilute to volume with RODI water. Invert to thoroughly mix. Final concentration = 10 and 1 mg L^{-1} Al and Mn, respectively. Store in polyethylene containers. Prepare fresh weekly. Store in the refrigerator. Allow to equilibrate to room temperature before use.

6.6.4 MCSS Blank: In 1-L volumetric flask, mix 500 RODI water, 33.3 mL concentrated HCl, and 14.91 g KCl and dilute to volume with RODI water. Invert to thoroughly mix. Store in polyethylene containers. Prepare fresh weekly. Store in the refrigerator. Allow to equilibrate to room temperature before use.

6.7 Argon gas, purity 99.9%

7. Procedure

Extraction of Al and Mn

7.1 Weigh 2.5 g of <2-mm, air-dry soil to the nearest mg and place in a labeled extraction tube (ET). If sample is fine-grind, weigh 1.25 g to the nearest mg. If sample is moist, weigh enough soil to achieve $\approx$ 2.5 or 1.25 g, respectively, of air-dry soil. Prepare one quality control sample per 24 samples.

7.2 Place labeled ET on extractor and connect to corresponding tared extraction tube (TET$_{KCl}$) with rubber tubing.

7.3 Use a dispenser and add 10 mL of 1 N KCl to the ET. All soil should be wetted and no air bubbles. Shaking, swirling, or stirring may be required to wet organic samples.

7.4 Secure reservoir tube (RT) to top of ET tube and let stand for 30 min. Extract at 30-min rate the KCl soultion until 2 mL of this solution remains above soil level. Turn off extractor. Do not let soil dry.

7.5 Add 45 mL of KCl solution to the RT if sample weight is 2.5 g. Add 17.5 mL of KCl solution to the RT if sample weight is 1.25 g. Set extractor for 45-min extraction. Extractor will turn off automatically.

7.6 Remove RT from top of extractor. Carefully remove TET$_{KCl}$. Leave the rubber tubing on the ET. Weigh each TET$_{KCl}$ containing the KCl extract to the nearest mg.

7.7 Mix the extract in each TET$_{KCl}$ by manually shaking. Fill a disposable tube with extract solution and discard the excess properly. This solution is reserved for extractable Al and Mn analyses. If extracts are not be determined immediately after collection, then store samples at 4°C.

Dilution of Extracts

7.8 Dilute samples (1:5 dilution). Dilute 1 part KCl sample extract with 4 parts 0.5 N HCl. Use Vortexer to mix sample.

7.9 Place the diluted sample solutions into test tubes and place in the sample holder of the sample changer.

ICP-AES Set-up and Operation

7.10 Refer to the manufacturer's manual for operation of the ICP-AES. The following parameters are only very general guidelines for instrument conditions for the analytes.

Parameter	Value
View	Radial
Wavelength Al	396.153
Wavelength Mn	257.610
Background Correction	"ON" 2-Point
Gas Flow	
Torch Gas	15.0 L min^{-1}
Auxiliary Gas Flow	0.5 L min^{-1}
Nebulizer Flow	0.80 L min^{-1}
Power	1450 W
Peristaltic Pump	
Sample Flow Rate	1.25 mL min^{-1}
Flush Rate	35 s
Relaxation Time	15 s
Pump Tubing Type	Solvflex black/black, 2-stop, 0.030 ID for sample
	Solvflex red/red, 2-stop, 0.045 ID for rinse

ICP-AES Calibration and Analysis

7.11 Calibrate the instrument by using the MCSS. The data system will then associate the concentrations with the instrument responses for each MCSS. Rejection criteria for MCSS, if $R^2 < 0.99$.

7.12 If sample exceeds calibration standard, the sample is diluted 1:10 with 1 N KCl followed by 1:5 0.5 HCl.

7.13 Perform one quality control (QC) (Low MCSS) every 12 samples. If reading is not within 10%, the instrument is re-calibrated and QC re-analyzed.

7.14 Record analyte readings to 0.01 mg L^{-1}.

8. Calculations

The instrument readings are the analyte concentration (mg L^{-1} Al and Mn). Use these values to calculate the analyte concentration in meq 100 g^{-1} and mg kg^{-1} for Al and Mn, respectively.

8.1 Al (meq 100 g^{-1}) = [A x [(B$_1$ - B$_2$)/B$_3$] x C$_1$ x C$_2$ x R x 100]/[1000 x E x F]

where:
A   = Al concentration in extract (mg L^{-1})
B$_1$ = Weight of extraction syringe and extract (g)
B$_2$ = Weight of tared extraction syringe (g)
B$_3$ = Density of 1 N KCl at 20°C (1.0412 g mL^{-1})
C$_1$ = Dilution, required (1:5)
C$_2$ = Dilution, if performed
100 = Conversion factor (100-g basis)
R   = Air-dry/oven-dry ratio (procedure 3D1) or field-moist/oven-dry ratio (procedure 3D2)
1000 = mL L^{-1}
E   = Soil sample weight (g)
F   = Equivalent weight (Al^{3+} = 8.99 mg meq^{-1})

8.2 Mn (mg kg^{-1}) = [A x [(B$_1$ - B$_2$)/B$_3$] x C$_1$ x C$_2$ x R x 1000]/[1000 x E]

where:

A = Mn concentration in extract (mg L^{-1})
B_1 = Weight of extraction syringe and extract (g)
B_2 = Weight of tared extraction syringe (g)
B_3 = Density of 1 N KCl at 20°C (1.0412 g mL^{-1})
C_1 = Dilution ratio, required (1:5)
C_2 = Dilution ratio, if needed
R = Air-dry/oven-dry ratio (procedure 3D1) or field-moist/oven-dry ratio (procedure 3D2)
1000 = Conversion factor in numerator to kg-basis
1000 = Factor in denominator (mL L^{-1})
E = Soil sample weight (g)

9. Report

Report KCl extractable Al to the nearest 0.1 meq 100 g^{-1} (cmol(+) kg^{-1}) and Mn to the nearest mg kg^{-1}.

10. Precision and Accuracy

Precision and accuracy data are available from the SSL upon request.

11. References

Gambrell, R. P. 1996. Manganese. p. 665-682. *In* D.L. Sparks (ed.) Methods of soil Analysis . Part 3. Chemical Methods. . Soil Sci. Am. Book Series No. 5. ASA and SSA, Madison WI.

Holmgren, George G.S., R.L. Juve, and R.C. Geschwender. 1977. A mechanically controlled variable rate leaching device. Soil Sci. Soc. Am. J. 41:1207-1208.

Bertsch, P.M., and P.R. Bloom. 1996. Aluminum. p. 517-530. *In* D.L. Sparks (ed.) Methods of soil analysis. Part 3. Chemical methods. No. 5. ASA and SSSA, Madison, WI.

Lee, R., B.W. Bache, M.J. Wilson, and G.S. Sharp. 1985. Aluminum release in relation to the determination of cation exchange capacity of some podzolized New Zealand soils. J. Soil Sci. 36:239-253.

Thomas, G.W. 1982. Exchangeable cations. p. 159-165. *In* A. Klute (ed.) Methods of soil analysis. Part 2. Chemical and microbiological properties. 2nd ed. Agron. Monogr. 9. ASA and SSSA, Madison, WI.

Ratios and Estimates Related to Ion Exchange and Extractable Cations (4B4)
Sum of Extractable Bases (4B4a)
Sum of Extractable Bases by NH₄OAc, pH 7 (4B4a1)
Sum of Extractable Bases by NH₄OAc, pH 7, Calculated (4B4a1a)

Sum the NH₄OAc, pH 7 extractable bases (Ca^{2+}, Mg^{2+}, K^+, and Na^+) (4B4a1a) obtained by procedure 4B1a1 and analyzed in procedures 4B1a1b1-4, respectively. This value is reported as meq 100 g^{-1} (cmol (+) kg^{-1}).

Ratios and Estimates Related to Ion Exchange and Extractable Cations (4B4)
Sum of Extractable Bases (4B4a)
Sum of Extractable Bases by NH₄Cl (4B4a2)
Sum of Extractable Bases by NH₄Cl, Calculated (4B4a2a)

Sum the NH₄Cl extractable bases (Ca^{2+}, Mg^{2+}, K^+, and Na^+) (4B4a2a) obtained by procedure 4B1b1 and analyzed in procedures 4B1b1b1-4, respectively. This value is reported as meq 100 g^{-1} (cmol (+) kg^{-1}).

Ratios and Estimates Related to Ion Exchange and Extractable Cations (4B4)
Cation Exchange Capacity (CEC) (4B4b)
CEC-8.2 (Sum of Cations) (4B4b1)
CEC-8.2, Calculated (4B4b1a)
CEC-8.2, Not Calculated (4B4b1b)

Calculate the CEC-8.2 (4B4b1a) by adding the sum of the NH_4OAc extractable bases (4B4a1) plus the $BaCl_2$-TEA extractable acidity (4B2a1a1 or 4B2b1a1). This value is reported as meq 100 g^{-1} (cmol (+) kg^{-1}). Cation summation is the basis for this procedure. The CEC-8.2 minus the CEC-7 is considered the pH dependent charge from pH 7.0 to pH 8.2. The CEC-8.2 is not calculated (4B4b1b) if significant quantities of soluble salts or carbonates are present in the soil. CEC-8.2 is calculated as follows:

CEC-8.2 = NH_4OAc Bases + $BaCl_2$-TEA Acidity

Ratios and Estimates Related to Ion Exchange and Extractable Cations (4B4)
Cation Exchange Capacity (CEC) (4B4b)
Effective Cation Exchange Capacity (ECEC) (4B4b2)
Sum of NH_4OAc Extractable Bases + 1 N KCl Extractable Aluminum, Calculated (4B4b2a)
Sum of NH_4OAc Extractable Bases + 1 N KCl Extractable Aluminum, Not Calculated (4B4b2b)

Calculate the ECEC (4B4b2a) by adding the sum of the NH_4OAc extractable bases (4B4a1) plus the 1 N KCl extractable Al (4B3a1a1). This value is reported as meq 100 g^{-1} (cmol (+) kg^{-1}). The ECEC is not calculated (4B4b2b) if significant quantities of soluble salts or carbonates are present in the soil. ECEC by NH_4OAc extractable bases and 1 N KCl Al is calculated as follows:

ECEC = NH_4OAc Bases + 1 N KCl Al

Ratios and Estimates Related to Ion Exchange and Extractable Cations (4B4)
Base Saturation (4B4c)
Base Saturation by NH_4OAc, pH 7 (CEC-7) (4B4c1)
Base Saturation by CEC-7, Calculated (4B4c1a)
Base Saturation by CEC-7, Set to 100% (4B4c1b)

Calculate the base saturation (4B4c1a) by dividing by the sum of NH_4OAc extractatable (4B4a1) bases by CEC-7 (4B1a1a1a1) and multiplying by 100. This value is reported as meq 100 g^{-1} (cmol (+) kg^{-1}). If a soil has significant quantities of soluble salts or carbonates, this value is set to 100% (4B4c1b). In soil taxonomy, base saturation determined by CEC-7 is used in mollic, umbric, and eutro-dystro criteria (Soil Survey Staff, 1999). Calculate base saturation by CEC-7 as follows:

Base Saturation (%) = (NH_4OAc Bases/CEC-7) x 100

References

Soil Survey Staff. 1999. Soil taxonomy: A basic system of soil classification for making and interpreting soil surveys, USDA-NRCS Agric. Handb. 436. 2nd ed. U.S. Govt. Print. Office, Washington, DC.

Ratios and Estimates Related to Ion Exchange and Extractable Cations (4B4)
Base Saturation (4B4c)
Base Saturation by NH₄Cl (4B4c2)
Base Saturation by NH₄Cl, Calculated (4B4c2a)
Base Saturation by NH₄Cl, Set to 100% (4B4c2b)

Calculate the base saturation (4B4c2a) by dividing the sum of the NH_4Cl extractatable (4B4a2) bases by CEC by NH_4Cl (4B1b1a1a1) and multiplying by 100. This value is reported as meq 100 g^{-1} or cmol (+) kg^{-1}. If a soil has significant quantities of soluble salts or carbonates, this value is set to 100% (4B4c2b). Calculate base saturation by NH4Cl as follows:

Base Saturation (%) = (NH₄Cl Bases/CEC by NH₄Cl) x 100

Ratios and Estimates Related to Ion Exchange and Extractable Cations (4B4)
Base Saturation (4B4c)
Base Saturation by CEC-8.2 (Sum of Cations) (4B4c3)
Base Saturation by CEC-8.2, Calculated (4B4c3a)
Base Saturation by CEC-8.2, Not Calculated (4B4c3b)

Calculate the base saturation (4B4c3a) by dividing the sum of the NH₄OAc extractable bases (4B4a1) by CEC-8.2 (4B4b1a) and multiplying by 100. This value is reported as meq 100 g^{-1} or cmol (+) kg^{-1}. If a soil has signficant quantities of soluble salts or carbonates, this value is not calculated (4B4c3b). Calculate base saturation by CEC-8.2 (Sum of Cations) as follows:

Base Saturation (%) = [NH₄OAc Bases/(NH₄OAc Bases + BaCl₂-TEA Acidity)] x 100

References

Soil Survey Staff. 1999. Soil taxonomy: A basic system of soil classification for making and interpreting soil surveys, USDA-NRCS Agric. Handb. 436. 2nd ed. U.S. Govt. Print. Office, Washington, DC.

Ratios and Estimates Related to Ion Exhange and Extractable Cations (4B4)
Base Saturation (4B4c)
Base Saturation by Effective Cation Exchange Capacity (ECEC) (4B4c4)
Base Saturation by Sum of NH₄OAc Extractable Bases + 1 *N* KCl Extractable Aluminum, Calculated (4B4c4a)
Base Saturation by Sum of NH₄OAc Extractable Bases + 1 *N* KCl Extractable Aluminum, Not Calculated (4B4c4b)

Calculate the base saturation (4B4c4a) by dividing the sum of NH₄OAc extractable bases (4B4a1) by the ECEC (4B4b2a) and multiplying by 100. If a soil has signficant quantities of soluble salts or carbonates, this value is not calculated (4B4c4b). Calculate base saturation by ECEC as follows:

Base Saturation (%) = (NH$_4$OAc Bases/NH$_4$OAc Bases + 1 N KCl Al) x 100

Ratios and Estimates Related to Ion Exchange and Extractable Cations (4B4)
Aluminum Saturation (4B4d)
Aluminum Saturation by Effective Cation Exchange Capacity (ECEC) (4B4d1)
Aluminum Saturation by Sum of NH$_4$OAc Extractable Bases + 1 N KCl Extractable Aluminum, Calculated (4B4d1a)
Aluminum Saturation by Sum of NH$_4$OAc Extractable Bases + 1 N KCl Extractable Aluminum, Not Calculated (4B4d1b)

Calculate the Al saturation (4B4d1a) by dividing the 1 N KCl extractable Al (4B3a1a1) by ECEC (4B4b2a) and multiplying by 100. If a soil has signficant quantities of soluble salts or carbonates, this value is not calculated (4B4d1b). Calculate Al saturation as follows:

Al Saturation (%) = [1 N KCl Al/(NH$_4$OAc Bases + 1 N KCl Al)]

Ratios and Estimates Related to Ion Exchange and Extractable Cations (4B4)
Activity (4B4e)
CEC-7/clay (4B4e1)

Divide the CEC-7 (procedure 4B1a1a1a1) by clay 7 (procedure 3A1a). This ratio is reported as a dimensionless value. In the past, the ratios of CEC:clay have been reported as meq g^{-1}. For more detailed information on the application of this ratio, refer to Soil Survey Staff (1995, 1999).

Reaction (4C)
Soil Suspensions (4C1)
Electrode (4C1a)
Standard Glass Body Combination (4C1a1)
Digital pH/Ion Meter (4C1a1a)

Soil pH is one of the most frequently performed determinations, and one of the most indicative measurements of soil chemical properties (McLean, 1982). Soil pH tells more about a soil than merely indicating whether it is acidic or basic but also the availability of essential nutrients and toxicity of other elements can be estimated because of their known relationship with pH (Thomas, 1996). Soil pH is affected by many factors, e.g., nature and type of inorganic and organic matter; amount and type of exchangeable cations and anions; soil:solution ratio; salt or electrolyte content; and CO$_2$ content (McLean, 1982). The acidity, neutrality, or basicity of a soil influences the solubility of various compounds; the relative ion bonding to exchange sites; and microbial activities. Depending on the predominant clay type, the pH may be used as a relative indicator of base saturation (Mehlich, 1934). Soil pH is also a critical factor in the availability of most essential elements for plants.

The SSL performs several pH determinations. These methods include but are not limited to as follows: NaF (1 N pH 7.5 to 7.8) (4C1a1a1); saturated paste pH (4C1a1a2); oxidized pH (4C1a1a3); 1:1 water and 1:2 CaCl$_2$ (final solution: 0.01 M CaCl$_2$) (4C1a2a1-2, respectively); 1 N KCl, procedure 4C1a2a3; and organic materials, CaCl$_2$ (final solution $\approx$ 0.01 M CaCl$_2$) (4C1a1a4).

An increase in the soil:water ratio or the presence of salts generally results in a decrease in the soil pH. The soluble salt content of the soil can be overcome by using dilute salt solutions, e.g., CaCl$_2$ or KCl, instead of distilled water. The use of dilute salt solutions is a popular method for masking seasonal variation in soil pH. The pH readings are usually less with dilute salt solutions than with distilled water but may be equal to or greater in highly weathered tropical soils, i.e., soils with a high anion exchange capacity. When the pH values of various soils are compared, determination by the same method is important (Foth and Ellis, 1988).

The 1 N KCl is an index of soil acidity and is more popular in those regions with extremely acid soils and in which KCl is used as an extractant of exchangeable Al. The KCl pH indicates the pH at which Al is extracted. Similar to the 1:2 CaCl$_2$ pH, the 1 N KCl pH readings also tend to be uniform regardless of time of year.

The saturated paste pH is popular in regions with soils with soluble salts. The water content varies with soil water storage characteristics. The saturated paste pH may be more indicative of the saturated, irrigated soil pH than is the soil pH measurement at a constant soil:water ratio. The saturated paste pH is also that pH at which the saturation extract is removed for salt analysis, and hence, is the pH and the dilution at which the sodium adsorption ratio (SAR) is computed (procedure 4E4b).

The 1 N NaF pH may be used as an indicator that amorphous material dominates the soil exchange complex. The oxidized pH may be used to assess the activities of soil microorganisms. In soil taxonomy, the CaCl$_2$ pH is used to distinguish two family reaction classes in Histosols (Soil Survey Staff, 1999).

References

Foth, H.D., and B.G. Ellis. 1988. Soil fertility. John Wiley and Sons. New York, NY.

McLean, E.O. 1982. Soil pH and lime requirement. p. 199-224. *In* A.L Page, R.H. Miller, and D.R. Keeney (eds.) Methods of soil analysis. Part 2. Chemical and microbiological properties. 2nd ed. Agron. Monogr. 9. ASA and SSSA, Madison, WI.

Mehlich, A. 1943. The significance of percentage of base saturation and pH in relation to soil differences. Soil Sci. Soc. Am. Proc. 7:167-174.

Soil Survey Staff. 1999. Soil taxonomy: A basic system of soil classification for making and interpreting soil surveys, USDA-NRCS Agric. Handb. 436. 2nd ed. U.S. Govt. Print. Office, Washington, DC.

Thomas, G.W. 1996. Soil pH and soil acidity. p. 475-490. *In* D.L. Sparks (ed.) Methods of soil analysis. Part 3. Chemical methods. No. 5. ASA and SSSA, Madison, WI.

Reaction (4C)
Soil Suspensions (4C1)
Electrode (4C1a)
Standard Glass Body Combination (4C1a1)
Digital pH/Ion Meter (4C1a1a)
1 N NaF, pH 7.5 – 7.8 (4C1a1a1)
Air-Dry or Field-Moist, <2 mm (4C1a1a1a-b1)

1. Application

The action of NaF upon noncrystalline (amorphous) soil material releases hydroxide ions (OH$^-$) to the soil solution and increases the pH of the solution. The amount of amorphous material in the soil controls the release of OH$^-$ and the subsequent increase in pH (Fields and Perrott, 1966). The following reactions illustrate this action and form the basis of this procedure.

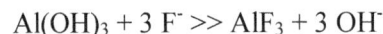

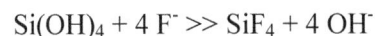

Al(OH)$_3$ + 3 F$^-$ >> AlF$_3$ + 3 OH$^-$

Si(OH)$_4$ + 4 F$^-$ >> SiF$_4$ + 4 OH$^-$

Most soils contain components that react with NaF and release OH⁻. However, a NaF pH $\geq$9.4 is a strong indicator that amorphous material dominates the soil exchange complex. Amorphous material is usually an early product of weathering of pyroclastic materials in a humid climate. Amorphous material appears to form in spodic horizons in the absence of pyroclastics.

2. Summary of Method

A 1-g sample is mixed with 50 mL of 1 N NaF and stirred for 2 min. While the sample is being stirred, the pH is read at exactly 2 min in the upper 1/3 of the suspension (4C1a1a1).

3. Interferences

The difference in the sediment and supernatant pH is called the suspension effect (McLean, 1982). To maintain uniformity in pH determination, measure the pH just above the soil sediment. Clays may clog the KCl junction and slow the electrode response. Clean the electrode by rinsing with distilled water and patting it dry with tissue. Wiping the electrode dry with a cloth, laboratory tissue, or similar material may cause electrode polarization. Use high purity (99%) NaF.

Soils with a 1:1 water pH >8.2 do not give a reliable NaF pH. Free carbonates in a soil result in a high NaF pH. In general, soils with a 1:1 water pH <7.0 are not affected.

4. Safety

The NaF is poisonous. Avoid eye contact and ingestion. Skin penetration and irritation are moderately hazardous. Do not eat or drink while using NaF. Thoroughly wash hands after use. Wear protective clothing, e.g., coats, aprons, and gloves, and eye protection, e.g., safety glasses or goggles, when using NaF. Use the fume hood when using NaF. Follow standard laboratory safety practices.

5. Equipment
5.1 Electronic balance, $\pm$1.0-mg sensitivity
5.2 Paper cup, 120 mL (4 fl. oz.), disposable, Solo Cup Co., No. 404
5.3 Titration beakers, polyethylene, 250 mL
5.4 Automatic titrator, Metrohm Titroprocessors, Control Units, Sample Changers, and Dosimats, Metrohm Ltd., Brinkmann Instruments, Inc.
5.5 Combination pH-reference electrode, Metrohm, Brinkmann Instruments, Inc.

6. Reagents
6.1 Reverse osmosis (RO) water, ASTM Type III grade of reagent water
6.2 Borax pH buffers, pH 4.00, pH 7.00, and pH 9.18, for electrode calibration, Beckman, Fullerton, CA.
6.3 Phenolphthalein
6.4 Sodium fluoride (NaF), 99% purity, EM Science
6.5 Sodium fluoride (NaF), 1.0 N solution. In a plastic bottle, add 400 g NaF in 8 L of distilled water. Let stand for 3 days. On the third day, after excess NaF has settled, measure 50 mL of the solution and read pH. The pH should be between 7.5 and 7.8. Add 3 to 5 drops 0.25% phenolphthalein and titrate to pink end point (pH 8.2 to 8.3). If pH is outside the 7.5 and 7.8 range, then adjust pH with either HF or NaOH. If solution has a pH >8.2 or if the titratable acidity is >0.25 meq L⁻¹, use another source of NaF.

7. Procedure

7.1 Weigh 1 g of <2-mm or fine-grind, air-dry soil to the nearest 1 mg and place in a 120-mL (4-oz) paper cup. If sample is moist, weigh enough soil to achieve $\approx$ 1 g of air-dry soil.

7.2 Calibrate the titrator with pH 4.00, 7.00, and 9.18 buffer solutions.

7.3 Sample stirring, waiting interval for readings, addition of NaF solution, pH readings, and rinsing of electrode are controlled by computer.

7.4 The general sequence used by the automated system is as follows:

7.4.1 The sample is lifted so that the pH electrode is positioned above the soil sediment. Stirring begins immediately and is maintained during each sample cycle.

7.4.2 A 50-mL solution is added to sample.

7.4.3 After 2 min, NaF pH is read. Record pH to the nearest 0.01 unit.

7.4.4 The sample is lowered, and the electrode and stirrer are rinsed with RO water.

7.4.5 The next sample is positioned for analysis.

7.4.6 The cycle is repeated until all samples have been analyzed.

7.5 Discard the solution and cup in safe containers. The paper cup with the NaF solution leaks in about 15 min.

8. Calculations
No calculations are required for this procedure.

9. Report
Report NaF pH to the nearest 0.1 pH unit.

10. Precision and Accuracy
Precision and accuracy data are available from the SSL upon request.

11. References
Fields, M., and K.W. Perrott. 1966. The nature of allophane in soils. Part 3. Rapid field and laboratory test for allophane. N.Z. J. Sci. 9:623-629.

McLean, E.O. 1982. Soil pH and lime requirement. p. 199-224. *In* A.L. Page, R.H. Miller, and D.R. Keeney (eds.) Methods of soil analysis. Part 2. Chemical and microbiological properties. 2nd ed. Agron. Monogr. 9. ASA and SSSA, Madison,

Reaction (4C)
Soil Suspensions (4C1)
Electrode (4C1a)
Standard Glass Body Combination (4C1a1)
Digital pH/Ion Meter (4C1a1a)
Saturated Paste pH (4C1a1a2)
Air-Dry, <2 mm (4C1a1a2a1)

1. Application
When making interpretations about the soil, the saturated paste pH is usually compared to the 1:1 water pH and the 1:2 $CaCl_2$ pH. The usual pH sequence is as follows: 1:1 water pH > 1:2 $CaCl_2$ pH > saturated paste pH. If saturated paste pH is > 1:2 $CaCl_2$ pH, the soil is not saline. If the saturated paste pH $\geq$ 1:1 water pH, the soil may be Na saturated and does not have free carbonates.

Because of the interrelations that exist among the various soil chemical determinations, the saturated paste pH value may be used as a means of cross-checking salinity data for internal consistency and reliability (U.S. Salinity Laboratory Staff, 1954). Some rules of thumb that apply to the saturated paste pH are as follows:

a. Soluble carbonates are present only if the pH is >9.

b. Soluble bicarbonate seldom >3 or 4 meq L^{-1}, if the pH is $\leq$7.

c. Soluble Ca^{2+} and Mg^{2+} seldom >2 meq L^{-1}, if the pH is >9.

d. Gypsiferous soils seldom have a pH >8.2.

2. Summary of Method

The saturated paste is prepared (4F2), and the pH of paste is measured with a calibrated combination electrode/digital pH meter (4C1a1a2).

3. Interferences

The difference in the sediment and supernatant pH is called the suspension effect (McLean, 1982). To maintain uniformity in pH determination, measure the pH just beneath the surface of saturated paste. Clays may clog the KCl junction and slow the electrode response. Clean the electrode by rinsing with RO water and patting it dry with tissue. Wiping the electrode dry with a cloth, laboratory tissue, or similar material may cause electrode polarization.

4. Safety

No significant hazards are associated with this procedure. Follow standard laboratory safety practices.

5. Equipment
5.1 Digital pH/ion meter, Accumet Model AR15, Fisher Scientific
5.2 Electrode, standard glass body combination, Accu-flow, Fisher Scientific

6. Reagents
6.1 Reverse osmosis (RO) water, ASTM Type III grade of reagent water
6.2 Borax pH Buffers, pH 4.00, pH 7.00 and pH 9.18, for electrode calibration, Beckman, Fullerton, CA.

7. Procedure

7.1 Prepare a saturated paste (procedure 4F2).

7.2 Calibrate the pH meter with pH 4.00, 7.00 and 9.18 buffer solutions.

7.3 After equipment calibration, gently wash the electrode with RO water. Dry the electrode. Do not wipe the electrode with a tissue as this may cause a static charge on the electrode.

7.4 Gently lower the electrode in the saturated paste until the KCl junction of the electrode is beneath the surface of saturated paste.

7.5 Allow the pH meter to stabilize before recording the pH. Record pH to the nearest 0.01 unit.

7.6 Gently raise the pH electrode from the paste and wash all particles adhering to the electrode with a stream of RO water.

8. Calculations

No calculations are required for this procedure.

9. Report

Report saturated paste pH to the nearest 0.1 pH unit.

10. Precision and Accuracy

Precision and accuracy data are available from the SSL upon request.

11. References

McLean, E.O. 1982. Soil pH and lime requirement. p. 199-224. *In* A.L. Page, R.H. Miller, and D.R. Keeney (eds.) Methods of soil analysis. Part 2. Chemical and microbiological properties. 2nd ed. Agron. Monogr. 9. ASA and SSSA, Madison, WI.

U.S. Salinity Laboratory Staff. 1954. L.A. Richards (ed.) Diagnosis and improvement of saline and alkali soils. USDA Handb. 60. U.S. Govt. Print. Office, Washington, DC.

Reaction (4C)
Soil Suspensions (4C1)
Electrode (4C1a)
Standard Glass Body Combination (4C1a1)
Digital pH/Ion Meter (4C1a1a)
Oxidized pH (4C1a1a3)

1. Application

Sulfidic material is waterlogged mineral, organic, or mixed soil material with a pH of 3.5 or higher, containing oxidizable sulfur compounds, and which if incubated as a 1-cm thick layer under moist, aerobic conditions (field capacity) at room temperature, shows a drop in pH of 0.5 or more units to a pH value of 4.0 or less (1:1 by weight in water or in a minimum of water to permit measurement) within 8 wk (Van Breemen, 1982; Soil Survey Staff, 1999). The intent of the method described herein is to determine if known or suspected sulfidic materials will oxidize to form a sulfuric horizon (Soil Survey Staff, 1999). Identification of H_2S in a soil by a "rotten-egg" smell or FeS in a saturated soil by its blue-black color indicates that sulfidic materials may be present. If such soils are drained and oxidized, the soil pH could drop to 3.5 or less, making the soil unsuitable for many uses. A field test for FeS is to add 1 N HCl and note the odor of H_2S.

2. Summary of Method

Transfer enough soil to fill a plastic cup one-half to two-thirds full. Add a little water if needed to make a slurry. Stir the slurry thoroughly to introduce air. Determine pH immediately. Place cup in a closed container with openings (inlet and outlet) providing humidified air. Keep at room temperature. After 24 h, open the container, stir the sample thoroughly, and determine the soil pH. Repeat the procedure for a minimum of 10 days until the pH reaches a steady state of ≤ 0.1 units over a 2-day period. Record daily pH readings (4C1a1a3).

3. Interferences

Samples should be shipped in watertight containers completely filled with water from the ambient soil solution to prevent potential oxidation of sulfides and reduction in soil pH.

Extended time in stirring of sample and/or in reading the pH may result in the introduction of sufficient O_2 into the mixture to change the pH reading. Quickly stirring the mixture and reading the pH reduce this error.

Clean the electrode by rinsing with reverse osmosis (RO) water and patting it dry with tissue. Wiping the electrode dry with a cloth, laboratory tissue, or similar material may cause electrode polarization.

4. Safety
No significant hazard has been identified with this procedure. Follow standard laboratory safety precautions.

5. Equipment
5.1 Cups, plastic

5.2 Closed container, with openings (inlet and outlet) providing for humidified airflow. A tube from the inlet of this closed container is connected to the outlet of a stoppered 2.5-L container full of RO water. A pump supplies air to a bubbling stone placed near the bottom of the 2.5-L container (Fig. 1).

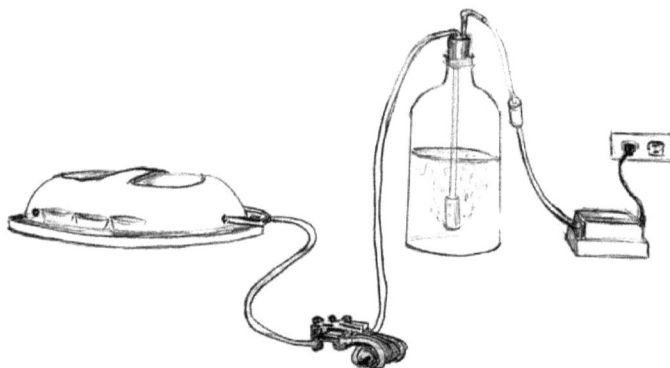

Fig. 1. Apparatus for measuring oxidized pH.

5.3 Digital pH/ion meter, Accumet Model AR15, Fisher Scientific

5.4 Electrode, standard glass body combination, Accu-flow, Fisher Scientific

6. Reagents
6.1 Reverse osmosis (RO) water, ASTM Type III grade of reagent water

6.2 Borax pH buffers, pH 4.00 and pH 7.00 for pH meter calibration, Beckman, Fullerton, CA.

7. Procedure

7.1 Calibrate the pH meter with pH 4.00 and pH 7.00 buffer solutions.

7.2 After equipment calibration, gently wash the electrode with RO water. Dry the electrode. Do not wipe the electrode with a tissue as this may cause a static charge on the electrode.

7.3 Transfer enough soil to fill a small plastic cup one-half to two-thirds full. Add a little water if needed to make a slurry. Stir the slurry thoroughly to introduce air.

7.4 Determine immediately the pH of sample by carefully placing the electrode into the soil mixture. Ensure that the KCl junction and sensor membrane are in contact with the mixture.

7.5 Allow the pH meter to stabilize before recording the pH. Record the pH to the nearest 0.01 pH unit.

7.6 After pH determination, immediately place sample in a closed container with openings (inlet and outlet) providing for humidified airflow. A tube from the inlet of this closed container is connected to the outlet of a stoppered 2.5-L container full of RO water. A pump supplies air to a bubbling stone placed near the bottom of the 2.5-L container. Keep at room temperature (20 to 25°C).

7.7 Stir the sample and record pH daily. Note any bubbling. After pH determination, immediately place sample back in closed container.

7.8 Record the pH for a minimum of 10 days until the change is ≤ 0.1 pH units for two days.

8. Calculations
No calculations are required for this procedure.

9. Report
Report the initial pH and the oxidized pH (end pH) to the nearest 0.1 pH unit.

10. Precision and Accuracy
Precision and accuracy data are available from the SSL upon request.

11. References
Van Breemen, N. 1982. Genesis, morphology, and classification of acid sulfate soils in Coastal Plains. p. 95-108. *In* J.A. Kittrick, D.S. Fanning, and L.R. Hossner (eds.) Acid sulfate weathering. Soil Soc. Am. Spec. Publ. No.10. ASA and SSSA, Madison, WI.

Soil Survey Staff. 1999. Soil taxonomy: A basic system of soil classification for making and interpreting soil surveys, USDA-NRCS Agric. Handb. 436. 2nd. ed. U.S. Govt. Print. Office, Washington, DC.

Reaction (4C)
Soil Suspensions (4C1)
Electrode (4C1a)
Standard Glass Body Combination (4C1a1)
Digital pH/Ion Meter (4C1a1a)
Organic Materials CaCl₂ pH, Final Solution ≈ 0.01 M CaCl₂ (4C1a1a4)

1. Application
This pH is used in soil taxonomy to distinguish two family reaction classes in Histosols (Soil Survey Staff, 1999). Dysic families have a pH <4.5 in 0.01 M $CaCl_2$ in all parts of the organic materials in the control section. Euic families have a pH >4.5 in 0.01 M $CaCl_2$ in some part of the control section.

2. Summary of Method
Place 2.5 mL (2.5 cm³) of the prepared sample in a 30-mL plastic container and add 4 mL of 0.015 M $CaCl_2$, i.e., yields final concentration of ≈ 0.01 M $CaCl_2$ with most packed, moist organic materials. Mix, cover, and allow to equilibrate at least 1 h. Uncover and measure pH with pH paper or pH meter (4C1a1a4).

3. Interferences
This test of organic soil material can be used in field offices. Since it is not practical in the field to base a determination on a dry sample weight, moist soil is used. The specific volume of moist material depends on how it is packed. Therefore, packing of material must be standardized in order to obtain comparable results by different soil scientists (Soil Survey Staff, 1999).

Clean the pH electrode by rinsing with reverse osmosis (RO) water and patting it dry with tissue. Wiping the electrode dry with a cloth, laboratory tissue, or similar material may cause electrode polarization.

4. Safety

No signifcant hazard has been identifed with this procedure. Follow standard laboratory safety precautions.

5. Equipment

5.1 Polycons, 30 mL

5.2 Digital pH/ion meter, Accumet Model AR15, Fisher Scientific

5.3 Electrode, standard glass body combination, Accu-flowt, Fisher Scientific

5.4 Half-syringe, 6 mL. Cut plastic syringe longitudinally to form a half-cylinder measuring device.

5.5 Metal spatula

6. Reagents

6.1 Reverse osmosis (RO) water, ASTM Type III grade of reagent water

6.2 Calcium chloride ($CaCl_2$), 0.015 M. Dissolve 1.10 g of $CaCl_2 \cdot 2H_2O$ in RO water and dilute to 500 mL.

6.3 Borax pH buffers, pH 4.00 and 7.00, for pH meter calibration, Beckman, Fullerton, CA.

7. Procedure

Sample Preparation

7.1 Prepare soil material. If the soil is dry, add water and let stand to saturate. Place 50 to 60 mL of a representative sample on a paper towel in a linear mound. Roll the towel around the sample and express water if necessary. Use additional paper towels as external blotters. Remove the sample and place on a fresh paper towel. The sample should be firm but saturated with water.

7.2 Use scissors to cut sample into 0.5- to 1.0-cm long segments.

7.3 Randomly select sample segments for determination of fiber (5C), solubility in pyrophosphate (5B), and pH (4C1a1a4).

pH Determination

7.4 Use a metal spatula to pack a half-syringe that is adjusted to the 5-mL mark or 2.5-mL (2.5-cm^3) volume with the moist sample.

7.5 Place 2.5 mL (2.5 cm^3) of the prepared sample in a 30-mL polycon and add 4 mL of 0.015 M $CaCl_2$, i.e. yields a final concentration of approximately 0.01 M $CaCl_2$ with most packed moist organic materials.

7.6 Mix, cover, and allow to equilibrate at least 1 h.

7.7 Uncover, mix again, immerse electrode, and measure pH. Rinse electrode with RO water.

7.8 Alternatively, place pH strip on top of sample so that it wets from the bottom. Close cover and allow to equilibrate approximately 5 min. Remove pH strip with tweezers. Use a wash bottle to gently wash soil from bottom of strip. Compare color of active segment (center) with reference segments and with pH scale on box to determine pH.

8. Calculations

No calculations are required for this procedure.

9. Report

Report the 0.01 M $CaCl_2$ pH to the nearest 0.1 pH unit.

10. Precision and Accuracy
Precision and accuracy data are available from the SSL upon request.

11. References
Soil Survey Staff. 1999. Soil taxonomy: A basic system of soil classification for making and interpreting soil surveys, USDA-NRCS Agric. Handb. 436. 2nd ed. U.S. Govt. Print. Office, Washington, DC.

Reaction (4C)
Soil Suspensions (4C1)
Electrode (4C1a)
Combination pH-Reference Electrode (4C1a2)
Automatic Titrator (4C1a2a)
1:1 Water pH (4C1a2a1)
Air-Dry or Field-Moist, <2 mm (4C1a2a1a-b1)
1:2 0.01 M CaCl$_2$ pH (4C1a2a2)
Air-Dry or Field Moist, < 2mm (4C1a2a2a-b1)

1. Application
The 1:1 water pH (4C1a2a1) and 1:2 0.01 M CaCl$_2$ pH (4C1a2a2) determinations are two commonly performed soil pH measurements. The CaCl$_2$ soil pH is generally less than the 1:1 water pH. The combination of exchange and hydrolysis in salt solutions (0.1 to 1 M) can lower the measured pH from 0.5 to 1.5 units, compared to the pH measured in RO water (Foth and Ellis, 1988).

In soil taxonomy, these pH values are used as a criterion for reaction classes (acid and nonacid) in families of Entisols and Aquepts (Soil Survey Staff, 1999). The acid class is <5.0 pH in 0.01 M CaCl$_2$ (2:1) or $\approx$ 5.5 in 1:1 water. The nonacid class is $\geq$5.0 pH in 0.01 M CaCl$_2$ (2:1).

2. Summary of Method
The pH is measured in soil-water (1:1) and soil-salt (1:2 CaCl$_2$) solutions. For convenience, the pH is initially measured in water and then measured in CaCl$_2$. With the addition of an equal volume of 0.02 M CaCl$_2$ to the soil suspension that was prepared for the water pH, the final soil-solution ratio is 1:2 0.01 M CaCl$_2$.

A 20-g soil sample is mixed with 20 mL of reverse osmosis (RO) water (1:1 w:v) with occasional stirring. The sample is allowed to stand 1 h with occasional stirring. The sample is stirred for 30 s, and the 1:1 water pH is measured. The 0.02 M CaCl$_2$ (20mL) is added to soil suspension, the sample is stirred, and the 1:2 0.01 M CaCl$_2$ pH is measured (4C1a2a2).

3. Interfernces
The pH will vary between the supernatant and soil sediment (McLean, 1982). Measure the pH just above the soil sediment to maintain uniformity. Clays may clog the KCl junction and slow the electrode response. Clean the electrode. Wiping the electrode dry with cloth, laboratory tissue or similar material may cause electrode polarization. Rinse the electrode with distilled water and pat dry.

Atmospheric CO$_2$ affects the pH of the soil:water mixture. Closed containers and nonporous materials will not allow equilibration with CO$_2$. At the time of pH determination, the partial pressure of CO$_2$ and the equilibrium point must be considered, if doing critical work.

4. Safety
No significant hazards are associated with the procedure. Follow standard laboratory safety practices.

5. Equipment

5.1 Measuring scoop, handmade, ≈ 20-g capability

5.2 Paper cup, 120 mL (4 fl. oz.), disposable, Solo Cup Co., No. 404

5.3 Dispenser, 0 to 30 mL, Repipet or equivalent

5.4 Beverage stirring sticks, wood

5.5 Titration beakers, polyethylene, 250 mL

5.6 Automatic titrator, Metrohm Titroprocessors, Control Units, Sample Changers, and Dosimats, Metrohm Ltd., Brinkmann Instruments, Inc.

5.7 Combination pH-reference electrode, Metrohm part no. 6.0210.100, Brinkmann Instruments, Inc.

6. Reagents

6.1 Reverse osmosis (RO) water, ASTM Type III grade of reagent water

6.2 Borax pH buffers, pH 4.00, pH 7.00, and pH 9.18, for electrode calibration, Beckman, Fullerton, CA.

6.3 Calcium chloride ($CaCl_2$), 0.02 M. Dissolve 23.52 g of $CaCl_2 \cdot 2H_2O$ in RO water and dilute to 8 L.

7. Procedue

7.1 Use a calibrated scoop to measure ≈ 20 g of <2-mm or fine-grind, air-dry soil. If sample is moist, use calibrated scoop to achieve ≈ 20 g of air-dry soil.

7.2 Place the sample in a 120-mL (4-oz) paper cup.

7.3 Dispense 20 mL of RO water into sample and stir.

7.4 Place paper cup with sample in 250-mL titration beaker, allow to stand for 1 h, stirring occasionally.

7.5 Load beakers into sample changer.

7.6 Calibrate the pH meter using the pH 9.18, 7.00 and pH 4.00 buffer solutions.

7.7 Sample stirring, waiting interval for readings, addition of $CaCl_2$ solution, pH readings, and rinsing of electrode are controlled by computer.

7.8 The general sequence used by the automated system is as follows:

7.8.1 The sample is lifted so that the pH electrode is positioned above the soil sediment.

7.8.2 The sample is stirred for 30 s.

7.8.3 After 1 min, 1:1 water pH is read. Record pH to the nearest 0.01 unit.

7.8.4 The 20 mL of 0.02 M $CaCl_2$ are added to sample. The sample is stirred for 30 s.

7.8.5 After 1 min, the 1:2 $CaCl_2$ pH is read. Record pH to the nearest 0.01 unit.

7.8.6 The sample is lowered, and the electrode and stirrer are rinsed with RO water.

7.8.7 The next sample is positioned for analysis.

7.8.8 The cycle is repeated until all samples have been analyzed.

8. Calculations
No calculations are required for this procedure.

9. Report
Report the 1:1 water pH and the 1:2 0.01 M CaCl$_2$ pH to the nearest 0.1 pH unit.

10. Precision and Accuracy
Precision and accuracy data are available from the SSL upon request.

11. References
Foth, H.D., and B.G. Ellis. 1988. Soil fertility. John Wiley and Sons. NY, NY.
McLean, E.O. 1982. Soil pH and lime requirement. p. 199-224. *In* A.L. Page, R.H. Miller, and D.R. Keeney (eds.) Methods of soil analysis. Part 2. Chemical and microbiological properties. 2nd ed. Agron. Monogr. 9. ASA and SSSA, Madison, WI.
Soil Survey Staff. 1999. Soil taxonomy: A basic system of soil classification for making and interpreting soil surveys, USDA-NRCS Agric. Handb. 436. 2nd ed. U.S. Govt. Print. Office, Washington, DC.

Reaction (4C)
Soil Suspensions (4C1)
Electrode (4C1a)
Combination pH-Reference Electrode (4C1a2)
Automatic Titrator (4C1a2a)
1 N KCl pH (4C1a2a3)
Air-Dry or Field-Moist, <2 mm (4C1a2a3a-b1)

1. Application
The 1 N KCl pH is an index of soil acidity. If KCl pH is <5, significant amounts of Al are expected in the solution, and if the pH is very much below 5, almost all the acidity is in the form of Al.

2. Summary of Method
A 20-g soil sample is mixed with 20 mL of 1 N KCl. The sample is allowed to stand for 1 h with occasional stirring. The sample is stirred for 30 s, and after 1 min, the KCl pH is read (4C1a2a3).

3. Interferences
The difference in the sediment and supernatant pH is called the suspension effect (McLean, 1982). To maintain uniformity in determination, measure the pH just above the soil sediment. Clays may clog the KCl junction and slow the electrode response. Clean the electrode by rinsing with reverse osmosis (RO) water and patting it dry with tissue. Wiping the electrode dry with a cloth, laboratory tissue, or similar material may cause electrode polarization.

4. Safety
No significant hazards are associated with the procedure. Follow standard laboratory safety practices.

5. Equipment
5.1 Measuring scoop, handmade, ≈ 20 g
5.2 Paper cup, 120 mL (4 fl. oz.), disposable, Solo Cup Co., No. 404
5.3 Dispenser, 0 to 20 mL, Repipet or equivalent
5.4 Beverage stirring sticks, wood, FSN 7340-00-753-5565
5.5 Titration beakers, polyethylene, 250 mL

5.6 Automatic titrator, Metrohm Titroprocessors, Control Units, Sample Changers, and Dosimats, Metrohm Ltd., Brinkmann Instruments, Inc.

5.7 Combination pH-reference electrode, Brinkmann Instruments, Inc.

6. Reagents

6.1 Reverse osmosis (RO) water, ASTM Type III grade of reagent water

6.2 Borax pH buffers, pH 4.00, 7.00, and 9.18 for electrode calibration, Beckman, Fullerton, CA.

6.3 Potassium chloride (KCl), 1.0 N. Dissolve 74.56 g of KCl in RO water. Dilute to 1 L.

7. Procedure

7.1 Use a calibrated scoop to measure ≈ 20 g of <2-mm or fine-grind, air-dry soil. If sample is moist, use calibrated scoop to achieve ≈ 20 g of air-dry soil.

7.2 Place the sample in a 120-mL (4-oz) paper cup.

7.3 Dispense 20 mL of 1 N KCl into sample and stir with wooden beverage stirrer.

7.4 Place paper cup with sample in 250-mL titration beaker, allow to stand 1 h, stirring occasionally.

7.5 Load beakers into sample changer table.

7.6 Calibrate the pH meter using the pH 4.00, 7.00, and 9.18 buffer solutions.

7.7 Sample stirring, waiting interval for reading, pH reading, and rinsing of electrode are controlled by computer.

7.8 The general sequence used by the automated system is as follows:

7.8.1 The sample is lifted so that pH electrode is positioned above the soil sediment.

7.8.2 The sample is stirred for 30 s.

7.8.3 After 1 min, the KCl pH is read. Record pH to the nearest 0.01 unit.

7.8.4 The sample is lowered, and the electrode and stirrer are rinsed with RO water.

7.8.5 The next sample is positioned for analysis.

7.8.6 The cycle is repeated until all samples have been analyzed.

8. Calculations
No calculations are required for this procedure.

9. Report
Report KCl pH to the nearest 0.1 pH unit.

10. Precision and Accuracy
Precision and accuracy data are available upon request from the SSL.

11. References

McLean, E.O. 1982. Soil pH and lime requirement. p. 199-224. *In* A.L. Page, R.H. Miller, and D.R. Keeney (eds.) Methods of soil analysis. Part 2. Chemical and microbiological properties. 2nd ed. Agron. Monogr. 9. ASA and SSSA, Madison, WI.

Soil Test Analyses (4D)

For more than 30 years, soil testing has been widely used as a basis for determining lime and fertilizer needs (Soil and Plant Analysis Council, 1999). In more recent years, some of these tests have been employed in more diverse agronomic and environmental uses (SERA-IEG, 2000). It is for these reasons that the SSL has expanded its suite of soil test analyses to more completely characterize the inorganic and organic N fractions and to provide a number P analyses for a broad spectrum of soil applications.

Methods development in soil P characterization (Bray and Kurtz, 1945; Olsen et al., 1954; Chang et al., 1957) has been instrumental in developing principles and understanding of the nature and behavior of P in soils (Olsen et al., 1982). Amounts, forms, and distribution of soil P vary with soil-forming factors (Walker, 1974; Stewart and Tiessen, 1987); level and kind of added P (Barrow, 1974; Tisdale et al., 1985; Sharpley et al., 1996); other soil and land management factors (Haynes, 1982; Sharpley, 1985); and soil P-sorption characteristics (Goldberg and Sposito, 1984; Van Riemsdijk et al., 1984; Polyzopoulos et al., 1985; Frossard et al., 1993). Knowledge of these factors and their impact upon the fate and transport of soil P has been used in developing soil P interpretations for such broad and diverse application as fertility, taxonomic classification, genesis and geomorphology models, and environmental studies (Burt et al., 2002).

To characterize the P in the soil system requires the selection of an appropriate method of determination. This selection is influenced by many factors, e.g., objectives of study; soil properties; sample condition or environment; accuracy; and reproducibility (Olsen and Sommers, 1982). Most soil P determinations have two phases, i.e., preparation of a solution that contains the soil P or fraction thereof and the quantitative determination of P in the solution. Most P analyses of soil solutions have been colorimetric procedures, as they are sensitive, reproducible, and lend themselves to automated analysis, accommodating water samples, digest solutions, and extracts (SERA-IEG 17, 2000). Inductively coupled plasma (ICP) spectrophotometry can also be used for P determination, with the popularity of this procedure increasing due to the use of multi-element soil extractants (SERA-IEG 17, 2000). Results from colorimetric analyses are not always comparable to those from ICP because ICP estimates the total amount of P in solution while the colorimetric procedures measure P that can react with the color developing reagent (SERA-IEG 17, 2000).

The selected colorimetric method for P determination depends on the concentration of solution P; concentration of interfering substances in the solution to be analyzed; and the particular acid system involved in the analytical procedure (Olsen and Sommers, 1982). The SSL determines a number of P analyses, mostly colorimetrically, with the exception of the multi-element extractant Mehlich No. 3 (4D6a1a1-18). These P analyses include but are not limited to as follows: anion-resin extractable (4D1a1a1a1-2); water soluble (4D2a1a1); Bray P-1 (4D3a1); Olsen sodium-bicarbonate (4D5a1); Mehlich No. 3 (4D6a1); citric acid soluble (4D7a1); and New Zealand P Retention (4D8a1). The procedures for total P analysis (4H1a1a1a1-20, 4H1b1a1a1-11) are described in another section of this manual entitled "Total Analysis".

Nitrogen is ubiquitous in the environment as it is continually cycled among plants, soil organisms, soil organic matter, water and the atomosphere (National Research Council, 1993). Nitrogen is one of the most important plant nutrients and forms some of the most mobile compounds in the soil-crop system, and as such, is commonly related to water-quality problems. Total N includes both organic and inorganic forms. The SSL procedure for total N (4H2a2) is described in another section of this manual entitled "Total Analysis".

Inorganic N in soils is predominately NO_3 and NH_4, with nitrite seldom found in detectable amounts except in neutral to alkalilne soils receiving NH_4 or NH_4-producing fertilizers (Maynard and Kalra, 1993; Mulvaney, 1996). There is considerable diversity among laboratories in the extraction and determination of NO_3 and NH_4 (Maynard and Kalra, 1993). Nitrate is water soluble and a number of soil solutions including water have been used as extractants, with the most common being KCl (refer to Maynard and Kalra, 1993; and Mulvaney, 1996 for review of extractants). The SSL determines inorganic N (nitrate-nitrite) by KCl extraction, with a flow injection automated ion analyzer used to measure the soluble inorganic nitrate (NO_3^-) (4D9a1a1-2). The nitrate is quantitatively reduced to nitrite by passage of the sample through a copperized cadmium column. The nitrite (reduced nitrate plus original nitrite) is then determined by diazotizing with sulfanilamide followed by coupling with N-1-naphthylethylenediamine dihydrochloride. The resulting water soluble dye has a magenta color which is read at 520 nm.

The concept of an organic N fraction that is readily mineralized has been used to assess soil N availability in cropland, forests, and waste-disposal sites (Campbell et al, 1993). Incubation-leaching techniques have been used to quantify the mineralizable pool of soil organic N. These techniques may be aerobic or anerobic. The SSL determines mineralizable N (N as NH_3) by anaerobic incubation (4D10a1a1). In addition, the SSL also determines mineralizable N after fumigation incubation (6B2a1a1), described in another section of this manual entitled "Soil Biological and Plant Analyses".

References

Barrow, N.J. 1974. Effect of previous additions of phosphate on phosphate adsorption by soils. Soil Sci. 118:82-89.

Bray, R.H., and L.T. Kurtz. 1945. Determination of total, organic, and available forms of phosphorus in soils. Soil Sci. 59:39-45.

Burt, R., M..D. Mays, E.C. Benham, and M.A. Wilson. 2002. Phosphorus characterization and correlation with properties of selected benchmark soils of the United States. Commun. Soil Sci. Plant Anal. 33:117-141.

Campbell, C.A., B.H. Ellert, and Y.W. Jame. 1993. Nitrogen mineralilzation potential in soils. p. 341-357. *In* Martin R. Carter (ed.) Soil sampling and methods of analysis. Can. Soc. Soil Sci. Lewis Publ., Boca Raton, FL.

Chang, S.C., and M.L. Jackson. 1957. Fractionation of soil phosphorus. Soil Sci. 84:133-144.

Frossard, E., C. Feller, H. Tiessen, J.W.B. Stewart, J.C. Fardeau, and J.L. Morel. 1993. Can an isotopic method allow for the determination of phosphate-fixing capacity of soils? Commun. Soil Sci. Plant Anal. 24:367-377.

Haynes, H.R. 1982. Effects of liming on phosphate availability in acid soils. Plant Soil. 63:289-308.

Goldberg, S., and G. Sposito. 1984. A chemical model of phosphate adsorption by soils. II. Noncalcareous soils. Soil Sci. Soc. Am. J. 48:779-783.

Maynard, D.G., and Y.P. Kalra. 1993. Nitrate and exchangeable ammonium nitrogen. p. 25-38. *In* Martin R. Carter (ed.) Soil sampling and methods of analysis. Can. Soc. Soil Sci. Lewis Publ., Boca Raton, FL.

Mulvaney, R.L. 1996. Nitrogen – Inorganic forms. p. 1123-1184. In *In* D.L. Sparks (ed.) Methods of soil analysis. Part 3. Chemical methods. No. 5. ASA and SSSA, Madison, WI.

National Research Council. 1993. Soil and water quality. An agenda for agriculture. Committee on long-range soil and water conservation. Natl. Acad. Press, Washington, DC

Olsen, S.R., and C.V. Cole, F.S. Watanabe, and L.A. Dean. 1954. Estimation of available phosphorus in soils by extraction with sodium bicarbonate. USDA Circ. 939. U.S. Govt. Print. Office, Washington, DC.

Olsen, S.R., and L.E. Sommers. 1982. Phosphorus. p. 403-430. *In* A.L. Page, R.H. Miller, and D.R. Kenney (eds.) Methods of soil analysis. Part 2. Chemical and microbiological properties. 2nd ed. Agron. Monogr. 9. ASA and SSSA, Madison, WI.

Polyzopoulos, N.A., V.Z. Keramidas, and H. Koisse. 1985. Phosphate sorption by some Alfisols as described by commonly used isotherms. Soil Sci. Soc. Am. J. 49:81-84.

SERA-IEG 17. 2000. Methods of phosphorus analysis for soils, sediments, residuals, and waters. *In* G. M. Pierzynski (ed.) Southern Cooperative Series Bull. No. XXX. USDA-CSREES Regional Committee: Minimizing Agricultural Phosphorus Losses for Protection of the Water Resource.

Sharpley, A.N. 1985. Depth of surface soil-runoff interaction as affected by rainfall, soil, slopoe, and management. Soil Sci. Soc. Am. J. 49:1010-1015.

Sharpley, A.N. 1996. Availability of residual phosphorus in manured soils. Soil Sci. Soc. Am. J. 60:1459-1466.

Stewart, J.W.B, and H. Tiessen. 1987. Dynamics of soil organic phosphorus. Biogeochemistry. 4:41-60.

Tisdale, S.L., W.L. Nelson, J.D. Beaton. 1985. Soil fertililty and fertilizers. 4th ed. Macmillan Publ. Co., New York, NY.

Van Riemsdijk, W.H., A.M.A. Van der Linden, and L.J.M. Boumans. 1984. Phosphate sorption by soils. III. The diffusioin-precipitation model tested for three acid sandy soils. Soil Sci. Soc. Am. J. 48:545-548.

Walker, T.W. 1974. Phosphorus as an index of soil development. Trans. Int. Congr. Soil Sci. 10:451-457.

Soil Tests (4D)
Anion Resin Extraction (4D1)
Two-Point Extraction (4D1a)
1-h, 24-h, 1 *M* NaCl (4D1a1a1)
UV-Visible Spectrophotometer, Dual-Beam (4D1a1a1a)
Phosphorus, Two points (4D1a1a1a1-2)
Air-dry or Field-Moist, <2 mm (4D1a1a1a1-2a-b1)

1. Application

Anion resins remove P from soils without chemical alterations and only minor pH changes. Amounts of P released from soil and adsorbed by resins have been used as a measure of available P; an assessment of the availability of residual phosphates; estimation of release characteristics and runoff P for agricultural land (Elrashidi, et al., 2003); and the buffer capacity of soils (Olsen and Sommers, 1982).

Plotting g log of extraction periods (0.25, 0.50, 1, 2, 4, 8, 24, 48 h) against amounts of P released (mg kg^{-1}) showed a linear relationship in 24 U.S. benchmark soils (Elrashidi et al., 2003). Two extraction periods (1 and 24 h) are sufficient to develop linear equations that predict P release characteristics (PRC), describing the whole relationship between the extraction time (1 min to 48 h) and amount of P released (mg kg^{-1}) for soils (Elrashidi et al., 2003). The method describes a two-point measurement (1 and 24 h extraction).

2. Summary of Method

A 2-g soil sample and 4-g resin bag are shaken with 100 mL of reverse osmosis deionized water for 1 h. Soil suspension is shaken again with another 4-g resin bag for 23 h. Phosphorus released from soil during shaking is adsorbed by resin. To remove P from resin, resin bags are shaken for 1 h in 1 *M* NaCl solution. Concentrated HCl is added to sample extracts. A 1-mL aliquot is diluted with 4 mL of ascorbic acid molybdate solution. Absorbance of the solution is read using a spectrophotometer at 882 nm. Data are reported as mg P kg^{-1} soil (4D1a1a1a1-2).

3. Interferences

The Mo blue methods, which are very sensitive for P, are based on the principle that in an acid molybdate solution containing orthophosphate ions, a phosphomolybdate complex forms that can be reduced by ascorbic acid, $SnCl_2$, and other reducing agents to a Mo color. The intensity of blue color varies with the P concentration but is also affected by other factors such as acidity, arsenates, silicates, and substances that influence the oxidation-reduction conditions of the system (Olsen and Sommers, 1982).

4. Safety

Wear protective clothing (coats, aprons, sleeve guards, and gloves) and eye protection (face shields, goggles, or safety glasses). When preparing reagents, exercise special care. Many metal salts are extremely toxic and may be fatal if ingested. Thoroughly wash hands after handling these metal salts. Restrict the use of concentrated H_2SO_4 and HCl to a fume hood. Use safety showers and eyewash stations to dilute spilled acids and bases. Use sodium bicarbonate and water to neutralize and dilute spilled acids.

5. Equipment

5.1 Electronic balance, ±1.0-mg sensitivity
5.2 Mechanical reciprocating shaker, 100 oscillations min^{-1}, 1 ½ in strokes, Eberbach 6000, Eberbach Corp., Ann Arbor, MI
5.3 Bottles, polyethylene, 250, 125, and 60 mL
5.4 Funnel, 60° angle, long stem, 50-mm diameter
5.5 Filter paper, Whatman 42, 150 mm
5.6 Cups, plastic
5.7 Dispenser, 50 mL
5.8 Pipettes, electronic digital, 2500 μL and 10 mL, with tips 2500 μL and 10 mL
5.9 Cuvettes, plastic, 4.5-mL, 1-cm light path, Daigger Scientific
5.10 Spectrophotometer, UV-Visible, Dual-View, Varian, Cary 50 Conc, Varian Australia Pty Ltd.
5.11 Computer, with Cary WinUV software, Varian Australia Pty Ltd., and printer

6. Reagents

6.1 Reverse osmosis deionized (RODI) water, ASTM Type I grade of reagent water
6.2 Anion exchange resin (AER), DOWEX Marathon, type II, 510-610 μm spherical beads (DOW Chemical Company), converted to bicarbonate form by soaking bags overnight in 1.0 M NaHCO$_3$ solution and washing out excess salt with RODI water. Store in RODI water in refrigerator.
6.3 Nitex nylon fabric, 300-μm pores, and nylon thread, Sefar America, Inc., for making and sewing resin bags
6.4 Hydrochloric acid (HCl), concentrated, 12 N, trace pure grade
6.5 NaCl solution. Dissolve 58.4 g NaCl in 1000 mL RODI water.
6.6 Sulfuric-tartarate-molybdate solution (STMS). Dissolve 60 g of ammonium molybdate tetrahydrate [(NH$_4$)$_6$Mo$_7$O$_{24}$·4H$_2$O] in 200 mL of boiling RODI water. Allow to cool to room temperature. Dissolve 1.455 g of antimony potassium tartarate (potassium antimony tartarate hemihydrate [K(SbO)C$_4$H$_4$O$_6$·1/2H$_2$O] in the ammonium molybdate solution. Slowly and carefully add 700 mL of concentrated H$_2$SO$_4$. Cool and dilute to 1 L with RODI water. Store in a dark bottle in the refrigerator.
6.7 Ascorbic acid solution. Dissolve 13.2 g of ascorbic acid in RODI water and dilute to 100 mL with RODI water. Make fresh daily.
6.8 Working ascorbic acid molybdate solution (WAMS). Mix 25 mL of STMS solution with 800 mL of RODI water. Add 10 mL of ascorbic acid solution and dilute to 1 L with RODI water. Allow to stand at least 1 h before using. Prepare fresh daily.
6.9 Working stock standard P solution (WSSPS), 100.0 mg P L^{-1}. In a 1-L volumetric flask, dissolve 0.4394 g primary standard grade anhydrous potassium dihydrogen phosphate (KH$_2$PO$_4$) that has been dried for 2 h at 110°C in about 800 mL 1.0 M NaCl. Dilute to volume with NaCl solution and invert to mix thoroughly. Store in a polyethylene bottle. Make fresh weekly. Store in a refrigerator.

6.10 Standard P calibration solutions (SPCS) or working standards, 4.0, 3.0, 2.0, 1.0, 0.5, 0.25, and 0.0 mg P L^{-1}. Prepare fresh weekly. Store in a refrigerator. Allow to equilibrate to room temperature before use. To seven 100-mL volumetric flasks add as follows:

6.10.1 4.0 mg P L^{-1} = 4.0 mL WSSPS
6.10.2 3.0 mg P L^{-1} = 3.0 mL WSSPS
6.10.3 2.0 mg P L^{-1} = 2.0 mL WSSPS
6.10.4 1.0 mg P L^{-1} = 1.0 mL WSSPS
6.10.5 0.5 mg P L^{-1} = 0.5 mL WSSPS
6.10.6 0.25 mg P L^{-1} = 0.25 mL WSSPS
6.10.7 0.0 mg P L^{-1} = 0.0 mL WSSPS

Add 70 mL of 1.0 *M* NaCl solution and 4.0 mL of concentrated HCl to each SPCS. Allow to cool and dilute to mark with 1.0 *M* NaCl solution. Invert to mix thoroughly.

6.11 Quality Control Samples: 0.1 mg L^{-1} solution made from SSPS; blanks; selected SPCS; and SSL standard.

7. Procedure

7.1 Weigh 2 g of <2-mm or fine-grind, air-dry soil to nearest mg on an electronic balance and place in a 250-mL polyethylene bottle. Add 100 mL RODI water to bottle. Use two replicates for each soil sample in addition to a control treatment where all steps of the extraction process are performed in the absence of soil.

7.2 Place 4-g resin bag in the soil suspension and control sample. Transfer sample to shaker. Shake for 1 h at 100 oscillations min^{-1} at room temperature (20°C ± 2°C). After shaking, remove P from resin as described in Section 7.4 and then proceed with Sections 7.5 – 7.14.

7.3 Place another 4-g resin bag in the soil suspension and control sample. Transfer sample to shaker. Shake for 23 h at 100 oscillations min^{-1} at room temperature (20°C ± 2°C). After shaking, remove P from resin as described in Section 7.4 and then proceed with Sections 7.5 – 7.14.

7.4 Remove P from resin by lifting resin bag out of soil suspension and rinsing with 5 mL RODI water to remove attached soil particles. Add RODI water to soil suspension. If necessary, keep soil suspension for the next extraction.

7.5 Place resin bag in a 125-mL polyethylene bottle containing 50 mL of 1.0 *M* NaCl solution.

7.6 Transfer bottle to shaker and shake for 1 h at 100 oscillations min^{-1} at room temperature (20 °C± 2°C).

7.7 Transfer extracting solutions to 60-mL polyethylene bottles. Filter if soil particles are observed in the extract. Add 2 mL concentrated HCl to each bottle. If extracts are not to be determined immediately after collection, then store samples at 4°C. Analyze samples within 72 h.

7.8 Use the pipette to transfer a 1-mL aliquot of the sample to a plastic cup. Also transfer a 1-mL aliquot of the each SPCS to a plastic cup. Use a clean pipette tip for each sample and SPCS.

7.9 Dispense 4 mL of the WAMS to sample aliquot and to each SPCS. Swirl to mix. The color reaction requires a minimum of 20 min before analyst records readings. Complete all readings within 2-h period since blue color may fade after this period.

7.10 Transfer sample extract and SPCS to cuvettes.

7.11 Set the spectrophotometer to read at 882 nm. Autozero with calibration blank.

7.12 Calibrate the instrument using the SPCS. The data system will then associate the concentrations with the instrument responses for each SPCS. Rejection criteria for SPCS, if R^2 <0.99.

7.13 Run samples using calibration curve. Sample concentration is calculated from the regression equation. Rejection criteria for batch, if blanks as samples >0.01: If SPCS as samples >±20%; and if SSL standard >± 20% mean. Record results to the nearest 0.01unit for the sample extract and each SPCS.

7.14 If samples are outside the calibration range, dilute sample extracts with extracting solution and re-analyze.

8. Calculations

Convert extract P (mg L^{-1}) to soil P (mg kg^{-1}) as follows:

Soil P (mg kg^{-1}) =[(A x B x C x R x 1000)/E]

where:
A   = Sample extract reading (mg L^{-1})
B   = Extract volume (L) (0.055)
C   = Dilution, if performed
R   = Air-dry/oven-dry ratio (procedure 3D1) or field-moist/oven-dry ratio (procedure 3D2)
1000 = Conversion factor to kg-basis
E   = Sample weight (g)

9. Report
Report data to the nearest 0.1 mg P kg^{-1} soil.

10. Precision and Accuracy
Precision and accuracy data are available from the SSL upon request.

11. References
Elrashidi, M.A., M.D. Mays, and P.E. Jones. 2003. A technique to estimate release characteristics and runoff phosphorus for agricultural land. Commun. Soil Sci. and Plant Anal. 34:1759-1790.

Olsen, S.R., and L.E. Sommers. 1982. Phosphorus. p. 403-430. *In* A.L. Page, R.H. Miller, and D.R. Keeney (eds.) Methods of soil analysis. Part 2. Chemical and microbiological properties. 2nd ed. Agron. Monogr. 9. ASA and SSSA, Madison, WI.

Soil Tests (4D)
Aqueous Extraction (4D2)
Single-Point Extraction (4D2a)
1:10, 30 min (4D2a1)
UV-Visible Spectrophotometer, Dual-Beam (4D2a1a)
Phosphorus (4D2a1a1)
Air-dry or Field-Moist, <2 mm (4D2a1a1a-b1)

1. Application

Phosphorus occurs in soil in both the solution and solid phase. These forms are well documented but questions still remain concerning the exact nature of the constituents and ionic forms found in water, soils, and sediments (National Research Council, 1993). These forms influence P availability in relation to root absorption and plant growth; runoff and water quality problems; and P loadings.

Water soluble P has been defined as P measured in water, dilute salt extracts (e.g., O.01 M CaCl$_2$), displaced soil solutions, or saturation paste extracts (Olsen and Sommers, 1982). Even though the water soluble fraction principally consists of inorganic orthophosphate ions, there is evidence that some organic P is also included (Rigler, 1968).

The water or dilute salt extracts represent an attempt to approximate the soil solution P concentration. As an index of P availability, the objectives of this method are (1) to determine the P concentration level in the soil extract that limits plant growth (Olsen and Sommers, 1982) and (2) to determine the composition of the soil solution so that the chemical environment of the plant roots may be defined in quantitative terms (Adams, 1974). The sum of water soluble P and pH 3 extractable P has also been defined as the available P in runoff (Jackson, 1958).

2. Summary of Method

A 2.5-g sample of <2-mm, air-dry soil is mechanically shaken for 30 min in 25-mL of reverse osmosis deionized water. The sample is then centrifuged until solution is free of soil mineral particles, and then filtered until clear extracts are obtained. Absorbance of the solution is read using a spectrophotometer at 882 nm. Data are reported as mg P kg^{-1} soil (4D2a1a1).

3. Interferences

The Mo blue methods, which are very sensitive for P, are based on the principle that in an acid molybdate solution containing orthophosphate ions, a phosphomolybdate complex forms that can be reduced by ascorbic acid, SnCl$_2$, and other reducing agents to a Mo color. The intensity of blue color varies with the P concentration but is also affected by other factors such as acidity, arsentes, silicates, and substances that influence the oxidation-reduction conditions of the system (Olsen and Sommers, 1982).

4. Safety

Wear protective clothing (coats, aprons, sleeve guards, and gloves) and eye protection (face shields, goggles, or safety glasses). When preparing reagents, exercise special care. Many metal salts are extremely toxic and may be fatal if ingested. Thoroughly wash hands after handling these metal salts. Restrict the use of concentrated H$_2$SO$_4$ and HCl to a fume hood. Use safety showers and eyewash stations to dilute spilled acids and bases. Use sodium bicarbonate and water to neutralize and dilute spilled acids.

5. Equipment
5.1 Electronic balance, ±1.0-mg sensitivity
5.2 Mechanical reciprocating shaker, 200 oscillations min^{-1}, 1 ½ in strokes, Eberbach 6000, Eberbach Corp., Ann Arbor, MI
5.3 Centrifuge tubes, 50-mL, polyethylene
5.4 Filter paper, Whatman No. 42, 150 mm
5.5 Funnel, 60° angle, long stem, 50-mm diameter

5.6 Centrifuge, Centra GP-8, Thermo IEC, Needham Heights, MA
5.7 Volumetric flasks, 2-L, 100-mL, and 25-mL
5.8 Bottles, plastic, dark, 2-L
5.9 Pipettes, electronic digital, 2500 µL and 10 mL, with tips 2500 µL and 10 mL
5.10 Cups, plastic
5.11 Cuvettes, plastic, 4.5-mL, 1-cm light path, Daigger Scientific
5.12 Dispenser, 30 mL or 10 mL
5.13 Spectrophotometer, UV-Visible, Varian, Cary 50 Conc, Varian Australia Pty Ltd.
5.14 Computer, with Cary WinUV software, Varian Australia Pty Ltd., and printer

6. Reagents

6.1 Reverse osmosis deionized (RODI) water, ASTM Type I grade of reagent water
6.2 Sulfuric acid (H_2SO_4), concentrated, 36 N, trace pure grade
6.3 Molybdate solution. Dissolve 12.0 g ammonium molybdate tetrahydrate [$(NH_4)_6Mo_7O_{24} \cdot 4H_2O$] in approximately 250 mL RODI water. Dissolve 0.2908 potassium antimony tartarate [$K(SbO)C_4H_4O_6 \cdot 1/2H_2O$] in 100 mL RODI water. Add these two dissolved reagents to 1-L 5 N H_2SO_4 (141 mL conc. H_2SO_4 diluted to 1-L), mix thoroughly, and dilute with RODI water to 2-L. Store in dark bottle and refrigerate (reagent A).
6.4 Ascorbic acid: Dissolve 2.112 g of ascorbic acid in 400 mL of reagent A and mix (reagent B). Prepare fresh daily.
6.5 Stock standard P solution (SSPS), 100.0 mg P L^{-1}. In a 1-L volumetric flask, dissolve 0.4394 g primary standard grade anhydrous potassium dihydrogen phosphate (KH_2PO_4) that has been dried for 2 h at 110°C in about 800 mL RODI water. Dilute to volume with RODI water and invert to thoroughly mix. Store in polyethylene containers. Make fresh weekly. Store in a refrigerator.
6.6 Working stock standard P solution (WSSPS), 10.0 mg P L^{-1}. In a 1-L volumetric flask, pipet 100 mL SSPS and in 700 mL RODI water. Dilute to volume with RODI water and invert to thoroughly mix. Make fresh daily.
6.7 Standard P calibration solutions (SPCS) or working standards, 0.8, 0.6, 0.4, 0.2, and 0.0 mg P L^{-1}. Make fresh daily. To five 25-mL volumetric flasks add as follows:

6.7.1 0.8 mg P L^{-1} = 2.0 mL WSSPS
6.7.2 0.6 mg P L^{-1} = 1.5 mL WSSPS
6.7.3 0.4 mg P L^{-1} = 1.0 mL WSSPS
6.7.4 0.2 mg P L^{-1} = 0.5 mL WSSPS
6.7.5 0.0 mg P L^{-1} = 0.0 mL WSSPS (blank)

Add 4 mL reagent B, dilute each SPCS to mark with RODI water (extracting solution), and invert to thoroughly mix.

6.8 Quality Control Samples: 0.1 mg L^{-1} solution made from SSPS; blanks; and selected SPCS. In addition, SSL soil standard is routinely included in a batch for quality control.

7. Procedure

7.1 Weigh 2.5 g of <2-mm or fine-grind, air-dry soil to nearest mg on an electronic balance and place into a 50-mL centrifuge flask. If sample is moist, weigh enough soil to achieve ≈ 2.5 g of air-dry soil.

7.2 Add 25.0 mL of RODI water to sample. Place the sample in shaker and shake for 30 min at 200 oscillations min^{-1} at room temperature (20°C± 2°C).

7.3 Remove the sample from the shaker. Centrifuge at 3000 rpm for 20 min, decant, filter, and collect extract in receiving cup. If extracts are not to be determined immediately after collection, then store samples at 4°C. Analyze samples within 24 h.

7.4 Pipet 2-mL of sample extract into plastic cup, add 4-mL reagent B and 19 mL RODI water. Color is stable for 24 h. Maximum intensity develops in 10 min.

7.5 Transfer SPCS and sample extracts into 4.5-mL cuvettes.

7.6 Set spectrophotometer to 882 nm. Autozero with calibration blank.

7.7 Calibrate the instrument using the SPCS. The data system will then associate the concentrations with the instrument responses for each SPCS. Rejection criteria for SPCS, if R^2 <0.99.

7.8 Run samples using calibration curve. Sample concentration is calculated from the regression equation. Rejection criteria for batch, if blanks as samples > 0.01; if SPCS as samples > ± 20%; and if SSL standard > ± 20% mean. Record results to the nearest 0.01 unit for the sample extract and each SPCS.

7.9 If samples are outside the calibration range, dilute samples with extracting solution and re-run.

8. Calculations

Convert extract P (mg L^{-1}) to soil P (mg kg^{-1}) as follows:

Soil P (mg kg^{-1}) = [(A x B x C$_1$ x C$_2$ x R x 1000)/E]

where:
A   = Sample extract reading (mg kg^{-1})
B   = Extract volume (L)
C$_1$ = Dilution, required
C$_2$ = Dilution, if performed
R   = Air-dry/oven-dry ratio (procedure 3D1) or field-moist/oven-dry ratio (procedure 3D2)
1000 = Conversion factor to kg-basis
E   = Sample weight (g)

9. Report
Report data to the nearest 0.1 mg P kg^{-1} soil.

10. Precision and Accuracy
Precision and accuracy data are available from the SSL upon request.

11. References

Adams, F. 1974. Soil solution. p. 441-481. *In* E.W. Carson (ed.) The plant root and its environment. Univ. Press of Virginia, Charlottesville, VA.
Jackson, M.L. 1958. Soil chemical analysis. Prentice-Hall, Inc., Englewood Cliffs, N.J.
National Research Council. 1993. Soil and water quality. An agenda for agriculture. Committee on long-range soil and water conservation. Natl. Acad. Press, Washington, DC.
Olsen, S.R., and L.E. Sommers. 1982. Phosphorus. p. 403-430. *In* A.L. Page, R.H. Miller, and D.R. Keeney (eds.) Methods of soil analysis. Part 2. Chemical and microbiological properties. 2nd ed. Agron. Monogr. 9. ASA and SSSA, Madison, WI.

Rigler, F.W. 1968. Further observations inconsistent with the hypothesis that the molybdenum blue method measures orthophosphate in lake water. Limnology and Oceanography 13:7-13.

Sharpley, A.N., and R.G. Menzel. 1987. The impact of soil and fertilizer phosphorus on the environment. Adv. Agron. 41:297-324.

U.S. Department of Interior, U.S. Geological Survey. 1993. Methods for determination of inorganic substances in water and fluvial sediments. Book 5. Chapter Al. Method I-2601-78.

U.S. Environmental Protection Agency. 1983. Methods for chemical analysis of water and wastes. EPA-600/4-79-020. Revised March, 1983, Method 365.1.

Soil Tests (4D)
Aqueous Extraction (4D2)
Single-Point Extraction (4D2a)
1:10, 30 min (4D2a1)
Flow Injection, Automated Ion Analyzer (4D2a1b)
Phosphorus (4D2a1b1)
Air-Dry or Field-Moist, < 2 mm (4D2a1b1a-b1)

1. Application

Phosphorus occurs in soil in both the solution and solid phase. These forms are well documented but questions still remain concerning the exact nature of the constituents and ionic forms found in water, soils, and sediments (National Research Council, 1993). These forms influence P availability in relation to root absorption and plant growth; runoff and water quality problems; and P loadings.

Water soluble P has been defined as P measured in water, dilute salt extracts (e.g., 0.01 M $CaCl_2$), displaced soil solutions, or saturation paste extracts (Olsen and Sommers, 1982). Even though the water soluble fraction principally consists of inorganic orthophosphate ions, there is evidence that some organic P is also included (Rigler, 1968).

The water or dilute salt extracts represent an attempt to approximate the soil solution P concentration. As an index of P availability, the objectives of this method are (1) to determine the P concentration level in the soil extract that limits plant growth (Olsen and Sommers, 1982) and (2) to determine the composition of the soil solution so that the chemical environment of the plant roots may be defined in quantitative terms (Adams, 1974). The sum of water soluble P and pH 3 extractable P has also been defined as the available P in runoff (Jackson, 1958).

2. Summary of Method

A 2.5-g sample of <2-mm, air-dry soil is mechanically shaken for 30 min in 25-mL of reverse osmosis deionized water (RODI). The sample is then centrifuged until solution is free of soil mineral particles, and then filtered until clear extracts are obtained.

A flow injection automated ion analyzer is used to measure the orthophosphate ion (PO_4^{3-}). This ion reacts with ammonium molybdate and antimony potassium tartrate under acidic conditions to form a complex. This complex is reduced with ascorbic acid to form a blue complex which absorbs light at 880 nm. Absorbance is proportional to the concentration of PO_4^{3-} in the sample. Data are reported as mg P kg^{-1} soil (4D2a1b1).

3. Interferences

Silica forms a pale blue complex which also absorbs at 880 nm. This interference is generally insignificant as a Si concentration of approximately 30 mg SiO_2 L^{-1} would be required to produce a 0.005 mg P L^{-1} positive error in orthophosphate (LACHAT, 1993).

Glassware contamination is a problem in low-level P determinations. Glassware should be washed with 1:1 HCl and rinsed with deionized water. Commercial detergents should rarely be needed but, if they are used, use P-free preparation for lab glassware (LACHAT, 1993).

Concentrations of ferric ion >50 mg L^{-1} will cause a negative error due to competition with the complex for the reducing agent ascorbic acid. Samples high in Fe can be pretreated with sodium bisulfite to eliminate this interference. Treatment with bisulfite will also remove the interference due to arsenates (LACHAT, 1993).

4. Safety

Wear protective clothing (coats, aprons, sleeve guards, and gloves) and eye protection (face shields, goggles, or safety glasses). When preparing reagents, exercise special care. Many metal salts are extremely toxic and may be fatal if ingested. Thoroughly wash hands after handling these metal salts. Restrict the use of concentrated H_2SO_4 and HCl to a fume hood. Use safety showers and eyewash stations to dilute spilled acids and bases. Use sodium bicarbonate and water to neutralize and dilute spilled acids.

5. Equipment

5.1 Electronic balance, ±1.0-mg sensitivity
5.2 Centrifuge tubes, 50-mL, polyethylene
5.3 Mechanical reciprocating shaker, 200 oscillations min^{-1}, 1 ½ in strokes, Eberbach 6000, Eberbach Corp., Ann Arbor, MI
5.4 Centrifuge, Centra GP-8, Thermo IEC, Needham Heights, MA
5.5 Filter paper, Whatman No. 42, 150 mm
5.6 Funnel, 60° angle, long stem, 50-mm diameter
5.7 Volumetric flasks, 1-L and 250-mL
5.8 Bottles, plastic, dark, 1-L
5.9 Cups, plastic
5.10 Flow Injection Automated Ion Analyzer, QuikChem AE, LACHAT Instruments, Milwaukee, WI, with computer and printer
5.11 XYZ Sampler, LACHAT Instruments, Milwaukee, WI
5.12 Reagent Pump, LACHAT Instruments, Milwaukee, WI
5.13 Automated Dilution Station, LACHAT Instruments, Milwaukee, WI
5.14 Sample Processing Module (SPM) or channel, QuikChem Method (10-115-01-1-A, orthophosphate in waters, 0.01 to 2.0 mg P L^{-1}), LACHAT Instruments, Milwaukee, WI
5.15 Computer, with QuikChem software, LACHAT Instruments, Milwaukee, WI, and printer
5.16 Pipettes, electronic digital, 2500 µL and 10 mL, with tips, 2500 µL and 10 mL
5.17 Vials, plastic, 25-mL (standards)
5.18 Culture tubes, glass, 10-mL (samples)
5.19 Dispenser, 30 mL or 10 mL

6. Reagents

6.1 Reverse osmosis deionized (RODI) water, ASTM Type I grade of reagent water
6.2 Helium, compressed gas
6.3 Sulfuric acid (H_2SO_4), concentrated, 36 N, trace pure grade
6.4 Stock ammonium molybdate solution. In 1-L volumetric flask dissolve 40.0 g ammonium molybdate tetrahydrate [$(NH_4)_6Mo_7O_{24}\cdot4H_2O$] in approximately 800 mL RODI water. Dilute to the mark with RODI water and invert to thoroughly mix. Stir for 4 h. Store in plastic and refrigerate.
6.5 Stock antimony potassium tartrate solution. In 1-L flask, dissolve 3.0 g antimony potassium tartrate (potassium antimony tartrate hemihydrate [$K(SbO)C_4H_4O_6\cdot1/2H_2O$] in approximately 800 mL RODI water. Dilute to the mark and invert to thoroughly mix. Store in dark bottle and refrigerate.
6.6 Molybdate color reagent. In 1 L volumetric flask, add about 500 mL RODI water, then add 35.0 mL concentrated sulfuric acid (CAUTION: The solution will get very hot!). Swirl to mix. When it can be comfortably handled, add 72 mL stock antimony potassium tartrate solution and 213 mL stock ammonium molybdate solution. Dilute to volume with RODI water and invert three times. Degas with helium ≈ 5 min.

6.7 Ascorbic acid reducing solution. In l-L volumetric flask, dissolve 60.0 g ascorbic acid in about 700 mL RODI water. Dilute to volume with RODI water and invert three times. Degas with helium ≈ 5 min. Optional: After dilution to volume and degassing, dissolve 1.0 g dodecyl sulfate ($CH_3(CH_2)_{11}OSO_3Na$). Prepare fresh weekly.

6.8 Sodium hydroxide - EDTA rinse. Dissolve 65 g sodium hydroxide (NaOH) and 6 g tetrasodium ethylenediamine tetraacetic acid (Na_4EDTA) in 1.0 L RODI water.

6.9 Stock standard P solution (SSPS), 100.0 mg P L^{-1} (ppm). In a 1-L volumetric flask, dissolve 0.4394 g primary standard grade anhydrous potassium dihydrogen phosphate (KH_2PO_4) that has been dried for 2 h at 110°C in about 800 mL RODI water. Dilute to volume and invert to thoroughly mix. Do not degas. Store in polyethylene containers. Make fresh weekly. Store in a refrigerator.

6.10 Working stock standard P solution (WSSPS), 10.0 mg P L^{-1}. In a 1-L volumetric flask, dilute 100.0 mL SSPS to mark with RODI water. Invert to thoroughly mix. Make fresh daily.

6.11 Standard P calibration solutions (SPCS) or working standards, 2.0, 1.0, 0.5, 0.20, 0.05, 0.01, and 0.00 mg P L^{-1}. Make fresh daily. To seven 250-mL volumetric flasks add as follows:

6.11.1 2.0 mg P L^{-1} = 50.0 mL WSSPS
6.11.2 1.0 mg P L^{-1} = 25.0 mL WSSPS
6.11.3 0.5 mg P L^{-1} = 12.5 ml WSSPS
6.11.4 0.20 mg P L^{-1} = 5.0 mL WSSPS
6.11.5 0.05 mg P L^{-1} = 1.25 mL WSSPS
6.11.6 0.01 mg P L^{-1} = 0.25 mL WSSPS
6.11.7 0.00 mg P L^{-1} = 0 mL WSSPS (blank)

Dilute each SPCS to the mark with RODI water and invert to thoroughly mix. Do not degas.

7. Procedure

7.1 Weigh 2.5 g of <2-mm or fine-grind, air-dry soil to nearest mg on an electronic balance and place into a 50-mL centrifuge tube. If sample is moist, weigh enough soil to achieve ≈ 2.5 g of air-dry soil.

7.2 Add 25.0 mL of RODI water to sample. Transfer the sample to the shaker. Shake for 30 min at 200 oscillations min^{-1} at room temperature (20°C± 2°C).

7.3 Remove the sample from the shaker. Centrifuge at 2000 rpm for 10 min decant, filter, and collect extract in receiving cup.

7.4 Transfer sample extracts into culture tubes and place in XYZ sample trays marked "Samples". If extracts are not to be determined immediately after collection, then store samples at 4°C. Analyze samples within 24 h.

7.5 Transfer SPCS standards into plastic vials and place in descending order in XYZ sample trays marked "Standards".

7.6 Refer to the operating and software reference manuals for LACHAT set-up and operation.

7.7 Turn main power switch "ON" and allow 15 min for heater module to warm up to 37°C.

7.8 On reagent pump, set speed to 35. Pump RODI water through system for 20 min.

7.9 On computer main menu, select "Methods" and then "Analysis Select and Download". On method list, select water soluble P method. System unit receives the downloaded method and initializes it.

7.10 Pump reagents into manifold. Continue this step and observe baseline. A good baseline needs to be smooth and at zero absorbance. Scatter is indicative of air bubbles and irregular reagent flow. Also observe for any back-pressure in manifold tubing.

7.11 On computer main menu, select "Samples", "Tray Definition and Submit", and then "Edit" to create new sample tray followed by "Submit" to run new sample tray.

7.12 Method parameters specific to water soluble P are defined within the "Method Definition" menu. Some of these parameters have been modified from the QuikChem Method 10-115-01-1-A, orthophosphate in waters (U.S. Environmental Protection Agency, 1983; LACHAT Instruments, 1993; U.S. Department of Interior, Geological Survey 1993). Modifications are primarily related to the criteria and strategies for calibration standards and to injection timing.

7.13 Some of the method parameters as they relate to calibration standards are as follows:

7.13.1 There are 7 calibration standards (2.00, 1.00, 0.50, 0.20, 0.05, 0.01, and 0.00 mg P L^{-1}) with a data format of ####.###, i.e., data rounded to 3 places.

7.13.2 The segments/boundaries for the calibration standards are A - D (2.0 to 0.20 mg P L^{-1}) and D - G (0.20 to 0.00 mg P L^{-1}).

7.13.3 The protocol (replications) for the calibration standards is as follows: AA BB CCC DDDD EEEE FFFF GG

7.13.4 The check standard is 2.0 mg P L^{-1}. Maximum number of consecutive trays between check standard is one; maximum number of consecutive samples between check standard is 60; and maximum elapse time between check standards is 2 h.

7.13.5 Calibration strategy for segments A - D and D - G are normal. The normal strategy requires a minimum correlation coefficient of 0.99. Both segments require a maximum standard deviation in slope of 50%. A calibration passes only when both criteria are met. Strategies are user designated. In addition, calibration strategies are based on the full chord. Chord 0 is full chord, and chord 1 - 5 are sections of peak from start of peak to end of peak.

7.13.6 The instrument is calibrated with the injection of SPCS. The data system then associates the concentrations with the instrument response for each SPCS.

7.14 Method parameters in relation to timing are as follows:

7.14.1 Cycle period: 40 s

7.14.2 Inject to start of peak period: 18 s. To see if peaks are being timed correctly, scan across correlation coefficients for all chords 1 - 5. The most peak area should be between chords 2 - 4 with the most signal-to-noise ratio in chords 1 and 5.

7.14.3 Inject to end of peak period: 52 s

7.14.4 Automatic timing, where standard assumptions are in effect; no manual timing.

7.15 Method parameters in relation to data presentation are as follows:

229

7.15.1 Top Scale Response: 0.50 abs

7.15.2 Bottom Scale Response: 0.00 abs

7.16 Method parameters in relation to data results are as follows:

7.16.1 Set Default Chord to 3. This change must be made to both the sample and the calibration RDF's.

7.17 Refer to the "Method Definition" for water soluble P for other method parameters not discussed here.

7.18 Run samples using calibration curve. Sample concentration is calculated from the regression equations. Report results to the nearest 0.01 unit for the sample extract and each SPCS.

7.19 If samples are outside calibration range, dilute samples with extracting solution and re-analyze.

7.20 Upon completion of run, place the transmission lines into the NaOH - EDTA solution. Pump the solution for approximately 5 min to remove any precipitated reaction products. Then place these lines in RODI water and pump for an additional 5 min and proceed with the normal "Shut-down" procedure.

8. Calculations

Convert extract P (mg L^{-1}) to soil P (mg kg^{-1}) as follows:

Soil P (mg kg^{-1}) = [(A x B x C x R x 1000/E]

where:
A   = Sample extract reading (mg L^{-1})
B   = Extract volume (L)
D   = Dilution, if performed
R   = Air-dry/oven-dry ratio (procedure 3D1) or field-moist/oven-dry ratio (procedure 3D2)
1000 = Conversion factor to kg-basis
E   = Sample weight (g)

9. Report
Report data to the nearest 0.1 mg P kg^{-1} soil.

10. Precision and Accuracy
Precision and accuracy data are available from the SSL upon request.

11. References
Adams, F. 1974. Soil solution. p. 441-481. *In* E.W. Carson (ed.) The plant root and its environment. Univ. Press of Virginia, Charlottesville.
Jackson, M.L. 1958. Soil chemical analysis. Prentice-Hall, Inc., Englewood Cliffs, NJ.
LACHAT Instruments. 1993. QuikChem method 10-115-01-1-A, orthophosphate in waters, 0.02 to 2.0 mg P L^{-1}. LACHAT Instruments, 6645 West Mill Rd., Milwaukee, WI.

National Research Council. 1993. Soil and water quality. An agenda for agriculture. Committee on long-range soil and water conservation. Natl. Acad. Press, Washington, DC.

Olsen, S.R., and L.E. Sommers. 1982. Phosphorus. p. 403-430. *In* A.L. Page, R.H. Miller, and D.R. Keeney (eds.) Methods of soil analysis. Part 2. Chemical and microbiological properties. 2nd ed. Agron. Monogr. 9. ASA and SSSA, Madison, WI.

Rigler, F.W. 1968. Further observations inconsistent with the hypothesis that the molybdenum blue method measures orthophosphate in lake water. Limnology and Oceanography 13:7-13.

Sharpley, A.N., and R.G. Menzel. 1987. The impact of soil and fertilizer phosphorus on the environment. Adv. Agron. 41:297-324.

Soil Tests (4D)
Bray P-1 Extraction (4D3)
UV-Visible Spectrophotometer, Dual-Beam (4D3a)
Phosphorus (4D3a1)
Air-Dry or Field-Moist, <2 mm (4D3a1a-b1)

1. Application

The Bray P-1 procedure is widely used as an index of available P in the soil. Bray and Krutz (1945) originally designed the Bray P-1 extractant to selectively remove a portion of the adsorbed form of P with the weak, acidified ammonium fluoride solution. Adsorbed phosphorus is the in the anion form adsorbed by different charged surface functional groups that have varying degrees of adsorption affinity. In general, this method has been most successful on acid soils (Olsen and Sommers, 1982). The acid solubilizes calcium and aluminum phosphates, and partially extracts iron phosphates compounds. The NH_4F complexes the aluminum in solution and limits readsorption of P on iron oxides (Kuo, 1996). The Bray P-1 has limited ability to extract P in calcareous soils due to the neutralization of the dilute acid by carbonates.

2. Summary of Method

A 2.5-g soil sample is shaken with 25 mL of Bray P-1 extracting solution for 15 min. The sample is centrifuged until solution is free of soil mineral particles, and then filtered until clear extracts are obtained. A 2-mL aliquot is diluted with 8-mL of ascorbic acid molybdate solution. Absorbance of the solution is read using a spectrophotometer at 882 nm. Data are reported as mg P kg^{-1} soil (4D3a1).

3. Interferences

Many procedures may be used to determine P. Studies have shown that incomplete or excessive extraction of P to be the most significant contributor to interlaboratory variation. The Bray P-1 procedure is sensitive to the soil/extractant ratio, shaking rate, and time. This extraction uses the ascorbic acid-potassium antimony-tartarate-molybdate method. The Fiske-Subbarrow method is less sensitive but has a wider range before dilution is required (North Central Regional Publication No. 221, 1988). For calcareous soils, the Olsen method is preferred. An alternative procedure for calcareous soils is to use the Bray P-1 extracting solution at a 1:50 soil:solution ratio. This procedure has been shown to be satisfactory for some calcareous soils (Smith et al., 1957; North Central Regional Publication No. 221, 1988).

4. Safety

Wear protective clothing (coats, aprons, sleeve guards, and gloves) and eye protection (face shields, goggles, or safety glasses). When preparing reagents, exercise special care. Many metal salts are extremely toxic and may be fatal if ingested. Thoroughly wash hands after handling these metal salts. Restrict the use of concentrated H_2SO_4 and HCl to a fume hood. Use safety showers and eyewash stations to dilute spilled acids and bases. Use sodium bicarbonate and water to neutralize and dilute spilled acids.

5. Equipment

5.1 Electronic balance, ±0.10-mg sensitivity
5.2 Mechanical reciprocating shaker, 200 oscillations min^{-1}, 1 ½ in strokes, Eberbach 6000, Eberbach Corp., Ann Arbor, MI
5.3 Centrifuge tubes, 50-mL, polyethylene
5.4 Funnel, 60° angle, long stem, 50-mm diameter
5.5 Filter paper, Whatman 42, ashless, 9-cm diameter
5.6 Centrifuge, Centra GP-8, Thermo IEC, Needham Heights, MA
5.7 Pipettes, electronic digital, 2500 μL and 10 mL, with tips 2500 μL and 10 mL
5.8 Cups, plastic
5.9 Cuvettes, plastic, 4.5-mL, 1-cm light path, Daigger Scientific
5.10 Dispenser, 30 mL or 10 mL
5.11 Spectrophotometer, UV-Visible, Dual-View, Varian, Cary 50 Conc, Varian Australia Pty Ltd.
5.12 Computer, with Cary WinUV software, Varian Australia Pty Ltd., and printer

6. Reagents

6.1 Reverse osmosis deionized (RODI) water, ASTM Type I grade of reagent water
6.2 Hydrochloric acid (HCl), concentrated, 12 N, trace pure grade
6.3 HCl, 1 N. Carefully add 83.33 mL of concentrated HCl to RODI water and dilute to 1-L volume.
6.4 Sulfuric acid (H_2SO_4), concentrated, 36 N
6.5 Bray No. 1 Extracting solution, 0.025 N HCl and 0.03 N NH$_4$F. Dissolve 8.88 g of NH$_4$F in 4 L RODI water. Add 200 mL of 1.0 N HCl and dilute to 8 L with RODI water. The solution pH should be 2.6 ± 0.5. Store in a polyethylene bottle.
6.6 Sulfuric-tartarate-molybdate solution (STMS). Dissolve 60 g of ammonium molybdate tetrahydrate [(NH$_4$)$_6$Mo$_7$O$_{24}$·4H$_2$O] in 200 mL of boiling RODI water. Allow to cool to room temperature. Dissolve 1.455 g of antimony potassium tartarate (potassium antimony tartarate hemihydrate [K(SbO)C$_4$H$_4$O$_6$·1/2H$_2$O] in the ammonium molybdate solution. Slowly and carefully add 700 mL of concentrated H_2SO_4. Cool and dilute to 1 L with RODI water. Store in the dark in the refrigerator.
6.7 Ascorbic acid solution. Dissolve 6.6 g of ascorbic acid in RODI water and dilute to 50 mL with RODI water. Make fresh daily.
6.8 Working ascorbic acid molybdate solution (WAMS). Mix 25 mL of STMS solution with 800 mL of RODI water. Add 10 mL of ascorbic acid solution and dilute to 1 L with RODI water. Allow to stand at least 1 h before using. Prepare fresh daily.
6.9 Working stock standard P solution (WSSPS), 100.0 mg P L^{-1}. In a 1-L volumetric flask, dissolve 0.4394 g primary standard grade anhydrous potassium dihydrogen phosphate (KH$_2$PO$_4$) that has been dried for 2 h at 110°C in about 800 mL extracting solution. Dilute to 1-L volume with extracting solution water and invert to thoroughly mix. Store in polyethylene containers. Make fresh weekly. Store in a refrigerator.
6.10 Standard P calibration solutions (SPCS) or working standards, 5.0, 4.0, 2.0, 0.8, 0.4, 0.0 mg P L^{-1}. Make fresh weekly. Store in a refrigerator. Allow to equilibrate to room temperature before use. To six 250-mL volumetric flasks add as follows:

6.10.1 5.0 mg P L^{-1} = 12.5 mL WSSPS

6.10.2 4.0 mg P L^{-1} = 10.0 mL WSSPS
6.10.3 2.0 mg P L^{-1} = 5.0 mL WSSPS
6.10.4 0.8 mg P L^{-1} = 2.0 mL WSSPS
6.10.5 0.4 mg P L^{-1} = 1.0 mL WSSPS
6.10.6 0.0 mg P L^{-1} = 0 mL WSSPS (blank)

Dilute each SPCS to the mark with extracting solution and invert to thoroughly mix.

6.11 Quality Control Samples: 0.1 mg L^{-1} solution made from SSPS; blanks; and selected SPCS. In addition, SSL soil standard and WEPAL ISE's (Wageningen Evaluating Programmes for Analytical Laboratories, International Soil Exchange) from The Netherlands are routinely included in a batch for quality control.

7. Procedure

7.1 Weigh 2.5 g of <2-mm or fine-grind, air-dry soil to nearest mg on an electronic balance and place into a 50-mL centrifuge tube. If sample is moist, weigh enough soil to achieve ≈ 2.5 g of air-dry soil.

7.2 Dispense 25.0 mL of extracting solution to tube.

7.3 Transfer the sample in the shaker. Shake for 15 min at 200 oscillations min^{-1} at room temperature (20°C).

7.4 Remove the sample from the shaker. Centrifuge at 2000 rpm for 10 min, decant, filter, and collect extract in receiving cup. If extracts are not to be determined immediately after collection, then store samples at 4°C. Analyze samples within 72 h.

7.5 Use the pipette to transfer a 2-mL aliquot of the sample to a plastic cup. Also transfer a 2-mL aliquot of each SPCS to a plastic cup. Use a clean pipette tip for each sample and SPCS.

7.6 Dispense 8 mL of the WAMS to sample aliquot and to each SPCS. Swirl to mix. The color reaction requires a minimum of 20 min before analyst records readings.

7.7 Transfer sample extract and SPCS to cuvettes.

7.8 Set the spectrophotometer to read at 882 nm. Autozero with calibration blank.

7.9 Calibrate the instrument using the SPCS. The data system will then associate the concentrations with the instrument responses for each SPCS. Rejection criteria for SPCS, if R^2 <0.99.

7.10 Run samples using calibration curve. Sample concentration is calculated from the regression equation. Rejection criteria for batch, if blanks as samples > 0.01; If SPCS as samples > ± 20%; if SSL standard > ± 20% mean; and if ISE > (3 x MAD), where MAD = median of absolute deviations. Record results to the nearest 0.01 unit for the sample extract and each SPCS.

7.11 If samples are outside the calibration range, dilute sample extracts with extracting solution and re-analyze.

8. Calculations

Convert the extract P (mg L^{-1}) to soil P (mg kg^{-1}) as follows:

Soil P (mg kg^{-1}) = =[(A x B x C x R x 1000)/E]

where:
A   = Sample extract reading (mg L^{-1})
B   = Extract volume (L)
C   = Dilution, if performed
R   = Air-dry/oven-dry ratio (procedure 3D1) or field-moist/oven-dry ratio (procedure 3D2)
1000 = Conversion factor to kg-basis
E   = Sample weight (g)

9. Report
Report data to the nearest 0.1 mg P kg^{-1} soil.

10. Precision and Accuracy
Precision and accuracy data are available from the SSL upon request.

11. References
Bray, R.H., and L.T. Kurtz. 1945. Determination of total, organic, and available forms of phosphorus in soils. Soil Sci. 59:39-45.

Kuo, S. 1996. Phosphorus. p. 869-919. *In* D.L. Sparks (ed.) Methods of soil analysis. Part 3. Chemical methods. No. 5. ASA and SSSA, Madison, WI.

North Central Regional Publication No. 221. 1988. Recommended chemical soil test procedures for the North Central region. Agric. Exp. Stn. of IL, IN, IA, KS, MI, MN, MS, NE, ND, OH, SD, WI, and USDA cooperating.

Olsen, S.R., and L.E. Sommers. 1982. Phosphorus. p. 403-430. *In* A.L. Page, R.H. Miller, and D.R. Keeney (eds.) Methods of soil analysis. Part 2. Chemical and microbiological properties. 2nd ed. Agron. Monogr. 9. ASA and SSSA, Madison, WI.

Smith, F.W., B.G. Ellis, and J. Grava. 1957. Use of acid-fluoride solutions for the extraction of available phosphorus in calcareous soils and in soils to which rock phosphate has been added. Soil Sci. Soc. Am. Proc. 21:400-404.

Soil Tests (4D)
Bray P-1 Extraction (4D3)
Flow-Injection, Automated Ion-Analyzer (4D3b)
Phosphorus (4D3b1)
Air-Dry or Field-Moist, <2 mm (4D3b1a-b1)

1. Application
The Bray P-1 procedure is widely used as an index of available P in the soil. Bray and Krutz (1945) originally designed the Bray P-1 extractant to selectively remove a portion of the adsorbed form of P with the weak, acidified ammonium fluoride solution. Adsorbed phosphorus is the in the anion form adsorbed by different charged surface functional groups that have varying degrees of adsorption affinity. In general, this method has been most successful on acid soils (Olsen and Sommers, 1982). The acid solubilizes calcium and aluminum phosphates, and partially extracts iron phosphates compounds. The NH$_4$F complexes the aluminum in solution and limits readsorption of P on iron oxides (Kuo, 1996). The Bray P-1 has limited ability to extract P in calcareous soils due to the neutralization of the dilute acid by carbonates. For most soils, Bray P-1 and Mehlich No. 3 are nearly comparable in their abilities to extract native P but exceed Olsen sodium-bicarbonate method by two- to three-fold, indicating that predictive models for Bray P-1, Mehlich No. 3, and Olsen sodium-bicarbonate are closely associated with pH buffering of extractant (acid versus alkaline) (Burt et al., 2002).

2. Summary of Method

A 2.5-g soil sample is mechanically shaken for 15 min in 25-mL of Bray P-1 extracting solution. The sample is then centrifuged until solution is free of soil mineral particles, and then filtered until clear extracts are obtained.

A flow injection automated ion analyzer is used to measure the orthophosphate ion (PO_4^{3-}). This ion reacts with ammonium molybdate and antimony potassium tartrate under acidic conditions to form a complex. This complex is reduced with ascorbic acid to form a blue complex which absorbs light at 660 nm. Absorbance is proportional to the concentration of PO_4^{3-} in the sample. Data are reported as mg P kg^{-1} soil (4D3b1).

3. Interferences

Silica forms a pale blue complex which also absorbs at 660 nm. This interference is generally insignificant as a silica concentration of approximately 4000 mg L^{-1} would be required to produce a 1 mg L^{-1} positive error in orthophosphate (LACHAT Instruments, 1989).

Concentrations of ferric iron greater than 50 mg L^{-1} will cause a negative error due to competition with the complex for the reducing agent ascorbic acid. Samples high in iron can be pretreated with sodium bisulfite to eliminate this interference. Treatment with bisulfite will also remove the interference due to arsenates (LACHAT Instruments, 1989).

The determination of phosphorus is sensitive to variations in acid concentrations in the sample since there is no buffer. With increasing acidity, the sensitivity of the method is reduced. Samples, standards, and blanks should be prepared in a similar matrix.

4. Safety

Wear protective clothing (coats, aprons, sleeve guards, and gloves) and eye protection (face shields, goggles, or safety glasses). When preparing reagents, exercise special care. Many metal salts are extremely toxic and may be fatal if ingested. Thoroughly wash hands after handling these metal salts. Restrict the use of concentrated HCl, NH_4F, and H_2SO_4 to a fume hood. Use safety showers and eyewash stations to dilute spilled acids and bases. Use sodium bicarbonate and water to neutralize and dilute spilled acids.

5. Equipment

5.1 Electronic balance, ±1.0-mg sensitivity

5.2 Centrifuge tubes, 50-mL, polyethylene

5.3 Mechanical reciprocating shaker, 200 oscillations min^{-1}, 1 ½ in strokes, Eberbach 6000, Eberbach Corp., Ann Arbor, MI

5.4 Centrifuge, Centra GP-8, Thermo IEC, Needham Heights, MA

5.5 Filter paper, Whatman No. 42, 150 mm

5.6 Funnel, 60° angle, long stem, 50-mm diameter

5.7 Volumetric flasks, 1-L and 250-mL

5.8 Bottles, plastic, dark, 1-L

5.9 Cups, plastic

5.10 Dispenser, 30 mL or 10 mL

5.11 Flow Injection Automated Ion Analyzer, QuikChem AE, LACHAT Instruments, Milwaukee, WI, with computer and printer

5.12 XYZ Sampler, LACHAT Instruments, Milwaukee, WI

5.13 Reagent Pump, LACHAT Instruments, Milwaukee, WI

5.14 Automated Dilution Station, LACHAT Instruments, Milwaukee, WI

5.15 Sample Processing Module (SPM) or channel, QuikChem Method (12-115-01-1-A, orthophosphate in waters, 0.4 to 20 mg P L^{-1}), LACHAT Instruments, Milwaukee, WI

5.16 Computer, with QuikChem software, LACHAT Instruments, Milwaukee, WI, and printer

5.17 Pipettes, electronic digital, 2500 µL and 10 mL, with tips 2500 µL and 10 mL

5.18 Vials, plastic, 25-mL (standards)

5.19 Culture tubes, glass, 10-mL (samples)

6. Reagents

6.1 Reverse osmosis deionized (RODI) water, ASTM Type I grade of reagent water

6.2 Helium, compressed gas

6.3 Hydrochloric acid (HCl), concentrated, 12 N, trace pure grade

6.4 Sulfuric acid (H_2SO_4), concentrated, 36 N, trace pure grade

6.5 HCl, 1 N. Carefully add 83.33 mL of concentrated HCl to RODI water and dilute to 1-L volume.

6.6 Bray No. 1 Extracting Solution. 0.025 M HCl, and 0.03 M NH₄F. Dissolve 8.88 g of NH₄F in 4 L RODI water. Add 200 mL of 1.0 N HCl and dilute to 8 L with RODI water. The solution pH should be 2.6 ± 0.05. Store in a polyethylene bottle.

6.7 Stock ammonium molybdate solution. In 1-L volumetric flask dissolve 40.0 g ammonium molybdate tetrahydrate [$(NH_4)_6Mo_7O_{24}\cdot4H_2O$] in approximately 800 mL RODI water. Dilute to the mark with RODI water and invert to thoroughly mix. Stir for 4 h. Store in plastic and refrigerate.

6.8 Stock antimony potassium tartrate solution. In 1-L flask, dissolve 3.0 g antimony potassium tartrate (potassium antimony tartrate hemihydrate [$K(SbO)C_4H_4O_6\cdot1/2H_2O$] in approximately 800 mL RODI water. Dilute to the mark and invert to thoroughly mix. Store in dark bottle and refrigerate.

6.9 Molybdate color reagent. In 1 L volumetric flask, add 72 mL stock antimony potassium tartrate solution and 213 mL stock ammonium molybdate solution. Dilute to volume with RODI water and invert to thoroughly mix. Degas with helium ≈ 15 min.

6.10 Ascorbic acid reducing solution. In 1-L volumetric flask, dissolve 60.0 g ascorbic acid in about 700 mL RODI. Dilute to volume with RODI water and invert to thoroughly mix. Degas with helium ≈ 5 min. After dilution to volume and degassing, dissolve 1.0 g dodecyl sulfate ($CH_3(CH_2)_{11}OSO_3Na$). Prepare fresh daily.

6.11 0.8 M H_2SO_4 Carrier. To 1-L container, add 44.4 mL concentrated H_2SO_4 and bring volume with RODI water. (CAUTION: The solution will get very hot!) Invert to thoroughly mix. Degas with helium ≈ 5 min.

6.12 Sodium hydroxide - EDTA rinse. Dissolve 65 g sodium hydroxide (NaOH) and 6 g tetrasodium ethylenediamine tetraacetic acid (Na_4EDTA) in 1.0 L RODI water.

6.13 Working stock standard P solution (WSSPS), 100.0 mg P L^{-1}. In a 1-L volumetric flask, dissolve 0.4394 g primary standard grade anhydrous potassium dihydrogen phosphate (KH_2PO_4) that has been dried for 2 h at 110°C in about 800 mL extracting solution. Dilute to 1-L volume with extracting solution and invert to thoroughly mix. Store in polyethylene containers. Make fresh weekly. Store in a refrigerator.

6.14 Standard P calibration solutions (SPCS) or working standards, 20.00, 12.00, 4.00, 0.800, and 0.000 mg P L^{-1} as PO_4^{3-}. Make fresh weekly. Store in refrigerator. Allow to equilibrate to room temperature before use. To five 250-mL volumetric flasks add as follows:

6.14.1 20.00 mg P L^{-1} = 50 mL WSSPS
6.14.2 12.00 mg P L^{-1} = 30 mL WSSPS
6.14.3 4.00 mg P L^{-1} = 10 ml WSSPS
6.14.4 0.80 mg P L^{-1} = 2 mL WSSPS
6.14.5 0.00 mg P L^{-1}= 0 mL WSSPS (blank)

Dilute each SPCS to the mark with extracting solution and invert to thoroughly mix. Do not degas.

7. Procedure

7.1 Weigh 2.5 g of <2-mm or fine-grind, air-dry soil to nearest mg on an electronic balance and place into a 50-mL centrifuge tube. If sample is moist, weigh enough soil to achieve ≈ 2.5 g of air-dry soil.

7.2 Dispense 25.0 mL of extracting solution to tube.

7.3 Transfer the sample to the shaker. Shake for 15 min at 200 oscillations min^{-1} at room temperature (20°C± 2°C).

7.4 Remove the sample from the shaker. Centrifuge at 2000 rpm for 10 min, decant, filter, and collect extract in receiving cup. If extracts are not to be determined immediately after collection, then store samples at 4°C. Analyze samples within 72 h.

7.5 Transfer sample extracts into culture tubes and place in XYZ sample trays marked "Samples".

7.6 Transfer SPCS standards into plastic vials and place in descending order in XYZ sample trays marked "Standards".

7.7 Refer to the operating and software reference manuals for LACHAT set-up and operation.

7.8 Turn main power switch "ON" and allow 15 min for heater module to warm up to 60°C.

7.9 On reagent pump, set speed to 35.

7.10 On computer main menu, select "Methods" and then "Analysis Select and Download". On method list, select Bray P-1 Method. System unit receives the downloaded method and initializes it.

7.11 Pump reagents into appropriate chambers of the manifold. Continue this step and observe baseline. A good baseline needs to be smooth and at zero absorbance. Scatter is indicative of air bubbles and irregular reagent flow. Also observe for any back-pressure in manifold tubing.

7.12 On computer main menu, select "Samples", "Tray Definition and Submit", and then "Edit" to create new sample tray followed by "Submit" to run new sample tray.

7.13 Method parameters specific to Bray P-1 are defined within the "Method Definition" menu. Some of these parameters have been modified from the QuikChem Method 12-115-01-1-A, orthophosphate in soils (U.S. Environmental Protection Agency, 1983; LACHAT Instruments, 1989; U.S. Department of Interior, Geological Survey, 1993). Modifications are primarily related to the criteria and strategies for calibration standards and to injection timing.

7.14 Some of the method parameters as they relate to calibration standards are as follows:

7.14.1 There are 5 calibration standards (20.00, 12.00, 4.00, 0.80, and 0.00 mg P L^{-1}) with a data format of ####.###, i.e., data rounded to 3 places.

7.14.2 The segments/boundaries for the calibration standards are A - C (20.0 to 4.0 mg P L^{-1}); C - E (4.0 to 0.0 mg P L^{-1}).

7.14.3 The protocol (replications) for the calibration standards is as follows: AA BB CC DDD EEE.

7.14.4 The check standard is 20.0 mg P L^{-1}. Maximum number of consecutive trays between check standard is one; maximum number of consecutive samples between check standard is 60; and maximum elapse time between check standards is 2 h.

7.14.5 Calibration strategy for segments A – C and C - E are normal. The normal strategy requires a minimum correlation coefficient of 0.99. Both segments require a maximum standard deviation in slope of 50%. A calibration passes only when both criteria are met. Strategies are user designated. In addition, calibration strategies are based on the full chord. Chord 0 is full chord, and chord 1 - 5 are sections of peak from start of peak to end of peak.

7.14.6 The instrument is calibrated with the injection of SPCS. The data system then associates the concentrations with the instrument response for each SPCS.

7.15 Method parameters in relation to timing are as follows:

7.15.1 Cycle period: 40 s

7.15.2 Inject to start of peak period: 18 s. To see if peaks are being timed correctly, scan across correlation coefficients for all chords 1 - 5. The most peak area should be between chords 2 - 4 with the most signal-to-noise ratio in chords 1 and 5.

7.15.3 Inject to end of peak period: 46 s

7.15.4 Automatic timing, where standard assumptions are in effect. Manual timing may be helpful in this method.

7.16 Method parameters in relation to data presentation are as follows:

7.16.1 Top Scale Response: 0.50 abs

7.16.2 Bottom Scale Response: 0.00 abs

7.17 Refer to the "Method Definition" for Bray P-1 for other method parameters not discussed here.

7.18 Run samples using calibration curve. Sample concentration is calculated from the regression equations. Report results to the nearest 0.01 unit for the sample extract and each SPCS.

7.19 If samples are outside calibration range, dilute samples with extracting solution and re-analyze.

7.20 Upon completion of run, place the transmission lines into the NaOH - EDTA solution. Pump the solution for approximately 5 min to remove any precipitated reaction products. Then place these lines in RODI water and pump for an additional 5 min and proceed with the normal "Shut-down" procedure.

8. Calculations

Convert extract P (mg L^{-1}) to soil P (mg kg^{-1}) as follows:

Soil P (mg kg^{-1}) = [(A x B x C x R x 1000)/E]

where:
A = Sample extract reading (mg L^{-1})
B = Extract volume (L)
C = Dilution, if performed
R = Air-dry/oven-dry ratio (procedure 3D1) or field-moist/oven-dry ratio (procedure 3D2)

1000 = Conversion factor to kg-basis
E   = Sample weight (g)

9. Report
Report data to the nearest 0.1 mg P kg^{-1} soil.

10. Precision and Accuracy
Precision and accuracy data are available from the SSL upon request.

11. References
Bray, R.H., and L.T. Kurtz. 1945. Determination of total, organic, and available forms of phosphorus in soils. Soil Sci. 59:39-45.

Burt, R., M..D. Mays, E.C. Benham, and M.A. Wilson. 2002. Phosphorus characterization and correlation with properties of selected benchmark soils of the United States. Commun. Soil Sci. Plant Anal. 33:117-141.

Kuo, S. 1996. Phosphorus. p. 869-919. *In* D.L. Sparks (ed.) Methods of soil analysis. Part 3. Chemical methods. No. 5. ASA and SSSA, Madison, WI.

LACHAT Instruments. 1989. QuikChem method 12-115-01-1-A, phosphorus as orthophosphate, 0.4 to 20 mg P L^{-1}. LACHAT Instruments, 6645 West Mill Rd., Milwaukee, WI.

Olsen, S.R., and L.E. Sommers. 1982. Phosphorus. p. 403-430. *In* A.L. Page, R.H. Miller, and D.R. Keeney (eds.) Methods of soil analysis. Part 2. Chemical and microbiological properties. 2nd ed. Agron. Monogr. 9. ASA and SSSA, Madison, WI.

U.S. Department of the Interior, U.S. Geological Survey. 1993. Methods for determination of inorganic substances in water and fluvial sediments. Book 5. Chapter Al. Method I-2601-78.

U.S. Environmental Protection Agency. 1983. Methods for chemical analysis of water and wastes. EPA-600/4-79-020. Revised March, 1983, Method 353.2.

Bray P-2 Extraction (4D4)
UV-Visible Spectrophotometer, Dual-Beam (4D4a)
Phosphorus (4D4a1)
Air-Dry or Field-Moist, <2 mm (4D4a1a-b1)

1. Application
The Bray P-2 procedure functions to extract a portion of the plant available P in the soil. It has a similar composition to Bray P-1 extraction solution. The difference is a slightly higher concentration of HCl (0.025 N to 0.1 N) in the Bray P-2. It was originally designated by Bray and Kurtz (1945) to extract the easily acid soluble P a well as a fraction of adsorbed phosphates. The HCl solublizes calcium and aluminum phosphates, and partially extracts iron phosphate compounds. The NH$_4$F complexes the aluminum in solution and limits readsorption of P on iron oxides (Kuo, 1996). The higher acid concentration of the Bray P-2 should allow greater extraction of P in calcareous soils compared to Bray P-1, but it is not as widely used by soil testing laboratories as Bray P-1.

2. Summary of Method
A 2.5-g soil sample is shaken with 25 mL of Bray P-2 extracting solution for 15 min. The sample is centrifuged until solution is free of soil mineral particles, and then filtered until clear extracts are obtained. A 2-mL aliquot is diluted with 8-mL of ascorbic acid molybdate solution. Absorbance of the solution is read using a spectrophotometer at 882 nm. Data are reported as mg P kg^{-1} soil (4D4a1).

3. Interferences

Many procedures may be used to determine P. Studies have shown that incomplete or excessive extraction of P to be the most significant contributor to interlaboratory variation. The Bray procedure is sensitive to the soil/extractant ratio, shaking rate, and time. This extraction uses the ascorbic acid-potassium antimony-tartarate-molybdate method. The Fiske-Subbarrow method is less sensitive but has a wider range before dilution is required (North Central Regional Publication No. 221, 1988). For calcareous soils, the Olsen method is preferred. An alternative procedure for calcareous soils is to use the Bray P-1 extracting solution at a 1:50 soil:solution ratio. This procedure has been shown to be satisfactory for some calcareous soils (Smith et al., 1957; North Central Regional Publication No. 221, 1988). The higher acid concentration of the Bray P-2 should allow greater extraction of P in calcareous soils compared to Bray P-1, but it is not as widely used by soil testing laboratories as Bray P-1.

4. Safety

Wear protective clothing (coats, aprons, sleeve guards, and gloves) and eye protection (face shields, goggles, or safety glasses). When preparing reagents, exercise special care. Many metal salts are extremely toxic and may be fatal if ingested. Thoroughly wash hands after handling these metal salts. Restrict the use of concentrated H_2SO_4 and HCl to a fume hood. Use safety showers and eyewash stations to dilute spilled acids and bases. Use sodium bicarbonate and water to neutralize and dilute spilled acids.

5. Equipment

5.1 Electronic balance, ±0.10-mg sensitivity
5.2 Mechanical reciprocating shaker, 200 oscillations min^{-1}, 1 ½ in strokes, Eberbach 6000, Eberbach Corp., Ann Arbor, MI
5.3 Centrifuge tubes, 50-mL, polyethylene
5.4 Funnel, 60° angle, long stem, 50-mm diameter
5.5 Filter paper, Whatman 42, ashless, 9-cm diameter
5.6 Centrifuge, Centra GP-8, Thermo IEC, Needham Heights, MA
5.7 Pipettes, electronic digital, 2500 μL and 10 mL, with tips 2500 μL and 10 mL
5.8 Cups, plastic
5.9 Cuvettes, plastic, 4.5-mL, 1-cm light path, Daigger Scientific
5.10 Dispenser, 30 mL or 10 mL
5.11 Spectrophotometer, UV-Visible, Dual-View, Varian, Cary 50 Conc, Varian Australia Pty Ltd.
5.12 Computer, with Cary WinUV software, Varian Australia Pty Ltd., and printer

6. Reagents

6.1 Reverse osmosis deionized (RODI) water, ASTM Type I grade of reagent water
6.2 Hydrochloric acid (HCl), concentrated, 12 *N*, trace pure grade
6.3 HCl, 1 *N*. Carefully add 83.33 mL of concentrated HCl to RODI water and dilute to 1-L volume.
6.4 Sulfuric acid (H_2SO_4), concentrated, 36 *N*
6.5 Bray No. 2 Extracting solution, 0.1 *N* HCl and 0.03 *N* NH_4F. Dissolve 8.88 g of NH_4F in 4 L RODI water. Add 800 mL of 1.0 *N* HCl and dilute to 8 L with RODI water. Store in a polyethylene bottle.
6.6 Sulfuric-tartarate-molybdate solution (STMS). Dissolve 60 g of ammonium molybdate tetrahydrate $[(NH_4)_6Mo_7O_{24} \cdot 4H_2O]$ in 200 mL of boiling RODI water. Allow to cool to room temperature. Dissolve 1.455 g of antimony potassium tartarate (potassium antimony tartarate hemihydrate $[K(SbO)C_4H_4O_6 \cdot 1/2H_2O]$ in the ammonium molybdate solution. Slowly and carefully add 700 mL of concentrated H_2SO_4. Cool and dilute to 1 L with RODI water. Store in the dark in the refrigerator.
6.7 Ascorbic acid solution. Dissolve 6.6 g of ascorbic acid in RODI water and dilute to 50 mL with RODI water. Make fresh daily.

6.8 Working ascorbic acid molybdate solution (WAMS). Mix 25 mL of STMS solution with 800 mL of RODI water. Add 10 mL of ascorbic acid solution and dilute to 1 L with RODI water. Allow to stand at least 1 h before using. Prepare fresh daily.

6.9 Working stock standard P solution (WSSPS), 100.0 mg P L^{-1}. In a 1-L volumetric flask, dissolve 0.4394 g primary standard grade anhydrous potassium dihydrogen phosphate (KH_2PO_4) that has been dried for 2 h at 110°C in about 800 mL extracting solution. Dilute to 1-L volume with extracting solution water and invert to thoroughly mix. Store in polyethylene containers. Make fresh weekly. Store in a refrigerator.

6.10 Standard P calibration solutions (SPCS) or working standards, 5.0, 4.0, 2.0, 0.8, 0.4, 0.0 mg P L^{-1}. Make fresh weekly. Store in a refrigerator. Allow to equilibrate to room temperature before use. To six 250-mL volumetric flasks add as follows:

6.10.1 5.0 mg P L^{-1} = 12.5 mL WSSPS
6.10.2 4.0 mg P L^{-1} = 10.0 mL WSSPS
6.10.3 2.0 mg P L^{-1} = 5.0 mL WSSPS
6.10.4 0.8 mg P L^{-1} = 2.0 mL WSSPS
6.10.5 0.4 mg P L^{-1} = 1.0 mL WSSPS
6.10.6 0.0 mg P L^{-1} = 0 mL WSSPS (blank)

Dilute each SPCS to the mark with extracting solution and invert to thoroughly mix.

6.11 Quality Control Samples: 0.1 mg L^{-1} solution made from SSPS; blanks; and selected SPCS. In addition, SSL soil standard and WEPAL ISE's (Wageningen Evaluating Programmes for Analytical Laboratories, International Soil Exchange) from The Netherlands are routinely included in a batch for quality control.

7. Procedure

7.1 Weigh 2.5 g of <2-mm or fine-grind, air-dry soil to nearest mg on an electronic balance and place into a 50-mL centrifuge tube. If sample is moist, weigh enough soil to achieve ≈ 2.5 g of air-dry soil.

7.2 Dispense 25.0 mL of extracting solution to tube.

7.3 Transfer the sample in the shaker. Shake for 15 min at 200 oscillations min^{-1} at room temperature (20°C).

7.4 Remove the sample from the shaker. Centrifuge at 2000 rpm for 10 min, decant, filter, and collect extract in receiving cup. If extracts are not to be determined immediately after collection, then store samples at 4°C. Analyze samples within 72 h.

7.5 Use the pipette to transfer a 2-mL aliquot of the sample to a plastic cup. Also transfer a 2-mL aliquot of each SPCS to a plastic cup. Use a clean pipette tip for each sample and SPCS.

7.6 Dispense 8 mL of the WAMS to sample aliquot and to each SPCS. Swirl to mix. The color reaction requires a minimum of 20 min before analyst records readings.

7.7 Transfer sample extract and SPCS to cuvettes.

7.8 Set the spectrophotometer to read at 882 nm. Autozero with calibration blank.

7.9 Calibrate the instrument using the SPCS. The data system will then associate the concentrations with the instrument responses for each SPCS. Rejection criteria for SPCS, if R^2 <0.99.

7.10 Run samples using calibration curve. Sample concentration is calculated from the regression equation. Rejection criteria for batch, if blanks as samples > 0.01; If SPCS as samples > ± 20%; if SSL standard > ± 20% mean; and if ISE > (3 x MAD), where MAD = median of absolute deviations. Record results to the nearest 0.01 unit for the sample extract and each SPCS.

7.11 If samples are outside the calibration range, dilute sample extracts with extracting solution and re-analyze.

9. Calculations

Convert the extract P (mg L^{-1}) to soil P (mg kg^{-1}) as follows:

Soil P (mg kg^{-1}) = [(A x B x C x R x 1000)/E]

where:
A   = Sample extract reading (mg L^{-1})
B   = Extract volume (L)
C   = Dilution, if performed
R   = Air-dry/oven-dry ratio (procedure 3D1) or field-moist/oven-dry ratio (procedure 3D2)
1000 = Conversion factor to kg-basis
E   = Sample weight (g)

9. Report
Report data to the nearest 0.1 mg P kg^{-1} soil.

10. Precision and Accuracy
Precision and accuracy data are available from the SSL upon request.

11. References
Bray, R.H., and L.T. Kurtz. 1945. Determination of total, organic, and available forms of phosphorus in soils. Soil Sci. 59:39-45.
Kuo, S. 1996. Phosphorus. p. 869-919. *In* D.L. Sparks (ed.) Methods of soil analysis. Part 3. Chemical methods. No. 5. ASA and SSSA, Madison, WI.
North Central Regional Publication No. 221. 1988. Recommended chemical soil test procedures for the North Central region. Agric. Exp. Stn. of IL, IN, IA, KS, MI, MN, MS, NE, ND, OH, SD, WI, and USDA cooperating.
Olsen, S.R., and L.E. Sommers. 1982. Phosphorus. p. 403-430. *In* A.L. Page, R.H. Miller, and D.R. Keeney (eds.) Methods of soil analysis. Part 2. Chemical and microbiological properties. 2nd ed. Agron. Monogr. 9. ASA and SSSA, Madison, WI.
Smith, F.W., B.G. Ellis, and J. Grava. 1957. Use of acid-fluoride solutions for the extraction of available phosphorus in calcareous soils and in soils to which rock phosphate has been added. Soil Sci. Soc. Am. Proc. 21:400-404.

Soil Tests (4D)
Olsen Sodium-Bicarbonate Extraction (4D5)
UV-Visible Spectrophotometer, Dual-Beam (4D5a)
Phosphorus (4D5a1)
Air-Dry or Field-Moist, <2 mm (4D5a1a-b1)

1. Application
Olsen extractant is 0.5 *M* sodium bicarbonate solution at pH 8.5. This extractant is most applicable to neutral to calcareous soils (Buurman et al., 1996). Solubility of Ca-phosphate in

calcareous, alkaline, or neutral soils in increased because of the precipitation of Ca^{++} as CaCO3 (Soil and Plant Analysis Council, 1999). Olsen extractant correlates with Mehlich No. 3 on calcareous soils ($R^2 = 0.918$), even though the quantity of Mehlich No. 3 extractable P is considerably higher (Soil and Plant Analysis Council, 1999). While Mehlich No. 3, Bray P-1, and Olsen sodium-bicarbonate are linearly related, relationships developed between some P tests (e.g., Olsen P, Mehlich No. 3) may have limited predictive capability with increasing soil P content (Burt et al. 2000).

2. Summary of Method

A 1.0-g soil sample is shaken with 20 mL of Olsen sodium-bicarbonate extracting solution for 30 min. The sample is centrifuged until solution is free of soil mineral particles, and then filtered until clear extracts are obtained. Dilute 5-mL of sample extract with 5-mL of color reagent. The absorbance of the solution is read using a spectrophotometer at 882 nm. Data are reported as mg P kg^{-1} soil (4D5a1).

3. Interferences

The Mo blue methods, which are very sensitive for P, are based on the principle that in an acid molybdate solution containing orthophosphate ions, a phosphomolybdate complex forms that can be reduced by ascorbic acid, $SnCl_2$, and other reducing agents to a Mo color. The intensity of blue color varies with the P concentration but is also affected by other factors such as acidity, arsenates, silicates, and substances that influence the oxidation-reduction conditions of the system (Olsen and Sommers, 1982).

4. Safety

Wear protective clothing (coats, aprons, sleeve guards, and gloves) and eye protection (face shields, goggles, or safety glasses). When preparing reagents, exercise special care. Many metal salts are extremely toxic and may be fatal if ingested. Thoroughly wash hands after handling these metal salts. Restrict the use of concentrated H_2SO_4 and HCl to a fume hood. Use safety showers and eyewash stations to dilute spilled acids and bases. Use sodium bicarbonate and water to neutralize and dilute spilled acids.

5. Equipment

5.1 Electronic balance, ±1.0-mg sensitivity
5.2 Mechanical reciprocating shaker, 200 oscillations min^{-1}, 1 ½ in strokes, Eberbach 6000, Eberbach Corp., Ann Arbor, MI
5.3 Centrifuge tube, 50 mL, polyethylene
5.4 Funnel, 60° angle, long stem, 50-mm diameter
5.5 Filter paper, Whatman 42, 150 mm
5.6 Centrifuge, Centra, GP-8, Thermo IEC, Thermo IEC, Needham Heights, MA.
5.7 Pipettes, electronic digital, 1000µL and 10 mL, with tips, 1000µL and 10 mL
5.8 Cups, plastic
5.9 Dispenser, 30 mL or 10 mL
5.10 Cuvettes, plastic, 4.5-mL, 1-cm light path, Daigger Scientific
5.11 Spectrophotometer, UV-Visible, Dual-View, Varian, Cary 50 Conc, Varian Australia Pty Ltd.
5.12 Computer, with Cary WinUV software, Varian Australia Pty Ltd., and printer

6. Reagents

6.1 Reverse osmosis deionized (RODI) water, ASTM Type I grade of reagent water
6.2 Sulfuric acid (H_2SO_4), concentrated, 36 N, trace pure grade
6.3 Sulfuric acid (H_2SO_4), 4 M. To 250-mL volumetric, carefully add 56 mL concentrated H_2SO_4 to 150 mL RODI water. Allow to cool. Make to final volume with RODI water. Invert to thoroughly mix.
6.4 NaOH, 1 M. Dissolve 4 g NaOH in 100 mL RODI water.

6.5 Olsen Sodium Bicarbonate Extracting solution (0.5 M NaHCO$_3$). To 6-L container, dissolve 252 g NaHCO$_3$ in RODI water. Adjust the pH to 8.5 with 1M NaOH. Make to final volume. Mix thoroughly. Check pH every day.

6.6 Ammonium molybdate, 4%. Dissolve 4 g of [(NH$_4$)$_6$Mo$_7$O$_{24}$·4H$_2$O] in 100-mL volumetric with RODI water. Dilute to volume with RODI water. Store in the dark in the refrigerator.

6.7 Potassium antimony – (III) oxide tartarate, 0.275%. Dissolve 0.275 g [K(SbO)C$_4$H$_4$O$_6$·1/2H$_2$O] in 100 mL RODI water.

6.8 Ascorbic acid, 1.75%. Dissolve 1.75 g ascorbic acid in 100 mL RODI water. Prepare fresh daily.

6.9 Color developing reagent. To a 500-mL bottle, add 50 mL 4 M H$_2$SO$_4$, 15 mL 4% ammonium molybdate, 30 mL 1.75% ascorbic acid, 5 mL 0.275% potassium antimony – (III) oxide tartarate, and 200 mL RODI water. Mix well after each addition. Prepare fresh daily.

6.10 Stock standard P solution (SSPS), 100.0 mg P L^{-1}. In a 1-L volumetric flask, dissolve 0.4394 g primary standard grade anhydrous potassium dihydrogen phosphate (KH$_2$PO$_4$) that has been dried for 2 h at 110°C in about 800 mL extracting solution. Dilute to 1-L volume with extracting solution and invert to thoroughly mix. Store in polyethylene containers. Make fresh weekly. Store in a refrigerator.

6.11 Working stock standard P solution (WSSPS), 4.0 mg P L^{-1}. Pipet 10 mL of 100 mg P L^{-1} SSPS to 250-mL volumetric flask. Dilute to 250-mL volume with extracting solution and invert to thoroughly mix. Make fresh weekly. Store in the refrigerator.

6.12 Standard P calibration solution (SPCS), or working standards, 2.0, 1.6, 1.2, 0.8, 0.4, and 0.0 mg P L^{-1}. Make fresh weekly. Store in the refrigerator. Allow to equilibrate to room temperature before use. To six 50-mL volumetric flasks, add as follows:

6.12.1 2.0 mg P L^{-1} = 25 mL WSSPS
6.12.2 1.6 mg P L^{-1} = 20 mL WSSPS
6.12.3 1.2 mg P L^{-1} = 15 mL WSSPS
6.12.4 0.8 mg P L^{-1} = 10 mL WSSPS
6.12.5 0.4 mg P L^{-1} = 5 mL WSSPS
6.12.6 0.0 mg P L^{-1} = 0 mL WSSPS (blank)

Dilute each SPCS to mark with extracting solution and invert to thoroughly mix.

6.13 Quality Control Samples: 0.1mg P L^{-1} solution made from SSPS; blanks; and selected SPCS. In addition, SSL soil standard and WEPAL ISE's (Wageningen Evaluating Programmes for Analytical Laboratories, International Soil Exchange) from The Netherlands are routinely included in a batch for quality control.

7. Procedure

7.1 Weigh 1.0 g of <2mm or fine-grind, air-dry soil to the nearest mg into a 50-mL centrifuge tube. If sample is moist, weigh enough soil to achieve ≈ 1.0 g of air-dry soil.

7.2 Dispense 20.0 mL of extracting solution to tube.

7.3 Transfer the sample to the shaker. Shake for 30 min at 200 oscillations min^{-1} at room temperature (20°C± 2°C).

7.4 Remove the sample from the shaker. Centrifuge at 2000 rpm for 10 min, decant, filter, and collect extract in receiving cup. If extracts are not to be determined immediately after collection, then store samples at 4°C. Analyze samples within 72 h.

7.5 Use the pipette to transfer a 5-mL aliquot of the sample to a plastic cup. Also transfer a 5-mL aliquot of each SPCS to a plastic cup. Use a clean pipette tip for each sample and SPCS.

7.6 Dispense 5 mL of color developing reagent to sample aliquot and to each SPCS. Swirl to mix. Do not place sample cups close together as carbon dioxide is released and solution will bubble. The color reaction requires a minimum of 20 min before analyst records readings. Allowing 1 h for color development usually improves results. Color will remain stable for 24 h.

7.7 Transfer sample extract and SPCS to cuvettes.

7.8 Set the spectrophotometer to read at 882 nm. Autozero with calibration blank.

7.9 Calibrate the instrument using the SPCS. The data system will then associate the concentrations with the instrument responses for each SPCS. Rejection criteria for SPCS, if R^2 <0.99.

7.10 Run samples using calibration curve. Sample concentration is calculated from the regression equation. Rejection criteria for batch, if blanks as samples > 0.01; if SPCS as samples > ± 20%; if SSL standard > ± 20% mean; and if ISE > (3 x MAD), where MAD = median of absolute deviations. Record results to the nearest 0.01 unit for the sample extract and each SPCS.

7.11 If samples are outside calibration range, dilute sample extracts with extracting solution and re-analyze.

8. Calculations

8.1 Convert the extract P (mg L^{-1}) to soil P (mg kg^{-1}) as follows:

Soil P (mg kg^{-1}) = [(A x B x C x R x 1000)/E]

where:
A = Sample extract reading (mg L^{-1})
B = Extract volume (L)
C = Dilution, if performed
R = Air-dry/oven-dry ratio (procedure 3D1) or field-moist/oven-dry ratio (procedure 3D2)
1000 = Conversion factor to kg-basis
E = Sample weight (g)

9. Report
Report data to the nearest 0.1 mg P kg^{-1} soil.

10. Precision and Accuracy
Precision and accuracy data are available from the SSL upon request.

11. References
Buurman, P, B. van Lagen, and E.J. Velthorst. 1996. Manual for soil and water analysis. Backhuys Publ., Leiden, The Netherlands.
Burt, R., M..D. Mays, E.C. Benham, and M.A. Wilson. 2002. Phosphorus characterization and correlation with properties of selected benchmark soils of the United States. Commun. Soil Sci. Plant Anal. 33:117-141.

Olsen, S.R., and L.E. Sommers. 1982. Phosphorus. p. 403-430. *In* A.L. Page, R.H. Miller, and D.R. Keeney (eds.) Methods of soil analysis. Part 2. Chemical and microbiological properties. 2nd ed. Agron. Monogr. 9. ASA and SSSA, Madison, WI.

Soil and Plant Analysis Council. 1999. Handbook on reference methods for soil analysis. Council on Soil Testing and Plant Analysis, CRC Press, Boca Raton, FL.

Soil Tests (4D)
Mehlich No. 3 Extraction (4D6)
UV-Visible Spectrophotometer, Dual-Beam (4D6a)
Phosphorus (4D6a1)
Air-Dry or Field-Moist, < 2 mm (4D6a1a-b1)

1. Application

Mehlich No. 3 was developed by Mehlich (1984) as a multielement soil extraction (Ca, Mg, K, Na, P). In the Mehlich No. 3 procedure, P is extracted by reaction with acetic acid and Fl compounds. Mehlich No. 3 is a used an index of available P in the soil. Extraction of P by Mehlich No. 3 is designed to be applicable across a wide range of soil properties ranging in reaction from acid to basic (Mehlich, 1984). Mehlich No. 3 correlates well with Bray P-1 on acid to neutral (R^2 = 0.966) but does not correlate with Bray P-1 on calcareous soils (Soil and Plant Analysis Council, 1999). Mehlich No. 3 correlates with Olsen extractant on calcareous soils (R^2= 0.918), even though the quantity of Mehlich No. 3 extractable P is considerably higher (Soil and Plant Analysis Council, 1999). The Mehlich No. 3 extractant is neutralized less by carbonate compounds in soil than the double acid (Mehlich No. 1) and the Bray P-1, and is less aggressive towards apatite or other Ca-phosphate than the double acid and Bray P-2 extractants (Tran and Simard, 1993). Mehlich No. 3 can also be used to extract Ca, Mg, K, and Na in a wide range of soils and correlates well with Mehlich No. 1, Mehlich No. 2, and NH_4OAc (Soil and Plant Analysis Council, 1999).

2. Summary of Method

A 2.5-g soil sample is shaken with 25 mL of Mehlich No. 3 extracting solution for 5 min. The sample is centrifuged until solution is free of soil mineral particles, and then filtered until clear extracts are obtained. Dilute 0.5-mL of sample extract with 13.5-mL of working solution. Absorbance of the solution is read using a spectrophotometer at 882 nm. Data are reported as mg P kg^{-1} soil (4D6a1).

3. Interferences

The Mo blue methods, which are very sensitive for P, are based on the principle that in an acid molybdate solution containing orthophosphate ions, a phosphomolybdate complex forms that can be reduced by ascorbic acid, $SnCl_2$, and other reducing agents to a Mo color, with ascorbic acid most commonly used by agricultural laboratories (Murphy and Riley, 1962). The intensity of blue color varies with the P concentration but is also affected by other factors such as acidity, arsenates, silicates, and substances that influence the oxidation-reduction conditions of the system (Olsen and Sommers, 1982).

4. Safety

Wear protective clothing (coats, aprons, sleeve guards, and gloves) and eye protection (face shields, goggles, or safety glasses). When preparing reagents, exercise special care. Many metal salts are extremely toxic and may be fatal if ingested. Thoroughly wash hands after handling these metal salts. Restrict the use of concentrated H_2SO_4 and HCl to a fume hood. Use safety showers and eyewash stations to dilute spilled acids and bases. Use sodium bicarbonate and water to neutralize and dilute spilled acids.

5. Equipment

5.1 Electronic balance, ±1.0-mg sensitivity

5.2 Mechanical reciprocating shaker, 200 oscillations min[-1], 1 ½ in strokes, Eberbach 6000, Eberbach Corp., Ann Arbor, MI

5.3 Centrifuge tubes, 50-mL, polyethylene

5.4 Funnel, 60° angle, long stem, 50-mm diameter

5.5 Filter paper, Whatman 42, 150 mm

5.6 Centrifuge, Centra, GP-8, Thermo IEC, Needham Heights, MA

5.7 Pipettes, electronic digital, 1000 μL and 10 mL, with tips, 1000 μL and 10 mL

5.8 Dispenser, 30 mL or 10 mL

5.9 Cups, plastic

5.10 Cuvettes, plastic, 4.5-mL, 1-cm light path, Daigger Scientific

5.11 Spectrophotometer, UV-Visible, Dual-View, Varian, Cary 50 Conc, Varian Australia Pty Ltd.

5.12 Computer with Cary WinUV software, Varian Australia Pty Ltd., and printer

6. Reagents

6.1 Reverse osmosis deionized (RODI) water, ASTM Type I grade of reagent water

6.2 Sulfuric acid (H_2SO_4), concentrated, 36 N, trace pure grade

6.3 Mehlich No. 3 Extracting solution (0.2 N CH_3COOH; 0.25 N NH_4NO_3; 0.015 N NH_4F; 0.13 N HNO_3; 0.001 M EDTA). Premixed Mehlich No. 3 Extractant, Special-20, Hawk Creek Laboratory, Rural Route 1, Box 686, Simpson Road, Glen Rock, PA, 17327.

6.4 Sulfuric-tartarate-molybdate solution (STMS). Dissolve 100 g of ammonium molybdate tetrahydrate [$(NH_4)_6Mo_7O_{24}\cdot4H_2O$] in 500 mL of RODI water. Dissolve 2.425 g of antimony potassium tartarate (potassium antimony tartarate hemihydrate [$K(SbO)C_4H_4O_6\cdot1/2H_2O$] in the ammonium molybdate solution. Slowly and carefully add 1400 mL of concentrated H_2SO_4 and mix well. Cool and dilute to 2 L with RODI water. Store in the dark in the refrigerator.

6.5 Ascorbic acid solution. Dissolve 8.8 g of ascorbic acid in RODI water and dilute to 100-mL with RODI water. Make fresh daily.

6.6 Working ascorbic acid molybdate solution (WAMS). Dilute 20 mL of STMS solution and 10 mL of the ascorbic acid solution with RODI water to make 1-L. Allow solution to come to room temperature before using. Prepare fresh daily.

6.7 Working stock standard P solution (WSSPS), 100.0 mg P L[-1]. In a 1-L volumetric flask, dissolve 0.4394 g primary standard grade anhydrous potassium dihydrogen phosphate (KH_2PO_4) (dried for 2 h at 110°C) in about 800 mL extracting solution. Dilute to 1-L volume with extracting solution and invert to thoroughly mix. Store in polyethylene containers. Make fresh weekly. Store in a refrigerator.

6.8 Standard P calibration solutions (SPCS) or working standards, 12.0, 10.0, 8.0, 4.0, 2.0, 1.0, 0.8, 0.4 and 0.0 mg P L[-1]. Prepare fresh weekly. Store in a refrigerator. Allow to equilibrate to room temperature before use. In nine 250-mL volumetric flasks add as follows:

6.8.1 12.0 mg P L[-1] = 30 mL WSSPS

6.8.2 10.0 mg P L[-1] = 25 mL WSSPS

6.8.3 8.0 mg P L[-1] = 20 mL WSSPS

6.8.4 4.0 mg P L[-1] = 10 mL WSSPS

6.8.5 2.0 mg P L[-1] = 5 mL WSSPS

6.8.6 1.0 mg P L[-1] = 2.5 mL WSSPS

6.8.7 0.8 mg P L[-1] = 2 mL WSSPS

6.8.8 0.4 mg P L[-1] = 1 mL WSSPS

6.8.9 0.0 mg P L[-1] = 0 mL WSSPS (blank)

Dilute each SPCS to the mark with extracting solution and invert to thoroughly mix.

6.9 Quality Control Samples: 0.1 mg P L^{-1} solution made from SSPS; blanks; and selected SPCS. In addition, SSL soil standard and WEPAL ISE's (Wageningen Evaluating Programmes for Analytical Laboratories, International Soil Exchange) from The Netherlands are routinely included in a batch for quality control.

7. Procedure

7.1 Weigh 2.5 g of <2-mm or fine-grind, air-dry soil to the nearest mg into a 50-mL centrifuge tube. If sample is moist, weigh enough soil to achieve ≈ 2.5 g of air-dry soil.

7.2 Dispense 25.0 mL of extracting solution to the tube.

7.3 Transfer the sample to the shaker. Shake for 5 min at 200 oscillations min^{-1} at room temperature (20°C± 2°C).

7.4 Remove the sample from the shaker. Centrifuge at 2000 rpm for 10 min, decant, filter, and collect extract in receiving cups. If extracts are not to be determined immediately after collection, then store samples at 4°C. Analyze samples within 72 h.

7.5 Use the pipette to transfer a 1 mL (or 0.5-mL) aliquot of the sample to a plastic cup. Also transfer a 1 mL (or 0.5-mL) aliquot of each SPCS to a plastic cup. Use a clean pipette tip for each sample and SPCS.

7.6 Dispense 27 mL (or 13.5 mL) of the WAMS to sample aliquot and to each SPCS. Swirl to mix. The color reaction requires a minimum of 20 min before analyst records readings. Allowing 1 h for color development usually improves results. Color will remain stable for 6 h.

7.7 Transfer sample extract and SPCS to cuvettes.

7.8 Set the spectrophotometer to read at 882 nm. Autozero with calibration blank.

7.9 Calibrate the instrument by using the SPCS. The data system will then associate the concentrations with the instrument responses for each SPCS. Rejection criteria for SPCS, if R^2 <0.99.

7.10 Run samples using calibration curve. Sample concentration is calculated from the regression equation. Rejection criteria for batch, if blanks as samples > 0.01; if SPCS as samples > ± 20%; if SSL standard > ± 20% mean; and if ISE > (3 x MAD), where MAD = median of absolute deviations. Record results to the nearest 0.01 unit for the sample extract and each SPCS.

7.11 If samples are outside calibration range, dilute sample extracts with extracting solution and re-analyze.

8. Calculation

8.1 Convert extract P (mg L^{-1}) to soil P (mg kg^{-1}) as follows:

Soil P (mg kg^{-1}) = [(A x B x C x R x 1000)/E]

where:
A   = Sample extract reading (mg L^{-1})
B   = Extract volume (L)
C   = Dilution, if performed

R   = Air-dry/oven-dry ratio (procedure 3D1) or field-moist/oven-dry ratio (procedure 3D2)
1000 = Conversion factor to kg-basis
E   = Sample weight (g)

9. Report
Report data to the nearest $0.1 mg\ P\ kg^{-1}$ soil.

10. Precision and Accuracy
Precision and accuracy data are available from the SSL upon request.

11. References

Mehlich, A. 1984. Mehlich 3 soil text extractant: A modification of Mehlich 2 extractant. Commun. Soil Sci. Plant Anal. 15:1409-1416.

Murphy, J., and J.R. Riley. 1962. A modified single solution method for the determination of phosphate in natural waters. Anal. Chem. Acta. 27:31-36.

Olsen, S.R., and L.E. Sommers. 1982. Phosphorus. p. 403-430. *In* A.L. Page, R.H. Miller, and D.R. Keeney (eds.) Methods of soil analysis. Part 2. Chemical and microbiological properties. 2nd ed. Agron. Monogr. 9. ASA and SSSA, Madison, WI.

Soil and Plant Analysis Council, Inc. 1999. Handbook on reference methods for soil analysis. Council on Soil Testing and Plant Analysis. CRC Press, Boca Raton, FL.

Tran, T.S., and R.R. Simard. 1993. Mehlich III-extractable elements. p. 43-50. *In* M.R. Carter (ed.) Soil sampling and methods of analysis. Can. Soc. Soil Sci. Lewis Publ., Boca Raton, FL.

Mehlich No. 3 Extraction (4D6)
Inductively Coupled Plasma Atomic Emission Spectrophotometer (4D6b)
Axial Mode (4D6b1)
Ultrasonic Nebulizer (4D6b1a)
Aluminum, Arsenic, Barium, Calcium, Cadmium, Cobalt, Chromium, Copper, Iron, Potassium Magnesium, Manganese, Sodium, Nickel, Phosphorus, Lead, Selenium, and Zinc (4D6b1a1-18)
Air-Dry or Field-Moist (4D6b1a1-18a-b1)

1. Application
Mehlich No. 3 was developed by Mehlich (1984) as a multielement soil extraction (Ca, Mg, K, Na, P). In the Mehlich No. 3 procedure, P is extracted by reaction with acetic acid and Fl compounds. Mehlich No. 3 is a used an index of available P in the soil. Extraction of P by Mehlich No. 3 is designed to be applicable across a wide range of soil properties ranging in reaction from acid to basic (Mehlich, 1984). Mehlich No. 3 correlates well with Bray P-1 on acid to neutral ($R^2 = 0.966$) but does not correlate with Bray P-1 on calcareous soils (Soil and Plant Analysis Council, 1999). Mehlich No. 3 correlates with Olsen extractant on calcareous soils ($R^2 = 0.918$), even though the quantity of Mehlich No. 3 extractable P is considerably higher (Soil and Plant Analysis Council, 1992). The Mehlich No. 3 extractant is neutralized less by carbonate compounds in soil than the double acid (Mehlich No. 1) and the Bray P-1, and is less aggressive towards apatite or other Ca-phosphate than the double acid and Bray P-2 extractants (Tran and Simard, 1993). Mehlich No. 3 can also be used to extract Ca, Mg, K, and Na in a wide range of soils and correlates well with Mehlich No. 1, Mehlich No. 2, and NH_4OAc (Soil and Plant Analysis Council, 1999). Additionally, Mehlich No. 3 can be used to extract Al, Cd, Cu, Fe, Mn, Ni, Pb, and Zn (Elrashidi et al., 2003).

2. Summary of Method
A 2.5-g soil sample is shaken with 25 mL of Mehlich No. 3 extracting solution for 5 min. The sample is centrifuged until solution is free of soil mineral particles, and then filtered until clear extracts are obtained. Calibration standards are prepared for elemental analysis. A

blank of Mehlich No. 3 is prepared. An inductively coupled plasma atomic emission spectrophotometer (ICP-AES) is used for analysis. The concentration of Al, As, Ba, Ca, Cd, Co, Cr, Cu, Fe, K, Mg, Mn, Na, Ni, P, Pb, Se, and Zn are determined by an inductively coupled plasma atomic emission spectrophotometer (ICP-AES) in radial mode. Data by this procedure (4D6b1a1-18) are reported as mg kg^{-1} soil.

3. Interferences

Spectral and matrix interferences exist. Interferences are corrected or minimized by using both an internal standard and inter-elemental correction factors. Also, careful selection of specific wavelengths for data reporting is important. Background corrections are made by ICP-AES software. Samples and standards are matrix-matched to help reduce interferences.

4. Safety

Wear protective clothing (coats, aprons, sleeve guards, and gloves) and eye protection (face shields, goggles, or safety glasses). When preparing reagents, exercise special care. Many metal salts are extremely toxic and may be fatal if ingested. Thoroughly wash hands after handling these metal salts. Restrict the use of concentrated acids to a fume hood. Use safety showers and eyewash stations to dilute spilled acids and bases. Use sodium bicarbonate and water to neutralize and dilute spilled acids.

5. Equipment
5.1 Electronic balance, ±1.0-mg sensitivity
5.2 Mechanical reciprocating shaker, 200 oscillations min^{-1}, 1 ½ in strokes, Eberbach 6000, Eberbach Corp., Ann Arbor, MI
5.3 Centrifuge tubes, 50-mL, polyethylene
5.4 Funnel, 60° angle, long stem, 50-mm diameter
5.5 Filter paper, Whatman 42, 150 mm
5.6 Centrifuge, Centra, GP-8, Thermo IEC, Needham Heights, MA
5.7 Pipettes, electronic digital, 1000 μL and 10 mL, with tips, 1000 μL and 10 mL
5.8 Dispenser, 30 mL or 10 mL
5.9 Cups, plastic
5.10 Inductively coupled plasma atomic emission spectrophotometer (ICP-AES), Perkin-Elmer Optima 3300 Dual View (DV), Perkin-Elmer Corp., Norwalk, CT.
5.11 RF generator, floor mounted power unit, 45 MHz free running, Perkin-Elmer Corp., Norwalk, CT.
5.12 Computer, with WinLab software ver. 4.1, Perkin-Elmer Corp., Norwalk, CT, and printer
5.13 Recirculating chiller, Neslab, CFT Series
5.14 Compressed gasses, Argon (minimum purity=99.996%) and Nitrogen (minimum purity=99.999%)
5.15 Autosampler, AS-90, Perkin-Elmer Corp., Norwalk, CT.
5.16 Quartz torch, Part No. N069-1662; alumina injector (2.0 mm id), Part No. N069-5362

6. Reagents
6.1 Reverse osmosis deionized (RODI) water, ASTM Type I grade of reagent water
6.2 Mehlich No. 3 Extracting solution (0.2 N CH$_3$COOH; 0.25 N NH$_4$NO$_3$; 0.015 N NH$_4$F; 0.13 N HNO$_3$; 0.001 M EDTA). Premixed Mehlich No. 3 Extractant, Special-20, Hawk Creek Laboratory, Rural Route 1, Box 686, Simpson Road, Glen Rock, PA, 17327.
6.3 Primary standards, 1000 mg L^{-1}: of Al, Fe, Ca, Mg, Na, K, Mn, P, As, Ba, Cd, Co, Cr, Cu, Ni, Pb, Se, and Zn, High Purity Standards, Charleston, SC.
6.4 Mixed Standard High A, 100, 100, 100, 20, 40, 20, and 10 mg L^{-1} of Ca, Mg, Al, K, Fe, Na, Cr, respectively: To a 500-mL volumetric, add 50, 50, 50, 10, 20, 10, and 5 mL of the 1000 mg L^{-1} Ca, Mg, Al, K, Fe, Na (Section 6.3) and dilute to volume with Mehlich No. 3 (Section 6.2) extracting solution. Invert to thoroughly mix. Store in a polyethylene container. Make fresh weekly. Store in a refrigerator.

6.5 Mixed Standard Medium A, 50, 50, 50, 10, 20, 10, and 5 mg L^{-1} of Ca, Mg, Al, K, Fe, Na, and Cr, respectively: To a 100-mL volumetric, add 50 mL of Mixed Standard High A and dilute to volume with Mehlich No. 3 extracting solution. Invert to thoroughly mix. Store in polyethylene container. Make fresh weekly. Store in a refrigerator.

6.6 Mixed Standard Low A, 5, 5, 5, 1, 2, 1, and 0.5 mg L^{-1} of Ca, Mg, Al, K, Fe, Na, and Cr, respectively: To a 100-mL volumetric, add 10 mL of Mixed Standard Medium A and dilute to volume with Mehlich No. 3 extracting solution. Invert to thoroughly mix. Store in polyethylene container. Make fresh weekly. Store in a refrigerator.

6.7 Mixed Standard Very Low A, 0.5, 0.5, 0.5, 0.1, 0.2, 0.1 and 0.05 mg L^{-1} of Ca, Mg, Al, K, Fe, Na, and Cr, respectively: To a 100-mL volumetric, add 10 mL of Mixed Standard Low A and dilute to volume with Mehlich No. 3 extracting solution. Invert to thoroughly mix. Store in a polyethylene container. Make fresh weekly. Store in a refrigerator.

6.8 Mixed Standard High B, 250, 10, 10, 10, 10, 10, 10, 10, 10, 10, 10 mg L^{-1} of P, As, Ba, Cd, Co, Cu, Mn, Ni, Se, Zn, and Pb, respectively: To a 100-mL volumetric, add 25, 1, 1, 1, 1, 1, 1, 1, 1, 1, and 1 mL of the 1000 mg L^{-1} of P, As, Ba, Cd, Co, Cu, Mn, Ni, Se, Zn, and Pb (Section 6.3) and dilute to volume with Mehlich No. 3 (Section 6.2) extracting solution. Invert to thoroughly mix. Store in a polyethylene container. Make fresh weekly. Store in a refrigerator.

6.10 Mixed Standard Medium B, 25, 1, 1, 1, 1, 1, 1, 1, 1, 1, 1 mg L^{-1} of P, As, Ba, Cd, Co, Cu, Mn, Ni, Pb, Se, and Zn, respectively: To a 100-mL volumetric, add 10 mL of Mixed Standard High B and dilute to volume with Mehlich No. 3 extracting solution. Invert to thoroughly mix. Store in polyethylene container. Make fresh weekly. Store in a refrigerator.

6.11 Mixed Standard Low B, 2.5, 0.1, 0.1, 0.1, 0.1, 0.1, 0.1, 0.1, 0.1, 0.1, 0.1 mg L^{-1} of P, As, Ba, Cd, Co, Cu, Mn, Ni, Pb, Se, and Zn, respectively: To a 100-mL volumetric, add 10 mL of Mixed Standard Medium B and dilute to volume with Mehlich No. 3 extracting solution. Invert to thoroughly mix. Store in polyethylene container. Make fresh weekly. Store in a refrigerator.

6.12 Mixed Standard Very Low B, 0.25, 0.01, 0.01, 0.01, 0.01, 0.01, 0.01, 0.01, 0.01, 0.01, 0.01 mg L^{-1} of P, As, Ba, Cd, Co, Cu, Mn, Ni, Pb, Se, and Zn, respectively: To a 100-mL volumetric, add 10 mL of Mixed Standard Low B and dilute to volume with Mehlich No. 3 extracting solution. Invert to thoroughly mix. Store in polyethylene container. Make fresh weekly. Store in a refrigerator.

6.13 Blanks: To a 100-mL volumetric, Mehlich No. 3 extracting solution. Invert to thoroughly mix. Store in polyethylene container. Make fresh weekly. Store in a refrigerator.

7. Procedure

Extraction of Al, As, Ba, Ca, Cd, Co, Cr, Cu, Fe, K, Mg, Mn, Na, Ni, P, Pb, Se, and Zn

7.1 Weigh 2.5 g of <2-mm or fine-grind, air-dry soil to the nearest mg into a 50-mL centrifuge tube. If sample is moist, weigh enough soil to achieve ≈ 2.5 g of air-dry soil.

7.2 Dispense 25.0 mL of extracting solution to the tube.

7.3 Transfer the sample to the shaker. Shake for 5 min at 200 oscillations min^{-1} at room temperature (20°C± 2°C).

7.4 Remove the sample from the shaker. Centrifuge at 2000 rpm for 10 min, decant, filter, and collect extract in receiving cups. If extracts are not to be determined immediately after collection, then store samples at 4°C. Analyze samples within 72 h.

ICP-AES Set-up and Operation

7.5 Use the ICP-AES in axial mode to analyze elements. Use ultrasonic nebulization of sample. No initial dilutions of samples are necessary prior to analysis. Perform instrument checks (Hg alignment; BEC and %RSD of 1 mg L^{-1} Mn solution) prior to analysis as discussed in operation manual of instrument. Check instrument alignment and gas pressures to obtain optimum readings with maximum signal to noise ratio.

Analyses are generally performed at two or more wavelengths for each element. The selected wavelengths are as follows (reported wavelength listed first and in boldface):

Element	Wavelength
	----------- nm ---------
Al	**308.215**, 396.153
Fe	**259.939**, 238.204
Ca	**315.887**, 317.932
Mg	**280.271**, 279.075
Na	**589.592**, 588.995
K	**766.490**
Mn	**260.570** 257.608
P	**178.221**, 214.915
As	**193.69**
Ba	**233.525** 455.507
Cd	**226.501**, 214.435
Co	**228.614**
Cr	**267.710** 205.558
Cu	**324.753**, 327.396
Ni	**232.003**, 231.604,
Pb	**220.353**, 216.998
Se	**196.026**
Zn	**213.857**, 206.197

7.6 Use the Mehlich No. 3 extracting solution to dilute those samples with concentrations greater than the high standard. Rerun all elements and use only the data needed from the diluted analysis.

7.7 Establish detection limits using the blank standard solution. The instrumental detection limits are calculated by using 3 times the standard deviation of 10 readings of the blank. These values establish the lower detection limits for each element. Analyzed values lower than the detection limits are reported as "ND" or non-detected. The fraction digested needs to be identified with each sample.

8. Calculations

The calculation of mg kg^{-1} of an element in the soil from mg L^{-1} in solution is as follows:

Analyte concentration in soil (mg kg^{-1}) = [A x B x C x R x 1000]/E

A = Sample extract reading (mg L^{-1})
B = Extract volume (L)
C = Dilution, if performed
R = Air-dry/oven-dry ratio (procedure 3D1) or field-moist/oven-dry ratio (3D2)
1000 = Conversion factor to kg-basis
E = Sample weight (g)

9. Report
Data are reported to the nearest 0.1mg kg^{-1}.

10. Precision and Accuracy

Precision and accuracy and precision data are available from the SSL upon request.

11. References

Elrashidi, M.A., M.D. Mays, and C.W. Lee. 2003. Assessment of Mehlich3 and ammonium bicarbonate–DTPA extraction for simultaneous measurement of fifteen elements in soils. Commun. Soil Sci. and Plant Anal. 34:2817-2838.

Mehlich, A. 1984. Mehlich 3 soil text extractant: A modification of Mehlich 2 extractant. Commun. Soil Sci. Plant Anal. 15:1409-1416.

Murphy, J., and J.R. Riley. 1962. A modified single solution method for the determination of phosphate in natural waters. Anal. Chem. Acta. 27:31-36.

Olsen, S.R., and L.E. Sommers. 1982. p. 403-430. In A.L. Page, R.H. Miller, and D.R. Keeney (eds.) Methods of soil analysis. Part 2. Chemical and microbiological properties. 2nd ed. Agron. Monogr. 9. ASA and SSSA, Madison, WI.

Soil and Plant Analysis Council, Inc. 1999. Handbook on reference methods for soil analysis. Council on Soil Testing and Plant Analysis. CRC Press, Boca Raton, FL.

Tran, T.S., and R.R. Simard. 1993. Mehlich III-extractable elements. p. 43-50. In M.R. Carter (ed.) Soil sampling and methods of analysis. Can. Soc. Soil Sci. Lewis Publ., Boca Raton, FL.

Soil Tests (4D)
Citric Acid Soluble (4D7)
UV-Visible Spectrophotometer, Dual-Beam (4D7a)
Phosphorus (4D7a1)
Air-Dry or Field-Moist, <2 mm (4D7a1a-b1)

1. Application

In soil taxonomy, citric acid soluble P_2O_5 is a criterion for distinguishing between mollic (<250 ppm P_2O_5) and anthropic epipedons (>250 ppm P_2O_5) (Soil Survey Staff, 1999). Additional data on anthropic epipedons from several parts of the world may permit improvements in this definition (Soil Survey Staff, 1999).

Phosphorus (citrate-soluble, Method 960.01) and phosphorus (citrate insoluble, Method 963.03) are recognized methods in the Official Methods of Analysis by the Association of Analytical Communities (AOAC) International (AOAC, 2000). The AOAC citrate-soluble P method considers the recovery of phosphite source materials as available phosphorus, even though the Association of American Plant Food Control Officials does not recognize phosphite as a source of available phosphorus. The procedure described herein is used by N.A.A.S. (England and Wales) and is based on the method developed by Dyer (1894).

2. Summary of Method

A sample is checked for $CaCO_3$ equivalent. Sufficient citric acid is added to sample to neutralize the $CaCO_3$ plus bring the solution concentration of citric acid to 1%. A 1:10 soil:solution is maintained for all samples. The sample is shaken for 16 h and filtered. Ammonium molybdate and stannous chloride are added. Absorbance is read using a spectrophotometer at 660 nm. Data are reported as mg P_2O_5 kg^{-1} soil (4D7a1).

3. Interferences

Unreacted carbonates interfere with the extraction of P_2O_5. Sufficient citric acid is added to sample to neutralize the $CaCO_3$. However, a high citrate level in sample may interfere with the molybdate blue test. If this occurs, the method can be modified by evaporating the extract and ashing in a muffle furnace to destroy the citric acid.

Positive interferences in the analytical determination of P_2O_5 are silica and arsenic, if the sample is heated. Negative interferences in the P_2O_5 determination are arsenate, fluoride, thorium, bismuth, sulfide, thiosulfate, thiocyanate, or excess molybdate. A concentration of Fe >1000 ppm interferes with P_2O_5 determination. Refer to Snell and Snell (1949) and Metson (1956) for additional information on interferences in the citric acid extraction of P_2O_5.

4. Safety

Wear protective clothing (coats, aprons, sleeve guards, and gloves) and eye protection (face shields, goggles, or safety glasses) when preparing reagents, especially concentrated acids and bases. Dispense concentrated acids and bases in fume hood. Use the safety showers and eyewash stations to dilute spilled acids and bases. Use sodium bicarbonate and water to neutralize and dilute spilled acids. Follow standard laboratory procedures.

5. Equipment

5.1 Electronic balance, ±0.10-mg sensitivity
5.2 Mechanical reciprocating shaker, 200 oscillations min^{-1}, 1 ½ strokes, Eberbach 6000, Eberbach Corp., Ann Arbor, MI
5.3 Centrifuge tubes, 50-mL, graduated, polyethylene
5.4 Bottles, with gas release caps
5.5 Filter paper, Whatman 42, 150 mm
5.6 Funnel, 60° angle, long-stem, 50-mm diameter
5.7 Pipettes, electronic digital, 1000 µL and 10 mL, with tips, 1000 µL and 10 mL
5.8 Dispenser, 30-mL or 10-mL
5.9 Cuvettes, plastic, 4.5-mL, 1-cm light path, Daigger Scientific
5.10 Spectrophotometer, UV-Visible, Dual-View, Varian, Cary 50 Conc, Varian Australia Pty Ltd.
5.11 Computer, with Cary WinUV software, Varian Australia Pty Ltd., and printer

6. Reagents

6.1 Reverse osmosis deionized (RODI) water
6.2 Hydrochloric acid (HCl), concentrated, 12 N, trace pure grade
6.3 Citric acid solution, 10%. In a 1-L volumetric, dissolve 100 g of anhydrous citric acid ($C_6H_8O_7$) in RODI water and bring to volume with RODI water.
6.4 Citric acid solution, 1%. Dilute 100.0 mL of 10% citric acid solution to 1-L with RODI water
6.5 Ammonium molybdate solution, 1.5%. Dissolve 15.0 g of ammonium molybdate [$(NH_4)_6MO_7O_{24} \cdot 4H_2O$] in 300 mL of distilled water. Transfer to a 1-L volumetric flask and carefully add 310 mL of concentrated HCl. Allow to cool. Make to 1-L volume with RODI water. Store in brown bottle in the dark in a refrigerator. Solution is stable for ≈ 3 months.
6.6 Stock stannous chloride solution (SSCS). Dissolve 10 g of stannous chloride ($SnCl_2 \cdot 2H_2O$) in 100 mL of concentrated HCl. Invert to mix thoroughly. Make fresh weekly. Store in a refrigerator.
6.7 Working stannous chloride solution (WSCS). Dilute 2 mL of SSCS with 100 mL of RODI water. Use immediately as solution is only stable for ≈ 4 h.
6.8 Stock standard P_2O_5 solution (SSPS), 250 mg L^{-1} P. Dissolve 1.099 g of potassium dihydrogen orthophosphate (KH_2PO_4) that has been dried for 2 h at 110°C with RODI water in 1-L volumetric flask. Add 5 ml of 2 N HCl. Make to 1-L volume with RODI water. Invert to mix thoroughly. Store in polyethylene bottles. Make fresh weekly. Store in a refrigerator.
6.9 Working stock standard P_2O_5 solution (WSSPS), 2.5 mg L^{-1} P. Pipet 10.0 mL of SSPS and dilute to 1-L in a volumetric flask with RODI water. Invert to mix thoroughly. Make fresh daily.
6.10 Standard P_2O_5 calibration solutions (SPCS). Pipet 0.0, 1.0, 2.0, 3.0, 4.0, and 5.0 mL of WSSPS into 50-mL tubes, add 1 ml of 1% citric acid solution, 4 mL of ammonium molybdate solution, and dilute to 25-mL with RODI water. Add 2 mL of the working

stannous chloride solution (WSCS). Final concentrations are 0.0, 0.10, 0.20, 0.30, 0.40, and 0.50 mg L^{-1}, respectively. Invert to mix thoroughly. Prepare fresh weekly. Store in a refrigerator. Allow to equilibrate to room temperature before use.

6.11 Quality Control Samples: 0.1 mg P L^{-1} solution from SSPS; blanks; selected SPCS; and SSL soil standard are routinely included in a batch for quality control.

7. Procedure

7.1 Weigh 3 g of <2-mm or fine-grind, air-dry soil to the nearest mg into a bottle with gas release tops. If sample is moist, weigh enough soil to achieve ≈ 3 g of air-dry soil. If the soil does not contain free carbonates, proceed to step 7.3.

7.2 If the soil contains free CaCO$_3$, refer to Table 1 to determine the amount of 10% citric acid solution required to neutralize the CaCO$_3$. Add required volume of 10% citric acid into a graduated cylinder and bring to a volume of 30-mL with RODI water. Add this solution to the soil. Place the bottle in a mechanical shaker and shake the bottle for 6 h at 200 oscillations min^{-1} at room temperature (20°C ± 2°C) to dissolve and neutralize the CaCO$_3$. Proceed to step 7.4.

Table 1. Volume of 10% citric acid (mL) required to decompose CaCO3 (%) and to bring to solution concentration to 1% in a final volume of 30 mL for 3-g sample.

%CC[1]	mL CA[2]	% CC	mL CA	% CC	mL CA	%CC	mL CA
0	3.0	16	9.1	32	15.3	48	21.4
1	3.4	17	9.5	33	15.7	49	21.8
2	3.8	18	9.9	34	16.1	50	22.2
3	4.2	19	10.3	35	16.4	51	22.6
4	4.5	20	10.7	36	16.8	52	23.0
5	4.9	21	11.1	37	17.2	53	23.3
6	5.3	22	11.4	38	17.6	54	23.7
7	5.7	23	11.8	39	18.0	55	24.1
8	6.1	24	12.2	40	18.4	56	24.5
9	6.5	25	12.6	41	18.7	57	24.9
10	6.8	26	13.0	42	19.1	58	25.3
11	7.2	27	13.4	43	19.5	59	25.6
12	7.6	28	13.7	44	19.9	60	26.0
13	8.0	29	14.1	45	20.3	61	26.4
14	8.4	30	14.5	46	20.7	62	26.8
15	8.8	31	14.9	47	21.0	63	27.2

[1]%CC = percent calcium carbonate in a sample
[2]CA = ml of 10% citric acid needed to be diluted to 30-ml volume with RODI water and added to sample

7.3 If the soil contains no free CaCO$_3$, add 30 mL of 1% citric acid solution to the sample.

7.4 Cap the bottles, place in a mechanical shaker and shake for 16 h at 200 oscillations min⁻ at room temperature (20°C ± 2°C).

7.5 Remove the sample from shaker and filter. Collect extract. If extracts are not to be determined immediately after collection, then store samples at 4° C. Analyze samples within 72 h.

7.6 Pipet 1 mL of sample extract into a plastic cup. Samples are treated the same as SPCS (Section 6.10). Add 4 mL of ammonium molybdate solution to all samples and SPCS, and diluted to 25-mL with RODI water. Add 2 mL working stannous chloride solution (WSCS). Swirl to mix and allow to stand 20 min for color development.

7.7 Transfer sample extract and SPCS to cuvettes.

7.8 Set the spectrophotometer to read at 660 nm. Autozero with calibration blank. A blank has all reagents contained in the sample extract except the soil.

7.9 Calibrate the instrument by using the SPCS. The data system will then associate the concentrations with the instrument responses for each SPCS. Rejection criteria for SPCS, if R^2 <0.99.

7.10 Run samples using calibration curve. Run samples using calibration curve. Sample concentration is calculated from the regression equation. Rejection criteria for batch, if blanks as samples > 0.01; if SPCS as samples > ± 20%; and if SSL standard > ± 20% mean. Record results to the nearest 0.01 unit for the sample extract and each SPCS.

7.11 If samples are outside calibration range, dilute sample extracts with extracting solution and re-analyze.

8. Calculations

Convert the extract P_2O_5 (mg L^{-1}) to soil P_2O_5 (mg kg^{-1}) as follows:

Soil P_2O_5 (mg kg^{-1}) = [A x B x C_1 x C_2 x R x 1000 x 2.29]/E

where:
A = P_2O_5 in sample extract (mg L^{-1})
B = Extract volume (L) (0.03)
C_1 = Automatic dilution (1:25)
C_2 = Dilution, if necessary
R = Air-dry/oven-dry ratio (procedure 3D1) or field-moist/oven-dry ratio (procedure 3D2)
1000 = Conversion factor to kg-basis
2.29 = Conversion factor P to P_2O_5
E = Sample weight (g)

9. Report
Report the 1% citrate acid extractable P_2O_5 in mg kg^{-1} to nearest whole number.

10. Precision and Accuracy
Precision and accuracy data are available from the SSL upon request.

11. References
American Association of Analytical Communities (AOAC), International. 2000. Phosphorus (Citrate-Soluble) in Fertilizers and Phosphorus (Citrate-Insoluble) in Fertilizers, Methods 960.01 and 963.03, respectively. Official Methods of Analysis.
Dyer. 1894. Trans. Chem. Soc. 65:115-167.
Metson, A.J. 1956. Methods of chemical analysis for soil survey samples. Soil Bur. Bull. No. 12. N.Z. Dept. Sci. and Industrial Res.
Snell, F.D., and C.T. Snell. 1949. Colorimetric methods of analysis. Phosphorus. Vol 2. 3rd ed. p.630-681. D. Van Nostrand Co., Inc.

Soil Survey Staff. 1999. Soil taxonomy. A basic system of soil classification for making and interpreting soil surveys. 2nd ed. USDA-NRCS. Govt. Print. Office, Washington DC.

Soil Tests (4D)
New Zealand P Retention (4D8)
UV-Visible Spectrophotometer, Dual-Beam (4D8a)
Phosphorus (4D8a1)
Air-Dry or Field-Moist, <2 mm (4D8a1a-b1)

1. Application

In soil taxonomy, the P retention of soil material is a criterion for andic soil properties (Soil Survey Staff, 1999). Andisols and other soils that contain large amounts of allophane and other amorphous minerals have capacities for binding P (Gebhardt and Coleman, 1984). The factors that affect soil P retention are not well understood. However, allophane and imogolite have been considered as major materials that contribute to P retention in Andisols (Wada, 1985). Phosphate retention is also called P adsorption, sorption, or fixation.

2. Summary of Method

A 5-g soil sample is shaken in a 25-mL aliquot of a 1000 mg L^{-1} P solution for 24 h. The mixture is centrifuged at 2000 rpm for 15 min. An aliquot of the supernatant is transferred to a colorimetric tube to which nitric vanadomolybdate acid reagent (NVAR) is added. Absorbance of the solution is read using a spectrophotometer at 466 nm. This absorbance correlates to the concentration of the non–adsorbed P that remains in the sample solution. The New Zealand P retention (Blakemore et al., 1987) is the initial P concentration minus the P remaining in the sample solution and is reported as percent P retained (4D8a1).

3. Interferences

No significant problems are known to affect the P retention measurement.

4. Safety

Wear protective clothing (coats, aprons, sleeve guards, and gloves) and eye protection (face shields, goggles, or safety glasses). When preparing reagents, exercise special care. Many metal salts are extremely toxic and may be fatal if ingested. Thoroughly wash hands after handling these metal salts. Restrict the use of concentrated HNO_3 to a fume hood. Use safety showers and eyewash stations to dilute spilled acids and bases. Use sodium bicarbonate and water to neutralize and dilute spilled acids.

5. Equipment
5.1 Electronic balance, ±1.0-mg sensitivity
5.2 Mechanical reciprocating shaker, 100 oscillations min^{-1}, 1 ½ in strokes, Eberbach 6000, Eberbach Corp., Ann Arbor, MI
5.3 Digital diluter/dispenser with syringes, 10000 and 1000 μL, gas tight, Microlab 500, Reno, NV
5.4 Centrifuge, Centra GP-8
5.5 Centrifuge tubes, 50 mL, polyethylene
5.6 Cups, plastic
5.7 Pipettes, electronic digital, 1000 μL and 10 mL, with tips, 1000 μL and 10 mL
5.8 Dispenser, 30 mL or 10 mL
5.9 Filter paper, Whatman 42, 150 mm
5.11 Cuvettes, plastic, 4.5-mL, 1-cm path, Daigger Scientific
5.12 Spectrophotometer, UV-Visible, Varian, Cary 50 Conc, Varian Australia Pty Ltd.
5.13 Computer, with Cary WinUV software, Varian Australia Pty Ltd., with printer

6. Reagents

6.1 Reverse osmosis deionized (RODI) water, ASTM Type I grade of reagent water

6.2 Superfloc 16, 0.2%, 2 g L^{-1} RODI water

6.3 Nitric acid (HNO$_3$), concentrated, 16 N

6.4 P retention solution (PRS), 1000 mg L^{-1} P. Dissolve 35.2 g of KH$_2$PO$_4$ that has been dried for 2 h at 110°C and 217.6 g of sodium acetate trihydrate (CH$_3$COONa·3H$_2$O) in RODI water. Add 92 mL of glacial acetic acid. Dilute to 8 L with RODI water. The solution pH should range between 4.55 and 4.65.

6.5 Nitric acid solution. Carefully and slowly dilute 200 mL of concentrated HNO$_3$ to 2 L of RODI water. Add the acid to the water.

6.6 Molybdate solution. Dissolve 32 g of ammonium molybdate [(NH$_4$)$_6$Mo$_7$O$_{24}$·4H$_2$O] in 50°C RODI water. Allow the solution to cool to room temperature and dilute to 2 L with RODI water.

6.7 Nitric vanadomolybdate acid reagent (NVAR), vanadate solution. Dissolve 1.6 g of NH$_4$VO$_3$ in 500 mL of boiling RODI water. Allow the solution to cool to room temperature. Carefully and slowly add 12 mL of concentrated HNO$_3$. Dilute to 2 L with RODI water. Mix the nitric acid solution with the vanadate solution and then add the molybdate solution. Mix well. Note: This solution needs to be made in this order.

6.8 Diluent for Standard P calibration solutions (DSPCS). Add 54.4 g of sodium acetate trihydrate (CH$_3$COONa·3H$_2$O) and 23 mL of glacial acetic acid and dilute to 2 L with RODI water.

6.9 Standard P calibration solutions (SPCS), 100, 80, 60, 40, 20, and 0% P retained. Make fresh weekly. Store in a refrigerator. Allow to equilibrate to room temperature before use. To six 100-mL volumetric flasks add the appropriate amount of PRS (1000 mg L^{-1}) and bring to volume with DSPCS solution as follows:

6.9.1 100% SPCS = (0 mg L^{-1}) = Bring to 100-mL volume with DSPCS solution.

6.9.2 80% SPCS = 1: 20 (200 mg L^{-1}) = Add 20 mL PRS and bring to 100-mL volume with DSPCS solution.

6.9.3 60% SPCS = 1:10 (400 mg L^{-1}) = Add 40 mL PRS and bring to 100-mL volume with DSPCS solution.

6.9.4 40% SPCS = 3:20 (600 mg L^{-1}) = Add 60 mL PRS and bring to 100-mL volume with DSPCS solution.

6.9.5 20% SPCS = 1:5 (800 mg L^{-1}) = Add 80 mL PRS and bring to 100-mL volume with DSPCS solution.

6.9.6 0% SPCS = 1:4 (1000 mg L^{-1}) = Add 100 mL PRS to a 100-mL volumetric.

6.10 Quality Control: SSL soil standard is routinely included in every batch of 24 samples for quality control.

7. Procedure

7.1 Weigh 5g of <2-mm or fine-grind, air-dry soil to the nearest mg into a 50-mL centrifuge tube. If sample is moist, weigh enough soil to achieve ≈ 5 g of air-dry soil. No fine-grind material is used for this procedure.

7.2 Use the dispenser to add 25.0 mL of P-retention solution to centrifuge tube.

7.3 Transfer the sample to the shaker. Shake for 24 h at 100 oscillations min^{-1} at room temperature (20°C± 2°C).

7.4 Remove sample from the shaker. Add 2 to 3 drops of Superfloc, 0.02% w/v to each tube. Centrifuge at 2000 rpm for 15 min, decant, filter, and collect extract in receiving cup. If extracts are not to be determined immediately after collection, then store samples at 4°C. Analyze samples within 72 h.

7.5 Use the digital diluter to add the nitric vanadomolybdate acid reagent (NVAR) to each sample supernatant and to each SPCS. To fill a 4.5-mL cuvette, use a dilution of 1:20 sample dilution.

7.6 The color reaction requires a minimum of 30 min before the analyst records readings.

7.7 Set the spectrophotometer to read at 466 nm. Auto zero with the calibration blank.

7.8 Calibrate the instrument using the SPCS. The data system will then associate the % P retained with the instrument responses. Rejection criteria for SPCS, if R^2 <0.99.

7.9 Run samples using calibration curve. Sample P retention is calculated from the regression equation, i.e., the absorbance is equated to the SPCS. Rejection criteria for batch, if blanks as samples > 0.01; if SPCS as samples > ± 20%; and if SSL standard > ± 20% mean. Record results to the nearest 0.01 unit for the sample extract and each SPCS.

7.10 If samples are outside calibration range, dilute sample extracts with extracting solution and re-analyze.

8. Calculations
None.

9. Report
Report the percent New Zealand P retention to the nearest whole number.

10. Precision and Accuracy
Precision and accuracy data are available from the SSL upon request.

11. References

Blakemore, L.C., P.L. Searle, and B.K. Daly. 1987. Methods for chemical analysis of soils. 43 p. NZ Bureau Scientific Report 80, NZ Soil Bur. Lower Hutt, NZ.

Gebhardt, H., and N.T. Coleman. 1984. Anion adsorption of allophanic tropical soils: III. Phosphate adsorption. p. 237-248. *In* K.H. Tan (ed.) Anodosols. Benchmark papers in soil science series. Van Nostrand Reinhold, Co., Melbourne, Australia.

Soil Survey Staff. 1999. Soil taxonomy. A basic system of soil classification for making and interpreting soil surveys. 2nd ed. USDA-NRCS. Govt. Print. Office, Washington DC.

Wada, K. 1985. The distinctive properties of Andosols. p. 173-229. *In* B.A. Stewart (ed.) Advances in Soil Science. Springer-Verlag, New York, NY.

Soil Tests (4D)
1 *M* KCl Extraction (4D9)
Cadmium-Copper Reduction (4D9a)
Sulfanilamide N-1-Naphthylethylenediamine Dihydrochloride (4D9a1)
Flow-Injection, Automated Ion Analyzer (4D9a1a)
Nitrate-Nitrite (4D9a1a1-2)
Air-Dry or Field-Moist, <2 mm (4D9a1a1-2a-b1)

1. Application
The inorganic combined N in soils is predominantly NH_4^+ and NO_3^- (Keeney and Nelson, 1982). Nitrogen in the form of ammonium ions and nitrate are of particular concern because they are very mobile forms of nitrogen and are most likely to be lost to the environment (National Research Council, 1993). All forms of nitrogen, however, are subject to transformation to ammonium ions and nitrate as part of the nitrogen cycle in agroecosystems and all can contribute to residual nitrogen and nitrogen losses to the environment (National Research Council, 1993).

2. Summary of Method
A 2.5-g soil sample is mechanically shaken for 30 min in 25 mL of 1 *M* KCl solution. The sample is then filtered through Whatman No. 42 filter paper. A flow injection automated ion analyzer is used to measure the soluble inorganic nitrate (NO_3^-). The nitrate is quantitatively reduced to nitrite by passage of the sample through a copperized cadmium column. The nitrite (reduced nitrate plus original nitrite) is then determined by diazotizing with sulfanilamide followed by coupling with N-1-naphthylethylenediamine dihydrochloride. The resulting water soluble dye has a magenta color which is read at 520 nm. Absorbance is proportional to the concentration of NO_3^- in the sample. Data are reported as mg N kg^{-1} soil as NO_3^- and/or NO_2^- (4D9a1a1-2, respectively).

3. Interferences
Low results can be obtained for samples that contain high concentration of Fe, Cu, or other metals. In this method, EDTA is added to the buffer to reduce this interference (LACHAT, 1993).

4. Safety
Wear protective clothing (coats, aprons, sleeve guards, and gloves) and eye protection (face shields, goggles, or safety glasses). When preparing reagents, exercise special care. Many metal salts are extremely toxic and may be fatal if ingested. Thoroughly wash hands after handling these metal salts. Restrict the use of NH_4OH and concentrated HCl to a fume hood. Use safety showers and eyewash stations to dilute spilled acids and bases. Use sodium bicarbonate and water to neutralize and dilute spilled acids. Cadmium is toxic and carcinogenic. Wear gloves and follow the precautions described on the Material Safety Data Sheet. If repacking the cadmium-copper reduction column, do all transfers over a special tray or beaker dedicated to this purpose. Preferably, send the cadmium-copper column to LACHAT for repacking.

5. Equipment
5.1 Electronic balance, ±1.0-mg sensitivity
5.2 Centrifuge tubes, 25 mL, polyethylene, Oak Ridge
5.3 Mechanical reciprocating shaker, 200 oscillations min^{-1}, 1 ½ in strokes, Eberbach 6000, Eberbach Corp., Ann Arbor, MI
5.4 Centrifuge, 50-mL, polyethylene
5.5 Pipettes, electronic digital, 2500 µL and 10 mL, with tips, 2500 µL and 10 mL
5.6 Filter paper, Whatman No. 42, 150 mm
5.7 Funnel, 60° angle, long stem, 50-mm diameter

5.8 Volumetric flasks, 1-L and 250-mL

5.9 Bottles, plastic, dark, 1-L

5.10 Cups, plastic

5.11 Dispenser, 30 mL or 10 mL

5.12 Flow Injection Automated Ion Analyzer, QuikChem AE, LACHAT Instruments, Milwaukee, WI, with computer and printer

5.13 XYZ Sampler, LACHAT Instruments, Milwaukee, WI

5.14 Reagent Pump, LACHAT Instruments, Milwaukee, WI

5.15 Automated Dilution Station, LACHAT Instruments, Milwaukee, WI

5.16 Sample Processing Module (SPM) or channel, QuikChem Method (12-107-04-1-B, nitrate/nitrite in 1 M KCl 0.02 to 20.0 mg N L^{-1}), LACHAT Instruments, Milwaukee, WI

5.17 Computer, with QuikChem software, LACHAT Instruments, Milwaukee, WI, and printer

5.18 Vials, plastic, 25-mL (standards)

5.19 Culture tubes, glass, 10-mL (samples)

6. Reagents

6.1 Reverse osmosis deionized (RODI) water, ASTM Type I grade of reagent water

6.2 Helium, compressed gas

6.3 15 M NaOH. In a 500-mL container, add 250 ml RODI water. Slowly add 300 g NaOH. (CAUTION: The solution will get very hot!) Swirl until dissolved. Cool and store in a plastic bottle. Used to adjust ammonium chloride buffer to pH 8.5 (reagent 6.4).

6.4 Ammonium chloride buffer, pH 8.5. In a hood, add 500 mL RODI to a 1-L volumetric flask. Add 105 mL concentrated HCl, 95 mL ammonium hydroxide (NH_4OH) and 1.0 g disodium EDTA. Dissolve and dilute to mark. Invert to mix. Degas with helium ≈ 5 min.

6.5 Sulfanilamide color reagent. To a 1-L volumetric flask, add 600 mL RODI H_2O followed by 100 mL 85 percent phosphoric acid (H_3PO_4), 40.0 g sulfanilamide, and 1.0 g N-1-naphthylethylenediamine dihydrochloride (NED). Shake to wet and stir to dissolve 20 min. Dilute to mark, invert to thoroughly mix. Degas with helium ≈ 5 min. Store in dark bottle and discard when pink.

6.6 1 M KCl extracting solution, carrier and standards diluent. Dissolve 74.5 g potassium chloride (KCl) in 800 mL RODI water. Dilute to mark and invert to thoroughly mix. The extracting solution is used also as the carrier and a component of the N standards. Degas with helium ≈ 5 min.

6.7 The following are standards preparation for a 1 channel system determining $NO_2^- + NO_3^-$ or NO_2^- and a 2 channel system where one channel is used for $NO_2^- + NO_3^-$ and the other channel is used for determining NO_2^-. For the 1 channel system, either NO_2^- or NO_3^- standards may be used. It is recommended that when running a 1 channel method for $NO_2^- + NO_3^-$ that NO_3^- standards are used. For the 2 channel system, the use of both $NO_2^- + NO_3^-$ standard sets are recommended.

6.7.1 Stock standard nitrate solution (SSNO$_3$S), 200.0 mg N L^{-1} as NO_3^- in 1 M KCl. In a 1-L volumetric flask, dissolve 1.444 g potassium nitrate (KNO_3) dried in an oven for 2 h at 110° C and 74.5 g KCl in 600 mL RODI water. Dilute to mark with RODI water and invert to thoroughly mix. Store in polyethylene containers. Make fresh weekly. Store in a refrigerator.

6.7.2 Working stock standard nitrate solution (WSSNO$_3$S), 20.0 mg N L^{-1} as NO_3^- in 1 M KCl. To a 1-L volumetric flask, add 100 ml SSNO$_3$S. Dilute to mark with 1 M KCl and invert to thoroughly mix. Make fresh daily.

6.7.3 Standard nitrate calibration standards (SNO$_3$CS) or working standards, 10.00, 2.00, 0.80, 0.08, and 0.00 mg N L^{-1} as NO_3^- in 1 M KCl. Make fresh daily. To five 250-mL volumetric flasks add as follows:

6.7.3.1 10.00 mg N L^{-1} = 125.0 mL WSSNO$_3$S

6.7.3.2 2.00 mg N L^{-1} = 25.0 mL WSSNO$_3$S

6.7.3.3 0.80 mg N L^{-1} = 10.0 mL WSSNO$_3$S
6.7.3.4 0.08 mg N L^{-1} = 1.00 mL WSSNO$_3$S
6.7.3.5 0.00 mg N L^{-1} = 0.0 mL WSSNO$_3$S (blank)

Dilute each SNO$_3$CS to the mark with 1 M KCl and invert to thoroughly mix. Do not degas.

6.7.4 Stock standard nitrite solution (SSNO$_2$S), 200.0 mg N L^{-1} as NO$_2^-$ in 1 M KCl. In a 1-L volumetric flask, dissolve 0.986 g sodium nitrite (NaNO$_2$) or 1.214 g potassium nitrite (KNO$_2$), and 74.5 g KCl in 800 mL RODI water. Dilute to mark with RODI water and invert to thoroughly mix. Store in polyethylene containers. Make fresh weekly. Store in a refrigerator.
6.7.5 Working stock standard nitrite solution (WSSNO$_2$S), 20.0 mg N L^{-1} as NO$_2^-$ in 1 M KCl. To a 1-L volumetric flask, add 100 ml SSNO$_2$S. Dilute to mark with 1 M KCl and invert to thoroughly mix. Make fresh daily.
6.7.6 Standard nitrite calibration standards (SNO$_2$CS) or working standards, 10.00, 2.00, 0.80, 0.08, and 0.00 mg N L^{-1} as NO$_3^-$ in 1 M KCl. Make fresh daily. To five 250-mL volumetric flasks add as follows:

6.7.6.1 10.00 mg N L^{-1} = 125.0 mL WSSNO$_2$S
6.7.6.2 2.00 mg N L^{-1} = 25.0 mL WSSNO$_2$S
6.7.6.3 0.80 mg N L^{-1} = 10.0 mL WSSNO$_2$S
6.7.6.4 0.08 mg N L^{-1} = 1.00 mL WSSNO$_2$S
6.7.6.5 0.00 mg N L^{-1} = 0.0 mL WSSNO$_2$S (blank)

Dilute each SNO$_2$CS to the mark with 1 M KCl and invert to thoroughly mix. Do not degas.

7. Procedure

Extraction
7.1 Weigh 2.5 g of <2-mm or fine-grind, air-dry soil to the nearest mg and place into a 50-mL centrifuge tube.   If sample is moist, weigh enough soil to achieve ≈ 2.5 g of air-dry soil.

7.2 Add ≈ 25 mL of 1 M KCl to sample. Transfer the sample to a shaker. Shake for 30 min at 200 oscillations min^{-1} at room temperature (20°C ± 2°C).

7.3 Remove the sample from the shaker. Decant, filter, and collect extract in receiving cups. If extracts are not to be determined immediately after collection, then store samples at 4°C. Analyze samples within 24 h.

Flow Injection Set-up and Operation
7.4 Transfer sample extracts into culture tubes and place in XYZ sample trays marked "Samples".

7.5 Transfer WNCS standards into plastic vials and place in descending order in XYZ sample trays marked "Standards".

7.6 Refer to the operating and software reference manuals for LACHAT for set-up and operation.

7.7 Turn main power switch "ON".

7.8 Before inserting cadmium-copper reduction column, pump all reagents into manifold. Turn pump off. On the column, disconnect the center tubing from one of the union connectors and immediately connect to the outlet tubing of the buffer mixing coil. Connect the open tubing on the column to the tee fitting where the color reagent is added. Do not let air enter the column. Return the pump to normal speed. If air is introduced accidentally, connect the column into the manifold, turn the pump on maximum, and tap column, working up the column until all air is removed. Average life of column is approximately 600 samples. Upon degradation, the column needs to be replaced.

7.9 On computer main menu, select "Methods" and then "Analysis Select and Download". On method list, select 1 M KCl N method. System unit receives the downloaded method and initializes it. On computer main menu, select "Samples", "Tray Definition and Submit", and then "Edit" to create new sample tray followed by "Submit" to run new sample tray.

7.10 Upon connection of cadmium column and downloading of 1 M KCl Method, continue to pump reagents into manifold. Continue this step and observe baseline. A good baseline needs to be smooth and at zero absorbance. Scatter is indicative of air bubbles and irregular reagent flow. Also observe for any back-pressure in manifold tubing.

7.11 Method parameters specific to 1 M KCl N are defined within the "Method Definition" menu. Some of these parameters have been modified from the QuikChem Method 12-107-04-1-B, nitrate-nitrite in 2 M (1 M) KCl soil extracts (U.S. Environmental Protection Agency, 1983; LACHAT Instruments, 1992; U.S. Department of Interior, Geological Survey, 1993).

7.12 Some of the method parameters as they relate to calibration standards (nitrate - nitrite) are as follows:

7.12.1 There are 5 calibration standards (10.00, 2.00, 0.80, 0.08, and 0.00 mg N L^{-1}), with a data format of ####.###, i.e., data rounded to 3 places.

7.12.2 The segments/boundaries for the calibration standards are A - C (10.0 to 0.8 mg N L^{-1}) and C - E (0.8 to 0.0 mg N L^{-1}).

7.12.3 The protocol (replications) for the calibration standards is as follows: AA BB CCC DDD EE.

7.12.4 Check standard is 10.0 mg N L^{-1}. Maximum number of consecutive trays between check standard is one; maximum number of consecutive samples between check standard is 60; and maximum elapse time between check standards is 2 h.

7.12.5 Calibration strategy for segments A - C and C - E are normal. The normal strategy requires a minimum correlation coefficient of 0.99. Both segments require a maximum standard deviation in slope of 50%. A calibration passes only when both criteria are met. Strategies are user designated. In addition, calibration strategies are based on the full chord. Chord 0 is full chord, and chord 1 - 5 are sections of peak from start of peak to end of peak.

7.12.6 The instrument is calibrated with the injection of SNO$_3$CS and/or SNO$_2$CS. The data system is calibrated with the injection of SNO$_3$CS and/or SNO$_2$CS. The data system then associates the concentrations with the instrument response for each SNO$_3$CS and/or SNO$_2$CS.

7.13 Method parameters in relation to timing are as follows:

7.13.1 Cycle period: 40 s

7.13.2 Inject to start of peak period: 27 s. To see if peaks are being timed correctly, scan across correlation coefficients for all chords 1 - 5. The most peak area is supposed to be in chords 2 - 4 with the most signal-to-noise ratio in chords 1 and 5.

7.13.3 Inject to end of peak period: 63 s

7.13.4 Automatic timing, where standard assumptions are in effect; no manual timing.

7.14 Method parameters in relation to data presentation are as follows:

7.14.1 Top Scale Response: 0.50 abs

7.14.2 Bottom Scale Response: 0.00 abs

7.15 Refer to the "Method Definition" for 1 M KCl nitrate for other method parameters not discussed here.

7.16 Run samples using calibration curve. Sample concentration is calculated from the regression equations. Report results to the nearest 0.01 unit for the sample extract and each SPCS.

7.17 If samples are outside calibration range, dilute samples with extracting solution and re-analyze.

7.18 Upon completion of run, place the transmission lines into RODI water and pump for approximately 20 min and proceed with the normal "Shut-down" procedure.

7.19 The accumulation of 1 M KCl in the manifold tubing and the fittings may cause clogs over time. Upon completion of analysis, the valves and fittings need to be washed with RODI water. Some fittings may need to be soaked overnight or placed in a sonic bath for 10 to 15 min to remove any KCl accumulations.

8. Calculations

Convert extract N (mg L^{-1}) to soil N (mg kg^{-1}) as follows:

Soil N = [(A x B x C x R x 1000)/E]

where:
A   = Sample extract reading (mg N L^{-1})
B   = Extract volume (L)
C   = Dilution, if performed
R   = Air-dry/oven-dry ratio (procedure 3D1) or field-moist/oven-dry ratio (procedure 3D2)
1000 = Conversion factor to kg-basis
E   = Sample weight (g)

9. Report
Report data to the nearest 0.1mg N kg^{-1} soil as NO$_3^-$ and/or NO$_2^-$.

10. Precision and Accuracy
Precision and accuracy data are available from the SSL upon request.

11. References
Keeney, D.R., and D.W. Nelson. 1982. p. 643-698. Nitrogen - inorganic forms. *In* A.L. Page, R.H. Miller, and D.R. Keeney (eds.) Methods of soil analysis. Part 2. Chemical and microbiological properties. 2nd ed. Agron. Monogr. 9. ASA and SSSA, Madison, WI.

LACHAT Instruments. 1992. QuikChem method 12-107-04-1-B, nitrate in 2 M (1 M) KCl soil extracts, 0.02 to 20.0 mg N L^{-1}. LACHAT Instruments, 6645 West Mill Rd., Milwaukee, WI.

National Research Council. 1993. Soil and water quality. An agenda for agriculture. Committee on long-range soil and water conservation. Natl. Acad. Press, Washington, DC.

U.S. Department of Interior, U.S. Geological Survey. 1993. Methods for determination of inorganic substances in water and fluvial sediments. Book 5. Chapter Al. Method I-2601-78.

U.S. Environmental Protection Agency. 1983. Methods for chemical analysis of water and wastes. EPA-600/4-79-020. Revised March, 1983, Method 353.2.

Soil Tests (4D)
Aerobic Incubation (4D10)
2 M KCl Extraction (4D10a)
Ammonia –Salicylate (4D10a1)
Flow Injection, Automated Ion Analyzer (4D10a1a)
N as NH$_3$ (Mineralizable N) (4D10a1a1)
Air-Dry or Field-Moist, <2 mm (4D10a1a1a-b1)

1. Application

The most satisfactory methods currently available for obtaining an index for the availability of soil N are those involving the estimation of the N formed when soil is incubated under conditions which promote mineralization of organic N by soil microorganisms (U.S. Environmental Protection Agency, 1992). The method described herein for estimating mineralizable N is one of anaerobic incubation and is suitable for routine analysis of soils. This method involves estimation of the ammonium produced by a 1-week period of incubation of soil at 40°C (Keeney and Bremner, 1966) under anaerobic conditions to provide an index of N availability.

2. Method Summary

An aliquot of air-dry homogenized soil is placed in a test tube with water, stoppered, and incubated at 40°C for 1 week. The contents are rinsed with 2 M KCl. A flow injection automated ion analyzer is used to measure the ammonium produced in the soil after incubation. Absorbance of the solution is read at 660 nm. Data are reported as mg N kg^{-1} soil as NH$_3$ by procedure 4D10a1a1.

3. Interferences

The temperature and incubation period must remain constant for all samples. The test can be performed on field-moist or air-dry soil samples. Soil extracts can contain sufficient concentrations of calcium and magnesium to cause precipitation during analysis. EDTA is added to eliminate this problem (Keeney and Nelson, 1982).

4. Safety

Wear protective clothing (coats, aprons, sleeve guards, and gloves) and eye protection (face shields, goggles, or safety glasses) when preparing reagents, especially concentrated acids and bases. Dispense concentrated acids and bases in a fume hood. Thoroughly wash hands after handling reagents. Use the safety showers and eyewash stations to dilute spilled acids and bases. Use sodium bicarbonate and water to neutralize and dilute spilled acids.

5. Equipment
5.1 Electronic balance, ±1.0-mg sensitivity
5.2 Test tubes, 16-mm x 150-mm
5.3 PVC stoppers

5.4 Incubator, Model 10-140, Quality Lab Inc., Chicago, IL
5.5 Centrifuge, 50-mL, polyethylene
5.6 Filter paper, Whatman No. 42, 150 mm
5.7 Funnel, 60° angle, long stem, 50-mm diameter
5.8 Flow Injection Automated Ion Analyzer, QuikChem AE, LACHAT Instruments, Milwaukee, WI
5.9 XYZ Sampler, LACHAT Instruments, Milwaukee, WI
5.10 Reagent Pump, LACHAT Instruments, Milwaukee, WI
5.11 Automated Dilution Station, LACHAT Instruments, Milwaukee, WI
5.12 Sample Processing Module (SPM) or channel, QuikChem Method (12-107-06-2-A, ammonia (salicylate) in 2 M KCl soil extracts, 0.05 to 10.0 mg P L^{-1}), LACHAT Instruments, Milwaukee, WI
5.13 Computer, with QuikChem software, LACHAT Instruments, Milwaukee, WI, and printer
5.14 Vials, plastic, 25-mL (standards)
5.15 Culture tubes, glass, 10-mL (samples)

6. Reagents

6.1 Reverse osmosis deionized (RODI) water, ASTM Type I grade of reagent water
6.2 Helium, compressed gas
6.3 Potassium chloride (KCl), 2 *M,* carrier and standards diluent. Dissolve 150 g KCl in RODI water and dilute to 1-L volume. Mix thoroughly. Degas with helium.
6.4 EDTA (ethylene tetraacetic acid disodium salt dihydrate), 6% solution. Dissolve 66 g EDTA in 900 mL RODI water. Dilute to 1-L and invert to mix thoroughly. Degas with helium.
6.5 NaOH, buffer. Dissolve 28.0 g NaOH and 50.0 g sodium phosphate dibasic heptahydrate ($Na_2HPO_4 \cdot 7H_2O$) in 900 mL RODI water. Dilute to 1-L and invert to mix thoroughly. Degas with helium.
6.6 Salicylate-Nitroprusside Color Reagent. Dissolve 150 g sodium salicylate [salicylic acid sodium salt ($C_6H_4(OH)(COO)Na$)], and 1.0 g sodium nitroprusside [sodium nitroferricyanide dihydrate ($Na_2Fe(CN)_5NO \cdot 2H_2O$)] in 800 mL RODI water. Dilute to 1-L and invert to mix thoroughly. Degas with helium. Store in dark bottle in a refrigerator.
6.7 Hypochlorite Reagent. In a 500-mL volumetric, dilute 250 mL of 5.25% sodium hypochlorite (NaOCl) to mark with RODI water. Invert to mix thoroughly. Degas with helium.
6.8 Stock standard N solution (SSNS), 100.0 mg N L^{-1}. In a 1-L volumetric flask, dissolve 150 g potassium chloride (KCl) and 0.3819 g of ammonium chloride (NH$_4$Cl) that has dried for 2 h at 110°C in about 800 mL RODI water. Dilute to volume with RODI water and invert to thoroughly mix. Do not degas with helium. Store in polyethylene containers. Make fresh weekly. Store in a refrigerator.
6.9 Standard N calibration solutions (SNCS) or working standards, 20.0, 8.00, 2.00, 0.50, 0.10, 0.00 mg N L^{-1}. Make fresh daily. To six 250-mL volumetric flasks add as follows:

6.9.1 20.00 mg P L^{-1} = 50.0 mL SSNS
6.9.2 8.00 mg P L^{-1} = 20.0 mL SSNS
6.9.3 2.00 mg P L^{-1} = 5.0 ml SSNS
6.9.4 0.50 mg P L^{-1} = 1.25 mL SSNS
6.9.5 0.10 mg P L^{-1} = 0.25 mL SSNS
6.9.6 0.00 mg P L^{-1} = 0.0 mL SSNS (blank)

Dilute to mark with 2 M KCl. Invert to mix thoroughly.

7. Procedure

Anaerobic Incubation of Soil Sample

7.1 Weigh 5 g of <2 mm, air-dry soil to the nearest mg into a 16-mm x 150-mm test tube. If soil is fine-grind, weigh 1.25 g. If sample is moist, weigh enough soil to achieve ≈ 5 or 1.25 g, respectively, of air-dry soil.

7.2 Add 12.5 ±1 mL of RODI water. Do not add ethanol to overcome any wetting difficulties as ethanol may act as an interference with microbial activity. Stopper the tube, shake, and place in a 40°C constant-temperature incubator for 7 days. Refer to the manufacturer's instructions for set-up and operation of the incubator.

7.3 At the end of 7 days, remove the tube and shake for 15 s.

7.4 Transfer the contents of the test tube to another test tube. Complete the transfer by rinsing the tube with 3 times with 4 ml of 2 M KCl, using a total of 12.5 ±1 mL of the KCl. Filter contents into a centrifuge tube. If extracts are not to be determined immediately after collection, then store samples at 4°C. Analyze samples within 24 h. Ammonia is volatile and will leave the sample slowly, even though contained in polyethylene bottles.

Flow Injection Set-up and Operation

7.5 Transfer sample extracts into culture tubes and place in XYZ sample trays marked "Samples".

7.6 Transfer WNCS standards into plastic vials and place in descending order in XYZ sample trays marked "Standards".

7.7 Refer to the operating and software reference manuals for LACHAT for set-up and operation.

7.8 Turn main power switch "ON" and allow 15 min for heater module to warm up to 60°C.

7.9 On reagent pump, set speed to 35.

7.10 On computer main menu, select "Methods" and then "Analysis Select and Download". On method list, select Mineralizable N Method. System unit receives the downloaded method and initializes it.

7.11 Pump reagents (6.4, 6.5, 6.6, 6.7) into appropriate chambers of the manifold. Continue this step and observe baseline. A good baseline needs to be smooth and at zero absorbance. Scatter is indicative of air bubbles and irregular reagent flow. Also observe for any back-pressure in manifold tubing.

7.12 On computer main menu, select "Samples", "Tray Definition and Submit", and then "Edit" to create new sample tray followed by "Submit" to run new sample tray.

7.13 Method parameters specific to mineralizable N are defined within the "Method Definition" menu. Some of these parameters have been modified from the QuikChem Method 12-107-06-2-A, ammonia (salicylate) in 2 M KCl soil extracts (U.S. Environmental Protection Agency, 1979; LACHAT Instruments, 2001). Modifications are primarily related to the criteria and strategies for calibration standards and to injection timing.

7.14 Some of the method parameters as they relate to calibration standards are as follows:

267

7.14.1 There are 6 calibration standards (20.00, 8.00, 2.00, 0.50, 0.10, and 0.00 mg P L^{-1}) with a data format of ####.###, i.e., data rounded to 3 places.

7.14.2 The segments/boundaries for the calibration standards are A - C (20.0 to 2.00 mg P L^{-1}); C - F (2.00 to 0.00 mg P L^{-1}).

7.14.3 The protocol (replications) for the calibration standards is as follows: AA BB CC DD EE FF.

7.14.4 The check standard is 20 mg P L^{-1}. Maximum number of consecutive trays between check standard is one; maximum number of consecutive samples between check standard is 60; and maximum elapse time between check standards is 2 h.

7.14.5 Calibration strategy for segments A – C and C - F are normal. The normal strategy requires a minimum correlation coefficient of 0.99. Both segments require a maximum standard deviation in slope of 50%. A calibration passes only when both criteria are met. Strategies are user designated. In addition, calibration strategies are based on the full chord. Chord 0 is full chord, and chord 1 - 5 are sections of peak from start of peak to end of peak.

7.14.6 The instrument is calibrated with the injection of SNCS. The data system the associates the concentrations with the instrument response for each SNCS.

7.15 Method parameters in relation to timing are as follows:

7.15.1 Cycle period: 40 s

7.15.2 Inject to start of peak period: 25 s. To see if peaks are being timed correctly, scan across correlation coefficients for all chords 1 - 5. The most peak area should be between chords 2 - 4 with the most signal-to-noise ratio in chords 1 and 5.

7.1.5.3 Inject to end of peak period: 46 s

7.15.4 Automatic timing, where standard assumptions are in effect. Manual timing may be helpful in this method.

7.16 Method parameters in relation to data presentation are as follows:

7.16.1 Top Scale Response: 0.50 abs

7.16.2 Bottom Scale Response: 0.00 abs

7.17 Refer to the "Method Definition" for Mineralizable N for other method parameters not discussed here.

7.18 Run samples using calibration curve. Sample concentration is calculated from the regression equations. Report results to the nearest 0.01 unit for the sample extract and each SNCS.

7.19 If samples are outside calibration range, dilute samples with extracting solution and re-analyze.

7.20 Upon completion of run, place the transmission lines into the 1 *M* HCl. Pump the solution for approximately 5 min to remove any precipitated reaction products. Then place these lines in

RODI water and pump for an additional 5 min and proceed with the normal "Shut-down" procedure.

8. Calculations

Convert extract N (mg L^{-1}) to soil N (mg kg^{-1}) as follows:

Soil N = [(A x B x C x R x 1000)/E]

where:
A   = Analyte reading (mg L^{-1})
B   = Extract volume (L)
C   = Dilution, if performed
R   = Air-dry/oven-dry ratio (procedure 3D1) or field-moist/oven-dry ratio (procedure 3D2)
1000 = Conversion factor to kg-basis
E   = Sample weight (g)

9. Report
Report data to the nearest mg N kg^{-1} soil as NH_3.

10. Precision and Accuracy
Precision and accuracy data are available from the SSL upon request.

11. References

Keeney, D.R., and J.M. Bremner. 1966. Comparison and evaluation of laboratory methods of obtaining an index of soil nitrogen availability. Agron. J. 58:498-503.

Keeney, D.R., and D.W. Nelson. 1982. Nitrogen – inorganic forms. p. 643-698. *In* A.L. Page, R.H. Miller, and D.R. Keeney (eds.) Methods of soil analysis. Part 2. Chemical and microbiological properties. 2nd ed. Agron. Monogr. 9. ASA and SSSA, Madison, WI.

U.S. Environmental Protection Agency. 1992. Handbook of laboratory methods for forest health monitoring. G.E. Byers, R.D. Van Remortel, T.E. Lewis, and M. Baldwin (eds.). Part III. Soil analytical laboratory. Section 10. Mineralizable N. U.S. Environmental Protection Agency, Office of Research and Development, Environmental Monitoring Systems Laboratory, Las Vegas, NV.

Carbonate and Gypsum (4E)
3 *N* HCl Treatment (4E1)
CO_2 Analysis (4E1a)
Manometer, Electronic (4E1a1)
Calcium Carbonates (4E1a1a1)
Air-Dry, <2 mm (4E1a1a1a1)
Air-Dry, <20 mm (4E1a1a1a2)

1. Application
The distribution and amount of $CaCO_3$ are important for fertility, erosion, available water-holding capacity, and genesis of the soil. Calcium carbonate provides a reactive surface for adsorption and precipitation reactions, e.g., phosphate, trace elements, and organic acids (Loeppert and Suarez, 1996; Amer et al., 1985; Talibudeen an Arambarri, 1964; Boischot et al., 1950). The determination of calcium carbonate ($CaCO_3$) equivalent is a criterion in soil taxonomy (Soil Survey Staff, 1999). Carbonate content of a soil is used to define carbonatic, particle-size, and calcareous soil classes and to define calcic and petrocalcic horizons (Soil Survey Staff, 1999). The formation of calcic and petrocalcic horizons has been related to a

variety of processes, some of which include translocation and net accumulation of pedogenic carbonates from a variety of sources as well as the alteration of lithogenic (inherited) carbonate to pedogenic carbonate (soil-formed carbonate through *in situ* dissolution and reprecipitation of carbonates) (Rabenhorst et al., 1991). The $CaCO_3$ equivalent is most commonly reported on the <2-mm base. However, in some soils with hard carbonate concretions, carbonates are determined on both the <2-mm (4E1a1a1a1) and the 2- to 20-mm basis (4E1a1a1a2).

2. Summary of Method

The amount of carbonate in the soil is measured by treating the samples with HCl. The evolved CO_2 is measured manometrically. The amount of carbonate is then calculated as percent $CaCO_3$.

3. Interferences

Chemical interference is the reaction by the acid with other carbonates, e.g., carbonates of Mg, Na, and K, that may be present in soil sample. The calculated $CaCO_3$ is only a semiquantitative measurement (Nelson, 1982).

Analytical interference may be caused by temperature changes within the reaction vessel. When sealing the vessel, the analyst should not hold the vessel any longer than necessary to tighten the cap. The internal pressure must be equalized with the atmosphere. After the septum has been pierced with a needle, $\approx$ 5 to 10 s are required to equalize the internal pressure of the bottle. With extensive use, the septa leak gas under pressure. The septa should be replaced at regular intervals. The analyst should not touch the glass of the vessel when reading the pressure.

4. Safety

Wear protective clothing (coats, aprons, sleeve guards, and gloves) and eye protection (face shields, goggles, or safety glasses) when handling acids. Thoroughly wash hands after handling acids. Use the fume hood when diluting concentrated HCl. Use the safety showers and eyewash stations to dilute spilled acids. Use sodium bicarbonate and water to neutralize and dilute spilled acids.

The gelatin capsule may leak acid while being filled. Keep other personnel away from the area when filling capsules.

High pressure may develop inside the bottle if there is a large amount of calcareous sample. Do not use more than 2 g of any sample in the bottles. If high pressure develops in the bottle, release the pressure by venting the gas with a syringe needle. Some bottles may break without shattering. Discard any bottle with hairline cracks or obvious defects.

5. Equipment

5.1 Electronic balance, ±0.10-mg sensitivity
5.2 Electronic balance, ±1mg-sensitivity
5.3 Threaded weighing bottles, wide-mouth, clear glass, standard, 120 mL (4 fl. oz.), 48-mm neck size. For best results, grind rim of bottle with 400-600 grit sandpaper on a flat glass plate.
5.4 Machined PVC caps for threaded 120-mL (4 fl. oz.) weighing bottles, 54-mm diameter with 12.7-mm diameter hole drilled in center, O-ring seal.
5.5 O-rings, 3.2 x 50.8 x 57.2 mm (1/8 x 2 x 2 1/4 in)
5.6 Flanged stopper No. 03-255-5, Fisher Scientific. Place in machined cap.
5.7 Manometer, hand-held gauge and differential pressure, PCL-200 Series, Omega Engineering, Stanford, CT.
5.8 Hypodermic needle, 25.4 mm (1 in), 23 gauge. Connect needle to pressure tubing on transducer.
5.9 Mechanical rotating shaker, 140 rpm, Eberbach 6140, Eberbach Corp., Ann Arbor, MI.

6. Reagents

6.1 Reverse osmosis deionized (RODI) water, ASTM Type I grade of reagent water

6.2 Methyl red indicator

6.3 Hydrochloric acid (HCl), concentrated, 12 N

6.4 HCl, 3 N. Dilute 500 mL of concentrated HCl with 1500 mL RODI water. Add a few crystals of methyl red indicator. Methyl red indicator will turn yellow if HCl is consumed by sample. If this reaction occurs, adjust the sample size (smaller).

6.5 Gelatin capsule, 10 mL, size 11, Torpac Inc., Fairfield, NJ.

6.6 Gylcerine, USP. Put the gylcerine in a small squeeze bottle and use as a lubricant for the O- rings.

6.7 $CaCO_3$, Ultrex, assay dried basis 100.01%,

7. Procedure

Manometer Calibration

7.1 Calibrate the manometer quarterly or whenever equipment changes (e.g., old rubber septum replaced). Calibrate by weighing three replicates of $CaCO_3$ standards (0, 0.025, 0.05, 0.1, 0.2, 0.3, 0.4, 0.5, 0.75 g). Weigh to the nearest 0.1mg. Dry the standard samples in the oven for 2 h at 110°C. Remove samples from oven, place in desiccator and cool to ambient temperature. Proceed as outlined in Sections 7.4 through 7.9.

<2-mm Basis

7.2 A $CaCl_2$ pH >6.95 is generally used as an indicator of the presence of carbonates. The presence of carbonates (effervescence with HCl) is also checked during lab preparation (procedure 1B1b2b5).

7.3 Weigh 0.5 to 2 g of fine-grind, air-dry soil sample to the nearest mg and place in a 120-mL, wide-mouth bottle. Run 3 blanks and a quality control check sample with every batch of 24 samples. The quality control check sample serves as a single point check. Vary the sample weight according to the $CaCO_3$ content based on effervescence of sample (procedure 1B1b2b5) as follows:

7.3.1 Use a 2-g sample weight if effervescence is None, Very Slight, or Slight.

7.3.2 Use a 1-g sample weight, if effervescence is Strong.

7.3.3 Use a 0.5-g sample weight, if effervescence is Violent.

7.4 Lubricate the O-ring of bottle cap with gylcerine from a squeeze bottle.

7.5 Dispense 10 mL of 3 N HCl into a gelatin capsule and carefully place the top on the capsule. The HCl may squirt or leak out of capsule. If this happens, discard the capsule.

7.6 Place the capsule in bottle and cap bottle immediately.

7.7 Release any pressure in the bottle by piercing the stopper with a hypodermic needle. Remove the needle after ≈ 5 to 10 s.

7.8 After 5 to 10 min, the HCl dissolves through the capsule. Shake the bottle at a rate of 140 rpm on the shaker for the first 10 min and last 10 min of a 1-h interval at room temperature (20°C ± 2°C). After this 1 h, measure the pressure in the bottle by piercing the stopper of the cap with a hypodermic needle connected to the manometer.

7.9 Auto-zero the manometer before taking readings. Record the manometer readings (mm Hg).

<20-mm Basis

7.10 Determine carbonate content of the 2- to 20-mm fraction on a fine-grind (<180 μm) air-dry sample by the above method.

7.11 The carbonate in the 2- to 20-mm fraction and in the <2-mm fraction are combined and converted to a <20-mm soil basis.

8. Calculations

8.1 Correct the manometer readings as follows:

$$CR = (MR - BR)$$

where:
CR = Corrected reading
MR = Manometer reading
BR = Blank reading

Three blanks are run with each batch of 24 samples. The average of three blanks is used as BR.

8.2 Calculate the regression equation for the corrected manometer readings. Use the $CaCO_3$ weights as the dependent variable (regressed or predicted values) and the corresponding manometer readings as the independent variable.

8.3 Use the corrected (CR) linear regression (slope, intercept) equation to estimate % $CaCO_3$ in the sample as follows:

8.4 CCE = [(CR x Slope + Intercept) /Sample Weight (g)] x AD/OD

where:
CCE   = Calcium Carbonate Equivalent (%) in <2-mm fraction or 2- to 20-mm fraction
CR   = Corrected manometer reading
AD/OD = Air-dry/oven-dry ratio (procedure 3D1)

8.5 Carbonate = (A x B) + [C x (1-B)]

where:
Carbonate = Carbonate as $CaCO_3$ on a <20-mm basis (%)
A   = $CaCO_3$ in <2-mm fraction (%)
B   = Weight of the <20-mm fraction minus the weight of the 2- to 20-mm fraction
             divided by the weight of <20-mm fraction (procedures 1B2b2f and 3A2).
C   = $CaCO_3$ in 2- to 20-mm fraction (%)

9. Report

Report $CaCO_3$ equivalent as a percentage of oven-dry soil to the nearest whole number.

10. Precision and Accuracy

Precision and accuracy data are available from the SSL upon request.

11. References

Amer, F.A., A. Mahmoud, and V. Sabel. 1985. Zeta potential and surface area of calcium carbonate as related to phosphate sorption. Soil Sci. Soc. Am. J. 49:1137-1142.

Boischot, P, M. Coppenet, and J. Hebert. 1950. Fixation de l'acide phosphorique sur le calcaire des sols. Plant Soil 2:311-322.

Loeppert, R. H., and D. L. Suarez. 1996. Carbonate and gypsum. p. 437-474. *In* D.L. Sparks (ed.) Methods of Soil Analysis. Part 3 – Chemical methods. Soil Sci. Am. Book Series No. 5. ASA and SSSA, Madison, WI.

Nelson, R.E. 1982. Carbonate and gypsum. p. 181-197. *In* A.L. Page, R.H. Miller, and D.R. Keeney (eds.) Methods of soil analysis. Part 2. Chemical and microbiological properties. 2nd ed. Agron. Monogr. 9. ASA and SSSA, Madison, WI.

Rabenhorst, M.C., L.T. West, and L.P. Wilding. 1991. Genesis of calcic and petrocalcic horizons in soils over carbonate rocks. p. 61-74. *In* W.D. Nettleton (ed.) Occurrence, characteristics, and genesis of carbonate, gypsum, and silica accumulations in soils. Soil Sci. Soc. Am. Spec. Publ. No. 26. ASA and SSSA, Madison, WI.

Soil Survey Staff. 1999. Soil taxonomy: A basic system of soil classification for making and interpreting soil surveys. USDA-NRCS Agric. Handb. 436. 2nd ed. U.S. Govt. Print. Office, Washington, DC.

Talibudeen, O., and P. Arambarri. 1964. The influence of the amount and the origin of calcium carbonates on the isotopically-exchangeable phosphate in calcareous soils. J. Agric. Sci. 62:93-97.

Carbonate and Gypsum (4E)
Aqueous Extraction (4E2)
Precipitation in Acetone (4E2a)
Conductivity Bridge (4E2a1)
Electrical Conductivity (4E2a1a)
Gypsum, Qualitative and Quantitative (4E2a1a1)
Air-Dry, < 2 mm (4E2a1a1a1)
Air-Dry, < 20 mm (4E2a1a1a2)

1. Application

If the electrical conductivity of a soil sample >0.50 dS cm^{-1} by procedure 4F1a1a1, gypsum content is determined. Gypsum content of a soil is a criterion for gypsic and petrogypsic horizons and for mineralogical class at the family level (Soil Survey Staff, 1999). Soil subsidence through solution and removal of gypsum can crack building foundations, break irrigation canals, and make roads uneven. Failure can be a problem in soils with as little as 1.5% gypsum (Nelson, 1982). The gypsum content in the soil may be used to determine if reclamation of sodic soils requires chemical amendments. Corrosion of concrete is also associated with soil gypsum.

Gypsum formation by precipitation of calcium sulfate ($CaSO_4$) is usually highest at the surface layers. Gypsum from deposits high in gypsum are usually highest in the lower part of the soil profile. However, leaching may disrupt this sequence. Gypsum is reported on both the <2- and the <20-mm base.

2. Summary of Method

A soil sample is mixed with water to dissolve gypsum. Acetone is added to a portion of the clear extract to precipitate the dissolved gypsum. After centrifuging, the gypsum is redissolved in water. The electrical conductivity (EC) of the solution is read. The EC reading is used to estimate the gypsum content in meq 100 g^{-1}.

In procedure 4E2a1a1, gypsum content (meq 100 g^{-1}) is converted to percent gypsum (uncorrected). The percent gypsum (uncorrected) is used to calculate percent gypsum (corrected). The percent gypsum (corrected) is used to correct the AD/OD (air-dry/oven-dry ratio). The AD/OD and corrected AD/OD are determined in procedures 3D1 and 3D3, respectively. The corrected AD/OD uses the correction for the crystal water of gypsum. Gypsum content on a <2-mm basis is reported in procedure 4E2a1a1a1.

Gypsum content may also be determined on the 2- to 20-mm fraction prepared by procedure 1B1b2f1a1. The gypsum determined on the 2- to 20-mm fraction and the gypsum determined on the fine earth (prepared by procedure 1B1b2d2) are combined and converted to a <20-mm soil basis. Gypsum on a <20-mm basis is reported in procedure 4E2a1a1a2.

3. Interferences

Loss of the precipitated gypsum is the most significant potential error. Care in handling the precipitated gypsum is required. Incomplete dissolution of gypsum is also possible. In soils with large gypsum crystals, use fine-ground samples to reduce sampling errors.

When present in sufficiently high concentrations, the sulfates of Na and K are also precipitated by acetone. The concentration limits for sulfates of Na and K are 50 and 10 meq L^{-1}, respectively.

4. Safety

Acetone is highly flammable. Avoid open flames and sparks. Use a nonsparking centrifuge. Standard laboratory equipment includes fire blankets and extinguishers for use when necessary. Proper use and appropriate load balance of the centrifuge is required. Follow standard laboratory safety precautions.

5. Equipment
5.1 Electronic balance, ±1.0-mg sensitivity
5.2 Bottles, 250 mL, with caps, Wheaton.
5.3 Mechanical reciprocating shaker, 200 oscillations min^{-1}, Eberbach 6000, Eberbach Corp., Ann Arbor, MI
5.4 Dispenser, 100 mL and 10 mL
5.5 Pipette, 20 mL, solvent, Pistolpet, Manostat
5.6 Pipette, 10 mL, electronic digital, with tips, polypropylene, 10 mL,
5.7 Centrifuge, Sorvall, GLC-1, General Laboratory Centrifuge
5.8 Centrifuge tubes, 15 mL, plain, conical
5.9 Thermometer, 0° to 100°C
5.10 Conductivity bridge and conductivity cell, Markson Model 1056, Amber Science, Eugene, OR
5.11 Filter paper, folded, 185-mm diameter, Whatman No. 2V
5.12 Funnel, 90 cm
5.13 Flask, Erhlenmeyer, 250 mL
5.14 Vortexer, mini, MV1, VWR Scientific Products

6. Reagents
6.1 Reverse osmosis (RO) water, ASTM Type III grade of reagent water
6.2 Potassium chloride (KCl), 0.010 N. Dry KCl overnight or at least 4 h in oven (105°C). Dissolve 0.7456 g of dry reagent grade KCl in RO water and bring to 1-L volume. Conductivity at 25°C is 1.41 mmhos cm^{-1}.
6.3 Acetone, purity 99%

7. Procedure

7.1 Weigh 5.0 g of fine-grind, air-dry soil into a 500-mL Wheaton bottle to the nearest mg. If a trace of gypsum is present, a 20-g sample size may be used. If the EC reading is >0.85 mmhos cm^{-1}, repeat the procedure using a smaller sample size (2.5, 1, 0.5. 0.25, or 0.1 g).

7.2 Use a dispenser to add 100 mL of RO water to sample and 1 blank.

7.3 Cap the bottle and shake at an oscillating rate of 200 oscillations min^{-1} for 30 min at room temperature (20°C ± 2°C).

7.4 Filter the suspension. The first few mL of filtrate is usually cloudy and should be discarded. Collect the clear filtrate in a 250-mL flask.

7.5 Pipet 5 mL of filtrate into 15-mL conical centrifuge tube.

7.6 Use a solvent dispenser to add 5 mL acetone.

7.7 Cap tube with a polyethylene stopper and mix.

7.8 Carefully release pressure within tube by loosening the stopper.

7.9 Let stand for at least 10 min to allow the precipitate to flocculate.

7.10 Use acetone (1 or 2 mL) to rinse stopper and inside rim of tube to prevent gypsum loss.

7.11 Remove stopper and centrifuge at 2200 rpm for 5 min.

7.12 Decant and discard supernatant. Invert and drain the tube on filter paper or on towel for 5 min.

7.13 Add 5 mL of acetone to the tube. Replace stopper. Use Vortexer to shake sample.

7.14 Carefully remove stopper and rinse it and the inside rim of tube with acetone (1 or 2 mL).

7.15 Centrifuge the sample tube at 2200 rpm for 5 min.

7.16 Discard supernatant. Drain tube upside down for 5 min.

7.17 Use a dispenser to add 10 mL of RO water to tube.

7.18 Stopper and shake with Vortexer until the precipitate dissolves.

7.19 Calibrate the EC meter and cell by drawing the 0.010 N KCl solution into the cell.

7.20 Flush the cell and fill with RO water. Digital reading should be 0.00.

7.21 Read the EC of dissolved precipitate by drawing up solution into cell and flush at least once.

7.22 If the EC reading is >0.85 mmhos cm^{-1}, repeat the procedure using a smaller sample weight.

8. Calculations

8.1 Calculate % Gypsum$_{uc}$ (Gypsum uncorrected) by using Table 1 to convert EC reading (mmhos/cm) to gypsum content (meq/100 g) and proceeding with the following the equation.

% Gypsum$_{uc}$ = [Gypsum x Water x 0.08609 x AD/OD]/[Sample Weight (g) x 5]

where:
% Gypsum$_{uc}$= % Gypsum in <2 mm fraction or 2 to 20-mm fraction

Gypsum   = Gypsum (meq L^{-1}). Refer to Table 1.
Water   = Volume RO water (100 mL) to dissolve gypsum
0.08609   = Conversion factor (gypsum % = meq 100 g^{-1} x 0.08609)
AD/OD   = Air-dry/oven-dry ratio, (procedure 3D1)
5   = Filtrate (5 mL)

8.2 Table 1 converts EC (mmhos cm^{-1}) to gypsum (meq L^{-1}) for the above calculations. Enter Table 1 using both the x and y axis for the EC reading to determine gypsum content (meq L^{-1}).

8.3 Alternatively to using Table 1, calculate % Gypsum$_{uc}$ from the following equation:

Result = (Exp (2.420384 + 1.1579713 x Log (EC – blank)) x Water x 0.08609 x ADOD)/(Sample Weight x 5)

8.4 The following equation for calculation of % Gypsum$_c$ (gypsum corrected) assumes the crystal-water content of gypsum is 19.42% (Nelson et al., 1978) as opposed to the theoretical water content (20.21%).

% Gypsum$_c$ = [% Gypsum$_{uc}$]/[1 + 0.001942 x % Gypsum$_{uc}$]

8.5 Use the % Gypsum$_{uc}$ to recalculate the AD/OD (procedure 3D1). The corrected AD/OD (procedure 6F3) uses the correction for the crystal water of gypsum.

8.6 Calculate gypsum on <20-mm basis (procedure 4E2a1a1) as follows:

(%) Gypsum = A x B + [C x (1 - B)]

where:
A = Gypsum (%) in <2-mm fraction
B = Weight of the <20-mm fraction minus the 20- to 2-mm fraction divided by the weight of the <20-mm fraction
C = Gypsum (%) in the 20- to 2-mm fraction

9. Report
Report gypsum as a percent to the nearest whole unit.

10. Precision and Accuracy
Precision and accuracy data are available from the SSL upon request.

11. References
Nelson, R.E. 1982. Carbonate and gypsum. p. 181-197. *In* A.L. Page, R.H. Miller, and D.R. Keeney (eds.) Methods of soil analysis. Part 2. Chemical and microbiological properties. 2nd ed. Agron. Monogr. 9. ASA and SSSA, Madison, WI.

Nelson, R.E., L.C. Klamath, and W.D. Nettleton. 1978. Determining soil gypsum content and expressing properties of gypsiferous soils. Soil Sci. Soc. Am. J. 42:659-661.

Soil Survey Staff. 1999. Soil taxonomy: A basic system of soil classification for making and interpreting soil surveys. USDA-NRCS Agric. Handb. 436. 2nd ed. U.S. Govt. Print. Office, Washington, DC.

Table 1. Convert EC reading (mmhos cm^{-1}) to gypsum content (meq L^{-1})

EC	0.00	0.01	0.02	0.03	0.04	0.05	0.06	0.07	0.08	0.09
0.0						0.40				
0.1	0.80	0.89	0.98	1.10	1.22	1.31	1.40	1.50	1.60	1.70
0.2	1.80	1.90	2.00	2.10	2.20	2.30	2.40	2.50	2.60	2.70
0.3	2.80	2.90	3.00	3.10	3.20	3.30	3.40	3.50	3.60	3.72
0.4	3.85	3.98	4.10	4.22	4.35	4.48	4.60	4.70	4.80	4.90
0.5	5.00	5.12	5.25	5.38	5.50	5.62	5.75	5.88	6.00	6.12
0.6	6.25	6.35	6.45	6.58	6.70	6.82	6.95	7.05	7.15	7.28
0.7	7.40	7.52	7.65	7.78	7.90	8.04	8.18	8.32	8.45	8.58
0.8	8.70	8.82	8.95	9.05	9.15	9.28	9.40	9.55	9.70	9.85
0.9	10.00	10.12	10.25	10.38	10.50	10.62	10.75	10.88	11.00	11.15
1.0	11.30									

Electrical Conductivity and Soluble Salts (4F)
Aqueous Extraction (4F1)
1:2 Extraction (4F1a)
Conductivity Bridge (4F1a1)
Electrical Conductivity (4F1a1a)
Salt Prediction (4F1a1a1)
Air-Dry, <2 mm (4F1a1a1a1)

1. Application

Salt prediction is used not only to predict which soils have measurable amounts of soluble salts but also to predict the quantity and the appropriate dilutions for salt analyses of those soils. If salt prediction or conductivity is <0.25 mmhos cm^{-1} (dS cm^{-1}) soils are considered nonsalty, and generally, no other salt analyses are performed on these soils by the SSL.

2. Summary of Method

A soil sample is mixed with water and allowed to stand overnight. The electrical conductivity (EC) of the mixture is measured using an electronic bridge. The EC by this method (4F1a1a1) is used to indicate the presence of soluble salts (U.S. Salinity Laboratory Staff, 1954).

3. Interferences

Reverse osmosis deionized water is used to zero and flush the conductivity cell. The extract temperature is assumed to be 25°C. If the temperature deviates significantly, a correction may be required.

Provide airtight storage of KCl solution and samples to prevent soil release of alkali-earth cations. Exposure to air can cause gains and losses of water and dissolved gases significantly affecting EC readings.

4. Safety

No significant hazards are associated with this procedure. Follow standard laboratory safety practices.

5. Equipment

5.1 Electronic balance, ±1.0-mg sensitivity
5.2 Conductivity bridge and conductivity cell, with automatic temperature adjustment, 25 ± 0.1°C, Markson Model 1056, Amber Science, Eugene, OR
5.3 Plastic cups, 30 mL (1 fl. oz.), with lids, Sweetheart Cup Co. Inc., Owings Mills, MD
5.4 Dispenser, Repipet or equivalent, 0 to 10 mL

6. Reagents

6.1 Reverse osmosis (RO) water, ASTM Type III grade of reagent water

6.2 Potassium chloride (KCl), 0.010 N. Dry KCl overnight in oven (110°C). Dissolve 0.7456 g of KCl in RODI water and bring to 1-L volume. Conductivity at 25°C is 1.412 mmhos cm^{-1}.

7. Procedure

7.1 Weigh 5.0 g of <2-mm, air-dry soil in a 30-mL (1-oz) condiment cup.

7.2 Add 10 mL of RO water to sample using a Repipet dispenser.

7.3 Swirl to mix, cap, and allow to stand overnight.

7.4 Standardize the conductivity bridge using RO water (blank) and 0.010 N KCl (1.41 mmhos cm^{-1}).

7.5 Read conductance of supernatant solution directly from the bridge.

7.6 Record conductance to 0.01 mmhos cm^{-1}.

8. Calculations

8.1 No calculations are required for this procedure.

8.2 Use the following relationship to estimate the total soluble cation or anion concentration (meq L^{-1}) in the soil.

EC (mmhos cm^{-1}) x 10 = Cation or Anion (meq L^{-1})

8.3 Use the following relationship to estimate the total soluble cation or anion concentration (meq g^{-1} oven-dry soil) in the soil.

EC (mmhos cm^{-1}) x 20 = Cation (meq g^{-1} soil)

EC (mmhos cm^{-1}) x 20 = Anion (meq g^{-1} soil)

9. Report

Report prediction conductance to the nearest 0.01 mmhos cm^{-1} (dS m^{-1}).

10. Precision and Accuracy

Precision and accuracy data are available from the SSL upon request.

11. References

U.S. Salinity Laboratory Staff. 1954. L.A. Richards (ed.) Diagnosis and improvement of saline and alkali soils. 160 p. USDA Handb. 60. U.S. Govt. Print. Office, Washington, DC.

Electrical Conductivity and Soluble Salts (4F)
Saturated Paste (4F2)

Salt-affected soils, i.e., excessive amounts of soluble salts and/or exchangeable sodium (ES), are common in, though not restricted to, arid and semi-arid regions. These soils are usually described and characterized in terms of the soluble salt concentrations, i.e., major dissolved inorganic solutes (Rhoades, 1982). Salt composition and distribution in the soil profile affect the plant response, i.e., osmotic stress, specific ion effects, and nutritional imbalances. Soil texture and plant species also are factors in this plant response to saline soils.

Traditionally, the classification of salt-affected soils has been based on the soluble salt concentrations in extracted soil solutions and on the exchangeable sodium percentage (ESP) in the associated soil (Bohn et al., 1979). In general, saline soils have been defined as having a salt content >0.1% or an EC $\geq$4 mmhos cm^{-1}; alkali soils have an ESP of $\geq$15%; and saline-alkali soils have properties of both saline and alkali soils (U.S. Salinity Laboratory Staff, 1954). In soil taxonomy, the ESP and the Na-adsorption ratio (SAR) have been used as criteria for natric horizons (Soil Survey Staff, 1999). The ESP and SAR are calculated in procedures 4F3a1 and 4F3b, respectively.

The measurable absolute and relative amounts of various solutes are influenced by the soil:water ratio at which the soil solution extract is made. Therefore, this ratio is standardized to obtain results that can be applied and interpreted universally. Soil salinity is conventionally defined and measured on aqueous extracts of saturated soil pastes (U.S. Salinity Laboratory Staff, 1954). This soil:water ratio is used because it is the lowest reproducible ratio at which the extract for analysis can be readily removed from the soil with common laboratory equipment, i.e., pressure or vacuum, and because this soil:water ratio is often related in a predictable manner to field soil water contents (Rhoades, 1982). Soil solutions obtained at lower soil moisture conditions are more labor intensive and require special equipment.

The SSL measures salinity on aqueous extracts of saturated soil pastes. The saturated paste is prepared (4F2), and the saturation percentage (SP) determined (4F2a1). The saturated paste extract is obtained with an automatic extractor (4F2c1). Electrical conductivity and soil resistivity of saturated paste are measured in procedures 4F2b1 and 4F2b2, respectively. The saturated paste pH is measured in procedure 4C1a1a2. The water-soluble cations of Ca^{2+}, Mg^{2+}, K^+, and Na^+ are measured by atomic absorption spectrophotometry in procedures 4F2c1a1-4, respectively. The water-soluble anions of Br, CH_3COO^-, Cl^-, F^-, NO_3^-, NO_2^-, PO_4^{3-}, and SO_4^{2-} are measured by ion chromatography in procedures 4F2c1b1a1-8, respectively. The carbonate and bicarbonate concentrations are determined by acid titration procedures 4F2c1c1a1-2, respectively. Estimated total salt is calculated in procedure 4F3c. The SSL also performs a salt prediction test (procedure 4F1a1a1) which is used not only to predict those soils that have measurable amounts of soluble salts but also to predict the quantity and the appropriate dilutions for salt analyses of those soils. If salt predictions or conductances are <0.25 mmhos cm^{-1}, soils are considered nonsalty, and generally, no other salt analyses are performed on these soils by the SSL.

The SP, i.e., the amount of moisture in the saturated paste, is an important measurement. An experienced analyst should be able to repeat the saturated paste preparation to an SP within 5%. The SP can be related directly to the field moisture range. Measurements on soils, over a considerable textural range, (U.S. Salinity Laboratory Staff, 1954) indicate the following general rules of thumb.

SP $\approx$ 4 x 15-bar water

SP $\approx$ 2 x upper end field soil moisture content

AWC $\approx$ SP/4

where:
Sp = Saturation percentage
AWC = Available water capacity

Therefore, at the upper (saturated) and lower (dry) ends of the field moisture range, the salt concentration of the soil solution is $\approx$ 4x and 2x the concentration in the saturation extract, respectively.

If the soil texture is known, and the 15-bar water content has been measured, the preceding SP relationships may be redefined (U.S. Salinity Laboratory Staff, 1954) as follows:

15 Bar Water %	Texture	Relationship
2.0 to 6.5	Coarse	SP $\approx$ 6 1/3 x 15 bar
6.6 to 15	Medium	SP $\approx$ 4 x 15 bar
>15	Fine	SP $\approx$ 3 1/4 x 15 bar
>15	Organic	SP $\approx$ 3 2/3 x 15 bar

The electrical conductivity of the saturated paste (EC_s) is measured and is commonly reported as resistivity (R_s). The EC_s measurement requires more time, i.e., preparation of saturated soil paste, than the R_s measurement. However, the EC_s is the easier measurement from which to make interpretations, i.e., EC_s is more closely related to plant response (U.S. Salinity Laboratory Staff, 1954). Furthermore, there is a limited correlation between EC_s and R_s, as the relationship is markedly influenced by variations in SP, salinity, and soil mineral conductivity. The EC_s has been related to R_s (U.S. Laboratory Staff, 1954) by the equation as follows:

$$EC_s \approx 0.25/R_s$$

where:
0.25 = Constant for Bureau of Soils electrode cup

Historically, the EC_s is adjusted to 60°F (15.5°C) basis before interpretative use. The EC_s and R_s increase $\approx$ 2% per °C. The SSL determines the EC_s and R_s in procedures 4F2b1 and 4F2b2, respectively. The unit EC x 10^3 is called mmhos cm^{-1}.

The EC_s (mmhos cm^{-1}) may be used to estimate the salt percentage (P_{sw}) in solution (U.S. Salinity Laboratory Staff, 1954) as follows:

$$P_{sw} \approx 0.064 \text{ x } EC_s \text{ (mmhos cm}^{-1})$$

The preceding equation may be used to estimate the salt percentage in the soil (P_{ss}) (U.S. Salinity Laboratory Staff, 1954) as follows:

$$P_{ss} \approx (P_{sw} \text{ x SP})/100$$

The EC_s (mmhos cm^{-1}) may be used to estimate the osmotic potential (OP) in atmospheres of a solution (U.S. Salinity Laboratory Staff, 1954) as follows:

$$OP \approx 0.36 \text{ x } EC_s \text{ (mmhos cm}^{-1})$$

The EC_s (mmhos cm^{-1}) may be used to estimate the total cation or anion concentration (meq L^{-1}) of the solution (U.S. Salinity Laboratory Staff, 1954) as follows:

Total cations $\approx$ 10 x EC_s (mmhos cm^{-1})

Total anions $\approx$ 10 x EC_s (mmhos cm^{-1})

where:
EC_s at 25°C

A means of cross-checking chemical analyses for consistency and reliability is provided by the interrelations that exist among the various soil chemical determinations (U.S. Salinity Laboratory Staff, 1954). The saturated paste pH is the apparent pH of the soil:water mixture and is a key indicator in many of these interrelations. The saturated paste pH is dependent upon the dissolved CO_2 concentration; moisture content of the mixture; exchangeable cation composition; soluble salt composition and concentration; and the presence and amount of gypsum and alkaline-earth carbonates. Some rules of thumb that apply to the saturated paste (U.S. Salinity Laboratory Staff, 1954) are as follows:

Total Cation and Anion Concentrations

Total cations ≈ Total anions, expressed on equivalent basis

pH and Ca and Mg Concentrations

Concentrations of Ca^{2+} and Mg^{2+} are seldom >2 meq L^{-1} at pH >9.

pH and Carbonate and Bicarbonate Concentrations

Carbonate concentration (meq L^{-1}) is measurable only if pH >9.

Bicarbonate concentration is rarely >10 meq L^{-1} in absence of carbonates.

Bicarbonate concentration is seldom >3 or 4 meq L^{-1} if pH <7.

Gypsum

Gypsum is rarely present if pH >8.2.

Gypsum has variable solubility in saline solutions (20 to 50 meq L^{-1}).

Check for the presence of gypsum if Ca concentration >20 meq L^{-1} and pH ≤8.2.

pH, ESP, and Alkaline-Earth Carbonates

Alkaline-earth CO_3^- and ESP ≥15 are indicated if pH ≥8.5.

ESP ≤15 may or may not be indicated if pH <8.5.

No alkaline-earth CO_3^- are indicated if pH <7.5.

pH and Exchangeable Acidity

Significant amounts of exchangeable acidity are indicated if pH <7.0.

The commonly determined soluble cations and anions in the saturation extract include calcium, magnesium, sodium, potassium, chloride, sulfate, nitrate, fluoride, carbonate, bicarbonate, and nitrite. The less commonly analyzed cations and anions include iron, aluminum, manganese, lithium, strontium, rubidium, cesium, hydronium, phosphate, borate, silicate, bromide, selenate, selenite, arsenate, and arsenite.

The effect of soluble cations upon the exchangeable cation determination is to increase the cation concentration in the extracting solution, i.e., NH_4OAc, buffered at pH 7.0 (procedure 4B1a1b1-4). The dissolution of salts by the extractant necessitates an independent determination of soluble cations and a correction to the exchangeable cations. Therefore, in soils with soluble salts or carbonates, the soluble cations (meq L^{-1} solution) must be measured separately and the results subtracted from the extractable bases for determination of exchangeable bases as follows:

Exchangeable = Extractable - Soluble

The presence of alkaline-earth carbonates prevents accurate determination of exchangeable Ca and Mg. Refer to procedures 4E3c1a1-4.

References

Bohn, H.L., McNeal, B.L., and O'Connor, G.A. 1979. Soil chemistry. John Wiley & Sons, New York, NY.

Rhoades, J.D. 1982. Soluble salts. p. 167-179. *In* A.L Page, R.H. Miller, and D.R. Keeney (eds.) Methods of soil analysis. Part 2. Chemical and microbiological properties. 2nd ed. Agron. Monogr. 9. ASA and SSSA, Madison, WI.

U.S. Salinity Laboratory Staff. 1954. L.A. Richards (ed.) Diagnosis and improvement of saline and alkali soils. 160 p. USDA Handb. 60. U.S. Govt. Print. Office, Washington, DC.

Electrical Conductivity and Soluble Salts (4F)
Saturated Paste (4F2)
Gravimetric (4F2a)
Water Percentage (4F2a1)
Air-Dry, <2 mm (4F2a1a1)

1. Application

The saturated soil paste is a particular mixture of soil and water, i.e. the soil paste glistens as it reflects light; flows slightly when the container is tipped; and slides freely and cleanly from a spatula except for those soils with high clay content. This soil:water ratio is used because it is the lowest reproducible ratio for which enough extract for analysis can be readily removed from the soil with pressure or vacuum, and because this ratio is often related in a predictable manner to the field soil water content (U.S. Salinity Laboratory Staff, 1954). Upon preparation of a saturated paste, an aqueous extract is obtained, which is used in a series of chemical analyses, e.g., electrical conductivity and concentrations of the major solutes.

2. Summary of Method

A saturated paste is prepared (4F2) by adding water to a soil sample while stirring the mixture until the soil paste meets the saturation criteria, i.e. the soil paste glistens as it reflects light; flows slightly when the container is tipped; and slides freely and cleanly from a spatula except for those soils with high clay content. The mixture is covered and allowed to stand overnight. The saturation criteria are then rechecked. If the mixture fails to meet these criteria, more water or soil is added until criteria are met. A saturated paste sub-sample is used to determine the moisture content, i.e., saturation percentage (SP) by procedure 4F2a1.

3. Interferences

Special precautions must be taken for peat and muck soils and very fine or very coarse-textured soils (Rhoades, 1982). Dry peat and muck soils, especially if coarse textured or woody, require an overnight wetting to obtain a definite end point for the saturated paste. After the first wetting, pastes of these soils usually stiffen and lose their glisten. However, upon adding water and remixing, the paste usually retains the saturated paste characteristics. With fine-textured soils, enough water should be added immediately, with a minimum of mixing, to bring the sample nearly to saturation. Care also should be taken not to overwet coarse-textured soils. The presence of free water on the surface of the paste after standing is an indication of oversaturation in the coarse-textured soils (Rhoades, 1982).

4. Safety
Use heat-resistant gloves to remove hot moisture cans from the oven. No other significant hazards are associated with this procedure. Follow standard laboratory safety practices.

5. Equipment
5.1 Aluminum cans, drying
5.2 Spatulas, stainless steel, hardwood handles
5.3 Electronic balance, ±1-mg sensitivity
5.4 Oven, thermostatically controlled, 110°C
5.5 Thermometer, 0° to 200°C
5.6 Plastic food containers, 1920 mL (16 fl. oz.) capacities with recessed lids, Sweetheart Products Group, Owings Mills, MD

6. Reagents
6.1 Reverse osmosis (RO) water, ASTM Type III grade of reagent water

7. Procedure

Saturated Soil Paste Preparation (4F2)
7.1 Place a <2-mm, air-dry, 250-g soil sample in the food container. This sample size is convenient to handle with the 1920-mL (16-oz) food containers and provides enough extract for most purposes. The sample size varies with the number of determinations to be made upon the paste or saturation extract.

7.2 Add enough RO water to bring the sample nearly to saturation. To reduce soil puddling and to obtain a more definite end point of the saturation criteria, mix with a minimum of stirring. Soils puddle easily when worked at a moisture content near field capacity. If the paste becomes too wet, add more dry soil.

7.3 Occasionally tap the container on the workbench to consolidate the soil:water mixture. At saturation, the soil paste glistens as it reflects light; flows slightly when the container is tipped; and slides freely and cleanly off the spatula except for those soils with high clay content.

7.4 Cover the container and allow the sample to stand overnight.

7.5 Recheck saturation criteria, i.e., ordinarily, free water should not collect on the soil surface; paste should not stiffen markedly; and paste should not lose its glisten upon standing.

7.6 If the paste does not meet the saturation criteria, remix the paste with more RO water or dry soil. Allow to stand for at least 4 h and recheck the saturation criteria.

Saturation Percentage Determination (4F2a1)
7.7 Tare a moisture can and cover. Label each moisture can with the appropriate sample number.

7.8 Add ≈ 20 to 40 g of the saturated soil paste to the moisture can.

7.9 Cover the can, weigh the can plus sample, and record the weight to the nearest mg.

7.10 Remove the can cover, place the can in a vented drying oven at 110°C, and leave in the oven overnight (12 to 16 h). A drying period of 24 h or longer is recommended. Do not place moist samples in the oven with other samples that are drying, unless these samples have been in the oven at least 12 to 16 h. Do not overcrowd the drying oven with samples.

7.11 Remove the cans from the oven and cover immediately. Allow the cans to cool for 1 h.

7.12 Weigh the oven-dry paste sample and record the weight. Before calculating the SP, subtract the tare weights from the saturated paste and oven-dry weights. Do not use the SP sub-sample for other analyses.

8. Calculations

$$SP = [(Wt_{SP} - Wt_{OD})/ (Wt_{OD})] \times 100$$

where:
SP = Saturation percentage
Wt_{SP} = Weight of saturated paste
Wt_{OD} = Weight of oven-dry soil

9. Report
Report the saturation percentage to the nearest 0.1%.

10. Precision and Accuracy
Precision and accuracy are available from the SSL upon request.

11. References
Rhoades, J.D. 1982. Soluble Salts. p. 167-179. *In* A.L. Page, R.H. Miller, and D.R. Keeney (eds.) Methods of soil analysis. Part 2. Chemical and microbiological properties. 2nd ed. Agron. Monogr. 9. ASA and SSSA, Madison, WI.

U.S. Salinity Laboratory Staff, 1954. L.A. Richards (ed.) Diagnosis and improvement of saline and alkali soils. 160 p. USDA Handb. 60, U.S. Govt. Print Office, Washington DC.

Electrical Conductivity and Soluble Salts (4F)
Saturated Paste (4F2)
Conductivity Bridge (4F2b)
Electrical Conductivity (4F2b1)
Air-Dry, <2 mm (4F2b1a1)

1. Application
The electrical conductivity of the saturation extract (EC_s) is used as a criterion for classifying a soil as saline. Other uses of this measurement include the estimation of the total cation concentration in the extract, salt percentage in solution (P_{sw}), salt percentage in soil (P_{ss}), and osmotic pressure (OP). The unit EC x 10^3 is called the mmhos cm^{-1}. For solutions with a low EC_s, i.e., dilute solutions, the EC_s (mmhos cm^{-1}) x 10 ≈ cation concentration (meq L^{-1}) (U.S. Salinity Laboratory Staff, 1954). The EC_s (mmhos cm^{-1}) x 0.064 ≈ (P_{sw}); the (P_{sw} x SP)/100 ≈ P_{ss}; and the EC_s (mmhos cm^{-1}) x 0.36 ≈ OP in atmospheres (U.S. Salinity Laboratory Staff, 1954).

2. Summary of Method
The EC_s of the saturation extract that is prepared in procedure 4F2 is measured using a conductivity cell and a direct reading digital bridge (4F2b1). The cell constant is set using a standard solution.

3. Interferences

Reverse osmosis deionized water is used to zero and flush the conductivity cell. The extract temperature is assumed to be 25°C. If the temperature deviates significantly, a correction may be required.

Provide airtight storage of KCl solution and samples to prevent soil release of alkali-earth cations. Exposure to air can cause gains and losses of water and dissolved gases significantly affecting EC readings.

4. Safety

No significant hazards are associated with this procedure. Follow standard laboratory safety procedures.

5. Equipment

5.1 Conductivity bridge and conductivity cell, with automatic temperature adjustment, 25± 0.1°C, Markson Model 1096, Amber Science, Eugene, OR

6. Reagents

6.1 Reverse osmosis (RODI) water, ASTM Type I grade of reagent water, for rinse and KCl preparation

6.2 Potassium chloride (KCl), 0.010 N. Dry KCl overnight in oven (110°C). Dissolve 0.7456 g of KCl in RODI water and bring to 1-L volume. Conductivity at 25°C is 1.412 mmhos cm^{-1}.

7. Procedure

7.1 Calibrate the conductivity meter and cell by drawing the 0.010 N KCl solution into the cell.

7.2 Set the meter to "D" Scale and adjust the digital reading to "1.41".

7.3 Flush the cell and fill with RODI water. Verify that digital reading is "0.00".

7.4 Read the electrical conductivity of saturation extract (EC$_s$) by drawing up the extract into the cell and flush at least once if the cell has not been dried. Draw up extract a second time. Reading is started on the "C" Scale. Higher readings may require the use of the "D" or "E" Scales.

7.5 When the reading has stabilized, record the EC$_s$. Rinse the cell with RODI water and ensure that the conductivity reading falls to zero.

8. Calculations

No calculations are required for this procedure.

9. Report

Report EC$_s$ to the nearest 0.01 mmhos cm^{-1} (dS m^{-1}).

10. Precision and Accuracy

Precision and accuracy data are available from the Soil Survey upon request.

11. References

U.S. Salinity Laboratory Staff. 1954. L.A. Richards (ed.) Diagnosis and improvement of saline and alkali soils. 160 p. USDA Handb. 60. U.S. Govt. Print. Office, Washington, DC.

Electrical Conductivity and Soluble Salts (4F)
Saturated Paste (4F2)
Conductivity Bridge (4F2b)
Resistivity (4F2b2)
Air-Dry, <2 mm (4F2b2a1)

1. Application

The resistivity of the soil paste is mainly used to estimate the salt content in the soil. The apparatus is simple and rugged, the measurements can be made quickly, and the results are reproducible. Many agencies use the Bureau of Soils electrode cup to estimate the soluble salt content in soils (Davis and Bryan, 1910; Soil Survey Staff, 1951).

There is no simple method to convert saturation extract electrical conductivity to soil paste resistivity or vice versa. There is a limited correlation between EC_s and R_s, as the relationship is markedly influenced by variations in SP, salinity, and soil mineral conductivity.

2. Summary of Method

A saturated paste that is prepared in procedure 4F2 is placed in an electrode cup. The resistance is measured (procedure 4F2b2). The temperature of the paste is measured. The resistance (ohms) is converted to a 60°F (15.5°C) basis using a fourth order equation (Benham, 2003).

3. Interferences

No significant interferences are known to affect the saturated paste resistivity measurement.

4. Safety

No significant hazards are associated with this procedure. Follow standard laboratory safety practices.

5. Equipment

5.1 Conductivity bridge, Standard Wheatstone, Model RC 16B2, Beckman Instruments, Inc.
5.2 Soil cup cell holder, soil cup Cel-M, Industrial Instruments, Inc.
5.3 Bureau of Soils electrode cup, cell constant is defined as 0.25.
5.4 Thermometer, 0 to 100°C

6. Reagents

No reagents or consumables are used in this procedure.

7. Procedure

7.1 Fill the electrode cup with the saturated paste that is prepared in procedure 4F2. Gently tap the cup to remove air bubbles. Level the soil paste by striking off the excess with a spatula.

7.2 Place the cup in the cell holder. Make sure that the surfaces of the cup and holder are clean and bright. Use steel wool or fine sandpaper to carefully clean the surfaces.

7.3 Set the conductivity meter to 1000 cycles s^{-1} and adjust the multiplier range and the dial control to obtain the most distinct butterfly pattern on the fluorescent tube.

7.4 Measure resistivity. Adjustment of the sensitivity control also may be necessary.

7.5 Record the resistivity.

7.6 Place a thermometer in the saturated paste. When the temperature is stabilized, record the temperature.

8. Calculations
Use Table 1 to convert measured resistance to specific resistance at 60°F (15.5°C).

Resistivity (ohms cm^{-1}) = ohms @ 60° F x electrode cup cell factor.

Alternatively, the following equation may be used reducing soil paste resistance readings to values at 60°F with final results reported to 4 significant figures.

$A = (-0.013840786 + 0.028627073B - 0.00037976971B^2 + 3.7891593^{e-06}B^3 - 1.2020657^{e-08}B^4)$ x C x D x E

Where:
A = Resistance (ohms) corrected to 60°F
B = Temperature (°F) at which the resistance was measured
C = Resistance (ohms) measured at temperature B
D = Electrode cup cell factor
E = Scale (range multiplier)

9. Report
Report saturated paste resistivity in units of ohms at 60°F (15.5°C) to the nearest whole number.

10. Precision and Accuracy
Precision and accuracy data are available from the SSL upon request.

11. References
Benham, E.C. 2003. Soil resistivity temperature correction. USDA-NRCS, Lincoln, NE.
Davis, R.O., and H. Bryan. 1910. The electrical bridge for the determination of soluble salts in soils. USDA, Bur. Soils Bull. 61, 36 pp.
Soil Survey Staff. 1951. Soil survey manual. USDA-SCS. Govt. Print. Office, Washington, DC.
Whitney, M., and T.H. Means. 1897. An electrical method of determining the soluble salt content of soils. USDA, Div. Soils Bul. 8, 30 pp., illus.

Table 1. Bureau of soils data for reducing soil paste resistance readings to values at 60°F (Whitney and Means, 1897).

Temp				Ohms					
°F.	1000	2000	3000	4000	5000	6000	7000	8000	9000
40	735	1470	2205	2940	3675	4410	5145	5880	6615
42	763	1526	2289	3052	3815	4578	5341	6104	6867
44	788	1576	2364	3152	3940	4728	5516	6304	7092
46	814	1628	2442	3256	4070	4884	5698	6512	7326
48	843	1686	2529	3372	4215	5058	5901	6744	7587
50	867	1734	2601	3468	4335	5202	6069	6936	7803
52	893	1786	2679	3572	4465	5358	6251	7144	8037
54	917	1834	2751	3668	4585	5502	6419	7336	8253
56	947	1894	2841	3788	4735	5682	6629	7576	8523
58	974	1948	2922	3896	4870	5844	6818	7792	8766
60	1000	2000	3000	4000	5000	6000	7000	8000	9000
62	1027	2054	3081	4108	5135	6162	7189	8216	9243
64	1054	2108	3162	4216	5270	6324	7378	8432	9486
66	1081	2162	3243	4324	5405	6486	7567	8648	9729
68	1110	2220	3330	4440	5550	6660	7770	8880	9990
70	1140	2280	3420	4560	5700	6840	7980	9120	10260
72	1170	2340	3510	4680	5850	7020	8190	9360	10530
74	1201	2402	3603	4804	6005	7206	8407	9608	10809
76	1230	2460	3690	4920	6150	7380	8610	9840	11070
78	1261	2522	3783	5044	6305	7566	8827	10088	11349
80	1294	2588	3882	5176	6470	7764	9058	10352	11646
82	1327	2654	3981	5308	6635	7962	9289	10616	11943
84	1359	2718	4077	5436	6795	8154	9513	10872	12231
86	1393	2786	4179	5572	6965	8358	9751	11144	12537
88	1427	2854	4281	5708	7135	8562	9989	11416	12843
90	1460	2920	4380	5840	7300	8760	10220	11680	13140
92	1495	2990	4485	5980	7475	8970	10465	11960	13455
94	1532	3064	4596	6128	7660	9192	10724	12256	13788
96	1570	3140	4710	6280	7850	9420	10990	12560	14130
98	1611	3222	4833	6444	8055	9666	11277	12888	14499

Electrical Conductivity and Soluble Salts (4F)
Saturated Paste (4F2)
Saturated Paste Extraction (4F2c)
Automatic Extractor (4F2c1)

1. Application
The saturated paste is operationally defined so that it may be reproduced by a trained analyst using limited equipment. The saturated paste extract derived from the saturated paste is an important aqueous solution because many soil properties have been related to the composition of the saturation extract, e.g., soluble salt composition and electrical conductivity. These soil properties or characteristics are related in turn to the plant response to salinity (U.S. Salinity Laboratory Staff, 1954).

2. Summary of Method
The saturated paste (prepared in 4F2) is transferred to a plastic filter funnel fitted with filter paper. The funnel is placed on a mechanical vacuum extractor (Holmgren et al., 1977), and the saturated paste is extracted (4F2c1). The extract is used in subsequent chemical analyses, e.g., water-soluble cations (procedure 4F2c1a1-4) and water-soluble anions (procedures 4F2c1b1a1-8 and 4F2c1c1a1-2).

3. Interferences
Some saturated pastes are difficult to extract, i.e., soil dispersion and puddling. Repeated extractions may be necessary to obtain sufficient extract. High speed centrifuging or filtration of the extract also may be necessary. If the extract is to be stored for an extended period, sodium hexametaphosphate may be added to prevent calcium carbonate precipitation in the extract.

4. Safety
No significant hazards are associated with this procedure. Follow standard laboratory procedures.

5. Equipment
5.1 Mechanical vacuum extractor, 24-place, SAMPLETEX, MAVCO Industries, Lincoln, NE
5.2 Paste extraction cups, 9-cm diameter, for mechanical vacuum extractors
5.3 Syringes, disposable, 60 mL, polypropylene, for extraction
5.4 Rubber tubing, 3.2 ID x 1.6 OD x 6.4 mm (1/8 ID x 1/16 OD x 1 in)
5.5 Polycons, Richards Mfg. Co.
5.6 Syringe filters, 0.45-µm diameter, Whatman, Cliffton, NJ
5.7 Filter paper, 3 and 9-cm diameter, Whatman No. 40 or equivalent

6. Reagents
6.1 Reverse osmosis (RO) water, ASTM Type III grade of reagent water

7. Procedure

7.1 Prepare the saturated paste extract cup to receive the saturated paste (procedure 4E3) by placing a 3-cm diameter filter paper circle over the center of the cup followed by two 9-cm diameter filter paper circles. Slightly moisten the filter paper to ensure that it remains in place. Ensure there is no trapped air between the two 9-cm filter paper circles.

7.2 Place the extraction syringe on the lower disk of the mechanical vacuum extractor.

7.3 Use a clamp to close rubber tubing on the bottom of the paste extraction cup. Carefully transfer the saturated paste into the extraction cup. Gently tap the cup to remove entrapped air in the paste. Place cups on the extractor, connect the syringe and remove the clamp.

7.4 When all cups are ready to extract, place a plastic cover over the extraction cup to retard evaporation.

7.5 Turn on the extractor. Set the extraction time to ≈ 1 h.

7.6 When the extractor stops, turn off the power.

7.7 If sufficient extract has been obtained, pull the plunger of the syringe down. Do not pull plunger from the barrel of the syringe. Carefully remove the syringe containing the extract. Leave the rubber tubing on the sample tube.

7.8 If insufficient extract has been obtained, re-extract by repositioning the extractor to its starting configuration. Remove excess air in the syringe and restart the extractor. Slowing the extraction time to an overnight extraction may be necessary to obtain sufficient extract. Alternate methods are to extract any "unused" saturated paste in a new extraction cup or to re-extract by removing the top "moist" paste from the extraction cup, mixing with any "unused" paste, and re-extracting with a clean extraction cup or centrifuging.

7.9 Filtering the saturation extract is recommended to prevent the development of microorganisms. Connect the syringe to a 0.45 μm diameter syringe filter and express the extract into a polycon. If extracts are not to be determined immediately after collection, then store samples at 4°C.

8. Calculations
No calculations are required for this procedure.

9. Report
None.

10. Precision and Accuracy
Precision and accuracy data are available from the SSL upon request.

11. References
Holmgren, George G.S., R.L. Juve, and R.C. Geschwender. 1977. A mechanically controlled variable rate leaching device. Soil Sci. Am. J. 41:1207-1208.
U.S. Salinity Laboratory Staff. 1954. L.A. Richards (ed.) Diagnosis and improvement of saline and alkali soils. 160 p. USDA Handb. 60. U.S. Govt. Print. Office, Washington, DC.

Electrical Conductivity and Soluble Salts (4F)
Saturated Paste (4F2)
Saturated Paste Extraction (4F2c)
Automatic Extractor (4F2c1)
Atomic Absorption Spectrophotometry (4F2c1a)
Calcium, Magnesium, Potassium, and Sodium (4F2c1a1-4)
Air-Dry, <2 mm (4F2c1a1-4a1)

1. Application
The commonly determined soluble cations are Ca^{2+}, Mg^{2+}, K^+, and Na^+. In soils with a low saturation pH, measurable amounts of Fe and Al may be present. Determination of soluble

cations is used to obtain the relations between total cation concentration and other properties of saline solutions such as electrical conductivity and osmotic pressure (U.S. Salinity Laboratory Staff, 1954). The relative concentrations of the various cations in the soil-water extracts also provide information on the composition of the exchangeable cations in the soil. Complete analyses of the soluble ions provide a means to determine total salt content of the soils and salt content at field moisture conditions.

2. Summary of Method

The saturation extract from procedure 4F2c1 is diluted with an ionization suppressant (La_2O_3). The analytes are measured by an atomic absorption spectrophotometer (AA). The data are automatically recorded by a computer and printer. The saturation extracted cations, Ca^{2+}, Mg^{2+}, K^+, and Na^+, are reported in meq L^{-1} (mmol (+) L^{-1}) in procedures 4F2c1a1-4, respectively.

3. Interferences

There are four types of interferences (matrix, spectral, chemical, and ionization) in the analysis of these cations. These interferences vary in importance, depending upon the particular analyte selected. Do not use borosilicate tubes because of potential leaching of analytes.

4. Safety

Wear protective clothing and eye protection. When preparing reagents, exercise special care. Restrict the use of concentrated HCl to a fume hood. Many metal salts are extremely toxic and may be fatal if ingested. Thoroughly wash hands after handling these metal salts.

Follow standard laboratory procedures when handling compressed gases. Gas cylinders should be chained or bolted in an upright position. Acetylene is highly flammable. Avoid open flames and sparks. Standard laboratory equipment includes fire blankets and extinguishers for use when necessary. Follow the manufacturer's safety precautions when using the AA.

5. Equipment

5.1 Electronic balance, ±1.0-mg sensitivity
5.2 Atomic absorption spectrophotometer (AAS), double-beam, AAnalyst 300, Perkin-Elmer Corp., Norwalk, CT
5.3 Autosampler, AS-90, Perkin-Elmer Corp., Norwalk, CT
5.4 Computer, with AA WinLab software, Perkin-Elmer Corp., Norwalk, CT, and printer
5.5 Single-stage regulator, acetylene
5.6 Digital diluter/dispenser, with syringes 10000 and 1000 μL, gas tight, MicroLab 500, Hamilton Co., Reno, NV
5.7 Plastic test tubes, 15-mL, 16 mm x 100, for sample dilution and sample changer
5.8 Containers, polyethylene
5.9 Peristaltic pump

6. Reagents

6.1 Reverse osmosis deionized (RODI) water, ASTM Type I grade of reagent water
6.2 Hydrochloric acid (HCl), concentrated 12 N
6.3 HCl, 1:1 HCl:RODI, 6 N. Carefully mix 1 part of concentrated HCl to 1 part RODI water.
6.4 Stock lanthanum ionization suppressant solution (SLISS), 65,000 mg L^{-1}. Wet 152.4 g of lanthanum oxide (La_2O_3) with 100 mL RODI water. Slowly and cautiously add 500 mL of 6 N HCl to dissolve the La_2O_3. Cooling the solution is necessary. Dilute to 2 L with RODI water. Filter solution. Store in polyethylene container.
6.5 Working lanthanum ionization suppressant solution (WLISS), 2000 mg L^{-1}. Dilute 61.5 mL of SLISS with 1800 mL of RODI water (1:10). Make up to volume with RODI water. Invert to mix thoroughly. Store in polyethylene container.

6.6   Primary stock standards solution (PSSS), high purity, 1000 mg L^{-1}: Ca, Mg, K, and Na.

6.7   Working stock mixed standards solution (WSMSS) for Ca, Mg, and K. In a 500-mL volumetric flask, add 250 mL Ca PSSS, 25 mL Mg PSSS, and 100 mL K PSSS = 500 mg L^{-1} Ca, 50 mg L^{-1} Mg, and 200 mg L^{-1} K. Dilute to volume with RODI water. Invert to thoroughly mix. Store in polyethylene containers. Prepare fresh weekly. Store in the refrigerator.

6.8   Mixed calibration standards solution (MCSS), High, Medium, Low, Very Low, and Blank as follows:

6.8.1   MCSS High Standard (1:100): Dilute WSMSS 1:100 with WLISS. Invert to mix thoroughly mix. Store in polyethylene containers. Prepare fresh weekly. Store in a refrigerator. Allow to equilibrate to room temperature before use. Final concentration = 5 mg L^{-1} Ca, 0.5 mg L^{-1} Mg, and 2 mg L^{-1} K.

6.8.2   MCSS Medium Standard (1:200): To a 100-mL volumetric flask, add 50 mL of WSMSS and bring to volume with RODI water. Dilute 1:100 with WLISS. Invert to thoroughly mix. Store in polyethylene containers. Prepare fresh weekly. Store in a refrigerator. Allow to equilibrate to room temperature before use. Final concentration =2.5 mg L^{-1} Ca, 0.25 mg L^{-1} Mg, and 1 mg L^{-1} K.

6.8.3   MCSS Low Standard (1:400): To a 100-mL volumetric flask, add 25 mL of WSMSS and bring to volume with RODI water. Dilute 1:100 with WLISS. Invert to mix thoroughly. Store in polyethylene containers. Prepare fresh weekly. Store in a refrigerator. Allow to equilibrate to room temperature before use. Final concentration = 1.25 mg L^{-1} Ca, 0.125 mg L^{-1} Mg, and 0.5 mg L^{-1} K.

6.8.4   MCSS Very Low Standard (1:600): To a 100-mL volumetric flask, add 16.65 mL of WSMSS and bring to volume with RODI water. Dilute 1:100 with WLISS. Invert to mix thoroughly. Store in polyethylene containers. Prepare fresh weekly. Store in a refrigerator. Allow to equilibrate to room temperature before use. Final concentration = 0.83 mg L^{-1} Ca, 0.08 mg L^{-1} Mg, and 0.33 mg L^{-1} K.

6.8.5   MCSS Blank = 0 mL of Ca, Mg, and K. Dilute RODI water 1:100 with WLISS.

6.9   Na Calibration Standards Solution (NaCSS), High, Medium, Low, and Very Low as follows:

6.9.1   NaCSS High Standard (1:100): Dilute Na PSMSS (1000 mg L^{-1}) 1:100 with WLISS. Invert to thoroughly mix. Store in polyethylene containers. Prepare fresh weekly. Store in a refrigerator. Allow to equilibrate to room temperature before use. Final concentration = 10 mg L^{-1} Na.

6.9.2   NaCSS Medium Standard (1:200): In a 50-mL volumetric, add 25 mL of Na PSMSS and bring to volume with RODI water. Dilute 1:100 with WLISS. Invert to mix thoroughly. Store in polyethylene containers. Prepare fresh weekly. Store in a refrigerator. Allow to equilibrate to room temperature before use. Final concentration = 5 mg L^{-1} Na.

6.9.3   NaCSS Low Standard (1:400): In a 50-mL volumetric flask, add 12.5 mL of PSMSS and bring to volume with RODI water. Dilute 1:100 with WLISS. Invert to thoroughly mix. Store in polyethylene containers. Prepare fresh weekly. Store in a refrigerator. Allow to equilibrate to room temperature before use. Final concentration = 2.5 mg L^{-1} Na.

6.9.4 NaCSS Very Low Standard (1:600) In a 50-mL volumetric flask, add 8.35 mL of PSMSS Na (1000 ppm) and bring to volume with RODI water. Dilute 1:100 with WLISS. Invert to thoroughly mix. Store in polyethylene containers. Prepare fresh weekly. Store in a refrigerator. Allow to equilibrate before use. Final concentration = 1.67 mg L^{-1} Na.

6.9.5 NaCSS Blank = 0 mL Na PSMSS. Dilute RODI water 1:100 with WLISS.

6.10 Compressed air with water and oil traps.

6.11 Acetylene gas, purity 99.6%.

7. Procedure

Dilution of Calibration Standards and Sample Extracts

7.1 The 10-mL syringe is for diluent (WLISS). The 1-mL syringe is for the MCSS and saturation sample extracts (procedure 4F2c1). Set the digital diluter at a 1:100 dilution. See regeants for the preparation of the MCSS and the NaCSS. Dilute the saturation extract sample with 100 parts of WLISS (1:100).

7.2 Dispense the diluted sample solutions into test tubes which have been placed in the sample holders of the sample changer.

AAS Set-up and Operation

7.3 Refer to the manufacturer's manual for operation of the AAS. The following are only very general guidelines for instrument conditions for the various analytes.

Analyte	Conc. (mg L^{-1})	Burner & angle	Wavelength (nm)	Slit (mm)	Fuel/Oxidant (C_2H_2/Air)
Ca	5.0	10 cm @ 0°	422.7	0.7	1.5/10.0
Mg	0.5	10 cm @ 0°	285.2	0.7	1.5/10.0
K	2.0	10 cm @ 0°	766.5	0.7	1.5/10.0
Na	10.0	10 cm @ 30°	589.0	0.2	1.5/10.0

7.4 Use the computer and printer to set instrument parameters and to collect and record instrument readings.

AAS Calibration and Analysis

7.5 Calibrate the instrument by using the MCSS and NaCSS. The data system will then associate the concentrations with the instrument responses fore each MCSS. Rejection criteria for MCSS, if $R^2 < 0.99$.

7.6 If sample exceeds calibration standard, the sample is diluted 1:5, 1:20, 1:100, etc., with RODI water followed by 1:100 dilution with WLISS.

7.7 Perform one quality control (QC) (Low Standard) for every 12 samples. If reading is not within 10%, the instrument is re-calibrated and QC re-analyzed.

7.8 Record analyte readings to 0.01 mg L^{-1}.

8. Calculations

The instrument readings for analyte concentration are in mg L^{-1}. These analyte concentrations are converted to meq L^{-1} as follows:

Analyte Concentration in Soil (meq L^{-1}) = (A x B)/C

where:
A = Analyte (Ca^{2+}, Mg^{2+}, K$^+$, Na$^+$) concentration in extract (mg L^{-1})
B = Dilution ratio, if needed
C = Equivalent weight

where:
Ca^{+2} = 20.04 mg meq^{-1}
Mg^{+2} = 12.15 mg meq^{-1}
Na^{+1} = 22.99 mg meq^{-1}
K^{+1} = 39.10 mg meq^{-1}

9. Report

Report the saturation extraction cations of Ca^{2+}, Mg^{2+}, Na$^+$, and K$^+$ to the nearest 0.1 meq L^{-1} (mmol (+) L^{-1}).

10. Precision and Accuracy

Precision and accuracy data are available from the SSL upon request.

11. References

U.S. Salinity Laboratory Staff. 1954. L.A. Richards (ed.) Diagnosis and improvement of saline and alkali soils. 160 p. USDA Handb. 60. U.S. Govt. Print. Office, Washington, DC.

Electrical Conductivity and Soluble Salts (4F)
Saturated Paste (4F2)
Saturated Paste Extraction (4F2c)
Automatic Extractor (4F2c1)
Ion Chromatograph 4F2c1b
Conductivity Detector (4F2c1b1)
Self-Regeneration Suppressor (4F2c1b1a)
Bromide, Acetate, Chloride, Fluoride, Nitrate, Nitrite, Phosphate, and Sulfate (4F2c1b1a1-8)
Air-Dry, <2 mm (4F2c1b1a1-8a1)

1. Application

The soluble anions that are commonly determined in saline and alkali soils are carbonate, bicarbonate, sulfate, chloride, nitrate, nitrite, fluoride, phosphate, silicate, bromide, and borate (Khym, 1974; U.S. Salinity Laboratory Staff, 1954). Carbonate and bicarbonate are determined by titration. Phosphate, silicate, bromide, borate, and aluminate are found only occasionally in measurable amounts in soils. Chloride, sulfate, nitrate, fluoride, and nitrite are measured in solution by chromatography. In saline and alkali soils, carbonate, bicarbonate, sulfate, and chloride are the anions that are found in the greatest abundance. In general, soluble sulfate is usually more abundant than soluble chloride.

2. Summary of Method

The soil saturation extract is diluted according to its electrical conductivity (EC$_s$). The diluted sample is injected into the ion chromatograph, and the anions are separated. A conductivity detector is used to measure the anion species and content. Standard anion concentrations are used to calibrate the system. A calibration curve is determined, and the anion concentrations are calculated. A computer program automates these actions. The saturation extract anions, Br^-, CH_3COO^-, Cl^-, F^-, NO_3^-, NO_2^-, PO_4^{3-}, and SO_4^{2-} are reported in meq L^{-1} (mmol (-) L^{-1}) in procedures 4F2c1b1a1-8, respectively. This same method may also be used for water analysis.

3. Interferences
Some saturation extracts contain suspended solids. Filtering after dilution removes the particles. Saturation extracts of acid soils that contain Fe and/or Al may precipitate and clog the separator column. Saturation extracts of very high pH may contain organic material which may clog or poison the column. Low molecular weight organic anions will co-elute with inorganic anions from the column.

4. Safety
Wear protective clothing and safety glasses. When preparing reagents, exercise special care. Many metal salts are extremely toxic and may be fatal if ingested. Thoroughly wash hands after handling these metal salts. Follow the manufacturer's safety precautions when using the chromatograph.

5. Equipment
5.1 Ion chromatograph, double-column, conductivity detection, Dionex DX-120, Dionex Corp., Sunnyvale, CA
5.2 Guard column, IonPac AG14, 4 x 50 mm, Dionex Corp., Sunnyvale, CA
5.3 Analytical column, IonPac AS14, 4 x 250 mm, Dionex Corp., Sunnyvale, CA
5.4 Self-regeneration suppressor, ASRS – ULTRA, 4 mm, Dionex Corp., Sunnyvale, CA
5.5 Autosampler, AS40, Dionex Corp., Sunnyvale, CA
5.6 Computer with PeakNet software and printer
5.7 Digital diluter/dispenser, with syringes 10000 and 1000 μL, gas tight, MicroLab 500, Hamilton Co., Reno, NV
5.8 Poly-vials with caps, 5 mL, Dionex Corp., Sunnyvale, CA

6. Reagents
6.1 Reverse osmosis deionized filtered (RODI), ASTM Type I grade of reagent water
6.2 Helium gas
6.3 Eluent solution. Solution A: Mix 5.30 g of Na_2CO_3 with RODI water in a 500-mL volumetric and make to volume with RODI water. Solution B: Mix 4.20 g of $NaHCO_3$ with RODI water in a 500-mL volumetric and make to volume with RODI water. Add 35 mL of Solution A and 10 mL of Solution B in a 1-L volumetric and make to volume with RODI water for a final concentration is 0.024 M Na_2CO_3 and 0.025 M $NaHCO_3$. Make 2 L of eluent solution if chromatograph is to run overnight. Place eluent in eluent tank and degas with helium 15 min L^{-1} eluent (30 min total). Make fresh daily.
6.4 Primary stock standards solutions, ($PSSS_{1000}$), high purity, 1000 mg L^{-1}: Cl^-, SO_4^{2-}, F, NO_3^-, NO_2^-, Br^-, $CH3COO^-$, and PO_4^{3-}.
6.5 Primary stock standards solutions ($PSSS_{100}$), 100 mg L^{-1}: F^-, NO_3^-, NO_2^-, Br^-, and PO_4^{3-}. Prepare fresh weekly. In five 250-mL volumetric flasks add as follows:

6.5.1 F^- $PSSS_{100}$ = 25 mL F^- $PSSS_{1000}$
6.5.2 Br^- $PSSS_{100}$ = 25 mL of Br^- $PSSS_{1000}$
6.5.3 NO_2^- $PSSS_{100}$ = 25 mL NO_2^- $PSSS_{1000}$
6.5.4 NO_3^- $PSSS_{100}$ = 25 mL NO_3^- $PSSS_{1000}$
6.5.5 PO_4^{3-} $PSSS_{100}$ = 25 mL PO_4^{3-} $PSSS_{1000}$

Dilute each PSSS$_{100}$ to volume with RODI water and invert to thoroughly mix. Store in the refrigerator.

6.6 Mixed calibration standards solutions (MCSS), A, B, C, D, E, and Blank as follows:

6.6.1 MCSSA = In a 250-mL volumetric flask, add as follows

12.5 mL Cl$^-$ PSSS$_{1000}$ = 50 mg L^{-1}
25 mL SO$_4^2$ PSSS$_{1000}$ = 100 mg L^{-1}
12.5 mL F$^-$ PSSS$_{100}$ = 5 mg L^{-1}
20 mL NO$_3^-$ PSSS$_{100}$ = 8 mg L^{-1}
20 mL NO$_2^-$ PSSS$_{100}$ = 8 mg L^{-1}
22 mL Br$^-$ PSSS$_{100}$ = 8.8 mg L^{-1}
20 mL CH3COO$^-$ PSSS$_{1000}$ = 80 mg L^{-1}
24 mL PO$_4^{3-}$ PSSS$_{100}$ = 9.6 mg L^{-1}

Dilute to volume with RODI water and invert to thoroughly mix. Store in glass containers. Store in the refrigerator. Prepare fresh weekly.

6.6.2 MCSSB = In a 100-mL volumetric flask, add 40 mL MCSSA and dilute to volume with RODI water. Final concentration = 20, 40, 2, 3.2, 3.2, 3.52, 30.4, and 3.52 mg L^{-1} Cl$^-$, SO$_4^{2-}$, F$^-$, NO$_3^-$, NO$_2^-$, Br$^-$, CH3COO$^-$, and PO$_4^{3-}$, respectively. Invert to thoroughly mix. Store in glass containers. Prepare fresh weekly. Store in the refrigerator.

6.6.3 MCSSC = In a 100-mL volumetric flask, add 20 mL MCSSA and dilute to volume with RODI water. Final concentration = 10, 20, 1, 1.6, 1.6, 1.76, 15.2, and 1.76 mg L^{-1} Cl$^-$, SO$_4^{2-}$, F$^-$, NO$_3^-$, NO$_2^-$, Br$^-$, CH3COO$^-$, and PO$_4^{3-}$, respectively. Invert to thoroughly mix. Store in glass containers. Prepare fresh weekly. Store in the refrigerator.

6.6.4 MCSSD = In a 250-mL volumetric flask, add 25 mL MCSSA and dilute to volume with RODI water. Invert to thoroughly mix. Final concentration = 5, 10, 0.5, 0.8, 0.8, 0.88, 7.6, and 0.88 mg L^{-1} Cl$^-$, SO$_4^{2-}$, F$^-$, NO$_3^-$, NO$_2^-$, Br$^-$, CH3COO$^-$, and PO$_4^{3-}$, respectively. Store in glass containers. Prepare fresh weekly. Store in the refrigerator.

6.6.5 MCSSE = In a 250-mL volumetric flask, add 5 mL MCSSA and dilute to volume with RODI water. Final concentration = 1, 2, 0.1, 0.16, 0.16, 0.176, 1.52, and 0.176 mg L^{-1} Cl$^-$, SO$_4^{2-}$, F$^-$, NO$_3^-$, NO$_2^-$, Br$^-$, CH3COO$^-$, and PO$_4^{3-}$, respectively. Invert to thoroughly mix. Store in glass containers. Prepare fresh weekly. Store in the refrigerator.

6.6.6 MCSS Blank = 0 mL of Cl$^-$, SO$_4^{2-}$, F$^-$, NO$_3^-$, NO$_2^-$, Br$^-$, CH3COO$^-$, and PO$_4^{3-}$. Dilute RODI water to volume.

7. Procedure
Dilution of sample extracts
7.1 To estimate the total soluble anion concentration (meq L^{-1}), multiply the EC$_s$ (procedure 4F2b1) by 10. Subtract the CO$_3^{2-}$ and HCO$_3^-$ concentrations (procedures 4F2c1c1a1-2) from the total anion concentration. The remainder is the ≈ concentration (meq L^{-1}) of anions to be separated by ion chromatography.

Anion concentration (meq L^{-1}) = EC$_s$ x 10 - (HCO$_3^-$ + CO$_3^{2-}$)

7.2 Dilute the saturation extract with the RODI water as follows:

ECₛ (dS cm⁻¹)	Dilution Factor
0.00 to 0.55	4
0.56 to 0.65	5
0.66 to 0.75	6
0.76 to 0.85	7
0.86 to 0.95	8
0.96 to 1.05	9
1.06 to 1.20	10
1.21 to 1.40	15
1.41 to 1.50	25
1.51 to 1.60	30
1.61 to 1.80	40
1.81 to 2.00	50
2.01 to 2.30	60
2.31 to 2.60	70
2.61 to 3.10	80
3.11 to 3.55	90
3.56 to 4.05	100
4.06 to 4.60	120
4.61 to 5.20	140
5.21 to 5.85	150
5.86 to 6.55	160
6.56 to 7.30	180
7.31 to 8.00	200
8.01 to 9.00	225
9.01 to 10.00	240
10.01 to 11.50	270
11.51 to 13.00	280
13.01 to 14.50	300
14.51 to 16.00	320
16.01 to 17.00	360
17.01 to 18.00	400
18.01 to 20.00	450
20.01 to 21.00	480
21.01 to 23.00	500
23.01 to 24.00	540
24.01 to 25.00	560
25.01 to 27.00	600
27.01 to 28.00	640
28.01 to 30.00	680
30.01 to 32.00	700
32.01 to 33.00	720
33.01 to 36.00	800
36.01 to 40.00	900
40.01 to 44.00	1000

7.3 Place the MCSS (B, C, D, E, Blank) and diluted extract samples in the Poly-vials and cap with filtercaps.

Set-up and Operation of Ion Chromatograph (IC)

7.4 Refer to the Manufacturer's manual for the operation of chromatograph. Because any number of factors may cause a change in IC operating conditions, only a general set-up of the Dionex DX-120 ion chromatograph is presented. Individual analysts may modify some or all of the operating conditions to achieve satisfactory results. Ranges and/or (typical settings) are as follows:

Parameter	Range and/or (Typical Setting)
Calibration	Peak Height or (Area)
Flow Setting	0.5 to 4.5 mL min^{-1} (1.63 mL min^{-1})
Pressure	0 to 4000 psi (2200 to 2400 psi)
Detection	Suppressed conductivity
Total Conductivity	0 to 999.9 μS
Injection Volume	10 μL
Auto Offset	-999.9 to 999.9 μS (ON)

7.5 Load the sample holder cassettes with the capped samples, standards, and check samples.

7.6 Use the computer and printer to set instrument parameters and to collect and record instrument readings.

IC Calibration and Analysis

7.7 Calibrate the instrument by using the MCSS (B, C, D, E, Blank). The data system will then associate the concentrations with the instrument responses for each MCSS. Rejection criteria for MCSS, R^2 <0.99.

7.8 If samples are outside calibration, dilute sample extracts with RODI water solution and re-analyze.

7.9 Perform one quality control (QC) (Low Standard MCSS, Standard C) for every 12 samples. If reading is not within tolerance limits (10 to 15%, based on analyte), the instrument is re-calibrated and QC re-analyzed.

7.10 Record analyte readings to 0.01 mg L^{-1}.

8. Calculations

The instrument readings for analyte concentration are in mg L^{-1}. These analyte concentrations are converted to meq L^{-1} as follows:

Analyte Concentration in Soil (meq L^{-1}) = (A x B)/C

where:
A = Analyte (Br$^-$, CH$_3$COO$^-$, Cl$^-$, F$^-$, NO$_3^-$, NO$_2^-$, PO$_4^{3-}$, SO$_4^{2-}$) concentration in extract (mg L^{-1})
B = Dilution ratio, if needed
C = Equivalent weight

where:
Cl$^-$ = 35.45 mg meq^{-1}
SO$_4^{2-}$ = 48.03 mg meq^{-1}

F^- = 19.00 mg meq^{-1}
NO_3^- = 62.00 mg meq^{-1}
NO_2^- = 46.00 mg meq^{-1}
Br^- = 79.90 mg meq^{-1}
$CH3COO^-$ = 59.05 mg meq^{-1}
PO_4^{3-} = 31.66 mg meq^{-1}

9. Report

Report the saturation extraction anions (Cl^-, SO_4^{2-}, F^-, NO_3^-, NO_2^-, Br^-, $CH3COO^-$, and PO_4^{3-}) to the nearest 0.1 meq L^{-1} (mmol (-) L^{-1}).

10. Precision and Accuracy

Precision and accuracy data are available from the SSL upon request.

11. References

Khym, J.X. 1974. Analytical ion-exchange procedures in chemistry and biology: Theory, equipment, techniques. Prentice-Hall, Inc., Englewood Cliffs, NJ.

U.S. Salinity Laboratory Staff. 1954. L.A. Richards (ed.) Diagnosis and improvement of saline and alkali soils. USDA Agric. Handb. 60. U.S. Govt. Print. Office, Washington, DC.

Electrical Conductivity and Soluble Salts (4F)
Saturated Paste (4F2)
Saturated Paste Extraction (4F2c)
Automatic Extractor (4F2c1)
Automatic Titrator (4F2c1c)
Combination pH-Reference Electrode (4F2c1c1)
Acid Titration, H$_2$SO$_4$ (4F2c1c1a)
Carbonate and Bicarbonate (4F2c1c1a1-2)
Air-Dry, < 2 mm (4F2c1c1a1-2a1)

1. Application

The water soluble anions that usually are determined in saturation extracts are carbonate, bicarbonate, sulfate, chloride, nitrate, nitrite, fluoride, phosphate, silicate, bromide, and borate. Carbonate and bicarbonate are analyzed by titration. In saturation extracts, carbonate is measurable if the pH >9 (U.S. Salinity Laboratory Staff, 1954). The bicarbonate concentration is seldom >10 meq L^{-1} in the absence of carbonate anions (U.S. Salinity Laboratory Staff, 1954). The bicarbonate concentration at pH ≤7 seldom exceeds 3 or 4 meq L^{-1} (U.S. Salinity Laboratory Staff, 1954).

The total dissolved ion amounts generally increase with increasing soil moisture content. While some ions increase, some ions may decrease. The carbonate and bicarbonate anions are among those ions which are most dependent upon soil moisture. Therefore, in making interpretations about carbonate and bicarbonate in soil solution, there must be careful consideration about the chemistry of the soil and the soil solution.

2. Summary of Method

An aliquot of the saturation extract (procedure 4F2c) is titrated on an automatic titrator to pH 8.25 and pH 4.60 end points. The carbonate and bicarbonate are calculated from the titers, aliquot volume, blank titer, and acid normality (procedures 4F2c1c1a1-2, respectively). Carbonate and bicarbonate are reported in meq L^{-1} (mmol (+) L^{-1}).

3. Interfernces

Clean the electrode by rinsing with distilled water and patting it dry with tissue. Wiping the electrode dry with a cloth, laboratory tissue, or similar material may cause electrode polarization.

Slow electrode response time may cause over shooting the end point. A combination of slowing the buret speed and increasing the time delay may help. Cleaning the electrode with detergent may decrease the response time. If all else fails, changing the electrode generally solves the problem. Blanks may not titrate properly because some sources of reverse osmosis (RO) water have a low pH.

4. Safety

Wear protective clothing and eye protection. When preparing reagents, exercise care. Thoroughly wash hands after handling reagents. Restrict the use of concentrated H_2SO_4 to the fume hood. Use showers and eyewash stations to dilute spilled acids. Use sodium bicarbonate and water to neutralize and dilute spilled acids. Follow the manufacturer's safety precautions when operating the automatic titrator.

5. Equipment

5.1 Automatic titrator, Metrohm 670 Titroprocessor, with control unit, sample changer, and dispenser, Metrohm Ltd., Brinkmann Instruments, Inc.
5.2 Combination pH-reference electrode, Metrohm Ltd., Brinkmann Instruments, Inc.
5.3 Pipettes, electronic digital, 2500 µL and 10 mL, with tips, 2500 µL and 10 mL

6. Reagents

6.1 Reverse osmosis (RO) water, ASTM Type III grade of reagent water
6.2 Helium gas
6.3 Sulfuric acid (H_2SO_4), concentrated, 36 N
6.3 H_2SO_4, 0.0240 N standardized. Carefully dilute 2.67 mL of concentrated H_2SO_4 in 4 L of RODI degassed water ($\approx$ 15 min). Re-standardize the acid at regular intervals. Refer to the procedure for standardization of acids.
6.4 Borax pH buffers, pH 4.00, pH 7.00, and pH 9.18, for titrator calibration, Beckman, Fullerton, CA.

7. Procedure

7.1 Pipet 3 mL of the fresh saturation extract (procedure 4E3c) into a 250-mL titration beaker.

7.2 Add 72 mL of RO water into a titration beaker. Final volume is 75 mL for blanks and samples. Run 8 to 12 blanks of RO water through the titration procedure.

7.3 Refer to manufacturer's manual for operation of the automatic titrator.

7.4 Calibrate automatic titrator with 9.18, 7.00 and 4.00 pH buffers. Set-up the automatic titrator to set end point titration mode. The "Set" pH parameters are listed as follows:

Parameter	Value
Ep_1	pH 8.25
Dyn change pH	1.5 units
Drift	0.4 mV s^{-1}
Time delay	10 s
Ep_2	pH 4.60
Dyn change pH	1.5 units
Drift	0.4 mV s^{-1}

Temp   25°C
Stop Volume   35 mL

7.5 Place the 250-mL titration beakers in the sample changer.

7.6 Press "Start".

7.7 If the titrator is operating properly, no other analyst intervention is required. The titers and other titration parameters are recorded on the Titroprocessor printer.

8. Calculations

8.1 CO_3^{2-} (meq L^{-1}) = $(2\ T_1 \times N \times 1000)/$Aliquot

8.2 HCO_3^- (meq L^{-1}) = $[(T_2 + T_1) - \text{Blank} - (2 \times T_1) \times N \times 1000]/$Aliquot

where:
T_1	=	Titer of CO_3^{2-} (mL)
T_2	=	Titer of HCO_3^- (mL)
N	=	Normality of H_2SO_4
Blank	=	Average titer of blank solutions (mL)
Aliquot	=	Volume of saturation extract titrated (mL)
1000	=	Conversion factor to meq L^{-1}

9. Report
Report saturation extract CO_3^{2-} and HCO_3^- to the nearest 0.1 meq L^{-1} (mmol (-) L^{-1}).

10. Precision and Accuracy
Precision and accuracy data are available from the SSL upon request.

11. References
U.S. Salinity Laboratory Staff. 1954. L.A. Richards (ed.) Diagnosis and improvement of saline and alkali soils. USDA Handb. 60. U.S. Govt. Print. Office, Washington, DC.

Ratios and Estimates Related to Soluble Salts (4F3)
Exchangeable Sodium Percentage (ESP), NH₄OAc, pH 7.0(4F3a)
ESP, Calculated without Saturated Paste Extraction (4F3a1)

Compute the exchangeable sodium percentage (ESP) by dividing the exchangeable sodium (ES) by the CEC by NH₄OAc, pH 7.0 (CEC-7) and multiplying by 100 (procedure 4F3a1). The ES is calculated by subtracting the water soluble Na^+ determined in procedure 4E3c1a4 from the NH₄OAc extractable Na^+ determined in procedure 4B1a1b4 (U.S. Salinity Laboratory Staff, 1954). The CEC-7 is determined in procedure 4B1a1a1a1. In soil taxonomy, an ESP $\geq 15\%$ is a criterion for natric horizons (Soil Survey Staff, 1999). When the saturation extract is not prepared, the ESP is calculated as follows:

ESP = (ES/CEC-7) x 100

where:
ESP	=	Exchangeable sodium percentage
ES	=	Extractable sodium (NH₄OAc extractable Na^+, (cmol (+) kg^{-1})).
CEC-7	=	CEC by NH₄OAc, pH 7.0 (cmol (+) kg^{-1}).

Ratios and Estimates Related to Soluble Salts (4F3)
Exchangeable Sodium Percentage (ESP), NH₄OAc, pH 7.0(4F3a)
ESP, Calculated with Saturated Paste Extraction (4F3a2)

Exchangeable Na is computed with acceptable accuracy unless salt contents > 20 mmhos cm^{-1} (dS m^{-1}) at 25 C. Exchangeable Na equals extractable Na minus saturation extract Na multiplied by saturation percentage. Saturation percentage is the water percentage in the saturated paste divided by 1000. Exchangeable Na can be determined with greater accuracy than the other cations in the presence of gypsum or carbonates. If exchangeable K is negligible compared to exchangeable Ca and Mg then exchangeable Ca plus Mg equals CEC (NH₄OAc, pH 7.0) minus exchangeable Na. This approximation is suitably reproducible for comparison between soils and for soil classification. Exchangeable Ca can be computed in the same manner as exchangeable Na. Results are not so satisfactory for exchangeable Ca when computed in the presence of carbonates or large amounts of gypsum.

When the saturation extract is prepared, the SSL calculates the ESP by procedure 4F3a2 as follows:

ESP = 100 x [Na$_{ex}$ - (Na$_{ws}$ x (H₂O$_{ws}$/1000))]/CEC-7

where:
ESP   = Exchangeable sodium percentage
Na$_{ex}$ = Extractable Na (NH₄OAc extractable Na^{+}, (cmol (+) kg^{-1}))
Na$_{ws}$ = Water-soluble Na (mmol (+) L^{-1})
H₂O$_{ws}$ = Water saturation percentage
CEC-7   = CEC by NH₄OAc, pH 7.0 (cmol (+) kg^{-1})
1000   = Conversion factor to (cmol (+) kg^{-1})
100   = Conversion factor to percent

Ratios and Estimates Related to Soluble Salts (4F3)
Sodium Adsorption Ratio (SAR) (4F3b)

Compute the sodium adsorption ratio (SAR) by dividing the molar concentration of the monovalent cation Na^{+} by the square root of the molar concentration of the divalent cations Ca^{2+} and Mg^{2+} (U.S. Salinity Laboratory Staff, 1954). The water soluble Ca^{2+}, Mg^{2+}, and Na^{+} are determined in procedures 4F2c1a1, 4F2c1a2, and 4F2c1a4, respectively. The SAR was developed as a measurement of the quality of irrigation water, particularly when the water is used for irrigating soils that are salt or Na affected (U.S. Salinity Laboratory Staff, 1954). In soil taxonomy, a SAR $\geq$13 is a criterion for natric horizons (Soil Survey Staff, 1999). The SSL calculates the SAR by procedure 4F3b. The SAR is calculated as follows:

$$SAR = \frac{\left[Na^{+}\right]}{\sqrt{\frac{\left[Ca^{++}\right]+\left[Mg^{++}\right]}{2}}}$$

SAR   = Sodium Adsorption Ratio
Na^{+} = Water soluble Na^{+} (mmol (+) L^{-1}).
Ca^{2+} = Water soluble Ca^{2+} (mmol (+) L^{-1}).
Mg^{2+} = water soluble Mg^{2+} (mmol (+) L^{-1}).

References

Soil Survey Staff. 1999. Soil taxonomy: A basic system of soil classification for making and interpreting soil surveys. USDA-NRCS Agric. Handb. 436. 2nd ed. U.S. Govt. Print. Office, Washington, DC.

U.S. Salinity Laboratory Staff. 1954. L.A. Richards (ed.) Diagnosis and improvement of saline and alkali soils. USDA Agric. Handb. 60. U.S. Govt. Print. Office, Washington, DC.

Ratios and Estimates Related to Soluble Salts (4F3)
Estimated Total Salt (4F3c)

Use the charts and graphs available in U.S. Salinity Laboratory Staff (1954) to estimate total salt content (4F3c) from the electrical conductivity (EC_s) of the saturation extract (procedure 4F2b1). The essential relations are summarized in the equations as follows:

Total Salt in soil (ppm) = $(-4.2333 + (12.2347 \times EC_s) + (0.058 \times EC_s^2) - (0.0003 \times EC_s^3)) \times 0.000064 \times SP$

where:
EC_s = Electrical conductivity of saturation extract
SP = Saturation percentage of saturation paste

Previous equations used to estimate total salt content are as follows:

Log total salt in soil (ppm) = $0.81 + 1.08 \times$ Log EC_s (mmhos cm^{-1}) + Log SP

where:
EC_s = Electrical conductivity of saturation extract
SP = Saturation percentage of saturation paste

Total salt in soil (%) = Total salt (ppm) x 10^{-4}

These equations are applicable to saturation extracts with an EC_s <20 mmhos cm^{-1}. Deviations occur at higher salt concentrations.

References

Soil Survey Staff. 1999. Soil taxonomy: A basic system of soil classification for making and interpreting soil surveys. USDA-NRCS Agric. Handb. 436. 2nd ed. U.S. Govt. Print. Office, Washington, DC.

U.S. Salinity Laboratory Staff. 1954. L.A. Richards (ed.) Diagnosis and improvement of saline and alkali soils. 160 p. U.S. Dept. of Agric. Handb. 60. U.S. Govt. Print. Office, Washington, DC.

Selective Dissolutions (4G)

Background: Over the years, various terms have been used to describe broad groupings of soil components, e.g., *crystalline phyllosilicates, amorphous, poorly crystalline, paracrystalline, noncrystalline, allophane, imogolite,* and *short-range-order minerals (SROMs),* etc (Jackson, 1986). These groupings have been related, in part, to various laboratory analyses, and thereby, have been operationally defined quantitatively and semiquantitatively by these analyses (Jackson, 1986). Some of these analytical procedures include X-ray diffraction analysis and selective chemical dissolutions, e.g., dithionite-citrate, sodium pyrophosphate, and ammonium oxalate extractions. These terms have not been used consistently in the literature. In addition, there is not always a clear delineation between dissolution data, either conceptually or operationally. For more detailed discussion of these various soil terms and the application of selective chemical extractions, refer to Wada (1989) and Soil Survey Staff (1995).

Selective dissolution data have been used extensively in the study of the noncrystalline material content of soils and sediments. However, there are limitations in using these data. In general, there exists a continuum of crystalline order, ranging from no long-range order to paracrystalline to poorly crystalline to well crystalline (Follet et al., 1965). Selective dissolution data are necessary for independent determinations of various inorganic constituents of soils because of the difficulty with many physical analytical methods in estimating or even recognizing the presence of noncrystalline and paracrystalline free oxides or aluminosilicates mixed with crystalline soil components (Jackson et al., 1986). In general, the crystalline free oxides and phyllosilicates of soils can be identified qualitatively and estimated semiquantitatively by X-ray diffraction analysis. Those soils containing hydroxyls (-OH groups), e.g., kaolinite, gibbsite, and goethite, can sometimes be determined quantitatively by differential thermal analysis (DTA), differential scanning colorimetry (DCS), and thermogravimetric analysis (TGA). Refer to additional discussion on X-ray diffraction and thermal analysis in the mineralogy section (7) of this manual.

With selective chemical dissolution data, there are difficulties in the adequate assessment of the portion that is extracted by particular reagents, e.g., dithionite-citrate, sodium pyrophosphate, and ammonium oxalate. In principle, it cannot be expected that chemical methods are able to perfectly distinguish the degrees of crystallinity, and some caution is required in the interpretation of these analytical data (van Wambeke, 1992). Refer to (Wada, 1989) for the dissolution of Al, Fe, and Si in various clay constituents and organic complexes by treatment with different reagents. The SSL routinely performs three selective chemical dissolutions as follows: dithionite-citrate (4G1a1-3); ammonium oxalate (4G2a1a1-5); and sodium pyrophosphate (4G3a1-3).

Dithionite-Citrate Extraction: The original objectives of the dithionite-citrate extraction were to determine the free Fe oxides and to remove the amorphous coatings and crystals of free Fe oxide, acting as cementing agents, for subsequent physical and chemical analysis of soils, sediments, and clay minerals (Weaver et al., 1968; Jackson, 1969; Jackson et al., 1986). Dithionite-citrate extractable Fe (Fe_d) is considered a measure of "free iron" in soils and, as such, is pedogenically significant. Dithionite-citrate extractable Fe data are of interest in soil genesis-classification studies because of its increasing concentration with increasing weathering, and its effect on soil colors (Schwertmann, 1992). "Free iron" is also considered an important factor in P-fixation and soil aggregate stability.

Sodium Pyrophosphate Extraction: Sodium pyrophosphate extracting solutions tend to selectively extract mainly Fe and Al associated with organic compounds while the dithionite-citrate extractions tend to extract these compounds plus the free oxides (McKeague et al., 1971). At one time, sodium pyrophosphate extractable Fe and Al in conjunction with dithionite-citrate data were used to help identify translocated Al and Fe humus complexes in spodic horizons (Soil Survey Staff, 1975). Numerous evaluations of pyrophosphate extracts have indicated that the pyrophosphate extraction does not necessarily correlate with organic-bound Fe and Al (Schuppli et al., 1983; Kassim et al., 1984; Parfitt and Childs, 1988; Birkeland et al., 1989) as commonly thought (Schwertmann and Taylor, 1977; Parfitt and Childs, 1988). Pyrophosphate not only extracts organic-bound Fe but also peptizes solid particles of ferrihydrite and in some instances even goethite (Yuan et al., 1993). The use of sodium pyrophosphate extract data in conjunction with dithionite-citrate data as chemical requirements for spodic horizons have been replaced by other criteria (Soil Survey Staff, 1999) and, at one time, were referred to as spodic horizon criteria on the SSL data sheets.

Ammonium Oxalate Extraction: In general, ammonium oxalate allowed to react in darkness has been considered to be a selective dissolution for noncrystalline materials (McKeague and Day, 1966; Higashi and Ikeda, 1974; Fey and LeRoux, 1976; Schwertmann and Taylor, 1989; Hodges and Zelazny, 1980). The ammonium oxalate procedure removes most noncrystalline and paracrystalline materials (allophane and imogolite) from soils (Higashi and Ikeda, 1974; Hodges and Zelazny, 1980) as well as short-range-ordered oxides and hydroxides of Al, Fe, and Mn (Schwertmann, 1959, 1964; McKeague and Day, 1966; McKeague et al., 1971; Fey and LeRoux, 1976). In addition, this method is assumed to extract Al + Fe humus. Opaline Si is not dissolved by this method (Wada, 1977). This procedure has been reported to dissolve very little hematite and goethite and small amounts of magnetite (Baril and Bitton, 1969; McKeague et al., 1971; Walker, 1983). There have been conflicting data on the effect of this procedure on clay minerals, but in general, the ammonium oxalate treatment is considered to have very little effect on phyllosilicates (kaolinite, montmorillonite, illite) or gibbsite. The ammonium oxalate extraction is assumed to dissolve selectively "active" Al and Fe components that are present in noncrystalline materials as well as associated or independent poorly crystalline silica. The method also extracts allophane, imogolite, Al + Fe humus complexes, and amorphous or poorly crystallized oxides and hydroxides.

The intent of the ammonium oxalate procedure is to measure the quantities of poorly crystalline materials in the soil. At the present time, the ammonium oxalate extraction is considered the most precise chemical method for measuring these soil components. However, in principle it cannot be expected that chemical methods are able to perfectly distinguish degrees of crystallinity, and some caution is to be exercised in the interpretation of the analytical data (van Wambeke, 1992). A more reliable and accurate estimation of soil properties and a better understanding of noncrystallinity are provided when ammonium oxalate extraction is used in conjunction with other selective dissolution procedures, thermal techniques, and chemical tests (Jackson, 1986).

Application, Ratios, and Estimates: In a general way, the Fe_d is considered to be a measure of the total pedogenic Fe, e.g., goethite, hematite, lepidocrocite, and ferrihydrite, while the ammonium oxalate extractable Fe (Fe_o) (probably ferrihydrite) is a measure of the paracrystalline Fe (Birkeland et al., 1989). The Fe_o/Fe_d ratio is often calculated because it is considered an approximation of the relative proportion of ferrihydrite in soils (Schwertmann, 1985).

The Mn_d is considered the "easily reducible Mn". The Al_d and Al_o are pedogenically significant. The Al_d represents the Al substituted in Fe oxides, which can have an upper limit of thirty-three mole percent substitution (Schwertmann et al., 1977; Schwertmann and Taylor, 1989). The Al_o is generally an estimate of the total pedogenic Al in soils dominated by allophane, imogolite, and organically-bound Al (Wada, 1977; Childs et al., 1983). Unlike Fe_d, the Al_d extract is commonly less than the Al_o (Childs et al., 1983; Birkeland et al., 1989) and so does not necessarily represent the total pedogenic Al (Wada, 1977).

Allophane in soils can be estimated from the Al_o and Si_o and the pyrophosphate extractable Al (Al_p) (Parfitt and Henmi, 1982; Parfitt and Wilson, 1985; Parfitt, 1990). The Al_o represents the Al dissolved from allophane, imogolite, and Al-humus complexes, and the Al_p is the Al from the Al-humus complexes alone (Parfitt and Kimble, 1989). The Al_o minus the Al_p gives an estimate of the Al in allophane and imogolite, whereas the Si_o gives an estimate of the Si in allophane and imogolite. The $(Al_o - Al_p)/Si_o$ times the molar ratio (28/27) is an estimate of the Al/Si ratio of allophane and imogolite in the soil. The values of 28 and 27 represent the atomic weights of Si and Al, respectively.

Selective chemical dissolution data are used as taxonomic criteria for mineralogy classes, e.g., ammonium oxalate Fe and Si for the amorphic and ferrihydritic mineralogy classes and dithionite-citrate Fe for the ferritic mineralogy class. The optical density of ammonium oxalate extract (ODOE) of ≥ 0.25 is used as a chemical criterion for spodic materials (Soil Survey Staff, 1999). An increase in ODOE is used as an indicator of the accumulation of translocated organic materials in an illuvial horizon (Soil Survey Staff, 1999). Ammonium oxalate extractable Al_o plus 0.5 Fe_o is also used as a taxonomic criterion for andic soil properties (Soil Survey Staff, 1999). The weight of Fe atoms is approximately twice that of Al atoms. In evaluating the relative proportion of Fe and Al atoms solubilized by ammonium oxalate, the weight percent of Fe must be divided by two, i.e., $Al_o + \frac{1}{2} Fe_o$. Refer to the Soil Survey Staff (1995, 1999) for more detailed discussion of these taxonomic criteria.

References:

Baril, R., and G. Bitton. 1969. Teneurs elevees de fer libre et identification taxonomique de certain sols due Quebec contenant de la magnetite. Can. J. Soil Sci. 49:1-9.

Birkeland, P.W., R.M. Burke, and J.B. Benedict. 1989. Pedogenic gradients of iron and aluminum accumulation and phosphorus depletion in arctic and alpine soils as a function of time and climate. Quat. Res. 32:193-204.

Fey, M.V., and J. LeRoux. 1976. Quantitative determinations of allophane in soil clays. Proc. Int. Clay Conf. 1975 (Mexico City). 5:451-463. Appl. Publ. Ltd., Wilmette, IL.

Follet, E.A.C., W.J. McHardy, B.D. Mitchell, and B.F.L. Smith. 1965. Chemical dissolution techniques in the study of soil clays. I. & II. Clay Miner. 6:23-43.

Higashi, T., and H. Ikeda. 1974. Dissolution of allophane by acid oxalate solution. Clay Sci. 4:205-212.

Hodges, S.C., and L.W. Zelazny. 1980. Determination of noncrystalline soil components by weight difference after selective dissolution. Clays Clay Miner. 28:35-42.

Jackson, M.L. 1969. Soil chemical analysis. Adv. 2nd ed. Univ. Wisconsin, Madison, WI.

Jackson, M.L., C.H. Lim, and L.W. Zelazny. 1986. Oxides, hydroxides, and aluminosilicates. *In* A. Klute (ed.) Methods of soil analysis. Part 1. Physical and mineralogical methods. 2nd ed. Agron. 9:101-150.

Kassim, J.K., S.N. Gafoor, and W.A. Adams. 1984. Ferrihydrite in pyrophosphate extracts of podzol B horizons. Clay Miner. 19:99-106.

McKeague, J.A., J.E. Brydon, and N.M. Miles. 1971. Differentiation of forms of extractable iron and aluminum in soils. Soil Sci. Soc. Am. Proc. 35:33-38.

McKeague, J.A., and J.H. Day. 1966. Dithionite- and oxalate-extractable Fe and Al as aids in differentiating various classes of soils. Can. J. Soil Sci. 46:13-22.

Parfitt, R.L. 1990. Allophane in New Zealand - A review. Aust. J. Soil Res. 28:343-360.

Parfitt, R.L., and C.W. Childs. 1988. Estimation of forms of Fe and Al: a review of and analysis of contrasting soils by dissolution and Moessbauer methods. Aust. J. Soil. Res. 26:121-144.

Parfitt, R.L., and T. Henmi. 1982. Comparison of an oxalate-extraction method and an infrared spectroscopic method for determining allophane in soil clays. Soil Sci. Nutr. 28:183-190.

Parfitt, R.L., and J.M. Kimble. 1989. Conditions for formation of allophane in soils. Soil Sci. Soc. Am. J. 53:971-977.

Parfitt, R.L. and A.D. Wilson. 1985. Estimation of allophone and halloysite in three sequences of volcanic soils, New Zealand. Catena Suppl. 7:1-9.

Schuppli, P.A., G.J. Ross, and J.A. McKeague. 1983. The effective removal of suspended materials from pyrophosphate extracts of soils from tropical and temperate regions. Soil Sci. Soc. Am. J. 47:1026-1032.

Schwertmann, U. 1959. Die fraktionierte Extraktion der freien Eisenoxide in Boden, ihre mineralogischen Formen und ihre Entstehungsweisen. Z. Pflanzenernahr. Dueng. Bodenk. 84:194-204.

Schwertmann, U. 1964. The differentiation of iron oxides in soil by extraction with ammonium oxalate solution. Z. Pflanzenernaehr. Bodenkd. 105:194-202.

Schwertmann, U. 1985. The effect of pedogenic environments on iron oxide minerals. Adv. Soil Sci. i, 172-200.

Schwertmann, U. 1992. Relations between iron oxides, soil color, and soil formation. *In* J.M. Bigham and E.J. Ciolkosz (eds.). Soil color. SSSA Spec. Publ. No. 31. Soil Sci. Soc. Am., Madison, WI.

Schwertmann, U., R.W. Fitzpatrick, and J.LeRoux. 1977. Al substitution and differential disorder in soil hematites. Clays Clay Miner. 25:373-374.

Schwertmann, U., and R.M. Taylor. 1977. Iron oxides. *In* J.B. Dixon and S.B. Weed (eds.). Minerals in soil environment. Soil Sci. Soc. Am. No. 1:145-180.

Schwertmann, U., and R.M. Taylor. 1989. Iron oxides. In J.B. Dixon and S.B. Weed (eds.). Minerals in soil environment. 2nd ed. Soil Sci. Soc. Am. No. 1:370-438.

Survey Staff. 1975. Soil taxonomy: A basic system of soil classification for making and interpreting soil surveys, USDA-SCS Agric. Handb. No. 436. U.S. Govt. Print. Office, Washington, DC.

Soil Survey Staff. 1995. Soil survey laboratory information manual. Version No. 1.0. USDA-NRCS. Soil Survey Investigations Report No. 42. U.S. Govt. Print. Office, Washington, DC.

Soil Survey Staff. 1999. Soil taxonomy: A basic system of soil classification for making and interpreting soil surveys. USDA-NRCS Agric. Handb. 436. 2nd ed. U.S. Govt. Print. Office, Washington, DC.

van Wambeke, A. 1992. Soils of the tropics: properties and appraisal. McGraw-Hill, Inc., NY.

Wada, K. 1977. Allophane and imogolite. *In* J.B. Dixon and S.B. Weed (eds.). Minerals in soil environment. 1st ed. Soil Sci. Soc. Am. No. 1:603-638.

Wada, K. 1989. Allophane and imogolite. *In* J.B. Dixon and S.B. Weed (eds.). Minerals in soil environment. 2nd ed. Soil Sci. Soc. Am. No. 1: 1051-1087.

Walker, A.L. 1983. The effects of magnetite on oxalate- and dithionite-extractable iron. Soil Sci. Soc. Am. J. 47:1022-1025.

Weaver, R.M., J.K. Syers, and M.L. Jackson. 1968. Determination of silica in citrate-bicarbonate-dithionite extracts of soils. Soil Sci. Soc. Am. Proc. 32:497-501.

Yuan, G., L.M. Lavkulich, and C. Wang. 1993. A method for estimating organic-bound iron and aluminum contents in soils. Commun. Soil Sci. Plant Anal. 24(11&12), 1333-1343.

Selective Dissolutions (4G)
Dithionite-Citrate Extraction (4G1)
Atomic Absorption Spectrophotometer (4G1a)
Aluminum, Iron, Manganese (4G1a1-3)
Air-Dry or Field Moist, <2 mm (4G1a1-3a-b1)

1. Application

Dithionite-citrate (CD) is used as a selective dissolution extractant for organically complexed Fe and Al, noncrystalline hydrous oxides of Fe and Al, and amorphous aluminosilicates (Wada, 1989). The CD solution is a poor extractant of crystalline hydrous oxides of Al, allophane, and imogolite. The CD solution does not extract opal, Si, or other constituents of crystalline silicate minerals (Wada, 1989).

This extraction is also sometimes referred to as citrate-dithionite or sodium citrate-dithionite. The method (4G1a1-3) described herein is not the same extraction as described in the Soil Survey Investigations Report (SSIR) No. 1 (1972), method 6C3. This obsolete SSL method (6C3) incorporated sodium bicarbonate as a buffer (pH 7.3) in the dithionite-citrate method, resulting in a buffered neutral citrate-bicarbonate-dithionite (Aguilera and Jackson, 1953; Mehra and Jackson, 1960; and Jackson, 1969), commonly referred to as the CBD method.

2. Summary of Method

A soil sample is mixed with sodium dithionite, sodium citrate, and reverse osmosis deionized (RODI) water, and shaken overnight. Solution is centrifuged, and a clear extract obtained. The CD extract is diluted with RODI water. The analytes are measured by an atomic absorption spectrophotometer (AAS). The data are automatically recorded by a computer and printer. The AAS converts absorption to analyte concentration. The percent CD extractable Al, Fe, and Mn are reported in procedures 4G1a1-3, respectively.

3. Interferences

There are four types of interferences (matrix, spectral, chemical, and ionization) in the AA analyses of these elements. These interferences vary in importance, depending upon the particular analyte selected.

The redox potential of the extractant is dependent upon the pH of the extracting solution and the soil system. Sodium citrate complexes the reduced Fe and usually buffers the system to a pH of 6.5 to 7.3. Some soils may lower the pH, resulting in the precipitation of Fe sulfides.

Filtered extracts can yield different recoveries of Fe, Mn, and Al, relative to unfiltered extracts.

4. Safety

Wear protective clothing (coats, aprons, sleeve guards, and gloves); eye protection (face shields, goggles, or safety glasses); and a breathing filter when handling dry sodium dithionite. Sodium dithionite may spontaneously ignite if allowed to become moist, even by atmospheric moisture. Keep dithionite in a fume hood.

Follow standard laboratory practices when handling compressed gases. Gas cylinders should be chained or bolted in an upright position. Acetylene gas is highly flammable. Avoid open flames and sparks. Standard laboratory equipment includes fire blankets and extinguishers for use when necessary. Follow the manufacturer's safety precautions when using the AA.

5. Equipment

5.1 Electronic balance, ±1.0-mg sensitivity

5.2 Mechanical reciprocating shaker, 200 oscillations min^{-1}, 1 ½ in strokes, Eberbach 6000, Eberbach Corp., Ann Arbor, MI

5.3 Atomic absorption spectrophotometer (AAS), double-beam optical system, AAnalyst, 300, Perkin-Elmer Corp., Norwalk, CT, with computer and printer

5.4 Autosampler, AS-90, Perkin-Elmer Corp., Norwalk, CT

5.5 Peristaltic pump

5.6 Single-stage regulators, acetylene and nitrous oxide

5.7 Centrifuge, Centra, GP-8, Thermo IEC, Needham Heights, MA

5.8 Digital diluter/dispenser, with syringes 10000 and 1000 µL, gas tight, MicroLab 500, Hamilton Co., Reno, NV

5.9 Dispenser, 30 mL

5.10 Test tubes, 15-mL, 16 mm x 100, for sample dilution and sample changer

5.11 Containers, polypropylene

5.12 Volumetrics, Class A, 100, 250, and 1000-mL

5.13 Measuring scoop, handmade, 0.4 g calibrated

5.14 Centrifuge tubes, 50-mL

6. Reagents

6.1 Reverse osmosis deionized (RODI) water, ASTM Type I grade of reagent water

6.2 Sodium dithionite ($Na_2S_2O_4$), purified powder

6.3 Sodium citrate dihydrate ($Na_3C_6H_5O_7 \cdot 2H_2O$), crystal, reagent. Dissolve 336 g sodium citrate in approximately 1 L RODI water, followed by diluting to 2 L with RODI water. Final concentration is 0.57 M sodium citrate.

6.4 Sulfuric acid (H_2SO_4), concentrated.

6.5 Phosphoric acid (H_3PO_4), concentrated (85%). For Fe analysis, samples are diluted 1:50 prior to analysis. A diluting solution for Fe analysis (for a final concentration of 0.5% H_3PO_4 in samples) may be made by adding 6.12 mL of concentrated H_3PO_4 to 500 mL volume of RODI water, diluting to 1000 mL and mixing thoroughly. (Note: 1:5 sample dilutions for Al and Mn are in RODI water.)

6.6 Primary Stock Standard Solution (PSSS), high purity, 1000 mg L^{-1}: Fe, Mn, and Al.

6.7 Calibration standards for Fe (Section 6.8): To each 100 mL volume of blank, calibration, and quality control (QC) standards, add 25 mL of the following matrix matching mixture. This mixture is made by combining 20 mL Na citrate extracting solution, 0.21 mL H_2SO_4, and 6 mL H_3PO_4 and diluting to 250 mL volume with RODI water. Invert to mix thoroughly. (Note: Matrix of standards is prepared to match a 1:50 dilution of samples. Also, H_2SO_4 substitutes for dithionite).

6.8 Standard Fe calibration solutions (SFeCS) or working standards (25.0, 20.0, 15.0, 10.0, 5.0, 1.0, and 0.0 mg Fe L^{-1}) and QC (12.5 mg L^{-1}). Prepare fresh weekly. In seven 100-mL volumetric flasks add as follows:

6.8.1 25.0 mg Fe L^{-1} = 2.5 mL $PSSS_{Fe}$
6.8.2 20.0 mg Fe L^{-1} = 2.0 mL $PSSS_{Fe}$
6.8.3 15.0 mg Fe L^{-1} = 1.5 mL $PSSS_{Fe}$
6.8.4 10.0 mg Fe L^{-1} = 1.0 mL $PSSS_{Fe}$
6.8.5 5.0 mg Fe L^{-1} = 0.5 mL $PSSS_{Fe}$
6.8.6 1.0 mg Fe L^{-1} = 0.1 mL $PSSS_{Fe}$
6.8.7 0.0 mg Fe L^{-1} = 0.0 mL $PSSS_{Fe}$ (blank)
6.8.8 12.5 m Fe L^{-1} = 1.25 mL $PSSS_{Fe}$ (QC)

Fill to volume with RODI water and invert to mix thoroughly. After dissolution, transfer solution to a plastic bottle.

6.9 Calibration standards for Mn (Section 6.10): To each 100 mL volume of blank, calibration, and quality control (QC) standards, add 25 mL of the following matrix matching mixture. This mixture is made by combining 200 mL Na citrate extracting solution, 2.1 mL H_2SO_4, and diluting to 250-mL volume with RODI water. Invert to mix thoroughly. (Note: Matrix of standards is prepared to match a 1:5 dilution of samples. Also, H_2SO_4 substitutes for dithionite).

6.10 Standard Mn calibration solutions (SMnCS) or working standards (15.0, 10.0, 5.0, 2.5, 1.5, and 0.0 mg Mn L^{-1}) and QC (6.5 mg L^{-1}). Prepare fresh weekly. In six 100-mL volumetric flasks add as follows:

6.10.1 15.0 mg Mn L^{-1} = 1.5 mL $PSSS_{Mn}$
6.10.2 10.0 mg Mn L^{-1} = 1.0 mL $PSSS_{Mn}$
6.10.3 5.0 mg Mn L^{-1} = 0.5 mL $PSSS_{Mn}$
6.10.4 2.5 mg Mn L^{-1} = 0.25 mL $PSSS_{Mn}$
6.10.5 1.5 mg Mn L^{-1} = 0.15 mL $PSSS_{Mn}$
6.10.6 0.0 mg Mn L^{-1} = 0.0 mL $PSSS_{Mn}$ (blank)
6.10.7 6.5 mg Mn L^{-1} = 0.65 mL $PSSS_{Mn}$ (QC)

Fill to volume with RODI water and invert to mix thoroughly. After dissolution, transfer solution to a plastic bottle.

6.11 Calibration standards for Al (Section 6.12): To each 100 mL volume of blank, calibration, and quality control (QC) standards, add 25 mL of the following matrix matching mixture. This mixture is made by combining 200 mL Na citrate extracting solution, 2.1 mL H_2SO_4, and then diluting to 250-mL volume with RODI water. Invert to mix thoroughly. (Note: Matrix of standards is prepared to match a 1:5 dilution of samples (same as with Mn). Also, H_2SO_4 substitutes for dithionite).

6.12 Standard Al calibration solutions (SAlCS) or working standards (100.0, 80.0, 60.0, 40.0, 20.0, 10.0, and 0.0 mg Al L^{-1}) and QC (50.0 mg L^{-1}). Prepare fresh weekly. In seven 100-mL volumetric flasks add as follows:

6.12.1 100.0 mg Al L^{-1} = 10.0 mL $PSSS_{Al}$
6.12.2 80.0 mg Al L^{-1} = 8.0 mL $PSSS_{Al}$
6.12.3 60.0 mg Al L^{-1} = 6.0 mL $PSSS_{Al}$

6.12.4 40.0 mg Al L^{-1} = 4.0 mL PSSS$_{Al}$
6.12.5 20.0 mg Al L^{-1} = 2.0 mL PSSS$_{Al}$
6.12.6 10.0 mg Al L^{-1} = 1.0 mL PSSS$_{Al}$
6.12.7 0.0 mg Al L^{-1} = 0.0 mL PSSS$_{Al}$ (blank)
6.12.8 50.0 mg Al L^{-1} = 5.0 mL PSSS$_{Al}$ (QC)

Fill to volume with RODI water and invert to mix thoroughly. After dissolution, transfer solution to a plastic bottle.

6.13 Acetylene gas, purity 99.6%
6.14 Nitrous oxide, USP
6.15 Compressed air with water and oil traps

7. Procedure

Extraction of Al, Fe, and Mn

7.1 Weigh 0.75 g of <2-mm or fine-grind, air-dry soil sample to the nearest mg and place in an 50-mL centrifuge tube. If sample is moist, weigh enough soil to achieve ≈ 0.75 g of air-dry soil.

7.2 Add 0.4 g of sodium dithionite (use one calibrated scoop) and 25 mL of sodium citrate solution.

7.3 Cap tubes and shake briefly by hand to dislodge soil from tube bottom. Place tubes in rack.

7.4 Place rack in shaker and shake overnight (12 to 16 h) at 200 oscillations min^{-1} at room temperature (20°C ± 2°C).

7.5 Remove tubes from shaker and manually shake tubes to dislodge any soil from cap. Allow samples to sit overnight.

7.6 The following day, centrifuge at 4000 rpm for 15 min. The Fe, Mn, and Al are determined on the AA from a clear aliquot of solution.

Dilution of Sample Extracts

7.7 No ionization suppressant is required as the Na in the extractant is present in sufficient quantity. For a 1:50 dilution of samples for Fe analysis, use the H$_3$PO$_4$ diluting solution (see Section 6.5). The dilution of Fe results in a final solution concentration of 0.5% H$_3$PO$_4$. Dilute 1 part CD sample extract with 49 parts of H$_3$PO$_4$ diluting solution (1:50 dilution).

7.8 A 1:5 dilution in RODI water is used for Al and Mn. Dilute 1 part CD sample extract with 4 parts RODI water.

7.9 Dispense the diluted sample solutions into test tubes that have been placed in the holders of the sample changer.

AAS Set-up and Operation

7.10 Refer to the manufacturer's manual for operation of the AAS. The following are only very general guidelines for instrument conditions for the various analytes.

Analyte	Conc. (mg L^{-1})	Wavelength (nm)	Burner Head	Slit (mm)	Fuel/Oxidant
Fe	25.0	248.8	10-cm parallel	0.2	3.0 C_2H_2/15.7 Air
Mn	15.0	279.8	10-cm parallel	0.2	3.0 C_2H_2/15.7 Air
Al	100.0	309.3	5-cm parallel	0.7	8.5 C_2H_2/15.7 N_2O

Typical read delay is 3 s, and integration time is 3 s but can vary depending on soil type. Three replicates are average for each sample.

7.11 Use the computer and printer to set instrument parameters and to collect and record instrument readings.

7.12 The instrument readings are programmed to display analyte concentration in mg L^{-1} (ppm).

AAS Calibration and Analysis

7.13 Each element is analyzed during separate runs on the AAS. Use the calibration reagent blank and calibration standards to calibrate the AAS. Calibrations are linear with calculated intercept.

7.14 Use the QC after every 12th sample. It must pass within 15% to continue. If it fails, recalibrate and reread the QC. The QC is also read at the end of each run.

7.15 If samples are outside the calibration range, a serial dilution is performed. A 1:5 dilution of the sample using the calibration blank, followed by the typical dilution (1:5 dilution with RODI water for Al and Mn, and 1:50 dilution with the H_3PO_4 diluting solution for Fe). Maintain matrix match between standards and diluted samples by performing this extra dilution with calibration blank.

7.16 Record analyte readings to 0.01 unit.

8. Calculations

Convert analyte concentrations (mg L^{-1}) to percent in soil as follows:

Soil Fe, Al, Mn (%) = (A x B x C x R x 100)/ (E x 1000)

where:
A   = Sample extract reading (mg L^{-1})
B   = Extract Volume (L)
C   = Dilution, required
R   = Air-dry/oven-dry ratio (procedure 3D1) or field-moist/oven-dry ratio (procedure 3D2)
E   = Sample weight (g)
100 = Conversion factor to 100-g basis
1000 = mg g^{-1}

9. Report
Report percent CD extractable Al, Fe, and Mn to the nearest 0.1 of a percent.

10. Precision and Accuracy
Precision and accuracy data are available from the SSL upon request.

11. References

Wada, K. 1989. Allophane and imogolite. p. 1051-1087. *In* J.B. Dixon and S.B. Weed (eds.). Minerals in soil environment. 2nd ed. Soil Sci. Am. Book Series No. 1. ASA and SSSA, Madison, WI.

Soil Survey Staff. 1999. Soil taxonomy: A basic system of soil classification for making and interpreting soil surveys. USDA-NRCS Agric. Handb. 436. 2nd ed. U.S. Govt. Print. Office, Washington, DC.

Selective Dissolutions (4G)
Ammonium Oxalate Extraction (4G2)
Automatic Extractor (4G2a)
Inductively Coupled Plasma Atomic Emission Spectrophotometer (4G2a1)
Radial Mode (4G2a1a)
Aluminum, Iron, Manganese, Phosphorus, and Silicon (4G2a1a1-5)
Air-Dry or Field-Moist, <2 mm (4G2a1a1-5a-b1)
UV-Visible Spectrophotometer, Dual Beam (4G2a2)
Transmittance (4G2a2a)
Optical Density (4G2a2a1)
Air-Dry or Field-Moist, <2 mm (4G2a2a1a-b1)

1. Application

Ammonium oxalate is used as a selective dissolution extractant for organically complexed Fe and Al, noncrystalline hydrous oxides of Fe and Al, allophane, and amorphous aluminosilicates (Wada, 1989). Ammonium oxalate is a poor extractant of imogolite and layer silicates and does not extract crystalline hydrous oxides of Fe and Al, opal, or crystalline silicate (Wada, 1989). A more reliable and accurate estimation of soil properties and a better understanding of soil exchange complex is provided when ammonium oxalate extraction is used in conjunction with other selective dissolution procedures, thermal techniques, and chemical tests. In soil taxonomy, ammonium oxalate extractable Fe and Al are criteria for andic soil properties (Soil Survey Staff, 1999). This extraction is also sometimes referred to as acid ammonium oxalate, acid oxalate, oxalate-oxalic acid, or oxalic acid-ammonium oxalate.

2. Summary of Method

A soil sample is extracted with a mechanical vacuum extractor (Holmgren et al., 1977) in a 0.2 M ammonium oxalate solution buffered at pH 3.0 under darkness. The ammonium oxalate extract is weighed. The ammonium oxalate extract is diluted with reverse osmosis deionized water. The analytes are measured by a inductively coupled plasma atomic emission spectrophotometer (ICP-AES). Data are automatically recorded by a computer and printer. The ammonium oxalate extractable Al, Fe, Mn, P, and Si are reported in procedures 4G2a1a1-5, respectively. All these data are reported in percent except Mn and P, which are reported in mg kg^{-1}. In procedure 4G2a2a1, the optical density of the extract is measured with a UV spectrophotometer at 430 nm.

3. Interferences

There are four types of interferences (matrix, spectral, chemical, and ionization) in the ICP analyses of these elements. These interferences vary in importance, depending upon the particular analyte chosen.

The ammonium oxalate buffer extraction is sensitive to light, especially UV light. The exclusion of light reduces the dissolution effect of crystalline oxides and clay minerals. If the sample contains large amounts of amorphous material (>2% Al), an alternate method should be used, i.e., shaking with 0.275 M ammonium oxalate, pH 3.25, 1:100 soil:extractant.

4. Safety

Wear protective clothing and eye protection. When preparing reagents, exercise special care. Follow standard laboratory practices when handling compressed gases. Gas cylinders should be chained or bolted in an upright position. Follow the manufacturer's safety precautions when using the UV spectrophotometer and ICP.

5. Equipment

5.1 Electronic balance, ±1.0-mg sensitivity
5.2 Mechanical vacuum extractor, 24 place, SAMPLETEX, MAVCO Industries, Lincoln, NE
5.3 Tubes, 60-mL, polypropylene, for extraction (0.45-μm filter), reservoir, and tared extraction tubes
5.4 Rubber tubing, 3.2 ID x 1.6 OD x 6.4 mm, (1/8 ID x 1/16 OD x 1 in) for connecting syringe barrels
5.5 Dispenser, 30-mL
5.6 Pipettes, electronic digital, 10000 and 1000 μL, with tips 10000 and 1000 μL
5.7 Containers, polyethylene
5.8 Inductively coupled plasma atomic emission spectrophotometer (ICP-AES), dual-view, with high-solids nebulizer, alumina or quartz injector, Optima 4300 DV, Perkin-Elmer Corp., Norwalk, CT.
5.9 Automsampler, AS-93, Perkin-Elmer Corp., Norwalk, CT
5.10 Computer, with WinLab32^{TM} software, and printer
5.11 Single-stage regulator, high-purity, high-flow, argon
5.12 Test tubes, 15-mL, 16 mm x 100, for sample dilution and sample changer, Fisher Scientific
5.13 Vortexer, mini, MV1, VWR Scientific Products
5.14 Spectrophotometer, UV-Visible, Varian, Cary 50 Conc, Varian Australia Pty Ltd.
5.15 Computer with Cary WinUV software, Varian Australia Pty Ltd., and printer
5.16 Cuvettes, plastic, 4.5-mL, 1-cm light path, Daigger Scientific

6. Reagents

6.1 Reverse osmosis deionized (RODI) water, ASTM Type I grade of reagent water
6.2 Ammonium oxalate buffer solution, 0.2 M, pH 3.0. *Solution A* (base): Dissolve 284 g of $(NH_4)_2C_2O_4 \cdot 2H_2O$ in 10 L of DDI water. *Solution B* (acid): Dissolve 252 g of $H_2C_2O_4 \cdot H_2O$ in 10 L of DDI water. Mix 4 parts solution A with 3 parts solution B. Adjust ammonium oxalate solution pH by adding either acid or base solution. Store in a polypropylene bottle.
6.3 Borax pH buffers, pH 4.00, 7.00, and 9.18 for electrode calibration, Beckman, Fullerton, CA.
6.4 Primary Fe standard, 1000 mg L^{-1}. Certified Reference Solution, Fisher Chemical Scientific Co., Fairlawn, N.J.
6.5 Primary Al standard, 1000 mg L^{-1}. Certified Reference Solution, Fisher Chemical Scientific Co., Fairlawn, N.J.
6.6 Primary Si standard, 1000 mg L^{-1}. Certified Reference Solution, Fisher Chemical Scientific Co., Fairlawn, N.J.
6.7 Primary Mn standard, 1000 mg L^{-1}. Certified Reference Solution, Fisher Chemical Scientific Co., Fairlawn, N.J.
6.8 Primary P standard, 1000 mg L^{-1}. Certified Reference Solution, Fisher Chemical Scientific Co., Fairlawn, N.J.
6.9 High mixed calibration standard. Mix 60 mL of each primary standard (Si, Fe, and Al) with 10 mL of primary Mn standard and 20 mL of primary P standard in 1 L volumetric flask. Add 100 mL of 0.2 M ammonium oxalate solution and make to 1-L volume with RODI water. The elements are added in the order (Si, Fe, Al, Mn, P) to avoid element precipitation. Resulting solution contains 60 mg L^{-1} each of Si, Fe, and Al, 10 mg L^{-1}

Mn, and 20 mg L^{-1} P. Invert to mix thoroughly. Store in a polyethylene bottle. Make fresh weekly. Store in a refrigerator.

6.10 Medium mixed calibration standard. Mix 30 mL of each primary standard (Si, Fe, and Al) with 5 mL of primary Mn standard and 10 mL of primary P standard in 1 L volumetric flask. Add 100 mL of 0.2 M ammonium oxalate solution and make to 1-L volume with RODI water. The elements are added in the order (Si, Fe, Al, Mn, P) to avoid element precipitation. Resulting solution contains 30 mg L^{-1} each of Si, Fe, and Al, 5 mg L^{-1} Mn, and 10 mg L^{-1} P. Invert to mix thoroughly. Store in a polyethylene bottle. Make fresh weekly. Store in a refrigerator.

6.11 Low mixed calibration standard. Mix 10 mL of each primary standard (Si, Fe, and Al) with 2 mL of primary Mn standard, and 3 mL primary P standard in 1 L volumetric flask. Add 100 mL of 0.2 M ammonium oxalate solution and make to 1-L volume with RODI water. The elements are added in the order (Si, Fe, Al, Mn, P) to avoid element precipitation. Resulting solution contains 10 mg L^{-1} each of Si, Fe, and Al, 2 mg L^{-1} Mn, and 3 mg L^{-1} P. Invert to mix thoroughly. Store in a polyethylene bottle. Make fresh weekly. Store in a refrigerator.

6.12 Low Si calibration standard. Mix 5 mL of Si primary standard in 1-L volumetric flask. Add 100 mL of 0.2 M ammonium oxalate solution and make to 1-L volume with RODI water. Resulting solution contains 5 mg L^{-1} Si. Invert to mix thoroughly. Store in polyethylene bottle. Make fresh weekly. Store in a refrigerator.

6.13 Very Low Si calibration standard. Mix 2 mL of Si primary standard in 1-L volumetric flask. Add 100 mL of 0.2 M ammonium oxalate solution and make to 1-L volume with RODI water. Resulting solution contains 2 mg L^{-1} Si. Invert to mix thoroughly. Store in polyethylene bottle. Make fresh weekly. Store in a refrigerator.

6.14 Calibration reagent blank solution. Add 100 mL of 0.2 M ammonium oxalate solution and make to 1-L volume with RODI water. Store in polyethylene bottle. Make fresh weekly. Store in a refrigerator.

6.15 Argon gas, purity 99.9%

6.16 Nitrogen, purity 99.9%

7. Procedure

Extraction of Fe, Mn, Al, Si, and P

7.1 Weigh 0.5 g of <2-mm, air-dry or fine-grind soil the nearest mg and place in sample tube. If sample is moist, weigh enough soil to achieve ≈ 0.5 g. Prepare two reagent blanks (no sample in tube) per set of 48 samples.

7.2 Place labeled ET on extractor and connect to corresponding tared extraction tube (TET$_{Oxalate}$) with rubber tubing.

7.3 Use a dispenser to add 15.00 mL of ammonium oxalate buffer to the ET. Make sure that the sample is thoroughly wetted. During the addition, wash sides of the tube and wet the sample. Shaking, swirling, or stirring may be required to wet organic samples. Allow sample to stand for at least 30 min. Cover samples with black plastic bag to exclude light.

7.4 Secure reservoir tube (RT) to top of ET tube. Set extractor for 30-min extraction rate and extract until the ammonium oxalate buffer solution is at a 0.5 to 1.0-cm height above sample. Turn off extractor.

7.5 Add 35 mL of ammonium oxalate buffer to the RT.

7.6 Cover the extractor with a black plastic bag to exclude light. Adjust the extraction rate for a 12-h extraction.

7.7 After the extraction, shut off the extractor. Carefully remove $TET_{Oxalate}$. Leave the rubber tubing on the ET.

7.8 Weigh each syringe containing ammonium oxalate extract to the nearest mg.

7.9 Mix extract in each $TET_{Oxalate}$ by manually shaking. Fill a disposable tube with extract solution. This solution is reserved for determinations of Fe, Mn, Al, Si, and P. If optical density is to be measured, fill a disposable cuvette with extract solution. Discard excess solution properly. If extracts are not to be determined immediately after collection, then store samples at 4°C.

Determination of Optical Density of Extract

7.10 Place 4 mL of ammonium oxalate extract in disposable cuvette.

7.11 Place 4 mL of ammonium oxalate reagent blank in disposable cuvette.

7.12 On the spectrophotometer, select a 430-nm wavelength. Select normal slit width and height. Refer to manufacturer's manual for operation of the spectrophotometer.

7.13 Use the ammonium oxalate extract reagent blank to zero spectrophotometer.

7.14 Record optical density of ammonium oxalate extract to nearest 0.000 unit.

Dilution of Sample Extracts and Standards

7.15 Dilute ammonium oxalate extracts (1:10) with RODI water. Add 1 part ammonium oxalate sample extract with 9 parts dilution solution. Pipet 0.7 mL of extract and 6.3 mL RODI water. Vortex. Calibration reagent blanks and calibration standards are not diluted.

7.16 Dispense the diluted solutions into test tubes that have been placed in the sample holder of the sample changer.

ICP-AES Set-up and Operation

7.17 Refer to the manufacturer's manual for operation of the ICP-AES. The following parameters are only very general guidelines for instrument conditions for the various analytes.

Parameter	Value
Plasma	
Source Equilibration Delay	20 sec
Plasma Aerosol Type	Wet
Nebulizer Start-up Conditions	Gradual
Plasma	15 L min^{-1}
Auxiliary	0.5 L min^{-1}
Nebulizer	0.85 L min^{-1}
Power	1450 Watts
View Dist	15.0
Plasma View	Radial
Peristaltic Pump	
Sample Flow Rate	2.00 L min^{-1}
Sample Flush Time	35 sec

Wash Parameters	Between samples
Wash Rate	2.00 mL min^{-1}
Wash Time	30 sec

| Background Correction | 2 point (all elements) |

| Read Delay | 2 sec |

| Replicates | 2 |

Nebulizer pressure depends on the type of nebulizer that is being used, i.e., low flow nebulizer requires a higher nebulizer pressure whereas a higher flow nebulizer requires a lower nebulizer pressure. To check for correct nebulizer pressure, aspirate with 1000.0 mg L^{-1} yttrium. Adjust pressure to correct yttrium bullet.

7.18 Analyte data are reported at the following wavelengths.

Element	Wavelength
	----------- nm ---------
Fe	259.94
Al	308.22
Si	251.61
Mn	257.61
P	213.61

7.19 Use the computer and printer to set instrument parameters and to collect and record instrument readings. The instrument readings are programmed in mg L^{-1}.

ICP-AES Calibration and Analysis

7.20 Use a multipoint calibration for ICP-AES analysis of ammonium oxalate extracts. The ICP calibrates the blank first, low standard, medium standard, followed by the high standard. Prepare a quality control (QC) standard with analyte concentration between the high and low calibration standards. The ICP reads the QC after the high standard. If the QC falls within the range set by operator (±10%), the instrument proceeds to analyze the unknowns. If the QC is outside the range, the instrument restandardizes. The QC is analyzed approximately every 12 samples.

7.21 If sample exceeds calibration standard, dilute 1:5 (1 mL of sample extract with 4 mL 0.02 *M* ammonium oxalate extracting solution), followed by a 1:10 (1 mL of 1:5 solution with 9 mL RODI water). This makes for a 1:50 dilution.

7.22 Record analyte readings to the nearest 0.01 unit.

8. Calculations (Al, Fe, Si)

The instrument readings are the analyte concentration (mg L^{-1} Fe, Mn, Al, Si, and P). Use these values to calculate the analyte concentration in percent in the soil for Fe, Al, and Si and mg kg^{-1} for Mn and P as follows:

8.1 Soil Fe, Al, Si (%) = [A x [(B$_1$ − B$_2$)/B$_3$] x C$_1$ x C$_2$ x R x 100]/(E x 1000 x 1000)

where:
A   = Sample extract reading (mg L^{-1})

B_1 = Weight of syringe + extract (g)
B_2 = Tare weight of syringe (g)
B_3 = Density of 0.2 M ammonium oxalate solution at 20° C (1.007 g mL^{-1})
C_1 = Dilution, required
C_2 = Dilution, if performed
R   = Air-dry/oven-dry ratio (procedure 3D1) or field-moist/oven-dry ratio (procedure 3D2)
E   = Sample weight (g)
100 = Conversion factor to 100-g basis
1000 = Factor in denominator (mL L^{-1})
1000 = Factor in denominator (mg g^{-1})

8.2 Soil Mn, P (mg kg^{-1}) = [A x [(B_1 – B_2)/B_3] x C_1 x C_2 x R x 1000]/(E x 1000)

where:
A   = Sample extract reading (mg L^{-1})
B_1 = Weight of syringe + extract (g)
B_2 = Tare weight of syringe (g)
B_3 = Density of 0.2 M ammonium oxalate solution at 20° C (1.007 g mL^{-1})
C_1 = Dilution, required
C_2 = Dilution, if performed
R   = Air-dry/oven-dry ratio (procedure 3D1) or field-moist/oven-dry ratio (procedure 3D2)
1000 = Conversion factor in numerator to kg-basis
1000 = Factor in denominator (mL L^{-1})
E   = Sample weight (g)

9. Report
Report the percent ammonium oxalate extractable Al, Fe, and Si to the nearest 0.01%. Report the concentration of ammonium oxalate extractable Mn and P to the nearest mg kg^{-1} soil. Report the optical density of the ammonium oxalate extract to the nearest 0.01 unit.

10. Precision and Accuracy
Precision and accuracy data are available from the SSL upon request.

10. References
Aguilera, N.H., and M.L. Jackson. 1953. Iron oxide removal from soils and clays. Soil Sci. Soc. Am. Proc. 17:359-364.

Holmgren, G.G.S., R.L. Juve, and R.C. Geschwender. 1977. A mechanically controlled variable rate leaching device. Soil Sci. Am. J. 41:1207-1208.

Jackson, M.L. 1979. Soil chemical analysis-advance course. 2nd ed., 11th Printing. Published by author, Madison, WI.

Mehra, O.P., and M.L. Jackson. 1960. Iron oxide removal from soils and clays by a dithionite-citrate system buffered with sodium bicarbonate. Clays Clay Miner. 7:317-327.

Soil Survey Staff. 1999. Soil taxonomy. A basic system of soil classification for making and interpreting soil surveys. 2nd ed. USDA-NRCS. Govt. Print. Office, Washington DC.

Wada, K. 1989. Allophane and imogolite. p. 1051-1087. *In* J.B. Dixon and S.B. Weed (eds.). Minerals in soil environment. 2nd ed. Soil Sci. Soc. Am. Book Series No. 1. ASA and SSSA, Madison.

Selective Dissolutions (4G)
Ammonium Oxalate (4G2)
Ratios and Estimates Related to Ammonium Oxalate Extraction (4G2b)
Al + ½ Fe (4G2b1)

The ratio using ammonium oxalate extractable Al plus ½ Fe determined in procedure 4G2a1a1-5 are used as taxonomic criterion for andic soil properties. Refer to Soil Survey Staff (1995, 1999) for more detailed information on the application of this ratio.

References

Soil Survey Staff. 1995. Soil survey laboratory information manual. Version No. 1.0. USDA-NRCS. Soil Survey Investigations Report No. 42. U.S. Govt. Print. Office, Washington, DC.

Soil Survey Staff. 1999. Soil taxonomy: A basic system of soil classification for making and interpreting soil surveys. USDA-NRCS Agric. Handb. 436. 2nd ed. U.S. Govt. Print. Office, Washington, DC.

Selective Dissolutions (4G)
Sodium Pyrophosphate Extraction (4G3)
Atomic Absorption Spectrophotometry (4G3a)
Aluminum, Iron, and Manganese (4G3a1-3)
Air-Dry or Field-Moist, <2 mm (4G3a1-3a-b1)
Acid Digestion (4G3b)
$K_2Cr_2O_7$ + (H_2SO_4 + H_3PO_4 Digestion)(4G3b1)
CO_2 Evolution, Gravimetric (4G3b1a)
Organic Carbon (4G3b1a1)
Air-Dry or Field-Moist, <2 mm (4G3b1a1a-b1)

1. Application

Sodium pyrophosphate (0.1 M $Na_4P_2O_7$) is used as a selective dissolution extractant for organically complexed Fe and Al (Wada, 1989). The $Na_4P_2O_7$ solution is a poor extractant for allophane, imogolite, amorphous aluminosilicates, and noncrystalline hydrous oxides of Fe and Al. The $Na_4P_2O_7$ solution does not extract opal, crystalline silicates, layer silicates, and crystalline hydrous oxides of Fe and Al (Wada, 1989). Sodium pyrophosphate extractable organic C, Fe, and Al were former criteria for spodic placement in soil taxonomy (Soil Survey Staff, 1975).

2. Summary of Method

The soil sample is mixed with 0.1 M $Na_4P_2O_7$ and shaken overnight. The solution is then allowed to settle overnight before centrifuging and filtering to obtain a clear extract. The analytes (Al, Fe, Mn) are measured by an atomic absorption spectrophotometer (AAS). The data are automatically recorded by a computer and printer. The AAS converts absorption to analyte concentration. Percent sodium pyrophosphate extractable Al, Fe, and Mn are reported in procedures 4G3a1-3, respectively. The organic C in the sodium pyrophosphate extract is wet oxidized in a fume hood and gravimetrically measured in procedure 4G3b1a1.

3. Interferences

There are four types of interferences (matrix, spectral, chemical, and ionization) in the AA analyses of these elements. These interferences vary in importance, depending upon the particular analyte selected.

There are several problems with this procedure, especially the peptization and dispersion of microcrystalline iron oxide by pyrophosphate (Jeanroy and Guilet, 1981). The quantity of Fe extracted with pyrophosphate decreases with increasing centrifugation (McKeague and Schuppli, 1982); therefore uniform high-speed centrifugation or micropore filtration treatments are required (Schuppli et al., 1983; Loveland and Digby, 1984). Sodium pyrophosphate extraction works best at pH 10 (Loeppert and Inskeep, 1996). The concentration of $Na_4P_2O_7$ solution must be close to 0.1 M. Variable amounts of Fe, Al, Mn, and organic C may be extracted by varying the pyrophosphate concentration.

4. Safety

Wear protective clothing and eye protection. When preparing reagents, exercise special care. Restrict the use of concentrated HCl to a fume hood. Many metal salts are extremely toxic and may be fatal if ingested. Thoroughly wash hands after handling these metal salts.

Follow standard laboratory procedures when handling compressed gases. Gas cylinders should be chained or bolted in an upright position. Acetylene is highly flammable. Avoid open flames and sparks. Standard laboratory equipment includes fire blankets and extinguishers for use when necessary. Follow the manufacturer's safety precautions when using the atomic absorption spectrophotometer (AAS).

5. Equipment

5.1 Electronic balance, ±0.1-mg sensitivity
5.2 Mechanical reciprocating shaker, 200 oscillations min^{-1}, 1 ½ in strokes, Eberbach 6000, Eberbach Corp., Ann Arbor, MI
5.3 Atomic absorption spectrophotometer (AAS), double-beam optical system, AAnalyst 300, Perkin-Elmer Corp., Norwalk, CT
5.4 Autosampler, AS-90, Perkin-Elmer Corp., Norwalk, CT
5.5 Peristaltic pump
5.6 Single-stage regulators, acetylene and nitrous oxide
5.7 Centrifuge, Centra, GP-8, Thermo IEC, Needham Heights, MA
5.8 Digital diluter/dispenser, with syringes 10000 and 1000 µL, gas tight, MicroLab 500, Hamilton Co., Reno, NV
5.9 Dispenser, 40 mL
5.10 Test tubes, 15-mL, 16 mm x 100, for sample dilution and sample changer
5.11 Containers, polypropylene
5.12 Volumetrics, Class A, 100, 250, and 1000-mL
5.13 Centrifuge tubes, 50-mL
5.14 Funnel, 60° angle, long stem, 50-mm diameter
5.15 Filter paper, Whatman 42, 150 mm
5.16 Absorption bulb, Nesbitt with stopper
5.17 Absorption bulb, Stetser-Norton
5.18 Flask, boiling, round bottom, short neck
5.19 Condenser, Allihn
5.20 Funnel, separatory, cylindrical, open top, with stopcock
5.21 Tube, drying, Schwartz

6. Reagents

6.1 Reverse osmosis deionized (RODI) water
6.2 Hydrochloric acid (HCl), concentrated, 12 N
6.3 Sodium pyrophosphate solution, 0.1 M. Dissolve 446.05 g of $Na_4P_2O_7 \cdot 10H_2O$ in 10 L of RODI water. pH solution to 10.0 with either HCl or NaOH.
6.4 Primary Fe standard, 1000 mg L^{-1}. Certified Reference Solution, Fisher Chemical Scientific Co., Fairlawn, N.J.
6.5 Primary Al standard, 1000 mg L^{-1}. Certified Reference Solution, Fisher Chemical Scientific Co., Fairlawn, N.J.
6.6 Primary Mn standard, 1000 mg L^{-1}. Certified Reference Solution, Fisher Chemical Scientific Co., Fairlawn, N.J.
6.7 Mixed Calibration Standards (MCS), Fe, Al, Mn. To six 250-mL volumetrics, add the following amounts of Primary Standards (1000 mg L-1) as follows. The elements are added in the order (Fe, Al, Mn) to avoid element precipitation.

6.7.1 25, 100, and 15 mg L^{-1} Fe, Al, and Mn = 6.25, 25.00, and 3.75 mL Fe, Al, and Mn, respectively.

6.7.2 20, 80, and 10 mg L^{-1} Fe, Al, and Mn = 5.00, 20.00, and 2.50 mL Fe, Al, and Mn, respectively.

6.7.3 10, 40, and 5 mg L^{-1} Fe, Al, and Mn = 2.50, 10.00, and 1.25 mL Fe, Al, and Mn, respectively.

6.7.4 5, 20, and 2.50 mg L^{-1} Fe, Al, and Mn = 1.25, 5.00, and 0.625 mL Fe, Al, and Mn, respectively.

6.7.5 1, 10, and 1.5 mg L^{-1} Fe, Al, and Mn = 0.25, 2.50, 0.375 mL Fe, Al, and Mn, respectively.

6.7.6 0, 0, and 0 mg L^{-1} Fe, Al, and Mn = 0.0, 0.0, and 0.0 mL Fe, Al, and Mn, respectively.

Add 50 mL of 0.1 M Na$_4$P$_2$O$_7$ to each MCS and make to volume with RODI water. Final concentration of MCS is 0.02 M Na$_4$P$_2$O$_7$. Quality control (QC) is the MCS with 10, 40, and 5 mg L^{-1} Fe, Al, and Mn, respectively.

6.8 Potassium dichromate (K$_2$Cr$_2$O$_7$) reagent.
6.9 Potassium iodide solution. Dissolve 100 g of KI in 100 mL of RODI water.
6.10 Silver sulfate, saturate aqueous solution
6.11 Digestion acid mixture: Mix 600 mL of concentrated H$_2$SO$_4$ and 400 mL of 85% H$_3$PO$_4$
6.12 Indicarb or Mikohibite
6.13 Soda lime
6.14 Zinc granules, 300 mesh
6.15 Anhydrone
6.16 Acetylene gas, purity 99.6%
6.17 Nitrous oxide gas, compressed
6.18 Compressed air with water and oil traps

7. Procedure

Extraction of Al, Fe, and Mn

7.1 Weigh 0.5 g <2-mm or fine-grind, air-dry soil to the nearest mg sample and place in a 50-mL centrifuge tube. If sample is moist, weigh enough soil to achieve ≈ 0.5 g of air-dry soil.

7.2 Add 30-mL of 0.1 M Na$_4$P$_2$O$_7$, pH 10.0 solution to centrifuge tube.

7.3 Cap tube and shake briefly by hand to dislodge soil from tube bottom. Place tube in rack.

7.4 Place rack in shaker and shake overnight (12 to 16 h) at 200 oscillations min^{-1} at room temperature (20°C ± 2°C).

7.5 Remove tubes from shaker and manually shake tubes to dislodge any soil from cap. Allow samples to sit overnight.

7.6 Next day centrifuge sample at 4000 rpm for 15 min. The Fe, Mn, and Al are determined from a clear aliquot of solution. Filter if necessary.

Dilution of Sample Extracts and Standards

7.7 No ionization suppressant is required as the Na in the extractant is present in sufficient quantity. Dilute samples 1:5 with RODI water. Samples have a final concentration of 0.02 M $Na_4P_2O_7$.

7.8 Dispense the MCS and diluted sample solutions into test tubes that have been placed in the sample holder of the sample changer.

AAS Set-up and Operation

7.9 The following are only very general guidelines for instrument conditions for the various analytes.

Analyte	Conc. (mg L^{-1})	Wavelength (nm)	Burner Head	Slit (mm)	Fuel/Oxidant
Fe	25.0	248.3	10-cm parallel	0.2	3.0 C_2H_2/15.2 Air
Mn	15.0	279.8	10-cm parallel	0.2	3.0 C_2H_2/15.7 Air
Al	100.0	309.3	5-cm parallel	0.7	7.0 C_2H_2/3.5 N_2O

Typical read delay is 3 s and integration time is 3 s but can vary depending on soil type. Three replicates are averaged for each sample.

7.10 Use the computer and printer to set instrument parameters and to collect and record instrument readings.

7.11 If sample exceeds calibration standard, dilute the sample 1:10 with 0.1 M $Na_4P_2O_7$ and then 1:5 with RODI water.

AAS Calibration

7.12 Each element is analyzed during separate runs on the AAS. Use the calibration reagent blank and calibration standards to calibrate the AAS. Calibrations are linear with calculated intercept.

7.13 Use the QC after every 12th sample. It must pass within 15% to continue. If it fails, recalibrate and reread the QC. The QC is also read at the end of each run.

7.14 If samples are outside the calibration range, dilute

7.15 Record analyte readings to 0.01 unit.

Organic C Determination

7.16 Pipet 100 mL of the extract into a 100-ml flask.

7.17 Evaporate the extract to near dryness using a 50°C water bath and a gentle stream of clean, filtered air.

7.18 Construct the wet combustion apparatus. Refer to Fig. 1 for the apparatus for gravimetric organic C determination.

7.19 Add 1 to 2 g of potassium dichromate.

7.20 Wash the neck of the flask with 3 mL of RODI H_2O and connect to condenser.

7.21 Attach a weighed Nesbitt bulb to the system and open the valve at the top.

7.22 Pour 25 mL of digestion-acid mixture into the funnel. Add the mixture to the flask and immediately close the stopcock. Use the digestion-acid mixture to lubricate the stopcock.

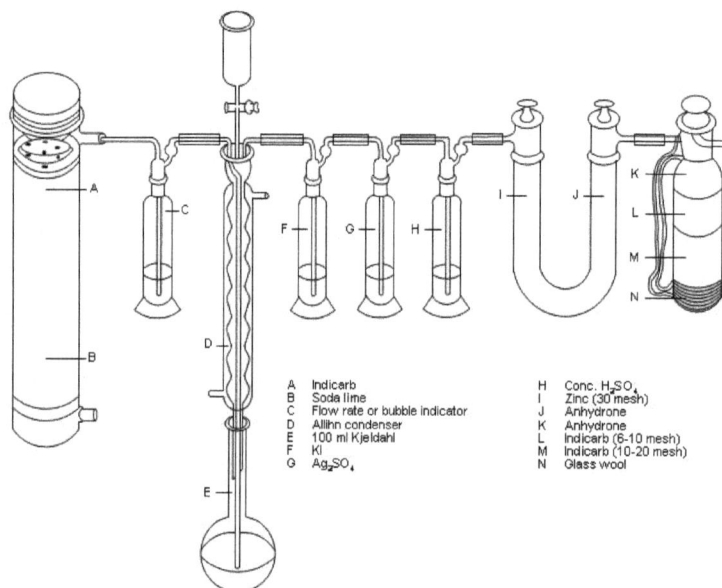

A	Indicarb	H	Conc. H_2SO_4
B	Soda lime	I	Zinc (30 mesh)
C	Flow rate or bubble indicator	J	Anhydrone
D	Allihn condenser	K	Anhydrone
E	100 ml Kjeldahl	L	Indicarb (6-10 mesh)
F	KI	M	Indicarb (10-20 mesh)
G	Ag_2SO_4	N	Glass wool

Figure 3. Apparatus for gravimetric organic carbon determinations of 0.1 M sodium pyrophosphate extracts.

7.23 The tip of the air-delivery tube should be $\approx$ 0.5 cm below the digestion-acid mixture. Adjust the flow of the "carrier stream" to maintain 1 to 2 bubble s^{-1} rate throughout the digestion. Apply suction on the outlet side of the Nesbitt bulb. Gentle air pressure and needle valve on the air-pressure line aids flow-adjustment.

7.24 With a gas flame or a variable power-heating mantle, gently heat the flask until the mixture boils ($\approx$ 3 to 4 min). Continue a gentle boiling for 10 min. Heating is too rapid if white fumes of SO_2 are visible above the second bulb of the reflux condenser.

7.25 Remove the heat and allow to aerate for 10 additional min at a rate of 6 to 8 bubbles s^{-1}.

7.26 Close the stopcock on the Nesbitt bulb, disconnect the bulb from the system, and weigh to the nearest 0.0001g.

8. Calculations

8.1 Soil Fe, Al, Mn (%) = (A x B x C_1 x C_2 x R x 100)/(E x 1000 x 1000)

where:
A = Analyte concentration reading (mg L^{-1})
B = Extract volume (L)
C_1 = Dilution, required
C_2 = Dilution, if performed

R = Air-dry/oven-dry ratio (procedure 3D1) or field-moist/oven-dry ratio (procedure 3D2)
E = Sample weight (g)
100 = Conversion factor to 100-g basis
1000 = Factor in denominator (mg g^{-1})
1000 = Factor in denominator (mL L^{-1})

8.2 Organic C (%) = [(Wt$_F$ - Wt$_I$) x 27.3 x Volume x R]/(Sample Weight (g) x 236.6)

where:
Wt$_F$ = Nesbitt bulb weight after digestion (g)
Wt$_I$ = Nesbitt bulb weight before digestion (g)
Volume = Extract volume digested (mL)
R = Air-dry/oven-dry ratio (procedure 3D1) or field-moist/oven-dry ratio (procedure 3D2)
27.3 = Conversion factor
236.6 = Total extract volume (mL)

9. Report

Report sodium pyrophosphate extractable Fe, Mn, and Al to the nearest 0.1 of a percent.

10. Precision and Accuracy

Precision and accuracy data are available from the SSL upon request.

11. References

Jeanroy, E., and B. Guillet. 1981. The occurrence of suspended ferruginous particles in pyrophosphate extracts of some soil horizons. Geoderma 26:95-105.

Loeppert, R. H., and W.P. Inskeep. 1996. Iron. p. 639-664. *In* D.L. Sparks (ed.) Methods of Soil Analysis. Part 3 – Chemical methods. Soil Sci. Am. Book Series No. 5. ASA and SSSA, Madison, WI.

Loveland, P.J., and P. Digby. 1984. The extraction of Fe and Al by 0.1*M* pyrophosphate solutions: A comparison of some techniques. J. Soil Sci. 35:243-250.

McKeague, J.A., and P.A. Schuppli. 1982. Changes in concentration of iron and aluminum in pyrophosphate extracts of soil and composition of sediment resulting from ultracentrifugation in relation to spodic horizon criteria. Soil Sci. 134:265-270.

Schuppli, P.A. P.A., G.J. Ross, and J.A. McKeague. 1983. The effective removal of suspended materials from pyrophosphate extracts of soils from tropical and temperate regions. Soil Sci. Soc. Am. J. 47:1026-1032.

Soil Survey Staff. 1975. Soil taxonomy: A basic system of soil classification for making and interpreting soil surveys. USDA-SCS Agric. Handb. 436. U.S. Govt. Print. Office, Washington, DC.

Wada, K. 1989. Allophane and imogolite. p. 1051-1087. *In* J.B. Dixon and S.B. Weed (eds.). Minerals in soil environment. 2nd ed. Soil Sci. Am. Book Series No. 1. ASA and SSSA, Madison, WI.

Total Analysis (4H)
Acid Digestion (4H1)
HNO₃ + HCl Digestion (4H1a)
Microwave (4H1a1)
Inductively Coupled Plasma Atomic Emission Spectrophotometer (4H1a1a)
Axial Mode (4H1a1a1)
Ultrasonic Nebulizer (4H1a1a1a)
Silver, Arsenic, Barium, Beryllium, Cadmium, Cobalt, Chromium, Copper, Manganese, Molybdenum, Nickel, Phosphorus, Lead, Antimony, Tin, Strontium, Thallium, Vanadium, Tungsten, and Zinc (4H1a1a1a1-20)
Air-Dry, <2 mm (4H1a1a1a1-20a1)

1. Application

The term of trace elements is widely applied to a variety of elements that are generally present in plants, soils, and water in low concentrations or what is termed background levels. Knowledge of these levels is important in understanding the consequences of increasing levels of trace elements in ecosystems (Tiller, 1989; Holmgren et al., 1992). These elements may become elevated in concentration due to natural (e.g., magmatic activity, mineral weathering, translocation through the soil or landscape) or through human-induced activities (e.g., pesticides, mining, smelting, manufacturing). The relative reactivity or bioavailability of these elements in soils is governed by a variety of chemical factors such as pH, redox potential, organic concentrations, and oxides (Pierzynski and Schwab, 1993; Gambrell, 1994; Keller and Vedy, 1994; Burt et al., 2002). Uses of elemental data in soil survey applications are broad and diverse, ranging from understanding natural (Wilcke and Amelung, 1996; Jersak et al., 1997) to human-induced distributions (Wilcke et al., 1998). Knowledge of the elemental amounts and distribution in soils and their relationships with other soil properties can enhance the understanding of the fate and transport of anthropogenic elements, thereby expanding the utility and application of soil survey knowledge in areas of environmental concern such as urban, mine spoil reclamation, smelter emissions, and agricultural waste applications (Burt et al., 2003).

2. Summary of Method

The approach of this digestion methodology is to maximize the extractable concentration of elements in digested soils while minimizing the matrix interferences such as found in digestion procedures that use HF acid. This method (4H1a1) follows EPA Method 3051A. A 500-mg <2-mm soil separate which has been air-dried and ground to < 200 mesh (75 μm) is weighed into a 100-ml Teflon (PFA) sample digestion vessel. To the vessel, 9.0 mL HNO₃ and 3.0 mL HCl are added. The vessel is inserted into a protection shield and covered, and placed into a rotor with temperature control. Following microwave digestion, the rotor and samples are cooled, and digestate quantitatively transferred into a 50-ml glass volumetric high purity reverse osmosis deionized water. The volumetrics are allowed to stand overnight, filled to volume, and samples transferred into appropriate acid-washed polypropylene containers for analysis. The concentration of Ag, As, Ba, Be, Cd, Co, Cr, Cu, Mn, Mo, Ni, P, Pb, Sb, Sn, Sr, Tl, V, W, and Zn are determined using an inductively coupled plasma atomic emission spectrophotometer (ICP-AES) in axial mode by procedures 4H1a1a1a1-20, respectively. Mercury is analyzed by a cold-vapor atomic absorption spectrophotometer (CVASS) (4H1a1c1), and As and Se are determined by flow through hydride-generation and atomic absorption spectrophotometer (HGAAS) (4H1a1b1a1-2), respectively.

3. Interferences

Organic constituents may contain metals and are difficult to digest if present in high concentrations. Certain elements are subject to volatile losses during digestion and transfer. Certain soil minerals (e.g., quartz, feldspars) are not soluble in HNO₃ + HCl.

Spectral and matrix interferences exist. Interferences are corrected or minimized by using both an internal standard and inter-elemental correction factors. Also, careful selection of specific wavelengths for data reporting is important. Background corrections are made by ICP software. Samples and standards are matrix-matched to help reduce interferences.

4. Safety
Wear protective clothing and eye protection. When preparing reagents, exercise special care. Restrict the use of concentrated acids to the fume hood. Wash hands thoroughly after handling reagents. Filling the digestion vessel to greater than 25 percent of the free volume or adding organic reagents or oxidizing agents to the cup may result in explosion of the digestion microwave system.

5. Equipment
5.1 Electronic balance, ±1.0-mg sensitivity
5.2 Pipet(s) capable of delivering 3 and 9 mL, Omnifit Corp. manufacturers variable volume, (10-ml maximum) pipettes suitable for HNO_3 and HCl delivery from 2.5 L bottles
5.3 Volumetric flasks, class A glass, 50 mL
5.4 Polypropylene bottles, 60 mL, with cap
5.5 Electronic balance, (±0.1 mg sensitivity)
5.6 Microwave oven, CEM Mars 5, 14 position-HP500 Plus vessel and rotor (vessels composed of PFA, sleeves composed of advanced composite)
5.7 Volumetrics, 500, 250, and 50-mL class A glass
5.8 Containers, 500-mL, polypropylene, with screw caps
5.9 Pipettes, electronic digital, 250 µL and 10 mL, Rainin Instrument Co., Woburn, MA
5.10 Inductively coupled plasma atomic emission spectrophotometer (ICP-AES), Perkin-Elmer Optima 3300 Dual View (DV), Perkin-Elmer Corp., Norwalk, CT
5.11 RF generator, floor mounted power unit, 45 MHz free running, Perkin-Elmer Corp., Norwalk, CT.
5.12 Computer, with WinLab software ver. 4.1, Perkin-Elmer Corp., Norwalk, CT, and printer
5.13 Recirculating chiller, Neslab, CFT Series
5.14 Compressed gasses, argon (minimum purity = 99.996%) and nitrogen (minimum purity = 99.999%)
5.15 Autosampler, AS-90, Perkin-Elmer Corp., Norwalk, CT.
5.16 Quartz torch, Part No. N069-1662; alumina injector (2.0 mm id), Part No. N069-5362
5.17 Ultrasonic nebulizer, Model U-5000AT+ , CETAC Corp., Omaha, NE
5.18 Peristaltic pump (for automatic injection of internal standard)

6. Reagents
6.1
6.2
6.3
6.4 Reverse osmosis deionized (RODI) water, ASTM Type I grade of reagent water
6.5 Concentrated hydrochloric acid (HCl), 12 N, trace pure grade
6.6 Concentrated nitric acid (HNO_3), 16 N, trace pure grade
6.7 Primary standards: 1000 mg L^{-1}, from High Purity Standards, Charleston, SC. Single elemental standards are manufactured in dilute HNO_3, HNO_3 + HF, or H_2O

7. Procedure

Microwave Acid Digestion

7.1 About 500 mg of fine-earth (<2-mm) or a specific particle size separate ground to <200-mesh (75 μm) is weighed to the nearest 0.1 mg in a 100-mL digestion vessel.

7.2 Note: If sample is principally composed of organic materials (organic C > 15%), perform a preliminary digestion in the muffle furnace in an digestion crucible: 250°C for 15 min, 450°C for 15 min, followed by 550°C for 1 h.

7.3 Pipet 9.0 mL HNO_3 and 3.0 mL HCl into the sample and allow to completely wet. Add acids in the fume hood. Allow acids to react and vent in uncovered vessels for about 30 min.

7.4 Place covered vessels in protective sleeve, cover and place into rotor.

7.5 Place digestion rotor in the microwave oven and insert the temperature probe into the reference vessel. Attach the probe cable into the fitting in the top of the microwave. Connect the pressure monitor to the vessel.

7.6 Microwave settings are as follows:

1200 watts at 100% power for 5.5 min until 175°C

Hold at 175°C for 4.5 min

Cool for 5 min

7.7 After cooling, disconnect temperature probe and pressure sensor from microwave.

7.8 Remove rotor from oven, and place in fume hood.

7.9 Open each vessel carefully and then quantitatively transfer contents of vessel to a 50-mL volumetric flask with RODI water.

7.10 Cap flask and mix well by inverting. Allow to stand overnight. Finish filling to volume with RODI water.

7.11 Decant contents into a labeled 60-mL polypropylene container.

7.12 Prepare working standards of a blank, reference soil sample from the SSL repository, NIST or other standard reference material, and blank by the same digestion method. Run two of these standards or blank with each set of 14 samples.

ICP-AES Calibration Standards, Set-Up, and Operation

7.13 A primary mixed calibration standard (PMCS) is prepared from the respective primary elemental standards (1000 mg L^{-1}) to a 500-mL final volume. From this PMCS, three working calibration standards (WCS) are prepared. In addition, a single element standard for Tl (STl_1) is prepared separately from a 1000 μg L^{-1} stock standard to a 50-mL final volume. Also, prior to diluting to volume, add 90 mL HNO_3 and 30 mL HCl to the PMCS and 9 mL HNO_3 and 3

326

mL HCl to the STl₁. Use RODI water to dilute to final volume for PMCS and STl₁. Invert to mix thoroughly. Store in polyethylene container in refrigerator. Make fresh on a routine basis. The amount of the primary standards (1000 mg L^{-1}) to make the PMCS, amount of the 1000 μg L^{-1} Tl stock standard to make the STl₁ and the final elemental concentrations of the PMCS and STl₁ are as follows:

Element	Concentration	Primary Standard Required
	-μg L^{-1}-	---- mL ----
As	2,000	1
Ni	4,000	2
P	60,000	30
Cr	4,000	2
Mn	40,000	20
Cu	20,000	10
Zn	20,000	10
Cd	2,000	1
Pb	4,000	2
Co	4,000	2
Ag	400	0.2
Ba	120,000	60
Be	600	0.3
Sb	400	0.2
Sr	60,000	30
Mo	400	0.2
V	20,000	10
Sn	8,000	4
Tl	500	25
W	2,000	1

7.14 The WCS are made from dilution of the PMCS with the exception of single element Tl (STl₂), which is made up separately. The three WCS (Low, Medium, and High) require 0.625, 6.25, and 62.5 ml PMCS diluted to 250-mL final volume, respectively. Also, prior to diluting to volume, add 44.89, 43.88, and 33.75 mL HNO₃ and 14.96, 14.63, and 11.25 mL HCl to the WCS (Low, Medium, High, respectively). The STl₂ requires 0.5, 5, and 50 mL of the STl₁ diluted to a 50-mL final volume. Also, prior to diluting to volume, add 8.91 and 8.1 mL HNO₃ and 2.97 and 2.7 mL HCl for the Low and Medium STl₂, respectively. Use RODI water to dilute to final volume for WCS and STl₂. Invert to mix thoroughly. Store in polyethylene container in a refrigerator. Make fresh on a routine basis. The elemental concentrations of the Low, Medium, and High WCS and the STl₂ are as follows:

Element	Concentration		
	Low	Medium	High
	μg L^{-1}		
Ni	10	100	1000
P	150	1500	15000
Cr	10	100	1000
Mn	100	1000	10000
Cu	50	500	5000
Zn	50	500	5000
Cd	5	50	500
Pb	10	100	1000
Co	10	100	1000
Ag	1	10	100
Ba	300	300	3000
Be	1.5	15	150
Sb	1.0	10	100
Sr	150	1500	15,000
Mo	1	10	100
V	50	500	5000
Sn	20	200	2000
Tl	5	50	500
W	5	50	500
As	5	50	500

7.15 Single element primary standards (1000 mg L^{-1}, Al, Fe, Mo, V, Mn) are required to create the inter-elemental correction (IEC) factors. These are prepared in the matrix of the digests and are combined into one solution for routine calibration. The single element IEC standards (SEIECS) are required to determine the IEC's. The mixed IEC standard (MIECS) is required for routine calibration. The SEIECS is based on a 50-mL final volume and the MIECS is based a 250-mL final volume. Use RODI water to dilute to final volume for SEIECS and MIECS. Invert to mix thoroughly. Store in polyethylene container in a refrigerator. Make fresh on routine basis. The amount of the primary standard (1000 mg L^{-1}) to make the SEIECS and MIECS solutions and the final elemental concentration of these IEC solutions are as follows:

Element	IEC Solution Concentration	Primary Standard Required	
		MIECS	SEIECS
	--mg L^{-1}--	mL	
Al	100	25.00	5.0
Fe	100	25.00	5.0
Mo	5	1.25	0.25
V	10	2.50	0.50
Mn	10	2.50	0.50

7.16 To MIECS, add 45 and 15 mL HNO_3 and HCl, respectively. To SEIECS, add 9 and 3 mL HNO_3 and HCl, respectively. Use RODI water to dilute to final volume for MIECS and SEIECS.

7.17 The elements chosen for IEC factors are based on established spectral interferences with chosen analyte wavelengths. The SEIECS should initially be prepared in separate 50-mL volumetrics for establishment of IEC factors and then prepared (MIECS) in a single 250-mL volumetric for routine analysis. IEC factors are established via a procedure in the WinLab software in which the amount of interference on the analyte (in $\mu g\ L^{-1}$) is measured for each mg L^{-1} of interferent concentration in the digest.

7.18 A 10 mg L^{-1} Lu internal standard (read at 291.138 nm) is added to the blank, all calibration standards, and samples. It is prepared by adding 5.0 mL Lu primary standard (1000 mg L^{-1}) and 10 ml conc. HNO_3 to 500-mL volumetric flask, and diluting to volume with RODI water.Internal standard is automatically injected via the peristaltic pump and mixing block.

7.19 Use the ICP-AES in axial mode and ultrasonic nebulization to analyze sample. Internal standard is added via an external peristaltic pump at 15% pump speed using 0.44 mm id. pump tubing. Internal standard and samples or standards are mixed via a mixing block and coil prior to entering the ultrasonic nebulizer. No initial dilutions of samples are necessary prior to analysis.Perform instrument checks (Hg alignment; BEC and %RSD of 1 mg L^{-1} Mn solution) prior to analysis as discussed in operation manual of instrument. Check instrument alignment and gas pressures to obtain optimum readings with maximum signal to noise ratio.

7.20 Analyses are generally performed at two or more wavelengths for each element. The selected wavelengths are as follows: (reported wavelength listed first and in boldface):

Element	Wavelength
	----------- nm ---------
Al	**237.312** 308.215
Fe	**302.107** 238.203
Mn	**260.570** 203.844
P	**178.221**, 213.620
Cr	**267.710**, 205.558
Cu	**324.753**, 327.396
Ni	**232.003**, 231.604,
Zn	**213.857**, 206.197
Cd	**228.802**, 226.501
Pb	**220.353**, 216.998
Co	**228.614**
Sb	**217.582**, 206.833
Sr	**460.733**, 407.771
Ba	**233.525**, 455.507
Be	**313.104**, 313.046
As	**188.979**
Ag	**328.068**, 338.287
Mo	**202.031**, 203.845
V	**292.402**, 310.230
Sn	**189.927**, 235.485
W	**207.912**, 224.876
Tl	**190.801**, 276.787
Lu	**291.138** (Internal Standard)
Al (IEC)	**237.312**
Fe (IEC)	**302.107**
Mo (IEC)	**202.031**
V (IEC)	**292.402**
Mn (IEC)	**260.568**

7.21 Use the blank standard solution to dilute those samples with concentrations greater than the high standard. Rerun all elements and use only the data needed from the diluted analysis.

7.22 Establish detection limits using the blank standard solution. The instrumental detection limits are calculated by using 3 times the standard deviation of 10 readings of the blank. These values establish the lower detection limits for each element. Analyzed values lower than the detection limits are reported as "ND" or non-detected.

7.23 The extract obtained in this procedure (4H1a1) is used in procedure 4H1a1c1 for Hg analysis and in procedures 4H1a1b1a1-2 for As and Se analysis, respectively.

8. Calculations

The calculation of mg kg^{-1} of an element in the soil from µg L^{-1} in solution is as follows:

Analyte concentration in soil (mg kg^{-1}) = [A x B x C x R x 1000]/E x 1000

A   = Sample extract reading (µg L^{-1})
B   = Extract volume (L)
C   = Dilution, if performed
R   = Air-dry/oven-dry ratio (procedure 3D1)
1000 = Conversion factor in numerator to kg-basis
E   = Sample weight (g)
1000 = Factor in denominator (µg mg^{-1})

9. Report

Analyses are generally performed at two or more wavelengths for each element, with the one selected wavelength for reporting purposes. The particle-size fraction digested needs to be identified with each sample. Data are reported to the nearest 0.01 mg kg^{-1}.

10. Precision and Accuracy

Precision and accuracy data are available from the SSL upon request.

10. References

Burt, R., M.A. Wilson, T.J. Keck, B.D. Dougherty, D.E. Strom, and J.A. Lindhal. 2002. Trace element speciation in selected smelter-contaminated soils in Anaconda and Deer Lodge Valley, Montana, USA. Adv. Environ. Res. 8:51-67.

Burt, R., M.A. Wilson, M.D. Mays, and C.W. Lee. 2003. Major and trace elements of selected pedons in the USA. J. Environ. Qual. 32:2109-2121.

Gambrell, R.P. 1994. Trace and toxic metals in wetlands–A review. J. Environ. Qual. 23:883-891.

Holmgren, G.G.S., M.W. Meyer, R.L. Chaney, and R.B. Daniels. 1993. Cadmium, lead, zinc, copper, and nickel in agricultural soils of the United States of America. J. Environ. Qual. 22:335-348.

Jersak, J., R. Amundson, and G. Brimhall, Jr. 1997. Trace metal geochemistry in Spodosols of the Northeastern United States. J. Environ. Qual. 26:551-521.

Keller, C., and J.C. Vedy. 1994. Distribution of copper and cadmium fractions in two forest soils. J. Environ. Qual. 23:987-999.

Pierzynski, G.M., and A. P. Schwab. 1993. Bioavailability of zinc, cadmium, and lead in a metal-contaminated alluvial soil. J. Environ. Qual. 22:247-254.

Tiller, K.G. 1989. Heavy metals in soils and their environmental significance. *In* B.A. Stewart (ed.) Adv. Soil Sci. 9:113-142.

Wilcke, W., and W. Amelung. 1996. Small-scale heterogeneity of aluminum and heavy metals in aggregates along a climatic transect. Soil Sci. Soc. Am. Proc. 60:1490-1495.

Wilcke, W., S. Muller, N. Kanchanakool, and W. Zech. 1998. Urban soil contamination in Bangkok: Heavy metal and aluminum partitioning in topsoils. Geoderma 86:211-228.

Total Analysis (4H)
Acid Digestion (4H1)
HNO₃ + HCl Digestion (4H1a)
Microwave (4H1a1)
HCl Digestion (4H1a1b)
Water Bath (4H1a1b1)
Flow Through Hydride-Generation and Atomic Absorption Spectrophotometer (4H1a1b1)
Arsenic and Selenium (4H1a1b1a1-2)
Air-Dry, <2 mm (4H1a1b1a1-2a1)

1. Application

Arsenic is an extremely toxic element that is occurs in both organic and inorganic forms in soils. It is typically found in low concentrations, but has been widely applied to soils as a component in pesticides and herbicides and also via industrial pollution and smelting operations. The element is used in drugs, soaps, dyes, and metals, though 90% of industrial As in the U.S. is used in wood preservatives (Pinsker, 2001). Concerns exist for both short-term (acute) and long-term (chronic) soil exposure. Primary route for exposure is via soil ingestion or inhalation of air-borne particles. Data from As measurements in groundwater by U.S. Geological Survey and Environmental Protection Agency has suggested that most As in groundwater is related to natural sources (Ryker, 2001), i.e., from mineral dissolution from minerals in geologic formations and soils. For example, As released via pyrite oxidation is in part responsible for groundwater As levels in Bangladesh ranging between 50 and 2,500 ug/L in many wells. Soil applied As is generally immobile, with soil chemistry similar to phosphorus. The element occurs as arsenate (As^{5+}) and arsenite (As^{3+}), and in soils, is in the form of the oxyanion, AsO_4^{3-}. The weathering of limestone and biological accumulation of the element by aquatic organisms is responsible for the high levels in wetland soils of Florida (Chen et. al., 2002)

Selenium is a naturally occurring element in rocks, but is especially concentrated in certain geologic formations, such as Mancos Shale in Colorado and Wyoming and in the shales of the Moreno and Kreyenhagen Formations of California (Martens and Suarez, 1997). Selenium occurs in four species (related to valance states): selenate (Se^{6+}), selenite (Se^{4+}), elemental Se (Se^0), and selenide (Se^{2-}). The bioavailability and toxicity is related to speciation. The oxidized species are more commonly found in soils and water. The element is important due to both deficiency (forages for animals) and toxicity (bioaccumulation) concerns (Huang and Fujii, 1996).

2. Summary of Method

A soil sample is digested with HNO₃ and HCl in a microwave oven (procedure 4H1a1). Following extraction, samples are diluted with water to a final 50-mL volume. A 6-mL aliquot of the digestate is combined with 6 mL concentrated H₂SO₄ and heated at 180 C for 5 min in the microwave oven to eliminate the HNO₃. Then, the extract is combined with 14 mL of water and 20 mL of concentrated HCl and boiled for 30 min. Sample extracts are allowed to cool. Potassium iodide is added as a pre-reduction step for As analysis, with a final concentration of 1% in analysis solutions. Solutions are allowed to stand 1 h before analyzed for total As and/or Se, using flow through hydride-generation and atomic absorption spectrophotometer (HGAAS). Under acidic conditions, sodium borohydride (NaBH₄) reduces As and Se to form gaseous products that can be detected by atomic adsorption. For example:

$$3NaBH_4 + 4H_2SeO_3 \rightarrow 4H_2Se_{(g)} + 3H_3BO_4 + 3NaOH$$

The data are automatically recorded by a computer and printer. The As and Se concentrations are reported as mg kg^{-1} in the soil by procedure 4H1a1b1a1-2, respectively.

3. Interferences

Oxidizing acids (e.g., nitric) can produce interferences, and inter-element interferences (e.g., Cu, Sn, Ni, Fe, Cr, Pb, Co) can affect determinations. Even small amounts of nitric acid produce suppressed, erratic absorbance signals and low recoveries, more so for As than Se. This interference can be effectively eliminated with either (1) addition of urea before the potassium iodide pre-reduction step, or (2) boiling the samples with H_2SO_4. Alternatively, if samples are boiled in HCl, the quantity of nitric acid can be sufficiently reduced to eliminate the need for urea. Hydride-forming elements may exist in more than one oxidation-state, affecting the signal. In general, As^{5+} methods produce a signal that may be 20 to 50% of that produced by As^{3+}. There are typically more inter-element interferences for As^{5+} methods than for As^{3+}. Inter-element inteferences can be reduced by using the lowest possible concentration of sodium borohydride. Best results can be obtained for difficult samples containing high concentrations of metals if the sodium borohydride concentration is reduced to 0.3% w/v.

4. Safety

Wear protective clothing (coats, aprons, sleeve guards, and gloves); eye protection (face shields, goggles, or safety glasses); and a breathing filter when handling As and Se solutions. These elements are extremely toxic.

Follow standard laboratory practices when handling compressed gases. Gas cylinders should be chained or bolted in an upright position. Acetylene gas is highly flammable. Avoid open flames and sparks. Standard laboratory equipment includes fire blankets and extinguishers for use when necessary. Follow the manufacturer's safety precautions when using the AA.

5. Equipment

5.1 Electronic balance, ±0.1-mg sensitivity
5.2 Atomic absorption spectrophotometer (AAS), PE Model Analyst 300, Perkin-Elmer Corp., Norwalk, CT
5.3 System 2 Electrodeless Discharge Lamp (EDL) Power Supply, with lamps for As and Se, Perkin-Elmer Corp., Norwalk, CT
5.4 Autosampler, Model 90A, Perkin-Elmer Corp., Norwalk, CT
5.5 WinLab Software, Ver. 4.1, Perkin-Elmer Corp., Norwalk, CT
5.6 Computer, Dell Optiplex GXM 333 MHz Pentium, Dell Computer Corp., 17 in color monitor
5.7 Printer, Hewlett-Packard LaserJet 880A
5.8 Single-stage regulator, acetylene service, part number E11-0-N511A, Air Products and Chemicals, Inc., Box 538, Allentown, PA
5.9 Double-stage regulator, argon service
5.10 Varian Vapor Generation Accessory, Model VGA-77
5.11 Tubes, 50-mL for calibration standards
5.12 Test tubes, 50-mL, Corning Pyrex, for sample digestion
5.13 Test tubes, 25-mL, 16 mm x 100, for sample dilution and autosampler
5.14 100, 250, and 1000-mL volumetrics, class A.
5.15 Containers, polypropylene and glass
5.16 Water bath, 95°C-capability

6. Reagents

6.1 Reverse osmosis deionized (RODI) water, ASTM Type I grade of reagent water
6.2 Sulfuric acid (H_2SO_4), 36 M, trace metal purity
6.3 Sodium hydroxide ($NaOH \cdot 3H_20$)

332

6.4 Sodium borohydride (NaBH$_4$) pellets

6.5 Hydrochloric acid (HCl), 12 M, trace metal purity, assay 35-38%

6.6 Acetylene gas, purity 99.6%.

6.7 Compressed air with water and oil traps

6.8 Compressed Argon

6.9 Primary standards, 1000 mg L^{-1} As and 1000 mg L^{-1} Se, High Purity Standards, Inc.

6.10 Primary mixed working standard (PMWS), 1000 µg L^{-1} As and Se: Add 500 mL RODI water to a 1-L volumetric. Add 1 mL of 1000 mg L^{-1} As, 1 mL of 1000 mg L^{-1} Se, and 10 mL concentrated HCl . Fill to volume with RODI water. Invert to mix thoroughly. Final concentration is ≈ 1% HCl. Store in polyethylene container in a refrigerator. Make fresh weekly.

6.11 Mixed Calibration Standards (MCS), 150, 100, 75, 50, 25, 12, and 0 µg L^{-1}: To seven 250-mL volumetrics, add PMWS and concentrated HCl as follows:

6.11.1 150 µg L^{-1} = 37.5 mL PMWS + 106 mL HCl

6.11.2 100 µg L^{-1} = 25.0 mL PMWS + 113 mL HCl

6.11.3 75 µg L^{-1} = 18.75 mL PMWS + 116 mL HCl

6.11.4 50 µg L^{-1} = 12.5 mL PMWS + 119 mL HCl

6.11.5 25 µg L^{-1} = 6.25 mL PMWS + 122 mL HCl

6.11.6 12 µg L^{-1} = 3.0 mL PMWS + 124 mL HCl

6.11.7 0 µg L^{-1} = 0 mL PMWS + 125 mL HCl

6.12 Bring to volume with RODI water and invert to mix thoroughly. Store in polyethylene container in a refrigerator. Standards will keep for 2 to 3 days with refrigeration. Quality Control (QC) check is MCS 100 µg L^{-1}. Final concentration of MCS and QC check is ≈ 6 M HCl.

6.13 Acid Carrier, 6 M HCl: Add 200 mL RODI water to a 500-mL volumetric. Carefully add 250 mL concentrated HCl and fill to volume with RODI water. Invert to mix thoroughly.

6.14 Reductant, 0.5% NaOH - 0.3% NaBH$_4$: Add 200 mL RODI water to a 500-mL volumetric. Always stabilize the solution by first adding the NaOH. Add 2.5 g NaOH and mix until dissolved. Add 1.5 g of NaBH$_4$, mix until dissolved, and fill to volume. Invert to mix thoroughly. Degas for 10 min. Make fresh daily.

6.15 Potassium iodide (KI) (10%) – ascorbic acid solution (10%): To a 500-mL volumetric, add 250 mL of RODI water. Dissolve 50 g of KI and 50 g of ascorbic acid and dilute to volume with RODI water. Invert to mix thoroughly. Make fresh daily. (Procedure requires 1-mL per sample.)

6.16 Diluent, 6 M HCl, for samples: Add 400 mL RODI water to a 1-L volumetric. Carefully add 500 mL concentrated HCl and fill to volume with RODI water. Invert to mix thoroughly.

6.17 QC Soil Standards: Loam C (certified reference material, purchased from High purity Standards) and NIST SRM 2710 (National Institute of Standards and Technology, Standard Reference Material), all prepared to < 200 mesh (75 µm).

7. Procedure

Digestion of acid (HNO$_3$ + HCl) extract

7.1 Prepare soil samples, soil standards, blanks, and spikes prepared in procedure 4H1a1 as follows: Pipette 6mL of acid (HNO$_3$ + HCl) extract and 6mL concentrated H$_2$SO$_4$ into a 50-mL Pyrex tube. Place in a glass beaker (4 tubes into each 250-mL beaker; maximum of 3 beakers, i.e., 12 samples, in microwave at one time). Heating program for microwave is as follows:

Parameter	Value
Stage	1
Max. Power (watts)	600
Heating Power (%)	100
Program Ramp (min)	5:00
Pressure (psi)	800
Temperature (°C)	180
Hold (min)	30:00

Place thermocouple (in thermowell) into one sample. Heat microwave until 180° C is reached for 5 minutes. At that time, fuming (loss of HNO_3) should cease. Allow samples to cool and vent in the microwave for 5 minutes prior to removal.

7.2 To the microwave digested soil samples and MCS, pipette 15 mL of RODI water + 20 mL of concentrated HCl. Include one QC soil standard and blank in each sample rack. Soil samples and QC soil standards have a final concentration of ≈ 6 M HCl.

7.3. Digest the open tubes (do not cover) in a sample rack by submerging up to the neck of the tubes in a water bath (95° C). Heat for 30 min and remove to cool.

7.4 Remove tubes, cool, fill to volume and cap. Mix well by inverting. (Note: digested soil samples and QC soil standards can be analyzed the same day, or placed into the refrigerator overnight for subsequent analysis the following day.) For As analysis, proceed to Section 7.5. For Se analysis, proceed to Section 7.7.

7.5 Arsenic (MCS): Pipette 36 mL of digested MCS + 4 mL of KI-ascorbic acid solution into test tubes that have been placed in the sample holder of the sample changer. Allow to stand 1 h before As analysis. Do not allow MCS to stand greater than 2 h before As analysis. For every sample extract rack, do a set of MCS.

7.6 Arsenic (sample extracts): Pipette 9 mL of sample extract + 1 mL of KI-ascorbic acid solution into test tubes that have been placed in the sample holder of the sample changer. Allow to stand 1 h before As analysis. Do not allow samples to stand longer than 2 h before As analysis.

7.7 Selenium (sample extracts and MCS): Pour sample extract and MCS (without KI-ascorbic acid solution) into test tubes that have been placed in the sample holder of the sample changer.

HGAAS Set-up and Operation

7.8 Each element is analyzed separately on the atomic absorption spectrometer. Refer to manufacturer's manual for operation of AAS. Connect EDL power supply to AAS. Allow EDL to warm up 20 to 30 minutes. Follow the manufacturer's operating procedures for AA EDL settings, warm-up, and adjustments to settings. Connect vapor generation accessory to AA. Follow the manufacturer's operating procedures for appropriate gas and liquid flow. Instrumental parameters for each element are as follows:

Element	Wavelength nm	Slit Width	EDL Current mA
Arsenic	193.7	0.7	380
Selenium	196.0	2.0	250

The flame is required only for heating the gas flow tube and is maintained as a low temperature as possible. Use an air/C_2H_2 mixture of 6.5 and 0.2 L min^{-1}.

7.9 For automated analysis (using the autosampler), the analysis is performed using atomic absorption with background correction, time average mode, with a read delay of 48 s, BOC time of 5 s, and read time of 2 s. Rinse for 30 s between samples. Reported values are the average of five replications.

7.10 Use the computer and printer to set instrument parameters and to collect and record instrument readings.

7.11 The instrument readings are programmed to display analyte concentration in μg L^{-1} (ppb).

HGAAS Calibration

7.12 Each element is analyzed during separate runs on the AA. Use the calibration reagent blank and calibration standards to calibrate the AAS. Detection limits are 5 and 10 μg L^{-1} for As and Se, respectively.

7.13 Use the QC standards after every 12th sample. The QC is 100 μg L^{-1} for As and Se, respectively, with $\pm$ 30% rejection criteria. If QC fails after three attempts, recalibrate and reread the QC. The QC is read at the end of each run.

7.14 If samples are outside the calibration range, a 1:5 serial dilution is performed using 6 *M* HCl. The QC soil standards (Loam C and SRM 2710) have As contents of 47±3 and 626±38, mg kg^{-1} soil, respectively, as consensus/certified values. These QC soil standards have automatic dilutions of 1:10 and 1:100, respectively.

Clean-up and Maintenance

7.15 Soak the absorption cell in dilute HNO_3 acid (0.1% w/v) for 30 min, rinse thoroughly with RODI water, and allow to dry.

7.16 If gas/liquid separator and tubing (including autosampler sipper) has been exposed to contamination with KI, pump a freshly prepared 1% sodium thiosulfate solution through the system for 10 min. The sodium thiosulfate solution must be removed by pumping RODI water through the system for 10 min.

7.17 If gas/liquid separator and tubing have not been exposed to contamination with KI, pump RODI water through the system for 10 min.

8. Calculations

Convert extract As and Se (μg L^{-1}) to soil As and Se (mg kg^{-1}) as follows:

Soil As (mg kg^{-1}) = [A x B x C x R x 1000]/E x 1000

where:
A   = Sample extract reading (μg L^{-1})
B   = Extract volume (L) (0.05)
C   = Dilution, if performed
1000 = Conversion factor in numerator to kg-basis
1000 = Factor in denominator (μg mg^{-1})
R   = Air-dry/oven-dry ratio (procedure 3D1)
E   = Sample weight (g) (0.5)

Soil Se (mg kg^{-1}) = [A x B x C x R x 1000]/E x 1000

where:
A   = Sample extract reading (μg L^{-1})
B   = Extract volume (L) (0.05)
C   = Dilution, if performed
1000 = Conversion factor in numerator to kg-basis
1000 = Factor in denominator (μg mg^{-1})
R   = Air-dry/oven-dry ratio (procedure 3D1)
E   = Sample weight (g) (0.5)

9. Report

Report As and Se to the nearest 0.1 mg kg^{-1} soil.

10. Precision and Accuracy

Precision and accuracy data are available from the SSL upon request.

11. References

Anawar, H.M., J. Akai, K.M.G. Mostofa, S. Safiullah, and S.M. Tareq. 2002. Arsenic poisoning in groundwater: Health risk and geochemical sources in Bangladesh. Environ. Int. 27:597-604.

Chen, M., L.Q. Ma, and W.G. Harris. 2002. Arsenic concentrations in Florida surface soils: influence of soil type and properties. Soil Sci. Soc. Am. J. 66:632-640.

Huang, P.M., and R. Fujii. 1996. Selenium and arsenic. p. 793-831. *In* D.L. Sparks (ed.) Methods of soil analysis. Part 3 – Chemical Analysis. Soil Sci. Soc. Am. Book Series No. 5. ASA and SSSA, Madison, WI.

Martens, D.A., and D.L. Suarez. 1997. Selenium speciation of marine shales, alluvial soils, and evaporation basin soils of California. J. Environ. Qual. 26:424-432.

Office of Environmental Health Assessment Services. 1999. Hazards of short-term exposure to arsenic contaminated soil. State of Washington. Olympia, WA.

Pinsker, L.M. 2001. Health hazards: Arsenic. Geotimes. Nov. p. 32-33.

Ryker, S.J. 2001. Mapping arsenic in groundwater: a real need, but a hard problem. Geotimes. Nov. p. 34-36.

Total Analysis (4H)
Acid Digestion (4H1)
HNO3 + HCl Digestion (4H1a)
Microwave (4H1a1)
Cold Vapor Atomic Absorption Spectrophotometer (4H1a1c)
Mercury (4H1a1c1)
Air-Dry, <2-mm (4H1a1c1a1)

1. Application

Mercury is highly toxic to both plants and animals, and enters the food chain primarily through atmospheric deposition (smelting, coal combustion, volcanic activity) and pesticide usage (Pais and Jones, 1997). Due to the absorption of Hg by both organic and inorganic soil components, many studies have been performed which have examined soil-Hg interactions (MacNaughton and James, 1974; Barrow and Cox, 1992; Yin et al., 1996) and ecosystem distributions (Hall et al., 1987; Inacio et al., 1998).

2. Summary of Method

Soil digests (HNO$_3$ + HCl) from procedure 4H1a1 are analyzed for Hg using cold-vapor atomic absorption spectroscopy (CVAAS). This method is based on absorption of radiation at 253.7 nm wavelength by Hg vapor. The digest is mixed with stannous chloride to reduce Hg to the elemental state. Using argon as a carrier gas, the solution is passed over a gas-liquid separator in a closed system to separate the gaseous Hg from solution. The Hg vapor passes through a cell positioned in the light path of an atomic absorption spectrophotometer. Absorbance (peak height) is measured as a function of mercury concentration. Mercury data are reported as μg kg-1 soil by procedure 4H1a1c1.

3. Interferences
Copper, chlorides, and certain volatile organic materials may be interferences.

4. Safety
Soil digests contain acid and must be handled appropriately. Procedure uses a primary Hg standard that is diluted for working standards. Use gloves and avoid skin contact. Gasses exhausting from the Hg analyzer cabinet, prior to passing through the Hg vapor trap may contain Hg vapor. Do not run the instrument unless the exhaust gas is properly scrubbed or removed.

5. Equipment
5.1 Cold-vapor atomic absorption spectrophotometer (CVAAS), CETAC M-6000A Mercury Analyzer, CETAC Corp., Omaha, NE
5.2 Autosampler, CETAC ASX-500 Model 510, CETAC Corp., Omaha, NE
5.3 Autodilutor Accessory, CETAC ADX-500, CETAC Corp., Omaha, NE
5.4 Nafion drying tube, CETAC Corp., Omaha, NE
5.5 Peristaltic Pump, CETAC Corp., Omaha, NE
5.6 Computer, Microsoft Windows 97, CETAC M-6000A Software, CETAC, Corp., Omaha, NE, and printer
5.7 Pipettors, electronic digital, Rainin Instrument Co., Woburn, MA, 2500 μL and 10 mL
5.8 Compressed argon gas
5.9 Calcium sulfate (anhydrous) or equivalent desiccant

6. Reagents
6.1 Reverse osmosis deionized (RODI) water, ASTM Type I grade of reagent water
6.2 Concentrated hydrochloric acid (HCl), 12 N. Use trace pure HCl.
6.3 Concentrated nitric acid (HNO$_3$), 16 N. Use trace pure HNO$_3$.
6.4 Primary standard: 1000 mg L^{-1} Hg, High-Purity Standards, Charleston, SC.
6.5 Stock standard, 500 μg L^{-1}: Add 0.25 mL primary standard to 500 mL volumetric and dilute to mark with RODI water and invert to mix thoroughly.
6.6 Standard Hg calibration solutions or working standards are 4.0, 3.0, 2.5, 1.0, 0.5, and 0.0 ug L^{-1}. To six 50-mL volumetric flasks add as follows:

6.6.1 0.0 ug L^{-1} = 9 mL HNO$_3$ + 3 mL HCl
6.6.2 0.5 ug L^{-1} = 0.05 mL stock standard + 9 mL HNO$_3$ + 3 mL HCl
6.6.3 1.0 ug L^{-1} = 0.1 mL stock standard + 9 mL HNO$_3$ + 3 mL HCl
6.6.4 2.5 ug L^{-1} = 0.25 mL stock standard + 9 mL HNO$_3$ + 3 mL HCl
6.6.5 3.0 ug L^{-1} = 0.3 mL stock standard + 9 mL HNO$_3$ + 3 mL HCl
6.6.6 4.0 ug L^{-1} = 0.4 mL stock standard + 9 mL HNO$_3$ + 3 mL HCl

Dilute each working standard to mark with RODI water and invert to mix thoroughly. These working standards are used for Normal and High Throughput Ranges.

6.7 Stannous chloride, reducing agent 10% stannous chloride solution (SnCl$_2$ in 7% HCl). Add 50 g of stannous chloride and 97.2 mL concentrated HCl to 500 mL volumetric and dilute to mark with RODI water. Invert to mix thoroughly.

6.8 Diluent: Add 90 mL concentrated HNO_3 and 30 mL concentrated HCl to 500 mL volumetric and dilute to mark with RODI water. Invert to mix thoroughly.

6.9 Rinse (5% HNO_3): Add 71.4 mL HNO_3 to 1-L volumetric and dilute to mark with RODI water. Invert to mix thoroughly.

6.10 Potassium permanganate, solid, crystalline, fills safety trap for Hg vapor exhaust.

6.11 Glass wool. Fine glass wool only.

7. Procedure

7.1. Oven radiator temperature must be at 125°C to maintain the actual gas temperature of 50° C. Argon gas carrier must be supplied at 100 psig (6.9 bar). Liquid flow is always set at fixed flow of 4.0 mL min^{-1} (sample) and 0.8 mL min^{-1} (reagent).

7.2. Turn mercury analyzer and mercury lamp for warm-up (90 min) prior to analysis. Ensure integrity of lamp (there is some loss of performance at 13 mA but replace at 15 mA).

7.3. Turn on computer and choose Worksheet Template appropriate to range of analysis. Three ranges of analysis (Highest Sensitivity, Normal, and High Throughput) have been developed on the CETAC Mercury Analyzer. Method parameters for each of these ranges are saved on a different Worksheet Template. General parameters are as follows:

High Throughput Range:
Sampling Times:
        Integration 1 s
        Read Delay 35 s

Auto-Adjust Integration
        Replicates 4

Instrument Control
        Gas Flow 300 mL min^{-1}

Autosampler Setup
        Sip Duration 20 s
        Rinse Time 20 s
        Repeats 1

Sample Matrix: Liquid
Reslope Frequency: 0
Reslope Standard: Standard No. 3 (check standard)
Detection Limit: 0.050 µg L^{-1}
Baseline Correction: 1 point

Normal Range:
Sampling Times:
        Integration 1 s
        Read Delay 50 s

Auto-Adjust Integration
       Replicates 4

Instrument Control
    Gas Flow 85 mL min^{-1}

Autosampler Setup
    Sip Duration 30 s
    Rinse Time 45 s
    Repeats 1

Sample Matrix: Liquid
Reslope Frequency: 0
Reslope Standard: Standard No. 3 (check standard)
Detection Limit: 0.015 µg L^{-1}
Baseline Correction: 1 point

Highest Sensitivity Range:
Sampling Times:
    Integration 1
    Read Delay 50

Auto-Adjust Integration
    Replicates 4

Instrument Control
    Gas Flow 40 mL min^{-1}

Autosampler Setup
    Sip Duration 60 s
    Rinse Time 140 s
    Repeats 1

Sample Matrix: Liquid
Reslope Frequency: 0
Reslope Standard: Standard No. 3 (check standard)
Detection Limit: 0.001 µg L^{-1}
Baseline Correction: 2 point

7.4. Prior to analysis ensure that the gas-liquid separator (GLS) post is fully wetted as follows:

7.4.1 Check the bottle supplying the ASX-500 rinse station is full of 5% HNO_3.

7.4.2 Use quick release mechanism and fully release clamp tension on lower two tube channels of peristaltic pump (drain channels).

7.4.3 Use sample and reagent tubes and pump 5% HNO_3 and stannous chloride reagent, respectively.

7.4.4 With drain pump tubes unclamped, GLS should begin to fill with 5% HNO_3.

7.4.5 Allow GLS to fill until liquid level reaches top of GLS center post or until gas bubble propels a meniscus upward to wet post all along its length, including apex.

7.4.6 Upon wetting, immediately reengage quick-release clamps on drain pump tubes.

7.4.7 Do not let liquid level overflow GLS into Nafion drying tube.

7.4.8 With drain tube clamps properly reengaged and pump running, liquid level normally stops rising and goes back down

7.4.9 Once GLS has emptied, leave pump running (keep liquid flowing)

7.5 Zero analyzer to compensate for baseline offsets (due to microscopic dust buildup on the optics, dirty sample windows, and thermal drift).

7.6 Define Time Profile using Signal Profile Chart (sample time, and 1st and 2nd baseline correction points) as follows:

7.6.1 Current sample time on chart will start at dark green vertical line and stop at light red vertical line.

7.6.2 Using mouse, define sample time by clicking on chart and dragging to right.

7.6.3 Integration times will automatically be recalculated based on new total sample time.

7.6.4 Current 1st baseline correction point will start at light blue vertical line and stop at dark red line.

7.6.5 Use mouse and shift key and change point times by clicking on chart and dragging to right.

7.6.6 To do this, press and hold shift key.

7.6.7 Click with left mouse button on left limit of part of signal to be used on baseline correction point.

7.6.8 Light blue vertical cursor line will appear.

7.6.9 Without releasing shift key or left mouse button, drag mouse to right until portion of signal to be used as baseline correction point is between light blue and dark red cursor lines.

7.6.10 Release shift and mouse key and chart will update baseline times in worksheet.

7.6.11 Current 2nd baseline correction point on chart will start at light green vertical line and stop at purple line.

7.6.12 Use mouse and control key, change point times by clicking on chart and dragging to right.

7.7 Enter sample numbers, final volume, and weights in "Labels".

7.8 Run calibration and analysis in "Analysis".

7.9 After completion of analytical run, run RODI water through sample sipper tube and stannous chloride reagent lines. Pump all lines dry after rinsing.

7.10 Perform shutdown: turn off mercury lamp, argon gas, ASX-500, pump, M-6000 main power, close the M-6000A Software, and turn off the computer.

8. Calculations

Analytical data is reported by the instrument in $\mu g\ mL^{-1}$ in solution. It is converted to $\mu g\ kg^{-1}$ in soil as follows:

$$Hg\ in\ soil\ (\mu g\ kg^{-1}) = [A \times B \times C \times R \times 1000]/E$$

where:
A = Sample extract reading $(\mu g\ L^{-1})$
B = Extract volume (mL)
C = Dilution, if performed
1000 = Conversion factor in numerator to kg-basis
R = Air-dry/oven-dry ratio (procedure 3D1)
E = Sample weight (g)

9. Report
Hg data are reported to the nearest $0.1\ \mu g\ kg^{-1}$.

10. Precision and Accuracy
Precision and accuracy data are available from the SSL upon request.

11. References

Barrow, N.J., and V.C. Cox. 1992. The effects of pH and chloride concentration on mercury sorption: I. By goethite. J. Soil Sci. 43:295-304.

Hall, A., A.C. Duarte, M.T. Caldeeira, and M.F. Lucas, M.F. 1987. Sources and sinks of mercury in the coastal lagoon of Aveiro, Portugal. Sci. Total Environ. 64:75-87.

Inacio, M.M., V. Pereira, and M.S. Pinto. 1998. Mercury contamination in sandy soils surrounding an industrial emission source (Estarreja, Portugal). Geoderma 85:325-339.

MacNaughton, M.G., and R.O. James. 1974. Adsorption of aqueous mercury (II) complexes at the oxide/water interface. J. Colloid Interface Sci. 47:431-440.

Pais, Istvan, and J.B. Jones, Jr. 1997. The handbook of trace elements. St. Lucie Press. Boca Raton, FL. 223 pp.

Yin, Y., H.E. Allen, Y. Li, C.P. Huang, and P.F. Sanders. 1996. Adsorption of mercury(II) by soil: Effects of pH, chloride, and organic matter. J.

Total Analysis (4H)
Acid Digestion (4H1)
HF + HNO₃ + HCl Digestion (4H1b)
Microwave (4H1b1)
Boric Acid (4H1b1a)
Inductively Coupled Plasma Atomic Emission Spectrophotometer (4H1b1a1a1-11)
Radial Mode (4H1b1a1a)
Aluminum, Calcium, Iron, Potassium, Magnesium, Manganese, Sodium, Phosphorus, Silicon, Zirconium, and Titanium (4H1b1a1a1-11)
Air-Dry, <2 mm (4H1b1a1a1-11a1)
Oven-Dry, <2 μm (4H1b1a1a1-11b2)

1. Application
Prior to the development of modern analytical techniques, e.g., X-ray diffraction and thermal analysis, identification of minerals was based on elemental analysis and optical properties (Washington, 1930; Bain and Smith, 1994). Chemical analysis is still essential to determine mineral structural formulas and to identify and quantify specific mineral species through elemental allocation to minerals. Many clay mineral groups are subdivided based on composition.

Analysis of the entire fine-earth (<2-mm) fraction or specific particle-size separates provides information on parent material uniformity, pedon development, and mineral weathering within or between pedons. This interpretation is determined from differences between horizons or pedons in elemental concentrations, elemental ratios such as Si/Al, Si/Al+Fe, or Ti/Zr, or from differences in total elemental concentrations compared to concentrations determined by selective dissolution techniques.

The inherent fertility of a soil derived from its parent material can be examined by determination of the basic cations relative to the Si or Al content. Phosphorus fertility of a soil and potential water quality problems can be better understood by measurements of total P, especially when compared to other P measurements, such as water-soluble or Bray-1 extractable P.

Hydrofluoric acid (HF) is efficient in the digestion and dissolution of silicate minerals for elemental dissolution (Bernas, 1968; Sawhney and Stilwell, 1994). HNO_3 + HCl aids in the digestion of soil components, especially the organic fraction. Procedure 4H1b1a is a digestion of 100 mg of dried clay suspension, the fine-earth (<2-mm) fraction, or other particle size separate with HF + HNO_3 + HCl. Samples are placed in Teflon digestion vessels and heated in a microwave. Elemental concentration of the digestate is determined using an inductively coupled plasma atomic emission spectrometer (ICP-AES). Procedure 4H1b1 follows EPA method 3052.

2. Summary of Method

A 250-mg <2-mm or other particle-size soil separate which has been oven-dried and ground to < 200 mesh (75 µm) is weighed into a 100-ml Teflon (PFA) sample digestion vessel. In addition, dried clay (<0.002 mm) may be used or a clay suspension (procedure 7A1a1) containing approximately 250 mg of clay material is pipeted into a digestion container and dried at 110°C. A equal amount of suspension is pipeted into a tared aluminum-weighing dish and dried at 110°C to obtain a dried sample weight. The P and Na content of the clay fraction is not measurable when the soil is dispersed in sodium hexametaphosphate (procedure 7A1a1).

To the vessel, 9.0 mL HNO_3, 3.0 mL HCl, and 4 mL HF are added. The vessel is inserted into a protection shield and covered, and placed into a rotor with temperature control. Following microwave digestion, the rotor and samples are cooled, and 20 ml of 4.5% boric acid solution is added (4H1b1a). The samples are then covered and heated in the microwave. The digestate then quantitatively transferred onto a 100-ml polypropylene volumetric with boric acid solution to achieve a final boric acid concentration of 2.1%. The volumetrics are allowed to stand overnight, filled to volume, and approximately 60 mL saved for analysis. The concentration of Al, Ca, Fe, K, Mg, Mn, Na, P, Si, Ti, and Zr are determined by ICP-AES by procedures 4H1b1a1a1-11, respectively.

3. Interferences

Insoluble fluorides of various metals may form. Formation of SiF_4 results in gaseous losses of Si, but additions of boric acid retards formation of this molecule as well as dissolves other metal fluorides. Spectral and matrix interferences exist. Careful selection of specific wavelengths for data reporting is important. Background corrections are made by ICP-AES software. Samples and standards are matrix matched to reduce interferences where possible.

4. Safety

Wear protective clothing and eye protection. When preparing reagents, exercise special care. Restrict the use of concentrated acids to the fume hood. Keep HF acid refrigerated and avoid contact with skin of all acids. Wash hands thoroughly after handling reagents. Filling the digestion vessel to greater than 25 percent of the free volume or adding organic reagents or oxidizing agents to the cup may result in explosion of the digestion microwave system.

5. Equipment

5.1 Electronic balance, ±1.0-mg sensitivity

5.2 Pipet(s) capable of delivering 3, 4, and 9 ml. Omnifit Corp. manufacturers variable volume, (10-ml maximum) pipettes suitable for HNO_3 and HCl delivery from 2.5 L bottles.

5.3 Volumetric flasks, nalgene, 100 mL

5.4 Polypropylene bottles, 60 mL, with cap

5.5 Electronic balance, (±0.1 mg sensitivity)

5.6 Microwave, CEM Mars 5, 14 position-HP500 Plus vessel and rotor (vessels composed of PFA, sleeves composed of advanced composite).

5.7 Desiccator

5.8 Disposable aluminum-weighing dishes

5.9 Volumetrics, 500-mL, polypropylene

5.10 Containers, 500-mL, polypropylene, with screw caps

5.11 Pipettes, electronic digital, 2500 μL and 10 mL, Rainin Instrument Co., Woburn, MA

5.12 Inductively coupled plasma atomic emission spectrophotometer (ICP-AES), Perkin-Elmer Optima 3300 Dual View (DV), Perkin-Elmer Corp., Norwalk, CT.

5.13 RF generator, floor mounted power unit, 45 MHz free running, Perkin-Elmer Corp., Norwalk, CT.

5.14 Computer, with WinLab software, ver. 4.1, Perkin-Elmer Corp., Norwalk, CT, and printer

5.15 Recirculating chiller, Neslab, CFT Series

5.16 Compressed gasses, argon (minimum purity = 99.996%) and nitrogen(minimum purity = 99.999%)

5.17 Autosampler, AS-90, Perkin-Elmer Corp., Norwalk, CT.

5.18 Quartz torch, Part No. N069-1662; alumina injector (2.0 mm id), Part No. N069-5362; Ryton

6. Reagents

6.1 Reverse osmosis deionized (RODI) water, ASTM Type I grade of reagent water

6.2 Calcium sulfate (anhydrous) or equivalent desiccant

6.3 Hydrofluoric acid (HF), 48%, low trace metal content

6.4 Concentrated hydrochloric acid (HCl), 12 N, trace pure grade

6.5 Concentrated nitric acid (HNO_3), 16 N, trace pure grade

6.6 Boric acid solution, 4.5 percent. Dissolve 45.0 g low trace metal, granular boric acid (H_3BO_3) in 1000 mL RODI water.

6.7 Boric acid solution, 1.9 percent. Dissolve 19.0 g low trace metal, granular boric acid (H_3BO_3) in 1000 mL RODI water.

6.8 Primary standards: 1000 mg L^{-1}, from High Purity Standards, Charleston, SC. Single elemental standards in dilute HNO_3, HNO_3 + HF, or H_2O

7. Procedure

Microwave Acid Digestion

7.1 Fine-earth (<2-mm) or a specific particle size separate ground to <200-mesh (75 μm) is used. A 250 mg sample is weighed to the nearest 0.1 mg in a 100-ml TFM vessel. If a clay suspension is used, proceed to section 7.2 to 7.5. If using dried specimen, proceed to section 7.6. (Note: samples are typically prepared in sets of 14, as the digestion rotor accommodates that number of samples; If sample is principally composed of organic materials (organic C > 15%), perform a preliminary digestion in the muffle furnace in an digestion crucible: 250°C for 15 min, 450°C for 15 min, followed by 550°C for 1 h.

7.2 Prepare Na-saturated clay as in procedure 7A1a1, (Preparation of Clay Suspension, Sections 7.8 to 7.19). Clay dispersion by this method eliminates quantitative analysis of Na and P in the clay due to dispersion by sodium hexametaphosphate. Digestion of the entire fine earth (<2-mm) fraction or any fraction not derived by dispersion with sodium hexametaphosphate (or other Na and P-containing dispersing agents) can be quantitatively analyzed for Na and P. Dispersion of clays and cleaning of test tubes and dishware should be with RODI water.

7.3 Pipet a known aliquot of clay suspension containing approximately 250 mg clay into a 100-mL TFM vessel. The volume of required suspension depends on the clay concentration in the suspension, but is generally from 6 to 10 mL. More dilute suspensions should be partially evaporated under a fume hood to concentrate the clay prior to transfer to the Teflon container.

7.4 Pipet a duplicate aliquot of suspension (as used in section 7.3) into a tared Al weighing dish, dry at 110°C, cool in a desiccator, and weigh to the nearest 0.1 mg. Use this value as the sample weight in the calculations.

7.5 Dry the Teflon container and clay suspension in an oven for 4 h or until the aqueous portion of the suspension is completely evaporated. Remove from oven and cool on the bench top or in a fume hood. Cooling in a desiccator is not required.

7.6 Pipet 9.0 mL HNO_3 and 3.0 mL HCl into the sample and allow to completely wet and then pipet 4 mL HF into sample. Add acids in the fume hood.

7.7 Place covered Teflon digestion vessels in protective sleeve, cover, and place into rotor.

7.8 Place digestion rotor in the microwave oven and insert the temperature probe into the reference vessel. Attach the probe cable into the fitting in the top of the microwave. Connect the pressure monitor to the vessel.

7.9 Microwave settings are as follows:

1200 Watts, 100% power for 10 min (350 psi) to approximately 180°C
1200 Watts, 70% power for 9.5 min (350 psi), holding at 180°C
Cool (Vent) for 15 min

7.10 After venting, disconnect temperature probe from microwave.

7.11 Remove rotor from oven and place in fume hood.

7.12 Open each vessel carefully, add 20 mL 4.5 percent boric acid (H_3BO_3) solution, then cover vessels and re-digest in microwave at: 1200 Watts, 100% power (350 psi) to 160°C, at 160°C for 10 min. Cool (vent) for 15 min.

7.13 Transfer contents of digestion vessel to a 100-mL nalgene volumetric flask and adjust to near volume with 1.9% boric acid, achieving a final concentration of 2.1 percent H_3BO_3.

7.14 Cap flask and mix well by inverting. Allow to stand overnight to dissolve any metal fluorides. Finish filling to volume with 1.9% boric acid.

7.15 Invert the volumetric flask to mix and decant approximately 60 mL into a labeled polypropylene container.

7.16 Prepare working standards of a blank, reference soil sample from the SSL repository, and a National Institute of Standards and Technology (NIST) standard reference or other reference material by the same digestion method. Run one of these standards with each set of 14 samples.

ICP-AES Calibration Standards, Set-up, and Operation

7.17 Instrument calibration standards for analysis are limited to specific combinations of elements because of chemical incompatibilities of certain elements. Each working standard is used in two concentrations, high and low. The concentrations of elements in the low standards (CALO, ALLO, and SILO) are 50 percent of the concentrations in the high standards (CAHI, ALHI, and SIHI). The amounts of primary standards (1000 mg L^{-1}) to make 500-mL volume of the low and high calibration standards, at the specified concentrations, for ICP-AES analysis are as follows:

7.17.1 CALO is 75, 25, 20, and 10 mg L^{-1} of Ca, K, Mg, and Mn, respectively. To a 500- mL volumetic flask, add 37.5, 12.5, 10.0, and 5.0 mL of the Ca, K, Mg, and Mn primary standards (1000 mg L^{-1}), respectively.

7.17.2 CAHI is 150, 50, 40, and 20 mg L^{-1} of Ca, K, Mg, and Mn, respectively. To a 500-mL volumetric flask, add 75.0, 25.0, 20.0, and 10.0 mL of the Ca, K, Mg, and Mn primary standards (1000 mg L^{-1}), respectively.

7.17.3 ALLO is 100, 75, 5, 5, and 25 mg L^{-1} of Al, Fe, Ti, Zr, and Na, respectively. To a 500-mL volumetric flask, add 50.0, 37.5, 2.5, 2.5, and 12.5 of the Al, Fe, Ti, Zr, and Na primary standards (1000 mg L^{-1}), respectively.

7.17.4 ALHI is 200, 150, 10, 10, and 50 mg L^{-1} of Al, Fe, Ti, Zr, and Na, respectively. To a 500-mL volumetric flask, add 100.0, 75.0, 5.0, 5.0, and 25.0 mL of the Al, Fe, Ti, Zr, and Na primary standards (1000 mg L^{-1}), respectively.

7.17.5 SILO is 225 and 5 mg L^{-1} of Si and P, respectively. To a 500-mL volumetric flask, add 112.5 and 2.5 mL of the Si and P primary standards (1000 mg L^{-1}), respectively.

7.17.6 SIHI is 450 and 10 mg L^{-1} of Si and P, respectivley. To a 500-mL volumetric flask, add 225.0 and 5.0 mL of the Si and P primary standards (1000 mg L^{-1}), respectivley.

7.18 To the calibration standards and a blank, also add the following chemicals: 20.0 mL HF; 45.0 mL HNO$_3$; 15.0 mL HCl; and 10.90 g granular Boric Acid. Make all standards and the blank to a final 500-mL volume with RODI water.

7.19 Use the ICP-AES in radial mode and analyze for the following elements: Fe, Mn, Al, Ca, Mg, Na, K, P, Si, Zr, and Ti. No initial dilutions of samples are necessary prior to analysis. Perform instrument checks (Hg alignment, BEC, %RSD of Mn solution) prior to analysis as discussed in operation manual of instrument. Check instrument alignment and gas pressures to obtain optimum readings with maximum signal to noise ratio.

7.20 Analyses are generally performed at two or more wavelengths for each element. The selected wavelengths are as follows: (reported wavelength listed first and in bold):

Element	Wavelength
	----------- nm ---------
Al	**308.215**, 396.157
Ca	**315.887**, 317.932
Fe	**259.939**, 238.205
K	**766.490**
Mg	**280.271**, 279.075
Mn	**257.610**, 260.570
Na	**589.592**, 588.995
P	**178.221**, 213.620
Si	**212.412**, 251.612
Ti	**334.940**, 368.522
Zr	**339.197**, 343.818

7.21 Use the blank standard solution to dilute those samples with concentrations greater than the high standard. Rerun all elements and use only the data needed from the diluted analysis.

7.22 Establish detection limits using the blank standard solution. These instrumental detection limits are calculated by using 3 times the standard deviation of 10 readings of the blank. These values establish the lower detection limits for each element. Analyzed values lower than the detection limits are set equal to zero.

8. Calculations

8.1 The calculation of mg kg^{-1} of an element in the soil from mg L^{-1} in solution is as follows:

Analyte concentration in soil (mg kg^{-1}) = [A x B x C x R x 1000]/E

A = Sample extract reading (mg L^{-1})
B = Extract volume (L)
C = Dilution, if performed
R = Air-dry/oven-dry ratio (procedure 3D1)
1000 = Conversion factor in numerator to kg-basis
E = Sample weight (g)

8.2 Data are recorded on an elemental basis. Often users request data in an oxide form. The factor for converting from an elemental form to an oxide form is based on the atomic weights of the element and oxygen. An example is as follows:

Atomic weight Si = 28.09
Atomic Weight O = 16.0
Molecular weight SiO$_2$= 60.09

Calculate percent Si in SiO$_2$ as follows:

Si (%) = (28.09/60.09) x 100 = 46.7 %

There is 46.7 percent Si in SiO$_2$. To convert from mg/kg Si to percent Si oxide (SiO$_2$) in the soil, divide by 10,000 to convert from mg kg^{-1} to %, then divide the percent Si by 0.467 or multiply by the inverse of this value. The element, oxide form, and the elemental percent in the oxide form are as follows:

Element	Oxide Form	Elemental %
Si	SiO$_2$	46.7
Al	Al$_2$O$_3$	52.9
Fe	Fe$_2$O$_3$	69.9
Mg	MgO	60.3
Mn	MnO	77.4
K	K$_2$O	83.0
Ti	TiO$_2$	59.9
Ca	CaO	71.5
Zr	ZrO$_2$	74.0
P	P$_2$O$_5$	43.6
Na	Na$_2$O	74.2

9. Report

Analyses are generally performed at two or more wavelengths for each element, with the one selected wavelength for reporting purposes. The particle-size fraction digested needs to be identified with each sample. Data are reported to the nearest mg kg^{-1}.

10. Precision and Accuracy

Precision and accuracy data are available from the SSL upon request.

11. References

Bain, D.C., and B.F.L. Smith. 1994. Chemical analysis. p. 300-332. *In* M.J. Wilson (ed.) Clay mineralogy: Spectroscopic and chemical determinative methods. Chapman and Hall, Inc. London, England.

Bernas, B. 1968. A new method for decomposition and comprehensive analysis of silicates by atomic absorption spectrometry. Anal. Chem. 40:1682-1686.

Sawhney, B.L., and D.E. Stilwell. 1994. Dissolution and elemental analysis of minerals, soils, and environmental samples. p. 49-82. *In* J.E. Amonette and L.W. Zelazny (eds.) Quantitative methods in soil mineralogy. Soil Sci. Soc. Am. Misc. Publ. SSSA, Madison, WI.

Washington, H.S. 1930. The chemical analysis of rocks. 4th ed. John Wiley and Sons, Inc., NY, NY.

Total Analysis (4H)
Dry Combustion (4H2)
Thermal Conductivity Detector (4H2a)
Total Carbon, Nitrogen, and Sulfur (4H2a1-3)
Air-Dry, <2 mm (4H2a1-3a1)

1. Application

Organic Matter and Organic Carbon: Soil organic matter has been defined as the organic fraction of the soil exclusive of undecayed plant and animal residues and has been used synonymously with "humus" (Soil Science Society of America, 1987). However, for laboratory analyses, the soil organic matter generally includes only those organic materials that accompany soil particles through a 2-mm sieve (Nelson and Sommers, 1982). The organic matter content influences many soil properties, e.g., water retention capacity; extractable bases; capacity to supply N, P, and micronutrients; stability of soil aggregates; and soil aeration (Nelson and Sommers, 1996).

Organic C is a major component of soil organic matter. Organic C consists of the cells of microorganisms; plant and animal residues at various stages of decomposition; stable "humus" synthesized from residues; and nearly inert and highly carbonized compounds, e.g., charcoal, graphite, and coal (Nelson and Sommers, 1982). As organic C is the major component of soil organic matter, a measurement of organic C can serve as an indirect determination of organic matter. Organic C determination is either by wet or dry combustion. In the past, the SSL used the wet combustion method, Walkley-Black modified acid-dichromate digestion, FeSO$_4$ titration, automatic titrator (6A1c, method obsolete, Soil Survey Staff, 1996).

Values for organic C are multiplied by the "Van Bemmelen factor" of 1.724 to calculate organic matter. This factor is based on the assumption that organic matter contains 58% organic C. The proportion of organic C in soil organic matter for a range of soils is highly variable. Any constant factor that is selected is only an approximation. Studies have indicated that subsoils have a higher factor than surface soils (Broadbent, 1953). Surface soils rarely have a factor <1.8 and usually range from 1.8 to 2.0. The subsoil factor may average ≈ 2.5. The preference is to report organic C rather than to convert the organic C to organic matter through use of an approximate correction factor.

The SSL also uses a direct determination of soil organic matter. The organic matter is destroyed, after which the loss in weight of the soil is taken as a measure of the organic matter content (5A). The percent organic matter lost on ignition (400°C) can be used in place of organic matter estimates by the Walkley-Black organic C method.

Total Carbon: Total C is the sum of organic and inorganic C. Most of the organic C is associated with the organic matter fraction, and the inorganic C is generally found with carbonate minerals. The organic C in mineral soils generally ranges from 0 to 12% (Nelson and Sommers, 1996).

Total C is quantified by two basic methods, i.e., wet or dry combustion. The SSL uses dry combustion (4H2a1). In total C determinations, all forms of C in a soil are converted to CO_2 followed by a quantification of the evolved CO_2. Total C can be used to estimate the organic C content of a soil. The difference between total and inorganic C is an estimate of the organic C. The inorganic C should be approximately equivalent to carbonate values measured by CO_2 evolution with strong acid (Nelson and Sommers, 1996). In the SSL procedure 4E1a1a1, the amount of carbonate in a soil is determined by treating a sample with HCl followed by manometrically measuring the evolved CO_2. The amount of carbonate is then calculated as a $CaCO_3$ equivalent basis. Organic C defines mineral and organic soils. In soil taxonomy, organic C is also used at lower taxonomic levels, e.g., ustollic and fluventic subgroups (Soil Survey Staff, 1999).

Total Nitrogen: Total N includes organic and inorganic forms. The total N content of the soil may range from <0.02% in subsoils, 2.5% in peats, and 0.06 to 0.5% in surface layers of many cultivated soil (Bremmer and Mulvaney, 1982). The total N data may be used to determine the soil C:N ratio, the soil potential to supply N for plant growth, and the N distribution in the soil profile. The C:N ratio generally ranges between 10 to 12. Variations in the C:N ratio may serve as an indicator of the amount of soil inorganic N. Uncultivated soils usually have higher C:N ratios than do cultivated soils.

Soils with large amounts of illites or vermiculties can "fix" significant amounts of N compared to those soils dominated by smectites or kaolinites (Young and Aldag, 1982; Nommik and Vahtras, 1982). Since the organic C of may soils diminishes with depth while the level of "fixed" N remains constant or increases, the C:N ratio narrows (Young and Aldag, 1982). The potential to "fix" N has important fertility implications as the "fixed" N is slowly available for plant growth.

Two methods of analysis of total N have gained acceptance for the determination of total N in soils. These include the Kjeldahl (1883) method, which is essentially a wet oxidation procedure and the Dumas (1831) method, which is fundamentally a dry oxidation (i.e., combustion) procedure (Bremmer, 1996). The SSL uses the combustion technique for analysis of total N (4H2a2).

Total Sulfur: Organic and inorganic S forms are found in soils, with the organic fraction accounting for >95% of the total S in most soils from humid and semi-humid (Tabatabai, 1996). Mineralization of organic S and its conversion to sulfate by chemical and biological activity may serve as a source of plant-available S. Total S typically ranges from 0.01 to 0.05% in most mineral soils. In organic soils, total S may be >0.05%. The proportion of organic and inorganic S in a soil sample varies widely according to soil type and depth of sampling (Tabatabai, 1996).

In well-drained, well-aerated soils, most of the inorganic S normally occurs as sulfate. In marine tidal flats, other anaerobic marine sediments, and mine spoils, there are usually large amounts of reduced S compounds that oxidize to sulfuric acid upon exposure to the air. In arid regions, significant amounts of inorganic S are found as sulfates such as gypsum and barite (Tabatabai, 1996).

The typical use of total S is an index of the total reserves of this element, which may be converted to plant-available S. The SSL uses the combustion technique for analysis of total S (4H2a3). Extractable sulfate S (SO_4^{2-}-S) is an index of readily plant-available S. Reagents that have been used for measuring SO_4^{2-}-S include water, hot water, ammonium acetate, sodium carbonate and other carbonates, ammonium chloride and other chlorides, potassium phosphate and other phosphate, and ammonium fluoride (Bray-1). Extractable SO_4^{2-}-S does not include the labile fraction of soil organic S that is mineralized during the growing season (Tabatabai, 1996). Extraction reagents for organic S include hydrogen peroxide, sodium bicarbonate, sodium hydroxide, sodium oxalate, sodium peroxide, and sodium pyrophosphate. There are other methods available for determination of S, especially for total S and SO_4^{2-}-S. The investigator may refer to the review by Beaton et al., (1968).

For detailed discussion of the application of total C, N, and S, refer to Soil Survey Staff (1995).

2. Summary of Method

An air-dry (80 mesh, <180 μm) sample is packed in a tin foil, weighed, and analyzed for total C, N, and S by an elemental analyzer (procedures 4H2a1-3, respectively). The elemental analyzer works according to the principle of catalytic tube combustion in an oxygenated CO_2 atmosphere and high temperature. The combustion gases are freed from foreign gases. The desired measuring components (N_2, CO_2, and SO_2) are separated from each other with the help of specific adsorption columns and determined in succession with a thermal conductivity detector, with helium as the flushing and carrier gas. Percent total C, N, and S are reported by methods 4H2a1-3a1, respectively.

3. Interferences

Contamination through body grease or perspiration must be avoided in sample packing. Substance loss after weighing should be avoided by exact folding of the sample into the tin foil. Air in the sample material should be minimized (falsifying the N value) by compressing the sample packing. Insufficient O_2 dosing reduces the catalysts, decreasing their effectiveness and durability. Burnt sample substance that remains in ash finger falsifies the results of subsequent samples. WO_3 is used as sample additive and combustion filling to aid combustion or bind interfering substances (alkaline or earth-alkaline elements, avoid non-volatile sulfates).

4. Safety

Exhaust gas pipes should lead into a ventilated fume hood. Aggressive combustible products should not be analyzed. Before working on electrical connections (adsorption columns) or before changing reaction tubes, the instrument must be cooled down and cooled off. Gloves and safety glasses should be worn at all times during operation and maintenance of instrument.

5. Equipment

5.1 Elemental analyzer with on-line electronic balance ($0.1\pm$ mg sensitivity) and automatic sample feeder, Elementar varioEL and Elementar varioEL III, Elementar Analysensysteme GmbH, Hanau-Germany, and combustibles (Elementar Americas, Inc., Mt. Laurel, NJ; Alpha Resources Inc., Stevensville, MI) as follows:

5.1.1 Quartz ash finger, quartz
5.1.2 Quartz bridge
5.1.3 Combustion tube
5.1.4 Reduction tube
5.1.5 Gas purification (U-tube, GL 18)
5.1.6 Support tube (65 mm)
5.1.7 Protective tube
5.1.8 O_2 lance (150 mm rapid N)
5.1.9 Tin boats (4 x 4 x 11 mm)
5.1.10 Tin foil cups
5.2 Computer, with varioEL software, Elementar Analysensysteme GmbH, Hanau-Germany, and printer

6. Reagents
6.1 Sulfanilic acid, calibration standard, 41.6% C, 4.1% H, 8.1 % N, 27.7% O, and 18.5% S
6.2 Copper sticks
6.3 Corundum balls, high purity, alumina spheres, 3 – 5 mm
6.4 Cerium dioxide, 1 – 2 mm
6.5 Tungsten oxide powder, sample additive
6.6 Tungsten trioxide granulate, combustion tube filling
6.7 Quartz wool
6.8 Silver wool
6.9 Phosphorus pentoxide, Sicapent, Elementar Americas, Inc., Mt. Laurel, NJ
6.10 Helium, carrier gas, 99.996% purity
6.11 Oxygen, combustion gas, 99.995% purity

7. Procedure

Elemental Analyzer Set-up and Operation
7.1 Refer to the manufacturer's manual for operation and maintenance of the elemental analyzer. Conditioning of the elemental analyzer and determination of factor and blank value limit are part of the daily measuring routine. The following are only very general guidelines for instrument parameters for the various analytes in the CNS mode.

Temperature
Furnace 1	1140° C
Furnace 2	850 ° C
Furnace 3	0 ° C
CO_2 Column	85 ° C
SO_2 Column	210 ° C
SO_2 Col. Standby	140 ° C

Timing
Flush	5 s
Oxygen Delay	10 s
Auto Zero Delay	30 s
Integrator Reset Delay	50 s
Peak Anticipation N	70 s
Peak Anticipation C	125 s
Peak Anticipation S	70 s

Integrated Reset Delay for S	60 s

Thresholds
N Peak	3 mV
C Peak	3 mV
S Peak	3 mV

O_2 Dosing
Index 1	150 s
Index 2	1 s
Index 3	90 s
Index 4	120 s
Index 5	180 s

Elemental Analyzer Calibration and Analysis

7.2 A calibration that covers the desired working range of each element is performed. The test substances sulfanilic acid with each given element content and different weights is analyzed. The PC program automatically computes the calibration function (linear, polynomial, or mixed). Calibration will typically remain stable for at least 6 months. Re-calibration is recommended when the daily factor is outside the range of 0.9 to 1.1 or if components that influence the results (e.g., detector or adsorption column) have been exchanged. Changing the desorption temperature of adsorption columns can also require a re-calibration. .

7.3 Add 0.100 g of tungsten oxide in tin foil and tare. An homogenized fine-grind, air-dry soil sample is then packed in this tin foil, weighed (0.100 to 0.05 g), and placed into the carousel of the automatic sample feeder of elemental analyzer. Sample weight is based on visual observation of the sample, related to element content, homogeneity, and combustion behavior of the sample. The sample weight is entered in the PC from an on-line electronic balance via an interface. A quality control (QC) sample is performed at a minimum of every 35 to 40 samples.

8. Calculations

$$C (\%) = C_i \times AD/OD$$

where:
C (%) = C (%), oven-dry basis
C_i = C (%) instrument
AD/OD = Air-dry/oven-dry ratio (procedure 3D1)

$$N (\%) = N_i \times AD/OD$$

where:
N (%) = N (%), oven-dry basis
N_i = N (%) instrument
AD/OD = Air-dry/oven-dry ratio (procedure 3D1)

$$S (\%) = S_i \times AD/OD$$

where:
S (%) = S (%) on oven-dry basis
S_i = S (%) instrument
AD/OD = Air-dry/oven-dry ratio (procedure 3D1)

9. Report
Report total C and S percentage to the nearest 0.01% and total N to the nearest 0.001%.

10. Precision and Accuracy
Precision and accuracy data are available from the SSL upon request.

11. References

Beaton, James D., G.R. Burns, and J. Platou. 1968. Determination of sulphur in soils and plant material. Tech. Bull. No. 14. The Sulfur Inst., Washington, DC.

Bremmer, J.M. 1996. Nitrogen – Total. p. 1085 – 1121. *In* D.L. Sparks (ed.) Methods of soil analysis. Part 3. Chemical methods. No. 5. ASA and SSSA, Madison, WI.

Bremmer, J.M., and C.S. Mulvaney. 1982. Nitrogen - Total. p. 595-624. *In* A.L. Page, R.H. Miller, and D.R. Keeney (eds.) Methods of soil analysis. Part 2. Chemical and microbiological properties. 2nd ed. Agron. Monogr. 9. ASA and SSSA, Madison, WI.

Broadbent, F.E. 1953. The soil organic fraction. Adv. Agron. 5:153-183.

Dumas, J.B.A. 1831. Procedes de l'analyse organique. Ann. Chim. Phys. 247:198-213.

Kjeldahl, J. 1883. Neue Methode zur Bestimmung des Stickstoffs in organischen Korpern. Z. Anal. Chem. 22:366-382.

Nelson, D.W., and L.E. Sommers. 1996. Total carbon, organic carbon, and organic matter. p. 961-1010. *In* D.L. Sparks (ed.) Methods of soil analysis. Part 3. Chemical methods. No. 5. ASA and SSSA, Madison, WI.

Nommik, H., and K. Vahtras. 1982. Retention and fixation of ammonium and ammonia in soils. P. 123-171. *In* F.J. Stevenson (ed.) Nitrogen in agricultural soils. Agronomy 22, ASA and SSSA, Madison, WI.

Soil Science Society of America. 1987. Glossary of soil science terms. Rev. ed. Soil Sci. Soc. Am., Madison, WI.

Soil Survey Staff. 1996. Soil survey laboratory methods manual. Version No. 3.0. USDA-NRCS. Soil Survey Investigations Report No. 42. U.S. Govt. Print. Office, Washington, DC.

Soil Survey Staff. 1999. Soil taxonomy: A basic system of soil classification for making and interpreting soil surveys. USDA-NRCS Agric. Handb. 436. 2nd ed. U.S. Govt. Print. Office, Washington, DC.

Tabatabai, M.A. 1996. Sulfur. p. 921-960. *In* D.L. Sparks (ed.) Methods of soil analysis. Part 3. Chemical methods. No. 5. ASA and SSSA, Madison, WI.

Young, J.L., and R.W. Aldag. 1982. Inorganic forms of nitrogen in soil. p. 43-66. *In* F.J. Stevenson (ed.) Nitrogen in agricultural soils. Agronomy 22. ASA and SSSA, Madison, WI.

Ground and Surface Water Analysis (4I)
Reaction (4I1)
Electrode (4I1a)
Standard Glass Body Combination (4I1a1)
Digital pH/Ion Meter (4I1a1a)
pH (4I1a1a1)

1. Application

The pH of a water sample is a commonly performed determination and one of the most indicative measurements of water chemical properties. The acidity, neutrality, or basicity is a key factor in the evaluation of water quality.

2. Summary of Method

The pH of the water sample is measured with a calibrated combination electrode/digital pH meter (procedure 4I1a1a1).

3. Interfernces

Water pH needs to be measured immediately upon arrival at the laboratory in order to maintain optimal preservation of sample (Velthorst, 1996).

4. Safety

No significant hazards are associated with the procedure. Follow standard laboratory safety practices.

5. Equipment

5.1 Syringe filters, 0.45-μm diameter, Whatman, Cliffton, NJ
5.2 Tubes, 50-mL, with caps
5.3 Digital pH/ion meter, Accumet Model AR 15, Fisher Scientific
5.4 Electrode, standard glass body combination, Accu-flow, Fisher Scientific

6. Reagents
6.1 Reverse osmosis (RO) water, ASTM Type III grade of reagent water
6.2 Borax pH buffers, pH 4.00, pH 7.00, and pH 9.18, for electrode calibration, Beckman, Fullerton, CA.

7. Procedure

7.1 Water sample is filtered into a 50-mL tube. If extracts are not to be determined immediately after collection, then store samples at 4° C. Analyze samples within 72 h.

7.2 Calibrate the pH meter with pH 4.00, 7.00, and 9.18 buffer solutions.

7.3 After equipment calibration, gently wash the electrode with RO water. Dry the electrode. Do not wipe the electrode with a tissue as this may cause a static charge on the electrode.

7.4 Gently lower the electrode in the water sample until the KCl junction of the electrode is beneath the water surface.

7.5 Allow the pH meter to stabilize before recording the pH. Record pH to the nearest 0.01 unit.

7.6 Gently raise the pH electrode and wash the electrode with a stream of RO water.

8. Calculations
No calculations are required for this procedure.

9. Report
Report the pH of the water sample to the nearest 0.1 pH unit.

10. Precision and Accuracy
Precision and accuracy data are available from the SSL upon request.

11. References
Velthorst, E.J. 1996. Water analysis. p. 121-242. *In* P. Buurman, B. van Lagen, and E.J. Velthorst (eds.) Manual for soil and water analysis. Backhuys Publ. Leiden.

Ground and Surface Water Analyses (4I)
Electrical Conductivity and Salts (4I2)
Conductivity Bridge and Cup (4I2a)
Electrical Conductivity (4I2a1)

1. Application
Measuring electrical conductivity (EC) and total dissolved salts (TDS) is straightforward, but it is not in soils because salinity is significantly affected by the prevailing moisture content. A primary source of salts is chemical weathering of the minerals present in soils and rocks, with the most important including dissolution, hydrolysis, carbonation, acidification, and oxidation-reduction (National Research Council, 1993). All of these reactions contribute to an increase in the dissolved mineral load in the soil solution and in waters.

2. Summary of Method
The electrical conductivity of the water sample is measured using an electronic bridge (4I2a1).

3. Interferences

Reverse osmosis water is used to zero and flush the conductivity cell. The extract temperature is assumed to be 25°C. If the temperature deviates significantly, a correction may be required.

Provide airtight storage of KCl solution and samples to prevent soil release of alkali-earth cations. Exposure to air can cause gains and losses of water and dissolved gases significantly affecting EC readings.

4. Safety

No significant hazards are associated with this procedure. Follow standard laboratory safety practices.

5. Equipment

5.1 Syringe filters, 0.45-μm diameter, Whatman, Cliffton, NJ

5.2 Tubes, 50-mL, with caps

5.3 Conductivity bridge and conductivity cell, with automatic temperature adjustment, 25 ± 0.1°C, Markson Model 1056, Amber Science, Eugene, OR

6. Reagents

6.1 Reverse osmosis (RO) water, ASTM Type III grade of reagent water

6.2 Potassium chloride (KCl), 0.010 N. Dry KCl overnight in oven (110°C). Dissolve 0.7456 g of KCl in RODI water and bring to 1-L volume. Conductivity at 25°C is 1.412 mmhos cm^{-1}.

7. Procedure

7.1 Water sample is filtered into a 50-mL tube and capped. If extracts are not to be determined immediately after collection, then store samples at 4° C. Analyze samples within 72 h.

7.2 Standardize the conductivity bridge using RO water (blank) and 0.010 N KCl (1.41 mmhos cm^{-1}).

7.3 Read conductance of water sample directly from the bridge.

7.4 Record conductance to 0.01 mmhos cm^{-1}.

8. Calculations

8.1 No calculations are required for this procedure.

8.2 Use the following relationship to estimate the total soluble cation or anion concentration (meq L^{-1}) in the water.

EC (mmhos cm^{-1}) x 10 = Cation or Anion (meq L^{-1})

9. Report

Report prediction conductance to the nearest 0.01 mmhos cm^{-1}

10. Precision and Accuracy

Precision and accuracy data are available from the SSL upon request.

11. References

National Research Council. 1993. Soil and water quality. An agenda for agriculture. Natl. Acad. Press, Washington, DC.

Ground and Surface Water Analyses (4I)
Electrical Conductivity and Salts (4I2)
Atomic Absorption Spectrophotometer (4I2b)
Calcium, Magnesium, Potassium, and Sodium (4I2b1-4)

1. Application

Nutrients (nitrogen and phosphorus), sediments, pesticides, salts, or trace elements in ground and surface water affect soil and water quality (National Research Council, 1993). This procedure is developed for the analysis of ground or surface water.

2. Summary of Method

The water sample is filtered and diluted with an ionization suppressant (La_2O_3). The analytes are measured by an atomic absorption spectrophotometer (AAS). The data are automatically recorded by a computer and printer. The saturation extracted cations, Ca^{2+}, Mg^{2+}, K^+, and Na^+ are reported in meq L^{-1} (mmol (+) L^{-1}) in procedures 4I2b1-4, respectively.

3. Interferences

There are four types of interferences (matrix, spectral, chemical, and ionization) in the analysis of these cations. These interferences vary in importance, depending upon the particular analyte selected. Do not use borosilicate tubes because of potential leaching of analytes.

4. Safety

Wear protective clothing and eye protection. When preparing reagents, exercise special care. Restrict the use of concentrated HCl to a fume hood. Many metal salts are extremely toxic and may be fatal if ingested. Thoroughly wash hands after handling these metal salts.

Follow standard laboratory procedures when handling compressed gases. Gas cylinders should be chained or bolted in an upright position. Acetylene is highly flammable. Avoid open flames and sparks. Standard laboratory equipment includes fire blankets and extinguishers for use when necessary. Follow the manufacturer's safety precautions when using the AAS.

5. Equipment

5.1 Electronic balance, ±1.0-mg sensitivity
5.2 Syringe filters, 0.45-μm diameter, Whatman , Cliffton, NJ
5.3 Tubes, 50-mL, with caps
5.4 Atomic absorption spectrophotometer (AAS), double-beam, AAnalyst 300, Perkin-Elmer Corp., Norwalk, CT
5.5 Autosampler, AS-90, Perkin-Elmer Corp., Norwalk, CT
5.6 Computer, with AA WinLab software, Perkin-Elmer Corp., Norwalk, CT, and printer
5.7 Single-stage regulator, acetylene
5.8 Digital diluter/dispenser, with syringes 10000 and 1000 μL, gas tight, MicroLab 500, Hamilton Co., Reno, NV
5.9 Plastic test tubes, 15-mL, 16 mm x 100, for sample dilution and sample changer
5.10 Containers, polyethylene
5.11 Peristaltic pump

6. Reagents

6.1 Reverse osmosis deionized (RODI) water, ASTM Type I grade of reagent water
6.2 Hydrochloric acid (HCl), concentrated 12 *N*

6.3   HCl, 1:1 HCl:RODI, 6 N. Carefully mix 1 part of concentrated HCl to 1 part RODI water.

6.4   Stock lanthanum ionization suppressant solution (SLISS), 65,000 mg L^{-1}. Wet 152.4 g of lanthanum oxide (La$_2$O$_3$) with 100 mL RODI water. Slowly and cautiously add 500 mL of 6 N HCl to dissolve the La$_2$O$_3$. Cooling the solution is necessary. Dilute to 2 L with RODI water. Filter solution. Store in polyethylene container.

6.5   Working lanthanum ionization suppressant solution (WLISS), 2000 mg L^{-1}. Dilute 61.5 mL of SLISS with 1800 mL of RODI water (1:10). Make to 2-L volume with RODI water. Invert to mix thoroughly. Store in polyethylene container.

6.6   Primary stock standards solution (PSSS), high purity, 1000 mg L^{-1}: Ca, Mg, K, and Na.

6.7   Working stock mixed standards solution (WSMSS) for Ca, Mg, and K. In a 500-mL volumetric flask, add 250 mL Ca PSSS, 25 mL Mg PSSS, and 100 mL K PSSS = 500 mg L^{-1} Ca, 50 mg L^{-1} Mg, and 200 mg L^{-1} K. Dilute to volume with RODI water. Invert to thoroughly mix. Store in polyethylene containers. Prepare fresh weekly. Store in the refrigerator.

6.8   Mixed calibration standards solution (MCSS), High, Medium, Low, Very Low, and Blank as follows:

6.8.1 MCSS High Standard (1:100): Dilute WSMSS 1:100 with WLISS. Invert to mix thoroughly mix. Store in polyethylene containers. Prepare fresh weekly. Store in a refrigerator. Allow to equilibrate to room temperature before use. Final concentration = 5 mg L^{-1} Ca, 0.5 mg L^{-1} Mg, and 2 mg L^{-1} K.

6.8.2 MCSS Medium Standard (1:200): To a 100-mL volumetric flask, add 50 mL of WSMSS and bring to volume with RODI water. Dilute 1:100 with WLISS. Invert to thoroughly mix. Store in polyethylene containers. Prepare fresh weekly. Store in a refrigerator. Allow to equilibrate to room temperature before use. Final concentration =2.5 mg L^{-1} Ca, 0.25 mg L^{-1} Mg, and 1 mg L^{-1} K.

6.8.3 MCSS Low Standard (1:400): To a 100-mL volumetric flask, add 25 mL of WSMSS and bring to volume with RODI water. Dilute 1:100 with WLISS. Invert to mix thoroughly. Store in polyethylene containers. Prepare fresh weekly. Store in a refrigerator. Allow to equilibrate to room temperature before use. Final concentration = 1.25 mg L^{-1} Ca, 0.125 mg L^{-1} Mg, and 0.5 mg L^{-1} K.

6.8.4 MCSS Very Low Standard (1:600): To a 100-mL volumetric flask, add 16.65 mL of WSMSS and bring to volume with RODI water. Dilute 1:100 with WLISS. Invert to mix thoroughly. Store in polyethylene containers. Prepare fresh weekly. Store in a refrigerator. Allow to equilibrate to room temperature before use. Final concentration = 0.83 mg L^{-1} Ca, 0.08 mg L^{-1} Mg, and 0.33 mg L^{-1} K.

6.8.5 MCSS Blank = 0 mL of Ca, Mg, and K. Dilute RODI water 1:100 with WLISS.

6.9   Na Calibration Standards Solution (NaCSS), High, Medium, Low, and Very Low as follows:

6.9.1 NaCSS High Standard (1:100): Dilute Na PSMSS (1000 mg L^{-1}) 1:100 with WLISS. Invert to thoroughly mix. Store in polyethylene containers. Prepare fresh weekly. Store in a refrigerator. Allow to equilibrate to room temperature before use. Final concentration = 10 mg L^{-1} Na.

6.9.2 NaCSS Medium Standard (1:200): In a 50-mL volumetric, add 25 mL of Na PSMSS and bring to volume with RODI water. Dilute 1:100 with WLISS. Invert to mix thoroughly.

Store in polyethylene containers. Prepare fresh weekly. Store in a refrigerator. Allow to equilibrate to room temperature before use. Final concentration = 5 mg L^{-1} Na.

6.9.3 NaCSS Low Standard (1:400): In a 50-mL volumetric flask, add 12.5 mL of PSMSS and bring to volume with RODI water. Dilute 1:100 with WLISS. Invert to thoroughly mix. Store in polyethylene containers. Prepare fresh weekly. Store in a refrigerator. Allow to equilibrate to room temperature before use. Final concentration = 2.5 mg L^{-1} Na.

6.9.4 NaCSS Very Low Standard (1:600) In a 50-mL volumetric flask, add 8.35 mL of PSMSS Na (1000 ppm) and bring to volume with RODI water. Dilute 1:100 with WLISS. Invert to thoroughly mix. Store in polyethylene containers. Prepare fresh weekly. Store in a refrigerator. Allow to equilibrate before use. Final concentration = 1.67 mg L^{-1} Na.

6.9.5 NaCSS Blank = 0 mL Na PSMSS. Dilute RODI water 1:100 with WLISS.

6.10 Compressed air with water and oil traps.
6.11 Acetylene gas, purity 99.6%.

7. Procedure

7.1 Water sample is filtered into a 50-mL tube and capped. If extracts are not to be determined immediately after collection, then store samples at 4° C. Analyze samples within 72 h.

Dilution of Calibration Standards and Sample Extracts
7.2 The 10-mL syringe is for diluent (WLISS). The 1-mL syringe is for the MCSS and water sample. Set the digital diluter at a 1:100 dilution. See Sections 6.8 and 6.9 for preparation of the MCSS and NaCSS. Dilute the saturation extract sample with 100 parts of WLISS (1:100).

7.3 Dispense the diluted sample solutions into test tubes which have been placed in the sample holders of the sample changer.

AAS Set-up and Operation
7.4 Refer to the manufacturer's manual for operation of the AAS. The following are only very general guidelines for instrument conditions for the various analytes.

Analyte	Conc. (mg L^{-1})	Burner & angle	Wavelength (nm)	Slit (mm)	Fuel/Oxidant (C_2H_2/Air)
Ca	5.0	10 cm @ 0°	422.7	0.7	1.5/0.0
Mg	0.5	10 cm @ 0°	285.2	0.7	1.5/10.0
K	2.0	10 cm @ 0°	766.5	0.7	1.5/10.0
Na	10.0	10 cm @ 30°	589.0	0.2	1.5/10.0

7.5 Use the computer and printer to set instrument parameters and to collect and record instrument readings.

AAS Calibration and Analysis
7.6 Calibrate the instrument by using the MCSS and NaCSS. The data system will then associate the concentrations with the instrument responses fore each MCSS and NaCSS. Rejection criteria for MCSS and NaCSS, if R^2 <0.99.

7.7 If sample exceeds calibration standard, the sample is diluted 1:5, 1:20, 1:100, etc., with RODI water followed by 1:100 dilution with WLISS.

7.8 Perform one quality control (QC) (Low Standard) for every 12 samples. If reading is not within 10%, the instrument is re-calibrated and QC re-analyzed.

7.9 Record analyte readings to 0.01 mg L^{-1}.

8. Calculations

The instrument readings for analyte concentration are in mg L^{-1}. These analyte concentrations are converted to meq L^{-1} as follows:

Analyte Concentration in Soil (meq L^{-1}) = (A x B)/C

where:
A = Analyte (Ca, Mg, K, Na) concentration in extract (mg L^{-1})
B = Dilution ratio, if needed
C = Equivalent weight

where:
Ca^{+2} = 20.04 mg meq^{-1}
Mg^{+2} = 12.15 mg meq^{-1}
K^{+1} = 39.10 mg meq^{-1}
Na^{+1} = 22.99 mg meq^{-1}

9. Report

Report the saturation extraction cations of Ca^{2+}, Mg^{2+}, K^+, and Na^+ to the nearest 0.1 meq L^{-1} (mmol (+) L^{-1}).

10. Precision and Accuracy

Precision and accuracy data are available from the SSL upon request.

11. References

National Research Council. 1993. Soil and water quality. An agenda for agriculture. Natl. Acad. Press, Washington, DC.

Ground and Surface Water Analyses (4I)
Electrical Conductivity and Salts (4I2)
Ion Chromatograph (4I2c)
Conductivity Detector (4I2c1)
Self-Regeneration Suppressor (4I2c1a)
Bromide, Acetate, Chloride, Fluoride, Nitrate, Nitrite, Phosphate, and Sulfate (4I2c1a1-8)

1. Application

Nutrients (nitrogen and phosphorus), sediments, pesticides, salts, or trace elements in ground and surface water affect soil and water quality (National Research Council, 1993). This procedure is developed for the analysis of ground or surface water.

2. Summary of Method

The water sample is filtered and is diluted according to its electrical conductivity (EC$_s$). The diluted sample is injected into the ion chromatograph, and the anions are separated. A conductivity detector is used to measure the anion species and content. Standard anion concentrations are used to calibrate the system. A calibration curve is determined, and the anion concentrations are calculated. A computer program automates these actions. The water anions, Br$^-$, CH$_3$COO$^-$, Cl$^-$, F$^-$, NO$_3^-$, NO$_2^-$, PO$_4^{3-}$, SO$_4^{2-}$ are reported in meq L^{-1} (mmol (-) L^{-1}) by procedure 4I2c1a1-8, respectively.

3. Interferences

Some water samples contain suspended solids and require filtering. Low molecular weight organic anions will co-elute with inorganic anions from the column.

4. Safety

Wear protective clothing and safety glasses. When preparing reagents, exercise special care. Many metal salts are extremely toxic and may be fatal if ingested. Thoroughly wash hands after handling these metal salts. Follow the manufacturer's safety precautions when using the chromatograph.

5. Equipment

5.1 Syringe filters, 0.45-μm diameter, Whatman, Cliffton, NJ
5.2 Tubes, 50-mL, with caps
5.3 Guard column, IonPac AG14, 4 x 50 mm, Dionex Corp., Sunnyvale, CA
5.4 Analytical column, IonPac AS14, 4 x 250 mm, Dionex Corp., Sunnyvale, CA
5.5 Self-regeneration suppressor, ASRS – ULTRA, 4 mm, Dionex Corp., Sunnyvale, CA
5.6 Autosampler, AS40, Dionex Corp., Sunnyvale, CA
5.7 Computer with PeakNet software and printer
5.8 Digital diluter/dispenser, with syringes 10000 and 1000 μL, gas tight, MicroLab 500, Hamilton Co., Reno, NV
5.9 Poly-vials with caps, 5 mL, Dionex Corp., Sunnyvale, CA

6. Reagents

6.1 Reverse osmosis deionized filtered (RODI), ASTM Type I grade of reagent water
6.2 Helium gas
6.3 Eluent solution. Solution A: Mix 5.30 g of Na$_2$CO$_3$ with RODI water in a 500-mL volumetric and make to volume with RODI water. Solution B: Mix 4.20 g of NaHCO$_3$ with RODI water in a 500-mL volumetric and make to volume with RODI water. Add 35 mL of Solution A and 10 mL of Solution B in a 1-L volumetric and make to volume with RODI water for a final concentration is 0.024 M Na$_2$CO$_3$ and 0.025 M NaHCO$_3$. Make 2 L of eluent solution if chromatograph is to run overnight. Place eluent in eluent tank and degas with helium 15 min L^{-1} eluent (30 min total). Make fresh daily.
6.4 Primary stock standards solutions, (PSSS$_{1000}$), high purity, 1000 mg L^{-1}: Cl$^-$, SO$_4^{2-}$, F$^-$, NO$_3^-$, NO$_2^-$, Br$^-$, CH3COO$^-$, and PO$_4^{3-}$.
6.5 Primary stock standards solutions (PSSS$_{100}$), 100 mg L^{-1}: F$^-$, NO$_3^-$, NO$_2^-$, Br$^-$, and PO$_4^{3-}$. Prepare fresh weekly. In five 250-mL volumetric flasks add as follows:

6.5.1 F$^-$ PSSS$_{100}$ = 25 mL F$^-$ PSSS$_{1000}$
6.5.2 Br$^-$ PSSS$_{100}$ = 25 mL Br$^-$ PSSS$_{1000}$
6.5.3 NO$_2^-$ PSSS$_{100}$ = 25 mL NO$_2^-$ PSSS$_{1000}$
6.5.4 NO$_3^-$ PSSS$_{100}$ = 25 mL NO$_3^-$ PSSS$_{1000}$
6.5.5 PO$_4^{3-}$ PSSS$_{100}$ = 25 mL PO$_4^{3-}$ PSSS$_{1000}$

Dilute each PSSS$_{100}$ to volume with RODI water and invert to thoroughly mix. Store in the refrigerator.

6.6 Mixed calibration standards solutions (MCSS), A, B, C, D, E, and Blank as follows:

6.6.1 MCSSA = In a 250-mL volumetric flask, add as follows

12.5 mL Cl⁻ $PSSS_{1000}$ = 50 mg L^{-1}
25 mL SO_4^{2-} $PSSS_{1000}$ = 100 mg L^{-1}
12.5 mL F⁻ $PSSS_{100}$ = 5 mg L^{-1}
20 mL NO_3^- $PSSS_{100}$ = 8 mg L^{-1}
20 mL NO_2^- $PSSS_{100}$ = 8 mg L^{-1}
22 mL Br⁻ $PSSS_{100}$ = 8.8 mg L^{-1}
20 mL $CH3COO^-$ $PSSS_{1000}$ = 80 mg L^{-1}
24 mL PO_4^{3-} $PSSS_{100}$ = 9.6 mg L^{-1}

Dilute to volume with RODI water and invert to thoroughly mix. Store in glass containers. Store in the refrigerator. Prepare fresh weekly.

6.6.2 MCSSB = In a 100-mL volumetric flask, add 40 mL MCSSA and dilute to volume with RODI water. Final concentration = 20, 40, 2, 3.2, 3.2, 3.52, 30.4, and 3.52 mg L^{-1} Cl⁻, SO_4^{2-}, F⁻, NO_3^-, NO_2^-, Br⁻, $CH3COO^-$, and PO_4^{3-}, respectively. Invert to thoroughly mix. Store in glass containers. Prepare fresh weekly. Store in the refrigerator.

6.6.3 MCSSC = In a 100-mL volumetric flask, add 20 mL MCSSA and dilute to volume with RODI water. Final concentration = 10, 20, 1, 1.6, 1.6, 1.76, 15.2, and 1.76 mg L^{-1} Cl⁻, SO_4^{2-}, F⁻, NO_3^-, NO_2^-, Br⁻, $CH3COO^-$, and PO_4^{3-}, respectively. Invert to thoroughly mix. Store in glass containers. Prepare fresh weekly. Store in the refrigerator.

6.6.4 MCSSD = In a 250-mL volumetric flask, add 25 mL MCSSA and dilute to volume with RODI water. Invert to thoroughly mix. Final concentration = 5, 10, 0.5, 0.8, 0.8, 0.88, 7.6, and 0.88 mg L^{-1} Cl⁻, SO_4^{2-}, F⁻, NO_3^-, NO_2^-, Br⁻, $CH3COO^-$, and PO_4^{3-}, respectively. Store in glass containers. Prepare fresh weekly. Store in the refrigerator.

6.6.5 MCSSE = In a 250-mL volumetric flask, add 5 mL MCSSA and dilute to volume with RODI water. Final concentration = 1, 2, 0.1, 0.16, 0.16, 0.176, 1.52, and 0.176 mg L^{-1} Cl⁻, SO_4^{2-}, F⁻, NO_3^-, NO_2^-, Br⁻, $CH3COO^-$, and PO_4^{3-}, respectively. Invert to thoroughly mix. Store in glass containers. Prepare fresh weekly. Store in the refrigerator.

6.6.6 MCSS Blank = 0 mL of Cl⁻, SO_4^{2-}, F⁻, NO_3^-, NO_2^-, Br⁻, $CH3COO^-$, and PO_4^{3-}. Dilute RODI water to volume.

7. Procedure

7.1 Water sample is filtered into a 50-mL tube and capped. If extracts are not to be determined immediately after collection, then store samples at 4° C. Analyze samples within 72 h.

Dilution of sample extracts

7.2 To estimate the total soluble anion concentration (meq L^{-1}), multiply the EC (procedure 4I2a1) by 10. Subtract the CO_3^{2-} and HCO_3^- concentrations (procedures 4I2d1a1-2) from the total anion concentration. The remainder is the ≈ concentration (meq L^{-1}) of anions to be separated by ion chromatography.

Anion concentration (meq L^{-1}) = EC_s x 10 - (HCO_3^- + CO_3^{2-})

7.2 Dilute the saturation extract with the RODI water as follows:

EC$_s$ (dS cm^{-1})	Dilution Factor
0.00 to 0.55	4
0.56 to 0.65	5
0.66 to 0.75	6
0.76 to 0.85	7
0.86 to 0.95	8
0.96 to 1.05	9
1.06 to 1.20	10
1.21 to 1.40	15
1.41 to 1.50	25
1.51 to 1.60	30
1.61 to 1.80	40
1.81 to 2.00	50
2.01 to 2.30	60
2.31 to 2.60	70
2.61 to 3.10	80
3.11 to 3.55	90
3.56 to 4.05	100
4.06 to 4.60	120
4.61 to 5.20	140
5.21 to 5.85	150
5.86 to 6.55	160
6.56 to 7.30	180
7.31 to 8.00	200
8.01 to 9.00	225
9.01 to 10.00	240
10.01 to 11.50	270
11.51 to 13.00	280
13.01 to 14.50	300
14.51 to 16.00	320
16.01 to 17.00	360
17.01 to 18.00	400
18.01 to 20.00	450
20.01 to 21.00	480
21.01 to 23.00	500
23.01 to 24.00	540
24.01 to 25.00	560
25.01 to 27.00	600
27.01 to 28.00	640
28.01 to 30.00	680
30.01 to 32.00	700
32.01 to 33.00	720
33.01 to 36.00	800
36.01 to 40.00	900
40.01 to 44.00	1000

7.3 Place the MCSS (B, C, D, E, Blank) and diluted extract samples in the Poly-vials and cap with filtercaps.

Set-up and Operation of Ion Chromatograph (IC)

7.4 Refer to the Manufacturer's manual for the operation of chromatograph. Because any number of factors may cause a change in IC operating conditions, only a general set-up of the Dionex DX-120 ion chromatograph is presented. Individual analysts may modify some or all of the operating conditions to achieve satisfactory results. Ranges and/or (typical settings) are as follows:

Parameter	Range and/or (Typical Setting)
Calibration	Peak Height or (Area)
Flow Setting	0.5 to 4.5 mL min^{-1} (1.63 mL min^{-1})
Pressure	0 to 4000 psi (2200 to 2400 psi)
Detection	Suppressed conductivity
Total Conductivity	0 to 999.9 μS
Injection Volume	10 μL
Auto Offset	-999.9 to 999.9 μS (ON)

7.5 Load the sample holder cassettes with the capped samples, standards, and check samples.

7.6 Use the computer and printer to set instrument parameters and to collect and record instrument readings.

IC Calibration and Analysis

7.7 Calibrate the instrument by using the MCSS (B, C, D, E, Blank). The data system will then associate the concentrations with the instrument responses for each MCSS. Rejection criteria for MCSS, R^2 <0.99.

7.8 If samples are outside calibration, dilute sample extracts with RODI water solution and re-analyze.

7.9 Perform one quality control (QC) (Low Standard MCSS, Standard C) for every 12 samples. If reading is not within tolerance limits (10 to 15%, based on analyte), the instrument is re-calibrated and QC re-analyzed.

7.10 Record analyte readings to 0.01 mg L^{-1}.

8. Calculations

The instrument readings for analyte concentration are in mg L^{-1}. These analyte concentrations are converted to meq L^{-1} as follows:

Analyte Concentration in Soil (meq L^{-1}) = (A x B)/C

where:
A = Analyte (Br$^-$, CH$_3$COO$^-$, Cl$^-$, F$^-$, NO$_3^-$, NO$_2^-$, PO$_4^{3-}$, SO$_4^{2-}$) concentration in extract (mg L^{-1})
B = Dilution ratio, if needed
C = Equivalent weight

where:
Cl$^-$ = 35.45 mg meq^{-1}
SO$_4^{2-}$ = 48.03 mg meq^{-1}

F^- = 19.00 mg meq^{-1}
NO_3^- = 62.00 mg meq^{-1}
NO_2^- = 46.00 mg meq^{-1}
Br^- = 79.90 mg meq^{-1}
$CH3COO^-$ = 59.05 mg meq^{-1}
PO_4^{3-} = 31.66 mg meq^{-1}

9. Report

Report the saturation extraction anions (Br^-, CH_3COO^-, Cl^-, F^-, NO_3^-, NO_2^-, PO_4^{3-}, SO_4^{2}) to the nearest 0.1 meq L^{-1} (mmol (-) L^{-1}).

10. Precision and Accuracy

Precision and accuracy data are available from the SSL upon request.

11. References

National Research Council. 1993. Soil and water quality. An agenda for agriculture. Natl. Acad. Press, Washington, DC.

Ground and Surface Water Analyses (4I)
Electrical Conductivity and Salts (4I2)
Automatic Titrator (4I2d)
Combination pH-Reference Electrode (4I2d1)
Acid Titration, H_2SO_4 (4I2d1a)
Carbonate and Bicarbonate (4I2d1a1-2)

1. Application

Nutrients (nitrogen and phosphorus), sediments, pesticides, salts, or trace elements in ground and surface water affect soil and water quality (National Research Council, 1993). This procedure is developed for the analysis of ground or surface water.

2. Summary of Method

The water sample is filtered and aliquot titrated on an automatic titrator to pH 8.25 and pH 4.60 end points. The carbonate and bicarbonate are calculated from the titers, aliquot volume, blank titer, and acid normality. Carbonate and bicarbonate are reported in meq L^{-1} (mmol (-) L^{-1}) by procedures 4I2d1a1-2, respectively.

3. Interferences

Clean the electrode by rinsing with distilled water and patting it dry with tissue. Wiping the electrode dry with a cloth, laboratory tissue, or similar material may cause electrode polarization.

Slow electrode response time may cause over shooting the end point. A combination of slowing the buret speed and increasing the time delay may help. Cleaning the electrode with detergent may decrease the response time. If all else fails, changing the electrode generally solves the problem. Blanks may not titrate properly because some sources of reverse osmosis (RO) water have a low pH.

4. Safety

Wear protective clothing and eye protection. When preparing reagents, exercise care. Thoroughly wash hands after handling reagents. Restrict the use of concentrated H_2SO_4 to the fume hood. Use showers and eyewash stations to dilute spilled acids. Use sodium bicarbonate and water to neutralize and dilute spilled acids. Follow the manufacturer's safety precautions when operating the automatic titrator.

5. Equipment

5.1 Syringe filters, 0.45-μm diameter, Whatman, Cliffton, NJ

5.2 Tubes, 50-mL, with caps

5.3 Automatic titrator, Metrohm 670 Titroprocessor, with control unit, sample changer, and dispenser, Metrohm Ltd., Brinkmann Instruments, Inc.

5.4 Combination pH-reference electrode, Metrohm Ltd., Brinkmann Instruments, Inc.

5.5 Pipettes, electronic digital, 2500 μL and 10 mL, with tips, 2500 μL and 10 mL

6. Reagents

6.1 Reverse osmosis (RO) water, ASTM Type III grade of reagent water

6.4 Helium gas

6.3 Sulfuric acid (H_2SO_4), concentrated, 36 N

6.5 H_2SO_4, 0.0240 N standardized. Carefully dilute 2.67 mL of concentrated H_2SO_4 in 4 L of RODI degassed water ($\approx$ 15 min). Re-standardize the acid at regular intervals. Refer to the procedure for standardization of acids.

6.4 Borax pH buffers, pH 4.00, pH 7.00, and pH 9.18, for titrator calibration, Beckman, Fullerton, CA.

7. Procedure

7.1 Water sample is filtered into a 50-mL tube and capped. If extracts are not to be determined immediately after collection, then store samples at 4° C. Analyze samples within 72 h.

7.2 Pipet 3 mL of the water sample into a 250-mL titration beaker.

7.3 Add 72 mL of RO water into a titration beaker. Final volume is 75 mL for blanks and samples. Run 8 to 12 blanks of RO water through the titration procedure.

7.4 Refer to manufacturer's manual for operation of the automatic titrator.

7.5 Calibrate automatic titrator with 9.18, 7.00 and 4.00 pH buffers. Set-up the automatic titrator to set end point titration mode. The "Set" pH parameters are listed as follows:

Parameter	Value
Ep_1	pH 8.25
Dyn change pH	1.5 units
Drift	0.4 mV s^{-1}
Time delay	10 s
Ep_2	pH 4.60
Dyn change pH	1.5 units
Drift	0.4 mV s^{-1}
Temp	25°C
Stop Volume	35 mL

7.6 Place the 250-mL titration beakers in the sample changer.

7.7 Press "Start".

7.8 If the titrator is operating properly, no other analyst intervention is required. The titers and other titration parameters are recorded on the Titroprocessor printer.

8. Calculations

8.1 CO_3^{2-} (meq L^{-1}) $= (2 \ T_1 \times N \times 1000)/Aliquot$

8.2 HCO_3^- (meq L^{-1}) $= [(T_2 + T_1) - Blank - (2 \times T_1) \times N \times 1000]/Aliquot$

where:
T_1 = Titer of CO_3^{2-} (mL)
T_2 = Titer of HCO_3^- (mL)
N = Normality of H_2SO_4
Blank   = Average titer of blank solutions (mL)
Aliquot = Volume of saturation extract titrated (mL)
1000   = Conversion factor to meq L^{-1}

9. Report
Report saturation extract CO_3^{2-} and HCO_3^- to the nearest 0.1 meq L^{-1} (mmol (-) L^{-1}).

10. Precision and Accuracy
Precision and accuracy data are available from the SSL upon request.

11. References
National Research Council. 1993. Soil and water quality. An agenda for agriculture. Natl. Acad. Press, Washington, DC.

Ground and Surface Water Analyses (4I)
Total Analysis (4I3)
Inductively Coupled Plasma Atomic Emission Spectrophotometer (4I3a)
Axial Mode (4I3a1)
Ultrasonic Nebulizer (4I3a1a)
Aluminum, Iron, Manganese, Phosphorus, and Silicon (4I3d1a1-5)

1. Application
Nutrients (nitrogen and phosphorus), sediments, pesticides, salts, or trace elements in ground and surface water affect soil and water quality (National Research Council, 1993). This procedure is developed for the analysis of ground or surface water. This procedure is developed for the analysis of the elemental content of ground or surface water.

2. Summary of Method
The water is filtered and acidified with HCl. Two calibration standards plus a blank are prepared for elemental analysis. An inductively coupled plasma atomic emission spectrophotometer (ICP-AES) in axial mode is used to determine the concentration of Al, Fe, Mn, P, and Si (mg L^{-1}) by procedures 4I3d1a1-5, respectively.

3. Interferences
Spectral and matrix interferences exist. Interferences are corrected or minimized by using an internal standard. Also, careful selection of specific wavelengths for data reporting is important. Samples and standards are matrix matched to help reduce interferences.

4. Safety
Wear protective clothing and eye protection. When preparing reagents, exercise special care. Restrict the use of concentrated acids to the fume hood. Wash hands thoroughly after handling reagents.

5. Equipment
5.1 Syringe filters, 0.45-μm diameter, Whatman, Cliffton, NJ
5.2 Tubes, 50-mL, with caps
5.3 Volumetrics, 500-mL and 200 ml, class A glass
5.4 Containers, 500-mL, polypropylene, with screw caps
5.5 Pipettors, electronic digital, Rainin Instrument Co., Woburn, MA, 2500 μL and 10 mL
5.6 Inductively coupled plasma atomic emission spectrophotometer (ICP-AES), Perkin-Elmer Optima 3300 Dual View (DV), Perkin-Elmer Corp., Norwalk, CT.
5.7 Computer, with WinLab software, ver. 4.1, Perkin-Elmer Corp., Norwalk, CT, and printer
5.8 Compressed gasses, Argon (minimum purity=99.996%) and Nitrogen (minimum purity=99.999%)
5.9 Autosampler, AS-90, Perkin-Elmer Corp., Norwalk, CT.
5.10 Quartz torch, alumina injector (2.0 mm id), Ultrasonic Nebulizer, Model U-5000AT+, CETAC Corp., Omaha, NE

6. Reagents
6.1 Deionized, reverse osmosis (RODI) water, ASTM Type 1 grade of reagent water
6.2 Concentrated hydrochloric acid (HCl), 12 N. Use trace-pure grade that contains low levels of impurities.
6.3 Primary standards:1000 mg L^{-1}, High Purity Standards, Charleston, SC. Single elemental standards are manufactured in dilute HNO_3, HNO_3 + HF, or H_2O
6.4 Internal standard: HNO_3, trace pure

7. Procedure

7.1 Water sample is filtered into a 50-mL tube and capped. If extracts are not to be determined immediately after collection, then store samples at 4° C. Analyze samples within 72 h.

7.2 The working calibration standards (WCS) are made from dilution of the primary standards. The low and high WCS for Al and Si are prepared as follows:

7.2.1 ALLO is 0.5, 2.5, and 2.5 mg L^{-1} of Al, Fe, and Mn, respectively. To a 500-mL volumetric flask, add 0.25, 1.25, and 1.25 mL of the Al, Fe, and Mn primary standard (1000 mg L^{-1}), respectively.
7.2.2 ALHI is is 1.0, 5.0, and 5.0 mg L^{-1} of Al, Fe, and Mn, respectively. To a 500-mL volumetric flask, add 0.50, 2.50, and 2.50 mL of the Al, Fe, and Mn primary standard (1000 mg L^{-1}), respectively.
7.2.3 SILO is 0.5 and 0.5 mg L^{-1} of Si and P, respectively. To a 500-mL volumetric flask, add 0.25 and 0.25 mL of the Si and P primary standard (1000 mg L^{-1}), respectively.
7.2.4 SIHI is 1.0 and 1.0 mg L^{-1} of Si and P, respectively. To a 500-mL volumetric flask, add 0.50 and 0.50 mL of the Si and P primary standard (1000 mg L^{-1}), respectively.

7.3 Samples are treated with 0.5 mL HCl for each 10 mL water.

7.4 A 10 mg L^{-1} Lu internal standard (read at 291.138 nm) is added to the blank, all calibration standards, and samples. It is prepared by adding 5.0 mL Lu primary standard (1000 mg L^{-1}) and 10 mL conc. HNO_3 to a 500 ml volumetric flask, and diluting to volume with RODI water.

7.5 Use the ICP-AES spectrophotometer in axial mode to analyze elements. Use ultrasonic nebulization of sample. Internal standard is added via an external peristaltic pump at 15% pump speed using 0.44 mm id. pump tubing. Internal standard and samples or standards are mixed via a mixing block and coil prior to entering the ultrasonic nebulizer. Typically, no initial dilutions of samples because of high concentrations are necessary prior to analysis. Perform instrument checks (Hg alignment; BEC and %RSD of 1 mg L^{-1} Mn solution) prior to analysis as discussed in operation manual of instrument. Check instrument alignment and gas pressures to obtain optimum readings with maximum signal to noise ratio.

7.6 Analyses are generally performed at two or more wavelengths for each element. The selected wavelengths are as follows: (reported wavelength listed first and in boldface):

Element	Wavelength
	----------- nm ---------
Mn	**260.570**, 257.610, 403.075
P	**178.221**, 213.620, 214.910
Al	**308.215**, 167.022, 396.153
Fe	**238.203**, 239.562, 259.939
Si	**251.611**, 212.412, 288.158
Lu	**291.138** (Internal Standard)

7.7 Use the blank standard solution to dilute those samples with concentrations greater than the high standard. Rerun all elements and use only the data needed from the diluted analysis.

7.8 Establish detection limits using the blank standard solution. The instrumental detection limits are calculated by using 3 times the standard deviation of 10 readings of the blank. These values establish the lower detection limits for each element. Analyzed values lower than the detection limits are reported as "ND" or non-detected.

8. Calculations

With the HCl treatment (0.5 mL per 10 mL water) in the calculations, the concentrations are then reported directly, unless additional dilutions are performed because of high analyte concentrations.

9. Report

Data are reported to the nearest 0.01 mg L^{-1}.

10. Precision and Accuracy

Precision and accuracy data are available from the SSL upon request.

11. References

National Research Council. 1993. Soil and Water Quality. An agenda for agriculture. Natl. Acad. Press, Washington, DC.

ANALYSIS OF ORGANIC SOILS OR MATERIALS (5)

Mineral Content (5A)

1. Application

The mineral content is the plant ash and soil particles that remain after organic matter removal. The percentage of organic matter lost on ignition can be used to define organic soils in place of organic matter estimates by the Walkley-Black organic C method (6A1c, method obsolete, Soil Survey Staff, 1996). The determination of organic matter by loss on ignition is a taxonomic criterion for organic soil materials (Soil Survey Staff, 1999). Organic C data by Walkley-Black are generally considered invalid if organic C is >8 %.

2. Summary of Method

Dry sample overnight at 110°C in moisture can. Cool and weigh. Place sample in a cold muffle furnace and raise the temperature to 400°C. Heat sample overnight (16 h), cool, and weigh. The ratio of the weights (400°C/110°C) is the mineral content percentage (procedure 5A).

3. Interferences

The sample must be placed in a cold muffle furnace to prevent rapid combustion and sample splattering.

4. Safety

Use caution when the muffle furnace is hot. Wear protective clothing and goggles. Handle the heated material with tongs.

5. Equipment

5.1 Metal weighing tins
5.2 Oven, 110°C
5.3 Muffle furnace, 400°C
5.4 Electronic Balance, ±0.01-g sensitivity

6. Reagents

None.

7. Procedure

7.1 Place a 10- to 15-g sample in a tared weighing tin.

7.2 Dry sample at 110°C overnight.

7.3 Remove sample from oven, cap, and cool in a desiccator.

7.4 When cool, record weight to nearest 0.01 g.

7.5 Place sample and weighing tin in a cold muffle furnace. Raise temperature to 400°C. Heat overnight (16 h).

7.6 Remove sample from oven, cap, and cool in a desiccator.

7.7 When cool, record sample weight to nearest 0.01 g.

8. Calculations

8.1 Mineral Content (%) = $(R_W/OD_W) \times 100$

where:
R_W = Residue weight after ignition
OD_W = Oven-dry soil weight

Organic matter percent can then be calculated as follows:

8.2 Organic Content (%) = 100 - Mineral Content (%)

9. Report

Report mineral content to the nearest whole percent.

10. Accuracy and Precision

Precision and accuracy data are available from the SSL upon request.

11. References

Soil Survey Staff. 1996. Soil survey laboratory methods manual. Version No. 3.0. USDA-NRCS. Soil Survey Investigations Report No. 42. U.S. Govt. Print. Office, Washington, DC.

Soil Survey Staff. 1999. Soil taxonomy: A basic system of soil classification for making and interpreting soil surveys, USDA-NRCS Agric. Handb. 436. 2nd ed. U.S. Govt. Print. Office, Washington, DC.

Pyrophosphate Color (5B)

1. Application

Decomposed organic materials are soluble in sodium pyrophosphate. The combination of organic matter and sodium pyrophosphate form a solution color which correlates with the decomposition state of the organic materials. Dark colors are associated with sapric materials and light colors with fibric materials (Soil Survey Staff, 1999).

2. Summary of Method

Organic material is combined with sodium pyrophosphate. After standing, the color is evaluated by moistening a chromatographic strip in the solution and comparing the color with standard Munsell color charts (5B).

3. Interferences

This test of organic soil material can be used in field offices. Since it is not practical in the field to base a determination on a dry sample weight, moist soil is used. The specific volume of moist material depends on how it is packed. Therefore, packing of material must be standardized in order to obtain comparable results by different soil scientists (Soil Survey Staff, 1999).

4. Safety

Use caution when handling sodium pyrophosphate.

5. Equipment

5.1 Polycons, 30 mL, Richards Mfg. Co.
5.2 Chromatographic paper, Schleicher and Schuell no. 470 A-3.
5.3 Munsell Color Book, 10YR and 7.5YR pages.

5.4 Half-syringe, 6 mL. Cut plastic syringe longitudinally to form a half-cylinder measuring device.
5.5 Scissors
5.6 Paper towel
5.7 Tweezers
5.8 Metal spatula

6. Reagents
6.1 Sodium pyrophosphate ($Na_4P_2O_7 \cdot 10H_2O$)
6.2 Reverse osmosis (RO) water

7. Procedure

Sample Preparation
7.1 Prepare soil material. If the soil is dry, add water and let stand to saturate. Place 50 to 60 mL of a representative sample on a paper towel in a linear mound. Roll the towel around the sample and express water if necessary. Use additional paper towels as external blotters. Remove the sample and place on a fresh paper towel. The sample should be firm but saturated with water.

7.2 Use scissors to cut sample into 5- to 10-mm long segments.

7.3 Randomly select sample segments for determination of fiber (5C), solubility in pyrophosphate (5B), and pH (4C1a1a4).

Pyrophosphate
7.4 Dissolve 1 g (heaping 1/8 tsp) of sodium pyrophosphate in 4 mL of water in a 30-mL polycon container. Allow to equilibrate for 5 min.

7.5 Use a metal spatula to pack a half-syringe that is adjusted to the 5-mL mark or 2.5-mL (2.5-cm^3) volume with the moist sample.

7.6 Transfer soil material cleanly into the container that holds the pyrophosphate solution.

7.7 Mix thoroughly using a wooden stirrer or metal spatula. Cover and let stand overnight.

7.8 Mix sample again next morning.

7.9 Use tweezers to insert a strip of chromatographic paper vertically into the sample to a 1-cm depth. Let stand until the paper strip has wetted to a 2-cm height above slurry surface. Generally, sample needs to stand $\approx$ 5 min but may stand longer if cover is closed. Remove the paper strip with tweezers. Cut strip and leave in the slurry that portion to which the soil adheres.

7.10 Place the strip on a piece of blotting paper and press gently with tweezers to make even contact.

7.11 Remove paper strip with tweezers and compare color of the strip to Munsell color charts.

8. Calculations
No calculations.

9. Report
Report color using Munsell color notation.

10. Precision and Accuracy

Precision and accuracy data are not available for this procedure. Experienced analysts can usually reproduce results within ± 1 color chip.

11. References

Soil Survey Staff. 1999. Soil taxonomy: A basic system of soil classification for making and interpreting soil surveys, USDA-NRCS Agric. Handb. 436. 2nd ed. U.S. Govt. Print. Office, Washington, DC.

Fiber Volume (5C)

1. Application

The water-dispersed fiber volume is a method to characterize the physical decomposition state of organic materials. The decomposition state of organic matter is used in soil taxonomy to define sapric, hemic, and fibric organic materials (Soil Survey Staff, 1999). Sapric material passes through a 100-mesh sieve (0.15-mm openings). Fibers are retained on the sieve. As defined in soil taxonomy, organic materials that are >2 mm in cross section and that are too firm to be readily crushed between thumb and fingers are excluded from fiber.

2. Summary of Method

The sample is prepared to a standard water content. The unrubbed fiber content is determined in a series of three steps designed to remove the sapric material by increasingly vigorous treatments. The rubbed fiber content is determined by rubbing the sample between the thumb and fingers. The percent unrubbed fiber after each step and the final unrubbed and rubbed fiber are reported (5C).

3. Interferences

This test of organic soil material can be used in field offices. Since it is not practical in the field to base a determination on a dry sample weight, moist soil is used. The specific volume of moist material depends on how it is packed. Therefore, packing of material must be standardized in order to obtain comparable results by different soil scientists (Soil Survey Staff, 1999).

4. Safety

Use caution when using electrical equipment.

5. Equipment

5.1 Half-syringe, 6 mL. Cut plastic syringe longitudinally to form a half-cylinder measuring device.
5.2 Sieve, 100 mesh, 7.6-cm diameter
5.3 Eggbeater
5.4 Microscope or hand lens
5.5 Electric mixer, Hamilton Beach no. 35
5.6 Scissors
5.7 Paper towel
5.8 Metal spatula

6. Reagents

Reverse osmosis (RO) water

7. Procedure

Sample Preparation
7.1 Prepare soil material. If the soil is dry, add water and allow to stand until saturated. Place 50 to 60 mL of a representative sample on a paper towel in a linear mound. Roll the towel around the sample and gently squeeze to express water if necessary. Use additional paper towels as external blotters. Remove the sample and place on a fresh paper towel. The sample should be firm but saturated with water.

7.2 Use scissors to cut sample into 0.5- to 1.0-cm length segments.

7.3 Randomly select sample segments for determination of fiber (procedure 5C), solubility in pyrophosphate (procedure 5B), and pH (procedure 4C1a1a4).

Unrubbed Fiber: Overview
7.4 The unrubbed fiber procedure involves a series of three steps designed to disperse sapric material by increasingly vigorous treatments. All three steps may not be necessary. Following each step that is performed, the percentage estimate of sapric material remaining is visually determined under a microscope or hand lens. Categories used to estimate the remaining sapric component are as follows:

7.4.1 Clean (<1% sapric)
7.4.2 Nearly clean (1 to 10% sapric)
7.4.3 Some sapric (10 to 30% sapric)
7.4.4 Sapric (>30% sapric)

Unrubbed Fiber: Part 1
7.5 Use a metal spatula to pack a half-syringe that is adjusted to the 5-mL mark or 2.5-mL (2.5 cm^3) volume with the moist sample.

7.6 Transfer all the soil material to a 100-mesh sieve and wash under a stream of tap water, adjusted to deliver 200 to 300 mL in 5 s. Wash sample until the water passing through the sieve appears clean. To more clearly determine the end point, catch the effluent in a white plastic container. Periodically empty the container until the effluent runs nearly clean.

7.7 Examine the sample under a microscope or hand lens to determine if sample is free of sapric material.

7.8 If sapric material is >10%, proceed to Unrubbed Fiber, Part 2. If sapric material is <10%, wash the residue to one side of the screen and blot from underneath with absorbent tissue to withdraw water and proceed on as follows with Unrubbed Fiber, Part 1.

7.9 Repack the residue into a half-syringe and blot again with absorbent tissue. The moisture content should be ≈ that of the original sample.

7.10 Measure the volume by withdrawing the plunger and reading the value on the syringe scale. Record as a percentage of the initial 2.5-mL (2.5 cm^3) volume.

7.11 Proceed with the Rubbed Fiber determination.

Unrubbed Fiber: Part 2
7.12 Transfer the residue obtained in Unrubbed Fiber, Part 1 to a 500-mL plastic container and fill about half full with water.

7.13 Stir vigorously with an eggbeater for 1 min.

7.14 Transfer to the 100-mesh sieve and repeat procedural steps in Unrubbed Fiber, Part 1 beginning with Section 7.9. If sapric material is >10%, proceed to Unrubbed Fiber, Part 3.

Unrubbed Fiber: Part 3
7.15 Transfer residue left from Unrubbed Fiber, Part 2 to an electric mixer container (malt mixer or blender) and fill to about two-thirds with water.

7.16 Mix for 1 min.

7.17 Transfer to a 100-mesh sieve and repeat Unrubbed Fiber Part beginning with Section 7.6, the washing procedure.

7.18 Examine the residue under a microscope or hand lens and estimate the percentage of sapric material, if any.

7.19 Record the kind of fiber observed. Typical fibers are herbaceous, woody, and diatomaceous.

7.20 Blot the sample and measure the residue volume.

7.21 Proceed with the Rubbed Fiber determination.

Rubbed Fiber
7.22 Transfer the residue from the unrubbed fiber treatment to the 100-mesh sieve.

7.23 Rub sample between thumb and fingers under a stream of tap water, adjusted to deliver 150 to 200 mL in 5 s, until water passing through the sieve is clean. Clean rubbed fibers roll between the thumb and fingers rather than slide or smear.

7.24 Blot sample and measure volume in half-syringe.

8. Calculations

Fiber volume (%) = Reading on half-syringe (mL) x 20

where:
Fiber volume = Rubbed + unrubbed fiber

9. Report
Record the percentage of unrubbed fiber after each completed step. Report the final unrubbed and the rubbed fiber to the nearest whole percent and report fiber type.

10. Precision and Accuracy
No precision and accuracy data are available for this procedure.

11. References
Soil Survey Staff. 1999. Soil taxonomy: A basic system of soil classification for making and interpreting soil surveys, USDA-NRCS Agric. Handb. 436. 2nd ed. U.S. Govt. Print. Office, Washington, DC.

Melanic Index (5D)

1. Application

Melanic and fulvic Andisols have high contents of humus, related to their soil color reflecting pedogenic processes (Honna et al., 1988). Typically, Melanic Andisols are formed under grassland ecosystems, with humus dominated by A type humic acid (highest degree of humification), whereas Fulvic Andisols are found under forest ecosystems, with humus characterized by the high ratio of fulvic acid to humic acid (low degree of humification, e.g., P or B type humic acid) (Honna et al., 1988). The organic matter thought to result from large amounts of gramineous vegetation can be distinguished from organic matter formed under forest vegetation by the melanic index (Soil Survey Staff, 1999).

2. Summary of Method

A 0.5-g soil sample is mechanically shaken for 1 h in 25 mL of 0.5% NaOH solution. One drop of 0.2% superfloc solution (flocculation aid) is added to sample and then mechanically shaken for 10 min. Either a 1 or 0.5 mL extract (<10% or >10% organic C, respectively) is pipetted into a test tube, followed by the addition of 20 mL of 0.1% NaOH solution and mixing thoroughly. Absorbance of the solution is read using a spectrophotometer at 450 and 520 nm, respectively, within 3 h after extraction. Melanic Index is calculated by dividing the absorbance at 450 nm by the absorbance at 520 nm (procedure 5D).

3. Interferences

No known interferences.

4. Safety

No significant hazards are associated with this procedure. Follow standard laboratory safety practices.

5. Equipment

5.1 Electronic balance, ±1.0-mg sensitivity
5.2 Mechanical reciprocating shaker, 200 oscillations min^{-1}, 1 ½ in strokes, Eberbach 6000, Eberbach Corp., Ann Arbor, MI
5.3 Centrifuge tubes, 50-mL polypropylene
5.4 Centrifuge, Centra, GP-8, Thermo IEC, Needham Heights, MA
5.5 Pipettes, electronic digital, 1000 µL and 10 mL, with tips, 1000 µL and 10 mL
5.6 Dispenser, 30 mL or 10 mL
5.7 Cuvettes, plastic, 4.5-mL, 1-cm light path, Daigger Scientific
5.8 Spectrophotometer, UV-visible, Varian, Cary 50 Conc
5.9 Computer, Microsoft Windows 98
5.10 Printer

6. Reagents

6.1 Reverse osmosis deionized (RODI) water, ASTM Type I grade of reagent water
6.2 NaOH, 0.5% and 0.1%
6.3 Superfloc 16, 0.2% (2 g L^{-1}) in RODI water.

7. Procedure

7.1 Weigh 0.5 g of <2mm or fine-grind air-dry soil to the nearest 1.0 mg into a 50-mL centrifuge tube. If sample is moist, weigh enough soil to achieve ≈ 0.5 of air-dry soil.

7.2 Dispense 25 mL of 0.5% NaOH solution to the tube.

7.3 Transfer the sample to the shaker. Shake for 1 h at 200 oscillations min⁻¹ at room temperature.

7.4 Remove the sample from the shaker. Add one drop of 0.2% Superfloc 16 solution and centrifuge at 4000 rpm for 10 min.

7.5 Use the pipette to transfer either a 1 or 0.5 mL extract (<10% or >10% organic C, respectively) into test tube.

7.6 Add 20 mL of 0.1% NaOH solution and mix thoroughly.

7.7 Set the spectrophotometer at 450 nm. Read absorbance.

7.8 Set the spectrophotometer at 520 nm. Read absorbance.

8. Calculations

Melanic Index is calculated as follows:

Absorbance at 450 nm/Absorbance at 520 nm

9. Report
Report Melanic Index.

10. Precision and Accuracy
Precision and accuracy data are available from the SSL upon request.

11. References
Honna, T., S. Yamamoto, and K. Matsui. 1988. A simple procedure to determine melanic index that is useful for differentiating melanic from fulvic Andisols. Pedologist 32:69-78.

Soil Survey Staff. 1999. Soil taxonomy: A basic system of soil classification for making and interpreting soil surveys, USDA-NRCS Agric. Handb. 436. 2nd ed. U.S. Govt. Print. Office, Washington, DC.

Ratios and Estimates Related to Organic Matter (5E)

On the SSL Primary Characterization Data Sheets, there are several ratios and estimates associated with organic matter, using either estimated or measured C values. For more detailed information on these ratios, their calculations and applications, refer to the SSIR No. 45, Soil Survey Information Manual (1995). Additional information on these ratios and estimates can also be obtained from the SSL upon request.

References
Soil Survey Staff. 1995. Soil survey laboratory information manual. Version No. 1.0. USDA-NRCS. Soil Survey Investigations Report No. 42. U.S. Govt. Print. Office, Washington, DC.

SOIL BIOLOGICAL AND PLANT ANALYSIS (6)

Soil is an ecosystem that contains a broad spectrum of biological components, representing many physiological types (Germida, 1996). Soil biota is critical to soil quality, affecting nutrient cycling, soil stability and erosion, water quality and quantity, and plant health

(USDA-NRCS, 2004). Many of this biota (e.g., fungi, bacteria, earthworms, protozoa, arthropods, and nematodes) and their relationship to soil health are discussed in the USDA-NRCS Soil Biology Primer (Tugel and Lewandowski, 2001). Also refer to Reeder et al. (2001) for information on root biomass and microbial biomass.

The SSL routinely performs several biological analyses as follows: root and plant (above-ground) biomass and nutrient cycling (6C); organic carbon extractions (6A1a1, 6A2a1); separation and total analysis of the soil organic matter fraction (6A4); and microbial biomass characterization (6B). Additionally, carbonates are determined by acid decomposition and CO_2 analysis by gas chromatography (6A3a1a1). This method is more commonly used in soil biochemical and biology studies, where organic C in soils with carbonates may be more precisely determined by subtracting the total carbonate (inorganic C) from total C.

References

Germida, J.J. 1986. Cultural methods for soil microorganisms. p. 263-275. *In* Martin R. Carter (ed.) Soil sampling and methods of analysis. Can. Soc. Soil Sci. Lewis Publ., Boca Raton, FL.

Reeder, J.D., C.D. Franks, and D.G. Milchunas. 2001. Root biomass and microbial biomass. p. 139-166. *In* R.F. Follett, J.M. Kimble, and R. Lal (eds.) The potential of U.S. grazing lands to sequester carbon and mitigate the greenhouse effect. Lewis Publ., Boca Raton, FL.

Tugel, A.J., and A.M. Lewandowski (eds.) 2001. Soil biology primer. Soil Quality – Soil Biology Technical Note No. 1. Available: http://soils.usda.gov/sqi/soil biology primer.htm1.

USDA-NRCS. 2004. Soil biology and land management. Soil Quality – Soil Biology Technical Note No. 4. Available: http://soils.usda.gov/sqi/soil_quality/soil_biology_landmgt.html.

Analyses (6A)
Sample Preparation (1B3b2b1)
0.5 *M* K₂SO₄ Extraction + Heating with Disodium Bicinchoninic Reagent (6A1)
UV-Visible Spectrophotometer, Dual Beam (6A1a)
Hot Water Extractable Organic Carbon (6A1a1)
Air-Dry, <2 mm (6A1a1a1)

1. Application

Hot water soluble soil carbohydrates are thought to be primarily extra-cellular polysaccharides of microbial origin. They help bind soil particles together into stable aggregates. Water stable aggregates reduce soil loss through erosion, increase organic matter, and nutrient content. They also occur as part of the fast or labile organic carbon pool in soils. This labile pool contains the most available carbon for plant, animal and microbial use. The hot water soluble organic C makes up from 4 to 10% of the microbial biomass C determined by chloroform fumigation. It also makes up about 6 to 8% of the total carbohydrate content in the soil. This pool is the most easily depleted of the three organic C pools (Joergensen et al., 1996; Haynes and Francis, 1993).

2. Summary of Method

Water is added to a 10-g soil sample and autoclaved at 1 h at 121°C. Extractable carbohydrates are measured by adding disodium bicinchoninic (BCA) reagent to a 0.5 *M* K₂SO₄ soil extract, heating to 60° C for 2 h, cooling, and reading the absorbance at 562 nm using a spectrophotometer. Glucose is used as a standard and results expressed as glucose-C. Data are reported as mg glucose equivalent-carbon kg^{-1} soil by procedure 6A1a1.

3. Interferences

Carbohydrates from non-microbial sources can be avoided by using the BCA reagent that is selective for microbial carbohydrates. 0.5 M K_2SO_4 extracts of soil are usually supersaturated with $CaSO_4$, with the excess $CaSO_4$ precipitating during storage, especially if samples are frozen and during heating of the extract with BCA reagent (Joergensen et al., 1996). Sodium hexmetaphosphate is added to buffer to prevent $CaSO_4$ precipitation (Joergensen et al., 1996). Aspartic acid is used to chelate Cu^{2+}, preventing undesirable oxidation and so improving precision (Sinner and Puls, 1978).

4. Safety

Wear safety glasses when preparing solutions and handling soil extracts. Use oven mitts, tongs, and other devices to avoid contact with hot water and instruments used.

5. Equipment

5.1 Electronic balance, ±1.0-mg sensitivity
5.2 Steam sterilizer, 121°C-capability
5.3 Erlenmeyer flasks, 125-mL
5.4 Filter paper, Whatman 42, 150 mm
5.5 Pipette, electronic digital, 10000 μL, with tips, 10000 μL
5.6 Test tubes, 10-mL
5.7 Hot water bath, 60°C capability
5.8 Cuvettes, plastic, 4.5-mL, 1-cm light path, Daigger Scientific
5.9 Spectrophotometer, UV-Visible, Varian, Cary 50 Conc, Varian Australia Pty Ltd.
5.10 Computer with Cary WinUV software, Varian Australia Pty Ltd., and printer

6. Reagents

6.1 Reverse osmosis deionized water (RODI), ASTM Type I grade of reagent water
6.2 0.5 M K_2SO_4 solution. In a 1-L volumetric, dissolve 87.135 g K_2SO_4 (dried for 2 h at 110°C) in RODI water. Dilute to volume and invert to mix thoroughly.
6.3 Stock standard glucose solution (SSGS), 20.0 mg glucose L^{-1}. In 1-L of 0.5 M K_2SO_4 solution, dissolve 0.02 g glucose (dextrose). Dilute to volume and invert to thoroughly mix. Store in polyethylene containers. Make fresh weekly. Store in a refrigerator.
6.4 Solution 1: Add 4 g Na_2CO_3, 4 g sodium hexametaphosphate $[(NaPO_3)]_6$, and 0.2 g DL-aspartic acid in 100 mL RODI water. pH solution to 11.25 with NaOH.
6.5 Solution 2: Dissolve 0.48 g bicinchoninic acid in 12 mL RODI water (0.1 M).
6.6 Solution 3: Dissolve 1 g $CuSO_4$ in 25 mL RODI water (0.25 M).
6.7 Disodium bicinchoninic (BCA) reagent: Mix 100 mL Solution 1, 12 mL Solution 2, and 1.8 mL of Solution 3 = BCA reagent. Store in polyethylene containers. Make fresh daily. Store in a refrigerator.
6.8 Standard glucose working solutions (SGWS), 10.0, 5.0, 2.5, 1.25, 0.75, and 0.375 mg glucose L^{-1}. In six test tubes, add 5 mL RODI water. Perform six serial dilutions. Begin as follows: add 5 mL of SSGS to Tube 1 and shake (10.0 mg glucose L^{-1}) and extract 5 mL from Tube 1, add to Tube 2, and shake (5.0 mg glucose L^{-1}). Proceed to make all six SGWS.
6.9 Standard glucose calibration solutions (SGCS). Add 2 mL of each SGWS to a separate test tube, followed by 2 mL BCA reagent. Blank = 2 mL RODI water and 2 mL BCA reagent.

7. Procedure

7.1 Weigh 10 g of <2-mm (sieved), air-dry soil to the nearest mg and place into a 125-mL Erlenmeyer flask If soil is highly organic, weigh 2 g of fine-grind material to the nearest mg.

7.2 Add RODI water to soil at a 1:4 ratio (10 g to 40 mL water or 2 g soil to 8 mL water).

7.3 Autoclave 1 h at 121°C and 15 psi. Cool and filter.

7.4 Pipette 2 mL of each sample extract, 0.5 mL K_2SO_4, and 2 mL disodium BCA reagent into test tubes.

7.5 Place all tubes (SGCS, blank, and samples) in hot water bath for 2 h at 60°C.

7.6 Allow to cool and transfer sample extract and SGCS to cuvettes.

7.7 Set the spectrophotometer to read at 562 nm. Autozero with calibration blank.

7.8 Calibrate the instrument by using the SGCS. The data system will then associate the concentrations with the instrument responses for each SGCS. Rejection criteria for SPCS, if R^2 <0.99.

7.9 Run samples using calibration curve. Sample concentration is calculated from the regression equation.

7.10 If samples are outside calibration range, dilute sample extracts with extracting solution and re-analyze.

8. Calculations

Convert extract glucose equivalent (mg L^{-1}) to glucose equivalent-carbon in the soil (mg kg^{-1}) as follows:

Soil glucose-C (mg kg^{-1}) = [(A x B x C x R x 0.40 x 1000)/E]

where:
A   = Sample reading (mg L^{-1})
B   = Extract volume (L)
C   = Dilution, if performed
R   = Air-dry/oven dry ratio (procedure 3D1)
0.40 = Mass fraction C in glucose
1000 = Conversion factor to kg-basis
E   = Sample weight (g)

9. Report
Report data as to the nearest 0.1 mg glucose equivalent-carbon kg^{-1} soil.

10. Precision and Accuracy
Precision and accuracy data are available from the SSL upon request.

11. References
Haynes, R.J., and G.S. Francis. 1993. Changes in microbial biomass C, soil carbohydrate composition and aggregate stability induced by growth of selected crop and forage species under field conditions. J. Soil Sci. 44:665-675.

Joergensen, R.G., T. Mueller and V.Wolters.1996. Total carbohydrates of the soil microbial biomass in 0.5 M K_2SO_4 soil extracts. Soil Biol. Biochem. 28:9:1147-1153.

Sinner, M., and J. Puls. 1978. Non-corrosive dye reagent for detection of reducing sugars in borate complex ion-exchange chromatography. J.Chrom. 15.

Soil Analyses (6A)
Sample Preparation (1B3b2b1)
0.02 M KMnO$_4$ Extraction (6A2)
UV-Visible Spectrophotometer, Dual Beam (6A2a)
Active Carbon (6A2a1)
Air-Dry, <2 mm (6A2a1a1)

1. Application

This method, commonly called Weil Carbon (Weil et al., 2003), is designed to be a quick and easy field test for the assessment of active soil organic carbon. Following the principle of bleaching chemistry, potassium permanganate (KMnO$_4$) is used as an oxidizer of organic matter present in soil. The oxidized organic matter is considered to be associated with the active carbon pool (Blair et al.,1995). An active soil carbon index can be expressed as the quotient of active carbon to soil organic carbon (Blair et al., 2001). The stability of this index over time is considered to be a useful measure of soil quality (Islam and Weil, 1997).

2. Summary of Method

A 5-g sample is oxidized with 0.02M potassium permanganate diluted with reverse osmosis water. Sample is shaken for 2 min and allowed to stand undisturbed for 5 to10 minutes. A small aliquot of the supernatant is diluted with reverse osmosis water and the absorbance of the solution is read at 550 nm using a spectrophotometer. Extractable carbon by 0.02 M KMnO$_4$ is reported as mg C kg^{-1} soil by procedure 6A2a1.

3. Interferences

Chemical oxidation methods for the determination of labile soil carbon have a number of limitations: different soil samples may have variable amounts of readily oxidizable fractions, making standardization of any method a difficult task, with results influenced by the amount of C in the sample, MnO$_4^-$ concentration, and contact time (Blair et al., 1995).

4. Safety

Wear protective clothing (coats, aprons, and gloves) and eye protection (safety glasses and other devices as appropriate) while preparing reagents and performing procedure. When preparing reagents, exercise special care. Use a vented hood. Thoroughly wash hands after handling all chemicals.

Potassium permanganate is a strong oxidizer. Contact with other material may cause fire. Avoid contact with eyes, skin, and clothing. In case of fire, soak with water. In case of spill, sweep up and remove. Flush spill area with water. Inhalation of KMnO$_4$ dust may be severely damaging to respiratory passages and/or lungs. Contact with skin or eyes may cause severe irritation or burns. Substance is readily absorbed through the skin. See the Material Safety Data Sheet (MSDS) for further information regarding KMnO$_4$.

5. Equipment

5.1 Centrifuge tubes, 50-mL, graduated, polyethylene, with screw tops
5.2 Dropper pipette, 1-mL, graduated, disposable
5.3 Squirt bottle
5.4 Scoop, 5-g (or electronic balance, ±0.01-g sensitivity)
5.5 Petri dish, for sun-drying crumbled soil, if necessary
5.6 Tissues for cleaning cuvettes
5.7 Stopwatch, timer, or a watch with a second hand
5.8 Pipette, electronic digital, 10-mL
5.9 Volumetric flasks, 50, 100, and 200-mL, and 2-L, with stoppers
5.10 Spectrophotometer, UV-Visible, Varian, Cary 50 Conc, Varian Australia Pty Ltd.
5.11 Computer with Cary WinUV software, Varian Australia Pty Ltd., and printer
5.12 Cuvettes, plastic, 4.5-mL, 1-cm light path, Daigger Scientific

6. Reagents

6.1 Reverse osmosis deionized (RODI) water, ASTM Type I grade of reagent water

6.2 $CaCl_2 \cdot 2\,H_2O$, 0.1 M solution. Dissolve 29.40 g of $CaCl_2 \cdot 2\,H_2O$ in a 2-L volumetric flask with 1L of RODI. Bring to 2L volume. Store in a polyethylene bottle.

6.3 KOH, 0.1 M solution. Dissolve 0.561 g of potassium hydroxide with 50-mL of RODI water in a 100-mL volumetric flask. Bring to volume with RODI water. Invert to mix thoroughly.

6.4 Stock $KMnO_4$ solution, 0.2 M in 0.1 M $CaCl_2$ solution (pH 7.2). In a 200-mL volumetric, dissolve 6.32 g of potassium permanganate crystals in 100-mL of 0.1 M $CaCl_2$. Bring to volume with 0.1 M $CaCl_2$. Adjust solution pH to 7.2 with 0.1 M KOH (usually requires 1-2 drops of KOH). Solution is stable for approximately 3 days.

6.5 Standard $KMnO_4$ working (SKMnO₄WS) solutions, 0.04, 0.02, 0.01 and 0.005 M $KMnO_4$. Prepare fresh weekly. Store in a refrigerator. Allow to equilibrate to room temperature before use. In four 50-mL volumetric flasks add as follows: 10, 5, 2.5, and 1.25 mL of Reagent 6.4 (0.2 M $KMnO_4$, in 0.1 M $CaCl_2$ solution, pH 7.2). Bring to volume with RODI water. Invert to mix thoroughly.

6.6 Standard $KMnO_4$ calibration (SKMnO₄CS) solutions, 0.0004, 0.0002, 0.0001, 0.00005, and 0 M $KMnO_4$. To four 50-mL volumetric flasks, add 0.5-mL aliquots of the SKMnO₄WS. For the blank, add 5 mL 0.1 M $CaCl_2$. Make fresh daily. Bring to volume with RODI water. Invert to mix thoroughly.

7. Procedure

7.1 Add 15-mL RODI water to a clean, labeled centrifuge tube. Add 2 mL of Reagent 6.4 solution to the centrifuge tube, taking 1 mL twice and washing the pipette with the dilute solution in the tube. Add RODI water to make to 20-mL volume.

7.2 Weigh or scoop 5 g of <2-mm (sieved), air-dry soil to the nearest mg and add to the centrifuge tube and cap it. Avoid coarse fragments and plant material.

7.3 Shake vigorously for 2 min and allow to stand for 5 to 10 min to settle the soil. Do not disturb during settling period.

7.4 Add 45 mL of RODI water to a clean, labeled centrifuge tube. Transfer 0.5 mL of the supernatant solution into the tubes using a graduated 1-mL dropper pipette. Rinse the pipette with the dilute solution from the tube. Add RODI water to make to 50-mL volume. Invert to mix thoroughly. Repeat for each sample. Standards (SKMnO₄CS) have been treated in the same manner as described in Section 6.6.

7.5 Transfer sample extracts and SKMnO₄CS to cuvettes.

7.8 Set the spectrophotometer to read at 550 nm. Autozero with calibration blank.

7.9 Calibrate the instrument by using the calibration solutions. The data system will then associate the concentrations with the instrument responses for each calibration solution. Rejection criteria for calibration, if R^2 <0.99.

7.10 Run samples using calibration curve. Sample concentration is calculated from the regression equation. Record results to the nearest 0.01 unit for the sample extract and each calibration solution.

7.11 If some samples have zero absorbance, reweigh at smaller sample size (e.g., 2.5 g) and re-analyze. Samples that have zero absorbance are those that have large amounts of active carbon.

8. Calculations

The bleaching of the pink KMnO4 color (reduction in absorbance) is proportional to the amount of oxidizable C in soil, i.e., the greater the KMnO4 color loss (the lower the absorbance reading), the greater the amount of oxidizable C in the soil (Weil et al., 2003). To estimate the amount of C oxidized, use the assumption of Blair et al. (1995) that 1 mol MnO_4 is consumed (reduced from Mn^{7+} to Mn^{2+}) in the oxidation of 0.75 mol (9000 mg) of C as follows:

$$KMnO_4 \; C \; (mg \; kg^{-1}) = [0.02 \; mol \; L^{-1} - A] \; x \; (9000 \; mg \; C \; mol^{-1}) \; x \; (0.02 \; L \; solution/0.005 \; kg)$$

where:
0.02 mol L^{-1} = initial solution concentration
A   = analyte reading (mol L^{-1})
9000   = mg C (0.75 mole) oxidized by 1 mole of MnO_4 changing from Mn^{7+} to Mn^{2+}
0.02 L   = volume of KMnO4 solution reacted
0.005   = kg of soil used

9. Report

Report data to the nearest 0.1 mg C kg^{-1} soil.

10. Precision and Accuracy

Precision and accuracy data are available from the SSL upon request.

11. References

Blair, G.J., R.Lefroy, and L. Lise. 1995. Soil carbon fractions based on their degree of oxidation, and the development of a carbon management index for agricultural systems. Australian J. Agric. Res. 46:1459-1466.

Blair, G.J., R. Lefroy, A.Whitbread, N. Blair, and A. Conteh. 2001. The development of the KMnO4 oxidation technique to determine labile carbon in soil and its use in a carbon management index. p. 323-337. *In* R. Lal. J. Kimble, R. Follet, and B. Stewart (eds.) Assessment methods for soil carbon. Lewis Publ. Boca Raton, FL.

Islam, K.R., and R.R. Weil. 2000. Soil quality indicator properties in mid-Atlantic soils as influenced by conservation management. J. Soil and Water Conserv. 55:69-78.

Weil, R. R., R. I. Kandikar, M.A. Stine, J. B. Gruver, and S. E. Samson-Liebig. 2003. Estimating active carbon for soil quality assessment: A simplified method for laboratory and field use. Am. J. Alternative Agric. 18(1):3-17.

Soil Analyses (6A)
Sample Preparation (1B3b2b1)
Acid Dissolution (6A3)
1 N HCl + $FeCl_2$ (6A3a)
CO_2 Analysis (6A3a1)
Gas Chromatography (6A3a1a)
Carbonates (6A3a1a1)
Air-Dry, <2 mm (6A3a1a1a1)

1. Application

Methods involving determination of CO_2 have usually been preferred for measuring soil carbonate (Loeppert and Suarez, 1996). CO_2 released can be measured gravimetrically (Allison, 1960; Alllison and Moodie, 1965), titrimetrically (Bundy and Bremmer, 1972), manometrically (Martin and Reeve, 1955; Presley, 1975), volumetrically (Dreimanis, 1962), spectrophotometrically by infrared spectroscopy, or by gas chromatography (Loeppert and Suarez, 1996). The SSL routinely determines the amount of carbonate in the soil by treating the

$CaCO_3$ with HCl, with the evolved CO_2 measured manometrically (procedures 4E1a1a1a1-2 for <2-mm and 2- to 20-mm bases, respectively). The method herein describes soil carbonate by acid decomposition and CO_2 analysis by gas chromatography. This method is more commonly used in soil biochemical and biology studies, where organic C in soils with carbonates may be more precisely determined by subtracting the total carbonates (inorganic C) from total C (procedure 4H2a1).

2. Summary of Method

Soil carbonate is determined by chromatographic analysis of CO_2 evolved upon acidification of soil in a closed system of known headspace. Ferrous iron ($FeCl_2$) is added to the acid as an anti-oxidant, and the dilute acid solution ($1N$ HCl) is chilled before addition to soil to minimize the decarboxylation of organic matter by the acid. Data are reported as mg CO_2-C per g of soil to the nearest 0.1 g (6A3a1a1). These data can be used to estimate soil organic carbon by subtracting (CO_2-C x 0.2727) from total carbon (4H2a1).

3. Interferences

It is essential that precautions be taken to ensure that there is no interference from organic matter oxidation (Loeppert and Suarez, 1996). This procedure may be more appropriate for soils with relatively low amounts of carbonates (<15%).

4. Safety

Wear protective clothing (coats, aprons, sleeve guards, and gloves) and eye protection (face shields, goggles, or safety glasses) when handling acids. Thoroughly wash hands after handling acids. Use the fume hood when diluting concentrated HCl. Use the safety showers and eyewash stations to dilute spilled acids. Use sodium bicarbonate and water to neutralize and dilute spilled acids.

5. Equipment

5.1 Canning jars, 1-qt (0.984-L), with lids fitted with gas-sampling septa
5.2 Syringe, 20-mL, with 18-gauge needle
5.3 Disposable syringe, 1-mL, for gas sampling
5.4 Needle, 18 or 20 gauge, for venting jars
5.5 Electronic balance, ±0.01-g sensitivity
5.6 Gas chromatograph (GC) with thermal conductivity detector (TCD)
5.7 Beaker, glass, 600-mL
5.8 Filter paper, Whatman 42, 150 mm
5.9 Stirrer

6. Reagents

6.1 Reverse osmosis deionized (RODI) water, ASTM Type I grade of reagent water
6.2 $1N$ HCl with 3.1 g $FeCl_2$ added per 100 mL of solution, chilled to 4 to 5°C. Prepare 500 mL of solution of as follows: weigh 15.5 g $FeCl_2$ into 600-mL beaker; add 20-mL RODI water; stir (medium speed) until all crystals have dissolved; filter solution to remove insoluble ferric iron particulates; transfer to 500-mL volumetric; add 41.7 mL concentrated HCl to aqueous $FeCl_2$; bring to volume with RODI water; and chill solution to 4 to 5 °C.

7. Procedure

7.1 Determine water content of soil sample so results may be expressed on an oven-dry soil basis. Refer to water retention methods (3C).

7.2 Weigh 5g of <2-mm (sieved), air-dry soil to the nearest 0.01 g into clean 1-qt jar.

7.3 Tightly seal jar with lid fitted with gas sampling septum. Also include one or more jars with no soil added to serve as reagent controls and CO_2 background indicators.

7.4 Add 20 ml of RODI water (room temperature) to jar through septum with 20-mL syringe, venting jar with 18 or 20 gauge needle.

7.5 Add 20 mL of chilled $1N$ HCl + $FeCl_2$ solution through septum as in Section 7.4 (do not vent), let stand 2 h, swirling soil-acid solution occasionally.

7.6 Take gas sample for gas chromatograph analysis using 1-mL syringe. Purge syringe several times before withdrawing final sample for analysis.

8. Calculations

Calculation of Soil Carbonate mass from CO_2% volumetric concentrations (Kettler and Doran, 1995).

8.1 Reaction equation:

$$CaCO_3 + 2HCl\ (aq) \longrightarrow CO_2 \uparrow + H_2O + CaCl_2\ (aq)$$

mg CO_2-C produced = mg CO_3-C dissolved by acid in soil

mg CO_2-C produced/OD = mg CO_3-C/OD

where:
OD = Oven-dry soil (g)

8.2 Net jar headspace
Volume of Empty 1-qt. jar = $978 cm^3$ (measured by H_2O volume displacement).

Soil Solid Volume (cm^3) = [g Moist Soil/(1 + %H_2O/100)]/(Particle Density)

= (5.00 g sample/(1 + %H_2O/100))/(2.65g cm^{-3})

where:
2.65 = assumed particle density (g cm^{-3})

Net jar Headspace (cm^3) = Empty Jar - Soil Solid Volume - Soil H_2O Volume – Liquid Volume

= 978 cm^3 - (Soil Solid Volume, cm^3) - (g Oven-dry Soil/1+ % H_2O/100) - (20 cm^3 H_2O + 20 cm^3 Acid)

= 978 cm^3 - 1.89 cm^3 - 40 cm^3

= 936.1 cm^3 (936 to 935.4 cm^3 for 3 - 30% H_2O)

where:
5.00 = g soil oven-dry basis, assumed

8.3 Gaseous CO_2-Carbon produced:

This step is important for soils with >15% carbonates.

mg CO_2-Carbon produced by soil carbonate decomposition and detected as CO_2 in vapor space

= 4.594 mg CO_2-Carbon / atm x ((Mole % CO_{2s} x P_{ts}, atm) - (Mole % CO_{2b} x P_{tb}, atm))

where:
P_{tb} = Total pressure in blank jar by electronic manometer
P_{ts} = Total pressure in sample jar by electronic manometer
Mole % CO_{2b} = Mole % CO_2 in blank jar by GC
Mole % CO_{2s} = Mole % CO_2 in sample jar by GC

8.4 CO_2-Carbon produced but dissolved in solution:

mg CO_2-Carbon produced by soil carbonate decomposition but dissolved in solution

= (0.163 mg CO_2-C / atm) x ((Mole % CO_{2s} x P_{ts}, atm) - (Mole % CO_{2b} x P_{tb}, atm))

Total mg CO_3-Carbon / OD = Total mg CO_2-Carbon / OD

= {[(0.163 mg CO_2-C / atm) x ((Mole % CO_{2s} x P_{ts}, atm) - (Mole % CO_{2b} x P_{tb}, atm))] + [4.594 mg CO_2-Carbon / atm x ((Mole % CO_{2s} x P_{ts}, atm) - (Mole % CO_{2b} x P_{tb}, atm))]}/{FM/FMOD)}

= {[4.757 mg CO_2-C / FM / atm] x [(Mole % CO_{2s} x Pts, atm) - (Mole % CO_{2b} x P_{tb}, atm)]} x (FMOD)

where:
OD   = Oven-dry soil (g)
FM   = Field-moist soil (g)
FMOD = Field-moist soil/oven-dry ratio (g/g) (procedure 3D2)

8.5 Convert mg CO_2-C per g of soil to percent CO_2-C in soil as follows:

(mg CO_2-C/g soil) (1g / 1000mg) (100g soil) = Percent CO_2-C in soil

9. Report
       Report % CO_2-C in soil to the nearest 0.1%. These data can be used to estimate soil organic carbon by subtracting (CO_2-C x 0.2727) from total carbon (4H2a1).

10. Precision and Accuracy
       Precision and accuracy data are available from the SSL upon request.

11. References
     Allison, L.E. 1960. Wet combustion apparatus and procedure for organic and inorganic carbon in soil. Soil Sci. Soc. Am. Proc. 24:36-40.
     Allison, L.E., and C.D. Moodie. 1965. Carbonate. p. 1379-1400. *In* C.A. Black, D.D. Evans, J.L. White, L.E. Ensminger, and F.E. Clark (eds.) Methods of soil analysis. Part 2. 2nd ed. Agron. Monogr. 9. ASA, CSSA, and SSSA, Madison, WI.
     Bundy, L.G., and J.M. Bremmer. 1972. A simple titrimetric method for determination of inorganic carbon in soils. Soil Sci. Soc. Am. Proc. 36:273-275.
     Dreimanis, A. 1962. Quantitative gasometric determination of calcite and dolomite by using Chittick apparatus. J. Sediment. Petrol. 32:520-529.
     Kettler, T., and J. Doran. 1995. Determination of soil carbonate concentration by acid decomposition and GC CO_2 analysis. USDA-ARS, University of Nebraska, Lincoln (unpublished).

Loeppert, R.H., and D.L. Suarez. 1996. Carbonate and gypsum. *In* D.L. Sparks (ed.) Methods of Soil Analysis. Part 3–Chemical methods. ASA and SSSA, Madison, WI

Martin, S.E., and R. Reeve. 1955. A rapid manometric method for determining soil carbonate. Soil Sci. 79:187-197.

Presley, B.J. 1975. A simple method for determining calcium carbonate in sediment samples. J. Sediment. Petrol. 45:745-746.

Particulate Organic Matter and C-Mineral (6A4)
Sample Preparation (1B3b2a1)
Total Analysis (6A4a)
Dry Combustion (6A4a1)
Thermal Conductivity Detector (6A4a1a)
Carbon, Nitrogen, Sulfur (6A4a1a1-3)
Air-Dry (6A4a1a1-3a)
>53 μm, Particulate Organic Matter (6A4a1a1-3a1)
<53 μm, C-Mineral (6A4a1a1-3a2)

1. Application

Particulate organic matter (POM) is a physical fraction of the soil >53μm in diameter (Elliott and Cambardella, 1991; Cambardella and Elliott 1992; Follett and Pruessner, 1997). Some researchers combine this fraction with the fast or labile pool. Others have described this pool as slow, decomposable, or stabilized organic matter (Cambardella and Elliott, 1992). To avoid confusion, this fraction may best described as representing an intermediate pool with regards to decomposition. This fraction is similar to various sieved and physical fractions such as the resistant plant material (RPM) (Jenkinson and Rayner, 1977), and size fractions (Gregorich et al., 1988), and variously determined light fractions of the soil organic matter (Strickland and Sollins, 1987; Hassink, 1995).

Under tillage, the POM fraction becomes depleted (Jenkinson and Rayner, 1977; Cambardella and Elliott, 1992). Reductions of more than 50% have been found in long-term tillage plots (20 yr.). Measurable reductions are believed to occur in the range of 1 to 5 years (Cambardella and Elliott, 1992).

When paired samples are selected either in time or between two tillage treatments a comparison can be made to determine the impact of the tillage practice. POM can be used in soil organic matter modeling, as a soil quality indicator and as an indicator of the SOM that can move into the active C pool.

Since the late 1970's several models have been developed to estimate the dynamics of organic matter in the soil. All of these models have at least two phases, slow and rapid. In measuring these two phases chemical fractionation (humic and fulvic acids) has been found to be less useful than physical fractionation (Hassink, 1995). Examples of some of these models can be found in Jenkinson and Rayner (1977), tests of the CENTURY Soil Organic Model, (Parton et al., 1987; Metherell et al., 1993; and Montavalli et al., 1994). A minimum data set for soil organic carbon is proposed by Gregorich et al (1994) that includes POM as one of the primary parameters.

2. Summary of Method

The procedural steps described herein encompasses the physical separation (1B3b2a) of the soil organic matter (<2 mm) into two fractions: (1) >53-μm, POM and (2) <53 μm, C-Mineral (C-Min) (Cambardella and Elliot, 1992; Follett and Pruessner, 1997) and the analysis of these two fractions for total C, N, and S by procedure 6A4a1a1-3, respectively. Typically, this procedure is determined on the A horizons (Soil Survey Staff, 1999) because detectable levels of both C and N are most likely to occur in this horizon.

3. Interferences

In some weathered soils there is approximately the same amount of C in both fractions. To date, no research has been done to establish the interpretation of this result. Charcoal in native sod that has been historically burned if residence time was to be determined from the two fractions, does not affect the POM determination and C and N analysis themselves.

4. Safety

Always wear safety glasses when working with glass containers.

5. Equipment

5.1 Pressure regulator for water, with stop cock attached to tubing
5.2 Sieve, 10 mesh, 2 mm
5.3 Sieve, 270 mesh, 53 μm
5.4 Mechanical reciprocating shaker, 200 oscillations min^{-1}, 1 ½ strokes, Eberbach 6000, Eberbach Corp., Ann Arbor, MI
5.5 Glass, Pyrex, pie or round cake pans
5.6 Evaporating crucibles
5.7 Drying Oven (110°C)

6. Reagents

6.1 Reverse osmosis deionized water (RODI), ASTM Type I grade of reagent water

7. Procedure

Physical Separation of Organic Matter (POM and C-Min)

7.1 Remove large pieces of organic matter, roots, plant residue, gravel, wood material from an air dry soil sample. Do not shake. Save the large fraction and weigh.

7.2 Weigh 10 g of <2-mm (sieved), air-dry soil to the nearest mg into a 125-mL Erlenmeyer flask. Add 30 mL of RODI water to sample.

7.3 If the soil has carbonates, take a sub-sample and measure the inorganic carbon following the gas chromatograph procedure (6A3a1a1).

7.4 Stopper sample tightly and shake for 15 h (overnight) at 200 oscillations min^{-1} at room temperature (20°C ± 2°C).

7.5 Sieve the soil through a 53-μm sieve. The POM fraction will remain on top of the sieve and the C-min will be a slurry that is collected in a pie pan underneath the sieve. Use the regulator that is attached to the RODI water to rinse the POM with a steady gentle stream of water. Keep rinsing until the water that comes through the sieve is clear. Capture all the soil slurry and water (C-Min fraction) that passes the sieve.

7.6 Label and tare a glass pie pan and an evaporating crucible to the nearest 0.01 g.

7.7 Transfer the POM into an evaporating crucible by rinsing the sieve, including the sides, with a small amount of RODI water.

7.8 Transfer the C-Min slurry into a glass pie pan for drying. Rinse the contents of the pie pan into the labeled glass pie pan.

7.9 Dry the two fractions (POM, C-Min) in an oven at 110 ° C oven. The C-Min fraction may require 48 h to dry, depending on how much water was used to rinse the sample.

7.10 Once the samples are dry, let them cool briefly and record the weight to the nearest 0.1 mg.

7.11 Transfer the entire contents for each fraction to an appropriately labeled scintillating vial.

Total Carbon, Nitrogen, and Sulfur Analysis
7.12 Determine total C, N, and S for POM and C-min fractions (6A4a1a1-3), using fine-grind samples ($\approx$ 180 μm). Refer to 4H2a1-3 for the remaining procedural steps for 6A4a1a1-3.

8. Calculations

Calculate POM-C and C-Min using soil bulk density values determined by procedures 4A or ASTM method D-2167 (American Society of Testing and Materials, 2004) and total C values determined by procedures 6A4a1a1.

8.1 (%Total C of POM Fraction/100) x (POM (g)/10) x FMOD = POM-C g/g soil

where:
FMOD = Field-moist/oven-dry ratio (procedure 3D2)

8.2 (%Total C of C-Min Fraction/100) x C-Min (g)/10) x FMOD = C-Min g/g soil

8.3 (% Total C of POM fraction/100) x [POM (g)/10] x Bulk Density x 100,000 x Depth = POM-C kg ha^{-1} at given depth interval

8.4 (% Total C of C-Min fraction/100) x [C-Min (g)/10] x Bulk Density x 100,000 x Depth = C-Min kg ha^{-1} at given interval depth.

Use the above equations for similar computations of N and S (6A4a1a2-3, respectively).

9. Report
Report POM-C and C-Min in kg ha^{-1} at a given depth interval (cm). Report separately similarly calculated values for N and S. Report the percent >2-mm fraction.

10. Precision and Accuracy
Precision and accuracy data are available from the SSL upon request.

11. References
American Society for Testing and Materials. 2004. Standard practice for density and unit weight of soil in place by the rubber balloon method. D 2167. Annual book of ASTM standards. Construction. Section 4. Soil and rock; dimension stone; geosynthesis. Vol. 04.08. ASTM, Philadelphia, PA.

Cambardella, C.A., and E.T. Elliott. 1992. Particulate soil organic-matter changes across a grassland cultivation sequence. Soil Sci. Soc. Am. J. 56:77-783.

Elliott, E.T., and C.A. Cambardella. 1991. Physical separation of soil organic matter. Agric. Ecosyst. Environ. 34:407-419.

Follett, R.F and E. Pruessner. 1997. POM procedure. USDA-ARS, Ft. Collins, CO.

Gregorich, E.G., M.R. Carter, D.A. Angers, C.M. Monreal, and B.H. Ellert. 1994. Towards a minimum data set to assess soil organic matter quality in agricultural soil. Can. J. Soil Sci. 74:367-385.

Gregorich, E.G., R.G. Kachanoski, and R.P. Voroney. 1988. Ultrasonic dispersion of aggregates: Distribution of organic matter in size fractions. Can. J. Soil. Sci. 68:395-403.

Hassink, J. 1995. Decomposition rate constants of size and density fractions of soil organic matter. Soil Sci. Soc. Am. J. 59:1631-1635.

Jenkinson, D.S., and J.H. Rayner. 1977. The turnover of soil organic matter in some of the Rothamsted classical experiments. Soil Sci. 123:298-305.

Metherell, A.K., L.A. Harding, C.V. Cove, and W.J. Parton. 1993. CENTURY soil organic matter model environment, technical documentation agroecosystem, Version 4.0. Great Plains System Res. Unit Tech. Rep. No. 4, USDA-ARS, Ft. Collins, CO.

Montavalli, P.P., C.Q. Palm, W.J. Parton, E.T. Elliott, and S.D. Frey. 1994. Comparison of laboratory and modeling simulation methods for estimating soil carbon pools in tropical forest soils. Soil Biol. Biochem. 26:935-944.

Parton, W.J., D.S. Ojima, C.V. Cole, and D.S. Schimel. 1994. A general model for soil organic matter dynamics: sensitivity to litter chemistry, texture and management. p. 147-167. *In* Quantitative modeling of soil forming processes, SSSA Spec. Publ. 39. SSSA, Madison, WI.

Soil Survey Staff. 1999. Soil taxonomy: A basic system of soil classification for making and interpreting soil surveys. USDA-NRCS Agric. Handb. 436. 2nd ed. U.S. Govt. Print. Office, Washington, DC.

Strickland, T.C., and P. Sollins. 1987. Improved method for separating light- and heavy-fraction organic material from soils. Soil Sci. Soc. Am. J. 51:1390-1393

Fumigation Incubation (6B)
Sample Preparation (1B3b1a1)
Gas Chromatography (6B1)
CO_2 Analysis (6B1a)
Microbial Biomass (6B1a1)
Field-Moist, <2 mm (6B1a1a1)
2 *M* KCl Extraction (6B2)
Automatic Extractor (6B2a)
Ammonia – Salicylate (6B2a1)
Flow Injection, Automated Ion Analyzer (6B2a1a)
N as NH_3 (Mineralizable Nitrogen) (6B2a1a1)
Field-Moist, <2 mm Fumigated and <2 mm Non-Fumigated (6B2a1a1a1-2)

1. Application

Soil microorganisms are an important component of soil organic matter. One of their functions is to breakdown non-living organic matter in the soil. A variety of methods exist to measure the biomass of living soil microbes. The method described herein by the chloroform fumigation incubation (Jenkinson and Powlson, 1976) with modifications, and CO_2 evolution measurement by gas chromatography. Mineralizable N may also be determined on microbial biomass.

2. Summary of Method

A freshly collected soil sample is weighed into two separate vials. One sample is fumigated using chloroform and the other is used as a control (non-fumigated). After fumigation both the fumigated and non-fumigated samples are brought up to 55% water filled pore space (WFPS) (Horwath and Paul, 1994). Both samples are placed in a sealed container and aerobically incubated for 20 days. During this incubation period it is assumed that normal respiration occurs in the control sample container. The fumigated sample having a large carbon source for food, supplied from the dead microorganisms, has a higher CO_2 production. At the end of 10 days respiration readings are taken on both the control and the fumigated sample to determine the amount of CO_2 evolved by gas chromatography. The CO_2 level of the control sample is also measured at the end of 20 days. CO_2 produced by biomass flush (g CO_2-C/g of soil) and soil microbial biomass (kg C/ha for a given depth interval) are reported by procedure 6B1a1. Mineralizable N by 2 *M* KCl extraction may also be determined on microbial biomass using a flow injection automated ion analyzer by procedure 6B2a1a. Mineralizable N is reported as mg N kg^{-1} soil as NH_3.

3. Interferences

The determination of CO_2 evolution by gas chromatography gives a rapid and accurate measurement and can be used in acidic soils. However, this technique is prone to error in neutral and alkaline soils (Martens, 1987), as accumulation of carbonate species in the soil solution can lead to lowered CO_2 determinations (Horwath and Paul, 1994).

4. Safety

All fumigation work needs to be conducted in an adequate fume hood because chloroform has carcinogenic-volatile properties. Never determine residual chloroform by sense of smell. Make sure the vacuum pump is maintained to ensure proper operations.

5. Equipment

5.1 Face Shield
5.2 Goggles
5.3 Rubber Apron
5.4 Rubber Gloves
5.5 Chloroform spill kit
5.6 Fume hood, 100-fpm face velocity
5.7 Vacuum chambers, fiberglass
5.8 Incubator, 25°C
5.9 Vacuum pump, 26 and 14 in Hg, organic/oil free
5.10 Mason jars, 1-qt, with lids and septa
5.11 Vials, 60-mL glass, with snap caps, for samples
5.12 Beakers, 100-mL
5.13 Refrigerator, for sample and titrate storage
5.14 Sieve, 10 mesh, 2 mm
5.15 Electronic balance, ±0.01-g sensitivity
5.16 Electronic balance, ±1.0-mg sensitivity
5.17 Oven, 110°C
5.18 Permanent marker
5.19 Paper Towels
5.20 Tongs, 12 in
5.21 Mechanical vacuum extractor, 24-place, SAMPLETEX, MAVCO Industries, Lincoln, NE
5.22 Tubes, 60-mL polypropylene, for extraction tubes
5.23 Rubber tubing, 3.2 ID x 1.6 OD x 6.4 mm (1/8 ID x 1/16 OD x 1 in) for connecting syringe barrels
5.24 Containers, polycon
5.25 Aluminum weighing pans, 60 mm diameter x 15 mm depth
5.26 Gas chromatograph (GC) with thermal conductivity detector (TCD)
5.27 GC syringes, 1-mL

Reagents

6.1 Reverse osmosis deionized (RODI), ASTM Type I grade of reagent water
6.2 Chloroform stabilized in amylene. Use purified chloroform within 3 weeks.
6.3 Helium, compressed gas

7. Procedure

7.1 Until sample preparation and analysis, keep soils moist and refrigerated.

7.2 Weigh soil to the nearest 0.01 g for bulk density determination. Remove a 15 to 20 g sample for water content. Sieve moist soil to <2 mm. Sub-sample (10 to 15 g) for post-sieve water content. Refrigerate samples until analyses can be performed. Dry post-sieve water content samples at 110°C overnight. Weigh samples the following day.

7.3 Prepare 1 sample for the Fumigated Day 10 (F 10) and 1 for the Non-Fumigated Day 20 (NF 20). Prepare replicates for each sample.

7.4 Mark the volume on the glass sample container corresponding to 20 mL (or more if a larger quantity of soil is needed).

7.5 Label the non-fumigated container clearly with permanent marker.

7.6 Use etched containers for the fumigated set, as chloroform can dissolve written labels.

7.7 Weigh enough moist soil (nearest mg) to achieve approximately 25 g (or 50 g) oven-dry soil into 60-mL glass vial. Use soil moisture content conversion. Adjust soil, by gently tapping against the counter, so that it is leveled off at the bulk density line. Carefully add RODI water with a dropper to bring the moisture up to 55% WFPS, using the moisture content conversion. Make the surface as uniformly moist as possible.

7.8 Cap the vials and refrigerate samples overnight to equilibrate.

Fumigated Samples
7.9 Line the vacuum chambers with wet paper towels to prevent desiccation.

7.10 Place the vacuum chambers in the fume hood.

7.11 Place beaker with 30 to 40 mL of stabilized chloroform in the chambers. Evacuate at 14" Hg to drive off the amylene stabilizer. A volume change will be visible, approximately 5 to 10 mL.

7.12 Place the sample vials in the vacuum chambers.

7.13 Place the pure chloroform into the pan of the vacuum chambers.

7.14 Fumigate samples for 24 h.

7.15 Evacuate the fumigated samples 4 times for approximately 15 min at 27" Hg to drive off the chloroform.

Fumigated and Non-fumigated Samples
7.16 Add 5 to 10 mL of RODI water to the bottom of the mason jar to prevent desiccation.

7.17 Seal mason jars securely with rings and lids. Lids must be airtight during the incubation.

7.18 Seal mason jars securely with rings and lids. Lids must be airtight during the incubation. Incubate samples at 25° C for 10 days.

Day 10 Samples
7.19 Remove mason jars from incubator. Proceed to Section 7.26 for analysis.

7.20 Remove all F 10 samples from the mason jars.

7.21 Cap F 10 samples and store at 4° C until they can be extracted for mineralizable N (2 M KCl). If a microbial inhibitor is used, samples can be stored in the refrigerator for up to two weeks before analysis. For longer periods they should be frozen. For extraction, proceed to Section 7.29.

7.22 Incubate NF 20 samples at 25° C for 10 days. Make sure the mason jar lids are still sealing. If not, replace with new lids

Day 20 Samples

7.23 Remove mason jars from incubator. Proceed to Section 7.26 for analysis.

7.24 Remove all NF 20 samples from the mason jars.

7.25 Cap samples and store at 4° C until they can be extracted for mineralizable N (procedure 6B3a1a1). If a microbial inhibitor is used, samples can be stored in the refrigerator for up to two weeks before analysis. For longer periods they should be frozen. For extraction, proceed to Section 7.29

Gas Chromatography

7.26 Measure the CO_2 accumulated in the headspace of the mason jars by gas chromatography.

7.27 Refer to manufacturer's manual for operation of the gas chromatograph.

7.28 Calibration curves and retention times for gas under analysis is established by analyzing the certified standard gas mixture (1% CO_2) by the procedure used for analysis of the sample. Flow rate is 30 mL min^{-1}. Monitor the baseline prior to analysis.

2 M KCl Extraction

7.29 Mix fumigated replicates together. Mix non-fumigated replicates together.

7.30 Weigh 5 g of moist soil to the nearest mg into 60-mL polypropylene extraction tubes. Tube will need to be tapped and rinsed with 2 M KCl in order to get the moist soil to bottom of tube.

7.31 Set-up vacuum extractors. Add 25 mL of 2 M KCl to extraction tubes. Extract for 1 h.

7.32 Following extraction, transfer contents of tubes into polycon containers. Proceed with determining mineralizable N. Also analyze N in reagent RODI water as blanks. Refer to 4D10a1a1 for the remaining procedural steps for 6B2a1a1.

8. Calculations

8.1 Calculate the bulk density of 25 g of <2-mm, field-moist soil in a 100-mL beaker manually compressed to 20 cm^3 volume.

$$Db_1 \text{ (g cm}^{-3}) = 25 \text{ g} /(1 + H_2O_f)/20 \text{ cm}^3$$

where:
Db_1 = Bulk density (g cm^{-3})
H_2O_f = Field water content (g g^{-1})

H_2O_f is determined by procedures 4B.

8.2 Calculate the gravimetric water content [Gravimetric $H_2O_{0.55}$ (g/g)] required for soil to be at 55% water filled pore space (WFPS), using calculated Db_1 in Section 8.1 and an assumed particle density of 2.65 g cm^{-3}:

Gravimetric $H_2O_{0.55}$ (g/g) = $\{0.55 \times [1 - (Db_1/2.65)]\}/Db_1$

where:
Gravimetric $H_2O_{0.55}$ (g/g) = 55% water filled pore space (WFPS)
2.65 = Assumed particle density (g cm^{-3})

8.3 Calculate the additional gravimetric water needed for soil to reach 55% WFPS:

H_2O_{add} = $\{[$Gravimetric H_2O 0.55 (g/g)$] - (H_2O_f)\} \times [W / (1 + H_2O_f)]$

where:
W = Weight of soil (g)
H_2O_{add} = Amount of water to add to reach 55% WFPS (g)

8.4 Determine the CO_2-Carbon produced by biomass flush during 10 day incubation.

8.4.1 At 10 days, determine the mole % concentration of CO_2 in head space of jar by GC analysis.

8.4.2 Calculate the CO_2-Carbon produced by biomass flush as follows:

CO_2-Carbon produced by biomass flush, g $BioCO_2$-Carbon / g FM Soil =

[(mole %CO_2, 10 days, fumigated - mole %CO_2, 10 days, non-fumigated) x (0.47 g CO_2-C)]/W

where:
FM = Field-moist soil (g)

8.5 Determine the difference between fumigated CO_2-Carbon produced during the 10 to 20 day incubation period and the non-fumigated CO_2-Carbon produced during the same period.

8.5.1 At 20 days, determine the mole % concentration of CO_2 in head space of jar by GC analysis.

8.5.2 Calculate the (fumigated - non-fumigated) CO_2-Carbon produced during the 10 to 20 day incubation period as follows:

(fumigated - non-fumigated) CO_2-Carbon produced during the 10 to 20 day incubation period, g $BioCO_2$-Carbon / g FM Soil =

$\{[$(mole %CO_2, 20 days, fumigated - mole %CO_2, 10 days, fumigated) - (mole % CO_2, 20 days, non-fumigated - mole %CO_2, 10 days, non-fumigated)$] \times (0.47$ g CO_2-C$)\} / W$

8.6 Calculate the Soil Biomass Flush (kg CO_2-C/ha), using the bulk density value (Db_2) determined by procedure 4A or ASTM method D-2167:

Soil biomass flush (kg CO_2-Carbon / ha) = (g $BioCO_2$-Carbon / g FM Soil) x (g FM Soil / g OD soil) x (Db_2: g OD soil / cm^3 FM Soil) x (1 kg CO_2-Carbon / 1000 g CO_2-Carbon) x (100000000 cm^2 / ha) x (layer thickness, cm)

8.7 Calculate Soil Microbial Biomass (kg C/ha for a given depth interval):

Soil Biomass Flush (kg C/ha) / 0.41

$0.41 = K_c$, fraction of biomass C mineralized to CO_2 (Anderson and Domsch, 1978)

8.8 Calculate (fumigated – non-fumigated) mineralizable N (mg kg^{-1}):

Fumigated N – Non-fumigated N = $\{[(F_1-F_2) \times A \times B \times E \times 1000] / C\} - \{[(N_1-N_2) \times D \times F \times E \times 1000]\} / G\}$

where:
F_1 = Analyte reading, fumigated (mg L^{-1})
F_2 = Blank reading, reagent RODI Water, fumigated (mg L^{-1})
A = Extract volume, fumigated (L)
B = Dilution, fumigated (if performed)
C = Sample weight, fumigated (g)
N_1 = Analyte reading non-fumigated sample extract (mg L^{-1})
N_2 = Blank reading, reagent RODI water, non-fumigated (mg L^{-1})
D = Extract volume, non-fumigated (L)
F = Dilution, non-fumigated (if performed)
G = Sample weight, non-fumigated (g)
E = Field-moist/oven-dry ratio (procedure 3D2)
1000 = Conversion factor to kg-basis

9. Report

Report CO_2 produced by biomass flush (g CO_2-C/g of soil) and soil microbial biomass (kg C ha^{-1} for a given depth interval). Report the difference between mineralizable N of fumigated and non-fumigated to the nearest mg N kg^{-1} soil as NH_3.

10. Precision and Accuracy

Precision and accuracy data are available from the SSL upon request.

11. References

American Society for Testing and Materials. 2004. Standard practice for density and unit weight of soil in place by the rubber balloon method. D 2167. Annual book of ASTM standards. Construction. Section 4. Soil and rock; dimension stone; geosynthesis. Vol. 04.08. ASTM, Philadelphia, PA.

Anderson, J.P.E., and K.H. Domsch. 1978. Mineralization of bacteria and fungi in chloroform fumigated soils. Soil. Biol. Biochem. 10:207-213.

Horwath, W.R., and E.A. Paul. 1994. Microbial mass. p. 753-771. In R.W. Weaver, J.S. Angle, and P.S. Bottomley (eds.) Methods of soil analysis. Part 2. Microbiological and biochemical properties. p. 753-771. ASA and SSSA, Madison, WI.

Jenkinson, D.S., and D.S. Powlson. 1976. The effects of biocidal treatments on metabolism in soil – V.A. method for measuring soil biomass. Soil Biol. Biochem. 8:209-213.

Kettler, T., and J. Doran. 1997. Procedure for determination of soil microbial biomass (gas chromatography method). USDA-ARS, University of Nebraska, Lincoln (unpublished).

Martens, R. 1987. Estimation of microbial biomass in soil by the respiration method: Importance of soil pH and flushing methods of the respired CO_2. Soil Biol. Biochem. 19:77-81.

Plant Analyses (6C)
Sample Preparation (1B3b3a1a1, 1B3b3b1a1)
Root Biomass (6C1)
Plant (above-ground) Biomass (6C2)
Plant Nutrition (6C3)
Total Analysis (6C3a)
Dry Combustion (6C3a1)
Thermal Conductivity Detector (6C3a1a)
Carbon, Nitrogen, and Sulfur (6C3a1a1-3)
Dry (50°C), Roots (6C3a1a1-3a1)
Dry (50°C), Plant Material (above-ground) (6C3a1a1-3a2)

1. Application

Root biomass in the upper 4 inches of the soil is an input value for the Revised Universal Soil Loss Equation (RUSLE) (Renard et al., 1997). The mass, size and distribution of roots in the near surface is among the most important factors in determining the resistance of the topsoil to water and wind erosion. Root biomass is also one of the major Carbon pools found in soil. Commonly, root mass and plant residue in the soil form between 3,000 (annual crop) and 15,000 (perennial grasses) lbs/ac/yr soil biomass (Harwood, et al, 1998). Above-ground biomass (production) represents annual yield and can be measured following the protocols found in the National Range and Pasture Handbook (USDA-NRCS, 1997). Root biomass represents biomass from more than one year.

The development of new roots and ultimately the decomposition of roots within the soil is a major contributor to the Soil Organic Carbon (SOC) pool. In this way, plant roots also contribute to the fertility of soils by slowly releasing macro- and micro-nutrients back into the soil.

Root biomass and SOC help bind the soil together by forming aggregates and granular structure. This improves the tilth as well as the erosion resistance of soil. Depending upon the root turnover rate (known for some species), climate, and residue decomposition rate (known for some areas, based on climate and soil moisture status) the amount of Carbon stored in the soil can be determined from the root biomass, plant residue and SOC.

Root biomass is frequently used to calculate root/shoot ratios in order to evaluate the health and vigor of plants, and determine the success of establishment of seeded plants at the 4-leaf stage.

Dried roots can be fine-grind and total C, N, P, and S can be determined. The C/N ratio can also be determined, which is typically different from the C/N ratio of the above ground plant material. Low levels of N in the soil will promote root growth over top growth (Bedunah and Sosebee, 1995). The C/N ratio of roots, plant residue in the soil and SOC each contribute to the residue decomposition rate for soils. Low C/N values lead to more rapid decomposition, high C/N levels slow decomposition. The C/N ratio required for decomposition of plant residue, without a net tie-up of N, is approximately 25:1. Plant residue from young legumes commonly has a C/N ratio of 15:1. Plant residue from woody materials commonly is 400:1 (Harwood et al., 1998). The C/N ratio of soil microbes is quite variable but commonly falls between 15:1 and 3:1 (Paul and Clark, 1989).

Root biomass/horizon can be paired with the description of roots in each soil horizon (i.e. few fine, many very fine, etc.) in the pedon description and thus a qualitative estimate can be made of the mass in each size fraction of roots.

This automated method for determining root biomass also includes some plant residue. Woody material is removed and weighed separately.

Because root biomass determined in this manner includes plant residue it can be used to estimate the soil plant residue pool in most models (Jenkinson and Rayner, 1977; Metherell et al., 1993).

2. Summary of Method

The procedural steps described herein encompasses the physical separation of roots and plant residue from a soil sample using an automated root washer (1B3b3); these weights recorded for root (6C1) and plant biomass (6C2); and these fractions analyzed for total C, N, and S by procedure 6C3a1a1-3, respectively.

3. Interferences

The soil must be dispersed for successful separation of the roots and plant residue from the soil sample. Tap water rather than distilled water should be used to help avoid puddling and dispersion problems.

4. Safety

Do not touch moving parts of the root washer when it is in operation. Avoid electrical shock by ensuring that the electrical cord is dry, and prevent the formation of pools of water near the cord.

5. Equipment

5.1 Automated root washer (after Brown and Thilenius, 1976)
5.1.1 Root cages, basket sieves, with No. 30 mesh and 0.5 mm-diameter openings.
5.1.2 Garden hose
5.1.3 Sediment tank
5.2 Buckets
5.3 Analytical balance, ± 0.01 g sensitivity
5.4 Drying oven (60°C-capability)
5.5 Weighing dishes
5.6 Scintillating vials
5.7 Tweezers
5.8 Drying trays

6. Reagents

6.1 Tap water
6.2 Algaecide, Bath Clear

7. Procedure

Sample Preparation

7.1 Weigh approximately 200 g of field-moist soil to the nearest 0.01 g and record the weight.

7.2 Pour all of the weighed soil into a root cage and cap it.

7.3 Immerse cage in tap water until soil disperses (overnight if samples are cloddy).

Root Washing

7.4 Make sure that machine is level and that the sediment tank is under the drain.

7.5 Load the root cages containing the soil and root slurry into the rotation bars. Be sure to load them evenly. If not using all of the rotation bar slots, load into every other slot.

7.6 Fill the washing tank with water to the top of the bottom cage.

7.7 Add 10 drops of algaecide to the washing tank. Attach machine to water source.

7.8 Turn on the water at the faucet then turn on the machines spray nozzle. Do not start the machine with the lid open. Once the rotator has started, turn on spray nozzles.

7.9 Depending upon the number of samples let the machine run from 40 to 90 min. (Ex: 12 samples usually take about 60 min.)

Clean Up and Maintenance

7.10 Upon completion of sample washing, shut down the sprayer first then the rotator. Drain the machine first by opening the bottom plug. Make sure the sediment tank is under the drain. After the machine is drained, let the water in the sediment tank settle. Replace plug in the machine.

7.11 Drain the sediment tank water off. Collect the sediment out of the machine and the sediment tank and properly dispose of it.

7.12 Flush out all of the sediment in the machine over the sediment tank. Repeat procedure until the machine is completely clean.

7.13 Clean the entire area. Run water down the drain for about 30 min after everything is clean.

Root/Plant Material Separation and Drying

7.14 Air-dry roots and plant material at room temperature overnight while still in the sieve cages.

7.15 Remove the roots/plant residue in the cage by tapping them. Brush out any roots/plant residue that clings to the side of the sieve cages.

7.16 Add water to a tray of roots/plant material. Float off as much of the organic matter as possible by adding water to a tray roots/plant residue. Much of the organic fraction will be less dense than the sand particles that are not removed during root washing. Pour floating matter into root cage to trap roots/plant residue; avoid introduction of inorganic portion into cage.

7.17 If roots/plant material remain in the inorganic fraction, use tweezers to remove as much of it as possible and return it to the cage.

7.18 Air dry at room temperature overnight all material in cage. Next day, tap and brush the air-dry material into a tray.

7.19 Remove the woody material, dry at 50° C in an oven overnight, and record weight of woody material.

7.20 Separate plant residue from roots, dry at 50°C in an oven overnight, and record weights of plant residue and roots.

7.21 Place the roots and plant residue into separate scintillation vials.

Total Carbon, Nitrogen, and Sulfur Analysis

7.22 Determine total C, N, and S for roots and plant material (6C3a1a1-3), using fine-grind samples ($\approx 180~\mu m$). Refer to 4H2a1-3 for the remaining procedural steps for 6C3a1a1-3.

Separating Roots and Organic Matter Residue (picking)

7.23 Following initial air-drying, use tweezers and separate organic matter residue from roots using tweezers. Roots are usually light colored, and organic residue is usually darker colored.

7.24 Place the organic residue and roots on separate tared watch glasses and re-dry and weigh.

7.25 Record each individual weight for plant residue and roots. Subtract the tare weights and record the total weight of air-dry roots and the total weight of air-dry plant residue. Report separately root biomass and plant residue rather than just roots including some organic residue.

8. Calculations

Calculate root biomass using soil bulk density values determined by procedures (3B) described in this manual or ASTM method D-2167 (American Society for Testing and Materials, 2004).

Root biomass/ha for soil layer of given thickness (kg ha^{-1}) =

[Dry Roots (g)/Total sample weight (g) FM soil] x (Bulk density: g OD soil/cm^3 FM soil) x (g FM soil/g OD soil) x (1 kg/1000 g) x (100000000 cm^2/ha) x (Layer thickness, cm).

where:
OD = Oven-dry
FM = Field-moist

9. Report
Report root biomass as kg ha^{-1} at a given depth interval (cm). If plant residue was separated from roots, report each separately.

10. Precision and Accuracy
Precision and accuracy data are available from the SSL upon request.

11. References

American Society for Testing and Materials. 2004. Standard practice for density and unit weight of soil in place by the rubber balloon method. D 2167. Annual book of ASTM standards. Construction. Section 4. Soil and rock; dimension stone; geosynthesis. Vol. 04.08. ASTM, Philadelphia, PA.

Bedunah, D.J. and R.E. Sosebee (eds.). 1995. Wildland plants: Physiological ecology and developmental morphology. Soc. Range Mgt., Denver, CO.

Brown, G.R. and J.F. Thilenius. 1976. A low-cost machine for separation of roots from soil material. J. Range Mgt. 29:506-507

Fribourg, H.A. 1953. A rapid method for washing roots. Agron. J. 45:334-335.

Harwood, R.R., M.A. Cavigelli, S.R. Deming, L.A. Frost and L.K. Probyn (eds.). 1998. Michigan field crop ecology: Managing biological processes for productivity and environmental quality. Michigan State Univ. Ext. Bull. E-2646, 92 pp.

Jenkinson, D.S., and J.H. Rayner. 1977. The turnover of soil organic matter in some of the Rothamsted classical experiments. Soil Sci. 123:298-305.

Lauenroth, W.K. and W.C. Whitman. 1971. A rapid method for washing roots. J. Range Mgt. 24:308-309.

Metherell, A.K., L.A. Harding, C.V. Cove and, W.J. Parton. 1993. CENTURY soil organic matter model environment, technical documentation agroecosystem, Version 4.0. Great Plains System Res. Unit Tech. Rep. No. 4, USDA-ARS, Ft. Collins, CO.

Paul, E.A., and F.E. Clark. 1989. Soil Microbiology and Biochemistry. Academic Press, Inc., Harcourt Brace Jovanovich, Publ., San Diego, CA.

Renard, K.G., G.R. Foster, G.A. Weesies, D.K. McCool, and D.C. Yoder. 1997. Predicting soil erosion by water: A guide to conservation planning with the revised Universal Soil Loss Equation (RUSLE). Agric. Handb. 703, USDA-ARS, U.S. Govt. Print Office, Washington, DC. .

Sosebee, R.E. (ed.) 1977. Rangeland plant physiology. Range Sci. Series No. 4. Soc. Range Mgt., Denver, CO.

Thornley, J.H.M. 1995. Shoot:root allocation with respect to C, N and P: An investigation and comparison of resistance and teleonomic models. An. of Bot. 75:391-405.

USDA-NRCS, 1997. Inventory and monitoring grazing land resources. p 4i – 4ex21. *In* National range and pasture handbook, Chapter 4. U.S. Govt. Print. Office, Washington DC.

Ratios and Estimates Related to Biological Analyses (6E)

Estimates or calculated values associated with soil biological and plant analyses are described within the respective methods. Additional information on the reporting of these calculated values can be obtained from the SSL upon request.

MINERALOGY (7)

Instrumental Analyses (7A)

The physical and chemical properties of a soil are controlled to a very large degree by the soil minerals, especially by those minerals constituting the clay fraction (McBride, 1989; Whittig and Allardice, 1986). Positive identification of mineral species and quantitative estimation of their proportions in soils usually require the application of several complementary qualitative and quantitative analyses (Whittig and Allardice, 1986). Some of the semi-quantitative and quantitative procedures that have been performed by the SSL include X-ray diffraction (procedure 7A1a1) and thermal analysis (7A2a, 7A3a, and 7A4a). Other indirect, ancillary procedures to infer mineral composition include linear extensibility, elemental analysis, and CEC/clay ratios.

Analysis by x-ray diffraction facilitates identification of crystalline mineral components of soil and semi-quantitative estimates of relative amounts. It is commonly applied to the clay fraction in soils and to layer silicate (phyllosilicates) minerals in particular. Identification is by d-spacings (spatial distance between repeating planes of atoms) characteristic of a mineral, according to Bragg's Law. Because layer silicates structures are very similar from one mineral to another, except in the direction perpendicular to the layers, several treatments (cation saturation and heating) must be used to correctly identify the several minerals. In x-ray analysis of soils or clay samples, there are difficulties in evaluation of and compensation for the variations in chemical composition, crystal perfection, amorphous substances, and particle-size (Whittig and Allardice, 1986; Hughes et al., 1994). A more reliable and accurate estimation of mineral percentages is provided when x-ray diffraction analysis is used in conjunction with other methods, e.g., differential-thermal, surface-area, elemental analysis, and other species-specific chemical methods (Alexiades and Jackson, 1966; Karathanasis and Hajek, 1982).

Many soil constituents undergo thermal reactions upon heating that serve as diagnostic properties for qualitative and quantitative identification of these substances (Tan et al., 1986; Karathanasis and Harris, 1994). Thermogravimetric analysis (TGA) (procedure 7A2a) is a technique for determining weight loss of a sample when it is heated at a constant rate. The TGA is an outgrowth of dehydration curves that were used in early studies of various phyllosilicate clay minerals (Jackson, 1956). However, with TGA the sample weight is monitored continuously rather than measured at discrete intervals after periods of heating at a constant temperature (Wendlandt, 1986). The TGA measures only reactions that involve weight loss of the sample.

Differential scanning calorimetry (DSC) (procedure 7A4a) is a calorimetric technique that theoretically measures the amount of energy required to establish zero temperature difference between sample and reference material as the two are heated side by side at a controlled rate (Tan et al, 1986; Karathanasis and Harris, 1994). Most common DCS instruments have sample and reference pans heated in a single furnace, with the difference in temperature measured during various endothermic and exothermic reactions. This difference in temperature is then converted to a value equivalent to an enthalpy change (expressed in calories) using instrumental calibrations (Karathanasis and Harris, 1994).

The TGA and DSC are complementary methods available to the analyst. Many of the same clay mineral reactions, e.g., dehydroxylation, loss of surface adsorbed water, decomposition of carbonates, and oxidation, that are studied by DSC can also be studied by TGA. However, some transformation reactions, e.g., melting or structural reorganization (quartz alpha-beta transition), cannot be measured by TGA because no weight loss is involved (Karathanasis and Harris, 1994). The DSC procedures provide information about energy relationships in the structures and reactions of the solid phase, whereas TGA provides quantitative information about quantities of substances gained or lost by the solid phase during certain thermally driven reactions.

References

Alexiades, C.A., and M.L. Jackson. 1965. Quantitative clay mineralogical analysis of soils and sediments. Clays and Clay Miner. 14:35-52.

Jackson, M.L. Soil chemical analysis. Advan. course. M. L. Jackson, Madison, WI.

Karathanasis, A.D., and B.F. Hajek. 1982. Revised methods for rapid quantitative determination of minerals in soil clays. Soil Sci. Soc. Am. J. 46:419-425.

Karathanasis, A.D., and W.G. Harris. 1994. Quantitative thermal analysis of soil minerals. p. 360-411. *In* J.E. Amonette and L.W. Zelazny (eds.) Quantitative methods in soil mineralogy. Soil Sci. Soc. Am. Misc. Publ. SSSA, Madison, WI.

McBride, M.B. 1989. Surface chemistry of soil minerals. p. 35-88. *In* J.B. Dixon and S.B. Weed (eds.) Minerals in soil environments. Soil Sci. Soc. Am. Book Series No. 1. ASA and SSSA, Madison, WI.

Tan, K.H., and B.F. Hajek, and I. Barshad. 1986. Thermal analysis techniques. p. 151-183. *In* A. Klute (ed.) Methods of soil analysis. Part 1. Physical and mineralogical properties. 2nd ed. Agron. Monogr. 9. ASA and SSSA, Madison, WI.

Wendlandt, W.W. 1986. Thermal analysis. 3rd ed. John Wiley and Sons, New York.

Whittig, L.D., and W.R. Allardice. 1986. X-ray diffraction techniques. p. 331-362. *In* A. Klute (ed.) Methods of soil analysis. Part 1. Physical and mineralogical methods. 2nd ed. Agron. Monogr. 9. ASA and SSSA, Madison, WI.

Instrumental Analyses (7A)
X-Ray Diffractometer (7A1)
Thin Film on Glass, Resin Pretreatment II (7A1a)
Mg Room Temperature, Mg Glycerol Solvated, K 300°, K 500° C (7A1a1)

1. Application

Clay fractions of soils are commonly composed of mixtures of one or more phyllosilicate minerals together with primary minerals inherited directly from the parent material (Olson et al., 1999). Positive identification of mineral species and quantitative estimation of their proportions in these polycomponent systems usually require the application of several complementary qualitative and quantitative analyses (Whittig and Allardice, 1986; Amonette and Zelazny, 1994; Wilson, 1994; Moore and Reynolds, 1997). One of the most useful methods to identify and to make semiquantitative estimates of the crystalline mineral components of soil is x-ray diffraction analysis (Hughes et al., 1994; Kahle et al., 2002). Quantification of a mineral by x-ray diffraction requires attention to many details, including sample (slide) size relative to the incident x-ray beam, thickness and particle size uniformity of sample, and beam-sample orientation (Moore and Reynolds, 1997). More complex quantification procedures include using standard additions, full pattern fitting, and determining mineral intensity factors (Kahle et. al., 2002). At best, quantification can approach a precision of $\pm 5\%$ and an accuracy of ± 10 to 20% (Moore and Reynolds, 1997).

The operational strategy at the SSL and the preceding Lincoln SSL has been to base mineral quantification on first order peak intensities. Semi-quantitative interpretations have been held consistent over time (1964 to the present) by adjusting instrumental parameters (e.g., scan speed) to maintain a constant peak intensity for a in-house reference clay standard and subsequently soil samples. The intent is to keep interpretations consistent from sample to sample.

2. Summary of Method

Soils are dispersed and separated into fractions of interest. Sands and silts are mounted on glass slides as slurries, on a smear of Vaseline, or on double sticky tape for analysis. Clay suspensions are placed on glass slides to dry and to preferentially orient clay minerals. Most samples of soil clays contain fewer than 7 minerals that require identification. The soil clay minerals of greatest interest are phyllosilicates, e.g., kaolinite, mica (illite), smectite, vermiculite, hydroxy-interlayered vermiculite, smectite, hydroxy-interlayered smectite and chlorite.

Diffraction maxima (peaks) develop from the interaction of x-rays with planes of elements that repeat at a constant distance (d-spacing) through the crystal structure. Generally, no two minerals have exactly the same d-spacings in three dimensions and the angles at which diffraction occurs are distinctive for a particular mineral (Whittig and Allardice, 1986; Moore and Reynolds, 1997). Phyllosilicates (or layer silicate minerals) have very similar structures except in the direction perpendicular to the layers (*c*-dimension). Several treatments are needed to sort out which minerals are present. Glycerol is added to expand smectites. Ionic saturation and/or heat treatments are used to collapse some 2:1 layer silicates and dehydroxylate kaolinite, gibbsite, and goethite, eliminating characteristic peaks.

The crystal "d" spacings of minerals, i.e., the interval between repeating planes of atoms, can be calculated by Bragg's Law as follows:

$$n\lambda = 2d \sin \theta$$

where:
n = interger that denotes order of diffraction
λ = x-radiation wavelength (Angstroms, Å)
d = crystal "d" spacing (Å)
θ = angle of incidence

When n = 1, diffraction is of the first order. The wavelength of radiation from an X-ray tube is constant and characteristic for the target metal in the tube. Copper radiation (CuKα) with a wavelength of 1.54 Å (0.154 nm) is used at the SSL. Because of the similar structure of layer silicates commonly present in soil clays, several treatments that characteristically affect the "d" spacings are necessary to identify the clay components. At the SSL, four treatments are used, i.e., Mg^{2+} (room temperature); Mg^{2+}-glycerol (room temperature); K^+ (300°C); and K^+ (500°C).

Standard tables to convert θ or 2θ angles to crystal d-spacings are published in the U.S. Geological Survey Circular 29 (Switzer et al., 1948) and in other publications (Brown, 1980). Through the years hardware has been updated and the recording of data has evolved from a strip chart recorder through several kinds of electronic software. X-ray by this procedure (7A1a1) is semiquantitative.

3. Interferences

Interstratification of phyllosilicate minerals causes problems in identification. These interstratified mixtures, differences in crystal size, purity, chemical composition, atomic unit cell positions, and background or matrix interferences affect quantification (Moore and Reynolds, 1997; Kahle et al., 2002). No pretreatments other than ionic saturation and dispersion with sodium hexametaphosphate are used for separation and isolation of the clay fraction in the routine procedure. Impurities such as organic matter, carbonates and iron oxides may act as matrix interferences causing peak attenuation during X-ray analysis or may interfere with clay dispersion and separation. Pretreatments to remove these impurities serve to concentrate the crystalline clay fraction and may increase accuracy, but also potentially result in degradation of certain mineral species (e.g, smectites), as well as loss of precision (Hughes et al., 1994).

The separation (centrifuge) procedure used to isolate the clay fraction from the other size fractions of the soil skews the <2-μm clay suspension toward the fine clay, but it minimizes the inclusion of fine silt in the fraction. Sedimentation of the clay slurry on a glass slide tends to cause differential settling by particle size (i.e., increasing the relative intensity of finer clay minerals).

Dried clay may peel from the XRD slide. One remedy is to rewet the peeled clay on the slide with 1 drop of glue-water mixture (1:7). Other remedies are:

a. Place double sticky tape on the slide prior to re-wetting the dried clay with the glue-water mixture.

b. Dilute the suspension if thick.

c. Crush with ethanol and dry, and then add water to make a slurry slide.

d. Roughen the slide surface with a fine-grit sandpaper.

An optimum amount of glycerol on the slides is required to solvate the clay, i.e., to expand smectites to 18 Å. X-ray analysis should be performed 1 to 2 days after glycerol addition. If excess glycerol is applied to the slide and free glycerol remains on the surface, XRD peaks are attenuated. Some suggestions to dry the slides and achieve optimum glycerol solvation are as follows:

a. Use a chamber such as a desiccator (with no desiccant) to dry slide, especially when the clay is thin.

b. If the center of slide is whitish and dry, usually with thick clay, brush slide with glycerol or add an additional drop of glycerol.

4. Safety
Operate the centrifuge with caution. Keep the centrifuge lid closed when in operation. Ensure that all rotors and tubes are seated firmly in proper location. Use tongs and appropriate thermal protection when operating the muffle furnace. The diffraction unit presents an electrical and radiation hazard. Analysts must receive radiation safety training before operating the equipment. Employees must wear a radiation film badge while in the room when the diffraction unit is in operation.

5. Equipment
5.1 Teaspoon (5 g)
5.2 Dispenser, 5 mL, for sodium hexametaphosphate solution
5.3 Centrifuge, International No. 2, with No. 240 head and carriers for centrifuge tubes, International Equip. Co., Boston, MA
5.4 Centrifuge tubes, plastic, 100 mL, on which 10-cm solution depth is marked
5.5 Rubber stoppers, No. 6, for centrifuge tubes
5.6 Mechanical reciprocating shaker, 100 oscillations min^{-1}, 1 ½ in strokes, Eberbach 6000, Eberbach Corp., Ann Arbor, MI
5.7 Plastic cups, 60 mL (2 fl. oz.) with lids
5.8 Label printer
5.9 Hypodermic syringes, plastic, 12 mL, with tip caps
5.10 Screen, 80 mesh, copper
5.11 Dropper bottle, plastic, 30 mL (1 fl. oz.), for a 1:7 glycerol:water mixture

5.12 Muffle furnace

5.13 X-ray diffractometer, Bruker 5000-Dmatic, with X-Y autosampler that accomodates 66 samples or standards, Bruker AXS Inc., Madison, WI

5.14 Computer, Diffract^plus EVA software, release 2000, Bruker AXS Inc., Madison, WI, and printer

5.15 XRD slides, glass, 2.54 X 2.54 mm (frosted glass slides used for K-treated samples)

5.16 XRD sample preparation board, wood, with 32 places for glass XRD slides

5.17 Slide holder.

5.18 Reference slides: quartz and clay from reference soil

6. Reagents

6.1 Reverse Osmosis (RO) water, ASTM Type III grade of reagent water.

6.2 Sodium hexametaphosphate solution. Dissolve 35.7 g of sodium hexametaphosphate $(NaPO_3)_6$ and 7.94 g of sodium carbonate (Na_2CO_3) in 1 L RO water.

6.3 Potassium chloride (KCl), 1.0 N. Dissolve 74.60 g KCl in 1 L RO water or 671.40 g KCl in 9 L RO water.

6.4 Magnesium chloride $(MgCl_2)$, 1.0 N. Dissolve 47.61 g $MgCl_2$ in 1 L RO water or 428.49 g $MgCl_2$ in 9 L RO water.

6.5 Glycerol:water mixture (1:7). Add 4 mL of glycerol to 28 mL RO water plus 2 drops of toluene.

6.6 Exchange resin, Rexyn 101 (H), analytical grade. Pretreatment of resin as follows:

6.6.1 Divide equally Rexyn 101 (H), approximately 250-g portions, into two 600-mL beakers labeled K and Mg and add appropriate salt solution (1.0 N KCl or 1.0 N MgCl$_2$). Cover resin with salt solution.

6.6.2 Stir, let settle for 10 min, decant clear solution, and add salt solution. Repeat 3 times. Leave resin covered in salt solution for 8 to 12 h.

6.6.3 Repeat step 6.6.2 on a second day. Resin is ready for syringes. Saturated resin not used initially for syringes can be saved for future use.

6.7 White glue, diluted 1:7 with RO water

7. Procedure

Preparation (Recharge) of Resin-Loaded Syringes

7.1 Place a small circle of 80-mesh screen in a 12-mL syringe and add 4 cm^3 of exchange resin from which salt solution has been drained. The procedure requires 2 Mg and 2 K slides for each sample, so two sets of syringes are prepared.

7.2 Saturate the resin in each of the syringes with 4 mL of the appropriate 1.0 N salt solution (MgCl$_2$ or KCl) and expel. Repeat saturation of resin. Individual steps follow.

7.3 Fill syringe completely with the salt solution and allow to equilibrate for 4 to 20 h.

7.4 Rinse syringe twice with 4 mL of RO water and rinse tip cap.

7.5 Completely fill syringe with RO water and allow to equilibrate for 4 to 20 h.

7.6 Rinse syringe twice with RO water.

7.7 Expel water, cap syringe, and store.

Preparation of Clay Suspension

7.8 Print a run sheets using LIMS. Each run consists of 8 samples with 4 treatments for each sample.

7.9 Label each 100ml plastic centrifuge tube with a sample number from the run sheet.

7.10 Place ≈ 5 g (1 tsp) of air-dry <2-mm soil in a 100-mL plastic centrifuge tube. If the sample appears to be primarily sand, use 10 g (2 tsp) of <2-mm soil to obtain sufficient clay.

7.11 Add 5 mL of sodium hexametaphosphate dispersion agent. If the soil contains gypsum or is primarily calcium carbonate, use 10 mL of sodium hexametaphosphate dispersing agent.

7.12 Fill tube to 9.5-cm height with RO water and close with a stopper.

7.13 Place the tubes in a mechanical shaker and shake overnight (at least 4 hours).

7.14 Remove stopper from tube and rinse stopper and sides of tube with enough water to bring the volume to the 10-cm mark.

7.15 Balance the pairs of tubes and place in centrifuge. Centrifuge at 750 rpm for 3.0 min.

7.16 If the clay is dispersed, carefully decant 30 mL of suspension into a labeled, 60-mL, plastic cup and cover with lid.

7.17 If the clay did not disperse after being shaken overnight, decant and discard the clear supernatant. Then add an additional 10mL sodium hexametaphosphate and sufficient RO water to bring the level up to 9.5 cm depth. Repeat Sections 7.13 to 7.16.

7.18 Clay suspension is used for X-ray diffraction analysis. It can be dried and used for elemental or thermal analysis.

Thin Film on Glass, Resin Pretreatment
7.19 The SSL uses sample boards that hold 32 slides each, i.e., 8 samples x 4 treatments. Place run number on the sample board. Prepare the sample board with glass XRD slides to receive the following 4 treatments per clay suspension sample.

Mg^{2+}- room temperature
Mg^{2+}- glycerol (room temperature)
K^+- 300°C (heated 2 h)
K^+- 500°C (heated 2 h)

7.20 Use a hypodermic syringe to place 6 drops of the glycerol:water mixture (1:7) on each Mg^{2+}-glycerol slide

7.21 Draw 3 to 4 mL of the clay suspension into the Mg syringe and invert back and forth to facilitate cation exchange.

7.22 Dispense 3 drops to clear the tip.

7.23 Dispense ≈ 0.3 mL (6 to 10 drops) to cover the Mg and mg-glycerol XRD slides. Similarly, use the K syringes to apply clay suspension to the frosted glass K-300 and K-500 slides. Draw RO water into each syringe and expel 3 times to remove all of the clay suspension, cap and store syringes. Recharge all syringes after 10 run boards.

7.24 When the clay suspension has dried, transfer the slides with the K^+-saturated clays to the muffle furnace. Heat for a minimum of 2 hours at 300°C, remove the K-300 batch of slides.

Set the temperature to 500 °C and heat slides for a minimum of 2 hours at 500°C. After slides are cool, return them to the run board.

X-ray Diffraction Operation

7.25 Complete X-ray analysis of the glycerol slide within 1 to 2 days after the slide dries. If this is not possible, add additional glycerol prior to run (e.g., add 6 drops of glycerol:water mixture to dry slide 24 h prior to x-ray analysis).

7.26 Place the tray with filled sample holders in the autosampler and execute the run. Use the following parameters:

CuKα radiation, λ = 1.54 Å (0.154 nm)
Scan range = 2° to 35°2θ
Generator settings = 40 kv, 30 ma
Divergence slit = 1°
Receiving slit = 0.2 mm

Step-size and scan-speed vary depending on intensity of X-rays generated. Settings should be adjusted to maintain the same peak intensities on the standard reference clay and quartz standard over the long term regardless of tube intensities.

7.27 In the laboratory information system (LIMS), create a batch file. Data in file is transferred to a job program on the x-ray computer software for data analysis. These data include project and sample identification. Include both the quartz and soil standard with each run.

7.28 Activate job program for analysis. The job stores raw data on the hard disk under the subdirectory designated by year, project type, project name.

7.29 Prepare and print a 4-color graphics chart. The four colors are blue (Mg^{2+}); green (Mg^{2+}-glycerol); pink (K^+ 300°C); and red (K^+ 500°C). File hard copies of detected peaks and graphics chart in pasteboard binders by state, county, and chronology.

7.30 Compare quartz and soil standard patterns electronically with previous runs to ensure peak intensity and positions have remained constant.

Interpretation of X-ray Diffraction Data

7.31 The angle in degrees two theta (2θ) measured in X-ray diffraction analyses is converted to angstroms (Å) using tables complied according to Bragg's Law. Refer to summary of method. Angstroms convert to nanometers (nm) by a factor of 0.1, e.g., 14 Å = 1.4 nm.

7.32 Use the following X-ray diffraction criteria to identify some common crystalline minerals. The reported "d" values are for 00*l* basal spacings. The Miller index (*hkl*)specifies a plane or crystal face which has some orientation to the three crystallographic axes of a, b, and c. The Miller index (00*l*) indicates a crystal face that is perpendicular to the a and b axes (Schultz, 1989). The following X-ray diffraction criteria also have some questions (Q) that may aid the analyst in interpreting the diffraction patterns. These questions are a suggested procedural approach to help the analyst identify the relative locations of a few peaks and to confirm key criteria. For a more complete list of d-spacings for confirmation or identification of a mineral consult the Mineral Powder Diffraction File – Data Book (JCPDS 1980).

X-Ray Diffraction Criteria

7.32.1 *Kaolinite and Halloysite*
a. Crystal structure missing at 500°C.
b. 7 Å (7.2 to 7.5 Å) with all other treatments

Q. Is there a 7 Å peak? Is it destroyed at 500°C? Kaolinite or Halloysite.

Q. Is the peak sharp and at ~ 7.1 Å (but absent at 500°C)? Kaolinite.

Q. Is the peak broad and at 7.2 to 7.5 Å (but absent at 500°C)? Halloysite.

7.32.2 *Mica (Illite)*
a. 10 Å with all treatments.
b. 10 Å with Mg^{2+}-saturation

Q. Is there a 10 Å peak with Mg^{2+}-saturation? Mica (Illite).

7.32.3 *Chlorite*
a. Crystal structure of Fe-chlorites destroyed at 650 to 700°C.
b. 14 Å with all other treatments.
c. 14 Å at 500°C.
d. Generally also has strong 7 Å peak.

Q. Is there a 14 Å peak when heated to 500°C? Chlorite.

7.32.4 *Vermiculite*
a. 14 Å with Mg^{2+}-saturation.
b. 14 Å with Mg^{2+}-glycerol solvation.
c. Nearly 10 Å with K^+ saturation.
d. 10 Å when K^+-saturated and heated to 300°C.

Q. Is there an enhanced 10 Å peak with K^+-saturation in comparison to Mg^{2+} saturation that cannot be attributed to smectitie? Vermiculite.

7.32.5 *Smectite*
a. 14 Å with Mg^{2+}-saturation
b. 12 to 12.5 Å with K^+- or Na^+-saturation.
c. 17 to 18 Å with Mg^{2+}-glycerol solvation.
d. 10 Å with K^+-saturation and heating to 300°C.

Q. Is there a 17 to 18 Å peak upon solvation? Smectite.

7.32.6 *Gibbsite*
a. Peak at 4.83 to 4.85 Å with Mg^{2+} and Mg^{2+}-glycerol but destroyed when heated to 300°C.

7.32.7 *Goethite*
a. Peak at 4.16 to 4.18 Å with Mg^{2+} and Mg^{2+}-glycerol but destroyed when heated to 300°C.

7.32.8 *Hydroxy-interlayed Vermiculite or Smectite*
a. Failure to completely collapse to 10 Å of smectite or vermiculite when K^+-saturated and heated to 300°C.

7.32.9 *Quartz*
a. Peaks at 4.27 Å and 3.34 Å with all treatments (only 3.34 if small amounts).

7.32.10 *Lepidocrocite*
a. Peak at 6.2 to 6.4 Å with Mg^{2+} and Mg^{2+}-glycerol but destroyed when heated to 300°C.

7.32.11 *Potassium Feldspar*
a. Peak at 3.24 Å with all treatments.

7.32.12 *Plagioclase Feldspar*
a. Twin peaks between 3.16 and 3.21 with all treatments.

7.32.13 *Calcite*
a. Peak at 3.035 Å with all treatments.

7.32.14 *Dolomite*
a. Peak at 2.88 to 2.89 Å with all treatments.

7.32.15 *Gypsum*
a. Peak at 7.56 Å with Mg^{2+} and Mg^{2+}-glycerol, but destroyed when heated to 300°C.

7.32.16 *Mixed Layer Vermiculite-Mica*
a. Randomly interstratified: Peak between 10 and 14 Å with Mg^{2+} that does not expand with Mg^{2+}-glycerol; peak collapses to 10 Å with K^+-saturation and heating to 300°C.
b. Regularly interstratified: A 24 Å peak (and higher orders); no change with Mg^{2+}-glycerol treatment; K^+ saturation and heating collapses vermiculite and a produces a 10 Å peak.

7.32.17 *Mixed Layer Smectite-Mica*
a. Randomly interstratified: Peak between 10 and 14 Å with Mg^{2+} that expands to 14-16 Å with Mg^{2+}-glycerol;Peak collapses to 10 Å with K^+-saturation and heating to 300°C.
b. Regularly interstratified: A small 24 Å peak and large peak at 12 Å with Mg^{2+}-saturation; expands to 28 Å with Mg^{2+}-glycerol treatment; K^+-saturation and heating collapses smectite, then produces a 10 Å peak.

7.32.18 *Mixed Layer Chlorite-Vermiculite*
Randomly Interstratified: Peak at 14 Å with Mg^{2+} and Mg^{2+}-glycerol; Peak collapses incompletely to between 10 and 14 Å with K^+-saturation and heating.
b. Regularly interstratified: A 28 Å peak (and higher orders) with Mg-saturation; no expansion with Mg^{2+}-glycerol treatment; K^+-saturation and heating to 500°C collapses vermiculite and a produces a 24 Å peak.

7.32.19 *Mixed Layer Chlorite-Smectite*
a. Randomly interstratified: Peak at 14 Å with Mg^{2+}-saturation; expands to higher spacings (≈16 Å) with Mg^{2+}-glycerol treatment; Peak collapses incompletely to between 10 and 14 Å with K^+-saturation and heating .

7.33 Use the X-ray diffraction criteria, i.e., diagnostic basal 00*l* spacings (Å), in Table 1 for identification and ready reference of some common crystalline minerals as affected by differentiating sample treatments.

7.34 Preferential orientation of clay mineral samples enhances diffraction from the basal (00*l*) spacing and tends to minimize the number and intensity of peaks from diffraction by other *hkl* planes. With preferential orientation, second, third, and fourth order peaks may be recorded in addition to the basal first order peaks. Groups of associated peaks that differ by order of diffraction are as follows:

7.34.1 *Smectite (Mg^{2+}-glycerol):*
a. 17 to 18 Å.
b. 8.5 to 9 Å (weak).

7.34.2 *Chlorite, vermiculite, and smectite:*
a. 14, 7, 4.7, and 3.5 Å.
b. 7, 4.7, and 3.5 Å weak for smectite.
(Note: High Fe substitution in the chlorite structure results in a decrease in the peak intensity of odd numbered orders (e.g., 14 and 4.7 Å) and increase in peak intensity of even number orders (7 and 3.5 Å)).

7.34.3 *Mica:*
a. 10, 5 (weak in biotites and moderate in muscovites), and 3.3 Å.

7.34.4 *Kaolinite:*
a. 7 and 3.5 Å.

7.35 The differentiation of kaolinite and halloysite in a sample can be aided by the use of formamide (Churchman et al., 1984). The intercalation and expansion of halloysite to a d-spacing of $\approx$ 10.4 Å is relatively rapid (20 to 30 min), whereas kaolinite expansion requires $\approx$ 4 h upon treatment. The procedure is as follows:

7.35.1 Lightly spray formamide as an aerosol on the dried Mg^{2+}-saturated slide.

7.35.2 Wait 15 min but not more than 1 h and X-ray approximately 7.6 to 13.5° 2θ (d = 11.6 to 6.55 Å).

7.35.3 Halloysite will expand to $\approx$ 10.4 Å, whereas kaolinite will remain unchanged.

7.35.4 Heating the sample to 110°C for 15 min will collapse the halloysite to $\approx$ 7 Å.

7.35.5 The total amount of kaolinite and halloysite can be determined by thermal analysis. The intensity ratio of the 10.4 to 7.2 Å peaks of the formamide-treated sample can be used to determine the relative percentage of halloysite and kaolinite.

7.35.6 The total amount of kaolinite and halloysite can be determined by thermal analysis. The intensity ratio of the 10.4 to 7.2 Å peaks of the formamide-treated sample can be used to determine the relative percentage of halloysite and kaolinite.

8. Calculations

X-ray diffraction produces peaks on a chart that corresponds to 2θ angle on a goniometer. Standard tables to convert θ or 2θ to crystal "d" spacings are published in the U.S. Geological Survey Circular 29 (Switzer et al., 1948) and in other publications (Brown, 1980). The crystal "d" spacings of minerals, i.e., the interval between repeating planes of atoms, can be calculated by Bragg's Law. Refer to summary of method.

9. Report

From the "Detected Peaks File" and graphics chart, identify the minerals present according to the registered "d" spacings. As a first approximation, use the following peak intensities, i.e., peak heights above background in counts s[-1], to assign each layer silicate mineral to one of the 5 semiquantitative classes.

Class	Peak Height above Background (counts sec^{-1})
5 (Very Large)	>1800
4 (Large)	1120 to 1800
3 (Medium)	360 to 1120
2 (Small)	110 to 360
1 (Very Small)	<110

Adjust class placement to reflect area under the curve if peak is broad relative to peak height or if thermal, elemental, clay activity data, or other evidence warrant class adjustment. If there are no peaks or no evidence of crystalline components, place the sample in NX class (noncrystalline). If there are only 1 to 3 very small (class1) peaks, also indicate NX to infer a major noncrystalline component.

10. Precision and Accuracy
X-ray by procedure 7A1a1 is semi-quantitative. Precision and accuracy data are available from the SSL upon request.

11. References

Amonette, J.E., and L.W. Zelazny. 1994. Quantitative methods in soil mineralogy. Soil Sci. Soc. Am. Misc. Publ. SSSA. Madison, WI.

Brown, G. 1980. Appendix I (Tables for the determination of d in Å from 2θ for the Kα and Kβ radiations of copper, cobalt, and iron). p. 439-475. *In* G.W. Brindley and G. Brown (eds.) Crystal structures of clay minerals and their x-ray identification. Mineralogical Soc. Monograph No. 5. Mineralogical Soc. Great Britain.

Churchman, G.J., J.S. Whitton, G.G.C. Claridge, and B.K.G. Theng. 1984. Intercalation method using formamide for differentiating halloysite from kaolinite. Clays and Clay Minerals 32:241-248.

Hughes, R.E., D.M. Moore, and H.D. Glass. 1994. Qualitative and quantitative analysis of clay minerals in soils. p. 330-359. *In* J.E. Amonette and L.W. Zelazny (eds.) Quantitative methods in soil mineralogy. Soil Sci. Soc. Am. Misc. Publ. SSSA. Madison, WI.

JCPDS. 1980. Mineral powder diffraction file – Data book. International Centre for Diffraction Data, Swarthmore, PA.

Kahle, M., M.Kleber, and R. Jahn. 2002. Review of XRD-based quantitative analyses of clay minerals in soils: The suitability of mineral intensity factors. Geoderma 109:191-205.

Moore, D.M., and R.C. Reynolds, Jr. 1997. X-ray diffraction and the identification and analysis of clay minerals. Oxford Univ. Press. NY, York, NY.

Olson, C.G., M.L. Thompson, and M.A. Wilson. 1999. Section F. Soil Mineralogy, Chapter 2. Phyllosilicates. P. F-77 to F-13. *In* M.E. Sumner (ed.) Handbook of soil science. CRC Press, Boca Raton, FL.

Schulze, D.G. 1989. An introduction to soil mineralogy. p. 1-34. *In* J.B. Dixon and S.B. Weed (eds.) Minerals in soil environments. Soil Sci. Soc. Am. Book Series No. 1. SSSA. Madison, WI.

Switzer, G., J.M. Axelrod, M.L. Lindberg, and E.S. Larsen 3d. 1948. U.S. Dept. Interior. Geological Survey. Circular 29. Washington, DC.

Whittig, L.D., and W.R. Allardice. 1986. X-ray diffraction techniques. *In* A. Klute (ed.) Methods of soil analysis. Part 1. Physical and mineralogical methods. 2nd ed. Agronomy 9:331-362.

Wilson, M.J. 1994. Clay mineralogy: Spectroscopic and chemical determinative methods. Chapman and Hall, NY.

Table 1. X-ray diffraction parameters of common soil minerals.

Mineral	Na$^+$	Mg^{2+}	Mg^{2+} Gly	K$^+$	K$^+$ 300°C	K$^+$ 500°C	K$^+$ 700°C
	00*l* diffraction spacing in angstroms						
Kaolinite	7	7	7	7	7	**LD**[1]	LD
Halloysite	7B[2]	**7B**	7B	7B	7B	**LD**	LD
Mica (Illite)	10	10	**10**	10	10	10	10
Chlorite	14*[3]	14*	14*	14*	14*	**14***	T[4]
Vermiculite	14	14	**14 10**		10	10	10
Smectite	12.5	14	**18**	12.5	10	10	10
Gibbsite	4.85	**4.85**	4.85	4.85	**LD**	LD	LD
Goethite	4.18	**4.18**	4.18	4.18	**LD**	LD	LD
Lepidocrocite	6.24	6.24	6.24	6.24	**LD**	LD	LD
Interlayer	10-14	10-14	10-18	10-14	10-14	10-14	10-14
Quartz	**3.34** and **4.27** for all treatments						
Calcite	**3.035** for all treatments						
Dolomite	**2.886** for all treatments						

[1] LD = Lattice destroyed
[2] B = Broad peak is common
[3] * = Sometimes <14 Å
[4] T = Temperature of decomposition varies with chemical composition, particle-size, and heating conditions.

Instrumental Analyses (7A)
Thermogravimetric Analysis (7A2)
Thermal Analyzer (7A2a)
Differential Thermal Analysis (7A3)
Thermal Analyzer (7A3a)

1. Application

Thermal analysis defines a group of analyses that determine some physical parameter, e.g., energy, weight, or evolved substances, as a dynamic function of temperature (Tan et al., 1986; Karathanasis and Harris, 1994). Thermogravimetric analysis (TGA) is a technique for determining weight loss of a sample as it is being heated at a controlled rate. The weight changes are recorded as a function of temperature, i.e., a thermogravimetric curve, and provide quantitative information about substances under investigation, e.g., gibbsite ($Al(OH)_3$), kaolinite ($Al_2Si_2O_5(OH)_4$), goethite (FeOOH) and 2:1 expandable minerals (smectite and vermiculite).

2. Summary of Method

A 5- to 10-mg sample of soil clay or fine earth (finely ground) is placed in a platinum sample pan, and the pan is placed in the TGA balance. The instrument records the initial sample weight. The sample is then heated from a temperature of 30 to 900°C at a rate of 20°C min^{-1} in a flowing N_2 atmosphere. The computer collects weight changes as a function of temperature and records a thermogravimetric curve. Gibbsite and kaolinite are quantified by calculating the weight loss between approximately 250 to 350°C and 450 to 550°C, respectively, and then relating these data to the theoretical weight loss of pure gibbsite or kaolinite. The weight loss is due to dehydroxylation, i.e., loss of crystal lattice hydroxyl ions. Though not presently performed by the SSL, quantification of the 2:1 expandable minerals (smectite + vermiculite) is related to weight loss at <250°C, i.e., loss of adsorbed water (Karanthasis and Hajek, 1982a; Karanthasis and Hajek, 1982b; Tan et al., 1986). At this low temperature, adsorbed water is proportional to the specific surface area of the sample (Jackson, 1956; Mackenzie, 1970; Tan and Hajek, 1977; Karathanasis and Hajek, 1982b). In the absence of gibbsite, goethite, a Fe oxyhydroxide, can be quantified based on the characteristic weight loss of 10.1 to 11.2% between 300 and 400°C (Karathanasis and Harris, 1994). Recent work in the SSL has found good agreement between gypsum quantification using dissolution procedures and thermal analysis. Gypsum has a weight loss of 20.9% between 100 and 350°C (Karathanasis and Harris, 1994). The TGA procedure (7A2a) is especially useful for soils with a large percentage (>20%) of gypsum. Burt et al. (2001) had good agreement between total Mg analysis and TGA quantification (12.9% weight loss between 600-650°C) of serpentine minerals in ultramafic-derived soils in Oregon.

3. Interferences

Organic matter is objectionable because it has a weight loss by dehydrogenation and by oxidation to CO_2 between 300 to 900°C (Tan, et al., 1986). Analysis in an inert N_2 atmosphere helps to alleviate this problem, but samples with significant organic matter should be pretreated with H_2O_2 (procedure 3A1a1). Mineral salts that contain water of crystallization also may interfere. Samples should be washed free of any soluble salts. In some cases, weight loss from gibbsite and goethite overlap and prevent quantitative interpretation. These samples can be deferrated (procedure 4F1a1) to eliminate goethite.

A representative soil sample is important as sample size is small (<10 mg). Avoid large aggregates in sample, the presence of which may cause thermal interferences, i.e., differential kinetics of gas diffusion through the sample and physical movement of sample in a reaction.

In general, the same reactions that interfere with DSC/DTA also interfere with TGA determinations of kaolinite, gibbsite, and 2:1 expandable minerals. However, TGA is more sensitive to small water losses at slow rates, whereas DSC/DTA is more sensitive to large water losses at rapid rates (Tan, et al., 1986). This sensitivity difference may help to explain why quantification of kaolinite and gibbsite in TGA vs. DSC/DTA often are not equivalent, i.e., TGA estimates tend to be greater than the corresponding DSC/DTA estimates. In TGA, there is a greater probability of measuring water losses in specific temperature regimes that are not specifically associated with dehydroxylation reactions of interest. This problem is particularly apparent with illitic samples, which characteristically contain more "structural" water than ideal structural formulae would indicate (Rouston, et al., 1972; Weaver and Pollard, 1973).

Even though it is well established that various minerals lose the major portion of their crystal lattice water in different temperature ranges (Tan et al., 1986), there are overlaps in weight loss regions (WLR) of minerals which interfere in the identification and measurement of the minerals of interest. The goethite WLR (250 to 400°C) overlaps the gibbsite WLR (250 to 350°C) (Mackenzie and Berggen, 1970). The illite WLR (550 to 600°C) overlaps the high end of the kaolinite WLR (450 to 550°C) (Mackenzie and Caillere, 1975). The WLR of hydroxy-Al interlayers in hydroxy-Al interlayered vermiculite (HIV) (400 to 450°C) overlaps the low end of the kaolinite WLR (450 to 550°C), especially in the poorly crystalline kaolinites (Mackenzie and Caillere, 1976). Similarly, the dehydroxylation of nontronites, Fe-rich dioctahedral smectites, (450 to 500°C) may interfere with kaolinite identification and measurement (Mackenzie and Caillere, 1975).

4. Safety
Secure high pressure N_2 tanks and handle with care. When changing the tanks, protect valves with covers. Do not program the analyzer for >950°C because it may present a safety hazard during sample analysis and cleaning cycles. Always use high quality purge gases with the TGA. Minimum purity of 99.9% is recommended. Handle hot furnace with care.

5. Equipment
5.1 Thermal analyzer, TGA 51, TA Instruments, New Castle, DE
5.2 Thermal analyzer operating system software, Thermal Analyst 2100, Version 8.10B, TA Instruments, New Castle, DE
5.3 Computer Data analysis software, TGA Standard Data Analysis Version 5.1, TA Instruments, New Castle, DE
5.4 Computer, IBM-PC 386, TA Instruments Operating System, Version, 8.10B, and printer
5.5 Thermal analyzer instrument controller (MIM), TA Instruments, New Castle, DE
5.6 N_2 gas, 99.99% purity
5.7 Two-stage gas regulators, 50 psi maximum outlet pressure
5.8 Forceps, flat-tipped
5.9 Weighing spatula
5.10 Desiccator, glass
5.11 Mortar and pestle
5.12 Sieve, 80 mesh
5.13 Kaolinite, standard, poorly crystalline, Georgia Kaolinite, Clay Minerals Society, Source Clay Minerals Project, sample KGa-2.
5.14 Gibbsite, standard, Surinam Gibbsite, SSL, 67L022.

6. Reagents
6.1 Magnesium nitrate saturated solution [$Mg(NO_3)_2 \cdot 6H_2O$]
6.2 Ethanol

7. Procedure

Derive <2μm Clay Fractions
7.1 Prepare Na-saturated sample (created during dispersion and separation of clay fraction from <2-mm soil). Refer to procedure 7A1a1, sections 7.8 to 7.19.

7.2 Dry the clay suspension and transfer to mortar. Moisten sample with ethanol and grind with pestle to make a homogeneous slurry.

7.3 Air-dry sample using flowing air in hood. Lightly grind sample with pestle to make a homogeneous powder.

7.4 Sieve sample through an 80-mesh screen. Equilibrate sample overnight over a saturated magnesium nitrate solution (55% relative humidity) in a glass desiccator.

TGA Operation

7.5 Turn on power to controller, instrument and computer.

7.6 Turn on the N_2 purge gas and set to 100 cm^3 min^{-1} sample purge.

7.7 Place the empty platinum sample pan on quartz rod and slide into quartz furnace tube. Use the Auto Zero function in the TA 2100 software to zero the balance.

7.8 Remove the sample pan from the quartz rod. Weigh ≈ 10 mg of sample, i.e., <80-mesh fine-earth (<2 mm) soil fraction or derived <2-μm clay fraction, into tared sample pan. Refer to section on derived <2-μm clay fractions, Steps 7.1 to 7.4.

7.9 Use flat-tipped forceps to tap the sample pan against a hard surface several times to uniformly distribute the sample.

7.10 Carefully place sample pan on the quartz rod of the TGA microbalance.

7.11 The standard sample run heating program is as follows: Equilibrate at 40°C, ramp up at 20°C min^{-1} to 110°C, hold at 110°C for 2 minutes (to mark an oven-dry base weight), ramp up to an ending temperature of 900°C.

7.12 Immediately start the "Run" program.

7.13 At the end of sample run (≈ 45 min), allow the TGA furnace to cool to 350°C.

7.14 Slide quartz furnace tube out of furnace. Pull furnace from main TGA unit and allow to cool on counter top. (Caution: Very Hot!).

7.15 Place alternate furnace in TGA unit and prepare for next sample run.

7.16 To analyze data file for weight loss, enter Data Analysis menu of the Thermal Analysis 2100 System and select the TGA Standard Data Analysis software.

7.17 Display file and calculate weight loss in specific regions for selected minerals.

8. Calculations

The thermogravimetric curve is displayed on the computer monitor. The ordinate (Y) is expressed in a relative weight percentage, i.e., the initial sample weight is 100.0%. Use the computer to calculate the total change in sample weight (Δ Y), within the predetermined temperature range, as a sample weight percent.

8.1 % Kaolinite =
{[(Δ sample weight % 450-550°C)]/14} x 100

or (Δ sample weight % 450-550°C) x 7.14

where:
Δ sample weight = total change in sample weight expressed as relative percent
14 = percent weight of hydroxyl water lost from pure kaolinite during dehydoxylation

8.2 % Gibbsite =
{[(Δ sample weight % 250-350°C)]/34.6} x 100

or (Δ sample weight % 250-350°C) x 2.89

where:
Δ sample weight = total change in sample weight expressed as relative percent of the 110°C base weight.
34.6 = percent weight of hydroxyl water lost from pure gibbsite during dehydroxylation

If Fe oxides are removed prior to analysis to prevent the interference with gibbsite determination, the calculation is modified to account for weight loss due to deferration as follows:

8.3 % Gibbsite ={[Δ Sample weight % 250-350°C x (Wt_2/Wt_1)]/34.6} x 100

where:
Wt_1 = Weight before deferration
Wt_2 = Weight after deferration

The percent weights of hydroxyl water lost from kaolinite and gibbsite are derived from the following assumed dehydroxylation reactions.

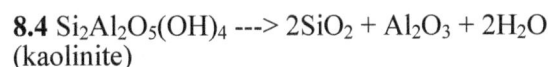

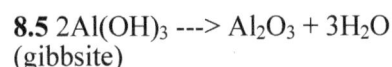

8.4 $Si_2Al_2O_5(OH)_4 \longrightarrow 2SiO_2 + Al_2O_3 + 2H_2O$
(kaolinite)

8.5 $2Al(OH)_3 \longrightarrow Al_2O_3 + 3H_2O$
(gibbsite)

Using kaolinite as an example, percent weight of hydroxyl water lost is calculated from the following formula weights.

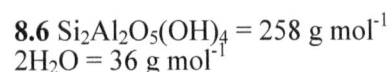

8.6 $Si_2Al_2O_5(OH)_4$ = 258 g mol^{-1}
$2H_2O$ = 36 g mol^{-1}

Percent weight of hydroxyl water lost = (36/258) x 100 = 14%

If serpentine minerals are present in the sample, TGA can be used to quantify these minerals (Burt et al., 2001) based on a the onset temperature of 600-650°C (Karathanasis and Harris, 1994) and a weight loss from 600 to 900°C (12.9%) based on the mineral structure $[Mg_3Si_2O_5(OH)_4]$:

8.7 % Serpentine minerals =
{[(Δ sample weight % 600-900°C)]/12.9} x 100

or (Δ sample weight % 600-900°C) x 7.75

Gypsum can be quantified based on a loss of 20.9% (Karathanasis and Harris, 1994) based on the weight loss in the region of 100 to 350°C:

8.8 % Gypsum =
{[(Δ sample weight % 100-350°C)]/20.9} x 100

or (Δ sample weight % 100-350°C) x 4.78

9. Report
Report percent gibbsite, kaolinite, gypsum, or antigorite to nearest whole number.

10. Precision and Accuracy
Precision and accuracy data are available from the SSL upon request.

11. References

Burt, R., M. Filmore, M.A. Wilson, E.R. Gross, R.W. Langridge, and D.A. Lammers. 2001. Soil Properties of selected pedons on ultramafic rocks in Klamath Mountains, Oregon. Commun. Soil Sci. Plant Anal. 32:2145-2175.

Jackson, M.L. 1956. Soil chemical analysis. Adv. course. M. L. Jackson, Madison, WI.

Karathanasis, A.D., and B.F. Hajek. 1982a. Revised methods for rapid quantitative determination of minerals in soil clays. Soil Sci. Soc. Am. J. 46:419-425.

Karathanasis, A.D., and B.F. Hajek. 1982b. Quantitative evaluation of water adsorption on soil clays. Soil Sci. Soc. Am. J. 46:1321-1325.

Karathanasis, A.D., and W.G. Harris. 1994. Quantitative thermal analysis of soil material. P. 260-411. *In* J.E. Amonette and L.W. Zelazny (eds.). Quantitative methods in soil mineralogy. Soil Sci. Soc. Am. Misc. Publ., SSSA, Madison, WI.

Mackenzie, R.C. 1970. Simple phyllosilicates based on gibbsite- and brucite-like sheets. p. 498-537. *In* R.C. Mackenzie (ed.) Differential thermal analysis. Vol. 1. Acad. Press, London.

Mackenzie, R.C., and G. Berggen. 1970. Oxides and hydroxides of higher-valency elements. p. 272-302. *In* R.C. Mackenzie (ed.) Differential thermal analysis. Acad. Press, NY.

Mackenzie, R.C., and S. Caillere. 1975. The thermal characteristics of soil minerals and the use of these characteristics in the qualitative and quantitative determination of clay minerals in soils. p. 529-571. *In* J.E. Gieseking (ed.) Soil components. Vol. 2. Inorganic components. Springer-Verlag, NY.

Mackenzie, R.C., and B.D. Mitchell. 1972. Soils. p. 267-297. *In* R.C. Mackenzie (ed.) Differential thermal analysis. Vol. 2. Acad. Press. London.

Rouston, R.C., J.A. Kittrick, and E.H. Hope. 1972. Interlayer hydration and the broadening of the 10A x-ray peak in illite. Soil Sci. 113:167-174.

Tan, K.H., and B.F. Hajek. 1977. Thermal analysis of soils. p. 865-884. *In* J.B. Dixon and S.B. Weed (eds.) Minerals in soil environments. 2nd ed. Soil Sci. Soc. Am. Book Series 1. ASA and SSSA, Madison, WI.

Tan, K.H., and B.F. Hajek, and I. Barshad. 1986. Thermal analysis techniques. p. 151-183. *In* A. Klute (ed.) Methods of soil analysis. Part 1. Physical and mineralogical properties. 2nd ed. Agron. Monogr. 9. ASA and SSSA, Madison, WI.

Weaver, C.E., and L.D. Pollard. 1973. The chemistry of clay minerals. Elsevier Sci. Publ. Co., Amsterdam.

Instrumental Analyses (7A)
Differential Scanning Calorimetry (7A4)
Thermal Analyzer (7A4a)

1. Application

Calorimetry measures specific heat or thermal capacity of a substance. Two separate types of differential scanning calorimetry (DSC) instruments have evolved over time. The term "DSC" is most appropriate for the power-compensated-type instrument in which the difference in the rate of heat flow between a sample and a reference pan is measured as materials are held isothermal to one another using separate furnaces (Karathanasis and Harris, 1994). The DSC therefore directly measures the magnitude of an energy change (ΔH, enthalpy or heat content) in a material undergoing an exothermic or endothermic reaction. Heat flow-type DSC instruments are more common and are similar in principal to differential thermal analyzers (DTA). The heat flow instruments have the sample and reference pans in a single furnace and monitor pan temperature from the conducting base. The difference in pan temperatures (ΔT) results from clay mineral decomposition reactions in the sample as the furnace temperature is increased. The configuration of this instrument results in a signal that is independent of the thermal properties of the sample and ΔT can be converted to a calorimetric value via instrument calibration (Karathanasis and Harris, 1994). DSC is commonly used to quantify gibbsite ($Al(OH)_3$) and kaolinite ($Al_2Si_2O_5(OH)_4$) in soils and clays by measuring the magnitude of their dehydroxylation endotherms which are between approximately 250 to 350°C and 450 to 550°C, respectively (Jackson, 1956; Mackenzie, 1970; Mackenzie and Berggen, 1970; Karathanasis and Hajek, 1982).

2. Summary of Method

An 8-mg sample of soil clay is weighed into an aluminum sample pan and placed in the DSC sample holder. The sample and reference pans are heated under flowing N_2 atmosphere from a temperature of 30 to 600°C at a rate of 10°C min^{-1}. Data are collected by the computer and a thermograph is plotted. Gibbsite and kaolinite are quantified by measuring the peak area of any endothermic reactions between 250 to 350°C and 450 to 550°C, respectively, and by calculating the ΔH of the reaction. These values are related to the measured enthalpies of standard mineral specimens (gibbsite and kaolinite). Percent kaolinite and gibbsite are reported by procedure 7A4a.

3. Interferences

Organic matter is objectionable because it produces irregular exothermic peaks in air or O_2, commonly between 300 to 500°C, which may obscure important reactions from the inorganic components of interest (Schnitzer and Kodama, 1977). Analysis in an inert N_2 atmosphere helps to alleviate this problem although thermal decomposition of organic matter is still observed. Pretreatment with H_2O_2 may be necessary for soils with significant amounts of organic matter. Mineral salts that contain water of crystallization also may be interferences. Samples should be washed free of any soluble salts.

Use a representative soil sample as sample size is small (<10 mg). Avoid large aggregates in sample, the presence of which may cause thermal interferences because of differential kinetics of gas diffusion through the sample and physical movement of sample in a reaction.

The dehydroxylation of goethite is between 250 to 400°C and may interfere with the identification and integration of the gibbsite endotherm (250 to 350°C) (Mackenzie and Berggen, 1970). The dehydroxylation of illite is between 550 to 600°C and partially overlaps the high end of the kaolinite endotherm (450 to 550°C), resulting in possible peak integrations (Mackenzie and Caillere, 1975). The dehydroxylation of hydroxy-Al interlayers in hydroxy-Al interlayered vermiculite (HIV) is between 400 to 450°C and may interfere with the low end of the kaolinite endotherm (450 to 550°C), especially in the poorly crystalline kaolinites (Mackenzie and Caillere, 1976). Similarly, the dehydroxylation of nontronites, Fe-rich dioctahedral smectites is between 450 to 500°C and may interfere with kaolinite identification and measurement (Mackenzie and Caillere, 1975).

4. Safety

Secure high pressure N_2 tanks and handle with care. When changing the tanks, valves should be protected with covers. Do not heat aluminum sample pans >600°C. Aluminum melts at 660°C, and the sample pans alloy with and destroy the DSC cell. Always use high quality purge gases with the DSC. Minimum purity of 99.9% is recommended.

5. Equipment

5.1 Thermal analyzer, DSC 910S, TA Instruments, New Castle, DE
5.2 Thermal analyzer operating system software, Thermal Analyst 2100, Version 8.10B, TA Instruments, New Castle, DE
5.3 Data analysis software, TGA Standard Data Analysis Version 4.0, TA Instruments, New Castle, DE
5.4 Computer, IBM-PC 386, TA Instruments Operating System, Version 8.10B
5.5 Thermal analyzer instrument controller (MIM), TA Instruments, New Castle, DE
5.6 Autosampler, 920 Auto DSC, TA Instruments, New Castle, DE
5.7 Printer, Hewlett Packard, HP-7440, 8-pen plotter
5.8 Two-stage gas regulators, 50 psi maximum outlet pressure
5.9 Electronic balance, ±0.1-mg sensitivity, Mettler AE160
5.10 Forceps, flat-tipped
5.11 Weighing spatula
5.12 Desiccator
5.13 Mortar and pestle
5.14 Sieve, 80 mesh
5.15 N_2 gas, 99.99% purity
5.16 Kaolinite, standard, poorly crystalline, Georgia Kaolinite, Clay Minerals Society, Source Clay Minerals Project, sample KGa-2.
5.17 Gibbsite, standard, Surinam Gibbsite, SSL 67L022.

6. Reagents

6.1 Magnesium nitrate saturated solution [$Mg(NO_3)_2 \cdot 6H_2O$]
6.2 Ethanol

7. Procedure

Derive <2μm Clay Fractions

7.1 Prepare Na-saturated clay as in procedure 7A1a1, preparation of clay suspension, sections 7.8 to 7.19.

7.2 Dry the clay suspension and transfer to mortar. Moisten sample with ethanol and grind with pestle to make an homogeneous slurry.

7.3 Air-dry sample using flowing air in hood. Lightly grind sample with pestle to make a homogeneous powder. Transfer to original container for storage until use.

7.4 Prior to analysis, sieve sample with 80-mesh screen. Equilibrate sample overnight over a saturated magnesium nitrate solution (55% relative humidity) in a glass desiccator.

DSC Operation

7.5 Set-up the instrument and calibrate. Refer to the manufacturer's manual for operation of the DSC. Samples can be analyzed singly or with the autosampler for multiple samples.

7.6 Weigh ≈ 8 mg of sample, i.e., <80-mesh fine-earth (<2mm) soil fraction or derived <2-μm clay fraction, into tared aluminum sample pan. Refer to section on derived <2-μm clay fractions, Steps 7.1 to 7.4.

7.7 Use flat-tipped forceps to remove aluminum sample pan from balance. Drop sample from a 4- to 5-mm height to uniformly distribute sample in pan. Return the sample pan with sample to the balance and record weight to nearest ±0.1 mg. This weight is entered into computer in appropriate menu.

7.8 Carefully place the aluminum sample pan in the center of DSC platinum sample side (front section) of sample holder.

7.9 Place empty aluminum sample pan in reference side (back section) of sample holder.

7.10 Cover the DSC cell.

7.11 The standard sample run heating program has a heating rate of $10°C\ min^{-1}$, a starting temperature of 30°C, and an ending temperature of 600°C.

7.12 Start the "Run" program.

7.13 When the run is complete, data are analyzed by entering the Data Analysis 2100 System and selecting the DSC Standard Data Analysis Program.

7.14 Display file and calculate joules g^{-1} for the mineral endotherm.

8. Calculations

The area under a curve representing an endothermic dehydroxylation reaction is proportional to the enthalpy (ΔH) of the reaction. The enthalpy is calculated with the DSC software per g of kaolinite or gibbsite (joules g^{-1}) as appropriate.

Analyze each of the standard clays on the DSC. Calculate the enthalpy per g for the endothermic reactions of the standard kaolinite and gibbsite (joules g^{-1}).

The purity of the standard clays is evaluated via TGA (7A2a). Adjust the DSC results of the standards using the purity measurements from TGA.

Determine the amount of kaolinite and gibbsite in soil samples by dividing the enthalpy of the sample (joules g^{-1}) by the enthalpy of the standard (joules g^{-1}). Multiply this result by 100 to express as a percentage.

9. Report
Report percent kaolinite and/or gibbsite to the nearest whole number.

10. Precision and Accuracy
Precision and accuracy data are available from the SSL upon request.

11. References
Karathanasis, A.D., and B.F. Hajek. 1982. Revised methods for rapid quantitative determination of minerals in soil clays. Soil Sci. Soc. Am. J. 46:419-425.

Karathanasis, A.D., and W.G. Harris. 1994. Quantitative thermal analysis of soil minerals. p. 360-411. *In* J.E. Amonette and L.W. Zelazny (eds.) Quantitative methods in soil mineralogy. Soil Sci. Soc. Am. Misc. Publ. SSSA. Madison, WI.

Mackenzie, R.C. 1970. Simple phyllosilicates based on gibbsite- and brucite-like sheets. p. 498-537. *In* R. C. Mackenzie (ed.) Differential thermal analysis. Vol. 1. Acad. Press, London.

Mackenzie, R.C., and G. Berggen. 1970. Oxides and hydroxides of higher-valency elements. p. 272-302. *In* R.C. Mackenzie (ed.) Differential thermal analysis. Acad. Press, NY.

Mackenzie, R.C., and S. Caillere. 1975. The thermal characteristics of soil minerals and the use of these characteristics in the qualitative and quantitative determination of clay minerals in soils. p. 529-571. *In* J.E. Gieseking (ed.) Soil components. Vol. 2. Inorganic components. Springer-Verlag, NY.

Mackenzie, R.C., and B.D. Mitchell. 1972. Soils. p. 267-297. *In* R.C. Mackenzie (ed.) Differential thermal analysis. Vol. 2. Acad. Press., London.

Rouston, R.C., J.A. Kittrick, and E.H. Hope. 1972. Interlayer hydration and the broadening of the 10A x-ray peak in illite. Soil Sci. 113:167-174.

Schnitzer, M., and H. Kodama. 1977. Reactions of minerals with soil humic substances. p. 741-770. *In* J.B. Dixon and S.B. Weed (eds.) Minerals in soil environments. 2nd ed. Soil Sci. Am. Book Series No. 1. ASA and SSSA, Madison, WI.

Tan, K.H., and B.F. Hajek. 1977. Thermal analysis of soils. p. 865-884. *In* J.B. Dixon and S.B. Weed (eds.) Minerals in soil environments. 2nd ed. Soil Sci. Am. Book Series No. 1. ASA and SSSA, Madison, WI.

Tan, K.H., and B.F. Hajek, and I. Barshad. 1986. Thermal analysis techniques. p. 151-183. *In* A. Klute (ed.) Methods of soil analysis. Part 1. Physical and mineralogical properties. 2nd ed. Agron. Monogr. 9. ASA and SSSA, Madison, WI.

Instrumental Analysis (7A)
Surface Area (7A5)
N$_2$ Adsorption (7A5a)
Brunauer, Emmett and Teller (BET) Theory (7A5a1)
Vacuum Degassing (7A5a1a)
Multi-point (7A5a1a1)
Air-Dry, <2 mm (7A5a1a1a1)
Single Point (7A5a1a2)
Air-Dry, <2-mm (7A5a1a2a1)

1. Application

Surface area influences many physical and chemical properties of materials, e.g., physical adsorption of molecules and the heat loss or gain that results from this adsorption, shrink-swell capacity, water retention and movement, cation exchange capacity, pesticide adsorption, and soil aggregation (Carter et al., 1986). In addition, many biological processes are closely related to specific surface. Soils vary widely in their relative surface area because of differences in mineralogical and organic composition and in their particle-size distribution. Specific surface area (SSA) is an operationally defined concept, dependent upon the measurement technique and sample preparation (Pennell, 2002).

The most common approach used to measure SSA is considered indirect, based on measurements of the adsorption or retention of probe molecules on a solid surface at monolayer coverage (Pennell, 2002). Two common methods of measuring SSA are by ethylene glycol monoethyl ether (EGME) and N$_2$-sorption, using the theory of Brunauer, Emmett and Teller (N$_2$-BET). N$_2$ as a nonpolar gas does not interact with or have access to interlayer crystallographic planes of expandable clay minerals and thus is considered to provide a measure of external surface area, whereas polar molecules such as EGME are known to penetrate the interlayer surfaces of expandable clay minerals and therefore have been used to provide a measure of total surface areas (internal + external surface area) (Pennell, 2002). Significant differences between these methods are most apparent in soils containing expandable clay minerals and soil organic matter (Chlou and Rutherford, 1993; Pennell et al.1995; de Jong,

1999; Quirk and Murray, 1999). In the past, the SSL determined surface area by glycerol retention (7D1, method obsolete, Soil Survey Staff, 1996) or EGME retention (7D2, method obsolete, Soil Survey Staff, 1996). The current method described herein is N_2–BET (multi-point).

2. Summary of Method

A <2-mm, air-dry soil sample is ground to pass a 0.25 mm (60 mesh) sieve and oven-dried (24 h, 110° C). Enough soil (typically 0.5 to 1 g) is added to weighed sample cell to achieve 2- to 50- m^2 total area. Soil is cleaned of contaminants, e.g., water and oils, by vacuum degassing at 10 millitorr for a minimum of 3 h at 110° C and then re-weighed to obtain degassed sample weight. The sample is brought to a constant temperature by means of an external bath (77° K) and then small amounts of gas (N_2) called the absorbate are admitted in steps to evacuate sample chamber. Gas molecules that stick to the surface of the solid (absorbent) are said to be adsorbed and tend to form a thin layer that covers the entire adsorbate surface. The number of molecules required to cover the adsorbent surface with a monolayer of adsorbed molecules, N_m, can be estimated based on the BET theory. Multiplying N_m by the cross-sectional area of an adsorbate molecule yields the sample's surface area. Specific surface area is reported in $m^2 g^{-1}$ by procedure 7A5a1a1 or 7A5a1a2.

3. Interferences

Organic material can coat or cover mineral surfaces, generally reducing SSA by N_2-BET. Removal of organic matter prior analysis will typically increase these values. Freeze-drying may provide SSA values that are more representative of field conditions, as air-drying may result in the collapse and shrinkage of soil humic acid, whereas freeze-drying maintains an intricate structural network more characteristic of a natural state (Pennell, 2002). Sample size will vary depending on the SSA of the solid.

4. Safety

The outer surfaces of the heating mantle may become hot during use. Do not hold hot heating mantles without wearing protective gloves. Never insert fingers inside the pocket to determine if the mantle is heating up. Do not place a hot heating mantle on a surface that is not heat-resistant. Switch off the heating mantle when not in use. Refer to the manufacturer's manual for safe operation of the surface area analyzer.

5. Equipment
5.1 Surface area analyzer, with vacuum degassing, Quantachrome, Nova 3000 Series, Boynton Beach, FL
5.2 Computer, with Nova, Enhanced Data Reduction Program, Version 2.13, Quantachrome, Boynton Beach, FL. and printer
5.3 Vacuum pump, capable of achieving 50 millitorr required, 10 millitorr recommended
5.4 Dewar flask with insulated lid for liquid N_2
5.5 Oven, 110° C
5.6 Sample cells, glass, with outside stem diameters of 6, 9, and 12 mm and internal diameters of 4, 7, and 10 mm, respectively
5.7 Glass filler rods, 6 mm x 268.5 mm and 6mm x 131.5 mm
5.8 Mortar and pestle

6. Reagents
6.1 N_2, high purity, 99.9%
6.2 Liquid N_2 (77°K)
6.3 Standard reference material, 31.16 $m^2 g^{-1}$ SSA, with ±2.03 $m^2 g^{-1}$ reproducibility, Quantachrome, Boynton Beach, FL

7. Procedure

Surface Area Analyzer Set-up and Operation

7.1 Set N_2-regulator at 10 PSIG (70 kPa).

7.2 Refer to the manufacturer's manual for the routine operation, manifold calibration, and sample cell calibration.

7.3 The dosing manifold is factory calibrated, and there is no need to repeat this calibration before every analysis. However, operator should check this calibration periodically (e.g. once every month), or if the operator feels that changes to the system may have altered the manifold volume.

7.4 There needs to be a sample cell calibration for each sample cell + filter rod + station combination and for each adsorbate/coolant combination. Once done, there is no need for further calibration for that particular combination. For most users, all standard (bulbless) cells can be considered equivalent (for each diameter), i.e., one cell/rod/station combination will suffice for a different cell (of the same diameter) with the same rode in the same station.

Sample Preparation and Vacuum Degas

7.5 Pulverize a <2-mm, air-dry soil sample with a mortar and pestle to break up any aggregates.

7.6 Weigh sample cell to the nearest 0.1 mg. Weigh enough soil (typically 0.5 to 1.0 g) into sample cell to achieve 2- to 50 m^2 total area (sample size will vary depending on the SSA of the soil). Oven-dry at 110°C for 24 h.

7.7 Remove sample cell with soil from oven and seal immediately. Allow to cool to touch.

7.8 Place the sample cell in the pouch of the heating mantle, set clamp in place, insert cell into fitting, tighten fitting, and loop elastic cords over hooks provided. Set the degas temperature at 110° C for a minimum of 3 h.

7.9 Upon completion of degassing, switch the mantle off. Allow sample to cool. Unload degasser when ready to analyze sample.

7.10 Remove cell and reweigh to obtain dry, degassed soil sample weight to the nearest 0.1 mg.

7.11 Use the Preset Analysis option on surface area analyzer allows user to preset and save before analysis the following: User ID, stations for analysis, setup files, cell numbers, sample ID numbers, and comments.

Analysis

7.12 Ensure dewar flask is filled to the red line (approximately 1 in from top) with liquid N_2. Allow 5 min for liquid N_2 to equilibrate for best results. If the liquid N_2 is still boiling heavily, then the dewar needs to be cleaned before filling it with liquid N_2. If boiling continues after a dry clean dewar is filled for 5 min, then replace the dewar. Residual boiling will require that the dewar be topped-off again. Ensure proper alignment of sample cells with dewar mouth. Foam cap is not required for BET analysis only. If however a long isotherm is required, then use the foam cap. Ensure that there is no condensed water visible in the stem of the sample cell.

7.13 Follow set-up instructions on the surface area analyzer (e.g., select cell for Stations, enter sample ID for Stations, enter dry degassed sample weight by measuring sample volume, etc) until prompted to proceed with analysis. Analytical data include Multi Point BET (adsorption), slope, intercept, correlation coefficient, BET C, total surface area in cell (m^2) and specific surface area ($m^2\ g^{-1}$). Set-up and analytical data are automatically recorded by computer and printer.

8. Calculations

The BET equation for determination of the surface area of solids (Quantachrome Corp., 2000) is as follows:

8.1 $1/[W((P0/P)-1)] = [1/(W_mC)] + [(C - 1)/W_mC] \times (P/P0)$

where:
W   = weight of gas adsorbed at a relative pressure of P/P0
W_m = weight of adsorbate constituting a monolayer of surface coverage
C   = BET C constant, related to energy of adsorption in first adsorbed layer and is indicative of the magnitude of absorbed/adsorbate interactions (typically, 50 to 250 for most solid surfaces)

The BET equation requires a linear plot of $1/[W((P0/P)-1)]$ versus P/P0 which for most solids, using nitrogen as adsorbate, is restricted to a limited region of the adsorption isotherm, usually in the P/P0 range of 0.05 to 0.35 (Quantachrome Corp., 2000). The standard multi-point BET procedure requires a minimum of three points in the appropriate relative pressure range. The weight of the monolayer of adsorbate W_m can then be obtained from the slope s and intercept i of the BET plot.

8.2 From the BET equation,

8.2.1 $s = (C - 1)/W_mC$

8.2.2 $i = 1/W_mC$.

8.3 Thus, the weight of the monolayer W_m can be obtained by combining these two equations as follows:

$W_m = 1/(s+i)$.

8.4 Total surface area S_t of the sample is calculated as follows:

$S_t = W_mNA_{cs}/M$

where:
N   = Avogadro's number (6.023×10^{23})
M   = Molecular weight of absorbate ($28.02\ g\ mol^{-1}$)
A_{cs} = Close-packed nitrogen monolayer at 77 K, the cross-sectional area for nitrogen = 16.2 Angstroms^2

8.6 Specific surface area S is calculated as follows:

$S = S_t/w$

where:
w = Degassed sample weight

9. Report
  Report specific surface area in $m^2 g^{-1}$ to the nearest 0.01 unit.

10. Precision and Accuracy

Precision and accuracy data are available from the SSL upon request.

11. References

Carter, D.L., M.M. Mortland, and W.D. Kemper. 1986. Specific surface. *In* A. Klute (ed.) Methods of Soil Analysis. Part 1. Physical and mineralogical methods. 2nd ed. Agron. Monogr. 9. Soil Sci. Soc. Am., Madison, WI.

Chlou, C.T., and D.W. Rutherford. 1993. Sorption of N2 and EGME vapors on some soils, clays, and mineral oxides and determination of sample surface areas by use of sorption data. Environ. Sci. Technol. 27:1587-1594.

de Jong, E. 1999. Comparison of three methods of measuring surface area of soils. Can. J. Soil Sci. 79:345-351.

Pennell, K.D. 2002. Specific surface area. *In* J.H. Dane and G.C. Topp (eds.) Methods of Soil Analysis. Part 4. Physical Methods. Soil. Sci. Soc. Am. Book Series No. 5, Madison, WI.

Pennell, K.D., S.A. Boyd, and L.M. Abriola. 1995. Surface area of soil organic matter reexamined. Soil Sci. Soc. Am. J. 59:1012-1018.

Quantachrome Corporation. 2000. Particle characterization technology. Quantachorme Corp., Boynton, FL.

Quirk, J.P., and R.S. Murray. 1999. Appraisal of the ethylene glycol monoethyl ether method for measuring hydratable surface area of clays and soils. Soil Sci. Soc. Am. J. 63:839-849.

Soil Survey Staff. 1996. Soil survey laboratory methods manual. Version No. 3.0. USDA-NRCS. Soil Survey Investigations Report No. 42. U.S. Govt. Print. Office, Washington, DC.

Optical Analyses[1] (7B)
Grain Studies (7B1)
Analysis and Interpretation (7B1a)

Minerals

Identification criteria: Important properties in grain identification are listed below in approximate order of ease and convenience of determination. Estimates of several of these properties often allow identification of a grain so that detailed or extremely accurate measurements are seldom necessary. In the finer soil separates, grain identification may be impossible because the grains may be too small or not in the right position to permit measurement of some properties, e.g., optic angle (2V) or optic sign. A process to help practice estimating properties is to crush, sieve, and mount a set of known minerals and to compare these known standards to unknowns.

Refractive index is the ratio of the speed of light in the medium (mineral) to the speed of light in a vacuum. It can be estimated by relief or can be accurately determined by using calibrated immersion liquids. When relief is used to estimate refractive index, the grain shape, color, and surface texture are considered, i.e., thin platy grains may be estimated low, whereas colored grains and grains with rough, hackly surface texture may be estimated high. Estimation is aided by comparing an unknown with known minerals.

---

[1] The discussion of identification and significance of minerals, microcrystalline aggregates, and amorphous substances in optical studies of grain mounts was from material after John G. Cady (1965), with permission, and modified by Warren C. Lynn, Research Soil Scientist, NRCS, Lincoln, NE.

Relief is an expression of the difference in refractive index between the grain and the mounting medium, and the greater the difference, the greater the relief. The analogy is to topographic relief. When viewed through the microscope, grains with high relief are distinct, whereas grains with low relief tend to fade into the background. The SSL selects a mounting medium with an index of refraction close to quartz, i.e., quartz has low relief. Most other minerals are identified by comparison.

Becke line is a bright halo of light that forms near the contact of the grain and the mounting medium because of the difference in refractive index between the two. As the plane of focus is moved upward through the grain, the Becke line appears to move into the component with the higher refractive index. In Petropoxy 154$_{TM}$, the Becke line moves away from potassium feldspar (index of refraction <1.54) but moves into mica (index of refraction >1.54).

Birefringence is the difference between the highest and lowest refractive index of the mineral. Accounting for grain thickness and orientation, the birefringence is estimated by interference color. Interference color is observed when an anisotropic mineral is viewed between crossed-polarized light. Several grains of the same species must be observed because the grains may not all lie in positions that show the extremes of refractive index. For example, mica has high birefringence but appears low when the platy mineral grain is perpendicular to the microscope axis because the refractive indices of the two crystallographic directions in the plane are similar. However, a mica grain viewed on edge in a thin section shows a high interference color. The carbonate minerals have extremely high birefringence (0.17 to 0.24). Most of the ferrogmagnesian minerals are intermediate (0.015 to 0.08). Orthoclase feldspar and apatite are low (0.008) and very low (0.005), respectively.

Color helps to discriminate among the heavy minerals. Pleochroism is the change in color or light absorption with stage rotation when the polarizer is inserted. Pleochroism is a good diagnostic characteristic for many colored minerals. Tourmaline, green beryl, and staurolite are examples of pleochroic minerals.

Shape, cleavage, and crystal form are characteristic or possibly unique for many minerals. Cleavage may be reflected in the external form of the grain or may appear as cracks within the grain that show as regularly repeated straight parallel lines or as sets of lines that intersect at definite repeated angles. The crystal shape may be different from the shape of the cleavage fragment. Plagioclase feldspars, kyanite, and the pyroxenes have strong cleavage. Zircon and rutile usually appear in crystal forms.

Extinction angle and character of extinction observed between crossed-polarized light are important criteria for some groups of minerals. To measure extinction angles, the grain must show its cleavage or crystal form. These angles may be different along different crystallographic axes. Some minerals have sharp, quick total extinction, whereas other minerals have more gradual extinction. In some minerals with high light dispersion, the interference color dims and changes at the extinction position.

Optic sign, optic angle, and sign of elongation are useful, if not essential, determinations but are often difficult, unless grains are large or in favorable orientation. Determination of optic sign requires that the grains show dim, low-order interference colors or show no extinction. Grains with bright colors and with sharp, quick extinction rarely provide usable interference figures.

Particular mineral species: The following are the outstanding diagnostic characteristics of the most commonly occurring minerals and single-particle grains in the sand and silt fractions of soils. The refractive indices that are provided are the intermediate values.

Quartz has irregular shapes. The refractive index of quartz (1.54) approximates that of the epoxy (Petropoxy 154$_{TM}$) mounting medium. The Becke line may be split into yellow and blue components. The interference colors are low order but are bright and warm. There is sharp extinction with a small angle of rotation, i.e., "blink extinction". Crystal forms are sometimes observed and usually indicate derivation from limestone or other low-temperature secondary origin.

Potassium feldspars: Orthoclase may resemble quartz, but the refractive index (1.52) and birefringence are lower than that of quartz. In addition, orthoclase may show cleavage. *Microcline* has a refractive index of 1.53. The Becke line moves away from the grain with upward focus. A twinning intergrowth produces a plaid or grid effect between crossed-polarized light that is characteristic of microcline. *Sanidine* has the same refractive index and birefringence as other potassium feldspars. Grains are usually clear and twinning is not evident. In sanidine, the 2V angle is low (12°) and characteristic. The 2V angle is the acute angle between two optic axes, or more simply, the optical axial angle.

Plagioclase feldspars have refractive indices that increase with an increase in the proportion of calcium. The refractive index of the sodium end-member albite (1.53) is lower than that of quartz, but the refractive index of the calcium end-member anorthite (1.58) is noticeably higher than that of quartz. Some *oligoclase* has the same refractive index as quartz which prevents distinctions by the Becke line. Plagioclase feldspars often show a type of twinning (defined as albite twinning) that appears as multiple alternating dark and light bands in crossed-polarized light. Cleavage is good in two directions parallel to (001) and (010), often producing lathlike or prismatic shapes.

Micas occur as platy grains that are often very thin. The plate view shows very low birefringence, whereas the edge view shows a very high birefringence. Plates are commonly equidimensional and may appear as hexagons or may have some 60° angles. *Biotite* is green to dark brown. Green grains may be confused with chlorite. Paler colors, a lowering of refractive index, and a distortion of the extinction and interference figure indicate weathering to *hydrobiotite, kaolinite,* or *vermiculite. Muscovite* is colorless. Muscovite has a moderate refractive index (1.59) in the plate view and an interference figure that shows a characteristic 2V angle of 30 to 40° which can be used as a standard for comparing 2V angles of other minerals.

Amphiboles are fibrous to platy or prismatic minerals with slightly inclined extinction, or occasionally with parallel extinction. Color and refractive index increase as the Fe content increases. Amphiboles have good cleavage at angles of ~ 56 and 124°. Refractive index of the group ranges from 1.61 to 1.73. *Hornblende* is the most common member of the amphiboles. Hornblende is slightly pleochroic; usually has a distinctive color close to olive-green; has inclined extinction; and is often used as an indicator of weathering.

Pyroxenes: Enstatite and aegerine-augite are prismatic and have parallel extinction. Aegerine-augite has unique and striking green-pink pleochroism. *Augite* and *diopside* have good cleavage at angles close to 90° and large extinction angles. Colors usually are shades of green, with interference colors of reds and blues. Refractive indices in the pyroxenes (1.65 to 1.79) are higher than for amphiboles.

Olivine is colorless to very pale green and usually irregular in shape (weak cleavage). Olivine has vivid, warm interference colors. Olivine is an easily weathered mineral and may have cracks or seams filled with serpentine or goethite. It is seldom identified in soils, but has been observed in certain soils from Hawaii.

Staurolite is pleochroic yellow to pale brown and sometimes contains holes, i.e., the "Swiss cheese" effect. The refractive index is ~ 1.74. Grains may have a foggy or milky appearance which may be caused by colloidal inclusions.

Epidote is a common heavy mineral, but the forms that occur in soils may be difficult to identify positively. Typical epidote is unmistakable with its high refractive index (1.72 to 1.76), strong birefringence, and a pleochroism that includes the pistachio-green color. It has typical interference colors of reds and yellows. Commonly, grains show an optic axis interference figure with a 2V angle that is nearly 90°. However, epidote is modified by weathering or metamorphism to colorless forms with lower birefringence and refractive index. *Zoisite* and *clinozoisite* in the epidote group are more common than some of the literature indicates. These minerals of the epidote group commonly appear as colorless, pale-green, or bluish-green, irregularly shaped or roughly platy grains with high refractive index (1.70 to 1.73). Most of these minerals show anomalous interference colors (bright pale blue) and no complete extinction and can be confused with several other minerals, e.g., kyanite and diopside. Zoisite has a distinctive deep blue interference color. Identification usually depends on determination of properties for many grains.

Kyanite is a common mineral but is seldom abundant. The pale blue color, the platy, angular cleavage flakes, the large cleavage angles, and the large extinction angles (30° extinction) usually can be observed and make identification easy.

Sillimanite and *andalusite* resemble each other. These minerals are fibrous to prismatic with parallel extinction. However, their signs of elongation are different. In addition, sillimanite is colorless, and andalusite commonly has a pink color.

Garnet is found in irregularly shaped, equidimensional grains that are isotropic and have high refractive index ($\geq$1.77). Garnet of the fine sand and silt size is often colorless. Pale pink or green colors are diagnostic in the larger grains.

Tourmaline has a refractive index of 1.62 to 1.66. Prismatic shape, strong pleochroism, and parallel extinction are characteristic. Some tourmaline is almost opaque when at right angles to the vibration plate of the polarizer.

Zircon occurs as tetragonal prisms with pyramidal ends. Zircon has very high refractive index (>1.9), parallel extinction, and bright, strong interference colors. Broken and rounded crystals frequently are found. Zircon crystals and grains are almost always clear and fresh appearing.

Sphene, in some forms, resembles zircon, but the crystal forms have oblique extinction. The common form of sphene, a rounded or subrounded grain, has a color change through ultrablue with crossed polarizers instead of extinction because of its high dispersion. Sphene is the only pale-colored or colorless high-index mineral that provides this effect. It is amber colored in reflected light. The refractive index of sphene is slightly lower than that of zircon, and the grains are often cloudy or rough-surfaced.

Rutile grains have prismatic shape. The refractive index and birefringence are extremely high (2.6 to 2.9). The interference colors usually are obscured by the brown, reddish-brown, or yellow colors of the mineral. Other TiO_2 minerals, *anatase* and *brookite*, also have very high refractive indices and brown colors and may be difficult to distinguish in small grains. The anatase and brookite usually occur as tabular or equidimensional grains.

Apatite is common in youthful soil materials. Apatite has a refractive index slightly <1.63 and a very low birefringence. Crystal shapes are common and may appear as prisms, and is often bullet shaped. Rounding by solution produces ovoid forms. Apatite is easily attacked by acid and may be lost in pretreatments.

Carbonates: Calcite, dolomite, and siderite, in their typical rhombohedral cleavage forms, are easily identified by their extremely high birefringence. In soils, these minerals have other forms, e.g., scales and chips; cements in aggregates; microcrystalline coatings or aggregates; and other fine-grained masses that are often mixed with clay and other minerals. The extreme birefringence is always the identification clue and is shown by the bright colors between crossed-polarized light and by the marked change in relief when the stage is rotated with one polarizer in. The microcrystalline aggregates produce a twinkling effect when rotated between crossed-polarized light. These three minerals have differences in their refractive indices which may be used to distinguish them. Siderite is the only one with both indices >Petropoxy 154$_{TM}$. It is more difficult to distinguish calcite from dolomite, and additional techniques such as staining or x-ray diffraction may be used.

Gypsum occurs in platy or prismatic flat grains with refractive index approximately equal to orthoclase. It usually has a brushed or "dirty" surface.

Opaque minerals, of which *magnetite* and *ilmenite* are the most common, are difficult to identify, especially when they are worn by transportation or otherwise affected by weathering. Observations of color and luster by reflected light, aided by crystal form if visible, are the best procedures. Magnetic separations help to confirm the presence of magnetite and ilmenite. Many grains that appear opaque by plain light can appear translucent if viewed between strong crossed-polarized light. Most grains that behave in this way are altered grains or aggregates and are not opaque minerals.

Microcrystalline Aggregates and Amorphous Substances

Identification criteria: Most microcrystalline aggregates have one striking characteristic feature, i.e., they show birefringence but do not have definite, sharp, complete extinction in crossed-polarized light. Extinction may occur as dark bands that sweep through the grain or parts of the grain when the stage is turned or may occur in patches of irregular size and shape. With a few exceptions, e.g., well-oriented mineral pseudomorphs and certain clay-skin fragments, some part of the grain is bright in all positions. Aggregates and altered grains should be examined with a variety of combinations of illumination and magnification in both plain and polarized lights. The principal properties that can be used to identify or at least characterize aggregates are discussed below.

Color, if brown to bright red, is usually related to Fe content and oxidation. Organic matter and Mn may contribute black and grayish-brown colors.

Refractive index is influenced by a number of factors, including elemental composition, atom packing, water content, porosity, and crystallinity. Amorphous (noncrystalline) substances have a single index of refraction, which may vary depending on chemical composition. For example, allophane has a refractive index of 1.47 to 1.49, but the apparent refractive index increases with increasing inclusion of ferrihydrite (noncrystalline Fe oxide) in the mineral.

Strength of birefringence is a clue to the identity of the minerals. Even though the individual units of the aggregate are small, birefringence can be estimated by interference color and brightness. Amorphous substances, having only a single index of refraction, exhibit no birefringence and are isotropic between crossed-polarized light.

Morphology may provide clues to the composition or origin of the aggregate. Some aggregates are pseudomorphs of primary mineral grains. Characteristics of the original minerals, i.e., cleavage traces, twining, or crystal form can still be observed. Morphology can sometimes be observed in completely altered grains, even in volcanic ash shards and basalt fragments. Other morphological characteristics may be observed in the individual units or in the overall structure, e.g., the units may be plates or needles, or there may be banding.

Particular species of microcrystalline aggregates and amorphous substances: For purposes of soil genesis studies, the aggregates that are present in sand or silt fractions are not of equal significance. Some are nuisances but must be accounted for, and others are particles with important diagnostic value. Useful differentiating criteria for some of the commonly occurring aggregate types are discussed below.

Rock fragments include chips of shale, schist, and fine-gained igneous rocks, e.g., rhyolite. Identification depends on the recognition of structure and individual components and the consideration of possible sources. Rock fragments are common in mountainous regions and are often hydrothermally-altered in the western U.S.

Clay aggregates may be present in a wide variety of forms. Silt and sand that are bound together into larger grains by a nearly isotropic brownish material usually indicate incomplete dispersion. Clay skins may resist dispersion and consequently may appear as fragments in grain mounts. Such fragments are usually brown or red and translucent with wavy extinction bands. Care is required to distinguish these fragments from weathered biotite. Clay aggregates may be mineral pseudomorphs. Kaolin pseudomorphs of feldspar commonly are found. Montmorillonite aggregates, pseudomorphic of basic rock minerals, have been observed. In this form, montmorillonite shows high birefringence and an extinction that is mottled or patchy on a small scale. Coarse kaolinite flakes, books, and vermicular aggregates resist dispersion and may be abundant in sand and silt. These particles may resemble muscovite, but they are cloudy; show no definite extinction; and have very low birefringence. Many cases of anomalously high cation exchange capacity (CEC) of sand and silt fractions that are calculated from whole soil CEC and from clay CEC and percent content, can be accounted for by the occurrence of these aggregates in the sand and silt fractions.

Volcanic glass is isotropic and has a low refractive index, lower than most of the silicate minerals. The refractive index ranges from 1.48 in the colorless siliceous glasses to as high as

1.56 in the green or brown glasses of basalt composition. Shapes vary, but the elongated, curved shard forms, often with bubbles, are common. This glassy material may adhere to or envelop other minerals. Particles may contain small crystals of feldspar or incipient crystals with needles and dendritic forms. The colorless siliceous types (acidic, pumiceous) are more common in soils, as the basic glasses weather easily. Acidic glasses are more commonly part of "ash falls", as the magma usually is gaseous and explosive when pressure is released. Basic glasses are more commonly associated with volcanic flow rocks which are usually not gaseous.

Allophane is present in many soils that are derived from volcanic ash. Allophane seldom can be identified directly, but its presence can be inferred when sand and silt are cemented into aggregates by isotropic material with low refractive index, especially if volcanic ash shards are also present.

Opal, an isotropic material, occurs as a cementing material and in separate grains, some of which are of organic origin, i.e., plant opal, sponge spicules, and diatoms. The refractive index is very low (<1.45), which is lower than the value for volcanic ash. Identification may depend in part on form and occurrence.

Iron oxides may occur as separate grains or as coatings, cementing agents, and mixtures with other minerals. Iron oxides impart brown and red colors and raise the refractive index in the mixtures. *Goethite* is yellow to brown. Associated red areas may be hematite. These red varieties have a refractive index and birefringence that are higher and seem to be better crystallized, often having a prismatic or fibrous habit. Aggregates have parallel extinction. In oriented aggregates, the interference colors often have a greenish cast. *Hematite* has higher refractive index than goethite and is granular rather than prismatic. Large grains of hematite are nearly opaque.

Gibbsite often occurs as separate, pure, crystal aggregates, either alone or inside altered mineral grains. The grains may appear to be well-crystallized single crystals, but close inspection in crossed-polarized light shows patchy, banded extinction, indicating intergrown aggregates. Gibbsite is colorless. The refractive index (1.56 to 1.58) and the birefringence are higher for gibbsite than the corresponding values for quartz. The bright interference colors and aggregate extinction are characteristic of gibbsite.

Chalcedony is a microcrystalline form of quartz that was formerly considered a distinct species. Chalcedony occurs as minute quartz crystals and exhibits aggregate structure with patchy extinction between crossed-polarized light. It may occur in nodules of limestone deposits and may be a pseudomorphic replacement in calcareous fossils. The refractive index is slightly lower than that of quartz, and the birefringence is lower than that of gibbsite. *Chert* is a massive form of chalcedony.

Glauconite occurs in aggregates of small micaceous grains with high birefringence. When fresh, glauconite is dark green and almost opaque, but it weathers to brown and more translucent forms. Glauconite is difficult to identify on optical evidence alone. Knowledge of source area or history are helpful in identification.

Titanium oxide aggregates have been tentatively identified in the heavy mineral separates of many soils. These bodies have an extremely high refractive index and high birefringence similar to rutile. The yellow to gray colors are similar to those of anatase. The TiO_2 aggregates are granular and rough-surfaced. This growth habit with the little spurs and projections suggests that TiO_2 aggregates may be secondary.

References

Cady, J.G. 1965. Petrographic microscope techniques. p. 604-631. *In* D.D. Evans, L.E. Ensminger, J.L. White, and F.E. Clark (eds.) Methods of soil analysis. Part 1. Physical and mineralogical properties, including statistics of measurement and sampling. 1st ed. Agron. Monogr. 9. ASA and SSSA, Madison, WI.

Optical Analyses (7B)
Grain Studies (7B1)
Separation by Heavy Liquids (7B1a1)

1. Application

The sand and silt fractions of most soils are dominated by quartz or by quartz and feldspars (Cady, 1965). These minerals have a relatively low specific gravity (2.57 to 2.76). The large numbers of "heavy" mineral grains (specific gravity >2.8 or 2.9) with a wide range of weatherability and diagnostic significance may be only a small percentage of the grains (Cady, 1965). However, these "heavy" minerals are often indicative of provenance, weathering intensities, and parent material uniformity (Cady, Wilding, and Drees, 1986).

2. Procedure

To study "heavy" minerals, a common practice is to concentrate these grains by specific-gravity separations in a heavy liquid. This procedure (7B1a1) is rarely used at the SSL but is done on occasion for special studies.

Micas are difficult to separate because of their shape and because a little weathering, especially in biotite, significantly decreases the specific gravity. These differences in density in biotite may be used to concentrate weathered biotite in its various stages of alteration .

Separation of grains by heavy liquids is most effective when grains are clean. Organic matter may prevent wetting and cause grains to clump or raft together. Light coatings may cause heavy grains to float, and iron-oxide coatings may increase specific gravity. In some kinds of materials, an additional technique is to separate and weigh the magnetic fraction, either before or after the heavy-liquid separation.

Concentrate the "heavy" minerals, i.e., specific gravity >2.8 or 2.9, by specific-gravity separations in a heavy liquid. The reagent of choice is sodium polytungstate (density 2.8 g^{-1} mL). Dilute the sodium polytungstate with distilled water to obtain required densities <2.8 g^{-1} mL. Use a specific gravity $\approx$ 2.5, to concentrate volcanic glass, plant opal, or sponge spicules. When using this liquid, avoid contact with skin and work in a well-ventilated area.

Separation by specific gravity alone in separatory funnels, cylinders or various kinds of tubes is usually adequate for grains >0.10 mm. Separation by centrifuging is required for grains <0.10 mm. Use pointed, 15-ml centrifuge tubes for these separations.

Decant the light minerals after inserting a smooth bulb glass rod to stop off the tapered end of centrifuge tube. Alternatively, remove the heavy minerals by gravity flow using a lower stopcock or maintain the heavy minerals in place by freezing the lower part of tube.

3. References

Cady, J.G. 1965. Petrographic microscope techniques. p. 604-631. *In* D.D. Evans, L.E. Ensminger, J.L. White, and F.E. Clark (eds.) Methods of soil analysis. Part 1. Physical and mineralogical properties, including statistics of measurement and sampling. 1st ed. Agron. Monogr. 9. ASA and SSSA, Madison, WI.

Cady, J.G., L.P. Wilding, and L.R. Drees. 1986. Petrographic microscope techniques. p. 198-204. *In* A. Klute (ed.) Methods of soil analysis. Part 1. Physical and mineralogical methods. 2nd ed. Agron. Monogr. 9. ASA and SSSA, Madison, WI.

Optical Analysis (7B)
Grain Mounts, Epoxy (7B1a2)

1. Application

Grain counts are used to identify and quantify minerals in the coarse silt and sand fractions of soils. Results are used to classify soil pedons in mineralogy families of soil taxonomy (Soil Survey Division Staff, 1999), to help determine substrate provenance of source materials, and to support or identify lithologic discontinuities.

2. Summary of Method

In particle-size analysis, soils are dispersed so that material <20 μm in diameter is separated by settling and decanting, and the sand and coarse silt fractions are separated by sieving. Refer to procedure for the separation by heavy liquids of the less abundant minerals with a specific gravity >2.8 or 2.9 (7B1a1).

Following sample selection, permanent mounts are prepared for the two most abundant particle-size fractions among the fine sand, very fine sand and coarse silt. The grains are mounted in a thermo-setting epoxy cement with a refractive index of 1.54. The grains are then identified and counted under a petrographic microscope.

A mineralogical analysis of a sand or silt fraction may be entirely qualitative, or it may be quantitative to different degrees (Cady, 1965). The SSL performs a quantitative analysis (7B1a2). Data are reported as a list of minerals and an estimated quantity of each mineral as a percentage of the grains counted in the designated fraction. The percentages of minerals are obtained by identifying and counting a minimum of 300 grains on regularly spaced line traverses that are 2 mm apart.

The identification procedures and reference data on minerals are described in references on sedimentary petrography (Krumbein and Pettijohn, 1938; Durell, 1948; Milner, 1962; Kerr, 1977; Deer et al., 1992) and optical crystallography (Bloss, 1961; Stoiber and Morse, 1972; Shelley, 1978; Klein and Hurlbut, 1985; and Drees and Ransom, 1994).

3. Interferences

The sample must be thoroughly mixed because the sub-sample on the slide is small. If grains are coated with clay or if aggregates of finer material remain in the fraction that is counted, results may be skewed. Variations in the time or temperature of heating the epoxy may result in either matrix stress or variation the refractive index of the epoxy. Do not use steel needles or spatulas because magnetic minerals may adhere to steel resulting in uneven distribution of grains on the slide.

4. Safety

Heat the epoxy in a fume hood. Use caution in handling hot glass slides. Use heat resistant gloves as needed. Immediately wash or remove any epoxy that comes in contact with the skin. Carefully handle slides and cover slips to avoid cuts.

5. Equipment

5.1 Petrographic microscope slides, precleaned, 27 x 46 mm
5.2 Cover slips, glass, 25 x 25 mm
5.3 Hot plate
5.4 Micro-spatula
5.5 Dissecting needle
5.6 Plywood covered with formica (6 x 8 x 1.25 cm)
5.7 Timer
5.8 Polarizing petrographic microscope
5.9 Tally counter
5.10 Set of 76-mm (3 in) sieves, square weave phosphor bronze wire cloth except 300 mesh which is twilled weave. U.S. series and Tyler Screen Scale equivalent designations are as follows:

Sand Size	Opening (mm)	U.S. No.	Tyler Mesh Size
VCS	1.0	18	16
CS	0.5	35	32
MS	0.25	60	60
FS	0.105	140	150
VFS	0.047	300	300

5.11 Oven, 110 C

6. Reagents
6.1 Petropoxy 154$_{TM}$ Resin and Curing Agent, Palouse Petro Products, 425 Sand Rd., Palouse, WA 99163
6.2 Index immersion oils
6.3 Reverse Osmosis (RO) water, ASTM Type III grade of reagent water

7. Procedure

Sample Selection and Grain Mount Preparation
7.1 Refer to the analysis request sheets. Record optical mineralogy requests in the LIMS Report "Optical Priority List." Note any special instructions by soil scientists. Sample selection depends on the purpose of analysis. In most work, e.g., checks on discontinuities or estimation of degree of weathering in different soil horizons, the study of those fractions that comprise a significant quantitative part of the soil is important. The SSL convention is to count the most abundant fraction, i.e., coarse silt (CSI), very fine sand (VFS) or fine sand (FS), especially if the fraction is clearly larger. This procedure works well in the establishment of mineralogy families for soil taxonomy (Soil Survey Staff, 1999). This procedure may result in different size fractions being counted for different horizons within a single pedon. If fractions are rather equal in abundance, the VFS is selected as it provides the widest range of information. The SSL does not count multiple fractions for a single sample, or count combined fractions, or present the data as weighted averages. If it is appropriate to count the same size fraction for each horizon within a pedon or project such as for a study of soil lithology, this request must be specified by the project coordinator.

7.2 Refer to the laboratory information system (LIMS) report "Optical Priority List" for daily work or check the sand box lids to determine which sands have been fractionated. Sands are fractionated during particle-size distribution analysis (PSDA) (procedure 3A1a). Fine sand, and very fine sand, fractions placed in gelatin capsules and stored in a labeled vial. Coarse silts are stored in aluminum pans. For projects, the LIMS report provides the following:

a. Sample numbers by project

b. Type of count requested (full grain count or glass count)

c. Percent of <2-mm fraction for CSI, VFS, or FS as determined by SSL PSDA.

7.3 If the particle-size section does not provide a sand and coarse silt separate, derive these fractions by repeated gravity sedimentation at 20 μm and sieving the 20-μm to 2.0-mm material as follows:

7.3.1 Disperse the sample in sodium hexametaphosphate as described in procedure 7A1a1.

7.3.2 Pour the soil suspension into a 200-mL beaker that has a line marked 5 cm above the bottom.

7.3.3 Add RO water to the beaker up to the 5-cm mark.

7.3.4 Stir the suspension and allow to settle 2.0 min. Use a stopwatch.

7.3.5 Decant and discard the suspension containing the clay and fine silt.

7.3.6 Repeat Steps 7.3.3 to 7.3.5 until the supernatant is clear or reasonably so.

7.3.7 Transfer the sediment to a drying dish and dry at 105-110°C.

7.3.8 Sieve the dried sample to isolate the individual fractions.

7.4 Review the PSDA data and select samples. Make grain mounts from the one or two most abundant fractions, preferably from the CSI, VFS, or FS. Record sample numbers and respective PSDA data.

7.5 Mix a small amount of Petropoxy 154$_{TM}$ resin and curing agent (1:10 ratio resin to curing agent) in a clean graduated plastic beaker that is provided with the reagents.

7.6 Prepare epoxy at least one day prior to use and refrigerate until needed.

7.7 Turn on hot plate and allow to equilibrate at 125°C for $\approx$ 1 h.

7.8 Remove mixture from refrigerator at least 40 min prior to use. If the petropoxy crystallizes, gently warm mixture until crystals dissolve.

7.9 At the base of the glass slides, record the grain size fraction (CSI, VFS, FS, etc.) and SSL LIMS sample number. An example is as follows:VFS 35126

7.10 Obtain sand vials and/or silt dishes. Arrange in an orderly manner. Work with 4-6 slides and samples at a time.

7.11 Remove lids from sand vials and place upside down in front of respective vials. Remove gelatin capsules (VFS or FS) from vial. Rotate capsule to mix contents and place in lid. Stir with a micro-spatula to mix coarse silts.

7.12 Use a small, rounded glass or plastic rod to drop petropoxy mixture on the upper middle of each slide. Use one drop of petropoxy for CSI or VFS and two drops for FS.

7.13 Use a micro-spatula to add the mixed grains to petropoxy. Use larger amounts for smaller fractions. The analyst's technique of adding the appropriate amount of petropoxy and of making grain counts on prepared slides develops with experience. Use a dissecting needle to slowly and carefully stir the grains into the petropoxy. Avoid introduction of air bubbles. Obvious air bubbles can be popped with the dissecting needle.

7.14 Gently place one cover slip (check to be certain) on the petropoxy. Avoid fingerprints. Allow the petropoxy to spread under the cover slip. Center the cover slip at top center of glass microscope slide so that there is a parallel, equidimensional border around the top and sides of slide.

7.15 To ensure the uniform distribution of grains and the removal of air bubbles, use a dissecting needle to gently tap or press down cover slip. If necessary, the analyst may need to recenter the cover slip. Be careful not to crack cover slip.

7.16 Align a batch of 4-6 slides in two rows on center of hot plate.

7.17 Set timer and heat slides at 125°C for 8 min. Time can be adjusted by experience. As a rule, when epoxy is set, it has cured to yield a refractive index of 1.540. Longer heating may result in a distortion of the optical characteristics of the petropoxy and a refractive index differing from 1.540.

7.18 As one batch of slides heats, prepare the next batch. After heating for 8 min, slide the glass slides off the hot plate onto the formica block. Allow to cool.

7.19 Examine the grain mount for quality. The epoxy medium should be isotropic. The presence of anisotropic stress lines around grains under X-Nicols may interfere with observation of optical properties. Remake any unsatisfactory grain mounts. Place satisfactory mounts in a microscope slide file box. Neatly record project number and grain mount positions on interior of box lid, and record box number(s) in the LIMS Sample Disposition files.

7.20 Return the petropoxy mixture to the refrigerator in order to extend shelf life of the mixture.

Observations of Grain Mount

7.21 Record raw grain count data in a logbook. Most grain counts are made with a 10X magnification ocular and either a 10X (for very fine or fine sand) or 25X (for coarse silt) magnification objective lens.

7.22 The first step is to seat the grain mount in the mechanical stage of the microscope and to survey the slide with a low-power magnification power (10X) to become familiar with the grain assemblage and to make a rough estimate of the relative abundance of minerals and other grains.

7.23 Initially, identify the most abundant minerals as they are probably the easiest to identify, and their elimination decreases the number of possibilities to consider in identifying the less common minerals. Furthermore, there are certain likely and unlikely assemblages of minerals, and an awareness of the overall types that are present gives clues to the minor species that may be expected.

7.24 Note the observed minerals by a two-letter code, e.g., QZ for quartz. Refer to the list of mineralogy codes, provided at the end of the mineralogy section (7).

7.25 Make grain counts in horizontal traverses across the grain mount. A 10X magnification objective is appropriate for FS and VFS. A 25X objective is appropriate for CSI.

7.26 To make a grain count, move the slide via the mechanical stage so that the left border of cover slip is in view and in the proximity of but not in the upper left corner. Place vertical scale on mechanical stage on an even number, e.g., 72 or 74 mm.

7.27 Set the rotating stage so that the horizontal movement of a grain, via the mechanical stage, parallels the horizontal cross-hair in the ocular.

7.28 List the most abundant grains and associated counter number in logbook. Mineral identification is facilitated by the familiarity with a few striking features and by the process of elimination.

7.29 Set counters to zero. Move the slide laterally one field width at a time. Identify and tally each grain that touches the horizontal cross-hair in each field of view until the right margin of cover slip is in view.

7.30 Translate the slide vertically a distance of 2 mm and run another traverse in the reverse direction.

7.31 Repeat process until the end of traverse in which 300 grains have been tallied. If there are only a few species, a counting of 300 grains provides a good indication of composition. As the number of species increases, the count should increase within limits of practicability. To count more than 1000 grains is seldom necessary.

7.32 The counting of complete traverses minimizes the effects of non-random distribution of grains on the slide. This non-random distribution of grains is usually most pronounced near the edges of the cover slip. If the entire slide has been traversed, and the total grain count is <300, reverse the direction of vertical translation and count traverses on odd-numbered settings, e.g., 81 or 79 mm.

7.33 Counting isotropic grains only (e.g., volcanic glass) can be done more rapidly using either of the following microscope configurations:

7.33.1 Positioning the polarizer slightly off the extinction or "blackout" position.

OR

7.33.2 With crossed Nicols and a gypsum plate, the outline of the grains is visible with the color of the grain being the same as the epoxy background.

7.34 When the count is complete, enter the raw data (project, sample number, fraction(s), minerals, and counts) into the SSL LIMS data base.

8. Calculations

Calculations are made by a computer program written to facilitate data entry and manipulation. Required inputs are as follows: project number; sample number; grain-size fractions; mineral identification; and number of grains counted per mineral.

Percentage of minerals (frequency per 100 grains) is calculated by the formula as follows:

8.1 Mineral frequency (%) = (Number of grains for a mineral x 100)/Total number of grains counted

9. Report

Report mineral contents to the nearest whole percentage of grains counted. These data are accurate number percentages for the size-fraction analyzed but may need to be recomputed to convert to weight percentages (Harris and Zelazny, 1985). Grain counts can deviate significantly from weight percentage due to platy grains and density variations. These data are reported on the mineralogy data page of the primary characterization data set. For each grain size counted, the mineral type and amount are recorded, i.e., quartz, 87% of fraction, is recorded as QZ87. The percentage of resistant minerals in each fraction is reported on the SSL datasheet. Refer to 7C1 for more information on resistant minerals.

10. Precision and Accuracy

Precision and accuracy data are available from the SSL upon request.

11. References

Bloss, D.F. 1961. An introduction to the methods of optical crystallography. Holt, Rinehart, and Winston, NY.

Cady, J.G. 1965. Petrographic microscope techniques. p. 604-631. *In* D.D. Evans, L.E. Ensminger, J.L. White, and F.E. Clark (eds.) Methods of soil analysis. Part 1. Physical and mineralogical properties, including statistics of measurement and sampling. 1st ed. Agron. Monogr. 9. ASA and SSSA, Madison, WI.

Deer, W.A., R.A. Howie, and J. Zussman. 1992. An introduction to the rock-forming minerals. 2nd ed. Longman Scientific and Technical, Essex, England.

Drees, L.R., and M.D. Ransom. 1994. Light microscope techniques in quantitative mineralogy. p.137-176. *In* J.E.Amonette and L.W. Zelazny (eds.) Quantitative methods in soil mineralogy. SSSA Misc. Publ. Soil Sci. Soc. Am., Madison, WI.

Durrell, Cordell. 1948. A key to common rock-forming minerals in thin section. W.H. Freeman and Co., Publ., San Francisco, CA.

Harris, W.G., and L.W. Zelazny. 1985. Criteria assessment for micaceous and illitic classes in Soil Taxonomy. p. 147-160. *In* J.A. Kittrick (ed.) Mineral classification of soils. SSSA Spec. Publ. No. 16. Madison, WI.

Kerr, P.F. 1977. Optical mineralogy. McGraw-Hill Book Co., Inc., NY.

Klein, Cornelis, and C.S. Hurlbut, Jr. 1985. Manual of mineralogy. John Wiley and Sons, NY.

Krumbein, W.C. and F.J. Pettijohn. 1938. Manual of sedimentary petrography. Appleton-Century-Crofts, NY.

Milner, H.B. 1962. Sedimentary petrography, 4th ed. The Macmillan Co., NY.

Shelley, David. 1978. Manual of optical mineralogy. Elsevier. North-Holland, Inc., NY.

Stoiber, R.E., and S.A. Morse. 1972. Microscopic identification of crystals. Ronald Press Company, NY.

Soil Survey Staff. 1999. Soil Taxonomy: A Basic System of Soil Classification for Making and interpreting Soil Surveys. 2nd ed. USDA-NRCS Agric. Handb. 436. U.S. Gov. Print. Office, Washington, DC.

Platy Grains (7B2)
Magnetic Separation (7B2a)

1. Application

A magnetic separator is used to separate magnetic or paramagnetic minerals from non-magnetic mineral grains from the fine earth fractions that range from 0.02 to 0.5 mm in size. In the SSL, the common application is to quantify the amount of platy grains (phyllosilicates) in micaceous or paramiceous soils. Ferrimagnetic magnetic grains such as magnetite and ilmenite typically are separated first with a hand magnet. The separator then concentrates grains of paramagnetic minerals such as biotite and muscovite from non-magnetic minerals such quartz and feldspar. This procedure (7B2a) is often used in combination with static tube separation (7B2b) or froth flotation (7B2c). Grains in each separate can be further analyzed by optical microscopy or x-ray diffraction.

2. Summary of Method

A magnetic separator applies a strong magnetic field along a shallow trough that slopes down from an entry point to an exit point. Grains travel the path under the force of gravity.

The trough is also tilted perpendicular to the travel path. The magnetic field draws paramagnetic grains up the tilt slope. A divider at midslope along the path separates the paramagnetic from non-magnetic grains. Grains exit the path into separate containers and the two components are weighed to obtain a relative percentage. Percent platy minerals of specific analyzed fraction are reported (7B2a).

3. Inerferences

Some mafic minerals are paramagnetic, and exit with the platy grains. The two groups may be separated by the static tube procedure (7B2b). Mafic minerals commonly are heavy minerals and may be separated from platy grains by density separation (7B1a2).

4. Safety

The magnetic field is not a direct biological health hazard. Keep mechanical watches away from the magnetic field. Keep iron/steel objects away from the magnetic field because of the strength of field. The separator will need to be turned off to retrieve them.

5. Equipment
5.1 300-mL fleaker, with a fill line marked at 5cm above bottom
5.2 Glass stirring rod, with a rubber policeman
5.3 Stop watch
5.4 Sonic probe, Vibra Cell Model V1A, Sonics and Materials, Inc., Danbury, CT
5.5 Magnetic separator; Frantz Isodynamic Magnetic Separator, Model L-1, S.G. Frantz, Inc, Trenton, NJ (Fig. 1)
5.6 Centrifuge tube, plastic, 100-mL
5.7 Beaker, pyrex, 150-200 mL, marked with a line 5-cm above bottom
5.8 Sieves:
5.8.1 0.5 - 2.0 mm = US 35 mesh sieve or Tyler 32 mesh sieve
5.8.2 0.1 - 0.5mm = US 140 mesh sieve or Tyler 150 mesh sieve
5.8.3 0.05 - 0.1mm = US 300 mesh sieve or Tyler 300 mesh sieve
5.8.4 0.02 - 0.05mm = Material that passes through the 300 mesh sieve
5.9 Mechanical reciprocating shaker, 100 oscillations min^{-1}, 1 ½ in strokes, Eberbach 6000, Eberbach Corp., Ann Arbor, MI
5.10 Oven, 110 C

6. Reagents
6.1 Reverse Osmosis (RO) water (ASTM type III grade)
6.2 Sodium hexametaphosphate solution. Dissolve 35.7 g of sodium hexametaphosphate $(NaPO_3)_6$ and 7.94 g sodium carbonate (Na_2CO_3) L^{-1}

7. Procedure

Separation of Target Fraction (0.02 to 2.0 mm)
(for magnetic separation, static tube separation, and/or froth flotation)
7.1 For magnetic separation alone or in combination with static tube separation, 5 grams of <2-mm soil is sufficient. If froth floatation is used in combination with magnetic separation, use 20 to 30 g of <2.0-mm sample.

7.2 Place sample into a 100-mL centrifuge tube for magnetic or tube methods, or in a 300-mL fleaker for froth flotation is intended. Add 10 to 60 mL of dispersing agent, as appropriate for sample weight.

7.3 Fill the centrifuge tube half full, or fleaker to the 5-cm mark with distilled water. Stopper and shake the solution overnight at 100 oscillations min^{-1}.

7.4 Transfer contents of the centrifuge tube to a 150-200 mL beaker. Add RO to the fleaker to the 5-cm mark, if needed.

7.5 With the stirring rod, stir the sample for 15 seconds and allow to stand for 2 minutes. Decant and discard the dispersant.

7.6 Repeat the settling process, usually 4 to 5 times, until the dispersant is clear. Sediment is the 0.02-2 mm fraction.

Sonic Cleaning of Sample
7.8 Add approximately 50 mL of RO water to the fleaker or beaker containing the 0.02-2.0 mm sample.

7.9 Turn on and tune the sonic cleaner for optimal performance. Set the % Duty cycle knob to 50; output control knob to 10; run time to 360 s; and turn on the pulser switch.

7.10 Lower the sonic probe horn 0.5 to 1.0 cm below the water level of the fleaker containing the washed sample and turn on the sonic probe.

7.11 Repeat the decanting process (see separation of working fraction) to remove fines (<0.02 mm) produced by the sonic probe treatment.

7.12 Oven-dry the sample at 110°C.

Particle-Size Separation
7.13 Dry sieve the .02-2mm fraction cleaned by sonication into the following subfractions: 0.5-2.0, 0.1-0.5, 0.05-0.1, and 0.02-0.05 mm.

7.14 Weigh and record dry weigh of each of the separates.

Procedure Options for Each Size Fraction
7.15 0.5-2mm fraction: Separate platy grains via magnetic separator only (Fig. 1). This fraction seems unsuited for static tube or froth floatation separation.

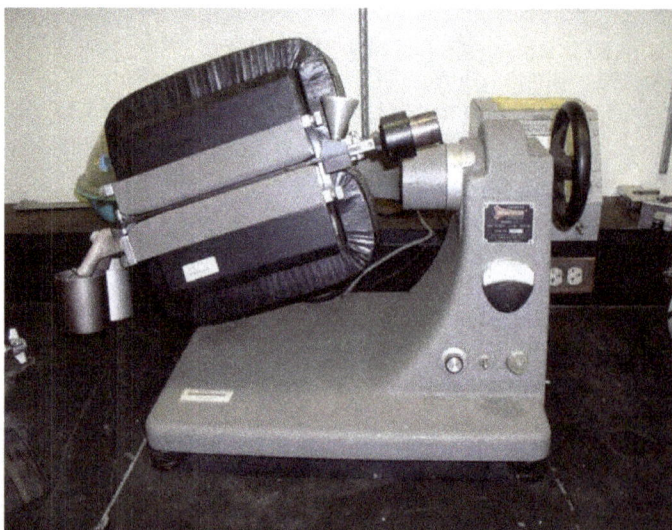

Fig. 1. Apparatus for Magnetic Separation

7.16 0.1-0.5, 0.05-0.1, and 0.02-0.05mm fractions: Separate by the following methods:

7.16.1 Magnetic separation (procedure 7B3a), or

7.16.2 Static tube separation (procedure 7B3b) followed by magnetic separation (procedure 7B3a) of residual fraction, or

7.16.3 Froth flotation separation (procedure 7B3c) followed by magnetic separation (procedure 7B3a) of residual fraction.

Magnetic Separation

7.17 Turn on the magnetic separator and adjust the machine to the high-high, positive position, and set the amp meter to 1.25 amperes.

7.18 Set the slope to about 20 degrees and the tilt to 15 degrees.

7.19 Introduce sample into the funnel and set the vibrator adjustment knob between 5 and 8. Observe the flow of the grains down the separator trough and adjust the vibrator setting accordingly.

7.20 Repeat the process using the non-magnetic portion 2-4 times or until the non-magnetic fraction is free of platy minerals.

7.21 Weigh and record the weight of the magnetic fraction as platy grains, and the non-magnetic fraction as residual grains. Calculate amount of platy grains as percent by weight of analyzed fraction.

8. Calculations

8.1 Percent platy grains = [100 x (weight of platy grains)]/sample weight)

8.2 Percent residual grains = [100 x (weight of residual grains)]/(sample weight)

8.3 Recovery = (weight of platy and residual grains)/(sample weight)

9. Report

Report platy grains as a percent of the specific particle size fraction analyzed, oven-dried soil weight.

10. Precision and Accuracy

Precision and accuracy data are not available for this method.

Platy Grains (7B2)
Static Tube Separation (7B2b)

1. Application

Static charge of mineral grains to glass and a magnetic separator are used to separate platy grains from non-platy grains in the 0.02 -2 mm fraction of soil. The separates are weighed to determine the quantity of platy minerals. The platy separates are examined by optical microscope and analyzed by X-ray diffraction to determine the kinds of minerals present.

2. Summary of Method

A sample of <2-mm soil is prepared according to the procedure described in 7B2a. A small portion of sample is introduced into the top of an inclined glass tube mounted on a vibrator. As the tube is rotated and vibrated, the platy grains adhere to the tube and the non-platy grains (residue) roll or slide through. The residue is run through a magnetic separator to separate the coarser platy grains that did not adhere to the glass tube. Percent platy minerals of specific analyzed fraction are reported (7B2b).

3. Interferences

Large platy grains tend to slide through the tube into the residue, especially if the plates are stacked into a book.

4. Safety

No known hazards exist.

5. Equipment

5.1 Glass tube, 1.5 cm inside diameter, 30 cm long vibrating mechanism
5.2 Receptacles to hold grains
5.3 Funnel or glassine paper or aluminum weighing dish
5.4 Camel's hair brush
5.5 Gelatin capsules.
5.6 Mechanical Vibrator

6. Reagents

None.

7. Procedure

7.1 Prepare sample (disperse, fractionate, and dry sample as described in Method 7B2a)

Static Charge Separation by Glass Tube

7.2 Set up vibrator as shown in Fig. 1.

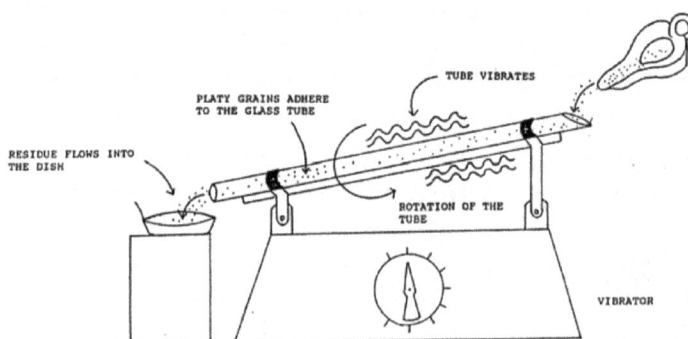

Fig. 1. Apparatus for Static Tube Separation

7.3 Weigh 0.1500 g of 0.02-2 mm material or particle-size separate onto a square of glassine paper and introduce into the upper end of the glass tube.

7.4 Turn on the vibrator until the material begins to flow. Rotate the tube slowly so the platy grains adhere to the tube wall. Adjust vibrator intensity and rotation to achieve a slow smooth flow rate.

7.5 When rounded grains have passed through the tube, remove the tube and hold it vertically over a tared weighing dish. Tap the tube to remove the platy grains. If grains remain, wash them out with RO water, dry, and weigh.

Note: The glass tube separation can be done by hand, without mechanical vibration in the field office to obtain a fair approximation of the platy grain component.

8. Calculations

8.1 Percent platy grains = [100 x (weight of platy grains)]/sample weight)

8.2 Percent residual grains = [100 x (weight of residual grains)]/(sample weight)

8.3 Recovery = (weight of platy and residual grains)/(sample weight)

9. Report
Report platy grains as a percent of the specific particle size fraction analyzed, oven-dried soil weight.

11. Precision and Accuracy
Precision and accuracy data are not available for this method.

Platy Grains (7B2)
Froth Flotation (7B2c)

1. Application
This method used with coarse silt, very fine, fine, and medium sand fractions of soils with significant amounts of platy minerals (mica, vermiculite, chlorite, and their pseudomorphically altered weathering products). It provides weight percent data on each of the fractions. Combined use of froth flotation and magnetic separation improve separation of platy and non-platy grains to better estimate weight percentages of components.

2. Summary of Method
Platy minerals (muscovite, biotite, vermiculite, and kaolinite) are floated off over the top of a container in an agitated aqueous suspension by action of a complexer and frother, adapted from procedure provided by Louis Schlesinger of the Minerals Research Laboratory School of engineering, North Carolina State University, in Asheville, NC. Percent platy minerals of specific analyzed fraction are reported (7B2c).

3. Interferences
There are no known interferences.

4. Safety
There are no known safety hazards.

5. Equipment

5.1 Modified 800 mL glass beaker for mixing container
5.2 Plastic bucket, 5 qt, for catch container
5.3 Mechanical mixer - 1 laboratory reagent mixer or a magnetic bar stirrer
5.4 Manual mixer - 1 glass rod pH meter
5.5 Oven, 110 °C
5.6 Wood tongue depressors or a similar spatula-like device
5.7 Syringe - 1 mL
5.8 Beaker - 800 mL aerator
5.9 Ring stand (to hold Aerator assembly)
5.10 Funnel and stand to hold funnel 300-mesh sieve
5.11 Glass rod with rubber policeman

6. Reagents

6.1 Reverse osmosis (RO) water, ASTM Type III grade of reagent water
6.2 Frother: Econofroth 910 frother, 100% solution (Mfg: Nottingham Company; P.O. Box 250049; Station N; 1303 Boyd Ave. NW; Atlanta, GA 303025; phone: 404-351-3501)
6.3 Promoter: Econofloat A-50 promoter, 5% solution (1mL/20mL or 50mL/1000mL); (Mfg: Nottingham Company)
6.4 NaOH (2.5% solution or 12.5 g per 500 g) 25 g/liter
6.5 H_2SO_4 , 0.9 N (2.5% solution or 12.5 g conc. H_2SO_4 per 500 g water)
6.6 Ethyl alcohol in wash bottle

7. Procedure

7.1 Prepare sample (disperse, fractionate, and dry sample as described in procedure 7B2a.

7.2 Set up apparatus as shown in Fig. 1.

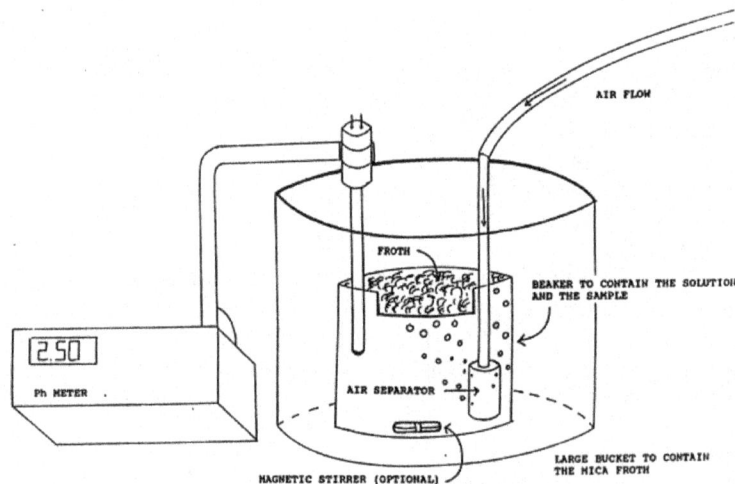

Fig. 1. Apparatus for Froth Flotation

7.3 Prepare 700 mL RO water, pH 2.5: Put 800-mL glass beaker on mechanical mixer and fill to 700 mL level. Adjust to pH 2.5 (not over 3.0) with 0.9 N H_2SO_4 solution and set the full beaker aside for use later.

7.4 Place 800 mL mixing container on mechanical mixer, and fill 3/4 full with RO water.

7.5 Add 5 g of sample to water and start a strong mixing action.

7.6 Adjust pH of sample and water to 2.5 with 0.9 N H_2SO_4 solution.

7.7 Add 0.25ml of Promoter.

7.8 Add 0.2ml of frother.

7.9 Continue to mix the sample solution for 2-5 minutes.

7.10 Check and readjust the pH level as necessary.

7.11 Place mixing container in plastic 5-qt catch bucket, lower aerator into solution, fill mixing container almost to top, using the water from section 7.3.

7.12 Turn off the mechanical mixer and turn on the air to the aerator to start the frothing action. The frothing foam should build to a point where the foam pours out of the modified mixing container. Add water from section 7.3 as needed to maintain foam overflow from the mixing container.

7.13 Use a glass stirring rod with a rubber policeman to mix the sample on the bottom of the mixing chamber in a grid like motion. As the platy minerals froth to the surface use the wood tongue depressor to rake them over the side of the mixing chamber.

7.14 After 1-5 minutes the amount of platy minerals floating to the surface should diminish. At this time turn off the air, thoroughly remix the sample with the glass stirring rod, and repeat the procedure starting at section 7.12.

7.15 After 2-5 minutes during the second run through section 7.12, carefully inspect the foam to see if platy minerals are still frothing to the surface. Continue until little or no platy minerals are frothing to the surface.

7.16 Transfer contents of the catch bucket with ethyl alcohol to the 300-mesh sieve and rinse the sample.

7.17 Transfer the rinsed grains to an aluminum dish, dry in an oven at 110°C, weigh and record the weight as platy grains.

7.18 Repeat steps 7.16 and 7.17 for the contents in the mixing container and record the weight as residual grains. <u>NOTE</u>: Delay weighing samples if magnetic separation (procedure 7B2a) will be done next.

8. Calculations

8.1 Percent platy grains = [100 x (weight of platy grains)]/sample weight)

8.2 Percent residual grains = [100 x (weight of residual grains)]/(sample weight)

8.3 Recovery = (weight of platy and residual grains)/(sample weight)

9. Report

Report platy grains as a percent of the specific particle size fraction analyzed, oven-dried soil weight.

10. Precision and Accuracy

Precision and accuracy data are not available for this method.

Ratios and Estimates Related to Optical Analysis (7C)
Total Resistant Minerals (7C1)

The sum of the grain-count percentages of resistant minerals are reported (7C1). For more detailed information on total resistant minerals, refer to the Soil Survey Staff (1995, 1999). Also refer to the list of mineralogy codes for resistant and weatherable minerals, following procedure 7C1.

References

Soil Survey Staff. 1995. Soil survey laboratory information manual. Version No. 1.0. USDA-NRCS. Soil Survey Investigations Report No. 42. U.S. Govt. Print. Office, Washington, DC.

Soil Survey Staff. 1999. Soil taxonomy: A basic system of soil classification for making and interpreting soil surveys, USDA-NRCS Agric. Handb. 436. 2nd ed. U.S. Govt. Print. Office, Washington, DC.

MINERALOGY CODES

Resistant Minerals

AE = Anatase
AG = Antigorite
AN = Andalusite
BY = Beryl
CD = Chalcedony (Chert, Flint, Jasper, Agate, Onyx)
CE = Cobalite
CH = Cliachite (Bauxite)
CN = Corundum
CR = Cristobalite
CT = Cassiterite
FE = Iron Oxides (Goethite, Magnetite, Hematite, Limonite)
GD = Gold
GE = Goethite
GI = Gibbsite
GN = Garnet
HE = Hematite
KK = Kaolinite
KY = Kyanite
LE = Lepidocrocite
LM = Limonite
LU = Leucoxene

MD = Resistant Mineraloids
MG = Magnetite
MH = Maghemite
MZ = Monazite
OP = Opaques
OR = Other Resistant Minerals
PN = Pollen
PY = Pyrophyllite
QC = Clay-Coated Quartz
QI = Iron-Coated Quartz
QZ = Quartz
RA = Resistant Aggregates
RE = Resistant Minerals
RU = Rutile
SA = Siliceous Aggregates
SL = Sillimanite
SN = Spinel
SO = Staurolite
SP = Sphene
TD = Tridymite
TM = Tourmaline
TP = Topaz
ZR = Zircon

Weatherable Minerals

AC = Actinolite
AF = Arfvedsonite
AH = Anthophyllite
AI = Aegerine-Augite
AL = Allophane
AM = Amphibole
AO = Aragonite
AP = Apatite
AR = Weatherable Aggregates
AU = Augite
AY = Anhydrite*
BA = Barite
BC = Biotite-Chlorite
BE = Boehmite
BK = Brookite
BR = Brucite
BT = Biotite
BZ = Bronzite
CA = Calcite*
CB = Carbonate Aggregates*
CC = Coal
CL = Chlorite
CM = Chlorite-Mica
CO = Collophane

CY = Chrysotile
CZ = Clinozoisite
DL = Dolomite
DP = Diopside
DU = Dumortierite
EN = Enstatite
EP = Epidote
FA = Andesite
FB = Albite
FC = Microcline
FD = Feldspar
FF = Foraminifera
FH = Anorthoclase
FK = Potassium Feldspar
FL = Labradorite
FM = Ferromagnesium Mineral
FN = Anorthite
FO = Oligoclase
FP = Plagioclase Feldspar
FR = Orthoclase
FS = Sanidine
FU = Fluorite
FZ = Feldspathoids
GG = Galena

GL = Glauconite
GO = Glaucophane
GY = Gypsum[*]
HA = Halite
HB = Hydrobiotite
HN = Hornblende
HS = Hydroxy-Interlayerd Smectite
HV = Hydroxy-Interlayered Vermiculite
ID = Iddingsite
IL = Illite (Hydromuscovite)
JO = Jarosite
KH = Halloysite
LA = Lamprobolite
LC = Analcime[*]
LI = Leucite
LO = Lepidomelane
LP = Lepidolite
LT = Lithiophorite
MC = Montmorillonite-Chlorite
ME = Magnesite
MI = Mica
ML = Melilite
MM = Montmorillonite-Mica
MR = Marcasite
MS = Muscovite
MT = Montmorillonite
MV = Montmorillonite-Vermiculite
NE = Nepheline
NJ = Natrojarosite
NX = Non-Crystalline
OV = Olivine

OW = Other Weatherable Minerals
PD = Piemontite
PG = Palygorskite
PI = Pyrite
PJ = Plumbjarosite
PK = Perovskite
PL = Phlogopite
PR = Pyroxene
PU = Pyrolusite
RB = Riebeckite (Blue Amphibole)
RO = Rhodocrosite
SC = Scapolite
SE = Sepiolite
SG = Sphalerite
SI = Siderite
SM = Smectite
SR = Sericite
ST = Stilbite
SU = Sulphur
TA = Talc
TE = Tremolite
TH = Thenardite[*]
VC = Vermiculite-Chlorite
VH = Vermiculite-Hydrobiotite
VI = Vivianite
VM = Vermiculite-Mica
VR = Vermiculite
WE = Weatherable Mineral
WV = Wavelite
ZE = Zeolite
ZO = Zoisite

Glass Count Minerals and Mineraloids

BG = Basic Glass
DI = Diatoms
GS = Glass
GA = Glass Aggregates
FG = Glass-Coated Feldspar
GC = Glass-Coated Grain
HG = Glass-Coated Hornblende
OG = Glass-Coated Opaque
QG = Glass-Coated Quartz

GM = Glassy Materials
OT = Other
PA = Palagonite
PO = Plant Opal
SS = Sponge Spicule

[*]Minerals not included as "weatherable minerals" as defined by Soil Taxonomy (1999) - "the intent is to include only those weatherable minerals that are unstable in a humid climate compared to other minerals such as quartz and 1:1 lattice clays, but are more resistant to weathering than calcite". This group of minerals is not part of the calculation for percent resistant minerals used in the siliceous family mineralogy class or percent weatherable minerals used as criteria for oxic horizon but are included in the calculation of "total resistant minerals" on the Soil Survey Laboratory mineralogy data sheet. Therefore, the value on the data sheet should be recalculated for strict use in Soil Taxonomy criteria if these minerals (e.g., calcite) are present in the grain count of a selected horizon.

SOIL SURVEY LABORATORY METHODS
(obsolete methods with old method codes)

This section of the Soil Survey Laboratory (SSL) Methods Manual describes the procedures that are no longer used at the SSL. These metods were described in earlier versions of Soil Survey Investigations Report (SSIR) No. 42 (1989, 1992, and 1996) and the SSIR No. 1, Procedures for Collecting Soil Samples and Methods of Analysis for Soil Survey (1972, 1982, and 1984). Some of these procedures are in the old format. Information is not available to describe some of these obsolete procedures in the same detail as used to describe the current methods in the laboratory. Also included in this section are examples of earlier versions of the Soil Survey Laboratory data sheet (Fig.1) and pedon description (Fig. 2). Figure 1 and 2 are located at the back of this section on obsolete methods.

Since the publication of the SSIR No. 42 (1996), there has been a significant increase in the number and kinds of methods performed at the SSL, resulting in a re-structuring of the method codes. All the methods described in this section of the manual carry the the old method codes. These old method codes have a maximum of four characters, e.g., 6A2b. These older codes are cross-referenced in a table preceding the descriptions of the obsolete SSL methods. This linkage between the method code systems is important to maintain as many older SSL data sheets and scientific publications report these older codes. While there may be slight differences between these cross-referenced methods, the intent here is to provide an historical linkage for the core SSL methods.

This section of the manual is divided into three parts as follows:

Part I, SSIR No. 42, Soil Survey Laboratory Methods Manual, Version 3.0 (1996)

Part II, SSIR No. 42, Soil Survey Laboratory Methods Manual, Versions 1.0 and 2.0 (1989, 1992, respectively)

Part III: SSIR No. 1, Procedures for Collecting Soil samples and Methods of Analysis for Soil Survey (1972, 1982, 1984)

METHODS OF ANALYSIS FOR SOILS, WATER, AND BIOLOGICAL MATERIALS

CROSS-REFERENCE BETWEEN CURRENT AND OBSOLETE METHODS

METHOD TITLE	CURRENT METHOD CODE	OBSOLETE METHOD CODE
SAMPLE COLLECTION AND PREPARATION	**1**	**1**
Field Sample Collection and Preparation	**1A**	**1A**
Site Selection	1A1	1A1
Geomorphic	1A1a	
Pedon	1A1b	1A2
Water	1A1c	
Biological	1A1d	
Laboratory Sample Collection and Preparation	**1B**	
Soils	1B1	
Samples	1B1a	
Bulk	1B1a1	
Natural Fabrics	1B1a2	
Clods	1B1a3	
Cores	1B1a4	
Preparation	1B1b	1B
Field-Moist Preparation	1B1b1	1B2
Whole Soil	1B1b1a	
Histosol Analysis	1B1b1a1	1B7
Biological Analysis	1B1b1a2	
<2-mm Fraction	1B1b1b	
Chemical and Selected Physical Analysis	1B1b1b1	
1500-kPa Water Content	1B1b1b2	
Field-Field-Moist/Air-dry Ratio	1B1b1b3	
<2-mm Fraction Sieved to < 0.425 mm	1B1b1c	
Atterberg Limits	1B1b1c1	
Air-Dry Preparation	1B1b2	1B1
Whole Soil	1B1b2a	
Aggregate Stability	1B1b2a1	
Repository Samples	1B1b2a2	
<2-mm Fraction	1B1b2b	
Standard Chemical, Physical, Mineralogical Analysis	1B1b2b1	
Salt Analysis	1B1b2b2	
1500-kPa Water Content	1B1b2b3	
Air-Dry/Oven-Dry Ratio	1B1b2b4	
Presence of Carbonates	1B1b2b5	
<2-mm Fraction Sieved to <0.425 mm	1B1b2c	
Atterberg Limits	1B1b2c1	
<2-mm Fraction Processed to ≈ 180 μm	1B1b2d	

CROSS-REFERENCE BETWEEN CURRENT AND OBSOLETE METHODS

METHOD TITLE	CURRENT METHOD CODE	OBSOLETE METHOD CODE
Total Carbon, Nitrogen, and Sulfur	1B1b2d1	
Carbonates and/or Gypsum	1B1b2d2	
Chemical Analysis of Organic Materials	1B1b2d3	
<2-mm Fraction Sieved to 75 μm	1B1b2e	
Major and/or Trace Elements	1B1b2e1	
Particles >2 mm	1B1b2f	
Particle-Size Analysis	1B1b2f1	
Particle-Size Analysis, Recorded	1B1b2f1a	1B1
2- to 20-mm Fraction Processed to <2-mm	1B1b2f1a1	1B5
Chemical, Physical, and Mineralogical Analysis	1B1b2f1a1a	
2- to 20-mm Fraction Processed to <2-mm and Recombined with <2-mm Fraction	1B1b2f1a2	
Chemical, Physical, and Mineralogical Analysis	1B1b2f1a2a	
Proportion and Particle-Size of Air-Dry Rock Fragments that Resist Abrupt Immersion in Tap Water	1B1b2f1a3	
Particle-Size Analysis, Not Recorded	1B1b2f1b	
Whole-Soil Processed to <2-mm	1B1b2f1b1	1B6
Chemical, Physical, and Mineralogical Analysis	1B1b2f1b1a	
Water	1B2	
Samples	1B2a	
Preparation	1B2b	
Biological Materials	1B3	
Samples	1B3a	
Biology Bulk Sample	1B3a1	
Microbial Biomass Sample	1B3a2	
Preparation	1B3b	
Field-Moist Preparation	1B3b1	
<2-mm Fractin	1B3b1a	
Microbial Biomass	1B3b1a1	
Air-Dry Preparation	1B3b2	
<2-mm Fraction Sieved <53 μm, with <53 μm (C-Mineral) and ≥ 53 μm (Particulate Organic Matter) Processed to ≈ 180 μm	1B3b2a	
Total Carbon, Nitrogen, and Sulfur Analysis	1B3b2a1	
<2-mm Fraction	1B3b2b	
Other Biological Analyses	1B3b2b1	
Dry (50°C) Preparation	1B3b3	
Roots	1B3b3a	
Root Biomass	1B3B3a1	
Roots Processed to ≈ 180 μm	1B3b3a1a	
Total Carbon, Nitrogen, and Sulfur Analysis	1B3b3a1a1	
Plants	1B3b3b	
Plant Biomass	1B3b3b1	
Plants Processed to ≈ 180 μm	1B3b3b1a	
Total Carbon, Nitrogen, and Sulfur Analysis	1B3b3b1a1	

CROSS-REFERENCE BETWEEN CURRENT AND OBSOLETE METHODS

METHOD TITLE	CURRENT METHOD CODE	OBSOLETE METHOD CODE
CONVENTIONS	**2**	**2**
Methods and Codes	**2A**	
Data Types	**2B**	
Particle-Size Fraction Base for Reporting Data	**2C**	**2A**
Particles <2 mm	2C1	2A1
Particles >2 mm	2C2	2A2
Sample Weight Base for Reporting Data	**2D**	
Air-Dry/Oven-Dry	2D1	
Field-Field-Moist/Oven-Dry	2D2	
Correction for Crystal Water	2D3	
Significant Figures and Rounding	**2E**	
Data Sheet Symbols	**2F**	**2B**
SOIL PHYSICAL AND FABRIC-RELATED ANALYSES	**3**	**3**
Particle-Size Distribution Analysis	**3A**	**3A**
Particles <2 mm	3A1	3A1
Pipet Analysis for routinely reported size fractions (1, 0.5, 0.25, 0.1, 0.047 mm, <20 µm, and <2 µm), with or without <0.2-µm fraction, with or without <2-µm carbonate clay	3A1a	3A1, 3A1b, 3A1d 3A2, 3A2b, 3A2d
Pipet Analysis with Standard Pretreatments and Dispersion	3A1a1	
Air-Dry	3A1a1a	
Field-Moist	3A1a1b	
Pipet Analysis with Carbonate Removal	3A1a2	
Air-Dry	3A1a2a	3A1e
Field-Moist	3A1a2b	3A2e
Pipet Analysis with Iron Removal	3A1a3	
Air-Dry	3A1a3a	3A1f
Field-Moist	3A1a3b	3A2f
Pipet Analysis with Silica Removal	3A1a4	
Air-Dry	3A1a4a	3A1g
Field-Moist	3A1a4b	3A2g
Pipet Analysis with Ultrasonic Dispersion	3A1a5	
Air-Dry	3A1a5a	3A1h
Field-Moist	3A1a5b	3A2h

CROSS-REFERENCE BETWEEN CURRENT AND OBSOLETE METHODS

METHOD TITLE	CURRENT METHOD CODE	OBSOLETE METHOD CODE
Pipet Analysis with Water Dispersion	3A1a6	
Air-Dry	3A1a6a	3A1c
Field-Moist	3A1a6b	3A2c
Particles > 2mm	3A2	3B
Weight Estimates	3A2a	3B1
By Field and Laboratory Weighing	3A2a1	3B1a
From Volume and Weight Estimates	3A2a2	3B1b
Volume Estimates	3A2b	3B2
Bulk Density	**3B**	**4A**
Saran-Coated Natural Clods	3B1	4A1
Field State	3B1a	4A1a
33 kPa	3B1b	4A1d
Oven-dry	3B1c	4A1h
Rewet	3B1d	4A1i
Reconstituted	3B2	
33 kPa	3B2a	
Oven-Dry	3B2b	
Compliant Cavity	3B3	4A
Field State	3B3a	4A5
Ring Excavation	3B4	
Field State	3B4a	
Frame Excavation	3B5	
Field State	3B5a	
Soil Cores	3B6	4A3
Field State	3B6a	4A3a
Water Retention	**3C**	**4B**
Pressure-Plate Extraction	3C1	4B1
6 10, 33, 100, or 200 kPa Water Retention	3C1a-e	
<2 mm Sieved, Air-Dry	3C1a-e1a	4B1a
Natural Clods	3C1a-d2	4B1c
Soil Cores	3C1a-d3	4B1d
Rewet	3C1c4	4B1e
Reconstituted	3C1c5	
Pressure-Membrane Extraction	3C2	4B2
1500 kPa Water Retention	3C2a	
<2 mm Sieved, Air-Dry	3C2a1a	4B2a
<2 mm Sieved, Field-Moist	3C2a1b	4B2b

CROSS-REFERENCE BETWEEN CURRENT AND OBSOLETE METHODS

METHOD TITLE	CURRENT METHOD CODE	OBSOLETE METHOD CODE
Water Retention at Field State	3C3	4B4
Ratios and Estimates Related to Particle Size Analysis, Bulk Density, and Water Retention	**3D**	
Air-Dry/Oven-Dry Ratio	3D1	4B5
Field-Moist/Oven-Dry Ratio	3D2	
Correction for Crystal Water	3D3	6S3
Coefficient of Linear Extensibility	3D4	4D
Air-Dry or Oven-Dry to 33-kPa Tension	3D4a	4D1
Water Retention Difference, Whole Soil	3D5	4C
Between 33-kPa and 1500-kPa Tension	3D5a	4C1
Between 10-kPa and 1500-kPa Tension	3D5b	4C2
Between 33-kPa and 1500-kPa (Air-Dry) Tension	3D5c	4C3
1500 kPa Water/Clay Ratio	3D6	8D1
Total Silt Fraction	3D7	
Total Sand Fraction	3D8	
2- to 5-mm Fraction	3D9	
5- to 20-mm Fraction	3D10	
20- to 75-mm Fraction	3D11	
0.1- to 75-mm Fraction	3D12	
>2-mm Fraction	3D13	
Micromorphology	**3E**	**4E**
Thin sections	3E1	4E1
Preparation	3E1a	4E1a
Interpretation	3E1b	4E1b
Wet Aggregate Stability	**3F**	**4G**
Wet Sieving	3F1	
Air-dry	3F1a	
2 to 1 mm	3F1a1	
2- to 0.5-mm Aggregates Retained	3F1a1a	4G1
Particle Density	**3G**	
Pycnometer Gas Displacement	3G1	
Oven-dry	3G1a	
<2 mm	3G1a1	
>2 mm	3G1a2	
Atterberg Limits	**3H**	**4F**
Liquid Limit	3H1	4F1
Air-Dry, <0.4 mm	3H1a1	
Field-Moist, <0.4 mm	3H1b1	
Plasticity Index	3H2	4F2
Air-Dry, <0.4 mm	3H2a1	
Field-Moist, <0.4 mm	3H2b1	

CROSS-REFERENCE BETWEEN CURRENT AND OBSOLETE METHODS

METHOD TITLE	CURRENT METHOD CODE	OBSOLETE METHOD CODE
SOIL AND WATER CHEMICAL EXTRACTIONS AND ANALYSES	**4**	
Acid Standardization	**4A**	
Ion Exchange and Extractable Cations	**4B**	**5A**
Displacement after Washing	4B1	
NH₄OAc, pH 7	4B1a	
Automatic Extractor	4B1a1	
2 *M* KCl Rinse	4B1a1a	
Steam Distillation	4B1a1a1	
HCl Titration	4B1a1a1a	
Cation Exchange Capacity (CEC-7)	4B1a1a1a1	5A8c
Air-Dry, <2 mm	4B1a1a1a1a1	
Field-Moist, <2 mm	4B1a1a1a1b1	
Atomic Absorption Spectrophotometer	4B1a1b	
Calcium, Magnesium, Potassium, and Sodium	4B1a1b1-4	6N2e, 6O2d, 6Q2b, 6P2b
Air-Dry, <2 mm	4B1a1b1-4a1	
Field-Moist, <2 mm	4B1a1b1-4b1	
NH₄Cl, Neutral Unbuffered	4B1b	
Automatic Extractor	4B1b1	
2M KCl Rinse	4B1b1a	
Steam Distillation	4B1b1a1	
HCl Titration	4B1b1a1a	
Cation Exchange Capacity	4B1b1a1a1	5A9c
Air-Dry, <2 mm	4B1b1a1a1a1	
Field-Moist, <2 mm	4B1b1a1a1b1	
Atomic Absorption Spectrophotometer	4B1a1b	
Calcium, Magnesium, Potassium, and Sodium	4B1b1b1-4	
Air-Dry, <2 mm	4B1b1b1-4a1	
Field-Moist, <2 mm	4B1b1b1-4b1	
BaCl₂-Triethanolamine, pH 8.2 Extraction	4B2	
Automatic Extractor	4B2a	
Automatic Titrator	4B2a1	
Back Titration with HCl	4B2a1a	
Extractable Acidity	4B2a1a1	6H5a
Air-Dry, <2 mm	4B2a1a1a1	
Field-Moist, <2 mm	4B2a1a1b1	

CROSS-REFERENCE BETWEEN CURRENT AND OBSOLETE METHODS

METHOD TITLE	CURRENT METHOD CODE	OBSOLETE METHOD CODE
Centrifuge	4B2b	
Automatic Titrator	4B2b1	
Back Titration with HCl	4B2b1a	
Extractable Acidity	4B2b1a1	
Air-Dry, <2 mm	4B2b1a1a1	
Field-Moist, <2 mm	4B2b1a1b1	
1 N KCl Extraction	4B3	
Automatic Extractor	4B3a	
Inductively Coupled Plasma Spectrophotometer	4B3a1	
Radial Mode	4B3a1a	
Al, Mn	4B3a1a1-2	6G9c, 6D3b
Air-Dry, <2 mm	4B3a1a1-2a1	
Field-Moist, <2 mm	4B3a1a1-2b1	
Ratios and Estimates Related to Ion Exchange and Extractable Cations	4B4	
Sum of Extractable Bases	4B4a	
Sum of Extractable Bases by NH$_4$OAc, pH 7	4B4a1	5B5a
Sum of Extractable Bases by NH$_4$OAc, pH 7, Calculated	4B4a1a	
Sum of Extractable Bases by NH$_4$Cl	4B4a2	
Sum of Extractable Bases by NH$_4$Cl, Calculated	4B4a2a	
Cation Exchange Capacity (CEC)	4B4b	
CEC-8.2 (Sum of Cations)	4B4b1	5A3a
CEC-8.2, Calculated	4B4b1a	
CEC-8.2, Not Calculated	4B4b1b	
Effective Cation Exchange Capacity (ECEC)	4B4b2	
Sum of NH$_4$OAc Extractable Bases + 1 N KCl Extractable Aluminum, Calculated	4B4b2a	5A3b
Sum of NH$_4$OAc Extractable Bases + 1 N KCl Extractable Aluminum, Not Calculated	4B4b2b	
Base Saturation	4B4c	
Base Saturation by NH$_4$OAc, pH 7 (CEC-7)	4B4c1	5C1
Base Saturation by CEC-7, Calculated	4B4c1a	
Base Saturation by CEC-7, Set to 100%	4B4c1b	
Base Saturation by NH$_4$Cl	4B4c2	
Base Saturation by NH$_4$Cl, Calculated	4B4c2a	
Base Saturation by NH$_4$Cl, Set to 100%	4B4c2b	
Base Saturation by CEC-8.2	4B4c3	5C3
Base Saturation by CEC-8.2, Calculated	4B4c3a	
Base Saturation by CEC-8.2, Not Calculated	4B4c3b	
Base Saturation by ECEC	4B4c4	
Base Saturation by Sum of NH$_4$OAc Extractable Bases + 1 N KCl Extractable Aluminum, Calculated	4B4c4a	
Base Saturation by Sum of NH$_4$OAc Extractable Bases + 1 N KCl Extractable Aluminum, Not Calculated	4B4c4b	

CROSS-REFERENCE BETWEEN CURRENT AND OBSOLETE METHODS

METHOD TITLE	CURRENT METHOD CODE	OBSOLETE METHOD CODE
Aluminum Saturation	4B4d	
Aluminum Saturation by ECEC	4B4d1	
Aluminum Saturation by Sum of NH$_4$OAc Extractable Bases + 1 N KCl Extractable Aluminum, Calculated	4B4d1a	
Aluminum Saturation by Sum of NH$_4$OAc Extractable Bases + 1 N KCl Extractable Aluminum, Not Calculated	4B4d1b	
Activity	4B4e	
1500 kPa water/CEC-7	4B4e1	8D1
Reaction	**4C**	
Soil Suspensions	4C1	
Electrode	4C1a	
Standard Glass Body Combination	4C1a1	
Digital pH/Ion Meter	4C1a1a	
1 N NaF, pH 7.5 – 7.8	4C1a1a1	8C1d
Air-Dry, <2 mm	4C1a1a1a1	
Field-Moist, <2 mm	4C1a1a1b1	
Saturated Paste pH	4C1a1a2	8C1b
Air-Dry, <2 mm	4C1a1a2a1	
Oxidized pH	4C1a1a3	8C1h
Organic Materials pH, Final Solution ≈ 0.01 M CaCl$_2$ pH	4C1a1a4	8C2a
Combination pH-reference electrode	4C1a2	
Automatic Titrator	4C1a2a	
1:1 water pH	4C1a2a1	8C1f
Air-Dry, <2 mm	4C1a2a1a1	
Field-Moist, <2 mm	4C1a2a1b1	
1:2 0.01 M CaCl$_2$ pH	4C1a2a2	8C1f
Air-Dry, <2 mm	4C1a2a2a1	
Field-Moist, <2 mm	4C1a2a2b1	
1 M KCl pH	4C1a2a3	8C1g
Air-Dry, <2 mm	4C1a2a3a1	
Field-Moist, <2 mm	4C1a2a3b1	
Soil Test Analyses	**4D**	
Anion Resin Extraction	4D1	
Two-Point Extraction	4D1a	
1 h, 24h, 1 M KCl	4D1a1a1	
UV-Visible Spectrophotometer, Dual-Beam	4D1a1a1a	
Phosphorus (2 points)	4D1a1a1a1-2	
Air-Dry, <2 mm	4D1a1a1a1-2a-b1	
Field-Moist, <2 mm	4D1a1a1a1-2b1	

CROSS-REFERENCE BETWEEN CURRENT AND OBSOLETE METHODS

METHOD TITLE	CURRENT METHOD CODE	OBSOLETE METHOD CODE
Aqueous Extraction	4D2	
Single-Point Extraction	4D2a	
1:10, 30 min	4D2a1	
UV-Visible Spectrophotometer, Dual Beam	4D2a1a	
Phosphorus	4D2a1a1	
Air-Dry, <2 mm	4D2a1a1a1	
Field-Moist, <2 mm	4D2a1a1b1	
Flow Injection, Automated Ion Analyzer	4D2a1b	
Phosphorus	4D2a1b1	6S7a
Air-Dry, <2 mm	4D2a1b1a1	
Field-Moist, <2 mm	4D2a1b1b1	
Bray P-1 Extraction	4D3	
UV-Visible Spectrophotometer, Dual Beam	4D3a	
Phosphorus	4D3a1	6S3
Air-Dry, <2 mm	4D3a1a1	
Field-Moist, <2 mm	4D3a1b1	
Flow Injection, Automated Ion Analyzer	4D3b	
Phosphorus	4D3b1	6S3b
Air-Dry, <2 mm	4D3b1a1	
Field-Moist, <2 mm	4D3b1b1	
Bray P-2 Extraction	4D4	
UV-Visible Spectrophotometer, Dual-Beam	4D4a	
Phosphorus	4D4a1	
Air-Dry, <2 mm	4D4a1a1	
Field-Moist, <2 mm	4D4a1b1	
Olsen Sodium-Bicarbonate Extraction	4D5	
UV-Visible Spectrophotometer, Dual Beam	4D5a	
Phosphorus	4D5a1	
Air-Dry, <2 mm	4D5a1a1	
Field-Moist, <2 mm	4D5a1b1	
Mehlich No. 3 Extraction	4D6	
UV-Visible Spectrophotometer, Dual Beam	4D6a	
Phosphorus	4D6a1	
Inductively Coupled Plasma Atomic Emission Spectrophotometer	4D6b	
Axial Mode	4D6b1	
Ultrasonic Nebulizer	4D6b1a	
Aluminum, Arsenic, Barium, Calcium, Cadmium, Cobalt, Chromium, Copper, Iron, Potassium, Magnesium, Manganese, Sodium, Nickel, Phosphorus, Lead, Selenium, and Zinc	4D6b1a1-18	
Air-Dry, <2 mm	4D6b1a1-18a1	
Field-Moist, <2 mm	4D6b1a1-18b1	

CROSS-REFERENCE BETWEEN CURRENT AND OBSOLETE METHODS

METHOD TITLE	CURRENT METHOD CODE	OBSOLETE METHOD CODE
Citric Acid Soluble	4D7	
UV-Visible Spectrophotometer, Dual Beam	4D7a	
Phosphorus	4D7a1	6S5
Air-Dry, <2 mm	4D7a1a1	
Field-Moist, < 2 mm	4D7a1b1	
New Zealand P Retention	4D8	
UV-Visible Spectrophotometer, Dual Beam	4D8a	
Phosphorus	4D8a1	6S4b
Air-Dry, <2 mm	4D8a1a1	
Field-Moist, <2 mm	4D8a1b1	
1 M KCl Extraction	4D9	
Cadmium-Copper Reduction	4D9a	
Sulfanilamide N-1-Naphthylethylenediamine Dihydrochloride	4D9a1	
Flow-Injection, Automated Ion Analyzer	4D9a1a	
Nitrate-Nitrite	4D9a1a1-2	6M2a
Air-Dry, <2 mm	4D9a1a1-2a1	
Field-Moist, <2 mm	4D9a1a1-2b1	
Field-Moist, <2 mm, Fumgiated and Non-fumigated	4D9a1a1-2b2-3	
Anaerobic Incubation	4D10	
2 M KCl Extraction	4D10a	
Ammonia – Salicylate	4D10a1	
Flow Injection, Automated Ion Analyzer	4D10a1a	
N as NH_3 – Mineralizable N	4D10a1a1	6B5a
Air-Dry, <2 mm	4D10a1a1a1	
Field-Moist, <2 mm	4D10a1a1b1	

Carbonate and Gypsum   4E

METHOD TITLE	CURRENT METHOD CODE	OBSOLETE METHOD CODE
3 N HCl Treatment	4E1	
CO_2 Analysis	4E1a	
Manometer, Electronic	4E1a1	
Carbonates	4E1a1a	
Calcium Carbonate	4E1a1a1	
Air-Dry, <2 mm	4E1a1a1a1	6E1g
Air-Dry, 2 to 20 mm	4E1a1a1a2	6E4
Aqueous Extraction	4E2a	
Conductivity Bridge	4E2a1	
Electrical Conductivity	4E2a1a	
Gypsum, qualitative and quantitative	4E2a1a	6F1a
Air-Dry, <2 mm	4E2a1a1a1	
Air-Dry, 2 to 20 mm	4E2a1a1a2	

CROSS-REFERENCE BETWEEN CURRENT AND OBSOLETE METHODS

METHOD TITLE	CURRENT METHOD CODE	OBSOLETE METHOD CODE
Electrical Conductivity and Soluble Salts	**4F**	
Aqueous Extraction	4F1	
1:2 Aqueous Extraction	4F1a	
Conductivity Bridge	4F1a1	
Electrical Conductivity	4F1a1a	
Salt Prediction	4F1a1a1	8I
Air-Dry, <2 mm	4F1a1a1a1	
Saturated Paste	4F2	8A
Gravimetric	4F2a	
Water Percentage	4F2a1	8A
Air-Dry, <2 mm	4F2a1a1	
Conductivity Bridge	4F2b	
Electrical Conductivity	4F2b1	8A3a
Air-Dry, <2 mm	4F2b1a1	
Resistivity	4F2b2	8E1
Air-Dry, <2 mm	4F2b2a1	
Saturated Paste Extraction	4F2c	8A3
Automatic Extractor	4F2c1	
Atomic Absorption Spectrophotometer	4F2c1a	
Calcium, Magnesium, Potassium, and Sodium	4F2c1a1-4	6N1b, 6O1b, 6Q1b, 6P1b
Air-Dry, <2 mm	4F2c1a1-4a1	
Ion Chromatograph	4F2c1b	
Conductivity Detector	4F2c1b1	
Self-Regeneration Suppressor,	4F2c1b1a	
*Bromide, Acetate, Chloride, Fluoride, Nitrate, Nitrite, **Phosphate**, and Sulfate	4F2c1b1a1-8	6K1d, 6U1b, 6M1d, 6W1d, 6L1d
Air-Dry, <2 mm	4F2c1b1a1-8a1	
Automatic Titrator	4F2c1c	
Combination pH-Reference Electrode	4F2c1c1	
Acid Titration, H_2SO_4	4F2c1c1a	
Carbonate and Bicarbonate	4F2c1c1a1-2	6I1b, 6J1b
Air-Dry, <2 mm	4F2c1c1a1-2a1	
Ratios and Estimates Related to Soluble Salts	4F3	
Exchangeable Sodium Percentage (ESP)	4F3a	
ESP, Calculated without Saturated Paste Extraction	4F3a1	5D2
ESP, Calculated with Saturated Paste Extraction	4F3a2	5D2
Sodium Adsorption Ratio (SAR)	4F3b	5E
Estimated Total Salt	4F3c	8D5

* Bromide, Acetate, and Phosphate – No previous obsolete method as cited in the SSIR No. 42 (1996).

CROSS-REFERENCE BETWEEN CURRENT AND OBSOLETE METHODS

METHOD TITLE	CURRENT METHOD CODE	OBSOLETE METHOD CODE
Selective Dissolutions	**4G**	
Dithionite-Citrate Extraction	4G1	
Atomic Absorption Spectrophotometer	4G1a	
Aluminum, Iron, and Manganese	4G1a1-3	6G7a, 6C2b, 6D2a
Air-Dry, <2 mm	4G1a1-3a1	
Field-Moist, <2 mm	4G1a1-3b1	
Ammonium Oxalate Extraction	4G2	
Automatic Extractor	4G2a	
Inductively Coupled Plasma Atomic Emission Spectrophotometer	4G2a1	
Radial Mode	4G2a1a	
Aluminum, Iron, Manganese, ***Phosphorus**, and Silicon	4G2a1a1-5	6G12b, 6C9b, 6D5b, 6V2b
Air-Dry, <2 mm	4G2a1a1-5a1	
Field-Moist, <2 mm	4G2a1a1-5b1	
UV-Visible Spectrophotometer, Dual Beam	4G2a2	
Transmittance	4G2a2a	
Optical Density	4G2a2a1	8J
Air-Dry, <2 mm	4G2a2a1a1	
Field-Moist, <2 mm	4G2a2a1b1	
Ratios and Estimates Related to Ammonium Oxalate Extraction	4G2b	
Al + ½ Fe	4G2b1	
Sodium Pyrophosphate Extraction	4G3	
Atomic Absorption Spectrophotometer	4G3a	
Aluminum, Iron, and Manganese	4G3a1-3	6G10a, 6C8a, 6D4a
Air-Dry, <2 mm	4G3a1-3a1	
Field-Moist, <2 mm	4G3a1-3b1	
Acid Digestion	4G3b	
$K_2Cr_2O_7 + (H_2SO_4 + H_3PO_4$ Digestion)	4G3b1	
Nesbitt Bulb	4G3b1	
CO_2 Evolution, Gravimetric	4G3b1a	
Organic Carbon	4G3b1a1	6A4a
Air-Dry, <2 mm	4G3b1a1a1	
Field-Moist, <2 mm	4G3b1a1b1	

*Phosphate – No previous obsolete method as cited in the SSIR No. 42 (1996).

CROSS-REFERENCE BETWEEN CURRENT AND OBSOLETE METHODS

METHOD TITLE	CURRENT METHOD CODE	OBSOLETE METHOD CODE
Total Analysis	**4H**	
Acid Digestion	4H1	
HNO_3 + HCl Digestion	4H1a	
Microwave	4H1a1	
Inductively Coupled Plasma Atomic Emission Spectrophotometer	4H1a1a	
Axial Mode	4H1a1a1	
Ultrasonic Nebulizer	4H1a1a1a	
Silver, Arsenic, Barium, Beryllium, Cadmium, Cobalt, Chromium, Copper, Manganese, Molybdenum, Nickel, Phosphorus, Lead, Antimony, Tin, Strontium, Thallium, Vanadium, Tungsten, Zinc	4H1a1a1a1-20	
Air-Dry, <2 mm	4H1a1a1-20a1	
HCl Digestion	4H1a1b	
Water Bath	4H1a1b1	
Flow through Hydride-Generation and Atomic Absorption Spectrophotometer	4H1a1b1	
Arsenic and Selenium	4H1a1b1a1-2	
Air-Dry, <2 mm	4H1a1b1a1-2a1	
Cold Vapor Atomic Absorption Spectrophotometer	4H1a1c	
Mercury	4H1a1c1	
Air-Dry, <2 mm	4H1a1c1a1	
HF + HNO_3 + HCl Digestion	4H1b	7C4a
Microwave	4H1b1	
Boric Acid	4H1b1a	
Inductively Coupled Plasma Atomic Emission Spectrophotometer	4H1b1a1	
Radial Mode	4H1b1a1a	
Aluminum, Calcium, Iron, Potassium, Magnesium, Manganese, Sodium, Phosphorus, Silicon, Zirconium, Titanium	4H1b1a1a1-11	6G11b, 6N5b, 6C7b, 6Q3b, 6O5b, 6D6a, 6P3b, 6S6a, 6V1b, 8K1a, 8O1a
Air-Dry, <2 mm	4H1b1a1a1-11a1	
Oven-Dry, <2 μm	4H1b1a1a1-11b2	

CROSS-REFERENCE BETWEEN CURRENT AND OBSOLETE METHODS

METHOD TITLE	CURRENT METHOD CODE	OBSOLETE METHOD CODE
Dry Combustion	4H2	
Thermal Conductivity Detector	4H2a	
Carbon, Nitrogen, and Sulfur	4H2a1-3	
Air-Dry, <2 mm	4H2a1-3a1	
Ground and Surface Water Analyses	**4I**	
Reaction	4I1	
Electrode	4I1a	
Standard Glass Body Combination	4I1a1	
Digital pH/Ion Meter	4I1a1a	
pH	4I1a1a1	
Electrical Conductivity and Salts	4I2	
Conductivity Bridge and Cup	4I2a	
Electrical Conductivity	4I2a1	
Atomic Absorption Spectrophotometer	4I2b	
Calcium, Magnesium, Potassium, and Sodium	4I2b1-4	
Ion Chromatograph	4I2c	
Conductivity Detector	4I2c1	
Self-Regeneration Suppressor	4I2c1a	
Bromide, Acetate, Chloride, Fluoride, Nitrate, Nitrite, Phosphate, and Sulfate	4I2c1a1-8	
Automatic Titrator	4I2d	
Combination pH-reference Electrode	4I2d1	
Acid Titration, H_2SO_4	4I2d1a	
Carbonate and Bicarbonate	4I2d1a1-2	
Total Analysis	4I3	
Inductively Coupled Plasma Atomic Emission Spectrophotometer	4I3a	
Radial Mode	4I3a1	
Ultrasonic Nebulizer	4I3a1a	
Aluminum, Iron, Manganese, Phosphorus, and Silicon	4I3a1a1-5	
ANALYSIS OF ORGANIC SOILS OR MATERIALS	**5**	
Mineral Content	**5A**	**8F1**
Pyrophosphate Color	**5B**	**8H**
Fiber Volume	**5C**	**8G1**

CROSS-REFERENCE BETWEEN CURRENT AND OBSOLETE METHODS

METHOD TITLE	CURRENT METHOD CODE	OBSOLETE METHOD CODE
Melanic Index	**5D**	
Ratios and Estimates Related to Organic Matter	**5E**	
SOIL BIOLOGICAL AND PLANT ANALYSES	**6**	
Soil Analyses	**6A**	
0.5 M K$_2$SO$_4$ Extraction + Heating with Disodium Bicinchoninic Reagent	6A1	
UV-Visible Spectrophotometer, Dual Beam	6A1a	
Hot Water Extractable Organic Carbon	6A1a1	
Air-Dry, <2 mm	6A1a1a1	
0.02 M KMnO$_4$ Extraction	6A2	
UV-Visible Spectrophotometer, Dual Beam	6A2a	
Active Carbon	6A2a1	
Air-Dry, <2 mm	6A2a1a1	
Acid Dissolution	6A3	
1 N HCl Dissolution + FeCl$_2$	6A3a	
Gas Chromatography	6A3a1	
CO$_2$ Analysis	6A3a1a	
Carbonates	6A3a1a1	
Air-Dry, <2 mm	6A3a1a1a1	
Particulate Organic Matter and Carbon-Mineral	6A4	
Total Analysis	6A4a	
Dry Combustion	6A4a1	
Thermal Conductivity Detector	6A4a1a	
Carbon, Nitrogen, and Sulfur	6A4a1a1-3	
Air-Dry, ≥53 μm (POM)	6A4a1a1-3a1	
Air-Dry, <53 μm (C-Min)	6A4a1a1-3a2	
Fumigation Incubation	**6B**	
Gas Chromatography	6B1	
CO$_2$ Analysis	6B1a	
Soil Microbial Biomass	6B1a1	
Field-Moist, Whole-Soil	6B1a1a1	

CROSS-REFERENCE BETWEEN CURRENT AND OBSOLETE METHODS

METHOD TITLE	CURRENT METHOD CODE	OBSOLETE METHOD CODE
2 *M* KCl Extraction	6B2	
Automatic Extractor	6B2a	
Ammonia – Salicylate	6B2a1	
Flow Injection, Automated Ion Analyzer	6B2a1a	
N as NH_3 – Mineralizable N	6B2a1a1	
Field-Moist, Whole-Soil Fumigated, Whole-Soil Non-Fumigated	6B2a1a1a1-2	
Plant Analyses	**6C**	
Root Biomass	6C1	
Plant (above-ground) Biomass	6C2	
Plant Nutrition	6C3	
Total Analysis	6C3a	
Dry Combustion	6C3a1	
Thermal Conductivity Detector	6C3a1a	
Carbon, Nitrogen, and Sulfur	6C3a1a1-3	
Dry (50°C), Roots	6C3a1a1-3a1	
Dry (50°C), Plants (above-ground)	6C3a1a1-3a2	
Ratios and Estimates Related to Biological Analyses	**6E**	
MINERALOGY	**7**	
Instrumental Analyses	**7A**	**7A**
X-Ray Diffraction	7A1	
Thin Film on Glass, Resin Pretreatment II	7A1a	
Mg Room Temperature, Mg Glycerol Solvated, K 300°C, K 500°C	7A1a1	7A2i
Thermogravimetric Analysis	7A2	7A4
Thermal Analyzer	7A2a	7A4c
Differential Scanning Colorimetry	7A3	7A6
Thermal Analyzer	7A3a	7A6b
Differential Thermal Analysis	7A4	7A3
Thermal Analyzer	7A4a	7A3c
Surface Area	7A5	
N_2 Adsorption	7A5a	
Brunauer, Emmett and Teller Theory	7A5a1	
Vacuum Degassing	7A5a1a	
Multi-Point	7A5a1a1	
Air-Dry, <2 mm	7A5a1a1a1	
Single Point	7A5a1a2	
Air-Dry, <2 mm	7A4a1a2a1	

CROSS-REFERENCE BETWEEN CURRENT AND OBSOLETE METHODS

METHOD TITLE	CURRENT METHOD CODE	OBSOLETE METHOD CODE
Optical Analyses	**7B**	**7B**
Grain Studies	7B1	7B1
Analysis and Interpretation	7B1a	7B1
Separation by Heavy Liquids	7B1a1	7B1
Grain Mounts, Epoxy	7B1a2	7B1a
Platy Grains	7B2	
Magnetic Separation	7B2a	
Static Tube Separation	7B2b	
Froth Flotation		
	7B2c	
Ratios and Estimates Related to Optical Analysis	**7C**	
Total Resistant Minerals	7C1	

PART I:

SSIR NO. 42, SOIL SURVEY LABORATORY METHODS MANUAL, VERSION 3.0 (1996)

ION ANALYSES (5)

Cation Exchange Capacity (5A)
NH₄OAc, pH 7.0 (5A8)
Automatic Extractor (CEC-7)
Steam Distillation
Kjeltec Auto 1035 Analyzer (5A8c)

1. Application

The CEC determined with 1 N NH₄OAc buffered at pH 7.0, is a commonly used method and has become a standard reference to which other methods are compared (Peech et al., 1947). The advantages of using this method are that the extractant is highly buffered so that the extraction is performed at a constant, known pH (7.0) and that the NH4$^+$ on the exchange complex is easily determined.

2. Summary of Method

Displacement after washing is the basis for this procedure. The CEC is determined by saturating the exchange sites with an index cation (NH4$^+$); washing the soil free of excess saturated salt; displacing the index cation (NH4$^+$) adsorbed by the soil; and measuring the amount of the index cation (NH4$^+$). A sample is leached using 1 N NH₄OAc and a mechanical vacuum extractor (Holmgren et al., 1977). The extract is weighed and saved for analyses of the cations. The NH₄$^+$ saturated soil is rinsed with ethanol to remove the NH₄$^+$ that was not adsorbed. Steam distillation and titration are used to determine the NH₄$^+$ adsorbed on the soil exchange complex. The CEC by NH₄OAc, pH 7 is reported in meq 100 g^{-1} oven-dry soil in procedure 5A8c.

3. Interferenes

Incomplete saturation of the soil with NH₄$^+$ and insufficient removal of NH₄$^+$ are the greatest interferences to this method. Ethanol removes some adsorbed NH₄$^+$ from the exchange sites of some soils. Isopropanol rinses has been used for some soils in which ethanol removes adsorbed NH₄$^+$. Soils that contain large amounts of vermiculite can irreversibly "fix" NH₄$^+$. Soils that contain large amounts of soluble carbonates can change the extractant pH and/or can contribute to erroneously high cation levels in the extract.

4. Safety

Wear protective clothing (coats, aprons, sleeve guards, and gloves) and eye protection (face shields, goggles, or safety glasses) when preparing reagents, especially concentrated acids and bases. Dispense concentrated acids and bases in a fume hood. Thoroughly wash hands after handling reagents. Use the safety showers and eyewash stations to dilute spilled acids and bases. Use sodium bicarbonate and water to neutralize and dilute spilled acids. Nessler's reagent contains mercury which is toxic. Proper disposal of the Nessler's reagent and clean-up of equipment in contact with the reagent is necessary.

Ethanol is flammable. Avoid open flames and sparks. Standard laboratory equipment includes fire blankets and extinguishers for use when necessary. Follow the manufacturer's safety precautions when using the vacuum extractor and the Kjeltec Auto 1035 Analyzer.

5. Equipment
5.1 Mechanical vacuum extractor, 24 place, Centurion International, Inc., Lincoln, NE
5.2 Mechanical vacuum extractor, Mavco Sampletex, 5300 N. 57th St., Lincoln, NE
5.3 Syringes, polypropylene, disposable, 60 mL, for extraction vessel, extractant reservoir and tared extraction syringe
5.4 Rubber tubing, 3.2 ID x 6.4 OD x 25.4 mm (1/8 ID x 1/4 OD x 1 in) for connecting syringe barrels.
5.5 Polycons, Richards Mfg. Co.
5.6 Kjeltec Auto 1035/1038 Sampler System, Tecator, Perstorp Analytical Inc.
5.7 Digestion tubes, straight neck, 250 mL
5.8 Analytical filter pulp, ash-free, Schleicher and Schuell, No. 289
5.9 Plunger, modified. Remove rubber and cut plastic protrusion from plunger end.
5.10 Electronic balance, ±1-mg sensitivity

6. Reagents
6.1 Distilled deionized (DDI) water
6.2 Ammonium acetate solution (NH_4OAc), 1 N, pH 7.0. Add 1026 mL of glacial acetic acid (CH_3COOH) to 15 L DDI water. Add 1224 mL of concentrated ammonium hydroxide (NH_4OH). Mix and cool. Dilute with DDI water to 18 L and adjust to pH 7.0 with CH_3COOH or NH_4OH.
6.3 Ethanol (CH_3CH_2OH), 95%, U.S.P.
6.4 Nessler's reagent. Add 4.56 g of potassium iodide (KI) to 30 mL DDI water. Add 5.68 g of mercuric iodide (HgI_2). Stir until dissolved. Dissolve 10 g of sodium hydroxide (NaOH) in 200 mL of DDI water. Transfer NaOH solution to a 250-mL volumetric flask and slowly add K-Hg-I solution. Dilute to volume with DDI water and thoroughly mix. Solution should not contain a precipitate. Solution can be used immediately. Store in brown bottle to protect from light.
6.5 Sodium chloride (NaCl), reagent, crystal.
6.6 Antifoam agent, slipicone release spray, Dow Chemical Corp. Alternatively, use n-octyl alcohol.
6.7 Boric acid, 4% (w:v), with bromcresol green-methyl red indicator (0.075 % bromcresol green and 0.05% methyl red), Chempure Brand
6.8 Hydrochloric acid (HCl), 0.05 N, standardized. Dilute 83 mL of concentrated HCl in 20 L of DDI water.
6.9 NaOH, 1 M. Add 500 mL of 50% NaOH solution to 8 L of DDI water. Dilute to 9 L with DDI water.

7. Procedure

Extraction of Bases
7.1 Prepare extraction vessel by tightly compressing a 1-g ball of filter pulp into the bottom of a syringe barrel with a modified plunger.

7.2 Weigh 2.50 g of <2-mm, air-dry soil and place in an extraction vessel. Weigh a smaller amount of sample, if the soil is highly organic. Prepare one quality control check sample per 48 samples.

7.3 Place extraction vessel on upper disk of the extractor and connect a tared extraction syringe. Use a 25.4-mm (1 in) length rubber tubing and insert the plunger in the slot of the stationary disk of the extractor.

7.4 Use a squeeze bottle to fill extraction vessel to the 20-mL mark with NH_4OAc solution ($\approx$ 10 mL). Thoroughly wet the sample. Let stand for at least 20 min.

7.5 Put reservoir tube on top of the extraction vessel. Rapidly extract the NH_4OAc solution to a 0.5- to 1.0-cm height above sample. Turn off extractor. Add $\approx$ 45 mL of NH_4OAc solution to the reservoir tube. Set extractor for an overnight (12 to 16 h) extraction.

7.6 Next morning turn off the extractor. Pull the plunger of the syringe down. Do not pull plunger from the barrel of the syringe. Carefully remove the syringe containing the extract. Leave the rubber tubing on the extraction vessel. Weigh each syringe containing the NH_4OAc extract to the nearest 0.01 g.

7.7 Mix the extract in each syringe by manually shaking. Fill a polycon with extract solution and discard the excess. The solution in the polycon is reserved for analyses of extracted cations (procedures 6N2, 6O2, 6P2, and 6Q2).

Removal of Excess Ammonium Acetate

7.8 Return the extractor to starting position. Attach syringe to the extraction vessel and rinse the sides of the extraction vessel with ethanol from a wash bottle. Fill the extraction vessel to the 20-mL mark with ethanol and let stand for 15 to 20 min.

7.9 Place reservoir tube on the extraction vessel. Rapidly extract the ethanol level to a 0.5- to 1.0-cm height above the sample. Turn off the extractor and add 55 to 60 mL of ethanol to the reservoir. Extract at a 45-min rate.

7.10 After the extractor has stopped, turn off the switch. Pull the plunger of the syringe down. Do not pull the plunger from the syringe barrel. Remove the syringe and discard the ethanol.

7.11 Repeat the ethanol wash.

7.12 After the second wash, collect a few drops of ethanol extract from the extraction vessel on a spot plate. Test for NH_4^+ by using Nessler's reagent. A yellow, red to reddish brown precipitate is a positive test. If the test is positive, repeat the ethanol wash and retest with Nessler's reagent. Repeat until a negative test is obtained.

Steam Distillation: Samples and Reagent Blanks

7.13 Remove the extraction vessel and transfer the sample to a 250-mL digestion tube. Add 6 to 7 g of NaCl to the digestion tube.

7.14 Perform the same transfer and addition of reagents for blanks as for samples.

7.15 Spray silicone antifoam agent (or 2 drops of n-octyl alcohol solution) into the digestion tubes for each of the samples and reagent blanks.

7.16 When using new reagents, e.g., boric acid, reagent blanks are distilled in 2 sets of 6, one set per Kjeltec machine. Each set of 6 is averaged and recorded on bench worksheet and manually set on each machine. During the steam distillation, the mean reagent blank titer is automatically subtracted from the sample titer.

7.17 On bench worksheet, record the normality of standardized acid, i.e., $\approx$ 0.05 N HCl.

7.18 Connect the tube to the distillation unit. Close the safety door. Distillation and titration are performed automatically. Record the titer in mL of titrant.

8. Calculations

CEC = [Titer x N x 100 x AD/OD]/[Sample Weight (g)]

where:
CEC   = Cation Exchange Capacity (meq 100 g^{-1})
Titer = Titer of sample (mL)
N = Normality of HCl titrant
100   = Conversion factor to 100-g basis
AD/OD= Air-dry/oven-dry ratio (procedure 4B5)

9. Report

Report CEC-7 in units of meq 100 g^{-1} of oven-dry soil to the nearest 0.1 meq 100 g^{-1}.

10. Precision

Precision data are not available for this procedure.

11. References

Holmgren, G.G.S., R.L. Juve, and R.C. Geschwender. 1977. A mechanically controlled variable rate leaching device. Soil Sci. Amer. J. 41:1207-1208.

Peech, M., L.T. Alexander, L.A. Dean, and J.F. Reed. 1947. Methods of soil analysis for soil fertility investigations. U.S. Dept. Agr. Circ. 757, 25 pp.

Cation Exchange Capacity (5A)
NH₄Cl (5A9)
Automatic Extractor
Steam Distillation (5A9c)
Kjeltec Auto 1035 Analyzer (5A9c)

1. Application

The CEC determined with a neutral unbuffered salt, e.g., 1 N NH₄Cl, is an estimate of the "effective" CEC (ECEC) of the soil (Peech et al., 1947). For a soil with a pH of <7.0, the ECEC value should be < CEC measured with a buffered solution at pH 7.0. The NH₄Cl CEC is ≈ equal to the NH₄OAc extractable bases plus the KCl extractable Al for noncalcareous soils.

2. Summary of Method

Displace

ment after washing is the basis for this procedure. The CEC is determined by saturating the exchange sites with an index cation ($NH4^+$); washing the soil free of excess saturated salt; displacing the index cation ($NH4^+$) adsorbed by the soil; and measuring the amount of the index cation ($NH4^+$). A sample is leached using 1 N NH₄Cl and a mechanical vacuum extractor (Holmgren et al., 1977). The extract is weighed and saved for analyses of the cations. The NH_4^+ saturated soil is rinsed with ethanol to remove the NH_4^+ that was not adsorbed. Steam distillation and titration are used to determine the NH_4^+ adsorbed on the soil exchange complex. The CEC by NH₄Cl is reported in meq 100 g^{-1} oven-dry soil in procedure 5A9c.

3. Interferences

Incomplete saturation of the soil with NH_4^+ and insufficient removal of NH_4^+ are the greatest interferences to this method. Ethanol removes some adsorbed NH_4^+ from the exchange sites of some soils. Isopropanol rinses have been used for some soils in which ethanol removes adsorbed NH_4^+. Soils that contain large amounts of vermiculite can irreversibly "fix" NH_4^+.

Soils that contain large amounts of soluble carbonates can change the extractant pH and/or can contribute to erroneously high cation levels in the extract.

4. Safety

Wear protective clothing (coats, aprons, sleeve guards, and gloves) and eye protection (face shields, goggles, or safety glasses) when preparing reagents, especially concentrated acids and bases. Dispense concentrated acids and bases in a fume hood. Nessler's reagent contains mercury which is toxic. Proper disposal of the Nessler's reagent and clean-up of equipment in contact with the reagent is necessary. Thoroughly wash hands after handling reagents. Use the safety showers and eyewash stations to dilute spilled acids and bases. Use sodium bicarbonate and water to neutralize and dilute spilled acids.

Ethanol is flammable. Avoid open flames and sparks. Standard laboratory equipment includes fire blankets and extinguishers for use when necessary. Follow the manufacturer's safety precautions when using the vacuum extractor and the Kjeltec Auto 1030 Analyzer.

5. Equipment

5.1 Mechanical vacuum extractor, 24 place, Centurion International, Inc., Lincoln, NE

5.2 Mechanical vacuum extractor, Mavco Sampletex, 5300 N. 57th St., Lincoln, NE

5.3 Syringes, polypropylene, disposable, 60 mL, for extraction vessel, extractant reservoir, and tared extraction syringe

5.4 Rubber tubing, 3.2 ID x 6.4 OD x 25.4 mm (1/8 ID x 1/4 OD x 1 in), for connecting syringe barrels.

5.5 Polycons, Richards Mfg. Co.

5.6 Kjeltec Auto 1035/1038 Sampler System, Tecator, Perstorp Analytical Inc.

5.7 Digestion tubes, straight neck, 250 mL

5.8 Analytical filter pulp, ash-free, Schleicher and Schuell, No. 289

5.9 Plunger, modified. Remove rubber and cut plastic protrusion from plunger end.

5.10 Electronic balance, ±1-mg sensitivity

6. Reagents

6.1 Distilled deionized (DDI) water

6.2 Ammonium chloride solution (NH_4Cl), 1 N. Dissolve 535 g of NH_4Cl reagent in DDI water and dilute to 10 L.

6.3 Ethanol (CH_3CH_2OH), 95%, U.S.P.

6.4 Nessler's reagent. Add 4.56 g of potassium iodide (KI) to 30 mL DDI water. Add 5.68 g of mercuric iodide (HgI_2). Stir until dissolved. Dissolve 10 g of sodium hydroxide (NaOH) in 200 mL DDI water. Transfer NaOH solution to a 250-mL volumetric flask and slowly add K-Hg-I solution. Dilute to volume with DDI water and thoroughly mix. Solution should not contain a precipitate. Solution can be used immediately. Store the reagent in a brown bottle to protect from light.

6.5 Sodium chloride (NaCl), reagent, crystal.

6.6 Antifoam agent, slipicone release spray, Dow Chemical Corp. Alternatively, use n-octyl alcohol

6.7 Boric acid, 4% (w:v), with bromcresol green-methyl red indicator (0.075 % bromcresol green and 0.05% methyl red), Chempure Brand

6.8 Hydrochloric acid (HCl), 0.05 N, standardized. Dilute 83 mL of concentrated HCl in 16 L of DDI water.

6.9 NaOH, 1 M. Add 500 mL of 50% NaOH solution to 8 L of DDI water. Dilute to 9 L with DDI water.

7. Procedure

Extraction of Bases

7.1 Prepare extraction vessel by tightly compressing a 1-g ball of filter pulp into the bottom of a syringe barrel with a modified plunger.

7.2 Weigh 2.50 g of <2-mm, air-dry soil and place in an extraction vessel. Weigh a smaller amount of sample, if the soil is highly organic. Prepare one quality control check sample per 48 samples.

7.3 Place extraction vessel on upper disk of the extractor and connect a tared extraction syringe. Use 25.4-mm (1 in) length rubber tubing and insert the plunger in the slot of the stationary disk of the extractor.

7.4 Use a squeeze bottle to fill extraction vessel to the 20-mL mark with NH_4Cl solution (≈ 10 mL). Thoroughly wet the sample. Let stand for at least 20 min.

7.5 Put reservoir tube on top of the extraction vessel. Rapidly extract the NH_4Cl solution to a 0.5- to 1.0-cm height above sample. Turn off extractor. Add ≈ 45 mL of NH_4Cl solution to the reservoir tube. Set extractor for an overnight (12 to 16 h) extraction. \

7.6 Next morning turn off the extractor. Pull the plunger of the syringe down. Do not pull plunger from the barrel of the syringe. Carefully remove the syringe containing the extract. Leave the rubber tubing on the extraction vessel. Weigh each syringe containing the NH_4Cl extract to the nearest 0.01 g.

7.7 Mix the extract in each syringe by manually shaking. Fill a polycon with extract solution and discard the excess. The solution in the polycon is reserved for analyses of extracted cations (procedures 6N2, 6O2, 6P2, and 6Q2).

Removal of Excess Ammonium Chloride

7.8 Return the extractor to starting position. Attach syringe to the extraction vessel and rinse the sides of the extraction vessel with ethanol from a wash bottle. Fill the extraction vessel to the 20-mL mark with ethanol and let stand for 15 to 20 min.

7.9 Place reservoir tube on the extraction vessel. Rapidly extract the ethanol level to a 0.5- to 1.0-cm height above the sample. Turn off the extractor and add 55 to 60 mL of ethanol to the reservoir. Extract at a 45-min rate.

7.10 After the extractor has stopped, turn off the switch. Pull the plunger of the syringe down. Do not pull the plunger from the syringe barrel. Remove the syringe and discard the ethanol.

7.11 Repeat the ethanol wash.

7.12 After the second wash, collect a few drops of ethanol extract from the extraction vessel on a spot plate. Test for NH_4^+ by using Nessler's reagent. A yellow, red to reddish brown precipitate is a positive test. If the test is positive, repeat the ethanol wash and retest with Nessler's reagent. Repeat until a negative test is obtained.

Steam Distillation: Samples and Reagent Blanks

7.13 Remove the extraction vessel and transfer the sample to a 250-mL digestion tube. Add 6 to 7 g of NaCl to the sample.

7.14 Perform the same transfer and addition of reagents for blanks as for samples.

7.15 Spray silicone antifoam agent (or 2 drops of n-octyl alcohol solution) into the digestion tubes for each of the samples and reagent blanks.

7.16 When using new reagents, e.g., boric acid, reagent blanks are distilled in 2 sets of 6, one set per Kjeltec machine. Each set of 6 is averaged and recorded on bench worksheet and manually set on each machine. During the steam distillation, the mean reagent blank titer is automatically subtracted from the sample titer.

7.17 On bench worksheet, record the normality of standardized acid, i.e., ≈ 0.05 N HCl.

7.18 Connect the tube to the distillation unit. Close the safety door. Distillation and titration are performed automatically. Record the titer in mL of titrant.

8. Calculations

CEC = [Titer x N x 100 x AD/OD]/[SampleWeight (g)]

where:
CEC = Cation Exchange Capacity (meq 100 g^{-1})
Titer = Titer of sample (mL)
N = Normality of HCl titrant
100 = Conversion factor to 100-g basis
AD/OD= Air-dry/oven-dry ratio (procedure 4B5)

9. Report
Report neutral salts CEC in units of meq 100 g^{-1} of oven-dry soil to the nearest 0.1 meq 100 g^{-1}.

10. Precision
Precision data are not available for this procedure.

11. References
Holmgren, G.G.S., R.L. Juve, and R.C. Geschwender. 1977. A mechanically controlled variable rate leaching device. Soil Sci. Amer. J. 41:1207-1208.
Peech, M., L.T. Alexander, L.A. Dean, and J.F. Reed. 1947. Methods of soil analysis for soil fertility investigations. U.S. Dept. Agr. Circ. 757, 25 pp.

CHEMICAL ANALYSES (6)

Organic Carbon (6A)
Walkley-Black Modified Acid-Dichromate Organic Carbon (6A1)
FeSO4 Titration, Automatic Titrator
Metrohm 686 Titroprocessor (6A1C)

1. Application
Organic C by the Walkley-Black method is a wet combustion technique to estimate organic C. A correction factor is used to convert the Walkley-Black value to an organic matter content. A common value for the factor is 1.724 based upon the assumption that soil organic matter contains 58% organic C. A review of the literature reveals that the factor is highly variable, not only among soils but also between horizons in the same soil (Broadbent, 1953). In addition, a recovery factor is used because the Walkley-Black method does not completely oxidize all the organic C.

2. Summary of Method

The SSL uses the Walkley-Black modified acid-dichromate $FeSO_4$ titration organic carbon procedure. A sample is oxidized with 1 N potassium dichromate and concentrated sulfuric acid (1:2 volume ratio). After 30 min, the reaction is halted by dilution with water. The excess dichromate is potentiometrically back-titrated with ferrous sulfate. A blank is carried throughout the procedure to standardize the ferrous sulfate. Percent organic C is reported on an oven-dry soil basis.

3. Interferences

Dichromate methods that do not use additional heating do not give complete oxidation of organic matter. Even with heating, the recovery may not be complete. Walkley and Black (1934) determined an average recovery factor of 76%. Other studies have found recovery factors ranging from 60% to 86%. Thus, an average correction factor yields erroneous values for many soils. The Walkley-Black method is only an approximate or semiquantitative estimate of organic C.

Maintain the ratio of dichromate solution to concentrated H_2SO_4 at 1:2 to help maintain uniform heating of the mixture.

The presence of significant amounts of chloride in the soil results in a positive error. If the chloride in the soil is known, use the following correction factor (Walkley, 1947) for the organic C.

Organic C (%) = Apparent soil C % - (Soil Cl⁻ %)/12

The presence of significant amounts of ferrous ions results in a positive error (Walkley, 1947). The dichromate oxidizes ferrous to ferric iron.

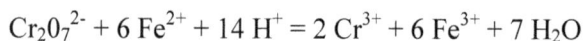

$$Cr_2O_7^{2-} + 6\ Fe^{2+} + 14\ H^+ = 2\ Cr^{3+} + 6\ Fe^{3+} + 7\ H_2O$$

The presence of manganese dioxide results in a negative error (Walkley, 1947). When heated in an acidic medium, the higher oxides of manganese, e.g., MnO_2, compete with dichromate for oxidizable substances.

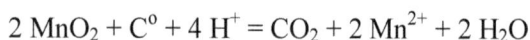

$$2\ MnO_2 + C^o + 4\ H^+ = CO_2 + 2\ Mn^{2+} + 2\ H_2O$$

All dichromate methods assume that the organic C in the soil has an average oxidation state of zero and an equivalent weight of 3 g per equivalent when reacting with dichromate. When the soil has carbonized material, e.g., charcoal, graphite, coal and soot, the Walkley-Black method gives low recovery of this material, i.e., recovery range is from 2 to 36%.

4. Safety

Wear protective clothing (coats, aprons, sleeve guards, and gloves) and eye protection (face shields, goggles, or safety glasses) when preparing acids and dichromate. Toxic chromyl chloride may be released from the sample, if high concentrations of chloride are present. Use the fume hood to contain the gases released by this procedure. Use the safety showers and eyewash stations to dilute spilled acids. Use sodium bicarbonate and water to neutralize and dilute spilled acids and dichromate. Follow the manufacturer's safety precautions when using the automatic titrator.

5. Equipment

5.1 Electronic balance, ±1-mg sensitivity
5.2 Titration beakers, borosilicate glass, 250 mL
5.3 Automatic dispenser, 5 to 20 mL, Oxford no. 470 or equivalent, for $K_2Cr_2O_7$, capable of volume adjustment to 10.00 ± 0.01 mL, 0.5% reproducibility.

5.4 Dispenser, Zippette 30 mL or equivalent, for concentrated H_2SO_4, Brinkmann Instruments, Inc.

5.5 Shaker, Eberbach 6000 power unit, fitted with spring holders for titration beakers, reciprocating speed of 60 to 260 epm, with 6040 utility box carrier and 6110 floor stand, Eberbach Corp., Ann Arbor, MI.

5.6 Automatic titrator, Metrohm 686 Titroprocessor Series 04, 664 Control Unit, 674 Sample Changer Series 5, and 665 Dosimat Series 14, Metrohm Ltd., Brinkmann Instruments, Inc.

5.7 Platinum electrode, Metrohm part no. 6.0412.000

6. Reagents

6.1 Distilled deionized (DDI) water

6.2 Potassium dichromate, 1.000 N, primary standard. Dissolve 49.035 g of $K_2Cr_2O_7$ reagent, dried @ 105°C, in 1-L volumetric flask with DDI water.

6.3 Sulfuric acid (H_2SO_4), concentrated, reagent

6.4 Ferrous sulfate, 1 N, acidic. Dissolve 1 kg of $FeSO_4 \cdot 7H_2O$ in 6 L of DDI water. Carefully add 640 mL of concentrated H_2SO_4 with stirring. Cool and dilute to 8 L with DDI water.

7. Procedure

Digestion of Organic C

7.1 Weigh 1.000 g air-dry soil and place in a titration beaker. If the sample contains >3% of organic C, use a smaller sample size. Refer to Table 1 for sample weight guide. If sample size is <0.5 g, use <80-mesh soil. If sample size is >0.5 g, use <2-mm soil.

7.2 With automatic dispenser, add 10.00 mL of $K_2Cr_2O_7$ solution to the titration beaker. Mix by swirling the sample.

7.3 Use the dispenser to carefully add 20 mL of concentrated H_2SO_4 to the beaker. Mix by swirling solution. Adjustment in the amount of $K_2Cr_2O_7$ added to sample requires appropriate adjustment in the amount of H_2SO_4 so that a 1:2 volume is maintained.

7.4 Place titration beaker on the reciprocating shaker and shake 1 min. If the dichromate-acid mixture turns a blue-green color, all the dichromate has been reduced. Add more dichromate and acid to maintain a 1:2 volume. Refer to Table 1 for dichromate:acid volumes.

Table 1. Digestion of organic C. Guide for sample weight and dichromate:acid volumes.

OC (%)	Sample (g)	$K_2Cr_2O_7$ (mL)	H_2SO_4 (mL)
0-3	1.000	10.00	20
3-6	0.500	10.00	20
3-6	1.000	20.00	40
6-12	0.500	20.00	40
12-24	0.250	20.00	40
24-50	0.100	30.00	60

7.5 Place the beaker on a heat resistant surface for 30 min.

7.6 Add≈ 180 mL DDI water to the beaker to stop the reaction.

Titration of Excess Dichromate

7.7 Titrate eight reagent blanks at the start of each batch to determine the normality of the ferrous sulfate. A blank is 10.00 mL $K_2Cr_2O_7$ plus H_2SO_4 without soil. The average titer is used for the blank titer value.

7.8 Place the appropriate blanks and samples in the sample holder magazines and place on the sample changer.

7.9 Refer to the manufacturer's instruction manual for operation of automatic titrator.

7.10 Set the endpoint to 700 mV. Set the controls of the 664 Control Unit to the appropriate settings.

7.11 Prime the buret with 50 mL of ferrous sulfate solution before starting the titrations.

7.12 When a long series of samples are being titrated, intersperse blank samples throughout the titrations. The blank titer drifts over time, mainly because of the temperature change of the solution. Any sample with a titer of less one milliter and/or endpoint of less than 620 millivolts should be reanalyzed.

7.13 Press "Start" on the titrator.

8. Calculations

OC (%) = [(Blank x Volume) - (10 x Titer) x 3 x 100 x AD/OD]/
       [Blank x Sample Weight (g) x 0.77 x 1000]

where:
OC (%)	= Organic C (%)
Blank	= Average titer of reagent blanks (mL)
Volume	= Volume of 1 N $K_2Cr_2O_4$ (mL)
Titer	= Titer of $FeSO_4$ (mL)
AD/OD	= Air-dry/oven-dry ratio (procedure 4B5)
3	= Equivalents per C (assumed)
1000	= Meq eq^{-1}
100	= Convert to 100-g basis
0.77	= Assumed C oxidation factor.

9. Report

Report organic C percentage to two decimal places, e.g., 0.95% OC, on an oven-dry basis.

10. Precision

Precision data are not available for this procedure. A quality control check sample is run in every batch of 20 samples. With 251 observations of the quality control check sample, the mean, standard deviation, and C.V. for organic carbon are 1.47, 0.025, and 1.7%, respectively.

11. References

Broadbent, F.E. 1953. The soil organic fraction. Adv. Agron. 5:153-183.

Walkley, A. 1946. A critical examination of a rapid method for determining organic carbon in soils: Effect of variations in digestion conditions and of inorganic soil constituents. Soil Sci. 63:251-263.

Walkley, A., and Black, I.A. 1934. An examination of the Degtjareff method for determining soil organic matter and a proposed modification of the chromic acid titration method. Soil Sci. 37:29-38.

Total Carbon (6A)
Dry Combustion (6A2)
LECO SC-444 Carbon Analyzer (6A2e)

1. Application

Total C in soils is the sum of organic and inorganic C. Most of the organic C is associated with the organic matter fraction, and the inorganic C is generally found with carbonate minerals. The organic C in mineral soils generally ranges from 0 to 12 percent.

Total C is quantified by two basic methods, i.e., wet or dry combustion. The SSL uses dry combustion. In total C determinations, all forms of C in a soil are converted to CO_2 followed by a quantification of the evolved CO_2. Total C can be used to estimate the organic C content of a soil. The difference between total and inorganic C is an estimate of the organic C. Organic C also can be determined directly (procedure 6A1c). The inorganic C should be equivalent to carbonate values measured by CO_2 evolution with strong acid (Nelson and Sommers, 1982).

Organic C defines mineral and organic soils. In *Soil Taxonomy*, organic C is also used at lower taxonomic levels, e.g., ustollic and fluventic subgroups (Soil Survey Staff, 1975).

2. Summary of Method

A fine-ground (<80-mesh) soil sample is oxidized at high temperatures. The released gases are scrubbed, and the CO_2 in the combustion gases is measured by using an infrared detector. The microprocessor formulates the analytical results (C_i) by combining the outputs of the infrared detector and the system ambient sensors with pre-programmed calibration, linearization and weight compensation factors. Percent total C is reported on an oven-dry soil basis.

3. Interferences

This procedure simultaneously measures inorganic and organic C. A high rate of combustion can oversaturate the carbon detection cell. The rate of combustion can be retarded by adding a solid/powder combustion controller.

4. Safety

Wear protective clothing and safety glasses. Magnesium perchlorate may form explosive mixtures. Magnesium perchlorate may contain traces of perchloric acid, which remain from manufacturer's operations. This acid is anhydrous because of the strong desiccating capability of the salt. Avoid prolonged contact with oxidizable material or material capable of forming unstable perchlorate esters or salts. Remove magnesium perchlorate by using an excess of water to thoroughly dilute the material.

The use of high temperatures in the oxidation of samples requires that extreme caution be used to prevent burns and fires. Follow standard laboratory procedures when handling compressed gases. Oxygen is highly flammable. Avoid open flames and sparks. Standard laboratory equipment includes fire blankets and extinguishers for use when necessary. Follow the manufacturer's safety precautions when using the carbon analyzer.

5. Equipment

5.1 Carbon analyzer, Leco Model SC-444, Sulfur and Carbon Analyzers, Leco Corp., St. Joseph, MI

5.2 Combustion boats, part no. 529-203, Leco Corp., St. Joseph, MI

5.3 Single-stage regulator, oxygen service, part no. E11-W-N115Box, Air Products and Chemicals, Inc., Box 538, Allentown, PA 18105

5.4 Electronic balance, ±1-mg sensitivity

6. Reagents
6.1 Anhydrous magnesium perchlorate, granular
6.2 Glass wool
6.3 Compressed oxygen, >99.5% @ 30 psi
6.4 Calcium carbonate, $CaCO_3$, reagent grade.
6.5 Solid/Powder Combustion Controller, part no. 501-426, Leco Corp., St. Joseph, MI
6.6 Soil Calibration Sample, part no. 502-062, Leco Corp., St. Joseph, MI

7. Procedure

7.1 Use a fine-ground 80-mesh, air-dry soil

7.2 Prepare instrument as outlined in the operator's instruction manual (Leco, 1994; Leco, 1993).

7.3 Methods are created with the method menu and stored in the instrument memory. System parameters are set as follows:

Furnace operating temperature: 1450°C
Lance delay: 20 s
Analysis time settings: 70 to 180 s
Comparator level settings: 0.1%

7.3 Condition instrument by analyzing a few soil samples, until readings are stable.

7.4 Calibrate instrument by analyzing at least three replicates of each calibration standard. Use the soil calibration standard for samples with less than three to four percent total carbon and calcium carbonate for samples with more than four percent total carbon. Weigh standards in a range from 0.2 to 0.7 g.

7.5 Load samples on autoload rack, place in the analyzer, and press analyze key.

7.6 Weigh 0.2 to 0.5 g sample in a tared combustion boat.

7.7 Load samples on autoload rack, place in the analyzer, and press analyze key.

7.8 If results exceed calibration range, reduced weight of sample. If carbon detection cell is saturated, add approximately 1 g of solid/powder combustion controller to sample.

7.9 Repack the reagent (anhydrous magnesium perchlorate) tubes whenever the reagent becomes caked or moist or the warning alarm displays.

8. Calculations

C (%) = C_i x AD/OD

where:
C (%) = C (%), oven-dry basis
C_i = C (%) instrument
AD/OD = air-dry/oven-dry ratio (procedure 4B5)

9. Report

Report total C percentage on an oven-dry basis to the nearest 0.1%.

10. Precision

A quality control check sample is included in every batch of ten samples. For 191 observations of calcium carbonate (actual total C = 12%), the mean, standard deviation, and C.V. for total carbon are 12.04, 0.31, and 2.5%, respectively. For 86 observations of soil calibration standard (reported total C = 0.77%), the mean, standard deviation, and C.V. for total carbon are 0.79, 0.02, and 2.2%, respectively.

11. References

Leco Corp. 1993. Sulfur and carbon in cements, soils, rock, ceramic and similar materials. Application Bulletin. Leco Corp., 3000 Lakeview Ave., St. Joseph, MI.

Leco Corp. 1994. Instruction Manual. SC-444 Sulfur and Carbon Analyzers. Leco Corp. , 3000 Lakeview Ave., St. Joseph, MI.

Nelson, D.W., and L.E. Sommers. 1982. Total carbon, organic carbon, and organic matter. *In* A.L. Page, R.H. Miller, and D.R. Keeney (eds.) Methods of soil analysis. Part 2. Chemical and microbiological properties. 2nd ed. Agronomy 9:539-579.

Soil Survey Staff. 1975. Soil taxonomy: A basic system of soil classification for making and interpreting soil surveys. USDA-SCS Agric. Handb. 436. U.S. Govt. Print. Office, Washington, DC.

Total Nitrogen (6B)
Dry Combustion (6B4)
LECO FP-428 Analyzer (6B4a)

1. Application

The total N content of the soil may range from <0.02% in subsoils, 2.5% in peats, and 0.06 to 0.5% in surface layers of many cultivated soils (Bremmer and Mulvaney, 1982). The total N data may be used to determine the soil C:N ratio, the soil potential to supply N for plant growth, and the N distribution in the soil profile. The C:N ratio generally ranges between 10 to 12. Variations in the C:N ratio may serve as an indicator of the amount of soil inorganic N. Uncultivated soils usually have higher C:N ratios than do cultivated soils.

Soils with large amounts of illites or vermiculites can "fix" significant amounts of N compared to those soils dominated by smectites or kaolinites (Young and Aldag, 1982; Nommik and Vahtras, 1982). Since the organic C of many soils diminishes with depth while the level of "fixed" N remains constant or increases, the C:N ratio narrows (Young and Aldag, 1982). The potential to "fix" N has important fertility implications as the "fixed" N is slowly available for plant growth.

2. Summary of Method

A soil sample is combusted at high temperature with oxygen to release NO_x. The gases released are scrubbed to remove interferences (e.g., CO_2 and H_2O), and the NO_x is reduced to N_2. The N_2 is measured by thermal conductivity detection and reported as percent N.

3. Interferences

The total N that is measured by the combustion method does not distinguish among the types of N that are present in the soil. The purity of the helium and oxygen gases used in the instrument may affect the results of the analysis. The highest purity gases available are required to assure low detection limits and consistent results.

4. Safety

Wear protective clothing (coats, aprons, sleeve guards, and gloves) and eye protection (goggles or safety glasses) when handling hot crucibles. Standard laboratory equipment includes fire blankets and extinguishers for use when necessary.

5. Equipment

5.1 Electronic balance, ±0.001-g sensitivity

6. Reagents

None

7. Procedure

7.1 Weigh 0.200 g of 80-mesh, air-dry soil into a tin foil cup.

7.2 Close the tin foil cup by twisting the top closed as to fit the sample holder.

7.3 Place the enclosed sample in the sample holder.

7.4 When all the samples are in the sample holder, place the sample holder on the instrument.

7.5 Refer to manufacturer's manual for operation and calibration of the LECO FP-438 Analyzer.

7.6 On the bench worksheet, record the percent N for the samples.

8. Calculations

N (%) = Instrument Reading x AD/OD

where:
AD/OD = Air-dry/Ovendry ratio (procedure 4B5)

9. Report

Report total N as a dimensionless value to the nearest 0.001 unit on an ovendry basis.

10. Precision

Precision data are not available for this procedure. For 105 observations of the quality control check sample for total N, the mean, standard deviation, and C.V. are 0.143, 0.004, and 2.7 percent, respectively.

11. References

Bremmer, J.M., and C.S. Mulvaney. 1982. Nitrogen - Total. *In* A.L. Page, R.H. Miller, and D.R. Keeney (eds.) Methods of soil analysis. Part 2. Chemical and microbiological properties. 2nd ed. Agronomy 9:595-624.

Nommik, H., and K. Vahtras. 1982. Retention and fixation of ammonium and ammonia in soils. *In* F.J. Stevenson (ed.) Nitrogen in agricultural soils. Agronomy 22:123-171.

Young, J.L., and R.W. Aldag. 1982. Inorganic forms of nitrogen in soil. *In* F.J. Stevenson (ed.) Nitrogen in agricultural soils. Agronomy 22:43-66.

Mineralizable Nitrogen (6B)
Steam Distillation (6B5)
Kjeltec Auto 1035 Sampler (6B5A)

1. Application

The most satisfactory methods currently available for obtaining an index for the availability of soil N are those involving the estimation of the N formed when soil is incubated under conditions which promote mineralization of organic N by soil microorganisms (Environmental Protection Agency, 1992). The method described herein for estimating mineralizable N is one of anaerobic incubation and is suitable for routine analysis of soils. This method involves estimation of the ammonium produced by a 1-week period of incubation of soil at 40°C (Keeney and Bremner, 1966) under anaerobic conditions to provide an index of N availability.

2. Method Summary

An aliquot of air-dry homogenized soil is placed in a test tube with water, stoppered, and incubated at 40°C for 1 week. The contents are transferred to a steam distillation, rinsed with 4 N KCl. The amount of ammonium-N is determined by steam distillation and titration for the KCl:soil mixture.

3. Interferences

There are no known interferences. The temperature and incubation period must remain constant for all samples. The test can be performed on field-moist or air-dry soil samples.

4. Safety

Wear protective clothing (coats, aprons, sleeve guards, and gloves) and eye protection (face shields, goggles, or safety glasses) when preparing reagents, especially concentrated acids and bases. Dispense concentrated acids and bases in a fume hood. Thoroughly wash hands after handling reagents. Use the safety showers and eyewash stations to dilute spilled acids and bases. Use sodium bicarbonate and water to neutralize and dilute spilled acids. Follow the manufacturer's safety precautions when using the incubator and Kjeltec Auto 1035 Analyzer.

5. Equipment
5.1 Electronic balance, ±1-mg sensitivity
5.2 Test tubes, 16-mm x 150-mm
5.3 PVC stoppers
5.4 Incubator, Model 10-140, Quality Lab Inc., Chicago, IL
5.5 Digestion tubes, straight neck, 250 mL
5.6 Kjeltec Auto 1035/1038 Sampler System, Tecator, Perstorp Analytical Inc.

6. Reagents
6.1 Distilled deionized (DDI) water
6.2 Potassium chloride (KCl), 4 N. Dissolve 298.24 g KCl in DDI water and dilute to 1-L volume.
6.3 Hydrochloric acid (HCl), 0.05 N, standardized. Dilute 83 mL of concentrated HCl in 20 L of DDI water.
6.4 Antifoam agent, slipicone release spray, Dow Chemical Corp. Alternatively, use n-octyl alcohol.
6.5 Boric acid, 4% (w:v), with bromcresol green-methyl red indicator (0.075 % bromcresol green and 0.05% methyl red), Chempure Brand

7. Procedure

Anaerobic Incubation of Soil Sample

7.1 Place 5.00 g of mineral soil (or 1.25 g of organic soil) into a 16-mm x 150-mm test tube. Record the soil sample weight to the nearest 0.00 g.

7.2 Add 12.5 ±1 mL of DDI water. Do not add ethanol to overcome any wetting difficulties as ethanol may act as an interference with microbial activity. Stopper the tube, shake, and place in a 40°C constant-temperature incubator for 7 days. Refer to the manufacturer's instructions for set-up and operation of the incubator.

7.3 At the end of 7 days, remove the tube and shake for 15 s.

7.4 Transfer the contents of the test tube to a 250-mL digestion tube. Complete the transfer by rinsing the tube with 3 times with 4 ml of 4 N KCl, using a total of 12.5 ±1 mL of the KCl.

Steam Distillation: Samples and Reagent Blanks

7.5 Remove the extraction vessel and transfer the sample to a 250-mL digestion tube.

7.6 Perform the same transfer and addition of reagents for blanks as for samples.

7.7 Spray silicone antifoam agent (or 2 drops of n-octyl alcohol solution) into the digestion tubes for each of the samples and reagent blanks.

7.8 When using new reagents, e.g., boric acid, reagent blanks are distilled in 2 sets of 6, one set per Kjeltec machine. Each set of 6 is averaged and recorded on bench worksheet and manually set on each machine. During the steam distillation, the mean reagent blank titer is automatically subtracted from the sample titer.

7.9 On bench worksheet, record the normality of standardized acid, i.e., ≈ 0.0500 N HCl.

7.10 Load samples in racks of 20. Distillation and titration are performed automatically. Record the titer in mL of titrant.

8. Calculations

N= (Titer x N x 100 x AD/OD)/ Sample Weight (g)

where:
N   = Mineralizable N (meq 100 g^{-1})
Titer   = Titer of sample (mL)
N = Normality of HCl titrant
100   = Conversion factor to 100-g basis
AD/OD = Air-dry/oven-dry ratio (procedure 4B5)

9. Report

Report mineralizable N in units of meq 100 g^{-1} of oven-dry soil to the nearest 0.001 meq 100 g^{-1}.

10. Precision

No precision data are available for this procedure.

11. References

Environmental Protection Agency. 1992. Handbook of laboratory methods for forest health monitoring. G.E. Byers, R.D. Van Remortel, T.E. Lewis, and M. Baldwin (eds.). Part III. Soil analytical laboratory. Section 10. Mineralizable N. U.S. Environmental Protection Agency, Office of Research and Development, Environmental Monitoring Systems Laboratory, Las Vegas, NV.

Keeney, D.R. and J.M. Bremner. 1966. Comparison and evaluation of laboratory methods of obtaining an index of soil nitrogen availability. Agron. J. 58:498-503.

Iron, Manganese, and Aluminum (6C, 6D, and 6G)
Dithionite-Citrate Extraction (6C2, 6D2, and 6G7)
Atomic Absorption Perkin-Elmer AA 5000
(6C2b, 6D2a, and 6G7a)

1. Application

Dithionite-citrate (CD) is used as a selective dissolution extractant for organically complexed Fe and Al, noncrystalline hydrous oxides of Fe and Al, and amorphous aluminosilicates (Wada, 1989). The CD solution is a poor extractant of crystalline hydrous oxides of Al, allophane, and imogolite. The CD solution does not extract opal, Si, or other constituents of crystalline silicate minerals (Wada, 1989). In *Soil Taxonomy*, the CD extractable Fe and Al are criteria for spodic placement (Soil Survey Staff, 1975).

2. Summary of Method

A soil sample is mixed with sodium dithionite, sodium citrate, and distilled deionized water, and shaken overnight. Superfloc 16 is added, and the mixture is made to volume. Solution is allowed to settle, and a clear extract is obtained. The CD extract is diluted with distilled deionized (DDI) water. The analytes are by an atomic absorption spectrophotometer (AA). The data are automatically recorded by a microcomputer and printer. The percent CD extractable Fe, Mn, and Al are reported in procedures 6C2b, 6D2a, and 6G7a, respectively.

3. Interferences

There are four types of interferences (matrix, spectral, chemical, and ionization) in the AA analyses of these elements. These interferences vary in importance, depending upon the particular analyze selected.

The redo potential of the extractant is dependent upon the pH of the extracting solution and the soil system. Sodium citrate complexes the reduced Fe and usually buffers the system to a pH of 6.5 to 7.3. Some soils may lower the pH, resulting in the precipitation of Fe sulfides. The SSL has not had significant problems with this interference.

Filtered extracts can yield different recoveries of Fe, Mn, and Al, relative to unfiltered extracts.

4. Safety

Wear protective clothing (coats, aprons, sleeve guards, and gloves); eye protection (face shields, goggles, or safety glasses); and a breathing filter when handling dry sodium dithionite. Sodium dithionite may spontaneously ignite if allowed to become moist, even by atmospheric moisture. Keep dithionite in a fume hood.

Follow standard laboratory practices when handling compressed gases. Gas cylinders should be chained or bolted in an upright position. Acetylene gas is highly flammable. Avoid open flames and sparks. Standard laboratory equipment includes fire blankets and extinguishers for use when necessary. Follow the manufacturer's safety precautions when using the AA.

5. Equipment

5.1 Electronic balance, ±1-mg sensitivity

5.2 Filter paper, pre-pleated, 185-mm diameter, Schleicher and Schuell

5.3 Atomic absorption spectrophotometer (AA), model 5000, Perkin-Elmer Corp., Norwalk, CT

5.4 Automatic burner control, model 5000, Perkin-Elmer Corp., Norwalk, CT

5.5 Autosampler, AS-50, Perkin-Elmer Corp., Norwalk, CT

5.6 Dot matrix printer, P-132, Interdigital Data Systems,Inc.

5.7 Single-stage regulator, acetylene service, part number E11-0-N511A, Air Products and Chemicals, Inc., Box 538, Allentown, PA

5.8 Digital diluter/dispenser, MicroLab 500, Hamilton Co., P.O. Box 10030, Reno, NV

5.9 Syringes, 10000 and 1000 μL, 1001 DX and 1010-TEL LL gas tight, Hamilton Co., P.O. Box 10030, Reno, NV

5.10 Test tubes, 15-mL, 16 mm x 100, for sample dilution and sample changer, Curtin Matheson Scientific, Inc., Houston, TX

5.11 Containers, polypropylene

6. Reagents

6.1 Distilled deionized (DDI) water

6.2 Sodium dithionite ($Na_2S_2O_4$), purified powder

6.3 Sodium citrate dihydrate ($Na_3C_6H_5O_7 \cdot 2H_2O$), crystal, reagent

6.4 Hydrochloric acid (HCl), concentrated 12 N

6.5 HCl, 1:1 HCl:DDI, 6 N. Carefully mix 1 part of concentrated HCl to 1 part DDI water.

6.6 HCl, 1% wt. Carefully dilute 25 mL of concentrated HCl to 1 L with DDI water.

6.7 Superfloc 16, 0.2% solution (w:v). Dissolve 2 g of Superfloc 16 in 1 liter of DDI water. Do not shake the mixture as this breaks the polymer chains of the Superfloc. Gently swirl the mixture occasionally over the several days that the solution requires to completely dissolve the Superfloc. Suggested source is American Cyanamid Co., P.O. Box 32787, Charlotte, NC.

6.8 Primary mixed standard, 4000 mg L^{-1} (4000 ppm) Fe, 600 mg L^{-1} (600 ppm) Mn, and 3000 mg L^{-1} (3000 ppm) Al. Dissolve 4.000 g of Fe wire, 0.6000 g of Mn metal powder, and 3.000 g of Al wire with 1:1 HCl in a glass beaker. When dissolved transfer to a 1-L volumetric flask and make to volume with 1% HCl solution. Store in a polypropylene bottle.

6.9 High calibration standard, 240 mg/8 oz (1012 ppm) Fe; 36 mg/8 oz (152 ppm) Mn; and 180 mg/8 oz (759 ppm) Al. Pipet 60 mL of primary mixed standard into 8 oz bottle. Add 20 g of sodium citrate dihydrate, 1.24 mL of concentrated H_2SO_4, and 2 mL of Superfloc 16 solution. In standards, the H_2SO_4 substitutes for the dithionite. Fill to 8-oz volume with DDI water and mix thoroughly. After dissolution, transfer solution to a plastic bottle.

6.10 Low calibration standard, 120 mg/8 oz (506 ppm) Fe; 18 mg/8 oz (76 ppm) Mn; and 90 mg/8 oz (380 ppm) Al. Pipet 30 mL of primary mixed standard into 8 oz bottle. Add 20 g of sodium citrate dihydrate, 1.24 mL of concentrated H_2SO_4, and 2 mL of Superfloc 16 solution. In standards and reagent blanks, the H_2SO_4 substitutes for the dithionite. Fill to 8-oz volume with DDI water and mix thoroughly. After dissolution, transfer solution to a plastic bottle.

6.11 Calibration reagent blank solution. Add 20 g of sodium citrate dihydrate, 1.24 mL of concentrated H_2SO_4, and 2 mL of Superfloc 16 solution. In standards and reagent blanks, the H_2SO_4 substitutes for the dithionite. Fill to 8-oz volume with DDI water and mix thoroughly. After dissolution, transfer solution to a plastic bottle.

6.12 Acetylene gas, purity 99.6%

6.13 Compressed air with water and oil traps

7. Procedure

Extraction of Fe, Mn, and Al

7.1 Weigh 4.00 g of <2-mm, air-dry soil sample and place in an 8-oz nursing bottle.

7.2 Add 2 g of sodium dithionite and 20 to 25 g of sodium citrate dihydrate.

7.3 Add DDI water to 4-oz level on bottle and securely stopper bottle.

7.4 Shake overnight (12 to 16 h) in a reciprocating shaker. After shaking, use a dispenser to add 2 ml of Superfloc 16 solution.

7.5 Fill bottle to 8-oz volume with DDI water. Stopper and shake thoroughly for ~ 15 s.

7.6 Allow to settle for at least 3 day (3 to 5 days typical). The Fe, Mn, and Al are determined from a clear aliquot of solution.

Dilution of Sample Extracts and Standards

7.7 No ionization suppressant is required as the Na in the extractant is present in sufficient quantity. Set the digital diluter at 66 for diluent and 35 for CD extracts, calibration reagent blanks, and calibration standards for a 1:20 dilution as follows:

7.8 Dilute 1 part CD sample extract with 19 parts of DDI water (1:20 dilution).

7.9 Dilute 1 part calibration reagent blank with 19 parts of DDI water (1:20 dilution).

7.10 Dilute 1 part low calibration standard with 19 parts of DDI water (1:20 dilution).

7.11 Dilute 1 part high calibration standard with 19 parts of DDI water (1:20 dilution).

7.12 Dispense the reagent blanks and calibration standards in polycons from which the solutions are transferred to test tubes. Dispense the diluted sample solutions into test tubes which have been placed in the sample holders of the sample changer.

AA Calibration

7.13 Use the calibration reagent blank and high calibration standard to calibrate the AA. The AA program requires a blank and a standard, in that order, to establish a single point calibration curve for element determination. Perform one calibration, i.e., blank plus standard, for every 12 samples.

7.14 Use the low calibration standard (120 mg/8 oz Fe; 18 mg/8 oz Mn; and 90 mg/8 oz Al) as a check sample. Use high calibration standard for Fe check sample and low calibration standard for Mn and Al check sample.

AA Set-up and Operation

7.15 Refer to the manufacturer's manual for operation of the AA. The following are only very general guidelines for instrument conditions for the various analytes.

Element	Wave-length (nm)	Burner head & Angle	Fuel/ Oxidant
Fe	248.5	5-cm parallel	$10\ C_2H_2/$ 25 Air
Al	309.35	5-cm parallel	$30\ C_2H_2/$ $17\ N_2O$
Mn	280.15	5-cm parallel	$10\ C_2H_2/$ 25 Air

Typical read delay is 6 s, and integration time is 8 s.

7.16 Use the microcomputer and printer to set instrument parameters and to collect and record instrument readings.

7.17 If sample exceeds calibration standard, dilute the sample (dilution ratio in calculation) with appropriate matrix and record dilution. Remember to keep the matrix the same after dilution by diluting with DDI water (1:20 dilution).

7.18 The instrument readings are usually programmed to display analyte concentration in mg/8 oz.

8. Calculations

Fe (%) = (Fe x DR x 100 x AD/OD)/(Sample x 1000)

Fe_2O_3 (%) = (Fe X DR x 1.43 x 100 x AD/OD)/(Sample x 1000)

Mn (%) = (Mn x DR x 100 x AD/OD)/(Sample x 1000)

Al (%) = (Al x DR x 100 x AD/OD)/(Sample x 1000)

where:
Fe   = mg/8 oz
Mn   = mg/8 oz
Al   = mg/8 oz
DR   = Dilution Ratio
Sample   = Sample weight (g)
1.43   = Conversion factor from Fe to Fe_2O_3
100   = Conversion factor to percent
AD/OD = Air-dry/oven-dry ratio (procedure 4B5)
1000   = Conversion factor (mg g^{-1})

9. Report
    Report percent CD extractable Fe, Mn, and Al on oven-dry soil basis to the nearest whole number.

10. Precision
    Precision data are not available for this procedure. A quality control check sample is run with every batch of samples. For the quality control check sample, the mean, standard deviation, and CV for Fe, Mn, and Al are as follows:

Element	Mean	n	Std. Dev.	C.V.
Fe	2.5	35	0.05	2.2%
Mn	0.01	19	0.00	0.0%
Al	0.26	33	0.01	5.4%

11. References

Wada, K. 1989. Allophane and imogolite. *In* J.B. Dixon and S.B. Weed (eds.). Minerals in soil environment. 2nd ed. Soil Sci. Soc. Amer. No. 1. p. 1051-1087.

Soil Survey Staff. 1975. Soil Taxonomy: A basic system of soil classification for making and interpreting soil surveys. USDA-SCS Agric. Handb. 436. U.S. Govt. Print. Office, Washington, DC.

Iron, Manganese, and Aluminum (6C, 6D, and 6G)
Dithionite-Citrate Extraction (6C2, 6D2, and 6G7)
Atomic Absorption
Thermo Jarrell Ash, Smith-Hietje 4000
(6C2c, 6D2b, and 6G7b)

1. Application

Dithionite-citrate (CD) is used as a selective dissolution extractant for organically complexed Fe and Al, noncrystalline hydrous oxides of Fe and Al, and amorphous aluminosilicates (Wada, 1989). The CD solution is a poor extractant of crystalline hydrous oxides of Al, allophane, and imogolite. The CD solution does not extract opal, Si, or other constituents of crystalline silicate minerals (Wada, 1989). In *Soil Taxonomy*, the CD extractable Fe and Al are criteria for spodic placement (Soil Survey Staff, 1975).

2. Summary of Method

A soil sample is mixed with sodium dithionite, sodium citrate, and distilled deionized water, and shaken overnight. Superfloc 16 is added, and the mixture is made to volume. Solution is allowed to settle, and a clear extract is obtained. The CD extract is diluted with distilled deionized (DDI) water. The analytes are measured by an atomic absorption spectrophotometer (AA). The data are automatically recorded by a microcomputer and printer. The percent CD extractable Fe, Mn, and Al are reported in procedures 6C2c, 6D2b, and 6G7b, respectively.

3. Interferences

There are four types of interferences (matrix, spectral, chemical, and ionization) in the AA analyses of these elements. These interferences vary in importance, depending upon the particular analyte selected.

The redox potential of the extractant is dependent upon the pH of the extracting solution and the soil system. Sodium citrate complexes the reduced Fe and usually buffers the system to a pH of 6.5 to 7.3. Some soils may lower the pH, resulting in the precipitation of Fe sulfides. The SSL has not had significant problems with this interference.

Filtered extracts can yield different recoveries of Fe, Mn, and Al, relative to unfiltered extracts.

4. Safety

Wear protective clothing (coats, aprons, sleeve guards, and gloves); eye protection (face shields, goggles, or safety glasses); and a breathing filter when handling dry sodium dithionite.

Sodium dithionite may spontaneously ignite if allowed to become moist, even by atmospheric moisture. Keep dithionite in a fume hood.

Follow standard laboratory practices when handling compressed gases. Gas cylinders should be chained or bolted in an upright position. Acetylene gas is highly flammable. Avoid open flames and sparks. Standard laboratory equipment includes fire blankets and extinguishers for use when necessary. Follow the manufacturer's safety precautions when using the AA.

5. Equipment

5.1 Electronic balance, ± 1-mg sensitivity
5.2 Filter paper, pre-pleated, 185-mm diameter, Schleicher and Schuell
5.3 Atomic absorption spectrophotometer (AA), Smith-Hieftje Model 4000, Thermo Jarrell Ash Corp., Franklin, MA
5.4 Autosampler, Model 150, Thermo Jarrell Ash Corp., Franklin, MA
5.5 ThermoSpec software, Version 3.01, Enable 4.0, DOS 5.0, Thermo Jarrell Ash Corp., Franklin, MA
5.6 Computer, CUi Advantage 486, Thermo Jarrell Ash Corp., Franklin, MA
5.7 Printer, NEC Pinwriter P3200
5.8 Single-stage regulator, acetylene service, part number E11-0-N511A, Air Products and Chemicals, Inc., Box 538, Allentown, PA
5.9 Digital diluter/dispenser, MicroLab 500, Hamilton Co., P.O. Box 10030, Reno, NV
5.10 Syringes, 10000 and 1000 μL, 1001 DX and 1010-TEL LL gas tight, Hamilton Co., P.O. Box 10030, Reno, NV
5.11 Test tubes, 15-mL, 16 mm x 100, for sample dilution and sample changer, Curtin Matheson Scientific, Inc., Houston, TX
5.12 Containers, polypropylene

6. Reagents

6.1 Distilled deionized (DDI) water
6.2 Sodium dithionite ($Na_2S_2O_4$), purified powder
6.3 Sodium citrate dihydrate ($Na_3C_6H_5O_7 \cdot 2H_2O$), crystal, reagent
6.4 Hydrochloric acid (HCl), concentrated 12 N
6.5 HCl, 1:1 HCl:DDI, 6 N. Carefully mix 1 part of concentrated HCl to 1 part DDI water.
6.6 HCl, 1% wt. Carefully dilute 25 mL of concentrated HCl to 1 L with DDI water.
6.7 Superfloc 16, 0.2% solution (w:v). Dissolve 2 g of Superfloc 16 in 1 liter of DDI water. Do not shake the mixture as this breaks the polymer chains of the Superfloc. Gently swirl the mixture occasionally over the several days that the solution requires to completely dissolve the Superfloc. Suggested source is American Cyanamid Co., P.O. Box 32787, Charlotte, NC.
6.8 Primary mixed standard, 4000 mg L^{-1} (4000 ppm) Fe, 600 mg L^{-1} (600 ppm) Mn, and 3000 mg L^{-1} (3000 ppm) Al. Dissolve 4.000 g of Fe wire, 0.6000 g of Mn metal powder, and 3.000 g of Al wire with 1:1 HCl in a glass beaker. When dissolved transfer to a 1-L volumetric flask and make to volume with 1% HCl solution. Store in a polypropylene bottle.
6.9 High calibration standard, 240 mg/8 oz (1012 ppm) Fe; 36 mg/8 oz (152 ppm) Mn; and 180 mg/8 oz (759 ppm) Al. Pipet 60 mL of primary mixed standard into 8 oz bottle. Add 20 g of sodium citrate dihydrate, 1.24 mL of concentrated H_2SO_4, and 2 mL of Superfloc 16 solution. In standards, the H_2SO_4 substitutes for the dithionite. Fill to 8-oz volume with DDI water and mix thoroughly. After dissolution, transfer solution to a plastic bottle.
6.10 Low calibration standard, 120 mg/8 oz (506 ppm) Fe; 18 mg/8 oz (76 ppm) Mn; and 90 mg/8 oz (380 ppm) Al. Pipet 30 mL of primary mixed standard into 8 oz bottle. Add 20 g of sodium citrate dihydrate, 1.24 mL of concentrated H_2SO_4, and 2 mL of Superfloc 16 solution. In standards and reagent blanks, the H_2SO_4 substitutes for the dithionite. Fill to

8-oz volume with DDI water and mix thoroughly. After dissolution, transfer solution to a plastic bottle.

6.11 Calibration reagent blank solution. Add 20 g of sodium citrate dihydrate, 1.24 mL of concentrated H_2SO_4, and 2 mL of Superfloc 16 solution. In standards and reagent blanks, the H_2SO_4 substitutes for the dithionite. Fill to 8-oz volume with DDI water and mix thoroughly. After dissolution, transfer solution to a plastic bottle.

6.12 Acetylene gas, purity 99.6%

6.13 Compressed air with water and oil traps

7. Procedure

Extraction of Fe, Mn, and Al

7.1 Weigh 4.00 g of <2-mm, air-dry soil sample and place in an 8-oz nursing bottle.

7.2 Add 2 g of sodium dithionite and 20 to 25 g of sodium citrate dihydrate.

7.3 Add DDI water to 4-oz level on bottle and securely stopper bottle.

7.4 Shake overnight (12 to 16 h) in a reciprocating shaker. After shaking, use a dispenser to add 2 ml of Superfloc 16 solution.

7.5 Fill bottle to 8-oz volume with DDI water. Stopper and shake thoroughly for ~ 15 s.

7.6 Allow to settle for at least 3 day (3 to 5 days typical). The Fe, Mn, and Al are determined from a clear aliquot of solution.

Dilution of Sample Extracts and Standards

7.7 No ionization suppressant is required as the Na in the extractant is present in sufficient quantity. Set the digital diluter at 66 for diluent and 35 for CD extracts, calibration reagent blanks, and calibration standards for a 1:20 dilution as follows:

7.8 Dilute 1 part CD sample extract with 19 parts of DDI water (1:20 dilution).

7.9 Dilute 1 part calibration reagent blank with 19 parts of DDI water (1:20 dilution).

7.10 Dilute 1 part low calibration standard with 19 parts of DDI water (1:20 dilution).

7.11 Dilute 1 part high calibration standard with 19 parts of DDI water (1:20 dilution).

7.12 Dispense the reagent blanks and calibration standards in polycons from which the solutions are transferred to test tubes. Dispense the diluted sample solutions into test tubes which have been placed in the sample holders of the sample changer.

AA Calibration

7.13 Use the calibration reagent blank and high calibration standard to calibrate the AA. The AA program requires a blank and a standard, in that order, to establish a single point calibration curve for element determination. Perform one calibration, i.e., blank plus standard, for every 12 samples.

7.14 Use the low calibration standard (120 mg/8 oz Fe; 18 mg/8 oz Mn; and 90 mg/8 oz Al) as a check sample. Use high calibration standard for Fe check sample and low calibration standard for Mn and Al check sample.

AA Set-up and Operation

7.15 Refer to manufacturer's manual for operation of the AA. The following are only very general guidelines for instrument conditions for the various analytes.

Element	Wave-length (nm)	Burner head & Angle	Fuel/ Oxidant
Fe	248.5	5-cm parallel	4 C_2H_2/ 16 Air
Al	309.35	5-cm parallel	20 C_2H_2/ 10 N_2O
Mn	280.15	5-cm parallel	4 C_2H_2/ 10 Air

Typical read delay is 6 s, and integration time is 8 s.

7.16 Use the microcomputer and printer to set instrument parameters and to collect and record instrument readings.

7.17 If sample exceeds calibration standard, dilute the sample (dilution ratio in calculation) with appropriate matrix and record dilution. Remember to keep the matrix the same after dilution by diluting with DDI water (1:20 dilution).

7.18 The instrument readings are usually programmed to display analyte concentration in mg/8 oz.

8. Calculations

Fe (%) = (Fe x DR x 100 x AD/OD)/ (Sample x 1000)

Fe_2O_3 (%) = (Fe X DR x 1.43 x 100 x AD/OD)/(Sample x 1000)

Mn (%)= (Mn x DR x 100 x AD/OD)/(Sample x 1000)

Al (%) = (Al x DR x 100 x AD/OD)/(Sample x 1000)

where:

Fe	= mg/8 oz
Mn	= mg/8 oz
Al	= mg/8 oz
DR	= Dilution Ratio
Sample	= Sample weight (g)
1.43	= Conversion factor from Fe to Fe_2O_3
100	= Conversion factor to percent
AD/OD	= Air-dry/oven-dry ratio (procedure 4B5)
1000	= Conversion factor (mg g^{-1})

9. Report

Report percent CD extractable Fe, Mn, and Al on oven-dry soil basis to the nearest whole number.

10. Precision
Precision data are not available for this procedure. A quality control check sample is run with every batch of samples.

11. References
Wada, K. 1989. Allophane and imogolite. *In* J.B. Dixon and S.B. Weed (eds.). Minerals in soil environment. 2nd ed. Soil Sci. Soc. Amer. No. 1. p. 1051-1087.

Soil Survey Staff. 1975. Soil Taxonomy: A basic system of soil classification for making and interpreting soil surveys. USDA-SCS Agric. Handb. 436. U.S. Govt. Print. Office, Washington, DC.

Iron, Manganese, Aluminum, Calcium, Magnesium, Sodium, Potassium, Phosphorus, Silicon, Zirconium, Copper, Zinc, Titanium, Cadmium, Lead, Nickel, Chromium, and Cobalt
HF Plus Aqua Regia (HF + HNO₃ + HCl) Dissolution
Inductively Coupled Plasma Spectrometry
Thermo Jarrell Ash ICAP 61E Optima 3300 DV
(6C7b, 6D6a, 6G11b, 6N5b, 6O5b, 6P3b, 6Q3b, 6S6a, 6V1b, 8K1a, 8L1a, 8M1a, 8N1a, 8O1a, 8P1a, 8Q1a, and 8R1a)

1. Application
This procedure is an integral part of total analysis (7C4a) and represents the spectroscopic analysis of elements in the digestate.

2. Summary of Method
High and low calibration standards are prepared for Ca, K, Mg, Mn, Cu, Zn, Cd, Pb, Co (mixed standards CALO and CAHI); Al, Fe, Ti, Zr, Na, (mixed standards ALLO and ALHI); and Si, P, Se, As (mixed standards SILO and SIHI). A blank of HF, HNO_3, HCl, and H_3BO_3 is prepared. A Thermo Jarrell Ash ICAP 61E spectrometer is used for analysis. The concentration of Fe, Mn, Al, Ca, Mg, Na, K, P, Si, Zr, Cu, Zn, Ti, Cd, Pb, Cr, and Co are determined by ICP analysis by procedures 6C7b, 6D6a, 6G11b, 6N5b, 6O5b, 6P3b, 6Q3b, 6S6a, 6V1b, 8K1a, 8L1a, 8M1a, 8N1a, 8O1a, 8P1a, 8Q1a, and 8R1a.

3. Interferences
None.

4. Safety
Wear protective clothing and eye protection. When preparing reagents, exercise special care. Restrict the use of concentrated acids to the fume hood. Keep HF acid refrigerated and avoid contact with skin of all acids. Wash hands thoroughly after handling reagents.

5. Equipment
5.1 Volumetrics, 500-mL, polypropylene
5.2 Containers, 500-mL, polypropylene, with screw caps
5.3 Pipettors, electronic digital, Rainin Instrument Co., Woburn, MA, 2500 μL and 10 mL
5.4 Inductively coupled plasma spectrometer, ICAP-61E, Thermo Jarrell Ash Corp. , Franklin, MA.
5.5 RF generator, floor mounted power unit, 45 MHz free running, Perkin-Elmer Corp., Norwalk, CT.
5.6 Computer, AT&T 386 Starstation, Model CPU-G72, and printer, NEC Pinwriter, P2200XE, Dot Matrix
5.7 ThermoSpec software, Thermo Jarrell Ash Corp., Franklin, MA
5.8 Line conditioner, Unity/1, Model UT8K, Best Power Technology, Inc., Necedah, WI
5.9 Compressed argon gas

5.10 Autosampler, Thermo Jarrell Ash Corp., Franklin, MA.

5.11 High flow torch, Part No. 126440-01; Saffire (HF-resistant) tip, Part No. 127190-00; Polypropylene spray chamber, Part 131129-00, Thermo Jarrell Ash Corp., Franklin, MA

6. Reagents

6.1 Deionized distilled (DDI) water

6.2 Hydrofluoric acid (HF), 48%, low trace metal content

6.3 Concentrated hydrochloric acid (HCl), 12 *N*. Use instrumental grade which contains low levels of impurities.

6.4 Concentrated nitric acid (HNO$_3$), 16 *N*. Use instrumental grade which contains low levels of impurities.

6.5 Boric acid solution. Dissolve 25.0 g low trace metal, granular boric acid (H$_3$BO$_3$) in 1000 mL DDI water.

6.6 Standards, 1000 ppm, suitable for atomic absorption spectroscopy for all elements.

7. Procedure

7.1 Instrument calibration standards for analysis are limited to specific combinations of elements because of chemical incompatibilities of certain elements. Specific combinations of elements in calibration standards are based on suggestion by Thermo Jarrell Ash (TJA), Inc. Each working standard is used in two concentrations, high and low. The concentrations of elements in the low standards (CALO, ALLO, and SILO) are 50 percent of the concentrations of elements in the low standards (CALO, ALLO, and SILO). Refer to Tables 1-3 for the amounts of primary standards (1000 ppm) to make 500-mL volume of the low and high calibration standards, at the specified concentrations, for ICP analysis.

Table 1. Calilbration standards for CALO and CAHI[1]

Element	Concentration	Concentration	Primary Std. Required For	Primary Std. Required For
	CALO	CAHI	CALO	CAHI
	ppm	ppm	mL	mL
Ca	75	150	37.5	75.0
K	25	50	12.5	25.0
Mg	20	40	10.0	20.0
Mn	10	20	5.0	10.0
Cu	5	10	2.5	5.0
Zn	5	10	2.5	5.0
Cd	5	10	2.5	5.0
Pb	5	10	2.5	5.0

[1]All calibration standards based on 500-ml final volume

Table 2. Calibration standards for ALLO and ALHI[1]

Element	Concentration	Concentration	Primary Std. Required For	Primary Std. Required For
	ALLO	AlHI	ALLO	AlHI
	ppm	ppm	mL	ML
Al	100	200	50.0	100.0
Fe	75	150	37.5	75.0
Ti	5	10	2.5	5.0
Zr	5	10	2.5	5.0
Na	25	50	12.5	25.0

[1]All calibration standards based on 500-ml final volume

Table 3. Calibration standards for SILO and SIHI[1]

Element	Concentration	Concentration	Primry Std. Required For	Primary Std. Required For
	SILO	SIHI	SILO	SIHI
	ppm	Ppm	mL	Ml
Si	225	450	112.5	225.0
P	5	10	2.5	5.0
Ti	5	10	2.5	5.0
Se	5	10	2.5	5.0
As	5	10	2.5	5.0

[1]All calibration standards based on 500-ml final volume

7.2 To the calibration standards and a blank, also add the following chemicals: 25.0 mL HF; 3.75 mL HNO_3; 1.25 mL HCl; and 12.5 g granular Boric Acid. Make all standards and the blank to a final 500-mL volume with DDI water.

7.3 Use the TJA ICAP 61E spectrophotometer and analyze for the following elements: Fe, Mn, Al, Ca, Mg, Na, K, P, Si, Zr, Cu, Zn, As, Ti, Se, Cd, and Pb. No initial dilutions of samples are necessary prior to analysis. Use polypropylene spray chamber and HF resistant torch on ICP. Check instrument alignment and gas pressures to obtain optimum readings with maximum signal to noise ratio. The torch tip used for HF digestions should not be run dry or used with RF powers exceeding 1350. The HF torch tip should only be used with high flow torch.

7.4 Use the HF blank standard solution to dilute those samples with concentrations greater than the high standard. Rerun all elements and use only the data needed from the diluted analysis.

7.5 Run the detection limits using the blank standard solution. These values establish the lower detection limits for each element. Analyzed values lower than the detection limits are set equal to zero.

7.6 When ICP analyses are completed, transfer data from the hard drive storage to a 3.5 inch floppy disk as an ASCII file via the "Report Writer" in the TJA software Thermospec, Version 5.06. These data are imported into a LOTUS 123, Version 3.1 spreadsheet for data analysis. Refer to procedure 7C4a.

8. Calculations
Refer to procedure 7C4a.

9. Report
Refer to procedure 7C4a.

10. Precision
No precision data are yet available for this procedure.

11. References
Refer to digestion procedure.

Organic Carbon, Iron, Manganese, and Aluminum (6A, 6C, 6D, and 6G)
Sodium Pyrophosphate Extraction (6A4)
CO$_2$ Evolution Gravimetric (6A4a)
Sodium Pyrophosphate Extraction (6C8, 6D4, and 6G10)
Atomic Absorption
Perkin-Elmer 5000 AA (6C8a, 6D4a, and 6G10a)

1. Application
Sodium pyrophosphate (0.1 M Na$_4$P$_2$O$_7$) is used as a selective dissolution extractant for organically complexed Fe and Al (Wada, 1989). The Na$_4$P$_2$O$_7$ solution is a poor extractant for allophane, imogolite, amorphous aluminosilicates, and noncrystalline hydrous oxides of Fe and Al. The Na$_4$P$_2$O$_7$ solution does not extract opal, crystalline silicates, layer silicates, and crystalline hydrous oxides of Fe and Al (Wada, 1989). In *Soil Taxonomy*, sodium pyrophosphate extractable organic C, Fe, and Al are criteria for spodic placement (Soil Survey Staff, 1975).

2. Summary of Method
The soil sample is mixed with 0.1 M Na$_4$P$_2$O$_7$ and shaken overnight. Superfloc 16 is added, and the mixture is made to volume. The solution is allowed to settle and a clear extract is obtained. The Na$_4$P$_2$O$_7$ extracted solution is diluted with distilled deionized (DDI) water. The diluted extract is aspirated into an atomic absorption spectrophotometer (AA). The analyte is measured by absorption of the light from a hollow cathode lamp. An automatic sample changer is used to aspirate a series of samples. The AA converts absorption to analyte concentration. Percent sodium pyrophosphate extractable Fe, Mn, and Al are reported in procedures 6C8a, 6D4a, and 6G10a, respectively. The organic C in the sodium pyrophosphate extract is wet oxidized and gravimetrically measured in procedure 6A4a.

3. Interferences
There are four types of interferences (matrix, spectral, chemical, and ionization) in the AA analyses of these elements. These interferences vary in importance, depending upon the particular analyte selected.

The concentration of Na$_4$P$_2$O$_7$ solution must be close to 0.1 M. Variable amounts of Fe, Al, Mn, and organic C may be extracted by varying the pyrophosphate concentration.

4. Safety

Wear protective clothing and eye protection. When preparing reagents, exercise special care. Restrict the use of concentrated HCl to a fume hood. Many metal salts are extremely toxic and may be fatal if ingested. Thoroughly wash hands after handling these metal salts.

Follow standard laboratory procedures when handling compressed gases. Gas cylinders should be chained or bolted in an upright position. Acetylene is highly flammable. Avoid open flames and sparks. Standard laboratory equipment includes fire blankets and extinguishers for use when necessary. Follow the manufacturer's safety precautions when using the AA.

5. Equipment

5.1 Electronic balance, ±0.0001 g
5.2 Shaker, Eberbach 6000 power unit, reciprocating speed of 60 to 260 epm, with 6040 utility box carrier and 6110 floor stand, Eberbach Corp., Ann Arbor, MI
5.3 Nursing bottle, 240 mL (8 fl. oz.), graduated
5.4 Rubber stoppers, No. 2, to fit nursing bottles
5.5 Dispenser/diluters, Repipet, 0 to 10 mL, Labindustries, 1802 2nd St., Berkeley, CA
5.6 Atomic absorption spectrophotometer (AA), model 5000, Perkin-Elmer Corp., Norwalk, CT
5.7 Automatic burner control, model 5000, Perkin-Elmer Corp., Norwalk, CT
5.8 Autosampler, AS-50, Perkin-Elmer Corp., Norwalk, CT
5.9 Single-stage regulator, acetylene service, part number E11-0-N511A, Air Products and Chemicals, Inc., Box 538, Allentown, PA
5.10 Heated regulator, single-stage, nitrous oxide, stock number 808 8039, Airco Welding Products, P.O. Box 486, Union, NJ
5.11 Diluter/dispenser, Microlab 500, Catalogue No. 69052, Hamilton Co., Bonaduz, GR, Switzerland
5.12 Syringes, 10000 and 1000 μL, 1001 DX and 1010-TEL LL gas tight, Hamilton Co., P.O. Box 10030, Reno, NV, 89510
5.13 Test tubes, 15-mL, 16 mm x 100, for sample dilution and sample changer, Curtin Matheson Scientific, Inc., Houston, TX
5.14 Absorption bulb, Nesbitt with stopper
5.15 Absorption bulb, Stetser-Norton
5.16 Flask, boiling, round bottom, short neck
5.17 Condenser, Allihn
5.18 Funnel, separatory, cylindrical, open top, with stopcock
5.19 Tube, drying, Schwartz
5.20 Containers, polypropylene
5.21 Dot matrix printer, P-132, Interdigital Data Systems, Inc.

6. Reagents

6.1 Distilled deionized (DDI) water
6.2 Hydrochloric acid (HCl), concentrated, 12 N
6.3 HCl, 1:1 HCl:DDI, 6 N. Carefully mix 1 part of concentrated HCl to 1 part DDI H_2O.
6.4 HCl, 1% wt. Carefully dilute 25 mL of concentrated HCl to 1 L with DDI H_2O.
6.5 Superfloc 16, 0.2% solution (w:v). Dissolve 2 g of Superfloc 16 in 1 liter of DDI H_2O. Do not shake the mixture as this breaks the polymer chains of the Superfloc. Gently swirl the mixture occasionally over the several days that the solution requires to completely dissolve the Superfloc. Suggested source is American Cyanamid Co., P.O. Box 32787, Charlotte, NC
6.6 Sodium pyrophosphate solution, 0.1 M. Dissolve 800 g of $Na_4P_2O_7 \cdot H_2O$ in 16 L of DDI H_2O. Dilute to 18 L with DDI H_2O.
6.7 Primary mixed standard, 4000 mg L^{-1} (4000 ppm) Fe; 2000 mg L^{-1} (3000 ppm) Al; and 600 mg L^{-1} (600 ppm) Mn. Dissolve 4.000 g of Fe wire, 3.000 g of Al wire, and 0.600 g

of Mn metal powder in 1:1 HCl:DDI in a glass beaker. When dissolved transfer to a 1-L volumetric flask and fill with 1% HCl solution. Store in a polypropylene bottle.

6.8 High calibration mixed standards solution (HCMSS), 80 mg/8 oz (80 ppm) Fe; 12 mg/8 oz (12 ppm) Mn; and 60 mg/8 oz (60 ppm) Al. Pipet 20 mL of primary mixed standard into an 8-oz bottle. Add 10.55 g of $Na_4P_2O_7 \cdot H_2O$ and 3.5 mL of concentrated H_3PO_4. Dilute to 8 oz with DDI H_2O. Store in a polypropylene bottle.

6.9 Low calibration mixed standards solution (LCMSS), 40 mg/8 oz (40 ppm) Fe; 6 mg/8 oz (6 ppm) Mn; and 30 mg/8 oz (30 ppm) Al. Pipet 10 mL of primary mixed standard into an 8-oz bottle. Add 10.55 g of $Na_4P_2O_7 \cdot H_2O$ and 3.5 mL of concentrated H_3PO_4. Dilute to 8 oz with DDI H_2O. Store in a polypropylene bottle.

6.10 Calibration reagent blank solution (CRBS). Add 10.55 g of $Na_4P_2O_7 \cdot H_2O$ and 3.5 mL of concentrated H_3PO_4. Dilute to 8 oz with DDI H_2O. Store in a polypropylene bottle.

6.11 Potassium dichromate ($K_2Cr_2O_7$), reagent

6.12 Potassium iodide solution. Dissolve 100 g of KI in 100 mL of DDI H_2O.

6.13 Silver sulfate, saturated aqueous solution.

6.14 Digestion acid mixture. Mix 600 mL of concentrated H_2SO_4 and 400 mL of 85% H_3PO_4.

6.15 Indicarb or Mikohlbite

6.16 Soda lime

6.17 Zinc granules, 300 mesh

6.18 Anhydrone

6.19 Acetylene gas, purity 99.6%

6.20 Nitrous oxide gas, compressed

6.21 Compressed air with water and oil traps

7. Procedure

Extraction of Fe, Mn, and Al

7.1 Weigh 2.00 g of <2-mm, air-dry soil sample and place in an 8-oz nursing bottle.

7.2 Add 0.1 M $Na_4P_2O_7$ solution to 7-oz level on bottle and securely stopper bottle.

7.3 Shake overnight (12 to 16 h) in a reciprocating shaker. After shaking, use a dispenser to add 4 mL of Superfloc 16 solution.

7.4 Fill bottle to 8-oz volume with $Na_4P_2O_7$ solution.

7.5 Stopper and shake vigorously for $\approx$ 15 s.

7.6 Allow to settle for at least 3 days (4 to 6 days typical). The Fe, Mn, and Al are determined from a clear aliquot of solution.

Dilution of Sample Extracts and Standards

7.7 No ionization suppressant is required as the Na in the extractant is present in sufficient quantity. Set the digital diluter on Hamilton diluter to 66 for diluent and 35 for sodium pyrophosphate sample extracts, calibration reagent blanks, and calibration standards for a 1:20 dilution as follows:

7.8 Dilute 1 part sample sodium pyrophosphate sample extract with 19 parts of DDI H_2O (1:20 dilution).

7.9 Dilute 1 part CRBS with 19 parts of DDI H_2O (1:20 dilution).

7.10 Dilute 1 part LCMSS with 19 parts of DDI H_2O (1:20 dilution).

7.11 Dilute 1 part HCMSS with 19 parts of DDI H_2O (1:20 dilution).

7.12 Dispense the diluted solutions into test tubes which have been placed in the sample holder of the sample changer.

AA Calibration

7.13 Use the calibration reagent blank and high calibration standard to calibrate the AA. The AA requires a blank and a standard, in that order, to establish a single point calibration curve for element determination. Perform one calibration, i.e., blank plus standard, for every 12 samples.

7.14 Use the low calibration standard as a check sample.

AA Set-up and Operation

7.15 Refer to the manufacturer's manual for operation of the Perkin-Elmer 5000 AA. The following are only very general guidelines for instrument conditions for the various analytes.

Analyte	Wavelength nm
Fe	248.8
Mn	280.1
Al	309.3

Analyte	Burner Head
Fe	5-cm parallel
Mn	5-cm parallel
Al	5-cm parallel

Analyte	Fuel/Oxidant
Fe	10 C_2H_2/25 Air
Mn	10 C_2H_2/25 Air
Al	30 C_2H_2/17 Air

Typical read delay is 6 s, and the integration by peak area is 8 s.

7.16 Use the microcomputer and printer to set instrument parameters and to collect and record instrument readings.

7.17 If sample exceeds calibration standard, dilute the sample (dilution ratio in calculation) with appropriate matrix and record dilution. Remember to keep matrix the same after dilution by diluting with DDI H_2O (1:20 dilution).

Organic C Determination

7.18 Pipet 100 mL of the extract into a 100-ml flask.

7.19 Evaporate the extract to near dryness using a 50°C waterbath and a gentle stream of clean, filtered air.

7.20 Construct the wet combustion apparatus. Refer to Fig. 3 of the apparatus for gravimetric organic C determination.

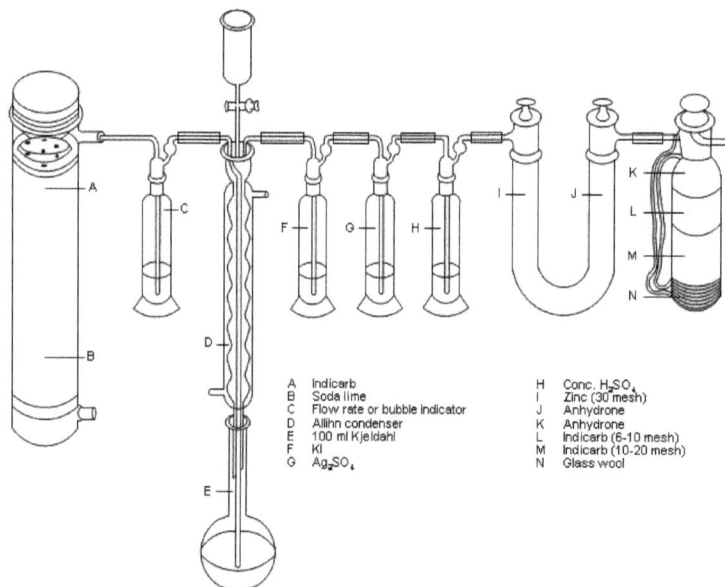

Figure 3. Apparatus for gravimetric organic carbon determinations of 0.1 M sodium pyrophosphate extracts.

A	Indicarb
B	Soda lime
C	Flow rate or bubble indicator
D	Allihn condenser
E	100 ml Kjeldahl
F	KI
G	Ag_2SO_4
H	Conc. H_2SO_4
I	Zinc (30 mesh)
J	Anhydrone
K	Anhydrone
L	Indicarb (6-10 mesh)
M	Indicarb (10-20 mesh)
N	Glass wool

7.21 Add 1 to 2 g of potassium dichromate.

7.22 Wash the neck of the flask with 3 mL of DDI H_2O and connect to condenser.

7.23 Attach a weighed Nesbitt bulb to the system and open the valve at the top.

7.24 Pour 25 mL of digestion-acid mixture into the funnel. Add the mixture to the flask and immediately close the stopcock. Use the digestion-acid mixture to lubricate the stopcock.

7.25 The tip of the air-delivery tube should be ≈ 0.5 cm below the digestion-acid mixture. Adjust the flow of the "carrier stream" to maintain 1 to 2 bubble s^{-1} rate throughout the digestion. Apply suction on the outlet side of the Nesbitt bulb. Gentle air pressure and needle valve on the air pressure line aids flow adjustment.

7.26 With a gas flame or a variable power heating mantle, gently heat the flask until the mixture boils (≈ 3 to 4 min). Continue a gentle boiling for 10 min. Heating is too rapid if white fumes of SO_2 are visible above the second bulb of the reflux condenser.

7.27 Remove the heat and allow to aerate for 10 additional min at a rate of 6 to 8 bubbles s^{-1}.

7.28 Close the stopcock on the Nesbitt bulb, disconnect the bulb from the system, and weigh to the nearest 0.0001 g.

8. Calculations

Analyte (%) = (AA x DR x AD/OD x 100)/(Sample Weight (g) x 1000)

where:

Analyte	= Fe, Mn, and Al	
AA	= Analyte concentration AA reading	
DR	= Dilution ratio of 1 if no additional dilution	
Sample	= Sample weight (g)	
100	= Factor to convert to percent	
1000	= Conversion factor (mg g^{-1})	
AD/OD	= Air-dry/oven-dry ratio	

Organic C (%) = [(Wt$_F$ - Wt$_I$) x 27.3 x Volume x AD/OD]/(Sample Weight (g) x 236.6)

where:

Wt$_F$	= Nesbitt bulb weight after digestion (g)
Wt$_I$	= Nesbitt bulb weight before digestion (g)
Volume	= Volume of extract digested (mL)
AD/OD	= Air-dry/oven-dry ratio (procedure 4B5)
27.3	= Conversion factor
236.6	= Total volume of extract (mL)

9. Report

Report percent sodium pyrophosphate extractable Fe, Mn, and Al on oven-dry soil basis to the nearest whole number.

10. Precision

Precision data are not available for this procedure.

11. References

Soil Survey Staff. 1975. Soil Taxonomy: A basic system of soil classification for making and interpreting soil surveys. USDA-SCS Agric. Handb. 436. U.S.Govt. Print. Office, Washington, DC.

Wada, K. 1989. Allophane and imogolite. *In* J.B. Dixon and S.B. Weed (eds.). Minerals in soil environment. 2nd ed. Soil Sci. Soc. Amer. No. 1. p. 1051-1087.

Iron, Manganese, Aluminum, Silicon, and Phosphate (6C, 6D, 6G,6V, 6S)
Ammonium Oxalate Extraction (6C9, 6D6, 6G12, 6V2, 6S8)
Inductively Coupled Plasma Spectrometry
Thermo Jarrell Ash, ICAP 61E
(6C9b, 6D5b, 6G12b, 6V2b, 6S8a)
Optical Density (8J)
(of ammonium oxalate extract)

1. Application

Oxalic acid-ammonium oxalate (acid oxalate) is used as a selective dissolution extractant for organically complexed Fe and Al, noncrystalline hydrous oxides of Fe and Al, allophane, and amorphous aluminosilicates (Wada, 1989). Acid oxalate is a poor extractant of imogolite and layer silicates and does not extract crystalline hydrous oxides of Fe and Al, opal, or crystalline silicate (Wada, 1989). A more reliable and accurate estimation of soil properties and a better understanding of soil exchange complex is provided when acid oxalate extraction is used in conjunction with other selective dissolution procedures, thermal techniques, and chemical tests. In *Soil Taxonomy*, acid oxalate extractable Fe and Al are criteria for andic soil properties (Soil Survey Staff, 1990).

2. Summary of Method

A soil sample is extracted with a mechanical vacuum extractor (Holmgren et al., 1977) in a 0.2 M acid oxalate solution buffered at pH 3.0 under darkness. The acid oxalate extract is weighed. The acid oxalate extract is diluted with 0.002 M DDBSA. The analytes are measured by a inductively coupled plasma emission spectrophotometer (ICP). Data are automatically recorded by a microcomputer and printer. The percent acid oxalate extractable Fe, Mn, Al, Si, and P are reported in procedures 6C9b, 6D5b, 6G12b, 6V2b, and 6S8a respectively. In procedure 8J, the optical density of the extract is measured with a UV spectrophotometer at 430 nm.

3. Interferences

There are four types of interferences (matrix, spectral, chemical, and ionization) in the ICP analyses of these elements. These interferences vary in importance, depending upon the particular analyte chosen.

The acid oxalate buffer extraction is sensitive to light, especially UV light. The exclusion of light reduces the dissolution effect of crystalline oxides and clay minerals. If the sample contains large amounts of amorphous material (>2% Al), an alternate method should be used, i.e., shaking with 0.275 M acid oxalate, pH 3.25, 1:100 soil:extractant.

4. Safety

Wear protective clothing and eye protection. When preparing reagents, exercise special care. Follow standard laboratory practices when handling compressed gases. Gas cylinders should be chained or bolted in an upright position. Follow the manufacturer's safety precautions when using the UV spectrophotometer and ICP.

5. Equipment

5.1 Electronic balance, ±1-mg sensitivity
5.2 Mechanical vacuum extractor, 24 place, Centurion International, Inc., Lincoln, NE
5.3 Mechanical vacuum extractor, Mavco Sampletex, 5300 N. 57th St., Lincoln, NE
5.4 Syringes, polypropylene, disposable, 60 mL, for extractant reservoir, extraction vessel, and tared extraction syringe
5.5 Rubber tubing, 3.2 ID x 6.4 OD x 25.4 mm, (1/8 ID x 1/4 OD x 1 in) for connecting syringe barrels
5.6 Pre-pulped tubes
5.7 Extraction vessel, 60-mL, 10u polypropylene, part no. 6986-6010, Whatman Inc., 9 Bridewell Place, Clifton, NJ
5.8 Disposable glass tubes
5.9 UV-visible spectrophotometer, Carey-50, Varian Instruments
5.10 Cuvettes, disposable, polystrene, 1-cm light path
5.11 Inductively coupled plasma spectrometer, ICAP 61E, Thermo Jarrell Ash Corp., Franklin, MA
5.12 Nebulizers, High-Solids, 41 psgi, Thermo Jarrell Ash Corp., Franklin, MA.
5.13 RF generator, floor mounted power unit, Model 7/90, Thermo Jarrell Ash Corp., Franklin, MA
5.14 Computer, AT&T 386 Starstation, Model CPU-G72, and printer, NEC Pinwriter, P2200XE, Dot Matrix
5.15 ThermoSpec software, Thermo Jarrell Ash Corp., Franklin, MA
5.16 Line conditioner, Unity/I, Model UT8K, Best Power Technology, Inc., Necedah, WI
5.17 Single-stage regulator, high-purity, high-flow, argon, product no. E11-X-N145DHF, Air Products and Chemicals, Inc., Box 538, Allentown, PA
5.18 Autosampler, Thermo Jarrell Ash Corp., Franklin, MA
5.19 Digital diluter/dispenser, MicroLab 500, Hamilton Co., P.O. Box 10030, Reno, NV
5.20 Syringes, 10000 and 1000 μL, 1001 DX and 1010-TEL LL gastight, Hamilton Co., P.O. Box 10030, Reno, NV

5.21 Test tubes, 15-mL, 16 mm x 100, for sample dilution and sample changer, Curtin Matheson Scientific, Inc., Houston, TX

5.22 Containers, polyproylene

6. Reagents

6.1 Distilled deionized (DDI) water

6.2 Acid oxalate buffer solution, 0.2 M, pH 3.0. *Solution A* (base): Dissolve 284 g of $(NH_4)_2C_2O_4 \cdot 2H_2O$ in 10 L of DDI water. *Solution B* (acid): Dissolve 252 g of $H_2C_2O_4 \cdot H_2O$ in 10 L of DDI water. Mix 4 parts solution A with 3 parts solution B. Adjust acid oxalate solution pH by adding either acid or base solution. Store in a polypropylene bottle.

6.3 pH buffers, pH 4.00 and 7.00, for electrode calibration

6.4 Primary Fe standard, 1000 ppm. Certified Reference Solution, Fisher Chemical Scientific Co., Fairlawn, N.J.

6.5 Primary Al standard, 1000 ppm. Certified Reference Solution, Fisher Chemical Scientific Co., Fairlawn, N.J.

6.6 Primary Si standard, 1000 ppm. Certified Reference Solution, Fisher Chemical Scientific Co., Fairlawn, N.J.

6.7 Primary Mn standard, 1000 ppm. Certified Reference Solution, Fisher Chemical Scientific Co., Fairlawn, N.J.

6.8 Primary P standard, 1000 ppm. Certified Reference Solution, Fisher Chemical Scientific Co., Fairlwan, N.J.

6.9 Dodecylbenzenesulfonic acid (DDBSA), Tech 97%, 0.1 M. Dissolve 32.2 g DDBSA in 1-L DDI water.

6.10 DDBSA solution. Dilution solution for acid oxalate extracts. Add 40.0 mL of 0.1 M DDBSA and make to 2-L volume with DDI water (0.002 M DDBSA solution).

6.11 High calibration standard. Mix 60 mL of each primary standard (Si, Fe, and Al) with 10 mL of primary Mn standard and 20 mL of primary P standard in 1 L volumetric flask. Add 50 mL of 0.4 M acid oxalate solution and 16.0 mL of 0.1 M DDBSA and make to 1-L volume with DDI water. The elements are added in the order (Si, Fe, Al, Mn, P) to avoid element precipitation. Resulting solution contains 60 ppm each of Si, Fe, and Al, 10 ppm Mn, and 20 ppm P. Store in a polypropylene bottle.

6.12 Medium calibration standard. Mix 30 mL of each primary standard (Si, Fe, and Al) with 5 mL of primary Mn standard and 10 mL of primary P standard in 1 L volumetric flask. Add 50 mL of 0.4 M acid oxalate solution and 16.0 mL of 0.1 M DDBSA and make to 1-L volume with DDI water. Resulting solution contains 30 ppm each of Si, Fe, and Al, 5 ppm Mn, and 10 ppm P. Store in a polypropylene bottle.

6.13 Low calibration standard. Mix 10 mL of each primary standard (Si, Fe, and Al) with 2 mL of primary Mn standard, and 3 mL primary P standard in 1 L volumetric flask. Add 30 mL of 0.4 M acid oxalate solution and 16.0 mL of 0.1 M DDBSA and make to 1-L volume with DDI water. Resulting solution contains 10 ppm each of Si, Fe, and Al, 2 ppm Mn, and 3 ppm P. Store in a polypropylene bottle.

6.14 Calibration reagent blank solution. Add 50 mL of 0.4 M acid oxalate solution and 16.0 mL of 0.1 M DDBSA and make to 1-L volume with DDI water.

6.15 Argon gas, purity 99.9%

7. Procedure

Extraction of Fe, Mn, Al, Si, and P

7.1 Prepare disposable sample tubes.

7.2 Weigh 0.500 g of <2-mm, air-dry soil and place in sample tube. Prepare two reagent blanks (no sample in tube) per set of 48 samples.

7.3 Place the sample tube on the upper disk of the extractor and connect a tared extraction syringe. Use 25.4-mm (1-in) length rubber tubing to insert the handle of the plunger in the slot of the stationary extractor disk.

7.4 Use a dispenser to add 15.00 mL of acid oxalate buffer to the sample tube. Make sure that the sample is thoroughly wetted. During the addition, wash sides of the tube and wet the sample. Shaking, swirling, or stirring may be required to wet organic samples. Allow sample to stand for at least 30 min.

7.5 Set extractor for 30-min extraction rate and extract until the acid oxalate buffer solution is at a 0.5 to 1.0-cm height above sample. Turn off extractor.

7.6 Put reservoir tube on top of the sample tube.

7.7 Add 35 mL of acid oxalate buffer to the reservoir tube.

7.8 Cover the extractor with a black plastic bag to exclude light. Adjust the extraction rate for a 12-h extraction.

7.9 After the extraction, shut off the extractor and pull plunger of syringe down. Do not remove the plunger from syringe barrel. Carefully remove the syringe with extract leaving the rubber tubing on the extraction vessel.

7.10 Weigh each syringe containing acid oxalate extract to the nearest 0.01 g.

7.11 Mix extract in each syringe by manually shaking. Fill a disposable tube with extract solution. This solution is reserved for determinations of Fe, Mn, Al, Si, and P. If optical density is to be measured, fill a disposable cuvette with extract solution. Discard excess solution.

Determination of Optical Density of Extract

7.12 Place 4 mL of acid oxalate extract in disposable cuvette.

7.13 Place 4 mL of acid oxalate reagent blank in disposable cuvette.

7.14 On Varian spectrophotometer, select a 430-nm wavelength. Select normal slit width and height. Refer to the manufacturer's manual for operation of the spectrophotometer.

7.15 Use the acid oxalate extract reagent blank to set spectrophotometer.

7.16 Record optical density of acid oxalate extract to nearest 0.000.

Dilution of Sample Extracts and Standards

7.17 Dilute acid oxalate extracts (1:10) with 0.002 M DDBSA solution. Add 1 part acid oxalate sample extract with 10 parts dilution solution.

7.18 Set the digital settings of the Hamilton diluter for a 1:10 dilution. Calibration reagent blanks and calibration standards are not diluted.

7.19 Dispense the diluted solutions into test tubes which have been placed in the sample holder of the sample changer.

ICP Calibration

7.20 Use a multipoint calibration for ICP analysis of acid oxalate extracts. The ICP calibrates the blank first, low standard, medium standard, followed by the high standard. Prepare a quality control (QC) standard with analyte concentration between the high and low calibration standards. The ICP reads the QC after the high standard. If the QC falls within the range set by operator, the instrument proceeds to analyze the unknowns. If the QC is outside the range, the instrument restandardizes. The QC is analyzed approximately every 12 samples.

ICP Set-up and Operation

7.21 Refer to the manufacturer's manual for operation of the ICP. The following parameters are only very general guidelines for instrument conditions for the various analytes.

Parameter	Value
Gas Flow	
Torch Gas	High Flow
Auxiliary Gas Flow	Medium 1.0 LPM
Nebulizer Pressure	41 psi
Power	
Approximate RF Power	1150
Peristaltic Pump	
Analysis Pump Rate	150 RPM
Flush Pump Rate	200 RPM
Relaxation Time	10 s
Pumping Tube Type	Silicone-Orange-3stop
Argon Flow Rate	2.0 LPM
Purged Optical Pathway Enclosure Purge	2.0 SLPM Air

Nebulizer pressure depends on the type of nebulizer that is being used, i.e., low flow nebulizer requires a higher nebulizer pressure whereas a higher flow nebulizer requires a lower nebulizer pressure. To check for correct nebulizer pressure, aspirate with 1000.0 ppm yttrium. Adjust pressure to correct yttrium bullet.

7.22 Analyte data are reported at the following wavelengths.

Analyte	Wavelength (nm)
Fe	259.940
Al	167.081
Si	251.611
Mn	257.610
P	178.28

7.23 Use the microcomputer and printer to set instrument parameters and to collect and record instrument readings. The instrument readings are usually programmed in ppm.

7.24 If sample exceeds calibration standard, dilute the sample (1:5) with 0,4 M acid oxalate solution (matrix) and then dilute (1:10) with the DDBSA solution.

8a. Calculations (Fe, Al, Si,)

Analyte (%) = [ICP x (Syr$_{fin}$ -Syr$_{init}$) x D.R. x AD/OD]/[Sample Weight (g) x 10000 x Density]

where:
ICP   = ICP analyte concentration (ppm)
Syr$_{fin}$ = Weight of syringe + extract (g)
Syr$_{init}$ = Tare weight of syringe (g)
D.R.   = Dilution ratio of samples over calibration range
Density   = Density of acid oxalate solution (1.007)
AD/OD   = Air-dry/oven-dry ratio (procedure 4B5)

8b. Calculations (Mn, P)

Analyte (ppm) = [ICP x (Syr$_{fin}$ -Syr$_{init}$) x D.R. x AD/OD]/[Sample Weight (g) x Density]

where:
ICP   = ICP analyte concentration (ppm)
Syr$_{fin}$ = Weight of syringe + extract (g)
Syr$_{init}$ = Tare weight of syringe (g)
D.R.   = Dilution ratio of samples over calibration range
Density   = Density of acid oxalate solution (1.007)
AD/OD   = Air-dry/oven-dry ratio (procedure 4B5)

9. Report
Report the percent acid oxalate extractable Fe, Al, and Si to the nearest 0.01%. Report the concentration of acid oxalate extractable Mn and P in ppm. Report the optical density of the acid oxalate extract to the nearest 0.001 unit.

10. Precision
Precision data are not available for this procedure.

11. References
Holmgren, G.G.S., R.L. Juve, and R.C. Geschwender. 1977. A mechanically controlled variable rate leaching device. Soil Sci. Amer. J. 41:1207-1208.

Soil Survey Staff. 1999. Soil Taxonomy. A Basic System of Soil Classification for Making and Interpreting Soil Surveys. 2nd ed. USDA-NRCS. Govt. Print.Office, Washington DC.

Wada, K. 1989. Allophane and imogolite. *In* J.B. Dixon and S.B. Weed (eds.). Minerals in soil environment. 2nd ed. Soil Sci. Soc. Amer. No. 1. p

Manganese and Aluminum (6D and 6G)
KCl, Automatic Extractor (6D3 and 6G9)
Inductively Coupled Plasma Spectrometry, Thermo Jarrell Ash, ICAP 61E (6D3b and 6G9c)

1. Application
The Al extracted by 1 N KCl approximates exchangeable Al and is a measure of the "active" acidity present in soils with a 1:1 water pH <5.5. Above pH 5.5, precipitation of Al occurs during analysis. This method does not measure the acidity component of hydronium ions (H$_3$O$^+$). If Al is present in measurable amounts, the hydronium is a minor component of the active acidity. Because the 1 N KCl extractant is an unbuffered salt and usually affects the soil pH one unit or less, the extraction is determined at or near the soil pH. The KCl

extractable Al is related to the immediate lime requirement and existing CEC of the soil. The "potential" acidity is better measured by the $BaCl_2$-TEA method (procedure 6H5a) (Thomas, 1982).

2. Summary of Method

A soil sample is leached with 1 N KCl using the mechanical vacuum extractor (Holmgren et al., 1977). The extract is weighed. The KCl extracted solution is diluted with distilled deionized water. The anlaytes are measured by inductively coupled plasma emission spectrophotometer (ICP). Data are automatically recorded by a microcomputer and printer. The Mn and Al are reported in mg kg^{-1} (ppm) and meq 100 g^{-1} ovendry soil in procedures 6D3b and 6G9c, respectively.

3. Interferences

There are four types of interferences (matrix, spectral, chemical, and ionization) in the ICP analyses of these cations. These interferences vary in importance, depending upon the particular analyte selected.

The soil:extractant ratio must remain constant. A soil:extractant ratio of 1:10 (w:v) for batch procedures is most commonly used. Using a leaching technique, a 1:20 (w:v) ratio gives comparable results. If the sample size is changed, the amount of extractable Al is changed. No other significant interferences have been identified for this procedure.

4. Safety

Wear protective clothing and eye protection. When preparing reagents, exercise special care. Follow standard laboratory practices when handling compressed gases. Gas cylinders should be chained or bolted in an upright position. Follow the manufacturer's safety precautions when using the ICP.

5. Equipment

5.1 Electronic balance, ±1-mg sensitivity
5.2 Mechanical vacuum extractor, 24 place, Centurion International, Inc., Lincoln, NE
5.3 Mechanical vacuum extractor, Mavco Sampletex, 5300 N. 57th St., Lincoln, NE
5.4 Syringes, polypropylene, disposable, 60 mL, for extraction vessel, extractant reservoir and tared extraction syringe
5.5 Rubber tubing, 3.2 ID x 6.4 OD x 25.4 mm, (1/8 ID x 1/4 OD x 1 in) for connecting syringe barrels
5.6 Analytical filter pulp, ash-free, Schleicher and Schuell, No. 289
5.7 Plunger, modified. Remove rubber and cut plastic protrusion from plunger end.
5.8 Wash bottle, 20 mL, to dispense KCl.
5.9 Polycons, Richards Mfg. Co.
5.10 Inductively coupled plasma spectrometer, ICAP 61E, Thermo Jarrell Ash Corp., Franklin, MA
5.11 RF generator, floor mounted power unit, Model 7/90, Thermo Jarrell Ash Corp., Franklin, MA
5.12 Computer, AT&T 386 Starstation, Model CPU-G72, and printer, NEC Pinwriter, P2200XE, Dot Matrix
5.13 ThermoSpec software, Thermo Jarrell Ash Corp., Franklin, MA
5.14 Line conditioner, Unity/I, Model UT8K, Best Power Technology, Inc., Necedah, WI
5.15 Single-stage regulator, high-purity, high-flow, argon, product no. E11-X-N145DHF, Air Products and Chemicals, Inc., Box 538, Allentown, PA
5.16 Autosampler, Thermo Jarrell Ash Corp., Franklin, MA
5.17 Nebulizers, Precision Glass, Type A, 2.3 mlpm, 35 psig, Precision Glass Co., 14775 Hinsdale Ave., Englewood, CO.
5.18 Digital diluter/dispenser, MicroLab 500, Hamilton Co., P.O. Box 10030, Reno, NV \

5.19 Syringes, 10000 and 1000 µL, 1001 DX and 1010-TEL LL gastight, Hamilton Co., P.O. Box 10030, Reno, NV

5.20 Test tubes, 15-mL, 16 mm x 100, for sample dilution and sample changer, Curtin Matheson Scientific, Inc., Houston, TX

5.21 Containers, polyproylene

6. Reagents

6.1 Distilled deionized (DDI) water

6.2 Potassium chloride solution (KCl), 1.0 *N*. Dissolve 1342 g of KCl reagent in 16 L DDI water. Allow solution to equilibrate to room temperature. Dilute to 18 L with DDI water. Use 1.0 *N* KCl for Al and Mn extraction.

6.3 Potassium chloride solution (KCl), 2.0 *N*. Dissolve 298.24 g of KCl reagent in 1.5 L DDI water. Allow solution to equilibrate to room temperature. Dilute to 2 L with DDI water. Use 2.0 *N* KCl for standards.

6.4 Primary Al standard, 1000 ppm (111 meq L^{-1}). Certified Reference Solution, Fisher Chemical Scientific Co., Fairlawn, N.J.

6.5 Primary Mn standard, 1000 ppm (250 meq L^{-1}). Certified Reference Solution, Fisher Chemical Scientific Co., Fairlawn, N.J.

6.6 High calibration Al and Mn standard, 10 meq L^{-1} and 8 ppm, respectively. Pipet 22.476 mL of primary Al standard into a 250-mL volumetric flask. Add 125.0 mL 2 *N* KCl solution to the flask and mix. Pipet 2.0 mL of primary Mn standard into flask and mix. Dilute to volume with DDI water.

6.7 Calibration reagent blank solution, 1.0 *N* KCl. Add 125 mL of 2.0 *N* KCl to a volumetric flask and make to 250-mL volume with DDI water. Store in polypropylene container.

6.8 Calibration Al and Mn check standard, 5 meq L^{-1} and 3.0 ppm, respectively. Pipet 11.24 mL of primary Al standard into a 250-mL volumetric flask. Add 125.0 mL 2 *N* KCl solution to the flask and mix. Pipet 0.75 mL of primary Mn standard into flask and mix. Dilute to volume with DDI water.

6.9 Dodecylbenzenesulfonic acid (DDBSA), tech 97%., 0.1 *M*. Dissolve 32.2 g DDBSA in 1-L DDI water.

6.10 DDBSA rinse solution. Dilute 40.0 mL 0.1 *M* DDBSA to 2-L volume with DDI water.

6.11 Argon gas, purity 99.9%

7. Procedure

Extraction of Al and Mn

7.1 Prepare extraction vessel by tightly compressing a 1-g ball of filter pulp into the bottom of a syringe barrel with a modified plunger.

7.2 Weigh exactly 2.50 g of <2-mm, air-dry soil and place in an extraction vessel. Prepare one quality control check sample per 48 samples.

7.3 Place the extraction vessel on the upper disk of the extractor and connect a tared extraction syringe. Use 25.4-mm (1-in) length rubber tubing to insert the handle of the plunger in the slot of the stationary extractor disk.

7.4 Use a squeeze bottle and fill extraction vessel to the 20-mL mark with 1.0 *N* KCl solution (≈ 10 mL). Make sure that the sample is thoroughly wetted. During the addition, wash sides of the tube and wet the sample. Shaking, swirling, or stirring may be required to wet organic samples. Allow sample to stand for at least 30 min.

7.5 Put reservoir tube on top of the extraction vessel. Set extractor for fast extraction rate and extract until the KCl solution is at a 0.5- to 1.0-cm height above sample. Turn off extractor.

7.6 Add 45 mL KCl solution to reservoir tube. Set extractor for 45-min extraction.

7.7 After the extraction, shut off extractor and pull plunger of syringe down. Do not remove the plunger from syringe barrel. Carefully remove the syringe with extract leaving the rubber tubing on the extraction vessel.

7.8 Weigh each syringe containing KCl extract to the nearest 0.01 g.

7.9 Mix extract in each syringe by manually shaking. Fill a polycon with extract solution and discard the excess. This solution is reserved for extractable Al and Mn analyses.

Dilution of Extracts and Standards

7.10 Set the digital settings at a 1:5 dilution for the KCl sample extracts, calibration reagent blanks, calibration standards, and calibration check standards as follows:

7.11 Dilute 1 part KCl sample extract with 4 parts of DDI water (1:5 dilution).

7.12 Dilute 1 part calibration reagent blank with 4 parts of DDI water (1:5 dilution).

7.13 Dilute 1 part calibration standard with 4 parts of DDI water (1:5 dilution).

7.14 Dilute 1 part calibration check standard with 4 parts of DDI water (1:5 dilution).

7.15 Dispense the diluted solutions into test tubes and place in the sample holder of the sample changer.

ICP Calibration

7.16 Use the calibration reagent blank (1.0 N KCl), high standard (10 meq L^{-1} Al and 8 ppm Mn), and the blank to calibrate the ICP.

7.17 Use the calibration check standard (5 meq L^{-1} Al and 3 ppm Mn) as a check sample. Perform a calibration check every 12 samples.

ICP Set-up and Operation

7.18 Refer to the manufacturer's manual for operation of the ICP. The following parameters are only very general guidelines for instrument conditions for the analytes.

Parameter	Value
Gas Flow	
Torch Gas	High Flow
Auxiliary Gas Flow	Medium 1.0 LPM
Nebulizer Pressure	32 PSI
RF Power	1150

Analyte	Wavelength (nm)	High/Low Offset	Peak Offset
Al	167.081	0/-21	-2
Mn	257.610	0/-18	0

7.19 Determine a set of 24 unknown samples for each successful calibration check.

7.20 Use the microcomputer and printer to set instrument parameters and to collect and record instrument readings.

7.21 If a sample exceeds the calibration standard, dilute the sample (1:5) as follows.

7.22 Analyze 4 quality control check sample for every 48 samples.

8. Calculations

8.1 The instrument readings are the analyte concentration (meq L^{-1} Al and ppm Mn) in undiluted extract. Use these values to calculate the analyte concentration on an oven-dry soil basis (meq 100 g^{-1}).

Analyte (meq 100 g^{-1}) = [ICP x ($Wt_{syr+ext}$ - Wt_{syr}) x D.R. x 100 x AD/OD]/[Sample Weight x 1.0412 x 1000]

where:
ICP   = ICP analyte reading
$Wt_{syr+ext}$ = Weight of extraction syringe & extract (g)
Wt_{syr} = Weight of tared extraction syringe (g)
D.R.   = Dilution ratio of samples over calibration range
1.0412   = Density of 1 N KCl @ 20°C
1000   = g L^{-1}
100   = Conversion factor (100-g basis)
AD/OD   = Air-dry/oven-dry ratio (procedure 4B5)

9. Report

Report KCl extractable Al and Mn in units of meq 100 g^{-1} of oven-dry soil to the nearest 0.01 meq 100 g^{-1}.

10. Precision

Precision data are not available for this procedure.

11. References

Holmgren, George G.S., R.L. Juve, and R.C. Geschwender. 1977. A mechanically controlled variable rate leaching device. Soil Sci. Amer. J. 41:1207-1208.

Thomas, G.W. 1982. Exchangeable cations. *In* A. Klute (ed.) Methods of soil analysis. Part 2. Chemical and microbiological properties. 2nd ed. Agronomy 9:159-165.

Manganese and Aluminum (6G)
KCl, Automatic Extractor (6G9)
Atomic Absorption
Thermo Jarrell Ash, Smith Hieftje 4000
(6D3c and 6G9d)

1. Application

The Al extracted by 1 N KCl approximates exchangeable Al and is a measure of the "active" acidity present in soils with a 1:1 water pH <5.5. Above pH 5.5, precipitation of Al occurs during analysis. This method does not measure the acidity component of hydronium ions (H_3O^+). If Al is present in measurable amounts, the hydronium is a minor component of the active acidity. Because the 1 N KCl extractant is an unbuffered salt and usually affects the soil pH one unit or less, the extraction is determined at or near the soil pH. The KCl extractable Al is related to the immediate lime requirement and existing CEC of the soil. The "potential" acidity is better measured by the $BaCl_2$-TEA method (procedure 6H5a) (Thomas, 1982).

2. Summary of Method

A soil sample is leached with 1 N KCl using the mechanical vacuum extractor (Holmgren et al., 1977). The extract is weighed. The KCl extract is diluted with distilled deionized (DDI) water. The analytes are measured by an atomic absorption spectrophotometer. The data are automatically recorded by a microcomputer and printer. The Al and Mn are reported in meq 100 g-1 and mg kg^{-1} (ppm) ovendry soil in procedure 6D3c and 6G9d.

3. Interferences

There are four types of interferences (matrix, spectral, chemical, and ionization) in the AA analyses of these cations. These interferences vary in importance, depending upon the particular analyte selected.

The soil:extractant ratio must remain constant. A soil:extractant ratio of 1:10 (w:v) for batch procedures is most commonly used. Using a leaching technique, a 1:20 (w:v) ratio gives comparable results. If the sample is changed, the amount of extractable Al is changed. No other significant interferences have been identified for this procedure.

4. Safety

Wear protective clothing and eye protection. When preparing reagents, exercise special care. Restrict the use of concentrated HCl to a fume hood. Many metal salts are extremely toxic and may be fatal if ingested. Follow standard laboratory practices when handling compressed gases. Gas cylinders should be chained or bolted in an upright position. Acetylene gas is highly flammable. Avoid open flames and sparks. Standard laboratory equipment includes fire blankets and extinguishers for use when necessary. Follow the manufacturer's safety precautions when using the AA.

5. Equipment

5.1 Electronic balance, ±1-mg sensitivity
5.2 Mechanical vacuum extractor, 24 place, Centurion International, Inc., Lincoln, NE
5.3 Mechanical vacuum extractor, Mavco Sampletex, 5300 N. 57th St., Lincoln, NE
5.4 Syringes, polypropylene, disposable, 60 mL, for extraction vessel, extractant reservoir, and tared extraction syringe
5.5 Rubber tubing, 3.2 ID x 6.4 OD x 25.4 mm, (1/8 ID x 1/4 OD x 1 in) for connecting syringe barrels
5.6 Analytical filter pulp, ash-free, Schleicher and Schuell, No. 289
5.7 Wash bottle, 20 mL, to dispense KCl.
5.8 Polycons, Richards Mfg. Co.
5.9 Atomic absorption spectrophotometer (AA), Smith-Hieftje Model 4000, Thermo Jarrell Ash Corp., Franklin, MA
5.10 Autosampler, Model 150, Thermo Jarrell Ash Corp., Franklin, MA
5.11 ThermoSpec software, Version 3.01, Enable 4.0, DOS 5.0, Thermo Jarrell Ash Corp., Franklin, MA
5.12 Computer, CUi Advantage 486, Thermo Jarrell Ash Corp., Franklin, MA
5.13 Printer, NEC Pinwriter P3200
5.14 Single-stage regulator, acetylene service, part number E11-0-N511A, Air Products and Chemicals, Inc., Box 538, Allentown, PA
5.15 Digital diluter/dispenser, MicroLab 500, Hamilton Co., P.O. Box 10030, Reno, NV
5.16 Syringes, 10000 and 1000 μL, 1001 DX and 1010-TEL LL gas tight, Hamilton Co., P.O. Box 10030, Reno, NV
5.17 Test tubes, 15-mL, 16 mm x 100, for sample dilution and sample changer, Curtin Matheson Scientific, Inc., Houston, TX

5.18 Containers, polypropylene

6. Reagents
6.1 Distilled deionized (DDI) water
6.2 Potassium chloride solution (KCl), 1.0 N. Dissolve 1342 g of KCl reagent in 16 L DDI water. Allow solution to equilibrate to room temperature. Dilute to 18 L with DDI water. Use 1.0 N KCl solution for Al extraction.
6.3 Potassium chloride solution (KCl), 2.0 N. Dissolve 298.24 g of KCl reagent in 1.5 L DDI water. Allow solution to equilibrate to room temperature. Dilute to 2 L with DDI water. Use 2.0 N KCl solution for standards.
6.4 Primary Al standard, 1000 ppm. Certified Reference Solution, Fisher Chemical Scientific Co., Fairlawn, N.J.
6.5 Primary Mn standard, 1000 ppm. Certified Reference Solution, Fisher Chemical Scientific Co., Fairlawn, N.J.
6.6 High calibration Al (4 meq L^{-1}) and Mn (3 ppm) standard. Mix 9 mL of primary Al standard with 125 mL 2 N KCl. Add 0.75 mL of primary Mn standard and make to 250-mL volume with DDI water.
6.7 Low calibration Al (0.2 meq L^{-1}) and Mn (0.2 ppm) standard. Mix 1.8 mL of primary Al standard with 125 mL 2 N KCl. Add 0.05 mL of primary Mn standard and make to 250-mL volume with DDI water.
6.8 Calibration Al (2 meq L^{-1}) and Mn (1.5 ppm) check standard. Mix 4.5 mL of primary Al standard with 125 mL 2 N KCl. Add 0.375 mL of primary Mn standard and make to 250-mL volume with DDI water.
6.9 Calibration reagent blank solution, 1.0 N KCl. Add 125 mL of 2.0 N KCl to a volumetric flask and make to 250-mL volume with DDI water.
6.10 Nitrous oxide gas, compressed
6.11 Acetylene gas, compressed, purity 99.6%
6.12 Compressed air with water and oil traps

7. Procedure

Extraction of Al
7.1 Prepare extraction vessel by tightly compressing a 1-g ball of filter pulp into the bottom of a syringe barrel with a modified plunger.

7.2 Weigh exactly 2.50 g of <2-mm, air-dry soil and place in an extraction vessel. Prepare one quality control check sample per 48 samples.

7.3 Place the extraction vessel on the upper disk of the extractor and connect a tared extraction syringe. Use 25.4-mm (1-in) length rubber tubing to insert the handle of the plunger in the slot of the stationary extractor disk.

7.4 Use a squeeze bottle and fill extraction vessel to the 20-mL mark with 1.0 N KCl solution ($\approx$10 mL). Make sure that the sample is thoroughly wetted. During the addition, wash sides of the tube and wet the sample. Shaking, swirling, or stirring may be required to wet organic samples. Allow sample to stand for at least 30 min.

7.5 Put reservoir tube on top of the extraction vessel. Set extractor for fast extraction rate and extract until the KCl solution is at a 0.5- to 1.0-cm height above sample. Turn off extractor.

7.6 Add 45 mL KCl solution to reservoir tube. Set extractor for 45-min extraction.

7.7 After the extraction, shut off extractor and pull plunger of syringe down. Do not remove the plunger from syringe barrel. Carefully remove the syringe with extract leaving the rubber tubing on the extraction vessel.

7.8 Weigh each syringe containing KCl extract to the nearest 0.01 g.

7.9 Mix extract in each syringe by manually shaking. Fill a polycon with extract solution and discard the excess. This solution is reserved for extractable Al and Mn analyses.

Dilution of Sample Extracts and Standards

7.10 No ionization suppressant is required as the K in the extractant is present in sufficient quantity. Set the digital settings at a 1:2 dilution for the KCl sample extracts, calibration reagent blanks, calibration standards, and calibration check standards as follows:

7.11 Dilute 1 part KCl sample extract with 1 part of DDI water (1:2 dilution).

7.12 Dilute 1 part calibration reagent blank with 1 part of DDI water (1:2 dilution).

7.13 Dilute 1 part calibration standard with 1 part of DDI water (1:2 dilution).

7.14 Dilute 1 part calibration check standard with 1 part of DDI water (1:2 dilution).

7.15 Dispense the diluted solutions into test tubes and place in the sample holder of the sample changer.

AA Calibration

7.16 Use the calibration reagent blank (1.0 N KCl), high standard (4 meq L^{-1} Al and 3 ppm Mn), and the low standard (2 meq L^{-1} Al and 1.5 ppm Mn) to calibrate the AA.

7.17 Use the calibration check standard (2 meq L^{-1} Al and 1.5 ppm Mn) as a check sample. Perform a calibration check every 12 samples.

AA Set-up and Operation

7.18 Refer to the manufacturer's manual for operation of the AA. The following parameters are only very general guidelines for instrument conditions for the analyte.

Element	Wave-length (nm)	Burner head & Angle	Fuel/ Oxidant
Al	309.3	5-cm parallel	20 C_2H_2/ 10 N_2O
Mn	280.1	5-cm parallel	4 C_2H_2/ 16 Air

Typical read delay is 6 s, and integration by peak area is 8 s.

7.19 Use the microcomputer and printer to set instrument parameters and to collect and record instrument readings.

7.20 If a sample exceeds the calibration standard, dilute the sample (dilution ratio in calculation) with appropriate matrix and record dilution. Remember to keep the matrix the same after dilution by diluting with DDI water (1:2 dilution).

7.21 Analyze one quality control check sample for every 48 samples.

8. Calculations

8.1 The instrument readings are the analyte concentration (meq L^{-1} Al) in undiluted extract. Use these values to calculate the analyte concentration on an oven-dry soil basis (meq 100 g^{-1}).

Al (meq 100 g^{-1}) = [AA$_{Al}$ x (Wt$_{syr+ext}$ - Wt$_{syr}$) x D.R. x 100 x AD/OD]/[Sample Weight (g) x 1.0412 x 1000]

where:

AA$_{Al}$ = AA Al reading (meq L^{-1})
Wt$_{syr+ext}$= Weight of extraction syringe and extract (g)
Wt$_{syr}$ = Weight of tared extraction syringe (g)
D.R. = Dilution ratio for samples over calibration range
1.0412 = Density of 1 N KCl @ 20°C
1000 = g L^{-1}
100 = Conversion factor (100-g basis)
AD/OD= Air-dry/oven-dry ratio (procedure 4B5)

9. Report
Report KCl extractable Al in units of meq 100 g^{-1} of oven-dry soil to the nearest 0.1 meq 100 g^{-1}.

10. Precision
Precision data are not available for this procedure. A quality control check sample is run with every batch of 48 samples.

11. References
Holmgren, George G.S., R.L. Juve, and R.C. Geschwender. 1977. A mechanically controlled variable rate leaching device. Soil Sci. Amer. J. 41:1207-1208.
Thomas, G.W. 1982. Exchangeable cations. *In* A. Klute (ed.) Methods of soil analysis. Part 2. Chemical and microbiological properties. 2nd ed. Agronomy 9:159-165.

Chloride, Sulfate, Nitrate, Fluoride, and Nitrite (6K, 6L, 6M, 6U, and 6W) Saturation Extract (6K1, 6L1, 6M1, 6U1, and 6W1) Chromatograph, Anion Suppressor Dionex 2110i Ion Chromatograph (6K1d, 6L1d, 6M1d, 6U1b, and 6W1b)

1. Application
The soluble anions that are commonly determined in saline and alkali soils are carbonate, bicarbonate, sulfate, chloride, nitrate, nitrite, fluoride, phosphate, silicate, and borate (Khym, 1974; U.S. Salinity Laboratory Staff, 1954). Carbonate and bicarbonate are determined by titration. Phosphate, silicate, and borate usually are not determined because they are found only occasionally in measurable amounts in soils. Chloride, sulfate, nitrate, fluoride, and nitrite are measured in solution by chromatography. In saline and alkali soils, carbonate, bicarbonate, sulfate, and chloride are the anions that are found in the greatest abundance. In general, soluble sulfate is usually more abundant than soluble chloride.

2. Summary of Method

The saturation extract is diluted according to its electrical conductivity (EC_s). The diluted sample is injected into the ion chromatograph, and the anions are separated. A conductivity detector is used to measure the anion. A chart recording is made of the chromatograph. Standard anions are used to calibrate the system. A calibration curve is determined, and the anion concentrations are calculated. A computer program automates these actions. The saturated extract anions, Cl^-, SO_4^{2-}, NO_3^-, F^-, and NO_2^- are reported in meq L^{-1} in procedures 6K1d, 6L1d, 6M1d, 6U1b, and 6W1b, respectively.

3. Interferences

Some saturation extracts contain suspended solids. Filtering after dilution removes the particles. Saturation extracts of acid soils that contain Fe and/or Al may precipitate and clog the separator column. Saturation extracts of very high pH may contain organic material which may clog or poison the column. Low molecular weight organic anions will co-elute with inorganic anions from the column.

4. Safety

Wear protective clothing and safety glasses. When preparing reagents, exercise special care. Many metal salts are extremely toxic and may be fatal if ingested. Thoroughly wash hands after handling these metal salts. Follow the manufacturer's safety precautions when using the chromatograph.

5. Equipment

5.1 Ion chromatograph, Series 2110i, dual-channel system
5.2 Analytical column, AS4A 4mm P/N 37041, Dionex Corp., 1228 Titan Way, Sunnyvale, CA
5.3 Guard column, AG4A 4mm P/N 37042, Dionex Corp., 1228 Titan Way, Sunnyvale, CA
5.4 Analytical pumps, Dionex Corp., 1228 Titan Way, Sunnyvale, CA
5.5 Automated sampler, Dionex Corp., 1228 Titan Way, Sunnyvale, CA
5.6 Conductivity detectors, Dionex Corp., 1228 Titan Way, Sunnyvale, CA
5.7 Anion self-regenerating suppressor (ASRS-1) with controller (SRC-1), Dionex Corp., 1228 Titan Way, Sunnyvale, CA
5.8 Computer interfaces, Dionex Corp., 1228 Titan Way, Sunnyvale, CA
5.9 Computer software, A1-450 Chromatography Software Program Release 3.32, Microsoft Windows Operating Environment; Dionex Corp., 1228 Titan Way, Sunnyvale, CA
5.10 Computer, DFI.
5.11 Printer, Epson, Fx-850
5.12 Poly-vials, 5 mL, P/N 038008, Dionex Corp., 1228 Titan Way, Sunnyvale, CA
5.13 Poly-vials, filtercaps, 5 mL, P/N 038009, Dionex Corp., 1228 Titan Way, Sunnyvale, CA
5.14 Digital diluter/dispenser, MicroLab 500, Hamilton C., P.O. Box 10030, Reno, NV
5.15 Syringes, gas tight, Hamilton 1001 DX and 1010-TEF LL, Hamilton Co., P.O. Box 10030, Reno, NV
5.16 Disposable 0.2-μm pore size, 25-mm filter assembly, Gelman Sciences, Inc., 674 South Wagner Road, Ann Arbor, MI 48106. Use for saturation extracts and standards.
5.17 Disposable 0.2-μm pore size, Ultipor N_{66} DFA3001NAEY, Pall Trinity Micro Corp., Cortland, NY. Use for filtering distilled deionized (DDI) water.

6. Reagents

6.1 Distilled deionized (DDI) filtered water
6.2 Sulfuric acid (H_2SO_4), concentrated, reagent
6.3 Toluene
6.4 Isopropanol to de-gas column
6.5 Stock $NaHCO_3$ solution, 0.480 M. Mix 40.34 g of dried $NaHCO_3$ with filtered DDI water and dilute to 1-L volume.

6.6 Stock Na$_2$CO$_3$ solution, 0.5040 M. Mix 53.42 g of dried Na$_2$CO$_3$ with filtered DDI water and dilute to 1-L volume.

6.7 Working eluent solution. Mix 100 mL of 0.5040 M NaHCO$_3$ and 100 mL of 0.4800 M Na$_2$CO$_3$ with filtered DDI water and dilute to 20-L volume. Add 8 drops of toluene to retard microbial growth.

6.8 Primary SO$_4^{2-}$ standard, 0.5 M (1.0 N). Mix 17.7560 g of Na$_2$SO$_4$ with filtered DDI water and dilute to 250-mL volume.

6.9 Primary Cl$^-$ standard, 1.0 M (1.0 N). Add 18.6392 g of KCl with filtered DDI water and dilute to 250-mL volume.

6.10 Primary F$^-$ standard, 0.125 M (0.125 N). Add 1.3122 g of NaF with filtered DDI water and dilute to 250-mL volume.

6.11 Primary NO$_3^-$ standard, 1.0 M (1.0 N). Add 25.2770 g of KNO$_3$ with filtered DDI water and dilute to 250-mL volume.

6.12 Primary mixed standard. Prepare 1 primary mixed standard by taking aliquots of each of the proceeding primary standards and diluting the combined aliquots to a 1-L volume with working eluent as follows:

Primary Standards	Aliquot	Final Volume w/Eluent	Concentration
	mL	mL	meq L^{-1}
Na$_2$SO$_4$	50	1000	50
KCl	10	1000	10
NaF	100	1000	12.5
KNO$_3$	30	1000	30

Add eight drops of toluene to primary mixed standard to retard microbial growth and store in a glass container.

6.13 Mixed calibration standards. Prepare 4 mixed calibration standards (0.5, 1.0, 3.0, and 7.0 readings) by taking aliquots of primary mixed standard and diluting each aliquot to 100-mL volume with working eluent as follows:

Primary Mixed Standards	Final Volume w/Eluent	Concentration			
		SO$_4^{2-}$	Cl$^-$	F$^-$	NO$_3^-$
mL	mL	meq L^{-1}			
0.5	100	0.25	0.05	0.0625	0.15
1.0	100	0.50	0.10	0.125	0.30
3.0	100	1.5	0.30	0.375	0.90
7.0	100	3.5	0.70	0.875	2.1

6.14 NaNO$_2$, Baker reagent grade, 99.5% purity

6.15 Primary NO$_2^-$ standard, 1 N (1000 meq L^{-1}). Mix 69.3568 g of reagent grade NaNO$_2$ with filtered DDI water and dilute to 1-L volume. Take 5 mL aliquot of primary NO$_2^-$ standard and dilute with 500 mL of filtered DDI water (10 meq L^{-1}). Add eight drops of toluene to primary NO$_2^-$ standard to retard microbial growth and store in a glass container.

6.16 NO_2^- calibration standards. Prepare 4 NO_2^- calibration standards (0.5, 1.0, 3.0, and 7.0 readings) by taking aliquots of primary NO_2^- standard (10 meq L^{-1}) and diluting each aliquot to 100-mL volume with working eluent as follows:

Primary Standard meq L^{-1}	Final Volume w/Eluent	NO_2^- Concentration
mL	mL	meq L^{-1}
0.5	100	0.5
1.0	100	1.0
3.0	100	3.0

7. Procedure

Dilution of extracts

7.1 To estimate the total soluble anion concentration (meq L^{-1}), multiply the EC_s (procedure 8A3a) by 10. Subtract the CO_3^{2-} and HCO_3^- concentrations (procedures 6I1b and 6J1b) from the total anion concentration. The remainder is the $\approx$ concentration (meq L^{-1}) of anions to be separated by ion chromatography.

$$\text{Anion concentration (meq } L^{-1}) = EC_s \times 10 - (HCO_3^- + CO_3^{2-})$$

7.2 Refer to Table 1 for dilution of saturation extract with the working eluent.

7.3 Place the diluted samples in the Poly-vials and cap with filtercaps.

7.4 Place the mixed calibration standards in the Poly-vials.

Set-up and Operation of Ion Chromatograph (IC)

7.5 Because any number of factors may cause a change in IC operating conditions, only a general set-up of the Dionex 2110i ion chromatograph is presented. Individual analysts may modify some or all of the operating conditions to achieve satisfactory results. Typical operation parameters are as follows:

Parameter	Range
Conductivity cell range	3 uS cm^{-1} full scale to 100 uS cm^{-1}
Auto offset	"On"
Analytical pump flow rate	2.0 to 2.5 mL min^{-1}
Low pressure limit	100 psi
High pressure limit	1200 psi
Regenerant flow rate	3 to 4 mL min^{-1}
Injector loop	0.50 mL
Air pressure	3 to 8 psi

7.6 Load the sample holder cassettes with the capped samples, standards, and check samples.

7.7 Refer to the manufacturer's manual for the operation of chromatograph.

8. Calculations

Calibration Calculations

8.1 Use the peak height of each anion standard to either construct a calibrated curve to plot anion concentration or use a least squares analysis to calculate anion concentration. The analytes are reported in meq L^{-1}.

8.2 *Calibration Curve*: Plot the peak height against the meq L^{-1} of each anion standard on graph paper. Construct the calibration curve by finding the "best" line that fits the plotted standards.

8.3 *Linear Squares Analysis*: Use a least squares criterion, i.e. best moving average. Refer to a statistical analysis book for additional information on least squares analysis. An example for the anion Cl^- is as follows:

Cl^- concentration (meq L^{-1}) = Y = 0.1 1.5 4.0
Peak height = X = 8.43 170.0 441.5
Number of standards = n = 3

$\sum Y_i = 5.6$ $\qquad\qquad$ $\sum X_i = 619.93$

$\sum Y_i/n = Y = 1.866$ $\qquad$ $\sum X_i/n = X = 206.6433$
$\sum X_i Y_i = 2021.843$ $\qquad$ $\sum X_i^2 = 223893.31$
$\sum X_i \sum Y_i = 3471.608$

$$b = \frac{\sum X_i Y_i - \sum X_i \sum Y_i/n}{\sum X_i^2 - (\sum X_i)^2/n} = \frac{2021.843 - 1157.2027}{223893.31 - 128104.4} = 0.0090265$$

b = slope of the line, i.e., the amount that Y changes when X changes by 1 unit.

The equation is as follows:

Y = Y + b (X - X)

Y = 1.866 + 0.0090265 (X) - 1.8653

Analyte Calculation

8.4 *Calibration curve*: Read the analyte concentration (meq L^{-1}) directly from the calibration curve.

Table 1. Dilution factor for saturated paste soil extracts based on EC readings.

EC_s (mmhos cm^{-1})	Dilution Factor
0.00 to 0.55	4
0.56 to 0.65	5
0.66 to 0.75	6
0.76 to 0.85	7
0.86 to 0.95	8
0.96 to 1.05	9
1.06 to 1.20	10
1.21 to 1.40	15
1.41 to 1.50	25
1.51 to 1.60	30
1.61 to 1.80	40
1.81 to 2.00	50
2.01 to 2.30	60
2.31 to 2.60	70
2.61 to 3.10	80
3.11 to 3.55	90
3.56 to 4.05	100
4.06 to 4.60	120
4.61 to 5.20	140
5.21 to 5.85	150
5.86 to 6.55	160
6.56 to 7.30	180
7.31 to 8.00	200
8.01 to 9.00	225
9.01 to 10.00	240
10.01 to 11.50	270
11.51 to 13.00	280
13.01 to 14.50	300
14.51 to 16.00	320
16.01 to 17.00	360
17.01 to 18.00	400
18.01 to 20.00	450
20.01 to 21.00	480
21.01 to 23.00	500
23.01 to 24.00	540
24.01 to 25.00	560
25.01 to 27.00	600
27.01 to 28.00	640
28.01 to 30.00	680
30.01 to 32.00	700
32.01 to 33.00	720
33.01 to 36.00	800
36.01 to 40.00	900
40.01 to 44.00	1000

8.5 *Linear regression*: Put the peak height in the preceding equation and solve for analyte concentration (meq L^{-1}). Thus, if sample extract has 204 peak height, the preceding equation is as follows:

$$Y = 1.866 + 0.0090265 (204) - 1.8653 = 1.84 \text{ meq } L^{-1}$$

8.6 Repeat the calibration set and analyte calculation for each anion.

8.7 The chromatograph software automatically calculates the analyte concentrations and prints a report of the results.

9. Report
Report the saturation extract anions in units of meq L^{-1} to the nearest 0.1 meq L^{-1}.

10. Precision
Precision data are not available for this procedure.

11. References
Khym, J.X. 1974. Analytical ion-exchange procedures in chemistry and biology: theory, equipment, techniques. Prentice-Hall, Inc., Englewood Cliffs, NJ.

U.S. Salinity Laboratory Staff. 1954. L.A. Richards (ed.) Diagnosis and improvement of saline and alkali soils. U.S. Dept. of Agric. Handb. 60. U.S. Govt. Print. Office, Washington, DC.

Calcium, Magnesium, Sodium, and Potassium
(6N, 6O, 6P, and 6Q)
Saturation Extraction
(6N1, 6O1, 6P1, and 6Q1)
Atomic Absorption
Perkin-Elmer AA 5000
(6N1b, 6O1b, 6P1b, and 6Q1b)

1. Application
The commonly determined soluble cations are Ca^{2+}, Mg^{2+}, Na^+, and K^+. In soils with a low saturation pH, measurable amounts of Fe and Al may be present. Determination of soluble cations is used to obtain the relations between total cation concentration and other properties of saline solutions such as electrical conductivity and osmotic pressure (U.S. Salinity Laboratory Staff, 1954). The relative concentrations of the various cations in the soil-water extracts also provide information on the composition of the exchangeable cations in the soil. Complete analyses of the soluble ions provide a means to determine total salt content of the soils and salt content at field-moisture conditions.

2. Summary of Method
The saturation extract from procedure 8A3a is diluted with an ionization suppressant ($LaCl_3$). The analytes are measured by an atomic absorption spectrophotometer (AA). The data are automatically recorded by a microcomputer and printer. The saturation extracted cations, Ca^{2+}, Mg^{2+}, Na^+, and K^+, are reported in meq L^{-1} in procedures 6N1b, 6O1b, 6P1b, and 6Q1b, respectively.

3. Interferences
There are four types of interferences (matrix, spectral, chemical, and ionization) in the analysis of these cations. These interferences vary in importance, depending upon the particular analyte selected.

4. Safety
Wear protective clothing and eye protection. When preparing reagents, exercise special care. Restrict the use of concentrated HCl to a fume hood. Many metal salts are extremely toxic and may be fatal if ingested. Thoroughly wash hands after handling these metal salts.

Follow standard laboratory procedures when handling compressed gases. Gas cylinders should be chained or bolted in an upright position. Acetylene is highly flammable. Avoid open flames and sparks. Standard laboratory equipment includes fire blankets and extinguishers for use when necessary. Follow the manufacturer's safety precautions when using the AA.

5. Equipment
5.1 Electronic balance, ±1-mg sensitivity
5.2 Filter paper, pre-pleated, 185-mm diameter, Schleicher and Schuell
5.3 Atomic absorption spectrophotometer (AA), model 5000, Perkin-Elmer Corp., Norwalk, CT
5.4 Automatic burner control, model 5000, Perkin-Elmer Corp., Norwalk, CT
5.5 Autosampler, AS-50, Perkin-Elmer Corp., Norwalk, CT
5.6 Dot matrix printer, P-132, Interdigital Data Systems,Inc.
5.7 Single-stage regulator, acetylene service, part number E11-0-N511A, Air Products and Chemicals, Inc., Box 538, Allentown, PA
5.8 Digital diluter/dispenser, MicroLab 500, Hamilton Co., P.O. Box 10030, Reno, NV
5.9 Syringes, 10000 and 1000 µL, 1001 DX and 1010-TEL LL gas tight, Hamilton Co., P.O. Box 10030, Reno, NV
5.10 Test tubes, 15-mL, 16 mm x 100, for sample dilution and sample changer, Curtin Matheson Scientific, Inc., Houston, TX
5.11 Containers, polypropylene

6. Reagents
6.1 Distilled deionized (DDI) water
6.2 Hydrochloric acid (HCl), concentrated 12 N
6.3 HCl, 1:1 HCl:DDI, 6 N. Carefully mix 1 part of concentrated HCl to 1 part DDI water.
6.4 HCl, 1% wt. Carefully dilute 25 mL of concentrated HCl to 1 L with DDI water.
6.5 NH_4OH, reagent grade, sp gr 0.90
6.6 Glacial acetic acid, 99.5%
6.7 Primary stock mixed standards solution (PSMSS). Dissolve 0.8759 g of oven-dry reagent grade calcium carbonate ($CaCO_3$) in a minimum of volume of 1:1 HCl:DDI. Add 0.2127 g of clean Mg ribbon dissolved in 1:1 HCl. Add 1.0956 g of dry reagent grade sodium chloride (NaCl) and 0.1864 g of dry reagent grade KCl. Transfer to a 250-mL volumetric and bring to volume with 1% HCl solution. Resulting solution contains 70 meq L^{-1} (1403 ppm) Ca; 70 meq L^{-1} (851 ppm) Mg; 75 meq L^{-1} (1724 ppm) Na; 10 meq L^{-1} (391 ppm) K. Store in a polypropylene container.
6.8 NH_4OAc solution, 1.0 N, pH 7.0, reagent blank. Mix 57 mL of glacial acetic acid in 600 mL DDI water. While stirring, carefully add 68 mL concentrated of NH_4OH. Cool and adjust pH to 7.0 using NH_4OH or acetic acid. Dilute to 1 L with DDI water. The NH_4OAc solution is used for extraction of cations (procedure 5A8c).
6.9 Working stock mixed standards solution (WSMSS). Dilute 20 mL of the PSMSS with 80 mL DDI water (1:5). Resulting solution contains 14 meq L^{-1} (281 ppm) Ca; 14 meq L^{-1} (170 ppm) Mg; 15 meq L^{-1} (345 ppm) Na; 2 meq L^{-1} (78 ppm) K. Store in a polypropylene container.
6.10 Stock lanthanum ionization suppressant solution, 65,000 ppm. Wet 152.4 g of lanthanum oxide (La_2O_3) with 100 mL DDI water. Slowly and cautiously add 500 mL of 6 N HCl to dissolve the La_2O_3. Cooling the solution is necessary. Dilute to 2 L with DDI water. Filter solution. Store in polypropylene container.

6.11 Lanthanum ionization suppressant solution, 6500 ppm. Dilute 200 mL of stock lanthanum ionization suppressant solution with 1800 mL of DDI water (1:10). Store in polypropylene container.

6.12 Dilute calibration mixed standards solution (DCMSS). Dilute 1 part of the WSMSS with 39 parts of the lanthanum solution (1:40). Resulting solution contains 0.35 meq L^{-1} (7 ppm) Ca; 0.35 meq L^{-1} (4 ppm) Mg; 0.375 meq L^{-1} (9 ppm) Na; 0.05 meq L^{-1} (2 ppm) K. Store in polypropylene container.

6.13 Dilute calibration reagent blank solution (DCRBS). Dilute 1 part of DDI water with 39 parts of the lanthanum solution (1:40). Store in polypropylene container.

6.14 Compressed air with water and oil traps.

6.15 Acetylene gas, purity 99.6%.

7. Procedure

Dilution of Sample Extracts and Standards

7.1 The 10-mL syringe is for diluent (lanthanum ionization suppressant solution). The 1-mL syringe is for saturation sample extracts (procedure 8A3a), calibration reagent blanks, and calibration standards. Set the digital diluter at 1:40 dilution for saturation sample extracts, reagent blanks, and calibration standards as follows:

7.2 Dilute 1 part saturation sample extract with 39 parts of lanthanum ionization suppressant solution (1:40 dilution).

7.3 Dilute 1 part WSMSS with 39 parts of lanthanum ionization suppressant solution (1:40 dilution). This dilution is the DCMSS. Refer to reagents section.

7.4 Dilute 1 part DDI water with 39 parts of lanthanum ionization suppressant solution (1:40 dilution). This dilution is the DCRBS. Refer to reagents section.

7.5 Dispense the diluted solutions into test tubes which have been placed in the sample holders of the sample changer.

AA Calibration

7.6 Use the DCRBS and the DCMSS to calibrate the AA. The AA program requires a blank and a standard, in that order, to establish a single point calibration curve for element determination. Perform one calibration, i.e., blank plus standard, for every 12 samples.

AA Set-up and Operation

7.7 Refer to the manufacturer's manual for operation of the AA. The following are only very general guidelines for instrument conditions for the various analytes.

Analyte	Conc. meq/L	Burner & Angle	Wave-length	Slit	Fuel/ Oxidant C2H2/ Air
Ca	14.00	50cm @ 0°	422.7	0.7	10/25
Mg	14.00	50 cm @ 30°	285.2	0.7	10/25
K	2.00	50 cm @ 0°	766.5	1.4	10/25
Na	15.00	50 cm @ 30°	589.0	0.4	10/25

516

7.8 Use the microcomputer and printer to set instrument parameters and to collect and record instrument readings.

7.9 If sample exceeds calibration standard, dilute the sample (dilution ratio in calculation) with appropriate matrix and record dilution. Remember to keep the matrix the same after dilution by diluting with the lanthanum ionization suppressant solution (1:40 dilution).

7.10 Analyze one quality control check sample for every 48 samples.

7.11 The instrument readings are usually programmed in meq L^{-1}. Record analyte readings to 0.01 meq L^{-1}.

8. Calculations

8.1 The instrument readings are the analyte concentration (meq L^{-1} cation) in undiluted extract. Use these values and dilution ratio (if any) and calculate the analyte concentration in meq L^{-1} cation.

Analyte Concentration in Soil (meq L^{-1}) = Analyte AA reading (meq L^{-1}) x Dilution ratio (if any)

9. Report
Report the saturation extraction cations of Ca^{2+}, Mg^{2+}, Na^+, and K^+ in units of meq L^{-1} to the nearest 0.1 meq L^{-1}.

10. Precision
Precision data are not available for this procedure.

11. References
U.S. Salinity Laboratory Staff. 1954. L.A. Richards (ed.) Diagnosis and improvement of saline and alkali soils. U.S. Dept. of Agric. Handb. 60. U.S. Govt. Print. Office, Washington, DC.

Calcium, Magnesium, Sodium, and Potassium
(6N, 6O, 6P, and 6Q)
Saturation Extraction
(6N1, 6O1, 6P1, and 6Q1)
Atomic Absorption
Thermo Jarrell Ash, Smith-Hieftje AA 4000
(6N1c, 6O1c, 6P1c, and 6Q1c)

1. Application
The commonly determined soluble cations are Ca^{2+}, Mg^{2+}, Na^+, and K^+. In soils with a low saturation pH, measurable amounts of Fe and Al may be present. Determination of soluble cations is used to obtain the relations between total cation concentration and other properties of saline solutions such as electrical conductivity and osmotic pressure (U.S. Salinity Laboratory Staff, 1954). The relative concentrations of the various cations in the soil-water extracts also provide information on the composition of the exchangeable cations in the soil. Complete analyses of the soluble ions provide a means to determine total salt content of the soils and salt content at field-moisture conditions.

2. Summary of Method
The saturation extract from procedure 8A3a is diluted with an ionization suppressant ($LaCl_3$). The analytes are measured by an atomic absorption spectrophotometer (AA). The

data are automatically recorded by a microcomputer and printer. The saturation extracted cations, Ca^{2+}, Mg^{2+}, Na^+, and K^+, are reported in meq L^{-1} in procedures 6N1c, 6O1c, 6P1c, and 6Q1c, respectively.

3. Interferences

There are four types of interferences (matrix, spectral, chemical, and ionization) in the analysis of these cations. These interferences vary in importance, depending upon the particular analyte selected.

4. Safety

Wear protective clothing and eye protection. When preparing reagents, exercise special care. Restrict the use of concentrated HCl to a fume hood. Many metal salts are extremely toxic and may be fatal if ingested. Thoroughly wash hands after handling these metal salts.

Follow standard laboratory procedures when handling compressed gases. Gas cylinders should be chained or bolted in an upright position. Acetylene is highly flammable. Avoid open flames and sparks. Standard laboratory equipment includes fire blankets and extinguishers for use when necessary. Follow the manufacturer's safety precautions when using the AA.

5. Equipment
5.1 Electronic balance, ±1-mg sensitivity
5.2 Filter paper, pre-pleated, 185-mm diameter, Schleicher and Schuell
5.3 Atomic absorption spectrophotometer (AA), Smith-Hieftje Model 4000, Thermo Jarrell Ash Corp., Franklin, MA
5.4 Autosampler, Model 150, Thermo Jarrell Ash Corp., Franklin, MA
5.5 ThermoSpec software, Version 3.01, Enable 4.0, DOS 5.0, Thermo Jarrell Ash Corp., Franklin, MA
5.6 Computer, CUi Advantage 486, Thermo Jarrell Ash Corp., Franklin, MA
5.7 Printer, NEC Pinwriter P3200
5.8 Single-stage regulator, acetylene service, part number E11-0-N511A, Air Products and Chemicals, Inc., Box 538, Allentown, PA
5.9 Digital diluter/dispenser, MicroLab 500, Hamilton Co., P.O. Box 10030, Reno, NV
5.10 Syringes, 10000 and 1000 μL, 1001 DX and 1010-TEL LL gas tight, Hamilton Co., P.O. Box 10030, Reno, NV
5.11 Test tubes, 15-mL, 16 mm x 100, for sample dilution and sample changer, Curtin Matheson Scientific, Inc., Houston, TX
5.12 Containers, polypropylene

6. Reagents
6.1 Distilled deionized (DDI) water
6.2 Hydrochloric acid (HCl), concentrated 12 N
6.3 HCl, 1:1 HCl:DDI, 6 N. Carefully mix 1 part of concentrated HCl to 1 part DDI water.
6.4 HCl, 1% wt. Carefully dilute 25 mL of concentrated HCl to 1 L with DDI water.
6.5 NH_4OH, reagent grade, sp gr 0.90
6.6 Glacial acetic acid, 99.5%
6.7 Primary stock mixed standards solution (PSMSS). Dissolve 0.8759 g of oven-dry reagent grade calcium carbonate ($CaCO_3$) in a minimum of volume of 1:1 HCl:DDI. Add 0.2127 g of clean Mg ribbon dissolved in 1:1 HCl. Add 1.0956 g of dry reagent grade sodium chloride (NaCl) and 0.1864 g of dry reagent grade KCl. Transfer to a 250-mL volumetric and bring to volume with 1% HCl solution. Resulting solution contains 70 meq L^{-1} (1403 ppm) Ca; 70 meq L^{-1} (851 ppm) Mg; 75 meq L^{-1} (1724 ppm) Na; 10 meq L^{-1} (391 ppm) K. Store in a polypropylene container.
6.8 NH_4OAc solution, 1.0 N, pH 7.0, reagent blank. Mix 57 mL of glacial acetic acid in 600 mL DDI water. While stirring, carefully add 68 mL concentrated of NH_4OH. Cool and

adjust pH to 7.0 using NH$_4$OH or acetic acid. Dilute to 1 L with DDI water. The NH$_4$OAc solution is used for extraction of cations (procedure 5A8c).

6.9 Working stock mixed standards solution (WSMSS). Dilute 20 mL of the PSMSS with 80 mL DDI water (1:5). Resulting solution contains 14 meq L^{-1} (281 ppm) Ca; 14 meq L^{-1} (170 ppm) Mg; 15 meq L^{-1} (345 ppm) Na; 2 meq L^{-1} (78 ppm) K. Store in a polypropylene container.

6.10 Stock lanthanum ionization suppressant solution, 65,000 ppm. Wet 152.4 g of lanthanum oxide (La$_2$O$_3$) with 100 mL DDI water. Slowly and cautiously add 500 mL of 6 N HCl to dissolve the La$_2$O$_3$. Cooling the solution is necessary. Dilute to 2 L with DDI water. Filter solution. Store in polypropylene container.

6.11 Lanthanum ionization suppressant solution, 6500 ppm. Dilute 200 mL of stock lanthanum ionization suppressant solution with 1800 mL of DDI water (1:10). Store in polypropylene container.

6.12 Dilute calibration mixed standards solution (DCMSS). Dilute 1 part of the WSMSS with 39 parts of the lanthanum solution (1:40). Resulting solution contains 0.35 meq L^{-1} (7 ppm) Ca; 0.35 meq L^{-1} (4 ppm) Mg; 0.375 meq L^{-1} (9 ppm) Na; 0.05 meq L^{-1} (2 ppm) K. Store in polypropylene container.

6.13 Dilute calibration reagent blank solution (DCRBS). Dilute 1 part of DDI water with 39 parts of the lanthanum solution (1:40). Store in polypropylene container.

6.14 Compressed air with water and oil traps.

6.15 Acetylene gas, purity 99.6%.

7. Procedure

Dilution of Sample Extracts and Standards

7.1 The 10-mL syringe is for diluent (lanthanum ionization suppressant solution). The 1-mL syringe is for saturation sample extracts (procedure 8A3a), calibration reagent blanks, and calibration standards. Set the digital diluter at 1:40 dilution for saturation sample extracts, reagent blanks, and calibration standards as follows:

7.2 Dilute 1 part saturation sample extract with 39 parts of lanthanum ionization suppressant solution (1:40 dilution).

7.3 Dilute 1 part WSMSS with 39 parts of lanthanum ionization suppressant solution (1:40 dilution). This dilution is the DCMSS. Refer to reagents section.

7.4 Dilute 1 part DDI water with 39 parts of lanthanum ionization suppressant solution (1:40 dilution). This dilution is the DCRBS. Refer to reagents section.

7.5 Dispense the diluted solutions into test tubes which have been placed in the sample holders of the sample changer.

AA Calibration

7.6 Use the DCRBS and the DCMSS to calibrate the AA. The AA program requires a blank and a standard, in that order, to establish a single point calibration curve for element determination. Perform one calibration, i.e., blank plus standard, for every 12 samples.

AA Set-up and Operation

7.7 Refer to the manufacturer's manual for operation of the AA. The following are only very general guidelines for instrument conditions for the various analytes.

Analyte	Conc. meq/L	Burner & Angle	Wave-length	Slit	Fuel/ Oxidant C2H2/ Air
Ca	14.00	50cm @ 0°	422.7	0.7	10/25
Mg	14.00	50 cm @ 30°	285.2	0.7	10/25
K	2.00	50 cm @ 0°	766.5	1.4	10/25
Na	15.00	50 cm @ 30°	589.0	0.4	10/25

7.8 Use the microcomputer and printer to set instrument parameters and to collect and record instrument readings.

7.9 If sample exceeds calibration standard, dilute the sample (dilution ratio in calculation) with appropriate matrix and record dilution. Remember to keep the matrix the same after dilution by diluting with the lanthanum ionization suppressant solution (1:40 dilution).

7.10 Analyze one quality control check sample for every 48 samples.

7.11 The instrument readings are usually programmed in meq L^{-1}. Record analyte readings to 0.01 meq L^{-1}.

8. Calculations

8.1 The instrument readings are the analyte concentration (meq L^{-1} cation) in undiluted extract. Use these values and dilution ratio (if any) and calculate the analyte concentration in meq L^{-1} cation.

Analyte Concentration in Soil (meq L^{-1}) = Analyte AA reading (meq L^{-1}) x Dilution ratio (if any)

9. Report
Report the saturation extraction cations of Ca^{2+}, Mg^{2+}, Na^{+}, and K^{+} in units of meq L^{-1} to the nearest 0.1 meq L^{-1}.

10. Precision
Precision data are not available for this procedure.

11. References
U.S. Salinity Laboratory Staff. 1954. L.A. Richards (ed.) Diagnosis and improvement of saline and alkali soils. U.S. Dept. of Agric. Handb. 60. U.S. Govt. Print. Office, Washington, DC.

Calcium, Magnesium, Sodium, and Potassium
(6N, 6O, 6P, and 6Q)
NH₄OAC Extraction
(6N2, 6O2, 6P2, and 6Q2)
Atomic Absorption
Perkin-Elmer AA 5000
(6N2e, 6O2d, 6P2b, and 6Q2b)

1. Application

The extractable bases (Ca^{2+}, Mg^{2+}, Na^+, and K^+) from the NH₄OAC extraction (procedure 5A8c) are generally assumed to be those exchangeable bases on the cation exchange sites of the soil. The abundance of these cations usually occurs in the sequence of $Ca^{2+} > Mg^{2+} > K^+ > Na^+$. Deviation from this usual order signals that some factor or factors, e.g., free $CaCO_3$ or gypsum, serpentine (high Mg^{2+}), or natric material (high Na^+), have altered the soil chemistry. The most doubtful cation extractions with this method are Ca^{2+} in the presence of free $CaCO_3$ or gypsum and K^+ in soils that are dominated by mica or vermiculite (Thomas, 1982).

2. Summary of Method

The NH₄OAc extract from procedure 5A8c is diluted with an ionization suppressant ($LaCl_3$). The analytes are measured by an atomic absorption spectrophotometer (AA). The analyte is measured by absorption of the light from a hollow cathode lamp. An automatic sample changer is used to aspirate a series of samples. The AA converts absorption to analyte concentration. The data are automatically recorded by a microcomputer and printer. The NH₄OAc extracted cations, Ca^{2+}, Mg^{2+}, Na^+, and K^+, are reported in meq 100 g^{-1} oven-dry soil in procedures 6N2e, 6O2d, 6P2b, and 6Q2b, respectively.

3. Interferences

There are four types of interferences (matrix, spectral, chemical, and ionization) in the analyses of these cations. These interferences vary in importance, depending upon the particular analyte selected.

4. Safety

Wear protective clothing and safety glasses. When preparing reagents, exercise special care. Restrict the use of concentrated HCl to a fume hood. Many metal salts are extremely toxic and may be fatal if ingested. Thoroughly wash hands after handling these metal salts.

Follow standard laboratory procedures when handling compressed gases. Gas cylinders should be chained or bolted in an upright position. Acetylene gas is highly flammable. Avoid open flames and sparks. Standard laboratory equipment includes fire blankets and extinguishers for use when necessary. Follow the manufacturer's safety precautions when using the AA.

5. Equipment

5.1 Electronic balance, ±1-mg sensitivity

5.2 Filter paper, pre-pleated, 185-mm diameter, Schleicher and Schuell

5.3 Atomic absorption spectrophotometer (AA), model 5000, Perkin-Elmer Corp., Norwalk, CT

5.4 Automatic burner control, model 5000, Perkin-Elmer Corp., Norwalk, CT

5.5 Autosampler, AS-50, Perkin-Elmer Corp., Norwalk, CT

5.6 Dot matrix printer, P-132, Interdigital Data Systems,Inc.

5.7 Single-stage regulator, acetylene service, part number E11-0-N511A, Air Products and Chemicals, Inc., Box 538, Allentown, PA

5.8 Digital diluter/dispenser, MicroLab 500, Hamilton Co., P.O. Box 10030, Reno, NV

5.9 Syringes, 10000 and 1000 µL, 1001 DX and 1010-TEL LL gas tight, Hamilton Co., P.O. Box 10030, Reno, NV

5.10 Test tubes, 15-mL, 16 mm x 100, for sample dilution and sample changer, Curtin Matheson Scientific, Inc., Houston, TX

5.11 Containers, polypropylene

6. Reagents

6.1 Distilled deionized (DDI) water

6.2 Hydrochloric acid (HCl), concentrated 12 N

6.3 HCl, 1:1 HCl:DDI, 6 N. Carefully mix 1 part of concentrated HCl to 1 part DDI water.

6.4 HCl, 1% wt. Carefully dilute 25 mL of concentrated HCl to 1 L with DDI water.

6.5 NH_4OH, reagent-grade, sp gr 0.90

6.6 Glacial acetic acid, 99.5%

6.7 Primary stock mixed standards solution (PSMSS). Dissolve 0.8759 g of ovendry reagent grade calcium carbonate ($CaCO_3$) in a minimum of volume of 1:1 concentrated HCl:DDI water. Add 0.2127 g of clean Mg ribbon dissolved in 1:1 HCl. Add 1.0956 g of dry reagent grade sodium chloride (NaCl) and 0.1864 g of dry reagent grade KCl. Transfer to a 250-mL volumetric and bring to volume with 1% HCl solution. Resulting solution contains 70 meq L^{-1} (1403 ppm) Ca; 70 meq L^{-1} (851 ppm) Mg; 75 meq L^{-1} (1724 ppm) Na; 10 meq L^{-1} (391 ppm) K. Store in a polypropylene container.

6.8 NH_4OAc solution, 1.0 N, pH 7.0, reagent blank. Mix 57 mL of glacial acetic acid in 600 mL of DDI water. While stirring, carefully add 68 mL of concentrated NH_4OH. Cool and adjust pH to 7.0 using NH_4OH or acetic acid. Dilute to 1 L with DDI water. The NH_4OAc solution is used for extraction of cations (procedure 5A8c).

6.9 Working stock mixed standards solution (WSMSS). Dilute 20 mL of the PSMSS with 80 mL DDI water (1:5). Resulting solution contains 14 meq L^{-1} (281 ppm) Ca; 14 meq L^{-1} (170 ppm) Mg; 15 meq L^{-1} (345 ppm Na); 2 meq L^{-1} (78 ppm) K. Store in a polypropylene container.

6.10 Stock lanthanum ionization suppressant solution, 65,000 ppm. Wet 152.4 g lanthanum oxide (La_2O_3) with 100 mL DDI water. Slowly and cautiously add 500 mL of 6 N HCl to dissolve the La_2O_3. Cooling the solution is necessary. Dilute to 2 L with DDI water. Filter solution. Store in polypropylene container.

6.11 Lanthanum ionization suppressant solution, 6500 ppm. Dilute 200 mL of stock lanthanum ionization suppressant solution with 1800 mL of DDI water (1:10). Store in polypropylene container.

6.12 Dilute calibration mixed standards solution (DCMSS). Dilute 1 part of the WSMSS with 39 parts of the lanthanum solution (1:40). Resulting solution contains 0.35 meq L^{-1} (7 ppm) Ca; 0.35 meq L^{-1} (4 ppm) Mg; 0.375 meq L^{-1} (9 ppm) Na; 0.05 meq L^{-1} (2 ppm) K. Store in polypropylene container.

6.13 Dilute calibration reagent blank solution (DCRBS). Dilute 1 part of DDI water with 39 parts of the lanthanum solution (1:40). Store in polypropylene container.

6.14 Compressed air with water and oil traps.

6.15 Acetylene gas, purity 99.6%.

7. Procedure

Dilution of Sample Extracts and Standards

7.1 The 10-mL syringe is for diluent (lanthanum ionization suppressant solution). The 1-mL syringe is for NH_4OAc sample extracts (procedure 5A8c), calibration reagent blanks, and calibration standards. Set the digital diluter at a 1:40 dilution for the NH_4OAc sample extracts, reagent blanks, and calibration standards as follows:

7.2 Dilute 1 part NH_4OAc sample extract with 39 parts of lanthanum ionization suppressant solution (1:40 dilution).

7.3 Dilute 1 part WSMSS with 39 parts of lanthanum ionization suppressant solution (1:40 dilution). This is the DCMSS. Refer to reagents section.

7.4 Dilute 1 part DDI water with 39 parts of lanthanum ionization suppressant solution (1:40 dilution). This is the DCRBS. Refer to reagents section.

7.5 Dispense the diluted solutions into test tubes which have been placed in the sample holders of the sample changer.

AA Calibration

7.6 Use the DCRBS and the DCMSS to calibrate the AA. The AA program requires a blank and a standard, in that order, to establish a single point calibration curve for element determination. Perform one calibration, i.e., blank plus standard, for every 12 samples.

AA Set-up and Operation

7.7 Refer to the manufacturer's manual for operation of the AA. The following are only very general guidelines for instrument conditions for the various analytes.

Analyte	Conc. meq/L	Burner & Angle	Wave-length	Slit	Fuel/ Oxidant C_2H_2/Air
Ca	14.00	50cm @ 0°	422.7	0.7	10/25
Mg	14.00	50 cm @ 30°	285.2	0.7	10/25
K	2.00	50 cm @ 0°	766.5	1.4	10/25
Na	15.00	50 cm @ 30°	589.0	0.4	10/25

7.8 Use the microcomputer and printer to set instrument parameters and to collect and record instrument readings.

7.9 If sample exceeds calibration standard, dilute the sample (dilution ratio in calculation) with appropriate matrix and record dilution. Remember to keep the matrix the same after dilution by diluting with the lanthanum ionization suppressant solution (1:40 dilution).

7.10 Analyze one quality control check sample for every 48 samples.

7.11 The instrument readings are usually programmed in meq L^{-1}. Record analyte readings to 0.01 meq L^{-1}.

8. Calculations

8.1 The instrument readings are the analyte concentration (meq L^{-1} cation) in undiluted extract. Use these values and calculate the analyte concentration on an ovendry soil basis (meq 100 g^{-1}).

Analyte Concentration in Soil (meq 100 g^{-1}) = (A x B x C x E)/(10 x D)

where:
A = Analyte concentration in extract (meq L^{-1})

B = Extract volume (mL). Refer to procedure 5A8c.

= Weight of extract in syringe (g)/Density of 1 N NH$_4$OAc (1.0124 g cm^{-3})

C = Dilution ratio, if needed
D = Soil sample weight (g)
E = AD/OD ratio (procedure 4B5)

9. Report

Report the extractable Ca^{2+}, Mg^{2+}, Na$^+$, and K$^+$ in units of meq 100 g^{-1} of oven-dry soil to the nearest 0.1 meq 100 g^{-1}.

10. Precision

Precision data are not available for this procedure. A quality control check sample is run with every batch of 48 samples. The number of observations, mean, standard deviation, and C.V. for the quality control check sample are as follows:

Cation	n	Mean	Std. Dev.	C.V.
Ca	85	18.4	0.95	5.1%
Mg	84	7.5	0.23	3.1%
K	81	2.04	0.10	

11. References

Thomas, G.W. 1982. Exchangeable cations. *In* A.L. Page, R.H. Miller, and D.R. Keeney (eds.) Methods of soil analysis. Part 2. Chemical and microbiological properties. 2nd ed. Agronomy 9:159-165.

Calcium, Magnesium, Sodium, and Potassium (6N, 6O, 6P, and 6Q)
NH$_4$OAC Extraction
(6N2, 6O2, 6P2, and 6Q2)
Atomic Absorption
Thermo Jarrell Ash, Smith-Hieftje 4000
(6N2f, 6O2e, 6P2c, and 6Q2c)

1. Application

The extractable bases (Ca^{2+}, Mg^{2+}, Na$^+$, and K$^+$) from the NH$_4$OAC extraction (procedure 5A8c) are generally assumed to be those exchangeable bases on the cation exchange sites of the soil. The abundance of these cations usually occurs in the sequence of Ca^{2+} > Mg^{2+} > K$^+$ > Na$^+$. Deviation from this usual order signals that some factor or factors, e.g., free CaCO$_3$ or gypsum, serpentine (high Mg^{2+}), or natric material (high Na$^+$), have altered the soil chemistry. The most doubtful cation extractions with this method are Ca^{2+} in the presence of free CaCO$_3$ or gypsum and K$^+$ in soils that are dominated by mica or vermiculite (Thomas, 1982).

2. Summary of Method

The NH$_4$OAc extract from procedure 5A8c is diluted with an ionization suppressant (LaCl$_3$). The analytes are measured by an atomic absorption spectrophotometer (AA). The data are automatically recorded by a microcomputer and printer. The NH$_4$OAc extracted cations, Ca^{2+}, Mg^{2+}, Na$^+$, and K$^+$, are reported in meq 100 g^{-1} oven-dry soil in procedures 6N2f, 6O2e, 6P2c, and 6Q2c, respectively.

3. Interferences

There are four types of interferences (matrix, spectral, chemical, and ionization) in the analyses of these cations. These interferences vary in importance, depending upon the particular analyte selected.

4. Safety

Wear protective clothing and safety glasses. When preparing reagents, exercise special care. Restrict the use of concentrated HCl to a fume hood. Many metal salts are extremely toxic and may be fatal if ingested. Thoroughly wash hands after handling these metal salts.

Follow standard laboratory procedures when handling compressed gases. Gas cylinders should be chained or bolted in an upright position. Acetylene gas is highly flammable. Avoid open flames and sparks. Standard laboratory equipment includes fire blankets and extinguishers for use when necessary. Follow the manufacturer's safety precautions when using the AA.

5. Equipment

5.1 Electronic balance, ±1-mg sensitivity
5.2 Filter paper, pre-pleated, 185-mm diameter, Schleicher and Schuell
5.3 Atomic absorption spectrophotometer (AA), Smith-Hieftje Model 4000, Thermo Jarrell Ash Corp., Franklin, MA
5.4 Autosampler, Model 150, Thermo Jarrell Ash Corp., Franklin, MA
5.5 ThermoSpec software, Version 3.01, Enable 4.0, DOS 5.0, Thermo Jarrell Ash Corp., Franklin, MA
5.6 Computer, CUi Advantage 486, Thermo Jarrell Ash Corp., Franklin, MA
5.7 Printer, NEC Pinwriter P3200
5.8 Single-stage regulator, acetylene service, part number E11-0-N511A, Air Products and Chemicals, Inc., Box 538, Allentown, PA
5.9 Digital diluter/dispenser, MicroLab 500, Hamilton Co., P.O. Box 10030, Reno, NV
5.10 Syringes, 10000 and 1000 µL, 1001 DX and 1010-TEL LL gas tight, Hamilton Co., P.O. Box 10030, Reno, NV
5.11 Test tubes, 15-mL, 16 mm x 100, for sample dilution and sample changer, Curtin Matheson Scientific, Inc., Houston, TX
5.12 Containers, polypropylene

6. Reagents

6.1 Distilled deionized (DDI) water
6.2 Hydrochloric acid (HCl), concentrated 12 N
6.3 HCl, 1:1 HCl:DDI, 6 N. Carefully mix 1 part of concentrated HCl to 1 part DDI water.
6.4 HCl, 1% wt. Carefully dilute 25 mL of concentrated HCl to 1 L with DDI water.
6.5 NH_4OH, reagent-grade, sp gr 0.90
6.6 Glacial acetic acid, 99.5%
6.7 Primary stock mixed standards solution (PSMSS). Dissolve 0.8759 g of ovendry reagent grade calcium carbonate ($CaCO_3$) in a minimum of volume of 1:1 concentrated HCl:DDI water. Add 0.2127 g of clean Mg ribbon dissolved in 1:1 HCl. Add 1.0956 g of dry reagent grade sodium chloride (NaCl) and 0.1864 g of dry reagent grade KCl. Transfer to a 250-mL volumetric and bring to volume with 1% HCl solution. Resulting solution contains 70 meq L^{-1} (1403 ppm) Ca; 70 meq L^{-1} (851 ppm) Mg; 75 meq L^{-1} (1724 ppm) Na; 10 meq L^{-1} (391 ppm) K. Store in a polypropylene container.
6.8 NH_4OAc solution, 1.0 N, pH 7.0, reagent blank. Mix 57 mL of glacial acetic acid in 600 mL of DDI water. While stirring, carefully add 68 mL of concentrated NH_4OH. Cool and adjust pH to 7.0 using NH_4OH or acetic acid. Dilute to 1 L with DDI water. The NH_4OAc solution is used for extraction of cations (procedure 5A8c).

6.9 Working stock mixed standards solution (WSMSS). Dilute 20 mL of the PSMSS with 80 mL DDI water (1:5). Resulting solution contains 14 meq L^{-1} (281 ppm) Ca; 14 meq L^{-1} (170 ppm) Mg; 15 meq L^{-1} (345 ppm Na); 2 meq L^{-1} (78 ppm) K. Store in a polypropylene container.

6.10 Stock lanthanum ionization suppressant solution, 65,000 ppm. Wet 152.4 g lanthanum oxide (La$_2$O$_3$) with 100 mL DDI water. Slowly and cautiously add 500 mL of 6 N HCl to dissolve the La$_2$O$_3$. Cooling the solution is necessary. Dilute to 2 L with DDI water. Filter solution. Store in polypropylene container.

6.11 Lanthanum ionization suppressant solution, 6500 ppm. Dilute 200 mL of stock lanthanum ionization suppressant solution with 1800 mL of DDI water (1:10). Store in polypropylene container.

6.12 Dilute calibration mixed standards solution (DCMSS). Dilute 1 part of the WSMSS with 39 parts of the lanthanum solution (1:40). Resulting solution contains 0.35 meq L^{-1} (7 ppm) Ca; 0.35 meq L^{-1} (4 ppm) Mg; 0.375 meq L^{-1} (9 ppm) Na; 0.05 meq L^{-1} (2 ppm) K. Store in polypropylene container.

6.13 Dilute calibration reagent blank solution (DCRBS). Dilute 1 part of DDI water with 39 parts of the lanthanum solution (1:40). Store in polypropylene container.

6.14 Compressed air with water and oil traps.

6.15 Acetylene gas, purity 99.6%.

7. Procedure

Dilution of Sample Extracts and Standards

7.1 The 10-mL syringe is for diluent (lanthanum ionization suppressant solution). The 1-mL syringe is for NH$_4$OAc sample extracts (procedure 5A8c), calibration reagent blanks, and calibration standards. Set the digital diluter at a 1:40 dilution for the NH$_4$OAc sample extracts, reagent blanks, and calibration standards as follows:

7.2 Dilute 1 part NH$_4$OAc sample extract with 39 parts of lanthanum ionization suppressant solution (1:40 dilution).

7.3 Dilute 1 part WSMSS with 39 parts of lanthanum ionization suppressant solution (1:40 dilution). This is the DCMSS. Refer to reagents section.

7.4 Dilute 1 part DDI water with 39 parts of lanthanum ionization suppressant solution (1:40 dilution). This is the DCRBS. Refer to reagents section.

7.5 Dispense the diluted solutions into test tubes which have been placed in the sample holders of the sample changer.

AA Calibration

7.6 Use the DCRBS and the DCMSS to calibrate the AA. The AA program requires a blank and a standard, in that order, to establish a single point calibration curve for element determination. Perform one calibration, i.e., blank plus standard, for every 12 samples.

AA Set-up and Operation

7.7 Refer to the manufacturer's manual for operation of the AA. The following are only very general guidelines for instrument conditions for the various analytes.

Analyte	Conc. meq/L	Burner & Angle	Wave-length	Slit	Fuel/ Oxidant C2H2/Air
Ca	14.00	50cm @ 0°	422.7	0.7	10/25
Mg	14.00	50 cm @ 30°	285.2	0.7	10/25
K	2.00	50 cm @ 0°	766.5	1.4	10/25
Na	15.00	50 cm @ 30°	589.0	0.4	10/25

7.8 Use the microcomputer and printer to set instrument parameters and to collect and record instrument readings.

7.9 If sample exceeds calibration standard, dilute the sample (dilution ratio in calculation) with appropriate matrix and record dilution. Remember to keep the matrix the same after dilution by diluting with the lanthanum ionization suppressant solution (1:40 dilution).

7.10 Analyze one quality control check sample for every 48 samples.

7.11 The instrument readings are usually programmed in meq L^{-1}. Record analyte readings to 0.01 meq L^{-1}.

8. Calculations

8.1 The instrument readings are the analyte concentration (meq L^{-1} cation) in undiluted extract. Use these values and calculate the analyte concentration on an ovendry soil basis (meq 100 g^{-1}).

Analyte Concentration in Soil (meq 100 g^{-1}) = (A x B x C x E)/(10 x D)

where:
A = Analyte concentration in extract (meq L^{-1})
B = Extract volume (mL). Refer to procedure 5A8c.

= Weight of extract in syringe (g)/Density of 1 N NH$_4$OAc (1.0124 g cm^{-3})

C = Dilution ratio, if needed
D = Soil sample weight (g)
E = AD/OD ratio (procedure 4B5)

9. Report
   Report the extractable Ca^{2+}, Mg^{2+}, Na^+, and K^+ in units of meq 100 g^{-1} of oven-dry soil to the nearest 0.1 meq 100 g^{-1}.

10. Precision
   Precision data are not available for this procedure. A quality control check sample is run with every batch of 48 samples.

11. References

Thomas, G.W. 1982. Exchangeable cations. *In* A.L. Page, R.H. Miller, and D.R. Keeney (eds.) Methods of soil analysis. Part 2. Chemical and microbiological properties. 2nd ed. Agronomy 9:159-165.

Total Sulfur (6R)
SO2 Evolution, Infrared (6R3)
LECO SC-444 Sulfur Analyzer (6R3c)

1. Application

Organic and inorganic S forms are found in soils, with the organic S fraction accounting for >95% of the total S in most soils from humid and semi-humid (Tabatabai, 1982). Mineralization of organic S and its conversion to sulfate by chemical and biological activity may serve as a source of plant available S. Total S typically ranges from 0.01 to 0.05% in most mineral soils. In organic soils, total S may be >0.05%.

In well-drained, well-aerated soils, most of the inorganic S normally occurs as sulfate. In marine tidal flats, other anaerobic marine sediments, and mine spoils, there are usually large amounts of reduced S compounds which oxidize to sulfuric acid upon exposure to the air. In arid regions, significant amounts of inorganic S are found as sulfates such gypsum and barite.

The typical use of total S is as an index of the total reserves of this element, which may be converted to plant available S. The SSL uses the combustion technique (LECO sulfur analyzer) for analysis of total S (procedure 6R3b). Extractable sulfate S (SO_4^2-S) is an index of readily plant-available S. Reagents that have been used for measuring SO_4^2-S include water, hot water, ammonium acetate, sodium carbonate and other carbonates, ammonium chloride and other chlorides, potassium phosphate and other phosphates, and ammonium fluoride (Bray-1). Extractable SO_4^2-S does not include the labile fraction of soil organic S that is mineralized during the growing season (Tabatabai, 1982). Extraction reagents for organic S include hydrogen peroxide, sodium bicarbonate, sodium hydroxide, sodium oxalate, sodium peroxide, and sodium pyrophosphate. There other methods available for determination of soil S, especially for total S and SO_4^2-S. The investigator may refer to the review by Beaton et al. (1968).

2. Summary of Method

A fine-ground (<80-mesh) soil sample is oxidized at high temperature. The gases released are scrubbed, and the SO_2 in the combustion gases are measured using an infrared detector. Percent S is reported on an oven-dry soil basis.

3. Interferences

No significant interferences are known to affect the oxidizable S measurement.

4. Safety

Wear protective clothing and safety glasses. Magnesium perchlorate may form explosive mixtures. Magnesium perchlorate may contain traces of perchloric acid, which remain from manufacturer's operations. This acid is anhydrous because of the strong desiccating capability of the salt. Avoid prolonged contact with oxidizable material or material capable of forming unstable perchlorate esters or salts. Remove magnesium perchlorate by using an excess of water to thoroughly dilute the material.

The use of high temperatures in the oxidation of samples requires that extreme caution be used to prevent burns and fires. Follow standard laboratory procedures when handling compressed gases. Oxygen is highly flammable. Avoid open flames and sparks. Standard laboratory equipment includes fire blankets and extinguishers for use when necessary. Follow the manufacturer's safety precautions when using the sulfur analyzer.

5. Equipment
5.1 Sulfur analyzer, Leco Model SC-444, Sulfur and Carbon Analyzers, Leco Corp., St. Joseph, MI
5.2 Combustion boats, part no. 529-203, Leco Corp., St. Joseph, MI
5.3 Single-stage regulator, Oxygen Service, Part No. E11-W-N115BOX, Air Products and Chemicals, Inc., Box 538, Allentown, PA 18105
5.4 Electronic balance, ±1-mg sensitivity

6. Reagents
6.1 Anhydrone, anhydrous magnesium perchlorate, granular
6.2 Glass wool
6.3 Compressed oxygen, >99.5% @ 30 psi
6.4 Calcium carbonate, $CaCO_3$, reagent grade.
6.5 Solid/Powder Combustion Controller, part no. 501-426, Leco Corp., St. Joseph, MI
6.6 Soil Calibration Sample, part no. 502-062, Leco Corp., St. Joseph, MI
6.7 Sulfur Calibration Sample, part no. 502-648, Leco Corp., St. Joseph, MI

7. Procedure
7.1 Use a fine-ground 80-mesh, air-dry soil

7.2 Prepare instrument as outlined in the operator's instruction manual (Leco, 1994; Leco, 1993).

7.3 Methods are created with the method menu and stored in the instrument memory. System parameters are set as follows:

Furnace operating temperature: 1450°C
Lance delay: 20 s
Analysis time settings: 120 to 300 s
Comparator level settings: 0.3%

7.3 Condition instrument by analyzing a few soil samples, until readings are stable.

7.4 Calibrate instrument by analyzing at least three replicates of each calibration standard. Use the soil calibration standard for samples with less than 0.01 percent TS and the sulfur standard for samples with more than 0.01 percent TS. Weigh standards in a range from 0.2 to 0.5 g.

7.5 Load samples on autoload rack, place in the analyzer, and press analyze key.

7.6 Weigh 0.2 to 0.5 g sample in a tared combustion boat. Add approximately 1 g of solid/powder combustion controller to sample.

7.7 Load samples on autoload rack, place in the analyzer, and press analyze key.

7.9 Repack the reagent (anhydrous magnesium perchlorate) tubes whenever the reagent becomes caked or moist or the warning alarm displays.

8. Calculations

$$S\ (\%) = S_i \times AD/OD$$

where:
$S\ (\%)$ = S (%) on oven-dry basis
S_i = S (%) instrument

AD/OD   = air-dry/oven-dry ratio (procedure 4B5)

9. Report
Report total S as a percentage of oven-dry weight to the nearest 0.1%.

10. Precision
Precision data are not available for this procedure. A quality control check sample is run in every batch of 12 samples. A blank (crucible only) and a rerun of one of the 12 samples (unknowns) also are run in every batch. For 27 observations of the quality control check sample, the mean, standard deviation, and C.V. for total S are 0.57, 0.02, and 4.3%, respectively.

11. References
Beaton, James D., G.R. Burns, and J. Platou. 1968. Determination of sulfur in soils and plant material. Tech. Bull. No. 14. The Sulfur Inst., 1725 K Street, N.W., Washington, D.C. 20006

Leco Corp. 1993. Sulfur and carbon in cements, soils, rock, ceramic and similar materials. Application Bulletin. Leco Corp., 3000 Lakeview Ave., St. Joseph, MI.

Leco Corp. 1994. Instruction Manual. SC-444 Sulfur and Carbon Analyzers. Leco Corp. , 3000 Lakeview Ave., St. Joseph, MI.

Tabatabai, M.A. 1982. Sulfur. *In* A.L. Page, R.H. Miller, and D.R. Keeney (eds.) Methods of soil analysis. Part 2. Chemical and microbiological properties. 2nd ed. Agronomy 9:501-538.

Bray P-1 Absorbed Phosphorus (6S)
Baush and Laumb, Spectrophotometer 20 (6S3)

1. Application
The Bray P-1 procedure is widely used as an index of available P in the soil. The selectivity of the Bray extractant is designed to remove the easily acid-soluble P, largely calcium phosphates, and a portion of the phosphates of Al and Fe (Bray and Kurtz, 1945; Olsen and Sommers, 1982). In general, this method has been most successful on acid soils (Bray and Kurtz, 1945; Olsen and Sommers, 1982).

2. Summary of Method
A 1-g soil sample is shaken with 10 mL of extracting solution for 15 min at 100 oscillations per min^{-1}. The solution is filtered. A 2-mL aliquot is transferred to a colorimetric tube to which 8-mL of ascorbic acid molybdate solution are added. The percent transmittance of the solution is read using a spectrophotometer. The Bray P-1 is reported in mg kg^{-1} (ppm) P.

3. Interferences
Many procedures may be used to determine P. Studies have shown that incomplete or excessive extraction of P to be the most significant contributor to interlaboratory variation. The Bray P-1 procedure is sensitive to the soil/extractant ratio, shaking rate, and time. This extraction uses the ascorbic acid-potassium antimonyl-tartrate-molybdate method. The Fiske-Subbarrow method is less sensitive but has a wider range before dilution is required (North Central Regional Publication No. 221, 1988). For calcareous soils, the Olsen method is preferred. An alternative procedure for calcareous soils is to use the Bray P-1 extracting solution at a 1:50 soil:solution ratio. This procedure has been shown to be satisfactory for some calcareous soils (North Central Regional Publication No. 221, 1988; Smith et al., 1957).

4. Safety

Wear protective clothing (coats, aprons, sleeve guards, and gloves) and eye protection (face shields, goggles, or safety glasses). When preparing reagents, exercise special care. Many metal salts are extremely toxic and may be fatal if ingested. Thoroughly wash hands after handling these metal salts. Restrict the use of concentrated H_2SO_4 and HCl to a fume hood. Use safety showers and eyewash stations to dilute spilled acids and bases. Use sodium bicarbonate and water to neutralize and dilute spilled acids.

5. Equipment

5.1 Electronic balance, ±0.01-g sensitivity
5.2 Shaker, Eberbach 6000 power unit, reciprocating speed of 60 to 260 oscillations min^{-1}, with 6040 utility box carrier and 6110 floor stand, Eberbach Corp., Ann Arbor, MI
5.3 Spectrophotometer 20, Baush and Laumb
5.4 Pipettors, electronic digital, Rainin Instrument Co., Woburn, MA, 2500 µL and 10 mL
5.5 Cuvettes, glass, 10 mL, 1-cm light path
5.5 Funnel, 60° angle, long stem, 50-mm diameter
5.7 Filter paper, quantitative, Whatman grade 2, 9-cm diameter
5.8 Erlenmeyer flasks, 50 mL
5.9 Centrifuge, high-speed, International Equipment Co., IECB-22M

6. Reagents

6.1 Distilled deionized (DDI) water
6.2 Hydrochloric acid (HCl), concentrated, 12 N
6.3 HCl, 1 N. Carefully add 83.33 mL of concentrated HCl to DDI water and dilute to 1-L volume.
6.4 Sulfuric acid (H_2SO_4), concentrated, 36 N
6.5 Bray No. 1 Extracting solution, 0.025 N HCl and 0.03 N NH_4F. Dissolve 8.88 g of NH_4F in 4 L DDI H_2O. Add 200 mL of 1.0 N HCl and dilute to 8 L with DDI water. The solution pH should be 2.6 ± 0.5. Store in a polyethylene bottle.
6.6 Stock standard P solution (SSPS), 100 ppm P. Add 0.2197 g of KH_2PO_4 in 25 mL of DDI water. Dilute to a final volume of 500 mL with extracting solution. Store in a refrigerator. Solution is stable to 1 yr.
6.7 Sulfuric-tartrate-molybdate solution (STMS). Dissolve 60 g of ammonium molybdate tetrahydrate [$(NH_4)_6Mo_7O_{24}\cdot4H_2O$] in 200 mL of boiling DDI water. Allow to cool to room temperature. Dissolve 1.455 g of antimony potassium tartrate (potassium antimonyl tartrate hemihydrate $K(SbO)C_4H_4O_6\cdot1/2H_2O$) in the ammonium molybdate solution. Slowly and carefully add 700 mL of concentrated H_2SO_4. Cool and dilute to 1 L with DDI water. Store in the dark in the refrigerator.
6.8 Ascorbic acid solution. Dissolve 33.0 g of ascorbic acid in DDI water and dilute to 250 mL with DDI water. Store in the dark in the refrigerator.
6.9 Working ascorbic acid molybdate solution (WAMS). Prepare fresh each day. Mix 25 mL of STMS solution with 800 mL of DDI water. Add 10 mL of ascorbic acid solution and dilute to 1 L with DDI water.
6.10 Standard P calibration solutions (SPCS), 0.2, 0.5, 1.0, 2.0, 3.0, 4.0, 5.0, and 6.0, 7.0, and 8.0 ppm. Dilute the SSPS with the extracting solution as follows: 0.2 ppm = 0.5:250; 0.5 ppm = 0.5:100; 1.0 ppm = 1:100; 2.0 ppm = 2:100; 3.0 ppm = 3:100; 4.0 ppm = 4:100; 5.0 ppm = 5:100; 6.0 ppm = 6:100; 7.0 ppm = 7:100; 8.0 ppm = 8:100.

7. Procedure

7.1 Weigh 1.00 g of air-dry soil into a 50-mL Erlenmeyer flask.

7.2 Dispense 10.0 mL of extracting solution to flask.

7.3 Securely place the flask in the shaker. Shake for 15 min at 100 oscillations min^{-1} at room temperature (20°C).

7.4 Remove the sample from the shaker. Decant, filter, and collect extract.

7.5 Centrifuging or repeated filtering may be necessary to obtain clear extracts. Decant into 13-mL centrifuge tube and centrifuge at 10000 RPM for 10 min.

7.6 Use the pipettor to transfer a 2-mL aliquot of the sample to a cuvette. Also transfer a 2-mL aliquot of each SPCS to a cuvette. Use a clean pipette tip for each sample and SPCS.

7.7 Dispense 8 mL of the WAMS to sample aliquot and to each SPCS (1:5 dilution).

7.8 The color reaction requires a minimum of 20 min before analyst records readings.

7.9 Set the spectrophotometer (red bulb) to read at 882 nm.

7.10 Set the 100% transmittance against the blank which has 8 mL of the WAMS solution and 2 mL of extracting solution.

8. Calculations

8.1 Transmittance of a solution is the fraction of incident radiation transmitted by the solution, i.e., $T = P/P_0$ and is often expressed as a percentage, i.e., $\%T = P/P_0 \times 100$. The absorbance of a solution is directly proportional to concentration and is defined by the equation, $A = -\log_{10} T$. These relationships are derived from Beer's law.

Calibration Calculations

8.2 Use transmission of each SPCS to either construct a calibrated curve to plot P or use a least squares analysis to calculate P. The P is reported in ppm.

8.3 *Calibration Curve*: Plot the transmittance against the ppm P of each SPCS on semilog graph paper or convert to absorbances and plot on linear graph paper. Construct the calibration curve by finding the "best" line that fits the plotted SPCS.

8.4 *Linear Squares Analysis*: Use a least squares criterion, i.e. best moving average. Refer to a statistical analysis book for additional information on least squares analysis. To facilitate data manipulation in a least squares analysis, the following standard curve is developed using the concentration of SPCS as a f[ln(%T)]. Final calculated analyte concentration with either $\log_{10}$ or ln base would be the same. Using the following example, calculate analyte concentration with P (ppm) in extract = Y variable and percent transmittance (% T) = the X variable. The X variable is the natural logarithm of T.

P (ppm)	T (%)
0	100
2	77
4	59
6	45
8	33
10	24

Number of standards = n = 6

$\sum Y_i = 30$ $\sum X_i = 23.5077$

$\sum Y_i/n = Y = 5$ $\sum X_i/n = X = 3.9180$
$\sum X_iY_i = 107.5902$ $\sum X_i^2 = 93.5185$
$\sum X_i \sum Y_i = 705.231$

$$b = \frac{\sum X_iY_i - \sum X_i\sum Y_i/n}{\sum X_i^2 - (\sum X_i)^2/n} = \frac{107.5902 - 117.5385}{93.5185 - 92.102} \quad -7.023$$

b = slope of the line, i.e., the amount that Y changes when X changes by 1 unit.

The equation is as follows:

$Y = Y + b (X - X)$

$Y = 5 - 7.023 (\ln(X) - 3.9180)$

Analyte Calculation

8.5 *Calibration Curve*: Read the P (ppm) directly from the calibration curve.

8.6 *Least Squares Analysis*: Put the ln(%T) in the preceding equation and solve for ppm P. Thus, if sample extract has 84% transmission, the preceding equation is as follows:

$Y = 5 - 7.023 \ln(84) + 27.516 = 1.40$ ppm

8.7 Convert the extract P (ppm) to soil P (ppm or lbs/A) as follows:

Soil P (ppm) = Extract P (ppm) x 10

Soil P (lbs/A) = Extract P (ppm) x 20

9. Report
Report the soil Bray P-1 mg kg^{-1} (ppm) to the nearest whole number.

10. Precision
Precision data are not available for this procedure.

11. References
Bray, R.H., and L.T. Kurtz. 1945. Determination of total, organic, and available forms of phosphorus in soils. Soil Sci. 59:39-45.

North Central Regional Publication No. 221. 1988. Recommended chemical soil test procedures for the North Central region. Agric. Exp. Stn. of IL, IN, IA, KS, MI, MN, MS, NE, ND, OH, SD, WI, and USDA cooperating.

Olsen, S.R., and L.E. Sommers. 1982. *In* A.L. Page, R.H. Miller, and D.R. Keeney (eds.) Methods of soil analysis. Part 2. Chemical and microbiological properties. 2nd ed. Agronomy 9:403-430.

Smith, F.W., B.G. Ellis, and J. Grava. 1957. Use of acid-fluoride solutions for the extraction of available phosphorus in calcareous soils and in soils to which rock phosphate has been added. Soil Sci. Soc. Am. Proc. 21:400-404.

New Zealand P Retention (6S)
UV-Visible Spectrophotometer (6S4)
Beckmann DU-7 (6S4b)

1. Application

In *Soil Taxonomy*, the P retention of soil material is a criterion for andic soil properties (Soil Survey Staff, 1990). Andisols and other soils that contain large amounts of allophane and other amorphous minerals have capacities for binding P (Gebhardt and Coleman, 1984). The factors that affect soil P retention are not well understood. However, allophane and imogolite have been considered as major materials that contribute to P retention in Andisols (Wada, 1985). Phosphate retention is also called P absorption, sorption, or fixation.

2. Summary of Method

A 5-g soil sample is shaken in a 1000-ppm P solution for 24 h. The mixture is centrifuged at 2000 rpm for 15 min. An aliquot of the supernatant is transferred to a colorimetric tube to which nitric vanadomolybdate acid reagent (NVAR) is added. The percent transmittance of the solution is read using a spectrophotometer. The New Zealand P retention is reported as percent P retained.

3. Interferences

No significant problems are known to affect the P retention measurement.

4. Safety

Wear protective clothing (coats, aprons, sleeve guards, and gloves) and eye protection (face shields, goggles, or safety glasses). When preparing reagents, exercise special care. Many metal salts are extremely toxic and may be fatal if ingested. Thoroughly wash hands after handling these metal salts. Restrict the use of concentrated HNO_3 to a fume hood. Use safety showers and eyewash stations to dilute spilled acids and bases. Use sodium bicarbonate and water to neutralize and dilute spilled acids.

5. Equipment

5.1 Electronic balance, ±0.01-g sensitivity
5.2 Shaker, Eberbach 6000 power unit, reciprocating speed of 60 to 260 epm, with 6040 utility box carrier and 6110 floor stand, Eberbach Corp., Ann Arbor, MI
5.3 Digital diluter/dispenser, Microlab 500, Hamilton Co., P.O. Box 10030, Reno, NV
5.4 Syringes, 10000 and 1000 µL, 1001 DX and 1010-TEL LL gas tight, Hamilton Co., P.O. Box 10030, Reno, NV
5.5 Diluter/dispenser, 25 mL
5.6 UV-Visible Spectrophotometer, DU-7, Beckmann Instruments Inc.
5.7 Cuvettes, Labcraft Brand, disposable, polystyrene, square-bottom, 4.5 mL, 12.5 mm x 12.5 mm x 46 mm, Curtin Matheson Scientific, Inc., Houston, TX
5.8 Centrifuge, International No. 2, Model V, with no. 250 A head, International Equip. Co., Boston, MA
5.9 Trunions, International no. 320, International Equip. Co., Boston, MA
5.10 Centrifuge tubes, 50 mL, Oak-Ridge, polyallomer, Nalgene 3119, Nalge Co., Box 20365, Rochester, NY
5.11 Plastic cups, 2 fl. oz.
5.12 Pipets, volumetric, class A, glass, various sizes of 1 to 20 mL

6. Reagents

6.1 Distilled deionized (DDI) water
6.2 Nitric acid (HNO_3), concentrated, 16 N
6.3 P retention solution, 1000 ppm P. Dissolve 35.2 g of KH_2PO_4 and 217.6 g of sodium acetate ($Na_2C_2H_3O_2 \cdot 3H_2O$) in DDI water. Add 92 mL of glacial acetic acid. Dilute to 8 L

with DDI water in a volumetric flask. The solution pH should range between 4.55 and 4.65.

6.4 Molybdate solution. Dissolve 16 g of ammonium molybdate (VI) $[(NH_4)_6Mo_7O_{24} \cdot 4H_2O]$ in 50°C DDI water. Allow the solution to cool to room temperature and dilute to 1 L with DDI water.

6.5 Nitric acid solution. Carefully and slowly dilute 100 mL of concentrated HNO_3 to 1 L of DDI water. Add the acid to the water.

6.6 Nitric vanadomolybdate acid reagent (NVAR), vanadate solution. Dissolve 0.8 g of NH_4VO_3 in 500 mL of boiling DDI water. Allow the solution to cool to room temperature. Carefully and slowly add 6 mL of concentrated HNO_3. Dilute to 1 L with DDI water. Mix the nitric acid solution with the vanadate solution and then add the molybdate solution. Mix well.

6.7 Stock P standard solution (SPSS), 4000 ppm P. Dissolve 17.6 g KH_2PO_4 in DDI water. Dilute to 1 L with DDI water.

6.8 Standard P calibration P solutions (SPCS), 100, 80, 60, 40, 20, and 0% P retained. Dilute the SPSS with a solution that contains 32.8 g of sodium acetate (CH_3COONa) and 23 mL of glacial acetic acid diluted to 2 L with DDI water as follows: 100% = DDI water (0 ppm); 80% = 1:20 (200 ppm); 60% = 1:10 (400 ppm); 40% = 3:20 (600 ppm); 20% = 1:5 (800 ppm); and 0% = 1:4 (1000 ppm). The percent amount refers to percent P retention.

7. Procedure

7.1 Weigh 5.00 g of air-dry soil into a 50-mL centrifuge tube.

7.2 Use the dispenser to add 25.0 mL of P-retention solution to centrifuge tube.

7.3 Cap centrifuge tube and place in shaker and shake for 24 h at room temperature (20°C).

7.4 Add 2 to 3 drops of Superfloc, 0.02% w/v to each tube.

7.5 Centrifuge sample at 2000 rpm for 15 min. Filter using a Milipore filter, if necessary.

7.6 Pour sample supernatant into plastic cup.

7.7 Use the digital diluter to add the nitric vanadomolybdate acid reagent (NVAR) to each sample supernatant and to each SPCS. To fill a 4.5-mL cuvette, use a dilution of 1:20 sample dilution.

7.8 The color reaction requires a minimum of 30 min before the analyst records readings.

7.9 Set the spectrophotometer to read at 466 nm. Auto zero using the DDI water (blank). A blank has all reagents contained in the sample extract except the soil.

7.10 Record the percent transmittance to the nearest 0.01 unit for the sample extract and each SPCS.

8. Calculations

8.1 Transmittance of a solution is the fraction of incident radiation transmitted by the solution, i.e., $T = P/P_0$ and is often expressed as a percentage, i.e., $T = P/P_0 \times 100$. The absorbance of a solution is directly proportional to concentration and is defined by the equation, $A = -\log_{10} T$. These relationships are derived from Beer's law.

Calibration Calculations

8.2 Use the transmittance of each SPCS to either construct a calibrated curve to plot P or use a least squares analysis to calculate P. The P is reported in percent retained.

8.3 *Calibration Curve*: Plot the transmittances against the ppm P of each SPCS on semilog graph paper or convert to absorbances and plot on linear graph paper. Construct the calibration curve by finding the "best" line that fits the plotted SPCS.

8.4 *Least Squares Analysis*: Use a least squares criterion, i.e. best moving average. Refer to a statistical analysis book for additional information on least squares analysis. To facilitate data manipulation in a least squares analysis, the following standard curve is developed using the concentration of SPCS as a f[ln(%T]. Final calculated analyte concentration with either $\log_{10}$ or ln base would be the same. Refer to procedure 6S3b for an example of least squares analysis.

Analyte Calculation

8.5 *Calibration Curve*: Read the percent P directly from the calibration curve.

8.6 *Least Squares Analysis*: Refer to procedure 6S3 for an example of least squares analysis.

9. Report

Report the percent New Zealand P retention to the nearest whole number.

10. Precision

Precision data are not available for this procedure.

11. References

Gebhardt, H., and N.T. Coleman. 1984. Anion adsorption of allophanic tropical soils: III. Phosphate adsorption. p. 237-248. *In* K.H. Tan (ed.) Anodosols. Benchmark papers in soil science series. Van Nostrand Reinhold, Co., Melbourne, Canada.

Soil Survey Staff. 1990. Keys to soil taxonomy. 4th ed. SMSS technical monograph no. 6. Blacksburg, VA.

Wada, K. 1985. The distinctive properties of Andosols. p. 173-229. *In* B.A. Stewart (ed.) Adv. Soil Sci. Springer-Verlag, NY.

Citric Acid Extractable Phosphorus (6S)
Beckmann DU-7, UV-Visible Spectrophotmeter (6S5)

1. Application

In *Soil Taxonomy*, citric acid soluble P_2O_5 is a criterion for distinguishing between mollic (<250 ppm P_2O_5) and anthropic epipedons (>250 ppm P_2O_5) (Soil Survey Staff, 1975). Additional data on anthropic epipedons from several parts of the world may permit improvements in this definition (Soil Survey Staff, 1994). The procedure 6S5 is used by N.A.A.S. (England and Wales) and is based on the method developed by Dyer (1894).

2. Summary of Method

A sample is checked for $CaCO_3$ equivalent. Sufficient citric acid is added to sample to neutralize the $CaCO_3$ plus bring the solution concentration of citric acid to 1%. A 1:10 soil:solution is maintained for all samples. The sample is shaken for 16 h and filtered. Ammonium molybdate and stannous chloride are added. The percent transmittance of the solution is read using a spectrophotometer. The 1% citric acid extractable P_2O_5 is reported in mg kg^{-1} (ppm).

3. Interferences

Unreacted carbonates interfere with the extraction of P_2O_5. Sufficient citric acid is added to sample to neutralize the $CaCO_3$. However, a high citrate level in sample may interfere with the molybdate blue test. If this occurs, the method can be modified by evaporating the extract and ashing in a muffle furnace to destroy the citric acid.

Positive interferences in the analytical determination of P_2O_5 are silica and arsenic, if the sample is heated. Negative interferences in the P_2O_5 determination are arsenate, fluoride, thorium, bismuth, sulfide, thiosulfate, thiocyanate, or excess molybdate. A concentration of Fe >1000 ppm interferes with P_2O_5 determination. Refer to Snell and Snell (1949) and Metson (1956) for additional information on interferences in the citric acid extraction of P_2O_5.

4. Safety

Wear protective clothing (coats, aprons, sleeve guards, and gloves) and eye protection (face shields, goggles, or safety glasses) when preparing reagents, especially concentrated acids and bases. Dispense concentrated acids and bases in fume hood. Use the safety showers and eyewash stations to dilute spilled acids and bases. Use sodium bicarbonate and water to neutralize and dilute spilled acids. Follow standard laboratory procedures.

5. Equipment

5.1 Electronic balance, ±0.01-g sensitivity

5.2 Shaker, Eberbach 6000 power unit, reciprocating speed of 60 to 260 oscillations min^{-1}, with 6040 utility box carrier and 6110 floor stand, Eberbach Corp., Ann Arbor, MI

5.3 Digital diluter/dispenser, product no. 100004, with hand probe and actuator, product no. 230700, Hamilton Co., P.O. Box 10030, Reno, NV

5.4 Syringes, 10000 and 1000 µL, 1001 DX and 1010-TEL LL gas tight, Hamilton Co., P.O. Box 10030, Reno, NV,

5.5 Centrifuge tubes, 50 mL, Oak-Ridge, polyallomer, Nalgene 3119, Nalge Co., Box 20365, Rochester, NY

5.6 Filter paper, quantitative, Whatman grade 2, 9-cm diameter

5.7 Funnel, 60° angle, long stem, 50-mm diameter

5.8 Erlenmeyer flasks, 50 ml

5.9 Bottles with gas release caps

5.10 Pipettors, electronic digital, Rainin Instrument Co., Woburn, MA, 2500 µL and 10 mL Digital Pipet, 10-ml

5.11 UV-visible spectrophotometer, DU-7, Beckmann Instruments, Inc.

5.12 Cuvettes, Labcraft Brand, disposable, polystyrene, square-bottom, 4.5 mL, 12.5 mm x 12.5 mm x 46 mm, Curtin Matheson Scientific, Inc., Houston, TX

6. Reagents

6.1 Distilled, dionized (DDI) water

6.2 Hydrochloric acid (HCl), concentrated, 12 N

6.3 Citric acid solution, 10%. Dissolve 100 g of anhydrous citric acid ($C_6H_8O_7$) in 1-L volumetric flask.

6.4 Citric acid solution, 1%. Dilute 100.0 ml of 10% citric acid solution to 1-L with DDI water

6.5 Ammonium molybdate solution, 1.5%. Dissolve 15.0 g of ammonium molybdate [$(NH_4)_6MO_7O_{24} \cdot 4H_2O$] in 300 mL of distilled water. Transfer to a 1-L volumetric flask and carefully add 310 mL of concentrated HCl. Allow to cool. Make to 1-L volume with DDI water. Store in brown bottle in the dark. Solution is stable for ~ 3 months.

6.6 Stock stannous chloride solution (SSCS). Dissolve 10 g of stannous chloride ($SnCl_2 \cdot 2H_2O$) in 100 mL of concentrated HCl.

6.7 Working stannous chloride solution (WSCS). Dilute 2 mL of SSCS with 100 mL of DDI water. Use immediately as solution is only stable for ~ 4 h.

6.8 Stock standard P_2O_5 solution (SSPS), 250 ppm P. Dissolve 1.099 g of potassium dihydrogen orthophosphate (KH_2PO_4) with DDI water in 1-L volumetric flask. Add 5 ml of 2 N HCL. Make to 1-L volume with DDI water.

6.9 Working stock standard P_2O_5 solution (WSSPS), 2.5 ppm P. Pipet 10.0 mL of SSPS and dilute to 1-L in a volumetric flask with DDI water.

6.10 Standard P_2O_5 calibration solutions (SPCS). Pipet 0, 1, 2, 3, 4, and 5 mL of WSSPS into 50-mL oakridge tubes. Add 1 ml of 1% citric acid solution. Continue color development as for samples. Distilled water may be used as a blank.

7. Procedure

7.1 Weigh 3.00 g of <2-mm, air-dry soil into a bottle gas release tops. If the soil does not contain free carbonates, proceed to step 7.3.

7.2 If the soil contains free $CaCO_3$, refer to Table 1 to determine the amount of 10% citric acid solution required to neutralize the $CaCO_3$. Add required mls of 10% citric acid into a graduated cylinder and bring to a volume of 30-ml with DDI water. Add this solution to the soil. Swirl the bottle over a period of 6 h at 100 oscillations min^{-1} dissolve and neutralize the $CaCO_3$. Proceed to step 7.4.

Table 1. Volume of 10% citric acid (mL) required to decompose CaCO3 (%) and to bring to solution concentration to 1% in a final volume of 30 mL for 3-g sample.

%CC[1]	mL CA[2]	% CC	mL CA	% CC	mL CA	%CC	mL CA
0	3.0	16	9.7	32	16.4	48	23.2
1	3.4	17	10.2	33		49	23.6
2	3.8	18	10.6	34		50	24.0
3	4.3	19	11.0	35	17.7	51	24.4
4	4.7	20	11.4	36	18.1	52	24.8
5	5.1	21	11.8	37	18.6	53	25.3
6	5.5	22	12.2	38	19.0	54	25.7
7	6.0	23	12.7	39	19.4	55	26.1
8	6.4	24	13.1	40	19.8	56	26.5
9	6.8	25	13.5	41	20.2	57	27.0
10	7.2	26	14.0	42	20.6	58	27.4
11	7.6	27	14.4	43	21.0	59	27.8
12	8.0	28	14.8	44	21.5	60	28.2
13	8.5	29	15.2	45	21.9	61	28.6
14	8.9	30	15.6	46	22.4	62	29.0
15	9.3	31	16.0	47	22.8	63	29.5

[1]%CC = percent calcium carbonate in a sample
[2]CA = ml of 10% citric acid needed to be diluted to 30-ml volume with RODI water and added to sample

7.3 If the soil contains no free $CaCO_3$, add 30 mL of 1% citric acid solution to the sample.

7.4 Cap the bottles, place in a shaker and shake for 16 h at 100 oscillations min^{-1}.

7.5 Remove the sample from shaker and filter.

7.6 Pipet 1 mL of sample extract into a 50-ml oakridge tube. Add 4mL of ammonium molybdate solution to all samples and standards. Bring up to 25 ml mark with DDI water. Add 2 ml stannous chloride. Shake to mix and allow to stand 20 min for color development.

7.7 Set the spectrophotometer to read at 660 nm. Set the zero against distilled water (blank). A blank has all reagents contained in the sample extract except the soil.

7.8 Record the percent transmittance to the nearest 0.01 unit for the sample extract and each SPCS.

8. Calculations

8.1 Transmittance of a solution is the fraction of incident radiation transmitted by the solution, i.e., $T = P/P_o$ and is often expressed as a percentage, i.e., $\%T = P/P_o \times 100$. The absorbance of a solution is directly proportional to concentration and is defined by the equation, $A = -\log_{10} T$. These relationships are derived from Beer's law.

Calibration Calculations

8.2 Use transmission of each SPCS to either construct a calibrated curve to plot P_2O_5 or use a least squares analysis to calculate P_2O_5. The P_2O_5 is reported in ppm.

8.3 *Calibration Curve*: Plot the transmittances against the ppm P_2O_5 of each SPCS on semilog graph paper or convert to absorbances and plot on linear graph paper. Construct the calibration curve by finding the "best" line that fits the plotted SPCS.

8.4 *Linear Squares Analysis*: Use a least squares criterion, i.e. best moving average. Refer to a statistical analysis book for additional information on least squares analysis. To facilitate data manipulation in a least squares analysis, the following SPCS curve is developed using the concentration of SPCS as a f[ln(%T]. Final calculated analyte concentration with either $\log_{10}$ or ln base would be the same. Refer to procedure 6S3 for an example of least squares analysis.

Analyte Calculation

8.5 *Calibration Curve*: Read the P_2O_5 (ppm) directly from the calibration curve.

8.6 *Least Squares Analysis*: Refer to procedure 6S3 for an example of least squares analysis.

8.7 Convert the extract P_2O_5 (ppm) to soil P_2O_5 (ppm or lbs/A) as follows:

Soil P_2O_5 = Extract P_2O_5 x DR x 100 x AD/OD/Sample Weight (g)

where:
Soil P_2O	= P_2O_5 in soil (ppm)
Extract P_2O_5	= P_2O_5 in extract (ppm)
DR	= Dilution ratio, if necessary, otherwise 1
100	= Conversion factor
AD/OD	= Air-dry/oven-dry ratio (procedure 4B5)

9. Report

Report the 1% citrate acid extractable P_2O_5 in mg kg^{-1} (ppm) to nearest whole number.

10. Precision

Precision data are not available for this procedure.

11. References

Dyer. 1894. Trans. Chem. Soc. 65:115-167.

Metson, A.J. 1956. Methods of chemical analysis for soil survey samples. Soil Bureau Bulletin no. 12. NZ Dept. Sci. and Industrial Res.

Snell, F.D., and C.T. Snell. 1949. Colorimetric methods of analysis. Phosphorus. Vol 2. 3rd ed. p.630-681. D. Van Nostrand Co., Inc.

Soil Survey Staff. 1975. Soil taxonomy: A basic system of soil classification for making and interpreting soil surveys. USDA-SCS Agric. Handb. 436. Appendix III: Tests of organic materials. p. 483. U.S. Govt. Print. Office. Washington, DC.

Soil Survey Staff. 1994. Keys to soil taxonomy. 6th ed. USDA-SCS. Govt. Print. Office, Washington DC.

MINERALOGY (7)

Instrumental Analyses (7A)
X-Ray Diffraction (7A2)
Phillips XRG-300 X-Ray Diffractometer
Thin Film on Glass, Resin Pretreatment II (7A2i)
(Mg Room Temp, Mg Glycerol Solvated, K 300°C, K 500°C)

1. Application

Clay fractions of soils are commonly composed of mixtures of one or more phyllosilicate minerals together with primary minerals inherited directly from the parent material (Whittig and Allardice, 1986). Positive identification of mineral species and quantitative estimation of their proportions in these polycomponent systems usually require the application of several complementary qualitative and quantitative analyses (Whittig and Allardice, 1986). One of the most useful methods to identify and to make semiquantitative estimates of the crystalline mineral components of soil is X-ray diffraction analysis.

The operational strategy at the SSL and the preceding Lincoln Soil Survey Laboratory has been to adjust instrumental parameters to keep peak intensity of a soil reference constant from 1964 to present through the evolution of instrumentation. The intent is to keep the same quantitative interpretations consistent from sample to sample.

2. Summary of Method

Soils are dispersed and separated into fractions of interest. Sands and silts are mounted on glass slides as slurries or on double sticky tape for analysis. Clay suspensions are placed on glass slides to dry and to preferentially orient the clay minerals. The soil clay minerals of greatest interest are phyllosilicates, e.g., kaolinite, mica (illite), smectite, vermiculite, hydroxy-interlayered vermiculite, and chlorite.

Generally, no two minerals have exactly the same interatomic distances in three dimensions and the angle at which diffraction occurs is distinctive for a particular mineral (Whittig and Allardice, 1986). These interatomic distances within a mineral crystal result in a unique array of diffraction maxima, which help to identify that mineral. When several minerals are present in a sample, species identification is usually accomplished most easily and positively by determining the interatomic spacings that give rise to the various maxima and by comparing these with known spacings of minerals (Whittig and Allardice, 1986).

X-ray diffraction produces peaks on a chart that correspond to 2θ angles on a goniometer. The angle of incidence of the goniometer is relative to the surface plane of the sample. Standard tables to convert θ or 2θ angles to crystal "d" spacings are published in the U.S. Geological Survey Circular 29 (Switzer et al., 1948) and in other publications (Brown, 1980). At the SSL, conversions are made by the analysis program on the Philips diffractometer, d-spacings are recorded on an IBM-compatible 486 DOS-based computer system, and hard copies are printed for interpretation and filing. The crystal "d" spacings of minerals, i.e., the interval between repeating planes of atoms, can be calculated by Bragg's Law as follows:

$$n\lambda = 2d \sin \theta$$

where:
n = order of diffraction (integer)
λ = x-radiation wavelength (Angstroms, A)
d = crystal "d" spacing (A)
θ = angle of incidence

When n = 1, diffraction is of the first order. The wavelength of radiation from an X-ray tube is constant and characteristic for the target metal in the tube. Copper radiation (CuKa) with a wavelength of 1.54 A (0.154 nm) is used at the SSL. Because of similar structures of layer silicates commonly present in soil clays, several treatments which characteristically affect the "d" spacings are necessary to identify components. At the SSL, four treatments are used, i.e., Mg^{2+} (room temperature); Mg^{2+}-glycerol (room temperature); K^+ (300°C); and K^+ (500°C).

3. Interferences

Intimate mixtures of similar phyllosilicate minerals on a fine scale cause problems in identification. The mixtures, differences in crystal size and purity, and background or matrix interferences affect quantification. No pretreatments other than dispersion with sodium hexametaphosphate are used for separation and isolation of the crystalline clay fraction. Impurities such as organic matter and iron oxides may act as matrix interferences causing peak attenuation during X-ray analysis or may interfere with clay dispersion and separation. The separation procedure to isolate the clay fraction from the other size fractions of the soil skews the <2-μm clay suspension toward the fine clay, but it minimizes the inclusion of fine silt in the fraction. Dried clay may peel from the XRD slide. One remedy is to rewet the peeled clay on the slide with 1 drop of glue-water mixture (1:7). Other remedies are as follows:

a. Place double sticky tape on the slide prior to adding the dried clay.

b. Dilute the suspension by half, if thick.

c. Crush with ethanol and dry, and then add water to make a slurry slide.

d. Roughen the slide surface with a fine grit sand paper.

Sufficient glycerol on the slides is required to solvate the clay, i.e., to expand smectites to 18 A. X-ray analysis should be performed 1 to 2 days after glycerol addition. If excess glycerol is applied to the slide and free glycerol remains on the surface, XRD peaks are attenuated. Some suggestions to dry the slides and achieve optimum glycerol solvation are as follows:

a. Use a desiccator to dry slide, usually when the clay is thin.

b. If the center of slide is whitish and dry, usually with thick clay, brush slide with glycerol or add an additional drop of glycerol.

4. Safety

Operate the centrifuge with caution. Keep the centrifuge lid closed when in operation. Ensure that all hangers and tubes are seated firmly in proper location. Use tongs and appropriate thermal protection when operating the muffle furnace. The diffraction unit presents an electrical and radiation hazard. Analysts must receive radiation safety training before operating the equipment. Employees must wear a radiation film badge while in the room when the diffraction unit is in operation.

5. Equipment
5.1 Teaspoon (5 g)
5.2 Dispenser, 5 mL, for sodium hexametaphosphate solution
5.3 Centrifuge, International No. 2, with No. 240 head and carriers for centrifuge tubes, International Equip. Co., Boston, MA
5.4 Centrifuge tubes, plastic, 100 mL, on which 10-cm solution depth is marked
5.5 Rubber stoppers, No. 6, for centrifuge tubes
5.6 Mechanical shaker, reciprocal, 120 oscillations min^{-1}
5.7 Plastic cups, 60 mL (2 fl. oz.) with lids
5.8 Label machine
5.9 Hypodermic syringes, plastic, 12 mL, with tip caps
5.10 Screen, 80 mesh, copper
5.11 Dropper bottle, plastic, 30 mL (1 fl. oz.), for a 1:7 glycerol:water mixture
5.12 Muffle furnace
5.13 X-ray diffractometer, Philips XRG-300, with PW-1170 automated sample changer
5.14 PC-APD, Philips, software for Automatic Powder Diffraction (PW-1877), Version 3.5
5.15 Computer, IBM-compatible 486, Gateway 2000 4D X2-66V
5.16 Printer, Hewlett Packard LaserJet IV
5.17 Plotter, Hewlett Packard 7550 Plus
5.18 XRD slides, glass, 14 x 19 mm
5.19 XRD sample preparation board, wood, with 32 places for glass XRD slides
5.20 Slide holder. Accepts 14 x 19 mm XRD glass slides. Modified so slide surfaces rest flush with surface of holder.
5.21 Magazine for slide holder, 35 positions
5.22 Reference slides: quartz and clay from reference soil

6. Reagents
6.1 Distilled deionized (DDI) water
6.2 Sodium hexametaphosphate solution. Dissolve 35.7 g of sodium hexametaphosphate $(NaPO_3)_6$ and 7.94 g of sodium carbonate (Na_2CO_3) in 1 L DDI water.
6.3 Potassium chloride (KCl), 1.0 N. Dissolve 74.60 g KCl in 1 L DDI water or 671.40 g KCl in 9 L DDI water.
6.4 Magnesium chloride ($MgCl_2$), 1.0 N. Dissolve 47.61 g $MgCl_2$ in 1 L DDI water or 428.49 g $MgCl_2$ in 9 L DDI water.
6.5 Glycerol:water mixture (1:7). Add 4 mL of glycerol to 28 mL DDI water plus 2 drops of toluene.
6.6 Exchange resin, Rexyn 101 (H), analytical grade. Pretreatment of resin as follows:
6.6.1 Divide equally Rexyn 101 (H), approximately 250-g portions, into two 600-mL beakers labelled K and Mg and add appropriate salt solution (1.0 N KCl or 1.0 N $MgCl_2$). Cover resin with salt solution.
6.6.2 Stir, let settle for 10 min, decant clear solution, and add salt solution. Repeat 3 times. Leave resin covered in salt solution for 8 to 12 h.
6.6.3 Repeat step 6.6.2 second day. Resin is ready for syringes. Saturated resin not used initially for syringes can be saved for future use.
6.7 White glue, diluted 1:7 with DDI water

7. Procedure

Preparation (Recharge) of Resin-Loaded Syringes

7.1 Place a small circle of 80-mesh screen in a 12-mL syringe and add 4 cm^3 of exchange resin from which salt solution has been drained. Our procedure requires each sample to have 2 Mg and 2 K slides prepared, so we produce our syringes in sets of two.

7.2 Saturate the resin in each of the four syringes with 4 mL of the appropriate 1.0 N salt solution (MgCl$_2$ or KCl) and expel. Repeat saturation of resin.

7.3 Fill syringe completely with the salt solution and allow to equilibrate for 4 to 20 h.

7.4 Rinse syringe twice with 4 mL of DDI water and rinse tip cap.

7.5 Completely fill syringe with DDI water and allow to equilibrate for 4 to 20 h.

7.6 Rinse syringe twice with DDI water.

7.7 Expel water, cap syringe, and store.

Preparation of Clay Suspension

7.8 Place ≈ 5 g (1 tsp) of air-dry <2-mm soil in a 100-mL plastic centrifuge tube. If the sample appears to be primarily sand, use 10 g (2 tsp) of <2-mm soil to obtain sufficient clay.

7.9 Add 5 mL of sodium hexametaphosphate solution. If the soil contains gypsum or is primarily calcium carbonate, use 10 mL of sodium hexametaphosphate dispersing agent.

7.10 Fill tube to 9.5-cm height with DDI water.

7.11 Place rubber stopper in tube and shake overnight in mechanical shaker.

7.12 Remove stopper from tube and rinse stopper and sides of tube with enough water to bring the volume to the 10-cm mark.

7.13 Balance the pairs of tubes and place in centrifuge. Centrifuge at 750 rpm for 3.0 min.

7.14 If the clay is dispersed, carefully decant 30 mL of suspension into a labelled, 60-mL, plastic cup. Place cap on cup.

7.15 If the clay did not disperse after being shaken overnight, remove the rubber stopper and carefully decant the clear supernatant liquid.

7.16 Add an additional 10 mL of sodium hexametaphosphate dispersing agent to sample and then add DDI water to 9.5-cm depth.

7.17 Stopper and shake overnight to disperse the clay. Rinse stopper and fill tube to 10-cm mark.

7.18 Centrifuge, decant, and store clay suspension.

7.19 Use the clay suspension for X-ray diffraction analysis and HF plus aqua regia dissolution analysis. Dry clay suspension for use in thermal analysis.

Thin Film on Glass, Resin Pretreatment

7.20 The SSL uses a sample board which holds 32 slides, i.e., 8 samples x 4 treatments. Prepare the sample board with glass XRD slides to receive the following 4 treatments per clay suspension sample.

Mg^{2+} - room temperature
Mg^{2+} - glycerol (room temperature)
K^+ - 300°C (heated for 2 h)
K^+ - 500°C (heated for 2 h)

7.21 Place one small drop of the glycerol:water mixture (1:7) on each Mg^{2+}-glycerol slide.

7.22 Draw 1 mL of <2-μm clay suspension into the resin-loaded syringe and invert back and forth to facilitate cation exchange.

7.23 Dispense 3 drops to clear the tip.

7.24 Dispense ≈ 0.1 mL (6 to 10 drops) to cover the appropriate XRD slide. Draw DDI water into the syringe and expel 3 times to remove all of the clay suspension. Recharge the syringe after 10 times of use.

7.25 When the clay suspension has dried, transfer the slides with the K^+-saturated clays to transite plates and heat for a minimum of 2 h in a muffle furnace.

7.26 Heat the following sample slides on the XRD sample board.

K^+-300°C - slides 3, 7, 11, 15, 19, 23, 27, and 31

K^+-500°C - slides 4, 8, 12, 16, 20, 24, 28, and 32

7.27 After heating, remove the transite plate from the furnace, cool to air temperature, and return slides to XRD sample board.

X-ray Diffraction Operation

7.28 The X-ray analysis of the glycerol slide must be done within 1 to 2 days after the slide dries. If this is not possible, skip Step 7.21 when slide is prepared. Add one small drop of glycerol:water mixture (1:7) to dry slide 24 h prior to X-ray analysis.

7.29 Transfer the slides (1 to 32) from XRD sample board to slide holders (1 to 32) and place in slots (1 to 32) in a magazine for the automated sample changer.

7.30 Analyze one reference soil sample in each run. Place this sample in slot 33.

7.31 Analyze one quartz standard for 2θ and intensity calibrations in each run. Place this sample in slot 34. Intensity is measured at peak maximum at or near 26.66° 2θ for 10 s.

7.32 The 32 samples from one XRD board constitute one run on the diffraction unit. Prepare a run sheet for samples on each XRD sample board. Refer to example run instruction (7.33). Refer to the manufacturer's manual for operation of the X-ray diffractometer.

7.33 Place the magazine in the automated sample changer. Confirm that the XRD shutter is off when changing magazines. Set the XRD unit parameters as follows:

CuKa radiation, λ: 1.54 A (0.154 nm)

Scan range: 2° to 34° 2θ
Generator settings: 40 kv, 20 ma
Divergence slit: 1°
Receiving slit: 0.2 mm
Monochrometer: Yes
Step size and scan speed vary depending on intensity of X-rays generated from tube. Adjust settings to maintain same long-term peak intensities on standard reference clay and quartz standard regardless of tube intensities.

7.34 Enter run instruction from the keyboard. Create a batch file for the automated run. File names specified are of the sample number. An example run instruction is as follows:

Batch File Name: Project number (e.g., CP95LA022)

Raw Data File Name: Run number

First Sample: 1

Last Sample: 33
(reference soil clay)

7.35 Activate program. The run stores raw data on the hard disk under the subdirectory designated by project type and year, e.g., CP95. Refer to example run instruction (7.34).

7.36 Print a hard copy of the "Detected Peaks File" for each sample and perform level 1 smoothing on diffraction patterns.

7.37 Prepare and print a 4-color graphics chart. The 4 colors are blue (Mg^{2+}); green (Mg^{2+}-glycerol); pink (K^+ 300°C); and red (K^+ 500°C). Stamp chart with label; enter run parameter information, and complete soil information, e.g., soil name, horizon designation, and depth. File hard copies of detected peaks and graphics chart in pasteboard binders by state, county, and chronology.

7.38 Record "d" spacing and intensity of quartz standard in the logbook. Record the peak intensities for designated peaks for the reference soil clay.

7.39 File the detected peaks printout and graph for the reference soil in the reference soil-clay folder.

Interpretation of X-ray Diffraction Data
7.40 The angle in degrees two theta (2θ) measured in X-ray diffraction analyses is converted to angstroms (Å) using tables complied according to Bragg's Law. Refer to summary of method. Angstroms convert to nanometers (nm) by a factor of 0.1, e.g., 14 A = 1.4 nm.

7.41 Use the following X-ray diffraction criteria to identify some common crystalline minerals. The reported "d" values are for 00*l* basal spacings. The Miller index (*hkl*) specifies a crystal face which has some orientation to the three crystallographic axes of a, b, and c. The Miller index (00*l*) indicates a crystal face that is parallel to the a and b axes, e.g., phyllosilicate minerals. The following X-ray diffraction criteria also has some questions (Q) that may aid the analyst in interpreting the diffraction patterns. These questions are a suggested procedural approach to help the analyst identify the relative locations of a few peaks and to confirm key criteria.

X-Ray Diffraction Criteria

1. Kaolinite and Halloysite
a. Crystal structure missing at 500°C.
b. 7 A (7.2 to 7.5 A) with all other treatments
Q. Is there a 7 A peak? Is it destroyed at 500°C? Kaolinite or Halloysite.
Q. Is the peak sharp and at ~ 7.1 A? Kaolinite.
Q. Is the peak broad and at 7.2 to 7.5 A? Halloysite.

2. Mica (Illite)
a. 10 A with all treatments.
b. 10 A with Mg^{2+}-saturation
Q. Is there a 10 A peak with Mg^{2+}-saturation? Mica (Illite).

3. Chlorite
a. Crystal structure of Fe-chlorites destroyed at 650 to 700°C.
b. 14 A with all other treatments.
c. 14 A at 500°C.
d. Generally also has strong 7 A peak.
Q. Is there a 14 A peak when heated to 500°C? Chlorite.

4. Vermiculite
a. 14 A with Mg^{2+}-saturation.
b. 14 A with Mg^{2+}-glycerol solvation.
c. Nearly 10 A with K^+ saturation.
d. 10 A when K^+-saturated and heated to 300°C.
Q. Is there an enhanced 10 A peak with K^+-saturation in comparison to Mg^{2+} saturation that cannot be attributed to smectitie? Vermiculite.

5. Smectite
a. 14 A with Mg^{2+}-saturation
b. 12 to 12.5 A with K^+- or Na^+-saturation.
c. 17 to 18 A with Mg^{2+}-glycerol solvation.
d. 10 A with K^+-saturation and heating to 300°C.
Q. Is there a 17 to 18 A peak upon solvation? Smectite.

6. Gibbsite
a. Peak at 4.83 A with Mg^{2+} and Mg^{2+}-glycerol but destroyed when heated to 300°C.

7. Goethite
a. Peak at 4.18 A with Mg^{2+} and Mg^{2+}-glycerol but destroyed when heated to 300°C.

8. Hydroxy-interlayed Vermiculite or Smectite
a. Incomplete collapse to 10 A of smectite or vermiculite when K^+-saturated and heated to 300°C.

9. Quartz
a. Peaks at 4.27 A and 3.34 A with all treatments (only 3.34 if small amounts).

10. Lepidocrocite
a. Peak at 6.2 to 6.4 A with Mg^{2+} and Mg^{2+}-glycerol but destroyed when heated to 300°C.

11. Potassium Feldspar
a. Peak at 3.24 A with all treatments.

12. Plagioclase Feldspar
a. Twin peaks between 3.16 and 3.21 with all treatments.

13. Calcite
a. Peak at 3.035 A with all treatments.

14. Dolomite
a. Peak at 2.88 to 2.89 A with all treatments.

15. Gypsum
a. Peak at 4.27 A with Mg^{2+} and Mg^{2+}-glycerol, but destroyed when heated to 300°C.

16. Mixed Layer Vermiculite-Mica
a. Peak at 11 to 13 A with Mg^{2+} that does not expand with Mg^{2+}-glycerol.
b. Peak collapses to 10 A with K^+-saturation and heating to 300°C.

17. Mixed Layer Smectite-Mica
a. Peak at 11 to 13 A with Mg^{2+} that expands to 14-16 A with Mg-glycerol.
b. Peak collapses to 10 A with K^+-saturation and heating to 300°C.

18. Mixed Layer Chlorite-Mica
a. Peak at 14 A with Mg^{2+} and Mg^{2+}-glycerol.
b. Peak collapses toward 10 A with K^+-saturation and heating to 300°C, and more completely with heating to 500°C, but never to 10 A.

19. Mixed Layer Chlorite-Smectite
a. Peak at 11 to 13 A with Mg^{2+}-saturation that expands to about 16 A with Mg^{2+}-glycerol.
b. Collapses to about 12 A with K^+-saturation and heating to 300°C and 500°C.

7.42 Use the X-ray diffraction criteria, i.e., diagnostic basal 00*l* spacings (A), in Table 1 for identification and ready reference of some common crystalline minerals as affected by differentiating sample treatments.

7.43 Preferential orientation of clay mineral samples enhances diffraction from the basal (00*l*) spacing and tends to minimize the number and intensity of peaks from diffraction by other *hkl* planes. With preferential orientation, second, third, and fourth order peaks may be recorded in addition to the basal first order peaks. Groups of associated peaks that differ by order of diffraction are as follows:

Smectite (Mg^{2+}-glycerol):
a. 17 to 18 A.
b. 8.5 to 9 A (weak).

Chlorite, vermiculite, and smectite:
a. 14, 7, 4.7, and 3.5 A.
b. 7, 4.7, and 3.5 A weak for smectite.

Mica:
a. 10, 5 (weak in biotites and moderate in muscovites), and 3.3 A.

Kaolinite:
a. 7 and 3.5 A.

7.44 The differentiation of kaolinite and halloysite in a sample can be aided by the use of formamide (Churchman et al., 1984). The intercalation and expansion of halloysite to a d-spacing of ≈ 10.4 A is relatively rapid (20 to 30 min), whereas kaolinite expansion requires ≈ 4 h upon treatment. The procedure is as follows:

a. Lightly spray formamide as an aerosol on the dried Mg^{2+}-saturated slide.

b. Wait 15 min but not more than 1 h and X-ray approximately 7.6 to 13.5° 2θ (d = 11.6 to 6.55 A).

c. Halloysite will expand to ≈ 10.4 A, whereas kaolinite will remain unchanged.

d. Heating the sample to 110°C for 15 min will collapse the halloysite to ≈ 7 A.

e. The total amount of kaolinite and halloysite can be determined by thermal analysis. The intensity ratio of the 10.4 to 7.2 A peaks of the formamide-treated sample can be used to determine the relative percentage of halloysite and kaolinite.

8. Calculations

X-ray diffraction produces peaks on a chart that corresponds to 2θ angle on a goniometer. Standard tables to convert θ or 2θ to crystal "d" spacings are published in the U.S. Geological Survey Circular 29 (Switzer et al., 1948) and in other publications (Brown, 1980). The crystal "d" spacings of minerals, i.e., the interval between repeating planes of atoms, can be calculated by Bragg's Law. Refer to summary of method.

Table 1. X-ray diffraction parameters of common soil clay minerals.

Mineral	Na$^+$	Mg^{2+}	Mg^{2+} Gly	K$^+$	K$^+$ 300°C	K$^+$ 500°C	K$^+$ 700°C
	00*l* diffraction spacing in angstroms						
Kaolinite	7	7	7	7	7	**LD**[1]	LD
Halloysite	7B[2]	**7B**	7B	7B	7B	**LD**	LD
Mica (Illite)	10	10	**10**	10	10	10	10
Chlorite	14*[3]	14*	14*	14*	14*	**14***	T[4]
Vermiculite	14	14	**14 10**		10	10	10
Smectite	12.5	14	**18**	12.5	10	10	10
Gibbsite	4.85	**4.85**	4.85	4.85	**LD**	LD	LD
Goethite	4.18	**4.18**	4.18	4.18	**LD**	LD	LD
Interlayer	10-14	10-14	10-18	10-14	10-14	10-14	10-14
Quartz	**3.14** and **4.27** for all treatments						
Calcite	**3.035** for all treatments						
Dolomite	**2.88** for all treatments						

[1] LD = Lattice destroyed
[2] B = Broad peak is common
[3] * = Sometimes <14A
[4] T = Temperature of decomposition varies with chemical composition, particle-size, and heating conditions.

9. Report

From the "Detected Peaks File" and graphics chart, identify the minerals present according to the registered "d" spacings. As a first approximation, use the following peak intensities, i.e., peak heights above background in counts s^{-1}, to assign each layer silicate mineral to one of the 5 semiquantitative classes.

Class	Peak Height above Background (counts sec^{-1})
5 (Very Large)	>1.88 X 10^3
4 (Large)	1.12 to 1.88 X 10^3
3 (Medium)	0.36 to 1.12 X 10^3
2 (Small)	0.11 to 0.36 X 10^3
1 (Very Small)	<0.11 X 10^3

Adjust class placement to reflect area under the curve if peak is broad relative to peak height or if thermal, elemental, clay activity data, or other evidence warrant class adjustment. If there are no peaks or no evidence of crystalline components, place the sample in NX class (noncrystalline).

10. Precision

Precision data are not available for this procedure. Procedure 7A2i (X-ray diffraction) is semiquantitative.

11. References

Brown, G. 1980. Appendix I (Tables for the determination of d in A from 2θ for the KA and KB radiations of copper, cobalt, and iron). *In* G.W. Brindley and G. Brown (eds.) Crystal structures of clay minerals and their x-ray identification. Mineralogical Soc. Monograph No. 5. Mineralogical Soc. Great Britain. pp 439-475.

Churchman, G.J., J.S. Whitton, G.G.C. Claridge, and B.K.G. Theng. 1984. Intercalation method using formamide for differentiating halloysite from kaolinite. Clays and Clay Minerals. 32:241-248.

Switzer, G., J.M. Axelrod, M.L. Lindberg, and E.S. Larsen 3d. 1948. U.S. Dept. Interior. Geological Survey. Circular 29. Washington, DC.

Whittig, L.D., and W.R. Allardice. 1986. X-ray diffraction techniques. *In* A. Klute (ed.) Methods of soil analysis. Part 1. Physical and mineralogical methods. 2nd ed. Agronomy 9:331-362.

Total Analysis (7C)
HF Plus Aqua Regia (HF + HNO₃ + HCl) Dissolution (7C4a)

1. Application

Prior to the development of modern analytical techniques, e.g., X-ray diffraction and thermal analysis, identification of minerals was based on elemental analysis and optical properties (Washington, 1930; Bain and Smith, 1994). Chemical analysis is still essential to determine mineral structural formulas and to identify and quantify specific mineral species through elemental allocation to minerals. Many clay mineral groups are subdivided based on composition.

Analysis of the entire fine earth (<2-mm) fraction or specific particle-size separates provides information on parent material uniformity, pedon development, and mineral weathering within or between pedons. This interpretation is determined from differences between horizons or pedons in elemental concentrations, elemental ratios such as Si/Al,

Si/Al+Fe, or Ti/Zr, or from differences in total elemental concentrations compared to concentrations determined by selective dissolution techniques.

The inherent fertility of a soil derived from its parent material can be examined by determination of the basic cations relative to the Si or Al content. Phosphorus fertility of a soil and potential water quality problems can be better understood by measurements of total P, especially when compared to other P measurements, such as water-soluble or Bray-extractable P.

Hydrofluoric acid (HF) is efficient in the digestion and dissolution of silicate minerals for elemental dissolution (Bernas, 1968; Sawhney and Stilwell, 1994). Aqua regia (HNO_3 and HCl) aids in digestion of soil components, especially the organic fraction. Procedure 7C4a is a digestion of 100 mg of dried clay suspension, the fine earth (<2-mm) fraction, or other particle size separate with HF and aqua regia. Closed digestion vessels (Parr Bombs) are heated in the oven at 110°C for at least six hours. Elemental concentration of the digestate is determined by inductively coupled plasma-atomic emission spectrometry (ICP-AES).

2. Summary of Method

A clay suspension (procedure 7A2i) containing approximately 100 mg of clay material is pipeted into a Teflon digestion container and dried at 110°C. A equal amount of suspension is pipeted into a tared aluminum-weighing dish and dried at 110°C to obtain a dried sample weight. An oven-dry 100-mg soil sample (<80 mesh) or a specific particle-size separate may be substituted for the clay suspension. The P and Na content of the clay fraction is not measurable when the soil is dispersed in sodium hexametaphosphate (procedure 7A2i). Total P and Na are measurable on the fine-earth fraction or other particle-size separates not dispersed in Na or P-containing reagents, and the analyses are included as a part of this procedure.

Following evaporation of the aqueous portion of the suspension, 0.75 mL HNO_3, 0.25 mL HCl, and 5 mL HF are added. The vessel is inserted into a stainless steel retainer vessel, heated, cooled, and 15 mL of 2.5 percent boric acid solution is added to neutralize the excess HF acid. The digestate is quantitatively transferred with boric acid solution, diluted to 100 mL, shaken, and allowed to stand overnight. Approximately 60 mL are saved for analysis. The concentration of Fe, Mn, Al, Ca, Mg, Na, K, P, Si, Zr, Cu, Zn, As, Ti, Se, Cd, and Pb are determined by ICP analysis in procedures 6C7b, 6D6a, 6G11b, 6N5b, 6O5b, 6P3b, 6Q3b, 6S6a, 6V1b, 8K1a, 8L1a, 8M1a, 8N1a, 8O1a, 8P1a, 8Q1a, and 8R1a, respectively. Data are reported in procedure 7C4a.

3. Interferences

Insoluble fluorides of various metals may form. Formation of SiF_4 results in gaseous losses of Si, but additions of boric acid retards formation of this molecule as well as dissolves other metal fluorides.

4. Safety

Wear protective clothing and eye protection. When preparing reagents, exercise special care. Restrict the use of concentrated acids to the fume hood. Keep HF acid refrigerated and avoid contact with skin of all acids. Wash hands thoroughly after handling reagents. Filling the Teflon cup of the acid digestion bomb to greater than 25 percent of the free volume or adding organic reagents or oxidizing agents to the cup may result in explosion of the digestion bomb.

5. Equipment

5.1 Pipet(s) capable of delivering 5, 0.75, and 0.25 mL
5.2 Volumetric flasks, nalgene, 100 mL
5.3 Polypropylene bottles, 60 mL, with cap
5.4 Electronic balance, ±0.1 mg sensitivity
5.5 Acid digestion bombs: 25-mL Teflon containers with stainless steel retainer vessels
5.6 Oven, 110°C

5.7 Desiccator with P_2O_5 drying agent

5.8 Disposable aluminum-weighing dishes

6. Reagents

6.1 Deionized distilled (DDI) water

6.2 Hydrofluoric acid (HF), 48%, low trace metal content

6.3 Concentrated hydrochloric acid (HCl), 12 *N*. Use instrumental grade reagents which contain low levels of impurities.

6.4 Concentrated nitric acid (HNO$_3$), 16 *N*. Use instrumental grade reagents which contain low levels of impurities.

6.5 Boric acid solution, 2.5 percent. Dissolve 25.0 g low trace metal, granular boric acid (H$_3$BO$_3$) in 1000 mL DDI water.

7. Procedure
HF plus Aqua Regia Dissolution

7.1 Prepare Na-saturated clay as in procedure 7A2i, Preparation of Clay Suspension, Steps 7.8 to 7.19. Clay dispersion by this method eliminates quantitative analysis of Na and P in the clay due to dispersion by sodium hexametaphosphate. Digestion of the entire fine earth (<2-mm) fraction or any fraction not derived by dispersion with sodium hexametaphosphate (or other Na and P-containing dispersing agents) can be quantitatively analyzed for Na and P. Dispersion of clays and cleaning of test tubes and dishware should be with DDI water.

7.2 Pipet a known aliquot of clay suspension containing approximately 100 mg clay into a 25-mL Teflon container. The milliliters of suspension required depends on the clay concentration of the suspension but is generally from 2 to 6 mL. More dilute suspensions should be partially evaporated under a fume hood to concentrate the clay prior to transfer to the Teflon container. Fine-earth (<2-mm) or a specific particle size separate ground to <80-mesh may be used instead of clay. Samples with greater than 3 percent organic C should be ashed in a muffle furnace at 400°C for 2 h prior to analysis to destroy the organic matter. Oven-dry the sample (110°C), cool over P_2O_5, and weigh to 100 ±0.1 mg. If a clay suspension is used, Steps 7.3 to 7.4 are performed. Proceed to Step 7.5 if using fine-earth or other oven-dried material.

7.3 Pipet a duplicate aliquot of suspension (as used in Step 7.2) into a tared Al weighing dish, dry at 110°C, cool in a desiccator with P_2O_5, and weigh to the nearest 0.1 mg. Use this value as the sample weight in the calculations.

7.4 Dry the Teflon container and clay suspension in an oven for 4 h or until the aqueous portion of the suspension is completely evaporated. Remove from oven and cool on the bench top or in a fume hood. Cooling in a desiccator is not required.

7.5 Pipet 0.75 mL HNO$_3$ and 0.25 mL HCl into the sample and allow to completely wet and then pipet 5 mL HF into sample.

7.6 Place covered Teflon container in stainless steel retainer vessel. Place sample in oven at 110°C for a minimum of 6 h. Samples can be left in the oven overnight at 110°C.

7.7 Remove samples from oven and cool for at least 4 h.

7.8 Under a hood, remove Teflon container from steel retainer vessel, open the Teflon container, and add 15 mL 2.5 percent boric acid solution.

7.9 Quantitatively transfer contents of Teflon container to a 100 mL nalgene volumetric flask and adjust to volume with 2.5 percent H$_3$BO$_3$.

7.10 Cap flask and mix well by inverting at least three times. Allow to stand overnight to dissolve any metal fluorides.

7.11 Invert the volumetric flask to mix and decant approximately 60 mL into a labeled polypropylene container.

7.12 Prepare working standards of a blank, a clay suspension from a SSL reference soil sample, and a National Institute of Standards and Technology (NIST) standard reference material by the same digestion method. Run one of these standards with each set of 20 samples.

7.13 Solutions and standards are analyzed by ICP spectrometry. Refer to procedures 6C7b, 6D6a, 6G11b, 6N5b, 6O5b, 6P3b, 6Q3b, 6S6a, 6V1b, 8K1a, 8L1a, 8M1a, 8N1a, 8O1a, 8P1a, 8Q1a, and 8R1a for analysis of Fe, Mn, Al, Ca, Mg, Na, K, P, Si, Zr, Cu, Zn, As, Ti, Se, Cd, and Pb, respectively.

8. Calculations

8.1 Data are transferred as an ASCII file from the ICP computer onto a 3.5-in floppy disk via "Report Writer" in the TJA software ThermoSpec, Version 5.06.

8.2 On a MS-DOS based PC computer, import the ASCII file of ICP data into the DOS editor and strip off unnecessary headers and data from standards. Save the file after editing, renaming using a format that can be imported into LOTUS, e.g., rename to .wk3 file for LOTUS 123, Version 3.1.

8.3 Import the file into an established total analysis spreadsheet in LOTUS 123. The spreadsheet has columns for sample number, soil fraction digested, soil weight, concentration of each element in ppm, and the calculated elemental percent. Each line of elemental data for a sample is imported as a single data string.

8.4 Parse the components of each data string into separate columns. Rearrange the data set in order to have all elemental values on a single line for a particular sample. Move the data into the correct columns of the spreadsheet.

8.5 Insert values for elements requiring dilution into the original line of sample data and replace all negative values with zero.

8.6 Input sample weights, or if possible, import sample weights (dried soil weights) from the ASCII file generated by computer attached to balance via RS-232.

8.7 Calculate the percent of an element in the soil from ppm in solution as shown in the Si example as follows:

Si (ppm) in solution = 75.2 ppm (75.2 µg/mL)
Volume extract   = 100 mL
Sample weight (110°C) = 100.0 mg

Calculate as follows:

% Si = 75.2µg mL^{-1} x 100 mL x (1 g/ 10^6 µg) x (1/0.1 g soil) x 100 = 7.52 %

8.8 The fraction digested needs to be identified with each sample. Use proper SSL database abbreviations.

8.9 Delete the Na and P data for clay samples dispersed in sodium hexametaphosphate.

8.10 Prepare the file to send to CMS. Save the file as an unformatted ASCII file using LOTUS.

8.11 Enter data for Si, Al, Fe, Mg, Mn, K, Ti, Ca, Zr, P, and Na into the SSL CMS database on a 110°C weight basis as percent of the element in the fraction digested. Data are converted to the oxide form on the data sheet.

8.12 The factor for converting from an elemental form to an oxide form is based on the atomic weights of the element and oxygen. An example is as follows:

Atomic weight Si = 28.09
Atomic Weight O = 16.0
Molecular weight SiO_2 = 60.09

Calculate percent Si in SiO_2 as follows:

Si (%) = (28.09/60.09) x 100 = 46.7 %

There is 46.7 percent Si in SiO_2. To convert from percent Si to percent Si oxide (SiO_2) in the soil, divide the percent Si by 0.467 or multiply by the inverse of this value. The following table lists the element, the oxide form, and the elemental percent in the oxide form.

Element Form	Oxide	Elemental %
Si	SiO_2	46.7
Al	Al_2O_3	52.9
Fe	Fe_2O_3	69.9
Mg	MgO	60.3
Mn	MnO	77.4
K	K_2O	83.0
Ti	TiO_2	59.9
Ca	CaO	71.5
Zr	ZrO_2	74.0
P	P_2O_5	43.6
Na	Na_2O	74.2

9. Report

Data are reported as percent to the nearest tenth for Fe, Al, Mg, Na, K, and Si; to the nearest hundredth for Mn, Ca, P, and Ti; and to the nearest thousandth for Zr. The remaining trace elements (Cu, Zn, As, Se, Cd, and Pb) are reported in mg kg^{-1} (ppm).

10. Precision

The mean, standard deviation, and C.V. are calculated for each element for both the NIST standard and the SSL reference standard.

11. References

Bain, D.C., and B.F.L. Smith. 1994. Chemical analysis. (In) M.J. Wilson ed., Clay Mineralogy: Spectroscopic and Chemical Determinative Methods. pp. 300-332. Chapman and Hall, Inc. London, England.

Bernas, B. 1968. A new method for decomposition and comprehensive analysis of silicates by atomic absorption spectrometry. Anal. Chem. 40:1682-1686.

Sawhney, B.L., and D.E. Stilwell. 1994. Dissolution and elemental analysis of minerals, soils, and environmental samples. (In) J.E. Amonette and L.W. Zelazny, eds., Quantitative methods in soil mineralogy. pp. 49-82. Soil Sci. Soc. Am. Miscellaneous Publ. Madison, WI.

Washington, H.S. 1930. The chemical analysis of rocks. 4th ed. John Wiley and Sons, Inc., New York, NY.

Surface Area (7D)
Ethylene Glycol Monoethyl Ether (EGME) Retention (7D2)

1. Application

Surface area determines many physical and chemical properties of materials. Water retention and movement, cation exchange capacity, pesticide adsorption, and many biological processes are closely related to specific surface (Carter et al., 1986). Soils vary widely in their reactive surface area because of differences in mineralogical and organic composition and in their particle-size distribution (Carter et al., 1965). Specific surface, defined as surface area per unit mass of soil, is usually expressed in units of $m^2 \, g^{-1}$ or $cm^2 \, g^{-1}$ soil. Specific surface has been measured for several clays, e.g., $810 \, m^2 \, g^{-1}$ for smectite and 20 to $40 \, m^2 \, g^{-1}$ for kaolinite and mica.

2. Summary of Method

Ethylene glycol monoethyl ether (EGME) retention is a surface-area determination. A soil sample is dried over phosphorus pentoxide (P_2O_5). The sample is saturated with EGME. A monomolecular layer of EGME is established by desorbing the EGME by vacuum over EGME-saturated $CaCl_2$. The solvate of $CaCl_2$ and EGME helps to maintain an EGME vapor pressure in the desiccator which results in the formation of a monomolecular layer of EGME on sample surfaces.

The weight of a monomolecular layer of EGME on the sample is determined by weighing the dried sample. EGME is determined by weighing the sample and sample plus EGME (Carter et al., 1965). The SSL determines EGME retention by procedure 7D2. The SSL reports EGME retention as mg EGME per g of soil to the nearest mg on a <2-mm base.

3. Interferences

The loss or contamination of sample and the variation in sample weight may cause erroneous results. Handle the weighing vessels with finger cots or tongs to prevent vessel contamination and the resulting weighing errors. High relative humidity in the laboratory may result in high moisture absorption by sample.

4. Safety

Wear protective clothing (e.g., coats, aprons, and gloves) and eye protection (e.g., face shields, goggles, or safety glasses) when handling reagents and working with vacuum desiccators. Follow standard laboratory safety procedures in handling reagents and vacuum devices. The P_2O_5 is corrosive and reacts violently with water. Use caution in cleaning P_2O_5 spills. The EGME is combustible and harmful is swallowed, inhaled, or absorbed through the skin. Keep samples and desiccators with EGME under fume hood at all times.

5. Equipment

5.1 Electronic balance, ±0.1-mg sensitivity, Mettler AE 160

5.2 Vacuum desiccator, 250 mm, Nalgene No. 5310, with desiccator plate, 230 mm

5.3 Laboratory vacuum or vacuum pump, 0.65 to 0.75 bars

5.4 EGME trap, anhydrous $CaCl_2$ in a large tube between desiccator and vacuum source

5.5 Syringe, polypropylene, 3 mL

5.6 Weighing bottle, cylindrical, low form, 50 x 30 mm

6. Reagents

6.1 Ethylene glycol monoethyl ether (EGME), reagent
6.2 Phosphorus pentoxide (P_2O_5), anhydrous
6.3 Calcium chloride ($CaCl_2$), pellets, 40 mesh, reagent grade

7. Procedure

7.1 Dry 3 to 5 g of <2-mm, air-dry soil in a weighing bottle in a vacuum desiccator over P_2O_5 for 2 days.

7.2 Prepare solvated $CaCl_2$ by weighing 100 g oven-dried $CaCl_2$, without cooling, into a large beaker. Add 20 g EGME and mix by stirring. Transfer to a desiccator in which EGME-saturated samples equilibrate.

7.3 Weigh the P_2O_5-dried soil sample to the nearest 0.1 mg. When working outside the desiccator, cover the sample to avoid moisture adsorption from the atmosphere.

7.4 Use a 3-mL syringe to saturate the soil with EGME. Add 5 drops in excess of saturation.

7.5 Place the uncovered, EGME-soil mixture in a vacuum desiccator over solvated $CaCl_2$. Use a laboratory vacuum of 0.65 to 0.75 bar pressure.

7.6 Loosely cover the tops of weighing bottles with a piece of aluminum foil that is smaller than the inside diameter of desiccator.

7.7 Apply suction for 16 to 24 h.

7.8 Carefully release the suction. Remove weighing bottles and weigh the EGME-soil mixture.

7.9 If a 3-g sample is used, the difference between the EGME-soil mixture and P_2O_5-dry soil is ≈ 10 mg EGME/g P_2O_5-dry soil. When this difference is <10 mg, reduce the vacuum time to 1 h day^{-1} and weigh twice daily.

7.10 Repeat the vacuum and weighing procedure until a constant weight is attained. Constant weight is defined as three successive daily weighings within 1 mg of EGME per gram P_2O_5-dry soil. When a constant weight is attained, make calculations.

8. Calculations

8.1 The EGME retention is calculated as follows:

Retention of EGME (mg g^{-1}) = (Wt$_1$ - Wt$_2$) x (1000/Wt$_3$)

where:
Wt$_1$ = Soil weight with monomolecular layer of EGME + Tare weight of bottle
Wt$_2$ = Soil weight after drying with P_2O_5 + Tare weight of bottle
Wt$_3$ = Soil weight after drying with P_2O_5 - Tare weight of bottle
1000 = Conversion factor (mg g^{-1})
The surface area in units of mg EGME per g of soil is converted to m^2 g^{-1}, the convention commonly used in clay mineralogy. The conversion is as follows:

Surface area (m^2 g^{-1}) = (EGME retention (mg g^{-1}))/0.286

where:

0.286 = Conversion factor (mg EGME m^{-2})

The constant, 0.286, is the amount of EGME (mg) that is required to cover a m^2 of clay surface with a monomolecular layer (Carter et al., 1986). This value is calculated from the measured value of 231.7 mg EGME per g of pure montmorillonite assumed to have 810 m^2 g^{-1} on the basis of other measurements.

9. Report
Report EGME as mg EGME per g of soil to the nearest mg.

10. Precision
Precision data are not available for this procedure. Two quality control checks, a high and a low standard, are routinely analyzed in EGME. The mean (mg EGME per g soil), standard deviation, and C.V. for the quality control check sample are as follows:

	Mean	n	Std. Dev.	C.V.
High Std	109.0	10	7.4	7.3
Low Std.	37.5	10	0.64	4.8

11. References
Carter. D.L., M.D. Heilman, and C.L. Gonzalez. 1965. Ethylene glycol monoethyl ether for determining surface area of silicate minerals. Soil Sci. 100:356-360.

Carter, D.L., M. M. Mortland, and W.D. Kemper. 1986. Sampling. *In* A. Klute (ed.) Methods of soil analysis. Part 1. Physical and mineralogical methods. 2nd ed. Agronomy 9:413-423.

MISCELLANEOUS (8)

Ratios and Estimates (8D)
To Noncarbonate Clay (8D2)

Divide the CEC-7 (procedure 5A8c), extractable Fe (procedure 6C2), or 15-bar water retention (procedure 4B2a or 4B2b) by the noncarbonate clay percentage. Noncarbonate clay is determined by subtracting the carbonate clay (procedure 3A1d or 3A2d) from total clay (procedure 3A1 or 3A2).

Ratios and Estimates (8D)
Ca to Mg (extractable) (8D3)

Divide extractable Ca^{2+} (procedure 6N2) by extractable Mg^{2+} (procedure 6O2).

Ratios and Estimates (8D)
Estimated Clay Percentage (8D4)

For most soils, clay percentage can be approximated as 2.5 x 15-bar water percentage (procedure 4B2a or 4B2b). Use caution in applying this factor to any particular situation, especially if organic matter or other amorphous material is present in significant quantities.

Ratios and Estimates (8D)
Estimated Total Salt (8D5)

Use the charts and graphs available in U.S. Salinity Laboratory Staff (1954) to estimate total salt content from the electrical conductivity (EC_s) of the saturation extract (procedure 8A3a). The essential relations are summarized in the equations as follows:

Log total salt in soil (ppm) = 0.81 + 1.08 x Log EC_s (mmhos cm^{-1}) + Log SP

where:
EC_s = Electrical conductivity of saturation extract
SP = Saturation percentage of saturation extract

Total salt in soil (%) = Total salt (ppm) x 10^{-4}

These equations are applicable to saturation extracts with an EC_s <20 mmhos cm^{-1}. Deviations occur at higher salt concentrations.

Ratios and Estimates (8D)
Iron Plus Aluminum, Pyrophosphate Extractable to Dithionite-Citrate Extractable (8D6)

Divide the sum of the pyrophosphate-extractable Fe plus Al (procedures 6C8a and 6G10a, respectively) by the sum of dithionite-citrate-extractable Fe plus Al (procedures 6C2 and 6G7, respectively). Pyrophosphate and dithionite-citrate extractable Fe and Al are former criteria for spodic placement (Soil Survey Staff, 1975).

Ratios and Estimates (8D)
Index of Accumulation (8D7)

Subtract 1/2 the clay percentage (procedure 3A1 or 3A2) of a subhorizon from the CEC at pH 8.2 (procedure 5A3a) and multiply the remainder by the thickness of subhorizon (cm). The combined index of accumulation of amorphous material is a former criterion for spodic placement (Soil Survey Staff, 1975).

References

Soil Survey Staff. 1975. Soil Taxonomy: A basic system of soil classification for making and interpreting soil surveys. USDA-SCS Agric. Handb. 436. U.S. Govt. Print. Office, Washington, DC.
U.S. Salinity Laboratory Staff. 1954. L.A. Richards (ed.) Diagnosis and improvement of saline and alkali soils. 160 p. U.S. Dept. of Agric. Handb. 60. U.S. Govt. Print. Office, Washington, DC.

Use Table 1 with the SSL preparation procedures 1B1, 1B2, 1B5, 1B6, and 1B7. Gravel codes are also defined in Table 1. In the "Code" column, "Char" refers to characterization sample. Laboratory preparation and >2-mm porosity are defined in footnotes on laboratory data sheet.

Table 1. Laboratory preparation codes and procedural summaries.

Code		Laboratory Preparation
Char	**>2mm**	
S	Blank	Weigh sample at field-moisture content and record weight. Air-dry, weigh, and record weight. Sieve >2-mm fractions, weigh, record weights, and discard. Report all analytical results on <2-mm basis. Refer to procedure 1B1, Standard Air-dry.
S	P	Lab preparation is same as S-blank. However, report clod parameters and Cm (correction factor for >2-mm content moist soil) on an whole-soil basis. Refer to procedure 1B1, Standard Air-dry.
N	Blank	Lab preparation is same as S-blank except do not record the weight of the >2-mm fraction. All analytical results are reported on a <2-mm basis. Refer to procedure 1B1, Standard Air-dry.
M	Blank	Lab preparation is same as S-blank except sieve <2-mm moist subsample for 15-bar moist analysis. Use <2-mm air-dry soil for all other analyses. Report all analytical results on <2-mm basis. Refer to procedure 1B2, Field-moist.
S	K	Lab preparation is same as S-blank except grind the 2- to 20-mm fraction to <2 mm and keep for CO_3 analyses, etc. Report the analytical results for the ground 2- to 20-mm fraction on a 2- to 20-mm basis and all other analytical results on a <2-mm basis. Refer to procedure 1B5, Coarse Fragments.
S	R	Lab preparation is same as S-blank except recombine the 2- to 20-mm fraction with the <2-mm fraction and grind the entire sample to <2 mm. Report all analytical results for ground sample on a <2-mm basis. Refer to procedure 1B5, Coarse Fragments
G	P	Weigh sample at field-moisture content and record weight. Air-dry, weigh, and record weight. Grind entire sample to <2 mm. Report all analytical results for ground sample on a whole-soil basis. Refer to procedure 1B6, Whole-soil.
W	P	Weigh sample at field-moisture content and record weight. Air-dry, weigh, and record weight. Sieve >2-mm fractions, weigh, and record weights. Recombine the >2-mm fractions with the <2-mm fraction and grind entire sample to <2 mm. Report all analytical results on a whole-soil basis. This procedure is no longer performed at the SSL.

Table 1. Laboratory preparation codes and procedural summaries (continued).

<u>Code</u>   <u>Laboratory Preparation</u>

<u>Char</u>   <u>>2mm</u>

Code Char	>2mm	Laboratory Preparation
H	**Blank**	Obtain a moist whole-soil subsample for Histosol analysis. Obtain a <2-mm moist subsample for 15-bar moist analysis. Weigh remaining sample at field-moisture content and record weight. Air-dry, weigh, and record weight. Sieve >2-mm fractions, weigh, record weights, and discard. Pulverize subsample of <2-mm air-dry soil to a <80-mesh size and use for lab analyses. Use <80-mesh air-dry for all analyses except AD/OD, 15-, 1/10- and 2-bar analyses. For the AD/OD, 15-, 1/10- and 2-bar analyses, use <2-mm air-dry soil. Use <2-mm moist subsample for 15-bar moist. Report all analytical results except fabric on a <2-mm basis. Refer to procedure 1B7, Organic Material.
A (L)	**Blank**	Lab preparation is same as N-blank except pulverize subsample of <2-mm air-dry soil to a <80-mesh size and use for lab analyses. Use <80-mesh air-dry for all analyses except AD/OD and 15-bar analyses. For the AD/OD and 15-bar analyses, use <2-mm air-dry soil. All analytical results are reported on a <2-mm basis. Refer to procedure 1B1, Standard Air-dry.

Gravel codes

P =   porous >2-mm material that is considered soil is used for clod or core measurements.
V =   volume estimate is used to calculate the weight percentage of a >2-mm fraction. If that fraction is porous (P), code the samples with "P" rather than with "V".

PART II:

SSIR NO. 42, SOIL SURVEY LABORATORY METHODS MANUAL, VERSIONS 1.0 - 2.0 (1989, 1992, respectively)

ION EXCHANGE ANALYSES (5)

Cation Exchange Capacity (5A)
NH₄Oac, pH 7.0 (5A8)
Automatic Extractor (CEC-7)
Steam Distillation (5A8b)

1. Application

The CEC determined with $1\ N$ NH$_4$OAc buffered at pH 7.0, is a commonly used method and has become a standard reference to which other methods are compared (Peech et al., 1947). The advantages of using this method are that the extractant is highly buffered so that the extraction is performed at a constant, known pH (7.0) and that the NH4$^+$ on the exchange complex is easily determined.

2. Summary of Method

Displacement after washing is the basis for this procedure. The CEC is determined by saturating the exchange sites with an index cation (NH4$^+$); washing the soil free of excess saturated salt; displacing the index cation (NH4$^+$) adsorbed by the soil; and measuring the amount of the index cation (NH4$^+$). A sample is leached using $1\ N$ NH$_4$OAc and a mechanical vacuum extractor (Holmgren et al., 1977). The extract is weighed and saved for analyses of the cations. The NH$_4$$^+$ saturated soil is rinsed with ethanol to remove the NH$_4$$^+$ that was not adsorbed. Steam distillation and titration are used to determine the NH$_4$$^+$ adsorbed on the soil exchange complex. The CEC by NH$_4$OAc, pH 7 is reported in meq/100 g oven-dry soil in procedure 5A8b (Soil Conservation Service, 1984).

3. Interferences

Incomplete saturation of the soil with NH$_4$$^+$ and insufficient removal of NH$_4$$^+$ are the greatest interferences to this method. Ethanol removes some adsorbed NH$_4$$^+$ from the exchange sites of some soils. Isopropanol rinses has been used for some soils in which ethanol removes adsorbed NH$_4$$^+$. Soils that contain large amounts of vermiculite can irreversibly "fix" NH$_4$$^+$. Soils that contain large amounts of soluble carbonates can change the extractant pH and/or can contribute to erroneously high cation levels in the extract.

4. Safety

Wear protective clothing (coats, aprons, sleeve guards, and gloves) and eye protection (face shields, goggles, or safety glasses) when preparing reagents, especially concentrated acids and bases. Dispense concentrated acids and bases in a fume hood. Thoroughly wash hands after handling reagents. Use the safety showers and eyewash stations to dilute spilled acids and bases. Use sodium bicarbonate and water to neutralize and dilute spilled acids.

Ethanol is flammable. Avoid open flames and sparks. Standard laboratory equipment includes fire blankets and extinguishers for use when necessary. Follow the manufacturer's safety precautions when using the vacuum extractor and the Kjeltec Auto 1030 Analyzer.

5. Equipment
5.1 Mechanical vacuum extractor, 24 place, Centurion International, Inc., Lincoln, NE
5.2 Syringes, polypropylene, disposable, 60 mL, for sample tube, extractant reservoir, and tared extraction syringe.
5.3 Rubber tubing, 3.2 ID x 6.4 OD x 25.4 mm (1/8 ID x 1/4 OD x 1 in) for connecting syringe barrels.
5.4 Polycons, Richards Mfg. Co.
5.5 Kjeltec Auto 1030 Analyzer, Tecator, Fisher Scientific Inc.
5.6 Digestion tubes, straight neck, 250 mL
5.7 Analytical filter pulp, Schleicher and Schuell, no. 289
5.8 Plunger, modified. Remove rubber and cut plastic protrusion from plunger end.
5.9 Electronic balance, ±1-mg sensitivity

6. Reagents
6.1 Distilled deionized (DDI) water
6.2 Ammonium acetate solution (NH$_4$OAc), 1 N, pH 7.0. Add 1026 mL of glacial acetic acid (CH$_3$COOH) to 15 L DDI water. Add 1224 mL of conc. ammonium hydroxide (NH$_4$OH). Mix and cool. Dilute with DDI water to 18 L and adjust to pH 7.0 with CH$_3$COOH or NH$_4$OH.
6.3 Ethanol (CH$_3$CH$_2$OH), 95%, U.S.P.
6.4 Nessler's reagent. Add 4.56 g of potassium iodide (KI) to 30 mL DDI water. Add 5.68 g of mercuric iodide (HgI$_2$). Stir until dissolved. Dissolve 10 g of sodium hydroxide (NaOH) in 200 mL of DDI water. Transfer NaOH solution to a 250-mL volumetric flask and slowly add K-Hg-I solution. Dilute to volume with DDI water and thoroughly mix. Solution should not contain a precipitate. Solution can be used immediately.
6.5 Sodium chloride (NaCl), reagent, crystal.
6.6 Antifoam agent, slipicone release spray, Dow Chemical Corp. Alternatively, mix equal parts of mineral oil and n-octyl alcohol.
6.7 Boric acid, 4% (w:v), with bromcresol green-methyl red indicator (0.075 % bromcresol green and 0.05% methyl red), Ricca Chemical Co.
6.8 Hydrochloric acid (HCl), 0.1 N, standardized. Dilute 148 mL of conc. HCl in 16 L of DDI water.
6.9 NaOH, 1 M. Add 500 mL of 50% NaOH solution to 8 L of DDI water. Dilute to 9 L with DDI water.

7. Procedure

Extraction of Bases
7.1 Prepare sample tube by tightly compressing a 1-g ball of filter pulp into the bottom of a syringe barrel with a modified plunger.

7.2 Weigh 2.50 g of <2-mm, air-dry soil and place in sample tube. Prepare one quality control check sample per 48 samples.

7.3 Place sample tube on upper disk of the extractor and connect a tared extraction syringe. Use 25.4-mm (1 in) length rubber tubing and insert the plunger in the slot of the stationary disk of the extractor.

7.4 Use a squeeze bottle to fill sample tube to the 20-mL mark with NH$_4$OAc solution ($\approx$ 10 mL). Thoroughly wet the sample. Let stand for at least 20 min.

7.5 Put reservoir tube on top of the sample tube. Rapidly extract the NH$_4$OAc solution to a 0.5- to 1.0-cm height above sample. Turn off extractor. Add ≈ 45 mL of NH$_4$OAc solution to the reservoir tube. Set extractor for an overnight (12 to 16 h) extraction.

7.6 Next morning turn off the extractor. Pull the plunger of the syringe down. Do not pull plunger from the barrel of the syringe. Carefully remove the syringe containing the extract. Leave the rubber tubing on the sample tube. Weigh each syringe containing the NH$_4$OAc extract to the nearest 0.01 g.

7.7 Mix the extract in each syringe by manually shaking. Fill a polycon with extract solution and discard the excess. The solution in the polycon is reserved for analyses of extracted cations (procedures 6N2e, 6O2d, 6P2b, and 6Q2b).

Removal of Excess Ammonium Acetate

7.8 Return the extractor to starting position. Attach syringe to the sample tube and rinse the sides of the sample tube with ethanol from a wash bottle. Fill the sample tube to the 20-mL mark with ethanol and let stand for 15 to 20 min.

7.9 Place reservoir tube on the sample tube. Rapidly extract the ethanol level to a 0.5- to 1.0-cm height above the sample. Turn off the extractor and add 55 to 60 mL of ethanol to the reservoir. Extract at a 45-min rate.

7.10 After the extractor has stopped, turn off the switch. Pull the plunger of the syringe down. Do not pull the plunger from the syringe barrel. Remove the syringe and discard the ethanol.

7.11 Repeat the ethanol wash.

7.12 After the second wash, collect a few drops of ethanol extract from the sample tube on a spot plate. Test for NH$_4^+$ by using Nessler's reagent. A yellow, red to reddish brown precipitate is a positive test. If the test is positive, repeat the ethanol wash and retest with Nessler's reagent. Repeat until a negative test is obtained.

Steam Distillation: Samples and Reagent Blanks

7.13 Remove the sample tube and transfer the sample with filter pulp to a 250-mL digestion tube. Add 6 to 7 g of NaCl to the digestion tube. Use a gentle flow of compressed air to blow the filter pulp and sample out of the syringe. Wash the tube with DDI water and use a rubber policeman to complete transfer. The amount of distilled water that is added depends on the amount that is required to complete the transfer of tube contents.

7.14 Perform the same transfer and addition of reagents for blanks as for samples.

7.15 Spray silicone antifoam agent (or 2 drops of octyl alcohol) into the digestion tubes for each of the samples and reagent blanks.

7.16 When using new reagents, e.g., boric acid, reagent blanks are distilled in 2 sets of 6, one set per Kjeltec machine. Each set of 6 is averaged and recorded on bench worksheet and manually set on each machine. During the steam distillation, the mean reagent blank titer is automatically subtracted from the sample titer.

7.17 On bench worksheet, record the normality of standardized acid, i.e., ≈ 0.1 N HCl.

7.18 Connect the tube to the distillation unit. Close the safety door. Distillation and titration are performed automatically. Record the titer in mL of titrant.

8. Calculations

CEC (meq/100g) = (Titer x N x 100 x AD/OD)/(Weight)

where
Titer   = Titer of sample (mL)
N = Normality of HCl titrant
Weight = Sample weight (g)
100   = Conversion factor to 100 g basis
AD/OD = Air-dry/oven-dry ratio (procedure 4B5)

9. Report

Report CEC-7 in units of meq/100 g of oven-dry soil to the nearest 0.1 meq/100 g.

10. Precision

Precision data are not available for this procedure. A quality control check sample is run with every batch of 48 samples. With 113 observations of the quality control check sample, the mean, standard deviation, and C.V. for the CEC are 27.1, 0.57, and 2.1%, respectively.

11. References

Holmgren, George G.S., R.L. Juve, and R.C. Geschwender. 1977. A mechanically controlled variable rate leaching device. Soil Sci. Amer. J. 41:1207-1208.

Peech, M., L.T. Alexander, L.A. Dean, and J.F. Reed. 1947. Methods of soil analysis for soil fertility investigations. U.S. Dept. Agr. Circ. 757, 25 pp.

Soil Conservation Service. 1984. Procedures for collecting soil samples and methods of analysis for soil survey. USDA-SCS Soil Surv. Invest. Rep. no. 1. U.S. Govt. Print. Office, Washington, DC.

NH_4Cl, pH 7.0 (5A9)
Steam Distillation (5A9b)

1. Application

The CEC determined with a neutral unbuffered salt, e.g., 1 N NH_4Cl, is an estimate of the "effective" CEC (ECEC) of the soil (Peech et al., 1947). For a soil with a pH of <7.0, the ECEC value should be < CEC measured with a buffered solution at pH 7.0. The NH_4Cl CEC is $\approx$ equal to the NH_4OAc extractable bases plus the KCl extractable Al for noncalcareous soils.

2. Summary of Method

Displacement after washing is the basis for this procedure. The CEC is determined by saturating the exchange sites with an index cation (NH_4^+); washing the soil free of excess saturated salt; displacing the index cation (NH_4^+) adsorbed by the soil; and measuring the amount of the index cation (NH_4^+). A sample is leached using 1 N NH_4Cl and a mechanical vacuum extractor (Holmgren et al., 1977). The extract is weighed and saved for analyses of the cations. The NH_4^+ saturated soil is rinsed with ethanol to remove the NH_4^+ that was not adsorbed. Steam distillation and titration are used to determine the NH_4^+ adsorbed on the soil exchange complex. The CEC by NH_4Cl is reported in meq/100 g oven-dry soil in procedure 5A9b (Soil Conservation Service, 1984).

3. Interferences

Incomplete saturation of the soil with NH_4^+ and insufficient removal of NH_4^+ are the greatest interferences to this method. Ethanol removes some adsorbed NH_4^+ from the exchange sites of some soils. Isopropanol rinses have been used for some soils in which ethanol removes adsorbed NH_4^+. Soils that contain large amounts of vermiculite can irreversibly "fix" NH_4^+. Soils that contain large amounts of soluble carbonates can change the extractant pH and/or can contribute to erroneously high cation levels in the extract.

4. Safety

Wear protective clothing (coats, aprons, sleeve guards, and gloves) and eye protection (face shields, goggles, or safety glasses) when preparing reagents, especially concentrated acids and bases. Dispense concentrated acids and bases in a fume hood. Thoroughly wash hands after handling reagents. Use the safety showers and eyewash stations to dilute spilled acids and bases. Use sodium bicarbonate and water to neutralize and dilute spilled acids.

Ethanol is flammable. Avoid open flames and sparks. Standard laboratory equipment includes fire blankets and extinguishers for use when necessary. Follow the manufacturer's safety precautions when using the vacuum extractor and the Kjeltec Auto 1030 Analyzer.

5. Equipment
5.1 Mechanical vacuum extractor, 24 place, Centurion International, Inc., Lincoln, NE
5.2 Syringes, polypropylene, disposable, 60 mL, for sample tube, extractant reservoir, and tared extraction syringe.
5.3 Rubber tubing, 3.2 ID x 6.4 OD x 25.4 mm (1/8 ID x 1/4 OD x 1 in), for connecting syringe barrels.
5.4 Polycons, Richards Mfg. Co.
5.5 Kjeltec Auto 1030 Analyzer, Tecator, Fisher Scientific Inc.
5.6 Digestion tubes, straight neck, 250 mL
5.7 Analytical filter pulp, Schleicher and Schuell, no. 289
5.8 Plunger, modified. Remove rubber and cut plastic protrusion from plunger end.
5.9 Electronic balance, ±1-mg sensitivity

6. Reagents
6.1 Distilled deionized (DDI) water
6.2 Ammonium chloride solution (NH_4Cl), 1 N. Dissolve 535 g of NH_4Cl reagent in DDI water and dilute to 10 L.
6.3 Ethanol (CH_3CH_2OH), 95%, U.S.P.
6.4 Nessler's reagent. Add 4.56 g of potassium iodide (KI) to 30 mL DDI water. Add 5.68 g of mercuric iodide (HgI_2). Stir until dissolved. Dissolve 10 g of sodium hydroxide (NaOH) in 200 mL DDI water. Transfer NaOH solution to a 250-mL volumetric flask and slowly add K-Hg-I solution. Dilute to volume with DDI water and thoroughly mix. Solution should not contain a precipitate. Solution can be used immediately.
6.5 Sodium chloride (NaCl), reagent, crystal.
6.6 Antifoam agent, slipicone release spray, Dow Chemical Corp. Alternatively, mix equal parts of mineral oil and n-octyl alcohol.
6.7 Boric acid, 4% (w:v), with bromcresol green-methyl red indicator (0.075 % bromcresol green and 0.05% methyl red), Ricca Chemical Co.
6.8 Hydrochloric acid (HCl), 0.1 N, standardized. Dilute 148 mL of conc. HCl in 16 L of DDI water.
6.9 NaOH, 1 M. Add 500 mL of 50% NaOH solution to 8 L of DDI water. Dilute to 9 L with DDI water.

7. Procedure

Extraction of Bases

7.1 Prepare sample tube by tightly compressing a 1-g ball of filter pulp into the bottom of a syringe barrel with a modified plunger.

7.2 Weigh 2.50 g of <2-mm, air-dry soil and place in sample tube. Prepare one quality control check sample per 48 samples.

7.3 Place sample tube on upper disk of the extractor and connect a tared extraction syringe. Use 25.4-mm (1 in) length rubber tubing and insert the plunger in the slot of the stationary disk of the extractor.

7.4 Use a squeeze bottle to fill sample tube to the 20-mL mark with NH_4Cl solution (~ 10 mL). Thoroughly wet the sample. Let stand for at least 20 min.

7.5 Put reservoir tube on top of the sample tube. Rapidly extract the NH_4Cl solution to a 0.5- to 1.0-cm height above sample. Turn off extractor. Add $\approx$ 45 mL of NH_4Cl solution to the reservoir tube. Set extractor for an overnight (12 to 16 h) extraction.

7.6 Next morning turn off the extractor. Pull the plunger of the syringe down. Do not pull plunger from the barrel of the syringe. Carefully remove the syringe containing the extract. Leave the rubber tubing on the sample tube. Weigh each syringe containing the NH_4Cl extract to the nearest 0.01 g.

7.7 Mix the extract in each syringe by manually shaking. Fill a polycon with extract solution and discard the excess. The solution in the polycon is reserved for analyses of extracted cations (procedures 6N2e, 6O2d, 6P2b, and 6Q2b).

Removal of Excess Ammonium Chloride

7.8 Return the extractor to starting position. Attach syringe to the sample tube and rinse the sides of the sample tube with ethanol from a wash bottle. Fill the sample tube to the 20-mL mark with ethanol and let stand for 15 to 20 min.

7.9 Place reservoir tube on the sample tube. Rapidly extract the ethanol level to a 0.5- to 1.0-cm height above the sample. Turn off the extractor and add 55 to 60 mL of ethanol to the reservoir. Extract at a 45-min rate.

7.10 After the extractor has stopped, turn off the switch. Pull the plunger of the syringe down. Do not pull the plunger from the syringe barrel. Remove the syringe and discard the ethanol.

7.11 Repeat the ethanol wash.

7.12 After the second wash, collect a few drops of ethanol extract from the sample tube on a spot plate. Test for NH_4^+ by using Nessler's reagent. A yellow, red to reddish brown precipitate is a positive test. If the test is positive, repeat the ethanol wash and retest with Nessler's reagent. Repeat until a negative test is obtained.

Steam Distillation: Samples and Reagent Blanks

7.13 Remove the sample tube and transfer the sample with filter pulp to a 250-mL digestion tube. Add 6 to 7 g of NaCl to the sample. Use a gentle flow of compressed air to blow the filter pulp and sample out of the syringe. Wash the tube with DDI water and use a rubber policeman to complete transfer. The amount of distilled water that is added depends on the amount that is required to complete the transfer of tube contents.

7.14 Perform the same transfer and addition of reagents for blanks as for samples.

7.15 Spray silicone antifoam agent (or 2 drops of octyl alcohol) into the digestion tubes for each of the samples and reagent blanks.

7.16 When using new reagents, e.g., boric acid, reagent blanks are distilled in 2 sets of 6, one set per Kjeltec machine. Each set of 6 is averaged and recorded on bench worksheet and manually set on each machine. During the steam distillation, the mean reagent blank titer is automatically subtracted from the sample titer.

7.17 On bench worksheet, record the normality of standardized acid, i.e., ≈ 0.1 N HCl.

7.18 Connect the tube to the distillation unit. Close the safety door. Distillation and titration are performed automatically. Record the titer in mL of titrant.

8. Calculations

CEC (meq/100g) = (Titer x N x 100 x AD/OD)/(Weight0

where
Titer = Titer of sample (mL)
N = Normality of HCl titrant
Weight = Sample weight (g)
100 = Conversion factor to 100 g basis
AD/OD = Air-dry/oven-dry ratio (procedure 4B5)

9. Report
Report neutral salts CEC in units of meq/100 g of oven-dry soil to the nearest 0.1 meq/100 g.

10. Precision
Precision data are not available for this procedure. A quality control check sample is run with every batch of 48 samples. With 19 observations of the quality control check sample, the mean, standard deviation, and C.V. for the CEC are 26.0, 0.37, and 1.4%, respectively.

11. References
Holmgren, George G.S., R.L. Juve, and R.C. Geschwender. 1977. A mechanically controlled variable rate leaching device. Soil Sci. Amer. J. 41:1207-1208.
Peech, M., L.T. Alexander, L.A. Dean, and J.F. Reed. 1947. Methods of soil analysis for soil fertility investigations. U.S. Dept. Agr. Circ. 757, 25 pp.

CHEMICAL ANALYSES (6)

Total Carbon (6A)
Dry Combustion (6A2)
LECO CR-12 Carbon Analyzer (6A2d)

1. Application
Total C in soils is the sum of organic and inorganic C. Most of the organic C is associated with the organic matter fraction, and the inorganic C is generally found with carbonate minerals. The organic C in mineral soils generally ranges from 0 to 12%.

Total C is quantified by two basic methods, i.e., wet or dry combustion. The SSL uses dry combustion. In total C determinations, all forms of C in a soil are converted to CO_2 followed by a quantification of the evolved CO_2. Total C can be used to estimate the organic C content of a soil. The difference between total and inorganic C is an estimate of the organic C. Organic C also can be determined directly (procedure 6A1c). The inorganic C should be equivalent to carbonate values measured by CO_2 evolution with strong acid (Nelson and Sommers, 1982).

Organic C defines mineral and organic soils. In *Soil Taxonomy*, organic C is also used at lower taxonomic levels, e.g., ustollic and fluventic subgroups (Soil Survey Staff, 1975).

2. Summary of Method

An 80-mesh soil sample is oxidized at high temperatures. The released gases are scrubbed, and the CO_2 in the combustion gases is measured by using an infrared detector. Percent total C is reported on an oven-dry soil basis.

3. Interferences

This procedure simultaneously measures inorganic and organic C.

4. Safety

Wear protective clothing and safety glasses. Magnesium perchlorate may form explosive mixtures. Magnesium perchlorate may contain traces of perchloric acid, which remain from manufacturer's operations. This acid is anhydrous because of the strong desiccating capability of the salt. Avoid prolonged contact with oxidizable material or material capable of forming unstable perchlorate esters or salts. Remove magnesium perchlorate by using an excess of water to thoroughly dilute the material.

The use of high temperatures in the oxidation of samples requires that extreme caution be used to prevent burns and fires. Follow standard laboratory procedures when handling compressed gases. Oxygen is highly flammable. Avoid open flames and sparks. Standard laboratory equipment includes fire blankets and extinguishers for use when necessary. Follow the manufacturer's safety precautions when using the carbon analyzer.

5. Equipment

5.1 Carbon analyzer, Leco Model CR-12 781-600 Carbon System, Leco Corp., St. Joseph, MI

5.2 Data transmit card, part no. 772-573, Leco Corp., St. Joseph, MI

5.3 Combustion boats, part no. 529-203, Leco Corp., St. Joseph, MI

5.4 Single-stage regulator, oxygen service, part no. E11-W-N115Box, Air Products and Chemicals, Inc., Box 538, Allentown, PA 18105

5.5 Electronic balance, ±1-mg sensitivity

6. Reagents

6.1 Anhydrone, anhydrous magnesium perchlorate, granular

6.2 Glass wool

6.3 Compressed oxygen, >99.5% @ 30 psi

6.4 Calcium carbonate, $CaCO_3$, reagent grade.

7. Procedure

7.1 Use a fine-ground 80-mesh, air-dry soil

7.2 Weigh sample in a tared combustion boat. The sample size is dependent upon the C content. The product of sample weight (g) multiplied by C percentage should not be >10%. In most cases, the sample size is 1.00 g, unless the C content is >10%.

7.3 Refer to the manufacturer's manual for operation of carbon analyzer.

7.4 Combust sample in an O_2 atmosphere in which the C is oxidized to CO_2. Moisture and dust are removed by the instrument, and the CO_2 gas is then measured by a solid state infrared detector. The microprocessor formulates the analytical results (C_i) by combining the outputs of the infrared detector and the system ambient sensors with pre-programmed calibration, linearization and weight compensation factors. Analytical results are displayed and printed on the control console.

8. Calculations

C (%) = C_i x AD/OD

where:
C (%) = C (%), oven-dry basis
C_i = C (%) instrument
AD/OD = air-dry/oven-dry ratio (procedure 4B5)

9. Report
Report total C percentage on an oven-dry basis to the nearest 0.1%.

10. Precision
Precision data are not available for this procedure. A quality control check sample is included in every batch of ten samples. For 41 observations of the quality control check sample, the mean, standard deviation, and C.V. for total carbon are 11.38, 0.062, and 5.5%, respectively.

11. References
Nelson, D.W., and L.E. Sommers. 1982. Total carbon, organic carbon, and organic matter. *In* A.L. Page, R.H. Miller, and D.R. Keeney (eds.) Methods of soil analysis. Part 2. Chemical and microbiological properties. 2nd ed. Agronomy 9:539-579.

Soil Survey Staff. 1975. Soil taxonomy: A basic system of soil classification for making and interpreting soil surveys. USDA-SCS Agric. Handb. 436. U.S. Govt. Print. Office, Washington, DC.

Nitrogen (6B)
Kjeldahl Digestion II (6B3)
Ammonia Steam Distillation, Automatic Titrator (6B3a)

1. Application
The total N content of the soil may range from <0.02% in subsoils, 2.5% in peats, and 0.06 to 0.5% in surface layers of many cultivated soils (Bremmer and Mulvaney, 1982). The total N data may be used to determine the soil C:N ratio, the soil potential to supply N for plant growth, and the N distribution in the soil profile. The C:N ratio generally ranges between 10 to 12. Variations in the C:N ratio may serve as an indicator of the amount of soil inorganic N. Uncultivated soils usually have higher C:N ratios than do cultivated soils.

Soils with large amounts of illites or vermiculites can "fix" significant amounts of N compared to those soils dominated by smectites or kaolinites (Young and Aldag, 1982; Nommik and Vahtras, 1982). Since the organic C of many soils diminishes with depth while the level of "fixed" N remains constant or increases, the C:N ratio narrows (Young and Aldag, 1982). The potential to "fix" N has important fertility implications as the "fixed" N is slowly available for plant growth.

2. Summary of Method
A soil sample is digested using the Kjeldahl technique. The digest is made alkaline, the steam is distilled to release NH_4^+-N, and the NH_4^+-N is complexed with boric acid. The complexed NH_4^+-N is titrated with HCl, and the total N is calculated against a reagent blank (Soil Conservation Service, 1984).

3. Interferences

The total N that is measured by the Kjeldahl method does not distinguish among the types of N that are present in the soil. Practically all of the N is measured, but some forms of N are not recovered. Generally, soils have small amounts of N in the nonrecoverable forms, i.e., NO_3^- and NO_2^-. Soils with significant amounts of NO_3^- or NO_2^- are usually saline. The anion analysis of the saturated paste extracts measures NO_3^- and NO_2^- (procedures 6M1c and 6W1a, respectively).

The most significant error in the Kjeldahl method is the heating of the digestion mixture over 400°C. Loss of N occurs when the temperature of the digestion is >400°C (Bremmer and Mulvaney, 1982).

4. Safety

Wear protective clothing (coats, aprons, sleeve guards and gloves) and eye protection (face shields, goggles, or safety glasses) when handling acids and bases. Use heat resistant gloves when handling hot digestion tubes during digestion and steam distillation. Use the provided safety showers and eyewash stations to dilute spilled acids and bases. Use sodium bicarbonate and water to neutralize and dilute spilled acids. Digestion blocks are used at high temperatures, i.e., 250 and 400°C. Standard laboratory equipment includes fire blankets and extinguishers for use when necessary. Use the fume hood and fume aspiration devices to control and dispose of the acid fumes when digesting samples.

Boric acid is toxic and must not be ingested. Hengar granules contain Se which is toxic. Concentrated H_2SO_4 reacts violently with water and must be handled with caution. The 50% NaOH solution is very corrosive. Follow prudent laboratory safety precautions when handling these chemicals. Follow the manufacturer's safety precautions when using the Kjeltec Auto 1030 Analyzer.

5. Equipment
5.1 Electronic balance, ±0.001-g sensitivity
5.2 Digestion tubes, 250 mL, with constricted neck, Ace Glass Co., Inc.
5.3 Digestion blocks, 250 and 400°C
5.4 Dispenser, Zippette, 30 mL or equivalent, for conc. sulfuric acid (H_2SO_4), Brinkmann Instruments Inc.
5.5 Kjeltec Auto 1030 Analyzer, Tecator, Fisher Scientific Inc.

6. Reagents
6.1 Distilled water
6.2 Distilled deionized (DDI) water
6.3 Hydrochloric acid (HCl), conc., 12 N
6.4 Sodium hydroxide (NaOH), 50% (w:v), reagent
6.5 Hengar granules (selenized)
6.6 Digestion salt mixture. Mix 1000 g of potassium sulfate powder, 55 g of ferrous sulfate powder (anhydrous), and 32 g of copper II sulfate powder (anhydrous) in a tumbling mill for at least 30 min.
6.7 Antifoam, silicone spray bottle, Slipicone release spray, Dow Chemical Corp.
6.8 Boric acid, 4% (w:v), with bromcresol green-methyl red (0.075% bromcresol green and 0.05% methyl red) indicator, Ricca Chemical Co.
6.9 HCl, 0.1 N, standardized. Dilute 148 mL of conc. HCl in 16 L of DDI water.
6.10 Sucrose

7. Procedure

Kjeldahl Digestion of Sample
7.1 Weigh 3.000 g of <2-mm, air-dry soil into a 250-mL digestion tube. Refer to Table 1 for sample size.

7.2 Prepare 3 to 5 reagent blanks in every batch of 20 analyses. Reagent blanks contain 0.5 g of sucrose plus all reagents used in sample analysis, i.e., 12 mL of H_2SO_4, 4.5 g of digestion salt mixture, and 1 or 2 Hengar granules. Samples do not receive the 0.5 g of sucrose. Reagent blanks are run as samples and are not automatically subtracted during distillation procedure.

7.3 Use a dispenser to add 5 mL of distilled water to sample tube. Shake the tube to wet the sample.

7.4 Use a dispenser to add 12 mL of conc. H_2SO_4 to sample.

7.5 Allow sample to stand overnight.

7.6 Use a calibrated scoop to add 4.5 g of digestion salt mixture to sample.

7.7 Add 1 or 2 Hengar granules to sample.

7.8 Preheat one digestion heating block to 250°C and the other to 400°C.

7.9 Place the tube in the 250°C block, attach a fume aspirator, and digest for at least 30 min.

7.10 Remove the tube, place in the 400°C block, and digest sample for 1 h.

7.11 Remove the tube, place on a cooling board, and allow sample to cool for at least 15 min.

7.12 Remove the aspirator. Add 50 mL of distilled water.

Table 1. Sample size for total N based on volume of titrant ($FeSO_4$) used in organic C analysis (procedure 6A1c).

Fe_2SO_4 (mL)	Sample Size (g)
>6.00	3.0
4.00 to 5.00	2.0
3.00 to 4.00	1.5
<2.00	1.0

If >10.00 mL of $K_2Cr_2O_7$ (procedure 6A1c) is used and/or sample size is <1.00 g, then divide the volume of $FeSO_4$ by 2 and then use Table 1. To obtain a representative sample, do not use a sample size <0.5 g for total N analysis.

Ammonia Steam Distillation, Automatic Titrator
7.13 Spray silicone antifoam solution (or 2 drops of octyl alcohol) into the digestion tube and connect to the distillation unit.

7.14 Close the safety door.

7.15 Distillation and titration are performed automatically.

7.16 On bench worksheet, record the mL that are titrated for samples and reagent blanks. On bench worksheet, also record the normality of standardized HCl.

8. Calculations

N (%) = [(Titer$_{sample}$ - Titer$_{blank}$) x N X 1.4 X AD/OD]/(Sample Weight)

where
Titer$_{sample}$ = Titer of sample (mL)
Titer$_{blank}$ = Average titer of reagent blank (mL)
N = Normality of HCl titrant solution
1.4 = Conversion factor
AD/OD = Air-dry/oven-dry ratio (procedure 4B5)

9. Report
Report total N as a dimensionless value to the nearest 0.001 unit on an oven-dry basis.

10. Precision
Precision data are not available for this procedure. For 105 observations of the quality control check sample, the mean, standard deviation, and C.V. for total N are 0.143, 0.004, and 2.7%, respectively.

11. References
Bremmer, J.M., and C.S. Mulvaney. 1982. Nitrogen - Total. *In* A.L. Page, R.H. Miller, and D.R. Keeney (eds.) Methods of soil analysis. Part 2. Chemical and microbiological properties. 2nd ed. Agronomy 9:595-624.

Nommik, H., and K. Vahtras. 1982. Retention and fixation of ammonium and ammonia in soils. *In* F.J. Stevenson (ed.) Nitrogen in agricultural soils. Agronomy 22:123-171.

Soil Conservation Service. 1984. Procedures for collecting soil samples and methods of analysis for soil survey. USDA-SCS Soil Surv. Invest. Rep. no. 1. U.S. Govt. Print. Office, Washington, DC.

Young, J.L., and R.W. Aldag. 1982. Inorganic forms of nitrogen in soil. *In* F.J. Stevenson (ed.) Nitrogen in agricultural soils. Agronomy 22:43-66.

Iron, Aluminum, and Potassium (6C, 6G, and 6Q)
HF Dissolution (6C7, 6G11, and 6Q3)
Atomic Absorption (6C7a, 6G11a and 6Q3a)

1. Application
Historically, elemental analysis was developed for the analysis of rocks and minerals (Washington, 1930). The elemental analysis of soils, sediments, and rocks necessitates their decomposition into soluble forms. Hydrofluoric acid (HF) is efficient in the digestion and dissolution of silicate minerals for elemental decomposition. Elemental concentrations of Fe, Al, and K are determined by atomic absorption using 100 mg of clay suspension contained in a closed vessel with boric acid (H_3BO_3) to neutralize excess acid (Berdanier, Lynn, and Threlkeld, 1978; Soil Conservation Service, 1984).

2. Summary of Method
To 100 mg of clay suspension (procedure 7A2i), 5 mL of HF acid are added. The solution is heated, cooled, and 2 to 3 g of H_3BO_3 are added to neutralize excess acid. The solution is diluted to 100 mL, allowed to stand overnight, and 20 mL are decanted (procedure 7C3). The concentrations of Fe, Al, and K are determined by atomic absorption (AA) in procedures 6C7a, 6G11a, and 6Q3a, respectively. Data are reported in procedure 7C3.

3. Interferences

There are four types of interferences (matrix, spectral, chemical, and ionization) in AA analyses of these cations. These interferences vary in importance, depending upon the particular analyte selected.

The stable matrix system (HBF_4-H_3BO_3-ionic constituents of silicates) provides a suitable salt-free single matrix that greatly diminishes the chemical ionization, matrix, and instrumental interferences for AA determinations. One of the principal advantages of this technique is that all elements may be determined from a single sample solution (Lim and Jackson, 1982).

4. Safety

There are no significant hazards to analyst by this procedure. Wear protective clothing, e.g., coats and aprons. Follow standard laboratory practices when handling compressed gases. Gas cylinders should be chained or bolted in an upright position. Acetylene gas is highly flammable. Follow the manufacturer's safety precautions when using the AA.

5. Equipment

5.1 Atomic Absorption spectrophotometer (AA), Perkin-Elmer Corp., Norwalk, CT
5.2 Autosampler, AS-50, Perkin-Elmer Corp., Norwalk, CT
5.3 Microcomputer, 7500 Professional Computer, Perkin-Elmer Corp., Norwalk, CT
5.4 Dot matrix printer, P-132, Interdigital Data Systems, Inc.
5.5 Digital diluter/dispenser, product no. 100004, with hand probe and actuator, product no. 230700, Hamilton Co., P.O. box 10030, Reno, NV, 89510
5.6 Syringes, 10000 and 1000 µL, 1001 DX an 10110-TEL LL gastight, Hamilton Co., P.O. Box 10030, Reno, NV, 89510
5.7 Centrifuge tubes, polystyrene, 15 mL, conical bottom, graduated, part no. 2087, for sample dilution and sample changer, Becton Dickinson Labware, Becton Dickinson and Co., 2 Bridgewater Lane, Lincoln, Park, NJ 07035
5.8 Containers, polypropylene or teflon

6. Reagents

6.1 Distilled Deionized (DDI) water
6.2 Sodium chloride (NaCl) solution, 1143 ppm Na. Dissolve 5.81 g of NaCl in 2 L of DDI water.
6.3 Boric acid, H_2BO_3
6.4 Hydrofluoric acid (HF) solution, 2.47 N. Fill a polyethylene volumetric flask 1/3 full with DDI water. In hood, slowly and carefully add 49.36 g of HF. Slowly and carefully add 20 g of H_2BO_3. Hot reaction. May not completely dissolve. Make to 1-L volume with DDI water. Store HF solution in refrigerator. Use HF solution as reagent blank.
6.5 Fe stock solution, 1000 ppm. Commercial. Weigh 1.0000 g of Fe wire, dissolve in HCl, and make to 1-L volume with DDI water. Store in polypropylene container.
6.6 Al stock solution, 1000 ppm. Commercial. Weigh 1.0000 g of Al wire, dissolve in HCl, and make to 1-L volume with DDI water. Store in polypropylene container.
6.7 K stock solution, 50 meq L^{-1}. Dissolve 3.7279 g of KCl in 1 L of DDI. Store in polypropylene container.
6.8 Fe standard, 200 ppm. To 50 ml of Fe stock solution, add 12.34 ml of HF solution and 5 g of H_2BO_3. Make to 250-ml volume with DDI water. Store in polypropylene container.
6.9 Al standard, 200 ppm. To 50 ml of Al stock solution add 12.34 ml of HF solution and 5 g of H_2BO_3. Store in polypropylene container.
6.10 K standard, 1 meq L^{-1}. Add 12.34 ml of HF solution and 5 g of H_2BO_3 to 10 ml of K stock solution. Store in polypropylene container.
6.11 NaCl solution (1143 ppm Na). Dissolve 2.54 g of NaCl in DDI and make to 1-L volume.

7. Procedure

Dilution of Sample Extracts and Standards

7.1 Set the digital settings at 60 for the diluent (NaCl solution) and 99 for the HF sample, calibration reagent blanks, and calibration standards.

7.2 Dilute 1 part HF sample with 7 parts of NaCl solution (1:7 dilution).

7.3 Dilute 1 part calibration reagent blank (HF solution) with 7 parts NaCl solution (1:7 dilution).

7.4 Dilute 1 part of each calibration standard (200 ppm Fe, 200 ppm Al, and 1 meq^{-1} K) with 7 parts of NaCl solution (1:7 dilution).

7.5 Dispense the diluted solutions into 15-mL conical polystyrene centrifuge tubes. Place tubes in carousels of the sample changer.

AA Calibration

7.6 Use calibration reagent blank (HF solution) and calibration standards to calibrate the AA. The AA program requires a blank and a standard, in that order, to establish a single point calibration curve for element determination. During AA determinations, perform one calibration, i.e., blank plus standard, for every 8 samples.

AA Operation

7.7 The following parameters are only very general guidelines for instrument conditions for the analyte.

Element	Wavelength	Angle	Fuel/Oxidant
Fe	302.1	Parallel	C_2H_2/Air 20/25
Al	308.2	Parallel	C_2H_2/N_2O 30/17
K	766.5	30°	C_2H_2/Air 20/25

7.8 Use the microcomputer and printer to set instrument parameters and to collect and record instrument readings.

7.9 If a sample exceeds the calibration standard, dilute the sample (dilution ratio in calculation) with appropriate matrix and record the dilution. Remember to keep the matrix the same after dilution.

8. Calculations

Calculations are reported in procedure 7C3.

9. Report

Report concentrations of Fe, Al, and K by atomic absorption. Elemental concentrations are converted to percent oxides. Data are reported in procedure 7C3.

10. Precision

Precision data are not available for this procedure.

11. References

Berdanier, C.R., W.C. Lynn, and G.W. Threlkeld. 1978. Illitic mineralogy in Soil Taxonomy: X-ray vs. total potassium. Agron. Absts. 1978 Annual Mtgs.

Lim, C.H., and M.L. Jackson. 1982. Dissolution for total elemental analysis. *In* A.L. Page, R.H. Miller, and D.R. Keeney (eds.) Methods of soil analysis. Part 2. Chemical and microbiological properties. 2nd ed. Agronomy 9:1-12.

Soil Conservation Service. 1984. Procedures for collecting soil samples and methods of analysis for soil survey. USDA-SCS Soil Surv. Invest. Rep. no. 1. U.S. Govt. Print. Office, Washington, DC.

Washington, H.S. 1930. The Chemical analysis of rocks. 4th ed. John Wiley and Sons, Inc., New York, NY.

Iron, Aluminum, and Silicon (6C, 6G, and 6V)
Ammonium Oxalate Extraction (6C6, 6G12, and 6V2)
Inductively Coupled Plasma Spectrometry (6C9a, 6G12a, 6V2a)
Optical Density (8J)
(of ammonium oxalate extract)

1. Application

Oxalic acid-ammonium oxalate (acid oxalate) is used as a selective dissolution extractant for organically complexed Fe and Al, noncrystalline hydrous oxides of Fe and Al, allophane, and amorphous aluminosilicates (Wada, 1989). Acid oxalate is a poor extractant of imogolite and layer silicates and does not extract crystalline hydrous oxides of Fe and Al, opal, or crystalline silicate (Wada, 1989). A more reliable and accurate estimation of soil properties and a better understanding of soil exchange complex is provided when acid oxalate extraction is used in conjunction with other selective dissolution procedures, thermal techniques, and chemical tests. In Soil Taxonomy, acid oxalate extractable Fe and Al are criteria for andic soil properties (Soil Survey Staff, 1990).

2. Summary of Method

A soil sample is extracted with a mechanical vacuum extractor (Holmgren et al., 1977) in a 0.2 M acid oxalate solution buffered at pH 3.0 under darkness. The acid oxalate extract is weighed. The acid oxalate extract is diluted with 0.1 N HCl. The diluted extract is vaporized and atomized by a inductively coupled plasma emission spectrophotometer (ICP). The atoms or ions of the analyte are energized in high temperatures, resulting in the movement of valence electrons to higher orbits from the nucleus. As the electrons fall back to a lower orbit, electromagnetic energy at a specific wavelength for a given atom is emitted in measurable amounts (Soltanpour et al., 1982). Data are automatically recorded by a microcomputer and printer. The percent acid oxalate extractable Fe, Al, and Si are reported in procedures 6C9a, 6G12a, and 6V2a, respectively (Soil Conservation Service, 1984). On a less routine basis, Mn is also measured. To date, however, a National Soil Survey Laboratory (NSSL) method code has not been assigned to the Mn determination by acid oxalate extraction. In procedure 8J, the optical density of the extract is measured with a UV spectrophotometer at 430 nm.

3. Interferences

There are four types of interferences (matrix, spectral, chemical, and ionization) in the ICP analyses of these elements. These interferences vary in importance, depending upon the particular analyte chosen.

The acid oxalate buffer extraction is sensitive to light, especially UV light. The exclusion of light reduces the dissolution effect of cystalline oxides and clay minerals. If the sample contains large amounts of amorphous material (>2% Al), an alternate method should be used, i.e., shaking with 0.275 M acid oxalate, pH 3.25, 1:100 soil:extractant.

4. Safety

Wear protective clothing and eye protection. When preparing reagents, exercise special care. Restrict the use of concentrated HCl to a fume hood. Follow standard laboratory practices when handling compressed gases. Gas cylinders should be chained or bolted in an upright position. Follow the manufacturer's safety precautions when using the UV spectrophotometer and ICP.

5. Equipment

5.1 Electronic balance, ±1-mg sensitivity
5.2 Mechanical vacuum extractor, 24 place, Centurion International, Inc., Lincoln, NE
5.3 Syringes, polypropylene, disposable, 60 mL, for sample tube, extractant reservoir, and tared extraction syringe
5.4 Rubber tubing, 3.2 ID x 6.4 OD x 25.4 mm (1/8 ID x 1/4 OD x 1 in) for connecting syringe barrels
5.5 Polycons, Richards Mfg. Co.
5.6 Plunger, modified. Remove rubber and cut plastic protrusion from plunger end.
5.7 UV-visible spectrophotometer, DU-7, Beckmann Instruments, Inc.
5.8 Cuvettes, disposable, polystrene, 1-cm light path
5.9 Inductively coupled plasma spectrophotometer (ICP), Perkin-Elmer model 6000
5.10 Autosampler, AS-50, Perkin-Elmer Corp., Norwalk, CT
5.11 Microcomputer, 7500 Professional Computer, Perkin-Elmer Corp., Norwalk, CT
5.12 Dot matrix printer, P-132, Interdigital Data Systems, Inc.
5.13 Single-stage regulator, high-purity, high-flow, argon, product no. E11-X-N145DHF, Air Products and Chemicals, Inc., Box 538, Allentown, PA 18105
5.14 Digital diluter/dispenser, product no. 100004, with hand probe and actuator, product no. 230700, Hamilton Co., P.O. Box 10030, Reno, NV, 89510
5.15 Syringes, 10000 and 1000 μL, 1001 DX and 1010-TEL LL gastight, Hamilton Co., P.O. Box 10030, Reno, NV, 89510
5.16 Centrifuge tubes, polystyrene, 15 mL, conical, graduated, part no. 2087, for sample dilution and sample changer, Becton Dickinson Labware, Becton Dickinson and Co., 2 Bridgewater Lane, Lincoln Park, NJ 07035
5.17 Containers, polyproylene

6. Reagents

6.1 Distilled deionized (DDI) water
6.2 Hydrochloric acid (HCl), conc. 12 N
6.3 HCl, 1:1 HCl:DDI, 6 N. Carefully mix 1 part of conc. HCl to 1 part DDI water.
6.4 HCl, 1% wt. Carefully dilute 25 mL of conc. HCl to 1 L with DDI water.
6.5 HCl, 0.1 N. Add 8.33 mL of conc. HCl to DDI water and make to 1-L volume.
6.6 Acid oxalate buffer solution, 0.2 M, pH 3.0. *Solution A* (base): Dissolve 284 g of $(NH_4)_2C_2O_4 \cdot H_2O$ in 10 L of DDI water. *Solution B* (acid): Dissolve 252 g of $H_2C_2O_4 \cdot H_2O$ in 10 L of DDI water. Mix 4 parts solution A with 3 parts solution B. Adjust acid oxalate solution pH by adding either acid or base solution. Store in a polypropylene bottle.
6.7 pH buffers, pH 4.00 and 7.00, for electrode calibration.
6.8 Primary Fe standard, 1000 ppm. Dissolve 1.000 g of Fe wire in a minimum volume of 1:1 HCl:DDI. Dilute to 1-L volume in a volumetric flask using 1% HCl. Store in a polypropylene bottle.
6.9 Primary Al standard, 1000 ppm. Dissolve 1.000 g of Al wire in a minimum volume of 1:1 HCl:DDI. Dilute to 1-L volume in a volumetric flask using 1% HCl water. Store in a polypropylene bottle.
6.10 Primary Si standard, 1000 ppm. Fuse 0.2139 g of SiO_2 with 2 g of Na_2CO_3 in a platinum crucible. Dissolve the melt with DDI water and transfer to a 100-mL volumetric flask. Dilute to 1-L volume with DDI water. Store in a polypropylene bottle.

6.11 Primary Mn standard, 1000 ppm. Dissolve 1.000 g of Mn wire in a minimum volume of 1:1 HCl:DDI. Dilute to 1-L volume in a volumetric flask using 1:1 HCl:DDI. Store in a polypropylene bottle.

6.12 High calibration standard. Mix 30 mL of each primary standard (Al, Fe, and Si) with 5 mL of primary Mn standard. Add 50 mL of 0.4 M acid oxalate solution, 20 mL of conc. HCl, and make to 1-L volume with DDI water. Resulting solution contains 5 ppm Mn and 30 ppm each of Al, Fe, and Si. Store in a polypropylene bottle.

6.13 Low calibration standard. Mix 10 mL of each primary standard (Al, Fe, and Si) with 2 mL of primary Mn standard. Add 30 mL of 0.4 M acid oxalate solution, 20 mL of conc. HCl, and make to 1-L volume with DDI water. Resulting solution contains 2 ppm Mn and 10 ppm each of Al, Fe, and Si. Store in a polypropylene bottle.

6.14 Calibration reagent blank solution. Add 30 mL of 0.4 M acid oxalate solution, 20 mL of conc. HCl, and make to 1-L volume with DDI water.

6.15 Argon gas, purity 99.9%

7. Procedure

Extraction of Fe, Al, Si, and Mn

7.1 Prepare sample tube by tightly compressing a 1-g ball of filter pulp into the bottom of a syringe barrel with a modified plunger.

7.2 Weigh 0.500 g of <2-mm, air-dry soil and place in sample tube. Prepare two reagent blanks (no sample in tube) per set of 48 samples.

7.3 Place the sample tube on the upper disk of the extractor and connect a tared extraction syringe. Use 25.4-mm (1-in) length rubber tubing to insert the handle of the plunger in the slot of the stationary extractor disk.

7.4 Use a dispenser to add 15.00 mL of acid oxalate buffer to the sample tube. Make sure that the sample is thoroughly wetted. During the addition, wash sides of the tube and wet the sample. Shaking, swirling, or stirring may be required to wet organic samples. Allow sample to stand for at least 30 min.

7.5 Set extractor for 30-min extraction rate and extract until the acid oxalate buffer solution is at a 0.5 to 1.0-cm height above sample. Turn off extractor.

7.6 Put reservoir tube on top of the sample tube.

7.7 Add 35 mL of acid oxalate buffer to the reservoir tube.

7.8 Cover the extractor with a black plastic bag to exclude light. Adjust the extraction rate for a 12-h extraction.

7.9 After the extraction, shut off the extractor and pull plunger of syringe down. Do not remove the plunger from syringe barrel. Carefully remove the syringe with extract leaving the rubber tubing on the sample tube.

7.10 Weigh each syringe containing acid oxalate extract to the nearest 0.01 g.

7.11 Mix extract in each syringe by manually shaking. Fill a polycon with extract solution. This solution is reserved for determinations of Fe, Mn, Al, and Si. If optical density is to be measured, fill a disposable cuvette with extract solution. Discard excess solution.

Determination of Optical Density of Extract

7.12 Place 4 mL of acid oxalate extract in disposable cuvette.

7.13 Place 4 mL of acid oxalate reagent blank in disposable cuvette.

7.14 On DU-7 spectrophotometer, select a 430-nm wavelength. Select normal slit width and height.

7.15 Use the acid oxalate extract reagent blank to set spectrophotometer.

7.16 Record optical density of acid oxalate extract to nearest 0.000.

Dilution of Sample Extracts and Standards

7.17 For better nebulization, add one drop of DDBSA solution to each tube (sample extracts, calibration standards, and reagent blanks) to reduce surface tensions. Add DDBSA to tube before the addition of diluted solution.

7.18 Set the digital settings of the Hamilton diluter at 63 for the diluent (0.1 N HCl) and 70 for the acid oxalate extracts for a 1:10 dilution. Calibration reagent blanks and calibration standards are not diluted.

7.19 Dilute 1 part acid oxalate sample extract with 10 parts of 0.1 N HCl (1:10 dilution).

7.20 Dispense the diluted solutions into 15-mL conical polystyrene centrifuge tubes which have been placed in carousels of the sample changer.

ICP Calibration

7.21 Use high calibration standard and calibration reagent blank to calibrate ICP. The ICP requires a standard and a blank, in that order, for calibration. During ICP determinations, perform one calibration, i.e., standard plus blank, for every 6 samples.

7.22 Use the low calibration standard as a check sample.

ICP Set-up and Operation

7.23 The following parameters are only very general guidelines for instrument conditions for the various analytes.

Parameter	Value
ICP power	1250 W
Plasma gas flow	Ar 12 L min^{-1}
Nebulizer gas flow	Ar 0.5 L min^{-1}
Auxiliary gas flow	Ar 0.05 to 1 L min^{-1}

Use a high solids nebulizer instead of the cross flow nebulizer.

7.24 Analyte data for some elements are reported at 2 wavelengths which serve as data checks.

Analyte	Wave-length (nm)	Low Stamdards (ppm)	High Standards (ppm)
Fe	238.204/ 239.562	30.00	10.00
Al	394.400/ 396.150	30.00	10.00
Si	212.412/ 251.611	30.00	10.00
Mn	257.610	5.00	2.00

7.25 Use the microcomputer and printer to set instrument parameters and to collect and record instrument readings. The instrument readings are usually programmed in ppm.

7.26 If sample exceeds calibration standard, dilute the sample (dilution ratio in calculation) with appropriate matrix and record dilution. Remember to keep the matrix the same after dilution by diluting with 0.1 N HCl at the 1:1 ratio.

8. Calculations

Analyte (%) = [ICP x (Syr$_{fin}$ -Syr$_{init}$) x DR x AD/OD]/[Sample x 10000 x Density]

where
ICP   = ICP analyte concentration (ppm)
Syr$_{fin}$ = Weight of syringe + extract (g)
Syr$_{init}$ = Tare weight of syringe (g)
DR   = Dilution ratio of samples over calibration range
Sample = Weight of sample (g)
Density = Density of acid oxalate solution (1.007)
AD/OD = Air-dry/oven-dry ratio (procedure 4B5)

9. Report

Report the percent acid oxalate extractable Fe, Al, and Si to the nearest 0.01%. Percent acid oxalate extractable is also reported. To date, however, no method code has been assigned to Mn determination by acid oxalate extraction. Report the optical density of the acid oxalate extract to the nearest 0.001 unit.

10. Precision

Precision data are not available for this procedure. The mean, standard deviation, and CV for Fe, Al, Si, and optical density for both the low and high standards are as follows:

High Std	n	Mean	Std. Dev.	C.V.
Optical Density	18	0.18	0.03	14.2
Fe	17	0.94	0.17	18.2
Al	17	2.6	0.19	7.6
Si	17	1.2	0.09	7.3

Low Std	N	Mean	Std. Dev.	C.V.
Optical Density	25	0.06	0.00	7.5
Fe	24	0.26	0.03	9.7
Al	25	0.17	0.01	8.4
Si	26	0.02	0.01	53.2

11. References

Holmgren, G.G.S., R.L. Juve, and R.C. Geschwender. 1977. A mechanically controlled variable rate leaching device. Soil Sci. Amer. J. 41:1207-1208.

Soil Conservation Service. 1984. Procedures for collecting soil samples and methods of analysis for soil survey. USDA-SCS Soil Surv. Invest. Rep. no. 1. U.S. Govt. Print. Office, Washington, DC.

Soil Survey Staff. 1990. Keys to soil taxonomy. 4th ed. SMSS technical monograph no. 6. Blacksburg, VA.

Soltanpour, P.N., J.B. Jones Jr., and S.M. Workman. 1982. *In* A. Klute (ed.) Methods of soil analysis. Part 2. Chemical and microbiological properties. 2nd ed. Agronomy 9:29-65.

Wada, K. 1989. Allophane and imogolite. *In* J.B. Dixon and S.B. Weed (eds.). Minerals in soil environment. 2nd ed. Soil Sci. Soc. Amer. No. 1. p. 1051-1087.

Manganese and Aluminum (6D and 6G)
1 *N* KCl Extractable, Automatic Extractor (6D3 and 6G9)
Inductively Coupled Plasma Spectrometry (6D3 and 6G9b)

1. Application

The Al extracted by 1 *N* KCl approximates exchangeable Al and is a measure of the "active" acidity present in soils with a 1:1 water pH <5.5. Above pH 5.5, precipitation of Al occurs during analysis. This method does not measure the acidity component of hydronium ions (H_3O^+). If Al is present in measurable amounts, the hydronium is a minor component of the active acidity. Because the 1 *N* KCl extractant is an unbuffered salt and usually affects the soil pH one unit or less, the extraction is determined at or near the soil pH. The KCl extractable Al is related to the immediate lime requirement and existing CEC of the soil. The "potential" acidity is better measured by the $BaCl_2$-TEA method (procedure 6H5a) (Thomas, 1982).

2. Summary of Method

A soil sample is leached with 1 *N* KCl using the mechanical vacuum extractor (Holmgren et al., 1977). The leachate is weighed. The KCl extracted solution is diluted with 0.5 *N* HCl. The diluted extract is vaporized and atomized by a inductively coupled plasma emission spectrophotometer (ICP). The atoms or ions of the analyte are energized in high temperatures, resulting in the movement of valence electrons to higher orbits from the nucleus. As the electrons fall back to a lower orbit, electromagnetic energy at a specific wavelength for a given atom is emitted in measurable amounts (Soltanpour et al., 1982). Data are automatically recorded by a microcomputer and printer. The Mn and Al are reported in meq/100 g ovendry soil in procedures 6D3 and 6G9b (Soil Conservation Service, 1984).

3. Interferences

There are four types of interferences (matrix, spectral, chemical, and ionization) in the ICP analyses of these cations. These interferences vary in importance, depending upon the particular analyte selected.

The soil:extractant ratio must remain constant. A soil:extractant ratio of 1:10 (w:v) for batch procedures is most commonly used. Using a leaching technique, a 1:20 (w:v) ratio gives comparable results. If the sample size is changed, the amount of extractable Al is changed. No other significant interferences have been identified for this procedure.

4. Safety

Wear protective clothing and eye protection. When preparing reagents, exercise special care. Restrict the use of concentrated HCl to a fume hood. Follow standard laboratory practices when handling compressed gases. Gas cylinders should be chained or bolted in an upright position. Follow the manufacturer's safety precautions when using the ICP.

5. Equipment

5.1 Electronic balance, ±1-mg sensitivity

5.2 Analytical filter pulp, Schleicher and Schuell, no. 289

5.3 Mechanical vacuum extractor, 24 place, Centurion International, Inc., Lincoln, NE

5.4 Syringes, polypropylene, disposable, 60 mL, for sample tube, extractant reservoir, and tared extraction syringe.

5.5 Rubber tubing, 3.2 ID x 6.4 OD x 25.4 mm (1/8 ID x 1/4 OD x 1 in), for connecting syringe barrels

5.6 Plunger, modified. Remove rubber and cut plastic protrusion from plunger end.

5.7 Wash bottle, 20 mL, to dispense KCl.

5.8 Polycons, Richards Mfg. Co.

5.9 Inductively coupled plasma (ICP) atomic emission spectrophotometer, Perkin-Elmer model 6000

5.10 Autosampler, AS-50, Perkin-Elmer Corp., Norwalk, CT

5.11 Microcomputer, 7500 Professional Computer, Perkin-Elmer Corp., Norwalk, CT

5.12 Dot matrix printer, P-132, Interdigital Data Systems, Inc.

5.13 High-purity, high-flow, single-stage regulator, argon, product no. E11-X-N145DHF, Air Products and Chemicals, Inc., Box 538, Allentown, PA 18105

5.14 Digital diluter/dispenser, product no. 100004, with hand probe and actuator, product no. 230700, Hamilton Co., P.O. Box 10030, Reno, NV, 89510

5.15 Syringes, 10000 and 1000 µL, 1001 DX and 1010-TEL LL gas tight, Hamilton Co., P.O. Box 10030, Reno, NV, 89510

6. Reagents

6.1 Distilled deionized (DDI) water

6.2 Hydrochloric acid (HCl), conc., 12 N

6.3 HCl, 1:1 HCl:DDI, 6 N. Carefully mix 1 part of conc. HCl to 1 part DDI water.

6.4 HCl, 1% wt. Carefully dilute 25 mL of conc. HCl to 1 L with DDI water.

6.5 HCl, 0.5 N. Add 1 part of conc. HCl to 24 parts DDI water (1:25 dilution).

6.6 Potassium chloride solution (KCl), 1.0 N. Dissolve 1342 g of KCl reagent in 16 L DD water. Allow solution to equilibrate to room temperature. Dilute to 18 L with DDI water. Use 1.0 N KCl for Al and Mn extraction.

6.7 Potassium chloride solution (KCl), 2.0 N. Dissolve 298.24 g of KCl reagent in 1.5 L DDI water. Allow solution to equilibrate to room temperature. Dilute to 2 L with DDI water. Use 2.0 N KCl for standards.

6.8 Primary Al standard, 2248.5 ppm (250 meq L^{-1}). Dissolve 2.2485 g of Al wire in a minimum volume of 1:1 HCl:DDI. This is a very slow reaction. Dilute to 1 L in a volumetric flask using 1% HCl solution. Store in polypropylene container.

6.9 Primary Mn standard, 1000 ppm (36 meq L^{-1}). Commercial. Dissolve 1.000 g of Mn metal in a minimum volume of 1:1 HCl:DDI. When dissolved, dilute to 1 L in a volumetric flask using 1% HCl solution. Store in a polypropylene container.

6.10 Mn standard, 250 ppm (9 meq L^{-1}). Mix 25 mL of primary Mn standard (1000 ppm) with 10 mL of 1:1 HCl:DDI and dilute to 100-mL volume with DDI water. Store in polypropylene bottle.

6.11 Calibration Al and Mn standard, 10 meq L^{-1} Al and 5 ppm Mn. Mix 10 mL of primary Al standard (250 meq L^{-1}) with 125 mL 2.0 N KCl solution. Add 5 mL of Mn standard (250 ppm). Make to 250-mL volume with DDI water. Store in polypropylene container.

6.12 Calibration Al and Mn check standard, 5 meq L^{-1} Al and 2 ppm Mn. Mix 5 mL of primary Al standard (250 meq L^{-1}) with 125 mL 2.0 N KCl solution. Add 2 mL of Mn standard (250 ppm). Store in polypropylene container.

6.13 Calibration reagent blank solution, 1.0 N KCl. Add 125 mL of 2.0 N KCl to a volumetric flask and make to 250-mL volume with DDI water. Store in polypropylene container.

6.14 Dodecylbenzenesulfonic acid (DDBSA), tech 97%. Working stock is 0.1 M. Dilute 25 mL of 0.1 M DDBSA to 1-L volume with DDI water.

6.15 Argon gas, purity 99.9%

7. Procedure

Extraction of Al and Mn

7.1 Prepare sample tube by tightly compressing a 1-g ball of filter pulp into the bottom of a syringe barrel with a modified plunger.

7.2 Weigh exactly 2.50 g of <2-mm, air-dry soil and place in sample tube. Prepare one quality control check sample per 48 samples.

7.3 Place the sample tube on the upper disk of the extractor and connect a tared extraction syringe. Use 25.4-mm (1-in) length rubber tubing to insert the handle of the plunger in the slot of the stationary extractor disk.

7.4 Use a squeeze bottle and fill sample tube to the 20-mL mark with 1.0 N KCl solution ($\approx$ 10 mL). Make sure that the sample is thoroughly wetted. During the addition, wash sides of the tube and wet the sample. Shaking, swirling, or stirring may be required to wet organic samples. Allow sample to stand for at least 30 min.

7.5 Put reservoir tube on top of the sample tube. Set extractor for fast extraction rate and extract until the KCl solution is at a 0.5- to 1.0-cm height above sample. Turn off extractor.

7.6 Add 45 mL KCl solution to reservoir tube. Set extractor for 45-min extraction.

7.7 After the extraction, shut off extractor and pull plunger of syringe down. Do not remove the plunger from syringe barrel. Carefully remove the syringe with extract leaving the rubber tubing on the sample tube.

7.8 Weigh each syringe containing KCl extract to the nearest 0.01 g.

7.9 Mix extract in each syringe by manually shaking. Fill a polycon with extract solution and discard the excess. This solution is reserved for extractable Al and Mn analyses.

Dilution of Extracts and Standards

7.10 For better neubilization, add one drop of DDBSA solution to KCl sample extracts, calibration reagent blanks and calibration standards to reduce surface tensions. Add DDBSA to tube before adding diluted solution.

7.11 Set the digital settings at 40 for the diluent (0.5 N HCl) and 99 for the KCl sample extracts, calibration reagent blanks, calibration standards, and calibration check standards for a 1:5 dilution as follows:

7.12 Dilute 1 part KCl sample extract with 4 parts of 0.5 N HCl (1:5 dilution).

7.13 Dilute 1 part calibration reagent blank with 4 parts of 0.5 N HCl (1:5 dilution).

7.14 Dilute 1 part calibration standard (10 meq L^{-1} Al and 5 ppm Mn) with 4 parts of 0.5 N HCl (1:5 dilution).

7.15 Dilute 1 part calibration check standard (5 meq L^{-1} Al and 2 ppm Mn) with 4 parts of 0.5 N HCl (1:5 dilution).

7.16 Dispense the diluted solutions into 15-mL conical polystyrene centrifuge tubes which have been placed in carousels of the sample changer.

ICP Calibration

7.17 Use calibration standard (10.00 meq L^{-1} Al and 5.00 ppm Mn,) and calibration reagent blank (1.0 N KCl) to calibrate ICP. The ICP requires a standard and a blank, in that order, for calibration. During ICP determinations, perform one calibration, i.e., standard plus blank, for every 6 samples.

7.18 Use the calibration check standard (5.00 meq L^{-1} Al and 2.00 ppm Mn) as a check sample.

ICP Set-up and Operation

7.19 The following parameters are only very general guidelines for instrument conditions for the analytes.

Parameter	Al	Fe
Plasma flow (ml Ar min-1)	12.0	12.0
Nebulizer flow (mL Ar min-1)	0.5	0.5
Auxiliary flow (mL Ar min-1)	0.5	0.5
Viewing Height (nm)	15.0	15.0

	Wavelength 1	
Wavelength (nm)	394.400	259.373
Scan speed (s)	2.0	1.0
Bkg. Correction (nm)	-0.069, +0.055	0.0

	Wavelength 2	
Wavelength (nm)	396.152	294.920
Scan speed (s)	2.0	2.0
Bkg. Correction (nm)	-0.069, +0.055	-0.046, +0.049

High solids nebulizer has a 20 s read delay.

7.20 Load sample tubes in the carousel so that the calibration standard, reagent blank, calibration check standard, and 6 unknown samples are determined in order. Determine a set of 24 unknown samples with each carousel.

7.21 Use the microcomputer and printer to set instrument parameters and to collect and record instrument readings.

7.22 If a sample exceeds the calibration standard, dilute the sample (dilution ratio in calculation) with appropriate matrix and record dilution. Remember to keep the matrix the same after dilution by diluting with the 0.5 N HCl solution (1:5 dilution).

7.23 Analyze one quality control check sample for every 48 samples.

8. Calculations

8.1 The instrument readings are the analyte concentration (meq L^{-1} Al and ppm Mn) in undiluted extract. Use these values to calculate the analyte concentration on an oven-dry soil basis (meq/100 g).

Analyte (meq/100 g) = [ICP x (Wt$_{syr+ext}$ - Wt$_{syr}$) x DR x 100 x AD/OD]/[Smp.Wt. x 1.0412 x 1000]

where
ICP   = ICP analyte reading
Wt$_{syr+ext}$ = Weight of extraction syringe & extract (g)
Wt$_{syr}$ = Weight of tared extraction syringe (g)
DR.   = Dilution ratio of samples over calibration range
Smp. Wt= Sample weight (g)
1.0412   = Density of 1 N KCl @ 20°C
1000   = g L^{-1}
100   = Conversion factor (100-g basis)
AD/OD = Air-dry/oven-dry ratio (procedure 4B5)

9. Report

Report KCl extractable Al and Mn in units of meq/100 g of oven-dry soil to the nearest 0.01 meq/100 g.

10. Precision

Precision data are not available for this procedure.

11. References

Holmgren, George G.S., R.L. Juve, and R.C. Geschwender. 1977. A mechanically controlled variable rate leaching device. Soil Sci. Amer. J. 41:1207-1208.

Soil Conservation Service. 1984. Procedures for collecting soil samples and methods of analysis for soil survey. USDA-SCS Soil Surv. Invest. Rep. no. 1. U.S. Govt. Print. Office, Washington, DC.

Soltanpour, P.N., J.B. Jones Jr., and S.M. Workman. 1982. *In* A. Klute (ed.) Methods of soil analysis. Part 2. Chemical and microbiological properties. 2nd ed. Agronomy 9:29-65.

Thomas, G.W. 1982. Exchangeable cations. *In* A. Klute (ed.) Methods of soil analysis. Part 2. Chemical and microbiological properties. 2nd ed. Agronomy 9:159-165.

Aluminum (6G)
KCl, Automatic Extractor (6G9)
Atomic Absorption (6G9a)

1. Application

The Al extracted by 1 N KCl approximates exchangeable Al and is a measure of the "active" acidity present in soils with a 1:1 water pH <5.5. Above pH 5.5, precipitation of Al occurs during analysis. This method does not measure the acidity component of hydronium ions (H_3O^+). If Al is present in measurable amounts, the hydronium is a minor component of the active acidity. Because the 1 N KCl extractant is an unbuffered salt and usually affects the soil pH one unit or less, the extraction is determined at or near the soil pH. The KCl extractable Al is related to the immediate lime requirement and existing CEC of the soil. The "potential" acidity is better measured by the $BaCl_2$-TEA method (procedure 6H5a) (Thomas, 1982).

2. Summary of Method

A soil sample is leached with 1 N KCl using the mechanical vacuum extractor (Holmgren et al., 1977). The leachate is weighed. The KCl extract is diluted with distilled deionized (DDI) water. The diluted extract is aspirated into an atomic absorption spectrophotometer (AA). The analyte is measured by absorption of the light from a hollow cathode lamp. An automatic sample changer is used to aspirate a series of samples. The AA converts absorption to analyte concentration. The data are automatically recorded by a microcomputer and printer. The Al is reported in meq/100 g ovendry soil in procedure 6G9a.

3. Interferences

There are four types of interferences (matrix, spectral, chemical, and ionization) in the AA analyses of these cations. These interferences vary in importance, depending upon the particular analyte selected.

The soil:extractant ratio must remain constant. A soil:extractant ratio of 1:10 (w:v) for batch procedures is most commonly used. Using a leaching technique, a 1:20 (w:v) ratio gives comparable results. If the sample is changed, the amount of extractable Al is changed. No other significant interferences have been identified for this procedure.

4. Safety

Wear protective clothing and eye protection. When preparing reagents, exercise special care. Restrict the use of concentrated HCl to a fume hood. Many metal salts are extremely toxic and may be fatal if ingested. Follow standard laboratory practices when handling compressed gases. Gas cylinders should be chained or bolted in an upright position. Acetylene gas is highly flammable. Avoid open flames and sparks. Standard laboratory equipment includes fire blankets and extinguishers for use when necessary. Follow the manufacturer's safety precautions when using the AA.

5. Equipment

5.1 Electronic balance, ±1-mg sensitivity
5.2 Mechanical vacuum extractor, 24 place, Centurion International, Inc., Lincoln, NE
5.3 Syringes, polypropylene, disposable, 60 mL, for sample tube, extractant reservoir, and tared extraction syringe.
5.4 Rubber tubing, 3.2 ID x 6.4 OD x 25.4 mm (1/8 ID x 1/4 OD x 1 in) for connecting syringe barrels
5.5 Wash bottle, 20 mL, to dispense KCl.
5.6 Plunger, modified. Remove rubber and cut plastic protrusion from plunger end.
5.7 Polycons, Richards Mfg. Co.
5.8 Atomic absorption spectrophotometer (AA), Perkin-Elmer model 5000
5.9 Autosampler, AS-50, Perkin-Elmer Corp., Norwalk, CT
5.10 Microcomputer, 7500 Professional Computer, Perkin-Elmer Corp., Norwalk, CT
5.11 Dot matrix printer, P-132, Interdigital Data Systems, Inc.
5.12 Single-stage regulator, acetylene service, part number E11-0-N511A, Air Products and Chemicals, Inc., Box 538, Allentown, PA 18105

5.13 Heated regulator, single-stage, nitrous oxide, stock number 808 8039, Airco Welding Products, P.O. Box 486, Union, NJ 07083

5.14 Digital diluter/dispenser, product no. 100004, with hand probe and actuator, product no. 230700, Hamilton Co., P.O. Box 10030, Reno, NV, 89510

5.15 Syringes, 10000 and 1000 μL, 1001 DX and 1010-TEL LL gas tight, Hamilton Co., P.O. Box 10030, Reno, NV, 89510

5.16 Centrifuge tubes, polystyrene, 15 mL, conical bottom, graduated, part no. 2087, for sample dilution and sample changer, Becton Dickinson Labware, Becton Dickinson and Co., 2 Bridgewater Lane, Lincoln Park, NJ 07035

5.17 Containers, polyproylene or teflon

6. Reagents

6.1 Distilled deionized (DDI) water

6.2 Hydrochloric acid (HCl), conc., 12 N

6.3 HCl, 1:1 HCl:DDI, 6 N. Carefully mix 1 part of conc. HCl to 1 part DDI water.

6.4 HCl, 1% wt. Carefully dilute 25 mL of conc. HCl to 1 L with DDI water.

6.5 Potassium chloride solution (KCl), 1.0 N. Dissolve 1342 g of KCl reagent in 16 L DDI water. Allow solution to equilibrate to room temperature. Dilute to 18 L with DDI water. Use 1.0 N KCl solution for Al extraction.

6.6 Potassium chloride solution (KCl), 2.0 N. Dissolve 298.24 g of KCl reagent in 1.5 L DDI water. Allow solution to equilibrate to room temperature. Dilute to 2 L with DDI water. Use 2.0 N KCl solution for standards.

6.7 Primary Al standard, 2248.5 ppm (250 meq L^{-1}). Dissolve 2.2485 g of Al wire in a minimum volume of 1:1 HCl:DDI. This is a very slow reaction. Dilute to 1 L in a volumetric flask using 1% HCl solution. Store in polypropylene bottle.

6.8 Calibration Al standard, 10 meq L^{-1}. Mix 10 mL of primary Al standard (250 meq L^{-1}) with 125 mL of 2.0 N KCl solution. Make to 250-mL volume with DDI water. Store in polypropylene bottle.

6.9 Calibration Al check standard, 5 meq L^{-1}. Mix 5 mL of primary Al standard (250 meq L^{-1}) with 125 mL of 2.0 N KCl solution. Make to 250-mL volume with DDI water. Store in polypropylene bottle.

6.10 Calibration reagent blank solution, 1.0 N KCl. Add 125 mL of 2.0 N KCl to a volumetric flask and make to 50-mL volume with DDI water. Store in polypropylene bottle.

6.11 Nitrous oxide gas, compressed

6.12 Acetylene gas, compressed, purity 99.6%

6.13 Compressed air with water and oil traps

7. Procedure

Extraction of Al

7.1 Prepare sample tube by tightly compressing a 1-g ball of filter pulp into the bottom of a syringe barrel with a modified plunger.

7.2 Weigh exactly 2.50 g of <2-mm, air-dry soil and place in sample tube. Prepare one quality control check sample per 48 samples.

7.3 Place the sample tube on the upper disk of the extractor and connect a tared extraction syringe. Use 25.4-mm (1-in) length rubber tubing to insert the handle of the plunger in the slot of the stationary extractor disk.

7.4 Use a squeeze bottle and fill sample tube to the 20-mL mark with 1.0 N KCl solution (~ 10 mL). Make sure that the sample is thoroughly wetted. During the addition, wash sides of the tube and wet the sample. Shaking, swirling, or stirring may be required to wet organic samples. Allow sample to stand for at least 30 min.

7.5 Put reservoir tube on top of the sample tube. Set extractor for fast extraction rate and extract until the KCl solution is at a 0.5- to 1.0-cm height above sample. Turn off extractor.

7.6 Add 45 mL KCl solution to reservoir tube. Set extractor for 45-min extraction.

7.7 After the extraction, shut off extractor and pull plunger of syringe down. Do not remove the plunger from syringe barrel. Carefully remove the syringe with extract leaving the rubber tubing on the sample tube.

7.8 Weigh each syringe containing KCl extract to the nearest 0.01 g.

7.9 Mix extract in each syringe by manually shaking. Fill a polycon with extract solution and discard the excess. This solution is reserved for extractable Al analysis.

Dilution of Sample Extracts and Standards

7.10 No ionization suppressant is required as the K in the extractant is present in sufficient quantity. Set the digital settings at 40 for the diluent (DDI water) and 99 for the KCl sample extracts, calibration reagent blanks, calibration standards, and calibration check standards for a 1:5 dilution as follows:

7.11 Dilute 1 part KCl sample extract with 4 parts of DDI water (1:5 dilution).

7.12 Dilute 1 part calibration reagent blank with 4 parts of DDI water (1:5 dilution).

7.13 Dilute 1 part calibration standard (10 meq L^{-1} Al) with 4 parts of DDI water (1:5 dilution).

7.14 Dilute 1 part calibration check standard (5 meq L^{-1} Al) with 4 parts of DDI water (1:5 dilution).

7.15 Dispense the diluted solutions into 15-mL conical polystyrene centrifuge tubes which are placed in carousels of the sample changer.

AA Calibration

7.16 Use calibration reagent blank (1.0 N KCl) and calibration standard (10 meq L^{-1} Al) to calibrate the AA. The AA program requires a blank and a standard, in that order, to establish a single point calibration curve for element determination. During AA determinations, perform one calibration, i.e., blank plus standard, for every 12 samples.

7.17 Use the calibration check standard (5 meq L^{-1} Al) as a check sample.

AA Set-up and Operation

7.18 The following parameters are only very general guidelines for instrument conditions for the analyte.

Element Head = Al
Wavelength (nm) = 309.3
Burner head & angle = 5 cm Parallel
Fuel/Oxidant (C_2H_2/N_2O) = 30/17
Typical read delay is 6 s, and integration by peak area is 8 s.

7.19 Use the microcomputer and printer to set instrument parameters and to collect and record instrument readings.

7.20 If a sample exceeds the calibration standard, dilute the sample (dilution ratio in calculation) with appropriate matrix and record dilution. Remember to keep the matrix the same after dilution by diluting with DDI water (1:5 dilution).

7.21 Analyze one quality control check sample for every 48 samples.

8. Calculations

8.1 The instrument readings are the analyte concentration (meq L^{-1} Al) in undiluted extract. Use these values to calculate the analyte concentration on an oven-dry soil basis (meq/100 g).

$$\text{Al (meq/100 g)} = [AA_{Al} \times (Wt_{syr+ext} - Wt_{syr}) \times DR \times 100 \times AD/OD]/[\text{Smp. Wt.} \times 1.0412 \times 1000]$$

where
AA_{Al} = AA Al reading (meq L^{-1})
$Wt_{syr+ext}$ = Weight of extraction syringe and extract (g)
Wt_{syr} = Weight of tared extraction syringe (g)
DR= Dilution ratio for samples over calibration range
Smp.Wt = Sample weight (g)
1.0412 = Density of 1 N KCl @ 20°C
1000 = g L^{-1}
100 = Conversion factor (100-g basis)
AD/OD = Air-dry/oven-dry ratio (procedure 4B5)

9. Report

Report KCl extractable Al in units of meq/100 g of oven-dry soil to the nearest 0.1 meq/100 g.

10. Precision

Precision data are not available for this procedure. A quality control check sample is run with every batch of 48 samples. For 21 observations of the quality control sample, the mean, standard deviation, and C.V. for extractable Al are 3.1, 0.18, and 5.7 %, respectively.

11. References

Holmgren, George G.S., R.L. Juve, and R.C. Geschwender. 1977. A mechanically controlled variable rate leaching device. Soil Sci. Amer. J. 41:1207-1208.
Thomas, G.W. 1982. Exchangeable cations. *In* A. Klute (ed.) Methods of soil analysis. Part 2. Chemical and microbiological properties. 2nd ed. Agronomy 9:159-165.

Chloride, Sulfate, Nitrate, Fluoride, and Nitrite (6K, 6L, 6M, 6U, and 6W) Saturation Extract (6K1, 6L1, 6M1, 6U1, and 6W1) Chromatograph (6K1c, 6L1c, 6M1c, 6U1a, and 6W1a)

1. Application

The soluble anions that are commonly determined in saline and alkali soils are carbonate, bicarbonate, sulfate, chloride, nitrate, nitrite, fluoride, phosphate, silicate, and borate (Khym, 1974; U.S. Salinity Laboratory Staff, 1954). Carbonate and bicarbonate are determined by titration in procedures 6I1b and 6J1b, respectively (Soil Conservation Service, 1984). Phosphate, silicate, and borate usually are not determined because they are found only occasionally in measurable amounts in soils. Chloride, sulfate, nitrate, fluoride, and nitrite are measured in solution in procedures 6K1c, 6L1c, 6M1c, 6U1a, and 6W1a, respectively (Soil

Conservation Service, 1984). In saline and alkali soils, carbonate, bicarbonate, sulfate, and chloride are the anions that are found in the greatest abundance. In general, soluble sulfate is usually more abundant than soluble chloride.

2. Summary of Method

The saturation extract is diluted according to its electrical conductivity (EC_s). The diluted sample is injected into the ion chromatograph, and the anions are separated. A conductivity detector is used to determine the anion. A chart recording is made of the chromatograph. Standard anions are used to calibrate the system. A calibration curve is determined, and the anion concentrations are calculated. The saturated extract anions, Cl^-, SO_4^{2-}, NO_3^-, F^-, and NO_2^- are reported in meq L^{-1} in procedures 6k1c, 6L1c, 6M1c, 6U1a, and 6W1a, respectively (Soil Conservation Service, 1984).

3. Interferences

Some saturation extracts contain suspended solids. Filtering after dilution removes the particles. Saturation extracts of acid soils that contain Fe and/or Al may precipitate and clog the separator column. Saturation extracts of very high pH may contain organic material which may clog or poison the column. Low molecular weight organic anions will co-elute with inorganic anions from the column.

4. Safety

Wear protective clothing and safety glasses. When preparing reagents, exercise special care. Many metal salts are extremely toxic and may be fatal if ingested. Thoroughly wash hands after handling these metal salts. Follow the manufacturer's safety precautions when using the chromatograph.

5. Equipment

5.1 Ion chromatograph, Series 2110i, with conductivity detector, Dionex Corp., 1228 Titan Way, Sunnyvale, CA 94086

5.2 HPIC AS3 analytical column, P/N 030985, Dionex Corp., 1228 Titan Way, Sunnyvale, CA 94086

5.3 HPIC AG3 analytical guard column, P/N 030986, Dionex Corp., 1228 Titan Way, Sunnyvale, CA 94086

5.4 Anion micro membrane suppressor, P/N 037072, Dionex Corp., 1228 Titan Way, Sunnyvale, CA 94086

5.5 Automated sampler, Dionex Corp., 1228 Titan Way, Sunnyvale, CA 94086

5.6 Poly-vials, 5 mL, P/N 038008, Dionex Corp., 1228 Titan Way, Sunnyvale, CA 94086

5.7 Poly-vials, filtercaps, 5 mL, P/N 038009, Dionex Corp., 1228 Titan Way, Sunnyvale, CA 94086

5.8 Chart recorder, Honeywell Corp., chart speed 0.5 cm min^{-1}, span 1000 mV F.S.

5.9 Digital diluter/dispenser, product number 100004, with hand probe and actuator, product number 230700, Hamilton Co., P.O. Box 10030, Reno, NV 89510

5.10 Syringes, gas tight, Hamilton 1001 DX and 1010-TEF LL, Hamilton Co., P.O. Box 10030, Reno, NV 89510

5.11 Syringes, disposable, polypropylene, 12 mL

5.12 Disposable 0.2-μm pore size, 25-mm filter assembly, Gelman Sciences, Inc., 674 South Wagner Road, Ann Arbor, MI 48106. Use for saturation extracts and standards.

5.13 Disposable 0.2-μm pore size, Ultipor N_{66} DFA3001NAEY, Pall Trinity Micro Corp., Cortland, NY 13045. Use for filtering distilled deionized (DDI) water.

6. Reagents

6.1 Distilled deionized (DDI) filtered water

6.2 Sulfuric acid (H_2SO_4), conc., reagent

6.3 Toluene

6.4 Isopropanol to de-gas column

6.5 Regenerant solution for membrane suppressor columns, 0.025 N H_2SO_4. Carefully mix 22.92 g of conc. H_2SO_4 with filtered DDI water and dilute to 18-L volume. Store in a clean glass carboy with a solid stopper. Cover the carboy top with aluminum foil to protect the contents from dust.

6.6 Stock $NaHCO_3$ solution, 0.480 M. Mix 40.34 g of dried $NaHCO_3$ with filtered DDI water and dilute to 1-L volume.

6.7 Stock Na_2CO_3 solution, 0.3838 M. Mix 40.68 g of dried Na_2CO_3 with filtered DDI water and dilute to 1-L volume.

6.8 Working eluent solution. Mix 112.5 mL of 0.480 M $NaHCO_3$ and 112.5 mL of 0.3838 M Na_2CO_3 with filtered DDI water and dilute to 18-L volume. Add 3 drops of toluene to retard microbial growth.

6.9 Primary SO_4^{2-} standard, 0.5 M (1.0 N). Mix 17.7560 g of Na_2SO_4 with filtered DDI water and dilute to 250-mL volume.

6.10 Primary Cl^- standard, 1.0 M (1.0 N). Add 18.6392 g of KCl with filtered DDI water and dilute to 250-mL volume.

6.11 Primary F^- standard, 0.125 M (0.125 N). Add 1.3122 g of NaF with filtered DDI water and dilute to 250-mL volume.

6.12 Primary NO_3^- standard, 1.0 M (1.0 N). Add 25.2770 g of KNO_3 with filtered DDI water and dilute to 250-mL volume.

6.13 Primary mixed standard. Prepare 1 primary mixed standard by taking aliquots of each of the proceeding primary standards and diluting the combined aliquots to a 1-L volume with working eluent as follows:

Primary Stds.	Aliquot	Final	Conc. Vol. w/Eluent
	mL	mL	Meq/L
Na_2SO_4	50	1000	50
KCl	10	1000	10
NaF	100	1000	12.5
KNO_3	30	1000	30

Add eight drops of toluene to primary mixed standard to retard microbial growth and store in a glass container.

6.14 Mixed calibration standards. Prepare 4 mixed calibration standards (0.5, 1.0, 3.0, and 7.0 readings) by taking aliquots of primary mixed standard and diluting each aliquot to 100-mL volume with working eluent as follows:

Primary Mixed Stds. (mL)	Final Vol. w/Eluent (mL)	SO_4^{2-} meq L^{-1}	Cl^- meq L^{-1}	F^- meq L^{-1}	NO_3^- meq L^{-1}
0.5	100	0.25	0.05	0.0625	0.15
1.0	100	0.50	0.10	0.125	0.30
3.0	100	1.5	0.30	0.375	0.90
7.0	100	3.5	0.70	0.875	2.1

6.15 $NaNO_2$, Baker reagent grade, 99.5% purity

6.16 Primary NO_2^- standard, 1 N (1000 meq L^{-1}). Mix 69.3568 g of reagent grade $NaNO_2$ with filtered DDI water and dilute to 1-L volume. Take 5 mL aliquot of primary NO_2^- standard and dilute with 500 mL of filtered DDI water (10 meq L^{-1}). Add eight drops of toluene to primary NO_2^- standard to retard microbial growth and store in a glass container.

6.17 NO_2^- calibration standards. Prepare 4 NO_2^- calibration standards (0.5, 1.0, 3.0, and 7.0 readings) by taking aliquots of primary NO_2^- standard (10 meq L^{-1}) and diluting each aliquot to 100-mL volume with working eluent as follows:

Primary Mixed Stds. (mL)	Final Vol. w/Eluent (mL)	NO_2^- meq L^{-1}
0.5	100	0.5
1.0	100	1.0
3.0	100	3.0
7.0	100	7.0

7. Procedure

Dilution of extracts

7.1 To estimate the total soluble anion concentration (meq L^{-1}), multiply the EC_s (procedure 8A3a) by 10. Subtract the CO_3^{2-} and HCO_3^- concentrations (procedures 6I1b and 6J1b) from the total anion concentration. The remainder is the $\approx$ concentration (meq L^{-1}) of anions to be separated by ion chromatography.

Anion concentration (meq L^{-1}) = EC_s x 10 - ($HCO_3^- + CO_3^{2-}$)

7.2 Dilute the saturation extract with the working eluent. Some typical dilutions are as follows:

EC_s (mmhos cm^{-1})	Dilution Factor
0.0 to 0.4	1:3
0.4 to 0.7	1:5
0.8 to 1.2	1:9
1.2 to 1.8	1:17
1.8 to 2.9	1:39
3.0 to 5.5	1:80
5.5 to 7.5	1:150
7.5 to 9.7	1:200
9.7 to 13.5	1:290
13.5 to 15.5	1:350
15.5 to 25.0	1:660
25.0 to 40.0	1:1100
40.0 to 55.0	1:2100
55.0 to 75.0	1:4800
+75.0	1:15500

7.3 Place the diluted samples in 12-mL syringes. Cap syringes to prevent evaporation or contamination.

7.4 Place the mixed calibration standards in 12-mL syringes.

Set-up and Operation of Ion Chromatograph (IC)

7.5 Because any number of factors may cause a change in IC operating conditions, only a general set-up of the Dionex 2110i ion chromatograph is presented. Individual analysts may modify some or all of the operating conditions to achieve satisfactory results. The uS cm^{-1} units are equivalent to mmhos cm^{-1}. Typical operation parameters are as follows:

Parameter	Range
Conductivity cell range	3 uS cm^{-1} full scale to 100 uS cm^{-1}
Auto Offset	"On"
Analytical pump flow rate	2.0 to 2.5 mL min^{-1}
Low pressure limit	200
High pressure limit	1000
Regenerant flow	3 to 4 mL min^{-1}
Injector loom	0.50 mL
Air pressure	3 to 6 psi
Chart recorder speed	0.5 cm min^{-1}
Chart recorder span	1000 mV full scale

7.6 Initial IC operation should be long enough to establish a stable baseline.

7.7 Inject the most concentrated standard. The IC adjustment may be necessary to obtain adequate stability, resolution, and reproducibility.

7.8 Inject standards in random order to detect if memory effects are evident.

7.9 Analyze blanks at frequent intervals.

7.10 The injection loop requires complete flushing, i.e., 3 to 5x the loop volume.

7.11 Inject samples, standards, and blanks in the IC after achievement of stability. The analyst may change the detector range to suit the sample.

7.12 The analyst records the detector range and peak height for each detected anion. The anion identity may be determined by comparison to standards. Peak height is determined from the baseline to the peak.

8. Calculations

Calibration Calculations

8.1 Use the peak height of each anion standard to either construct a calibrated curve to plot anion concentration or use a least squares analysis to calculate anion concentration. The analytes are reported in meq L^{-1}.

8.2 *Calibration Curve*: Plot the peak height against the meq L^{-1} of each anion standard on graph paper. Construct the calibration curve by finding the "best" line that fits the plotted standards.

8.3 *Linear Squares Analysis*: Use a least squares criterion, i.e. best moving average. Refer to a statistical analysis book for additional information on least squares analysis. An example for the anion Cl$^-$ is as follows:

Cl^- (meq L^{-1}) = Y = 0.1 1.5 4.0

Peak height = X = 8.43 170.0 441.5

Number of standards = n = 3

$\sum Y_i = 5.6$ $\sum X_i = 619.93$

$\sum Y_i/n = Y = 1.866$ $\sum X_i/n = X = 206.6433$

$\sum X_i Y_i = 2021.843$ $\sum X_i^2 = 223893.31$

$\sum X_i \sum Y_i = 3471.608$

$$b = \frac{\sum X_i Y_i - \sum X_i \sum Y_i/n}{\sum X_i^2 - (\sum X_i)^2/n} =$$

$$\frac{2021.843 - 1157.2027}{223893.31 - 128104.4} = 0.0090265$$

b = slope of the line, i.e., the amount that Y changes when X changes by 1 unit.

The equation is as follows:

$$Y = \overline{Y} + b(X - \overline{X})$$

Y = 1.866 + 0.0090265 (X) - 1.8653

Analyte Calculation
8.4 *Calibration curve*: Read the analyte concentration (meq L^{-1}) directly from the calibration curve.

8.5 *Linear regression*: Put the peak height in the preceding equation and solve for analyte concentration (meq L^{-1}). Thus, if sample extract has 204 peak height, the preceding equation is as follows:

Y = 1.866 + 0.0090265 (204) - 1.8653 = 1.84 meq L^{-1}

8.6 Repeat the calibration set and analyte calculation for each anion.

9. Report
Report the saturation extract anions in units of meq L^{-1} to the nearest 0.1 meq L^{-1}.

10. Precision
Precision data are not available for this procedure.

11. References
Khym, J.X. 1974. Analytical ion-exchange procedures in chemistry and biology: theory, equipment, techniques. Prentice-Hall, Inc., Englewood Cliffs, NJ.

Soil Conservation Service. 1984. Procedures for collecting soil samples and methods of analysis for soil survey. USDA-SCS Soil Surv. Invest. Rep. no. 1. U.S. Govt. Print. Office, Washington, DC.

U.S. Salinity Laboratory Staff. 1954. L.A. Richards (ed.) Diagnosis and improvement of saline and alkali soils. U.S. Dept. of Agric. Handb. 60. U.S. Govt. Print. Office, Washington, DC.

Total Sulfur (6R)
SO$_2$ Evolution, Infrared (6R3)
LECO SC-132 Sulfur Analyzer (6R3b)

1. Application

Organic and inorganic S forms are found in soils, with the organic S fraction accounting for >95% of the total S in most soils from humid and semi-humid (Tabatabai, 1982). Mineralization of organic S and its conversion to sulfate by chemical and biological activity may serve as a source of plant available S. Total S typically ranges from 0.01 to 0.05% in most mineral soils. In organic soils, total S may be >0.05%.

In well-drained, well-aerated soils, most of the inorganic S normally occurs as sulfate. In marine tidal flats, other anaerobic marine sediments, and mine spoils, there are usually large amounts of reduced S compounds which oxidize to sulfuric acid upon exposure to the air. In arid regions, significant amounts of inorganic S are found as sulfates such gypsum and barite.

The typical use of total S is as an index of the total reserves of this element, which may be converted to plant available S. The SSL uses the combustion technique (LECO sulfur analyzer) for analysis of total S (procedure 6R3b). Extractable sulfate S (SO$_4^{2}$-S) is an index of readily plant-available S. Reagents that have been used for measuring SO$_4^{2}$-S include water, hot water, ammonium acetate, sodium carbonate and other carbonates, ammonium chloride and other chlorides, potassium phosphate and other phosphates, and ammonium fluoride (Bray-1). Extractable SO$_4^{2}$-S does not include the labile fraction of soil organic S that is mineralized during the growing season (Tabatabai, 1982). Extraction reagents for organic S include hydrogen peroxide, sodium bicarbonate, sodium hydroxide, sodium oxalate, sodium peroxide, and sodium pyrophosphate. There other methods available for determination of soil S, especially for total S and SO$_4^{2}$-S. The investigator may refer to the review by Beaton et al. (1968).

2. Summary of Method

A fine-ground (<80-mesh) soil sample is oxidized at high temperature. The gases released are scrubbed, and the SO$_2$ in the combustion gases are measured using an infrared detector. Percent S is reported on an oven-dry soil basis.

3. Interferences

No significant interferences are known to affect the oxidizable S measurement.

4. Safety

Wear protective clothing and safety glasses. Magnesium perchlorate may form explosive mixtures. Magnesium perchlorate may contain traces of perchloric acid, which remain from manufacturer's operations. This acid is anhydrous because of the strong desiccating capability of the salt. Avoid prolonged contact with oxidizable material or material capable of forming unstable perchlorate esters or salts. Remove magnesium perchlorate by using an excess of water to thoroughly dilute the material.

The use of high temperatures in the oxidation of samples requires that extreme caution be used to prevent burns and fires. Follow standard laboratory procedures when handling compressed gases. Oxygen is highly flammable. Avoid open flames and sparks. Standard laboratory equipment includes fire blankets and extinguishers for use when necessary. Follow the manufacturer's safety precautions when using the sulfur analyzer.

5. Equipment
5.1 Sulfur analyzer, Leco Model SC-132 781-400 Sulfur System, Leco Corp., St. Joseph, MI
5.2 Data transmit card, part no. 772-573, Leco Corp., St. Joseph, MI
5.3 Combustion boats, part no. 529-203, Leco Corp., St. Joseph, MI
5.4 Single-stage regulator, Oxygen Service, Part No. E11-W-N115BOX, Air Products and Chemicals, Inc., Box 538, Allentown, PA 18105
5.5 Electronic balance, ±1-mg sensitivity

6. Reagents
6.1 Anhydrone, anhydrous magnesium perchlorate, granular
6.2 Glass wool
6.3 Compressed oxygen, >99.5% @ 30 psi

7. Procedure

7.1 Weigh an air-dry, fine-ground (<80-mesh) soil sample in a tared combustion boat. Sample size depends upon S content. The product of sample weight in g multiplied by the S percent must not be >2. In most cases, the sample size is 1.00 g, unless the S content is >2%.

7.2 Refer to the manufacturer's manual for operation of sulfur analyzer. An overview of the sulfur analyzer is as follows:

a. Samples are combusted in an O_2 atmosphere in which the S is oxidized to SO_2.

b. Moisture and dust are removed, and the SO_2 gas is then measured by a solid state infrared detector.

c. The microprocessor formulates the analysis results. The control console displays and prints results by combining the outputs of the infrared detector and system ambient sensors with pre-programmed calibration, linearization, and weight compensation factors.

8. Calculations

$$S (\%) = S_i \times AD/OD$$

where:
S (%) = S (%) on oven-dry basis
S_i = S (%) instrument
AD/OD = air-dry/oven-dry ratio (procedure 4B5)

9. Report
Report total S as a percentage of oven-dry weight to the nearest 0.1%.

10. Precision
Precision data are not available for this procedure. A quality control check sample is run in every batch of 12 samples. A blank (crucible only) and a rerun of one of the 12 samples (unknowns) also are run in every batch. For 27 observations of the quality control check sample, the mean, standard deviation, and C.V. for total S are 0.57, 0.02, and 4.3%, respectively.

11. References

Beaton, James D., G.R. Burns, and J. Platou. 1968. Determination of sulfur in soils and plant material. Tech. Bull. No. 14. The Sulfur Inst., 1725 K Street, N.W., Washington, D.C. 20006

Tabatabai, M.A. 1982. Sulfur. *In* A.L. Page, R.H. Miller, and D.R. Keeney (eds.) Methods of soil analysis. Part 2. Chemical and microbiological properties. 2nd ed. Agronomy 9:501-538.

Phosphorus (6S)
New Zealand P Retention (6S4)

1. Application

In *Soil Taxonomy*, the P retention of soil material is a criterion for andic soil properties (Soil Survey Staff, 1990). Andisols and other soils that contain large amounts of allophane and other amorphous minerals have capacities for binding P (Gebhardt and Coleman, 1984). The factors that affect soil P retention are not well understood. However, allophane and imogolite have been considered as major materials that contribute to P retention in Andisols (Wada, 1985). Phosphate retention is also called P absorption, sorption, or fixation.

2. Summary of Method

A 5-g soil sample is shaken in a 1000-ppm P solution for 24 h. The mixture is centrifuged at 2000 rpm for 15 min. An aliquot of the supernatant is transferred to a colorimetric tube to which nitric vanadomolybdate acid reagent (NVAR) is added. The percent transmittance of the solution is read using a colorimeter. The New Zealand P retention is reported as percent P retained.

3. Interferences

No significant problems are known to affect the P retention measurement.

4. Safety

Wear protective clothing (coats, aprons, sleeve guards, and gloves) and eye protection (face shields, goggles, or safety glasses). When preparing reagents, exercise special care. Many metal salts are extremely toxic and may be fatal if ingested. Thoroughly wash hands after handling these metal salts. Restrict the use of concentrated HNO_3 to a fume hood. Use safety showers and eyewash stations to dilute spilled acids and bases. Use sodium bicarbonate and water to neutralize and dilute spilled acids.

5. Equipment

5.1 Electronic balance, ±0.01-g sensitivity
5.2 Shaker, Eberbach 6000 power unit, reciprocating speed of 60 to 260 epm, with 6040 utility box carrier and 6110 floor stand, Eberbach Corp., Ann Arbor, MI
5.3 Digital diluter/dispenser, product no. 100004, with hand probe and actuator, product no. 230700, Hamilton Co., P.O. Box 10030, Reno, NV, 89510
5.4 Syringes, 10000 and 1000 µL, 1001 DX and 1010-TEL LL gas tight, Hamilton Co., P.O. Box 10030, Reno, NV, 89510
5.5 Diluter/dispenser, 25 mL
5.6 Calorimeter, Baush and Laumb
5.7 Calorimeter tubes, glass, 10 mL, 1-cm light path, Baush and Laumb
5.8 Centrifuge, International no. 2, Model V, with no. 250 A head, International Equip. Co., Boston, MA
5.9 Trunions, International no. 320, International Equip. Co., Boston, MA
5.10 Centrifuge tubes, 50 mL, Oak-Ridge, polyallomer, Nalgene 3119, Nalge Co., Box 20365, Rochester, NY 14602.
5.11 Plastic cups, 2 fl. oz.

5.12 Pipets, volumetric, class A, glass, various sizes of 1 to 20 mL

6. Reagents
6.1 Distilled deionized (DDI) water
6.2 Nitric acid (HNO_3), conc.
6.3 P retention solution, 1000 ppm P. Dissolve 8.80 g of KH_2PO_4 and 32.8 g of sodium acetate (CH_3COONa) in DDI water. Add 23 mL of glacial acetic acid. Dilute to 2 L with DDI water in a volumetric flask. The solution pH should range between 4.55 and 4.65.
6.4 Molybdate solution. Dissolve 16 g of ammonium molybdate (VI) [$(NH_4)_6MO_7O_{24} \cdot 4H_2O$] in 50°C DDI water. Allow the solution to cool to room temperature and dilute to 1 L with DDI water.
6.5 Nitric acid solution. Carefully and slowly dilute 100 mL of conc. HNO_3 to 1 L of DDI water. Add the acid to the water.
6.6 Nitric vanadomolybdate acid reagent (NVAR), vanadate solution. Dissolve 0.8 g of NH_4VO_3 in 500 mL of boiling DDI water. Allow the solution to cool to room temperature. Carefully and slowly add 6 mL of conc. HNO_3. Dilute to 1 L with DDI water. Mix the nitric acid solution with the vanadate solution and then add the molybdate solution. Mix well.
6.7 Stock P standard solution (SPSS), 4000 ppm P. Dissolve 17.6 g K_2HPO_4 in DDI water. Dilute to 1 L with DDI water.
6.8 Standard P calibration P solutions (SPCS), 100, 80, 60, 40, 20, and 0% P retained. Dilute the SPSS with a solution that contains 32.8 g of sodium acetate (CH_3COONa) and 23 mL of glacial acetic acid diluted to 2 L with DDI water as follows: 100% = DDI water (0 ppm); 80% = 1:20 (200 ppm); 60% = 1:10 (400 ppm); 40% = 3:20 (600 ppm); 20% = 1:5 (800 ppm); and 0% = 1:4 (1000 ppm). The percent amount refers to percent P retention.

7. Procedure

7.1 Weigh 5.00 g of air-dry soil into a 50-mL centrifuge tube.

7.2 Use the dispenser to add 25.0 mL of P-retention solution to centrifuge tube.

7.3 Cap centrifuge tube and place in shaker and shake for 24 h at room temperature (20°C).

7.4 Add 2 to 3 drops of Superfloc, 0.02% w/v to each tube.

7.5 Centrifuge sample at 2000 rpm for 15 min.

7.6 Pour sample supernatant into plastic cup.

7.7 Use the digital diluter to add the nitric vanadomolybdate acid reagent (NVAR) to each sample supernatant and to each SPCS. To fill a 10-mL Calorimeter tube, the diluter setting is 66 for diluent (NVAR) and 35 for sample (1:20 dilution).

7.8 The color reaction requires a minimum of 30 min before the analyst records readings.

7.9 Set the Calorimeter (blue bulb) to read at 466 nm. Set the zero against DDI water (blank). A blank has all reagents contained in the sample extract except the soil.

7.10 Record the percent transmittance to the nearest 0.01 unit for the sample extract and each SPCS.

8. Calculations

8.1 Transmittance of a solution is the fraction of incident radiation transmitted by the solution, i.e., $T = P/P_0$ and is often expressed as a percentage, i.e., $T = P/P_0$ x 100. The absorbance of a solution is directly proportional to concentration and is defined by the equation, $A = -\log_{10} T$. These relationships are derived from Beer's law.

Calibration Calculations

8.2 Use the transmittance of each SPCS to either construct a calibrated curve to plot P or use a least squares analysis to calculate P. The P is reported in percent retained.

8.3 *Calibration Curve*: Plot the transmittances against the ppm P of each SPCS on semilog graph paper or convert to absorbances and plot on linear graph paper. Construct the calibration curve by finding the "best" line that fits the plotted SPCS.

8.4 *Least Squares Analysis*: Use a least squares criterion, i.e. best moving average. Refer to a statistical analysis book for additional information on least squares analysis. To facilitate data manipulation in a least squares analysis, the following standard curve is developed using the concentration of SPCS as a f[ln(%T]. Final calculated analyte concentration with either $\log_{10}$ or ln base would be the same. Refer to procedure 6S3b for an example of least squares analysis.

Analyte Calculation

8.5 *Calibration Curve*: Read the percent P directly from the calibration curve.

8.6 *Least Squares Analysis*: Refer to procedure 6S3 for an example of least squares analysis.

9. Report

Report the percent New Zealand P retention to the nearest whole number.

10. Precision

Precision data are not available for this procedure.

11. References

Gebhardt, H., and N.T. Coleman. 1984. Anion adsorption of allophanic tropical soils: III. Phosphate adsorption. p. 237-248. *In* K.H. Tan (ed.) Anodosols. Benchmark papers in soil science series. Van Nostrand Reinhold, Co., Melbourne, Canada.

Soil Survey Staff. 1990. Keys to soil taxonomy. 4th ed. SMSS technical monograph no. 6. Blacksburg, VA.

Wada, K. 1985. The distinctive properties of Andosols. p. 173-229. *In* B.A. Stewart (ed.) Adv. Soil Sci. Springer-Verlag, NY.

MINERALOGY (7)

Instrumental Analyses (7A)
Thermal Gravimetric Analysis (7A4)
Perkin-Elmer 7 Series (7A4b)

1. Application

Thermal analysis defines a group of analyses that determine some physical parameter, e.g., energy, weight, or evolved substances, as a dynamic function of temperature (Tan et al., 1986). Thermogravimetric analysis (TGA) is a technique for determining weight loss of a sample as it is being heated at a controlled rate. The weight changes are recorded as a function of temperature, i.e., a thermogravimetric curve, and provide quantitative information about substances under investigation, e.g., gibbsite ($Al(OH)_3$), kaolinite ($Al_2Si_2O_5(OH)_4$), and 2:1 expandable minerals (smectite and vermiculite).

2. Summary of Method

A 5- to 10-mg sample of soil clay is weighed into a platinum sample pan and placed in the TGA balance. The instrument records the initial sample weight. The analyst zeros the balance. The sample is then heated from a temperature of 30 to 900°C at a rate of 20°C min^{-1} in a flowing N_2 atmosphere. The computer collects weight changes as a function of temperature and records a thermogravimetric curve. Gibbsite and kaolinite are quantified by noting the weight loss between 250 to 350°C and 450 to 550°C, respectively, and then relating these data to the theoretical weight loss of pure gibbsite or kaolinite (Soil Conservation Service, 1984). The weight loss is due to dehydroxylation, i.e., loss of crystal lattice water. Though not presently performed by the National Soil Survey Laboratory (NSSL), quantification of the 2:1 expandable minerals (smectite + vermiculite) is related to weight loss at <250°C, i.e., loss of adsorbed water (Karanthasis and Hajek, 1982; Tan et al., 1986). At this low temperature, adsorbed water is proportional to the specific area of the sample (Jackson, 1956; Karathanasis and Hajek, 1982; Mackenzie, 1970; Tan and Hajek, 1977).

3. Interferences

Organic matter is objectionable because it has a weight loss by dehydrogenation and by oxidation to CO_2 between 300 to 900°C (Tan, et al., 1986). Analysis in an inert N_2 atmosphere alleviates this problem. Mineral salts that contain water of crystallization also may be interferences. Samples should be washed free of any soluble salts.

A representative soil sample is important as sample size is small (<10 mg). Avoid large aggregates in sample, the presence of which may cause thermal interferences, i.e., differential kinetics of gas diffusion through the sample and physical movement of sample in a reaction.

In general, the same reactions that interfere with DSC/DTA also interfere with TGA determinations of kaolinite, gibbsite, and 2:1 expandable minerals. However, TGA is more sensitive to small water losses at slow rates, whereas DSC/DTA is more sensitive to large water losses at rapid rates (Tan, et al., 1986). This sensitivity difference may help to explain why kaolinite and gibbsite quantifications in TGA vs. DSC/DTA often are not equivalent, i.e., TGA estimates tend to be greater than the corresponding DSC/DTA estimates. In TGA, there is a greater probability of measuring water losses in specific temperature regimes that are not specifically associated with dehydroxylation reactions of interest. This problem is particularly apparent with illitic samples, which characteristically contain more "structural" water than ideal structural formulae would indicate (Rouston, et al., 1972; Weaver and Pollard, 1973).

Even though it is well established that various minerals lose the major portion of their crystal lattice water at different temperature ranges (Tan et al., 1986), there are overlaps in these weight loss regions (WLR) of minerals which interfere in the identification and measurement of the minerals of interest. The goethite WLR (250 to 400°C) overlaps the gibbsite WLR (250 to 350°C) (Mackenzie and Berggen, 1970). The illite WLR (550 to 600°C) overlaps the high end of the kaolinite WLR (450 to 550°C) (Mackenzie and Caillere, 1975). The WLR of hydroxy-Al interlayers in hydroxy-Al interlayered vermiculite (HIV) (400 to 450°C) overlaps the low end of the kaolinite WLR (450 to 550°C), especially in the poorly crystalline kaolinites (Mackenzie and Caillere, 1976). Similarly, the dehydroxylation of nontronites, Fe-rich dioctahedral smectites, (450 to 500°C) may interfere with kaolinite identification and measurement (Mackenzie and Caillere, 1975).

4. Safety

Secure high pressure N_2 tanks and handle with care. When changing the tanks, protect valves with covers. Do not program the analyzer for >950°C because it may present a safety hazard during sample analysis and cleaning cycles. Do not heat aluminum sample pans >600°C. Aluminum melts at 660°C, and the pans alloy with and destroy the sample holders.

Always use high quality purge gases with the TGA. Minimum purity of 99.9% is recommended.

5. Equipment
5.1 Thermal analysis system, Perkin-Elmer 7 series, 7500 computer, TAC7 instrument controllers
5.2 Thermogravimetric analyzer module, TGA7, Hewlett-Packard digital plotter
5.3 Pressure tanks (2), N_2, purity 99.99%
5.4 Two-stage gas regulators (2), 50 psi outlet pressure
5.5 One-stage gas regulator for compressed air
5.6 Electronic balance, ±0.1-mg sensitivity, Mettler AE160
5.7 Forceps, flat-tipped
5.8 Weighing spatula
5.9 Desiccator, glass
5.10 Mortar and pestle
5.11 Sieves, 100 mesh or 80 mesh
5.12 Kaolinite, standard, poorly crystalline, Georgia Kaolinite, Clay Minerals Society, Source Clay Minerals Project, sample KGa-2.
5.13 Gibbsite, standard, Surinam Gibbsite, National Soil Survey Laboratory (NSSL), 67L022.

6. Reagents
6.1 Magnesium nitrate saturated solution [$Mg(NO_3)_2 \cdot 6H_2O$]
6.2 Ethanol

7. Procedure

Derive <2µm Clay Fractions
7.1 Prepare Na-saturated clay as in procedure 7A2i, preparation of clay suspension, 7.8 to 7.19.

7.2 Dry the clay suspension and transfer to mortar. Moisten sample with ethanol and grind with pestle to make a homogeneous slurry.

7.3 Air-dry sample using flowing air in hood. Lightly grind sample with pestle to make a homogeneous powder.

7.4 Sieve sample with 80-mesh screen. Equilibrate sample overnight over a saturated magnesium nitrate solution (55% rh) in a glass desiccator.

TGA Operation
7.5 Set-up the instrument and calibrate.

7.6 Turn on the N_2 purge gases and set to 6 and 3.5 psi for balance and sample purge, respectively. The balance purge pressure should always be greater than the sample purge pressure.

7.7 Turn on compressed air and set to 25 psi.

7.8 Place the platinum sample pan in the balance stirrup. Use the computer to raise the furnace tube and to zero the balance. Lower the furnace tube.

7.9 Remove the sample pan from the stirrup. Weigh ≈ 5 mg of sample, i.e., <100-mesh whole-soil or derived <2-µm clay fraction, into tared sample pan. Refer to section on derived <2-µm clay fractions, 7.1 to 7.4.

7.10 Use flat-tipped forceps to remove the sample pan from the analytical balance. Tap the sample pan against a hard surface several times to uniformly distribute the sample.

7.11 Carefully place sample pan in the stirrup of the TGA microbalance.

7.12 The standard sample run heating program has a heating rate of $20°C$ min^{-1}, a starting temperature of $30°C$, and an ending temperature of $900°C$.

7.13 Raise the furnace tube and allow it to seat. Press "Read Weight" key (usually twice) until a relative weight percentage of 100.0% is displayed. The computer then reads the weight.

7.14 Immediately start the "Run" program.

7.15 At the end of the sample run ($\approx$ 45 min), remove the sample pan from the microbalance stirrup. The furnace tube is lowered automatically at the end of run.

7.16 To store data, enter the appropriate file name on the computer for the completed run. If data are not stored by appropriate file name, data are stored under a default file name of "gsav". Only four of these files can be saved at any one time, after which files are overwritten. Once a file is named, it cannot be changed.

8. Calculations

8.1 The thermogravimetric curve is displayed on the computer monitor. The ordinate (Y) is expressed in a relative weight percentage, i.e., the initial sample weight is 100.0%. Use the computer to calculate the total change in sample weight (Δ Y), within the predetermined temperature range, as a 8 sample weight percent.

% Kaolinite $= [(\Delta$ sample weight % $_{450-550°C})/14]$ x 100

14

where
Δ sample weight = total Δ in sample weight expressed as relative percent
14 = percent weight of hydroxyl water Lost from pure kaolinite

% Gibbsite $= [(\Delta$ sample weight % $_{250-350°C})/ 34.6]$ x 100

where
Δ sample weight = total Δ in sample weight expressed as relative percent
34.6 = percent weight of hydroxyl water lost from pure gibbsite

The percent weights of hydroxyl water lost from kaolinite and gibbsite are derived from the following assumed dehydroxylation reactions.

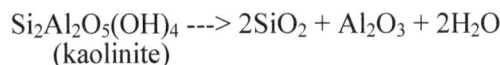

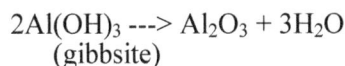

$Si_2Al_2O_5(OH)_4$ ---> $2SiO_2 + Al_2O_3 + 2H_2O$
  (kaolinite)

$2Al(OH)_3$ ---> $Al_2O_3 + 3H_2O$
  (gibbsite)

Using kaolinite as an example, percent weight of hydroxyl water lost is calculated from the following formula weights.

$Si_2Al_2O_5(OH)_4 = 258$ g mol^{-1}

$2H_2O = 36$ g mol^{-1}

Percent weight of hydroxyl water lost = (36/258) x 100 = 34.6%

9. Report

Report percent gibbsite and/or kaolinite to nearest whole number.

10. Precision

Precision data are not available for this procedure.

11. References

Jackson, M.L. 1956. Soil chemical analysis. Advan. course. M. L. Jackson, Madison, WI.

Karathanasis, A.D., and B.F. Hajek. 1982. Revised methods for rapid quantitative determination of minerals in soil clays. Soil Sci. Soc. Am. J. 46:419-425.

Mackenzie, R.C. 1970. Simple phyllosilicates based on gibbsite- and brucite-like sheets. p. 498-537. *In* R. C. Mackenzie (ed.) Differential thermal analysis. Vol. 1. Acad. Press, London.

Mackenzie, R.C., and G. Berggen. 1970. Oxides and hydroxides of higher-valency elements. p. 272-302. *In* R.C. Mackenzie (ed.) Differential thermal analysis. Acad. Press, NY.

Mackenzie, R.C., and S. Caillere. 1975. The thermal characteristics of soil minerals and the use of these characteristics in the qualitative and quantitative determination of clay minerals in soils. p. 529-571. *In* J.E. Gieseking (ed.) Soil components. Vol. 2. Inorganic components. Springer-Verlag, NY.

Mackenzie, R.C., and B.D. Mitchell. 1972. Soils. p. 267-297. *In* R.C. Mackenzie (ed.) Differential thermal analysis. Vol. 2. Acad. Press. London.

Rouston, R.C., J.A. Kittrick, and E.H. Hope. 1972. Interlayer hydration and the broadening of the 10A x-ray peak in illite. Soil Sci. 113:167-174.

Soil Conservation Service. 1984. Procedures for collecting soil samples and methods of analysis for soil survey. USDA-SCS Soil Surv. Invest. Rep. no. 1. U.S. Govt. Print. Office, Washington, DC.

Tan, K.H., and B.F. Hajek. 1977. Thermal analysis of soils. p. 865-884. *In* J.B. Dixon and S.B. Weed (eds.) Minerals in soil environments. 2nd ed. SSSA Book Series 1. SSSA, Madison, WI.

Tan, K.H., and B.F. Hajek, and I. Barshad. 1986. Thermal analysis techniques. p. 151-183. *In* A. Klute (ed.) Methods of soil analysis. Part 1. Physical and mineralogical properties. 2nd ed. Agronomy 9:151-183.

Weaver, C.E., and L.D. Pollard. 1973. The chemistry of clay minerals. Elsevier Sci. Publ. Co., Amsterdam.

Instrumental Analyses (7A)
Differential Scanning Calorimetry (7A6)

1. Application

Calorimetry measures specific heat or thermal capacity of a substance. Differential scanning calorimetry (DSC) is a calorimetric technique in which the rate of heat flow between a sample and a reference material is measured as materials are held isothermal to one another. The DSC directly measures the magnitude of an energy change (H, enthalpy or heat content) in a material undergoing an exothermic or endothermic reaction. DSC is commonly used to quantify gibbsite ($Al(OH)_3$) and kaolinite ($Al_2Si_2O_5(OH)_4$) in soils and clays by measuring the magnitude of their dehydroxylation endotherms which are between 250 to 350°C and 450 to 550°C, respectively (Karathanasis and Hajek, 1982; Jackson, 1956; Mackenzie and Berggen, 1970; Mackenzie, 1970).

602

2. Summary of Method

An 8 mg sample of soil clay is weighed into an aluminum sample pan and placed in the DSC sample holder. The sample and reference are heated under flowing N_2 atmosphere from a temperature of 30 to 600°C at a rate of 10°C min^{-1}. Data are collected by the computer and a thermogram is plotted. Gibbsite and kaolinite are quantified by measuring the peak area of any endothermic reactions between 250 to 350°C and 450 to 550°C, respectively, and by calculating the H of the reaction. These values are related to the values for the respective known quantities of the two minerals (gibbsite and kaolinite).

3. Interferences

Organic matter is objectionable because it produces irregular exothermic peaks in air or O_2, commonly between 300 to 500°C, which may obscure important reactions from the inorganic components of interest (Schnitzer and Kodama, 1977). Analysis in an inert N_2 atmosphere alleviates this problem. Mineral salts that contain water of crystallization also may be interferences. Samples should be washed free of any soluble salts.

Use a representative soil sample as sample size is small (<10 mg). Avoid large aggregates in sample, the presence of which may cause thermal interferences because of differential kinetics of gas diffusion through the sample and physical movement of sample in a reaction.

The dehydroxylation of goethite is between 250 to 400°C and may interfere with the identification and integration of the gibbsite endotherm (250 to 350°C) (Mackenzie and Berggen, 1970). The dehydroxylation of illite is between 550 to 600°C and partially overlaps the high end of the kaolinite endotherm (450 to 550°C), resulting in possible peak integrations (Mackenzie and Caillere, 1975). The dehydroxylation of hydroxy-Al interlayers in hydroxy-Al interlayered vermiculite (HIV) is between 400 to 450°C and may interfere with the low end of the kaolinite endotherm (450 to 550°C), especially in the poorly crystalline kaolinites (Mackenzie and Caillere, 1976). Similarly, the dehydroxylation of nontronites, Fe-rich dioctahedral smectites is between 450 to 500°C and may interfere with kaolinite identification and measurement (Mackenzie and Caillere, 1975).

4. Safety

Secure high pressure N_2 tanks and handle with care. When changing the tanks, valves should be protected with covers. Do not program the analyzer for >950°C because it may present a safety hazard during sample analysis and cleaning cycles. Do not heat aluminum sample pans >600°C. Aluminum melts at 660°C, and the sample pans alloy with an destroy the sample holders. Always use high quality purge gases with the DSC. Minimum purity of 99.9% is recommended.

5. Equipment

5.1 Thermal analysis system, Perkin-Elmer 7 series, 7500 computer, TAC7 instrument controllers

5.2 Differential scanning calorimeter module, DSC7, Hewlett-Packard digital plotter

5.3 Pressure tanks (2), N_2, purity 99.99%

5.4 Two-stage gas regulators (2), 50 psi outlet pressure

5.5 Electronic balance, ±0.1-mg sensitivity, Mettler AE160

5.6 Forceps, flat-tipped

5.7 Weighing spatula

5.8 Desiccator, glass

5.9 Mortar and pestle

5.10 Sieves, 100 mesh or 80 mesh

5.11 Kaolinite, standard, poorly crystalline, Georgia Kaolinite, Clay Minerals Society, Source Clay Minerals Project, sample KGa-2.

5.12 Gibbsite, standard, Surinam Gibbsite, National Soil Survey Laboratory (NSSL), 67L022.

6. Reagents

6.1 Magnesium nitrate saturated solution [Mg(NO$_3$)$_2$·6H$_2$O]

6.2 Ethanol

7. Procedure

Derive <2µm Clay Fractions

7.1 Prepare Na-saturated clay as in procedure 7A2i, preparation of clay suspension, 7.8 to 7.19.

7.2 Dry the clay suspension and transfer to mortar. Moisten sample with ethanol and grind with pestle to make a homogeneous slurry.

7.3 Air-dry sample using flowing air in hood. Lightly grind sample with pestle to make a homogeneous powder. Transfer to original container for storage until use.

7.4 Prior to TGA analysis, sieve sample with 80-mesh screen. Equilibrate sample overnight over a saturated magnesium nitrate solution (55% rh) in a glass desiccator.

DSC Operation

7.5 Set-up the instrument and calibrate.

7.6 Weigh ≈ 8 mg of sample, i.e., <100-mesh whole-soil or derived <2-µm clay fraction, into tared aluminum sample pan. Refer to section on derived <2-µm clay fractions, 7.1 to 7.4.

7.7 Use flat-tipped forceps to remove aluminum sample pan from balance. Drop sample from a 4- to 5-mm height to uniformly distribute sample in pan. Return the sample pan with sample to the balance and record weight to nearest ±0.1 mg. This weight is entered into computer in appropriate menu.

7.8 Carefully place aluminum sample pan in the center of DSC platinum sample side (left side) of sample holder. Place platinum two-hole lid on holder that covers the sample pan. Align lid holes with purge gas exit hole in DSC head.

7.9 Place empty aluminum sample pan in reference side (right side) of sample holder. Place remaining platinum two-hole lid on holder that covers the sample pan. Align lid holes as in previous step.

7.10 Close DSC head cover and lock.

7.11 The standard sample run heating program has a heating rate of 10°C min^{-1}, 5.3 min data delay, 5.0 min N$_2$ purge.

7.12 Start the "Run" program.

7.13 Observe the milliwatts (mW) readout on the computer display terminal and when reading stabilizes (≈ 5 to 10 s), remove the sample pan and sample from the sample side of sample holder. Do not disturb the reference side.

7.14 To store data, enter the appropriate file name on the computer for the completed run. If data are not stored by appropriate file name, data are stored under a default file name of "gsav". Only four of these files can be saved at any one time, after which files are overwritten. Once a file is named, it cannot be changed.

8. Calculations

The thermogram is displayed on the computer monitor. The area under the DSC curve is proportional to the enthalpy (H). Use the computer to calculate the H or enthalpy of reaction per g of kaolinite and/or gibbsite (joules g^{-1}) as appropriate.

8.1 % Kaolinite weight =   H/12.62

where:
12.62 = factor obtained from standard curve of kaolinite mixtures using China clay of undetermined purity

8.2 % Gibbsite weight =   H/15.03

where:
15.03 = factor obtained from standard curve of gibbsite values using deferrated Surinam gibbsite of undetermined purity

9. Report

Report percent kaolinite and/or gibbsite to the nearest whole number.

10. Precision

Precision data are not available for this procedure.

11. References

Karathanasis, A.D., and B.F. Hajek. 1982. Revised methods for rapid quantitative determination of minerals in soil clays. Soil Sci. Soc. Am. J. 46:419-425.

Mackenzie, R.C. 1970. Simple phyllosilicates based on gibbsite- and brucite-like sheets. p. 498-537. *In* R. C. Mackenzie (ed.) Differential thermal analysis. Vol. 1. Acad. Press, London.

Mackenzie, R.C., and G. Berggen. 1970. Oxides and hydroxides of higher-valency elements. p. 272-302. *In* R.C. Mackenzie (ed.) Differential thermal analysis. Acad. Press, NY.

Mackenzie, R.C., and S. Caillere. 1975. The thermal characteristics of soil minerals and the use of these characteristics in the qualitative and quantitative determination of clay minerals in soils. p. 529-571. *In* J.E. Gieseking (ed.) Soil components. Vol. 2. Inorganic components. Springer-Verlag, NY.

Mackenzie, R.C., and B.D. Mitchell. 1972. Soils. p. 267-297. *In* R.C. Mackenzie (ed.) Differential thermal analysis. Vol. 2. Acad. Press. London.

Rouston, R.C., J.A. Kittrick, and E.H. Hope. 1972. Interlayer hydration and the broadening of the 10A x-ray peak in illite. Soil Sci. 113:167-174.

Schnitzer, M., and H. Kodama. 1977. Reactions of minerals with soil humic substances. *In* J.B. Dixon and S.B. Weed (eds.) Minerals in soil environments. 2nd ed. SSSA Book Series 1. SSSA, Madison, WI.

Tan, K.H., and B.F. Hajek. 1977. Thermal analysis of soils. p. 865-884. *In* J.B. Dixon and S.B. Weed (eds.) Minerals in soil environments. 2nd ed. SSSA Book Series 1. SSSA, Madison, WI.

Tan, K.H., and B.F. Hajek, and I. Barshad. 1986. Thermal analysis techniques. p. 151-183. *In* A. Klute (ed.) Methods of soil analysis. Part 1. Physical and mineralogical properties. 2nd ed. Agronomy 9:151-183.

Total Analysis (7C)
HF Dissolution (7C3)

1. Application

Historically, elemental analysis was developed for the analysis of rocks and minerals (Washington, 1930). The elemental analysis of soils, sediments, and rocks necessitates their decomposition into soluble forms. Hydrofluoric acid (HF) is efficient in the digestion and dissolution of silicate minerals for elemental decomposition. Procedure 7C3 is an HF acid digestion. Elemental concentration is determined by atomic absorption using 100 mg of clay suspension contained in a closed vessel with boric acid (H_3BO_3) to neutralize excess acid (Berdanier, Lynn, and Threlkeld, 1978; Soil Conservation Service, 1984).

2. Summary of Method

To 100 mg of clay suspension (procedure 7A2i), 5 mL of HF acid are added. The solution is heated, cooled, and 2 to 3 g of H_3BO_3 are added to neutralize excess acid. The solution is diluted to 100 mL, allowed to stand overnight, and 20 mL are decanted. The concentrations of Fe, Al, and K are determined by atomic absorption in procedures 6C7a, 6G11a, and 6Q3a, respectively. Data are reported in procedure 7C3.

3. Interferences

Organic material may remain as a residue with this method.

4. Safety

Perform procedure in hood. Keep HF acid refrigerated and avoid contact with skin.

5. Equipment

5.1 Pipet, 5 mL,
5.2 Volumetric flask, nalgene, 100 mL
5.3 Polyethylene container, 25 mL, with cover
5.4 Electronic balance, ±0.1-mg sensitivity

6. Reagents

6.1 Distilled water
6.2 Hydrofluoric acid (HF), 48%,
6.3 Boric acid, (H_3BO_3), granular

7. Procedure

HF Dissolution

7.1 Prepare Na-saturated clay as in procedure 7A2i, preparation of clay suspension, 7.8 to 7.19.

7.2 Pipet 2 mL of clay suspension into a 25-mL Teflon cup and add 5 mL of HF. A 100 mg of 100-mesh whole-soil sample may be substituted for the clay suspension.

7.3 Pipet a duplicate sample into a weighing dish, dry at 105°C, and weigh. Use this sample for calculations.

7.4 Place covered Teflon cup in stainless steel retainer and tighten Teflon cap. Place sample in oven at 105°C for ≈ 4 h.

7.5 Turn off oven, open door and let stand overnight to cool.

7.6 Remove sample from oven.

7.7 Under a hood, remove Teflon cup from steel retainer vessel and add 2 to 3 g of H_3BO_3 acid.

7.8 Rinse contents of Teflon cup into a 100-mL nalgene volumetric flask and adjust to volume with distilled water. Allow to stand overnight.

7.9 Decant $\approx$ 20 mL into a 25-mL polyethylene container for elemental analysis by atomic absorption. Refer to procedures 7C7a, 6G11a, and 6Q3a.

8. Calculations

Use the MR 2.0 to perform calculations. Inputs are as follows: project number; sample number; tare value; tare + sample value; Al and Fe readings (mg L^{-1}); and K readings (meq L^{-1}). Review data for internal consistency. Request a rerun, if necessary, at this time. Store data on a data disk.

The following example illustrates the conversion calculations of atomic absorption readings or element concentrations of Fe, Al, and K to appropriate oxide forms. The concentrations of Fe and Al (mg L^{-1}), and K (meq L^{-1}) are converted to percent Fe_2O_3, Al_2O_3, and K_2O, respectively. Refer to procedure 4A5 for air-dry/oven-dry ratio (AD/OD).

Sample weight = S = 0.1071 g
Fe reading = [Fe] = 34.2 mg L^{-1}
Fe_2O_3 molecular weight = Fe_2O_3 = 159.70
Fe atomic weight = Fe = 55.85
Al reading = [Al] = 72.8 mg L^{-1}
Al_2O_3 molecular weight = Al_2O_3 = 101.94
Al atomic weight = Al = 26.98
K reading weight = [K] = 0.61 meq L^{-1}
K_2O molecular weight = K_2O = 94.19
K atomic weight = K = 39.10
K equivalent weight = 39 = 39
AD/OD = 1.024 = AD/OD
100 mL/1000 mL = dil. factor = 100/1000
1/1000 mg g^{-1} = conv. factor = 1/1000
100/1 P/100 pts. = conv. factor = 100/1

% Fe_2O_3 =
[Fe] x 1/S x 100/1000 x 1/1000 x 100/1 x AD/OD x Fe_2O_3/Fe =
34.2 x 1/0.1071 x 0.1 x 0.001 x 100 x 1.024 x 159.7/111.70
% Fe_2O_3 = 4.68

% Al_2O_3 =
[Al] x 1/S x 100/1000 x 1/1000 x 100/1 x AD/OD x Al_2O_3/Al =
72.8 x 1/0.1071 x 0.1 x 0.001 x 100 x 1.024 x 101.94/53.96
% Al_2O_3 = 13.15%

% K_2O =
[K] x 1/S x 100/1000 x 1/1000 x 100/1 x AD/OD x K_2O/K x 39 =
0.61 x 1/0.1071 x 0.1 x 0.001 x 100 x 1.024 x 94.19/78.2 x 39 =
% K_2O = 2.74

9. Report

Report data to nearest whole percent.

10. Precision

Precision data are not available for this procedure. A quality control check sample is routinely run in HF analyses. For 38 observations of the quality control check sample, the mean, standard deviation, and C.V. for percent Fe, Al_2O_3, and K_2O are as follows:

Analyte	Mean	Std. Dev.	C.V.
% Fe	2.8	0.40	14
% Al_2O_3	12.1	1.04	9
% K_2O	2.4	0.38	16

11. References

Berdanier, C.R., W.C. Lynn, and G.W. Threlkeld. 1978. Illitic mineralogy in Soil Taxonomy: X-ray vs. total potassium. Agron. Absts. 1978 Annual Mtgs.

Soil Conservation Service. 1984. Procedures for collecting soil samples and methods of analysis for soil survey. USDA-SCS Soil Surv. Invest. Rep. no. 1. U.S. Govt. Print. Office, Washington, DC.

Washington, H.S. 1930. The Chemical analysis of rocks. 4th ed. John Wiley and Sons, Inc., New York, NY.

Whittig, L.D., and W.R. Allardice. 1986. X-ray diffraction techniques. *In* A. Klute (ed.) Methods of soil analysis. Part 1. Physical and mineralogical methods. 2nd ed. Agronomy 9:331-362.

SSIR No. 1, PROCEDURES FOR COLLECTING SOIL SAMPLES AND METHODS OF ANALYSIS FOR SOIL SURVEY (1972, 1982, 1984)

SAMPLE COLLECTION AND PREPARATION (1)

Laboratory Preparation of Soil Samples (1B)
Carbonate-Containing Material (1B3)

Procedure

Prepare dialysis membrane sacks from 5 1/2-inch cellulose casing (Visking Company), using large rubber bands to tie the bottoms. Place the sample (as much as 6 kg if very gravelly and highly calcareous) in a dialysis membrane and add about 1 L pH 5, N NaOAc buffer. Tie the top of the dialysis membrane around a glass breather tube 4 in long and hang the assembly in a 60-L reservoir of buffer held in a 20-gal plastic garbage can. If carbonate is dissolving, knead the membrane to release bubbles of CO_2. When bubbles of CO_2 no longer form on kneading, open the dialysis membrane and use strong acid to check the coarser material for carbonate coatings (carbonate remains longer in the coarser material). When sample is free of carbonate, desalt it by dialysis against tap water flowing continuously through a large plastic garbage can. Check the ionic concentration inside the membrane by measuring conductivity of a small volume of the supernatant liquid poured out through the breather tube. Continue dialysis until the salt concentration is less than 10 meq/L.

The procedure used to dry the sample depends on whether the particles larger than 2 mm have been removed before buffer treatment. If they have been removed, withdraw excess water from the sample in the membrane with filter candles. Knead the membrane to mix the sample and place it in contact with ethanol to desiccate further. Remove the sample from the membrane and air-dry.

If the buffer-treated sample contains particles larger than 2 mm, wet sieve the sample through a 2-mm sieve. Then dry sieve the material remaining on the sieve (>2 mm) and add the <2-mm fraction from this sieving to the <2-mm fraction separated by the wet sieving. Remove most of the water from the <2-mm fraction with filter candles. Use ethanol to transfer the samples to shallow pans and dry. Ethanol prevents aggregation of clay into durable flakes during drying.

Discussion

The time required for carbonate removal varies greatly, depending on particle size, percentage and type of carbonate, and sample size. Samples from horizons strongly cemented by carbonate have required as long as 2 months. The concentration of alkaline-earth ions in the buffer greatly affects the rate of carbonate removal. Changing the buffer in the reservoir well before the buffer capacity has been exhausted, thereby keeping the alkaline-earth ion concentration low, increases the rate markedly. Desalting usually takes about 4 days.

For carbonate-cemented horizons, the whole sample, not just the <2-mm material, must be buffer treated. Furthermore, for horizons without carbonate cementation, buffer treatment of the whole sample has the advantage of washing the >2-mm skeletal material free of adhering fines and organic material. This problem is considered further in 1B4.

For very gravelly horizons, large samples (several kilograms) are necessary for buffer treatment because of the small amount of <2-mm material. Using large samples also increases precision of the >2-mm percentage.

References
Grossman and Millet (1961).

Carbonate-Indurated Material Containing Coarse Fragments (1B4)

Break the field sample to get several representative subsamples. Remove the carbonate from one subsample by acid treatment and separate the coarse fragments from the fine earth (1B3). Weigh the two fractions. Use the noncarbonate fine earth for the standard characterization and mineralogical measurements (sections 6-7).

Grind another subsample of the whole field sample to pass 80-mesh sieve. Determine the carbonate content (weight) of this whole ground subsample (6E).

These weights can be used to calculate the $CaCO_3$ percentage of the fine earth. Any analytical value based on the noncarbonate fine earth can be converted to the whole-soil basis as well as to the basis of the carbonate-containing fine earth.

PARTICLE-SIZE ANALYSIS (3)

Particles < 2mm (Pipet Method) (3A)

An automated balance system, consisting of a Radio Shack Model II microcomputer interfaced to a Mettler PL2000 electronic balance (for sand) and a Mettler AE160 electronic balance (for silt and clay), is used for determining, storing, and processing sample weights.

Air-Dry Samples (3A1)

Apparatus
Fleaker, 300 ml (tare to 1 mg).

Pasteur-Chamberlain filter candles, fineness "F".

Shaker, horizontal, 120 oscillations per minute.

Cylinders, 1000 ml.

Stirrer, motor-driven
Stirrer, hand. Fasten a circular piece of perforated plastic to one end of a brass rod.

Shaw pipet rack.

Pipets, 25 ml automatic (Lowy with overflow bulb).

Polyurethane foam, pipe-insulating cover.

Shaker with 1/2-in vertical and lateral movements and 500 oscillations per minute. Accommodates a nest of sieves.

Wide-mouth glass pill bottles with screw caps, 90 ml (tare to 1 mg).

Electronic balance (0.1-mg sensitivity).

Set of sieves. Square mesh woven phosphor bronze wire cloth. U.S. Series and Tyler Screen Scale equivalent designations as follows:

Sand Size	Opening (mm)	U.S. No.	Tyler Mesh Size
VCS	1.0	18	16
CS	0.5	35	32
MS	0.25	60	60
FS	0.105	140	150
VFS	0.047	300	300

Reagents

Hydrogen peroxide (H_2O_2), 30 to 35 percent.

Sodium hexametaphosphate $(NaPO_3)_6$. Dissolve 35.7 grams of $(NaPO_3)_6$ and 7.94 grams of Na_2CO_3 per liter of water.

Demineralized water.

Procedure

Removing organic matter.--Place about 10 air-dry soil containing no particles larger than 2 mm in a tared Fleaker. Add about 50-ml of demineralized water (referred to subsequently as water) and then add 5 ml of H_2O_2. Cover the fleaker with a watchglass. If a violent reaction occurs, repeat the cold H_2O_2 treatment periodically until no more frothing occurs. Heat the Fleaker to about 90°C on an electric hot plate. Add H_2O_2 in 5-ml quantities at 45-min intervals until the organic matter is destroyed, as determined visually. Continue heating for about 30 min to remove any excess H_2O_2.

Removing cementing agents (optional).--Treat the sample with about 200 ml of 1 N sodium acetate buffered at pH 5 to remove carbonates. When CO_2 bubbles are no longer evident, wash free of salts with a filter candle system. Highly calcareous samples may need a second treatment.

Remove siliceous cementing agents by soaking the sample overnight in 0.1 N NaOH. Iron oxide cementing agents are removed by shaking overnight in sodium dithionite (6C2). Wash free of salts with filter candle system before proceeding.

Removing dissolved mineral and organic components.--After the H_2O_2 treatment, place the Fleaker in a rack and add about 150 ml of water in a jet strong enough a short Pasteur-Chamberlain filter of "F" fineness. Five such washings and filterings are usually enough except for soils containing much coarse gypsum. Remove soil adhering to the filter by gentle back pressure; use finger as policeman. Dry the sample overnight in an oven at 105°C, cool in a desiccator, and weigh to the nearest milligram. Use the weight of the ovendry, H_2O_2-treated sample as the base weight for calculating percentages of the various fractions.

Dispersing the sample.--Add 10 ml of sodium hexametaphosphate dispersing agent to the Fleaker containing ovendry treated sample. Make the volume to approximately 200 ml. Stopper and shake overnight on a horizontal reciprocating shaker at 120 oscillations per minute.

Separating sands from silt and clay.--Wash the dispersed sample with water on a 300-mesh sieve. Silt and clay pass through the sieve into a 1-L cylinder. Use a clamp and stand to hold the sieve above the cylinder. Avoid using jets of water in washing the sample. Gently tap the sieve clamp with the side of the hand to facilitate sieving. Continue washing until the suspension volume in the cylinder is about 800 ml. Sand and some coarse silt remain on the sieve. It is important to wash all particles of less than 20µ diameter through the sieve. Remove the sieve from the holder, wash the sands into an evaporating dish with water, and dry at 105 to 110°C. Bring the silt and clay suspension in the cylinder to 1 L with water and cover with a watchglass.

Pipeting.--First pipet the <20µ fraction at a 10-cm depth. Vary sedimentation times according to temperature. Next, pipet the <2µ fraction after a predetermined setting time (usually 4 1/2 to 6 1/2 hr). Vary depth according to time and temperature. Use a Lowy 25-ml automatic pipet and regulate filling time to about 12 s. Before each pipeting, stir material in the sedimentation cylinder, and stir the suspension for 30 s with a hand stirrer, using an up-and-down motion. Note the time at completion of stirring. About 1 min before sedimentation is complete, lower the tip of the pipet slowly into the suspension to the proper depth with a Shaw pipet rack. At the appropriate time, fill the pipet and empty into a 90-ml, wide-mouth bottle. Rinse the pipet into the bottle once. Dry in an oven overnight at 105°C. Cool in a desiccator containing phosphorus pentoxide (P_2O_5). Weigh.

Sieving and weighing the sand fractions.--Transfer the dried sands to a nest of sieves. Shake for 3 min on a shaker that has 1/2-in vertical and lateral movements and oscillates at 500 strokes per minute. Record the weights of the individual sand fractions.

Calculations

Pipetted fractions:

Percentage of pipetted fractions = (A - B)KD

where
A = Weight (g) of pipetted fraction
B = Weight correction for dispersing agent (g)
K = 1000/(ml in pipet)
D = 100/(g of H_2O_2-treated ovendry total sample)

The <20-μ fraction minus the <2-μ fraction equals fine silt.

Sand fractions:
Percentage of sieved fractions = weight (g) of fraction on sieve times D.

Coarse silt fraction:
Obtain by difference. Subtract the sum of the percentages of sand plus the <20-μ fraction from 100.

References
Kilmer and Alexander (1949), Kilmer and Mullins (1954), Tyner (1939), and Grossman and Millet (1961).

Carbonate and Noncarbonate Clay I (3A1a)

Apparatus
Warburg manometer.
1/4-oz (5-ml) gelatin capsules.

30-ml plastic cups.

Reagents
Hydrochloric acid (HCl), 6 *N*.

Procedure
If carbonate is present, use the glass
bottle containing clay residue from regular pipet analysis and determine carbonate as in 6E1b. Use Warburg manometer.

Calculations

Cc = {[(A - B - C) x Factor]/D} x 100

where
Cc   = Carbonate clay (pct <2 mm)
A   = Upper reading
B   = Lower reading
C   = Blank
Factor = Factor derived from standard curve and includes pipette volume factor
D   = Total sample weight (3A1)

Nc = Total clay - Cc

where:
Nc= Noncarbonate clay (pct <2mm)
Cc= Carbonate clay (pct <2mm)

References
Shields and Meyer (1964).

Moist Samples (3A2)

If drying affects dispersion of treated sample, ovendrying may be avoided by removal of a pipet sample to estimate the total weight of the sample. Pipet 50 ml at a depth of 20 cm at time zero while the suspension is still turbulent. Use the ovendry weight of the aliquot to calculate the total weight of the <0.05-mm fraction. Add this weight to the total weight of the sands to obtain the total weight of the sample.

An optional procedure is to carefully weigh out two identical samples and pretreat to remove organic matter and dissolved mineral matter. The first sample is continued through the standard procedure, excluding ovendrying. The second sample is ovendried, weighed, and discarded. The ovendry weight of the second sample is substituted in the calculations for the first sample.

Carbonate and Noncarbonate Clay (3A2a)

Proceed as in 3A1a except use field-moist sample.

FABRIC-RELATED ANALYSES (4)

Bulk Density (4A)

Density is defined as mass per unit volume. Soil density as commonly used differs from most density measurements in that the volume of interparticle space is included but the mass of the liquid phase is excluded. Therefore, soil density has been called bulk density, Db, to distinguish it from the more usual density that is based on intraparticle volume only. Since the volume of a shrinking-swelling soil changes with a change in its water content, subscripts are added to designate the moisture condition when the measurement was made. Thus, Db_m is the bulk density of a moist sample, $Db_{1/3}$ is the bulk density of a clod sample equilibrated at 1/3-bar tension, and Db_d is the bulk density of a dry sample.

Saran-Coated Clods (4A1)

Reagents
Methyl ethyl ketone.

Dow Saran F310.--The saran resin dissolves readily in acetone or methyl ethyl ketone. In this method, methyl ethyl ketone is used as a solvent because it is less soluble in water than is acetone and there is less penetration of the Saran-solvent solution into a moist clod. However, acetone is adequate for a first (field) coat and is more readily available. Saran-solvent ratios of 1:4 to 1:7 are used, depending on the porosity of the soil to be coated.

Coating solution.--To prepare the solution, fill a weighted container with a solvent to about three-fourths its volume. From the weight of the solvent, calculate the weight of resin required to obtain a predetermined resin-solvent ratio and add to the solvent. Since the solvent is flammable and its vapors form explosive mixtures with air, mix the components with an air-powered or nonsparking electric stirrer under an exhaust hood. Information on the safe handling and use of methyl ethyl ketone is available in Chemical Safety Data Sheet SD-83, Manufacturing Chemists' Association, Inc., 1825 Connecticut Avenue NW, Washington, D.C. The threshold limits of methyl ethyl ketone are 200 ppm as given in OSHA standards, Part 2, Section 1910.93, table G1.

If a high-speed stirrer is used, the resin dissolves in about 1 hr. In the field, mix with a wooden stick. Metal cans (1 gal) are satisfactory containers for mixing and storing the plastic. Keep the containers tightly closed to prevent evaporation of the solvent.

Procedure

Collect natural clods (three per horizon) of about 100 to 200 cm^3 in volume (fist-sized). Remove a piece of soil larger than the clod from the face of a sampling pit with a spade. From this piece prepare a clod by directly cutting or breaking off protruding peaks and material sheared by the spade. If roots are present, they can be cut conveniently with scissors or side cutters. In some soils, clods can be removed directly from the face of a pit with a knife or spatula. No procedure for taking clod samples fits all soils; the procedure must be adjusted to meet the conditions in the field at the time of sampling.

The clods are tied with the fine copper wire or placed in hairnets and suspended from a rope or string, hung out like a clothesline. Moisten dry clods with a fine mist spray. The suspended clods are dipped by raising a container of the dipping mixture upward around each clod, so it is immersed momentarily. The saran-coated clods should be allowed to dry for 30 min or longer. Clods coated in this way can be transported to the laboratory and examined microscopically in an undisturbed state. For convenience, either of two concentrations of plastic solution is usually used--a 1:7 solution for most soil samples or a 1:4 solution for clods that have larger pores. If bulk density at field-moisture content is desired, store the clods in waterproof plastic bags as soon as the coating dries since the coating is permeable to water vapor. Although the coating keeps the clods intact, they may be crushed in transport unless they are packed in rigid containers.

In the laboratory, additional coatings of plastic are applied to make the clod waterproof and to prevent its disruption during wetting. Then weigh the clod, either in its natural moisture condition or in an adjusted moisture condition (e.g., 1/3-bar tension) in air and in water to obtain its volume by Archimedes' principle. Subsequent changes in moisture condition and volume of the soil sample can be followed by reweighing the coated clod in air and in water. Finally, weigh the ovendry clod in air and in water.

Be careful not to lose any soil material because the weight of material lost is calculated as soil moisture, and calculated bulk densities depend on the final ovendry weight of the clod.

Bulk-density values determined by this method are reported on the basis of fine-earth fabric. Weight and volume measurements are made on clod samples that may contain particles >2 mm; however, after the measurements are made, the weight and volume of the coarse fraction are subtracted. The remainder consists of the weight of <2-mm material and the volume of these fine-earth particles and the pore space associated with them.

Sometimes it is necessary to correct bulk density for weight and volume of the plastic coating. The coating has a density of about 1.3 g/cm^3 and it loses 10 to 20 percent of its air-dry weight on ovendrying at 105°C. Thus, the amount of correction becomes smaller as bulk density of the soil approaches the density of the coating and as moisture content of the soil approaches the weight loss of the coating.

Calculations

$$Db_{1/3} = \frac{wtclod_{od} - wt{>}2mm - tcoat_{od}}{volclod_{1/3} - vol{>}2mm - vol\ coat}$$

$$Db_{od} = \frac{wtclod_{od} - wt{>}2mm - wtcoat_{od}}{volclod_{od} - vol{>}2mm - vol\ coat}$$

$$W_{1/3} = \frac{wtclod_{1/3} - wtclod_{od} - (wtclod_{ad} - wtcoat_{od})}{wtclod_{od} - wt{>}2mm - wtcoat_{od}} \times 100$$

where
$Db_{1/3}$ = bulk density of <2-mm fabric at 1/3-bar tension in grams per cubic centimeter

Db_{od} = bulk density of <2-mm fabric at ovendryness in grams per cubic centimeter

$W_{1/3}$ = the weight percentage of water retained at 1/3-bar tension

wt clod$_{od}$ = weight of ovendry coated clod

wt clod$_{1/3}$ = weight of coated clod equilibrated at 1/3-bar tension

vol clod$_{od}$ = volume of ovendry coated clod

vol clod$_{1/3}$ = volume of coated clod equilibrated at 1/3-bar tension

vol >2 mm = volume of material >2 mm separated from clod after ovendrying

wt > 2 mm = weight of material > 2mm separated from after ovendrying

wt coat$_{ad}$ = weight of Saran coating before ovendrying

wt coat$_{od}$ = weight of Saran coating after ovendrying

vol coat = volume of Saran coating (estimated)

The field coat (initial coat) of plastic penetrates the clod to some extent. Weight of the field coat, estimated to 1.5 times the weight of each additional coat, is computed by:

$$Wtcoat_{init} = \frac{(Wtclod_B - Wtclod_A) \times 1.5}{3}$$

where

Wtcoat$_{init}$ = Weight of field (initial) coat

Wtclod$_A$ = Weight of clod with one coat of plastic

Wtclod$_B$ = Weight of clod with three additional coats of plastic

References
Brasher et al. (1966).

Air-Dry (Db$_d$) (4A1b)

After measuring field-state volume, place clods in a drying room kept at 90° F. Weigh a few clods each day until they reach a constant weight. Assume then that all the clods are airdry. Coat them again with Saran and measure "airdry" volume as described in 4A1a. Determine ovendry weight and calculate bulk density as described in 4A1.

30-cm Absorption (Db$_{30}$) (4A1c)

After measuring airdry volume, remove a patch of the Saran coating from one side of each clod. Next place the clods on a sand tension table with the exposed side in contact with very fine sand that has been equilibrated to 30-cm water tension. Again weigh a few clods each day until they reach constant weight and assume that all the clods are at 30-cm water tension. Most clods reach equilibrium in 7 to 10 days. Remove the clods from the tension table and coat with Saran until waterproof. Measure volume of the clods and calculate bulk density as described in 4A1.

1/3-Bar Desorption II (Db$_{1/3}$) (4A1e)

Cut a flat surface on the coated field-moist clods with a sharp knife or diamond saw. Seat the clods on saturated ceramic plates with the flat surface in contact with the plates. Place the plates in pans and add water to just cover the surface of the plates. After the clods become wet by capillary movement, place the plates in a pressure cooker and equilibrate at 1/3 bar. After equilibration, carefully remove the clods from the plates and dip in Saran until waterproof. Measure volume of the clods and calculate bulk density as described in 4A1.

1/3-Bar Desorption III (Db$_{1/3}$) (4A1f)

Proceed as in 4A1e except prewet the clods at 10-cm tension on porous bricks (cheesecloth layer between clod and brick) instead of saturating them on ceramic plates.

1/10-Bar Desorption (Db$_{1/10}$) (4A1g)

Proceed as in 4A1d, e, or f except make final desorption at 1/10 bar.

Parafffin-Coated Clods (4A2)
Ovendry (Db$_d$) (4A2a)

Ovendry the clods, coat with paraffin, and weigh in water and in air. Calculate bulk density as follows:

$$Db_d \text{ (g/cc)} = \frac{Wtair - Wt{>}2mm}{Wtair - WtH_2O - (Wt{>}2mm/2.65)}$$

where
Wtair = Weight in air
Wt>2mm = Weight of >2-mm fraction in clod
WtH_2O = Weight in water

Nonpolar-Liquid Saturated Clods (4A4)

Procedure

Place a natural clod in a nonpolar liquid of low viscosity, e.g., high-purity kerosene. Evacuate under vacuum until bubbles cease to appear and weigh the clod suspended in the nonpolar liquid. Remove the clod, place it on a sand table under 3-cm tension against the nonpolar liquid to drain off excess nonpolar liquid, and weigh it in air. The difference in weight of the clod in air and suspended in the nonpolar liquid divided by the density of the nonpolar liquid is the clod volume. Determine the ovendry weight and calculate bulk density as in 4A1. The difference between the clod's initial weight before immersion in the nonpolar liquid and its ovendry weight is the moisture content.

References
Rennie (1957).

FABRIC-RELATED ANALYSES (4)

Water Retention (4B)
Pressure-Plate Extraction (4B1)

After measuring the 1/3-bar volume (4A1d), the Saran coating is removed from the flat surface of the clods. The clods are allowed to air-dry (4 to 6 days) and then placed in the drying room 2 or 3 days. They are then placed on a tension table of very fine sand and equilibrated to 5-cm tension as in 4A1d. After about 2 weeks, some of the highly organic clods that have not rewetted are placed in a pan of free water overnight to make certain that wetting is complete. The clods are again desorbed to 1/3-bar as in 4A1d and volume measurements of the clod are made and bulk density is calculated as described in 4A1.

Soil Pieces (4B1b)

Procedure

Make desorption measurements of soil pieces concurrently with the sieved-sample measurements. Cover the sieved samples in retainer rings with small squares of industrial tissues (Kimwipes). Place the soil pieces (about 2.5 cm in diameter) on the tissues before adding water to the plate. Proceed as in 4B1a. If the soil pieces contain >2-mm material, wet sieve and weigh the ovendry >2-mm material. Report moisture content as percentage of ovendry weight of <2mm material.

References
Young (1962).

Sand-Table Absorption (4B3)

Saran-coated clods that have been equilibrated on a sand table to determine bulk density (4A) can also be used to determine water content at these tensions.

FABRIC-RELATED ANALYSES (4)

Micromorphology (4E)
Thin Sections (4E1)
Moved-Clay Percentage (4E1c)

Apparatus
Diamond tile saw.

Thin-section equipment.

Point-counting eyepiece.

Reagents
Aroclor 5460 (Monsanto).

Polyester resin.

Styrene.

Procedure
Impregnate an undisturbed field sample with Aroclor (4E1b). With a diamond saw cut the clods into pieces about 3 by 1 by 1 cm. Mount about 10 pieces side by side with polyester resin (use a little styrene) to form a block. Cut this assembly to form slices 3 by 1 by 1 cm. Slice all the field sample, composite, and withdraw subsamples of 10 to 15 slices. Stack these slices, tape them together, and mount in plastic (polyester resin plus styrene). Cut a section through the stack parallel to the direction of stacking and along the longer of the two remaining axes. Mount one such section from each stack on a glass slide and prepare a thin section.

To estimate the moved-clay volume insert a point-counting eyepiece into the microscope and run a transect along each strip. Keep the transect length and the number of fields in a transect constant. Count the number of points that fall on moved clay. Divide this number by the total number of points to get an estimate of the proportion of moved clay. To convert these volume estimates to weight estimates, multiply by the ratio of the bulk density of the moved clay to the bulk density of the appropriate dry fabric. Assume that the moved clay has a bulk density of 2.00 g per cubic centimeter.

References
Grossman (1964).

Scanning Electron Microscopy (4E2)

Electronically reproduced images of fabric surfaces can be obtained at magnifications ranging from 50 to 30,000 diameters. Depth of focus by this technique is large compared to that by light microscope. Stereoscopic pictures can be taken to give three-dimensional viewing.

Procedure
Take a sample of fabric up to 10 mm in diameter and 2 to 3 mm thick. Coat with a thin metallic layer and insert in the instrument. The image is displayed on a cathode ray tube.

ION EXCHANGE ANALYSES (5)

Cation Exchange Capacity (CEC) (5A)
NH₄OAc, pH 7.0 (Buchner funnel) (5A1)

Reagents
Ammonium acetate (NH$_4$OAc), 1 N, pH 7.0. Mix 68 ml ammonium hydroxide (NH$_4$OH), specific gravity 0.90, and 57 ml 99.5-percent acetic acid (CH$_3$COOH) per liter of solution desired. Cool, dilute to volume with water, and adjust to pH 7.0 with CH$_3$COOH or NH$_4$OH. Optionally prepare from NH$_4$OAc reagent salt and adjust pH.

Ethanol (CH$_3$CH$_2$OH), 95-percent, U.S.P.

Nessler's reagent (optional). Prepare according to Yuen and Pollard.

Procedure
Weigh 25 g airdry <2-mm soil (some early work was done with 50-g samples) into a 250-ml Erlenmeyer flask and add 35 to 50 ml NH$_4$OAc solution. Stopper, shake the flask for several minutes, and allow to stand overnight. Transfer contents of the flask to a Buchner funnel (Coors No. 1) fitted with moist Whatman No. 42 filter paper. Filter, using gentle suction if needed. Leach with 200 ml NH$_4$OAc, adding small amounts at a time so that leaching requires no less than 1 hour. Transfer leachate from suction flask to volumetric flask and retain for analysis of NH$_4$OAc-extractable cations (methods 6N2, 6O2, 6P2, 6Q2).

Add 95-percent ethanol in small amounts to the ammonium-saturated soil remaining on the Buchner funnel until the leachate gives a negative test for ammonia with Nessler's reagent or leach with 100 ml ethanol.

References
Peech et al. (1947) and Yuen and Pollard (1952).

Direct Distillation of Adsorbed Ammonia, Kjeldahl (5A1a)

Reagents
Sodium chloride (NaCl).

Antifoam mixture. Mix equal parts of mineral oil and n-octyl alcohol.

Sodium hydroxide (NaOH), 1 N.

Hydrochloric acid (HCl), 0.2 N, standardized.

Boric acid (H$_3$BO$_3$), 4-percent.

Mixed indicator. Mix 1.250 g methyl red and 0.825 g methylene blue in 1 liter 95-percent ethanol.

Brom cresol green, 0.1-percent, aqueous solution.

Procedure
Transfer the soil plus filter paper from method 5A1 to a Kjeldahl flask. Add 400 ml water and about 10 g NaCl, 5 drops antifoam mixture, a gram or two of granular zinc, and 40 ml 1 N NaOH. Connect the flask with the condenser and distill 200 ml into 50 ml 4-percent H$_3$BO$_3$ solution. Titrate the distillate to the first tinge of purple with 0.2 N HCl, using 10 drops mixed indicator and 2 drops brom cresol green.

Calculations

CEC(meq/100 g) $=$(A/B) x N x 100

where
A = Volume HCl (mL)
B = Sample weight (g)
N = Normality of acid

Report on ovendry basis.

References
Peech et al. (1947).

Displacement of Adsorbed Ammonia, Semimicro Kjeldahl (5A1b)

Reagents
Sodium chloride (NaCl), acidified, 10-percent. Dissolve 100 g NaCl, reagent-grade, ammonia-free, in 750 ml warm water; add 25 ml 2 N hydrochloric acid (HCl) and bring to 1000-ml volume.

Sodium hydroxide (NaOH), 1 N.

Boric acid (H_3BO_3), 2-percent.

Sulfuric acid (H_2SO_4), 0.01 N, standardized.

Ethanol, 95-percent.

Mixed indicator. Dissolve 0.1 g methyl red and 0.1 g brom cresol green in 250 ml ethanol.

Procedure
Leach soil from method 5A1 with 240 ml 10-percent acidified NaCl solution, using small increments. Drain completely between each increment. Transfer the leachate to a 250-ml volumetric flask and adjust volume to mark. Pipet a suitable aliquot of the leachate into a micro-Kjeldahl distillation flask and attach to steam-distillation apparatus. Start steam distillation and slowly add 10 ml 1 N NaOH. Catch distillate in a 250-ml Erlenmeyer flask containing 10 ml H_3BO_3 and 10 drops of mixed indicator. Distill for 5 minutes after H_3BO_3 turns green, lower receiving flask, and rinse condenser and outlet hose into receiving flask. Titrate the ammonia with 0.01 N H_2SO_4 to a red end point, using a blank for comparison.

Calculations

CEC(meq/100 g) = (A/B) x N x (C/D) x 100

where
A = Volume H_2SO_4 (mL)
B = Sample weight (g)
N = Normality of acid
C = Volume leachate (mL)
D = Volume aliquot (mL)

Report on ovendry basis.

NaOAc, pH 8.2 (5A2)
Centrifuge Method (5A2a)

Reagents
Sodium acetate (NaOAc), 1 N, pH 8.2.

Ethanol, 95-percent.

Ammonium acetate (NH_4OAc), 1 N, pH 7.0. Add 57 ml concentrated acetic acid and 68 ml concentrated NH_4OH, specific gravity 0.90, to about 800 ml water. Cool and dilute to 1 liter and adjust to pH 7.0 by adding more NH_4OH or acetic acid.

Procedure
Weigh 5-g samples to an accuracy of 1 percent and place in centrifuge tubes. Add 33 ml NaOAc, stopper the tubes, and shake for 5 minutes. Remove stopper and centrifuge until the supernatant liquid is clear (usually 5 min). Decant the supernatant liquid as completely as possible and discard. Repeat four times, discarding the supernatant liquid each time. After the last saturation, wash the rubber stoppers and use absorbent paper to remove any acetate crystals remaining on lip of centrifuge tube. Add about 30 ml ethanol to each tube, stopper, shake for 5 minutes, remove stopper, and centrifuge until the supernatant liquid is clear. Decant and discard the supernatant liquid. Continue washing until the electrical conductivity of the supernatant liquid from the last washing is between 55 and 40 μmho per centimeter. Optionally, decrease volume by about 5 ml each washing. Replace the absorbed sodium from the sample by extracting with three 30-ml portions of NH_4OAc solution. Dilute to 100 ml and determine the sodium concentration as described in 6P2a.

Calculations

CEC (meq/100 g) = (A/B) x dilution x 10

where
A = Na from curve (meq/L)
B = Sample weight (g)

Report on ovendry basis.

References
Richards (1954).

KOAc, pH 7.0 (5A4)

Procedure
Proceed as in 5A1 except substitute 1 N KOAc, pH 7.0, for NH_4OAc. Determine potassium with flame photometer.

BaCl$_2$, pH 8.2 (5A5)

Apparatus
Leaching tubes.

Flame photometer.

Reagents

Buffer solution. Barium chloride ($BaCl_2$), 0.5 N, and triethanolamine (TEA), 0.2 N. Adjust to pH 8.2 with HCl. Protect from CO_2 of the air by attaching a drying tube containing soda lime (sodium calcium hydrate) to the air intake.

Replacement solution. Barium chloride ($BaCl_2$), 0.5 N. Add 0.4 ml buffer solution per liter and mix. Protect from CO_2 with soda-lime tube.

Magnesium nitrate ($Mg(NO_3)_2$), 1 N.

Procedure

Transfer a 5-g sample to a leaching tube. For field-moist samples use a sample large enough to give an ovendry weight of about 5 g. Leach with 50 ml $BaCl_2$-TEA solution, controlling the leaching rate to give at least 4 hours of soil:solution contact time. Follow with 100 ml $BaCl_2$ replacement solution, controlling the leaching rate so that the soil and $BaCl_2$ solutions are in contact for a total of 20 to 24 hours. Rinse walls of leaching tube with 15 to 20 ml H_2O, collecting this washing with leachates from $BaCl_2$ solutions. Extractable acidity can be determined by using this solution (6H1a). Place leaching tube on a clean flask and wash with methanol until free of chloride ion. For many samples 100 ml methanol is enough, but more methanol may be needed for some soils, particularly those of heavy texture and containing large amounts of hydrous oxides. Leach with 100 ml 0.001 N $BaCl_2$ to remove methanol.

Disconnect leaching tube and flask, rinse underside of leaching tube, place over a 250-ml volumetric flask, and leach with 100 ml 1 N $Mg(NO_3)_2$ solution. Control leaching rate to give a soil-solution contact time of 16 hours or more. Rinse walls of leaching tube with 15 to 20 ml H_2O; collect rinse in the $Mg(NO_3)_2$ leachate. Make to volume.

Barium by Flame Photometry (5A5a)

Make standards in 1 N $Mg(NO_3)_2$. Determine barium by flame photometry at 489 mμ.

Calculations

CEC(meq/100 g) = (A/B) x dilution x 25

where
A = Ba from curve (meq/L)
B = Sample weight (g)

NH_4OAc, pH 7.0, Leaching Tube (5A6)

Apparatus
Allihn leaching tubes or 50-ml plastic syringe barrels.

Reagents
Same as in 5A1.

Procedure

Prepare the Allihn tubes by placing either filter paper (Reeve Angel No. 934 AH, 3-cm fiber glass) or filter paper pulp on the fritted glass plate. If the syringe barrel is used as a leaching tube, compress the filter paper pulp in the barrel bottom with the syringe plunger. Place a Gooch perforated plate over the filter paper to permit stirring the soil without damage to the filter. (This plate is not necessary if an adequate pulp pad is used.) Place 5 or 10 g soil and a teaspoon of Celite into the tubes. (Optionally place a layer of Celite under the soil.) Add 25 ml N NH_4OAc; stir and leach. Add an additional 25 ml N NH_4OAc and let stand overnight. Stop the leaching with a pinch clamp or by stoppering the leaching tube. Add the NH₄OAc directly to the leaching tube or use a constant level device (refer to Fig. 8, procedure 6N3). A volumetric flask can be substituted for the 250-ml Erlenmeyer flask and tubing.

Make the leachate to volume if a volumetric flask is used or, if tared suction flasks are used, make to the appropriate calibrated weight for 100 ml NH₄OAc. Set aside the leachate for further analysis. Add about 10 ml ethanol to the soil pad, stir, and leach. Leach with 100 ml ethanol and check for NH_4^+ in leachate. If NH_4^+ is present, leach with an additional 100 ml ethanol. Some soils, particularly those containing amorphous material, require as much as 400 ml ethanol to clear the ammonia from the leachate.

Direct Distillation (5A6a)

Transfer soil cake to Kjeldahl flask and determine ammonia as described in 5A1a.

NH₄Cl, pH 7.0, Mechanical Extraction (5A7)
Direct Distillation (5A7a)

Determine ammonia by Kjeldahl distillation as described in 5A1a.

NH₄OAc, pH 7.0, Automatic Extractor (5A8)
Direct Distillation (5A8a)

Reagents
Sodium chloride (NaCl).

Antifoam mixture. Mix equal parts of mineral oil and n-octyl alcohol.

Sodium hydroxide (NaOH), 1 N.

Hydrochloric acid (HCl), 0.2 N, standardized.

Boric acid (H_3BO_3), 4 percent.

Procedure
Transfer the soil plus filter pulp from methods 5A8 or 5A9 to a Kjeldahl flask. Add 400 ml water and about 10 g NaCl, 5 drops antifoam mixture, a gram or two of granular zinc, and 40 ml of 1 N NaOH. Connect the flask with the condenser and distill 140 ml into 50 ml of 4-percent H_3BO_3 solution in 250-ml titrator beaker. Titrate with automatic titrator to end point pH setting of 4.60.

Calculations

CEC(meq/100g) = (A/B) x N x 100

where
A = Volume HCl (mL)
B = Sample weight (g)
N = Normality of acid

Report on ovendry basis.

References
Peech et al. (1947).

NH₄Cl, pH 7.0, Automatic Extractor (5A9)
Direct Distillation (5A9a)

Determine ammonia by Kjeldahl distillation as described in 5A8a.

Extractable Bases (5B)
NH₄OAc, pH 7.0, Buchner Funnel (5B1)

Procedure
Analyze the NH_4OAc leachate from method 5A1 a for calcium, magnesium, sodium, and potassium (methods 6N2, 6O2, 6P2, 6Q2).

Uncorrected (extractable) (5B1a)

If a soil does not contain soluble salts, the extractable bases are presumed to equal the exchangeable bases. They are, however, reported as extractable bases.

Corrected (exchangeable) (5B1b)

If a soil contains soluble salts, estimate their amount from the saturation extract as follows. Multiply cation concentration in the saturation extract (meq/L) by the saturation percentage (divided by 1000) to convert to milliequivalents per 100 g. Subtract this quantity from the concentration of the extracted cation. This procedure is not valid for calcium and magnesium in the presence of carbonates or for calcium in the presence of gypsum because these salts are soluble in NH_4OAc.

References
Peech et al. (1947).

KCl-Triethanolamine Extraction, pH 8.2 (5B2)

Reagents
Buffer solution. Potassium chloride (KCl), 1.0 N, and triethanolamine (TEA), 0.2 N, pH 8.2.

Procedure
Proceed as in 5B1 except leach with 1 N KCl buffered at pH 8.2 with triethanolamine. Determine calcium by method 6N4, magnesium by 6O4.

References
North-Central Regional Research Committee (1955).

KCl-Triethanolamine Extraction, pH 8.2 (revised) (5B3)

Reagents
Buffer solution. Potassium chloride (KCl), 1.0 N, and triethanolamine (TEA), 0.2 N, pH 8.2.

Procedure
Weigh 10-g samples and transfer to 100-ml beakers. Add 40 ml buffer solution. Stir thoroughly at least three times over a period of not less than 1 hour. Filter the suspension and collect the leachate in a 100-ml volumetric flask.
Analyze the leachate for Ca and Mg by an appropriate procedure (6N4, 6O4).

Uncorrected (extractable) (5B3a)

If a soil does not contain soluble salts, the extractable bases are presumed to equal the exchangeable bases. They are, however, reported as extractable bases.

Corrected (exchangeable) (5B3b)

If a soil contains soluble salts, estimate their amounts from the saturation extract and correct as in 5B1b.

NH_4OAc, pH 7.0, Leaching Tube (5B4)

Analyze the NH_4OAc leachate from method 5A6 for Ca, Mg, Na, and K (methods 6N2, 6O2, 6P2, 6Q2).

Uncorrected (extractable) (5B4a)

If a soil does not contain soluble salts, the extractable bases are presumed to equal the exchangeable bases. They are, however, reported as extractable bases.

Corrected (exchangeable) (5B4b)

If a soil contains soluble salts, estimate their amounts from the saturation extract and correct as in 5B1b.

NH_4OAc, pH 7.0, Automatic Extractor (5B5)
Corrected (exchangeable) (5B5b)

If a soil contains soluble salts, estimate their amount from the saturation extract as follows. Multiply cation concentration in the saturation extract (meq/L) by the saturation percentage (divided by 1000) to convert to milliequivalents per 100 g. Subtract this quantity from the concentration of the extracted cation. This procedure is not valid for calcium and magnesium in the presence of carbonates that contain those elements, or for calcium in the presence of gypsum, because these compounds are soluble in NH_4OAc.

ION EXCHANGE ANALYSES (5)

Base Saturation (5C)
NaOAc, pH 8.2 (5C2)

Divide sum of NH_4OAc-extracted bases by the exchange capacity determined by method 5A2a.

Exchangeable Sodium Percentage (ESP (5D)
NaOAc, pH 8.2 (5D1)

Divide exchangeable sodium (meq/100 g) by the exchange capacity determined by method 5A2a.

Calcium Saturation (Exchangeable-Calcium Percentage) (5F)
NH$_4$OAc, pH 7.0 (5F1)

Divide the NH$_4$OAc-extracted calcium by the exchange capacity determined by procedure 5A1 or 5A6.

Organic Carbon (6A)

Determine carbon for each horizon that may contain organic matter. Report as carbon percentage by weight of <2-mm material.

To calculate total carbon per unit area, convert these weight percentages to volume percentages. Multiply each value by the bulk density Dbm, where m is usually 1/3 bar or 30 cm, and by the thickness (inches) of that horizon. If coarse fragments are present, further multiply by Cm (4A). Sum the organic-matter percentages and multiply by 0.254 to convert to kilograms of carbon per square meter.

Acid-Dichromate Digestion (6A1)
FeSO$_4$ Titration (6A1a)

Reagents
Potassium dichromate (K$_2$Cr$_2$O$_7$), 1.00 N (49.04 g per liter).

Ferrous sulfate, 1.0 N. Dissolve 280 g reagent-grade FeSO$_4$·7H$_2$O in water, add 80 ml concentrated H$_2$SO$_4$, cool, and dilute to 1 liter. Standardize this reagent each day by titrating against 10 ml N K$_2$Cr$_2$O$_7$ as directed.

Barium diphenylaminesulfonate indicator, 0.16 percent aqueous solution.

Orthophenanthroline-ferrous complex (optional), 0.025 M solution of one of the phenanthroline-ferrous complex indicators.

H$_2$SO$_4$, at least 96-percent.
Phosphoric acid (H$_3$PO$_4$), 86-percent.

Procedure
Transfer 1 g (0.5 g or less if high in organic matter) soil ground to pass an 80-mesh sieve to a 500-ml Erlenmeyer flask. Add 10 ml N K$_2$Cr$_2$O$_7$. Add 20 ml concentrated H$_2$SO$_4$ rapidly, directing the stream into the solution. Immediately swirl vigorously or place in wrist-action shaker for 1 minute. Let the flask stand on a sheet of asbestos for about 30 minutes. Add 200 ml water and 10 ml H$_3$PO$_4$. Add 0.5 ml barium diphenylaminesulfonate just before titrating. Titrate by adding FeSO$_4$ drop by drop to a light green end point. If more than 8 ml of the available 10 ml K$_2$Cr$_2$O$_7$ are reduced, repeat the determination, using less soil. If orthophenanthroline-ferrous complex is the indicator, it is not necessary to add H$_3$PO$_4$.

Calculations

Organic carbon (pct.) = ((A - B)/C) x N x (0.30/0.77)

where
A = Volume FeSO$_4$ blank (mL)
B = Volume FeSO$_4$ sample (mL)
C = Sample weight (g)
N = Normality of FeSO$_4$

0.77 = Recovery factor proposed by Walkley (1935)

Report on ovendry basis.

References
Peech et al. (1947) and Walkley (1935).

CO₂ Evolution, Gravimetric (6A1b)
Apparatus
See Fig. 4.

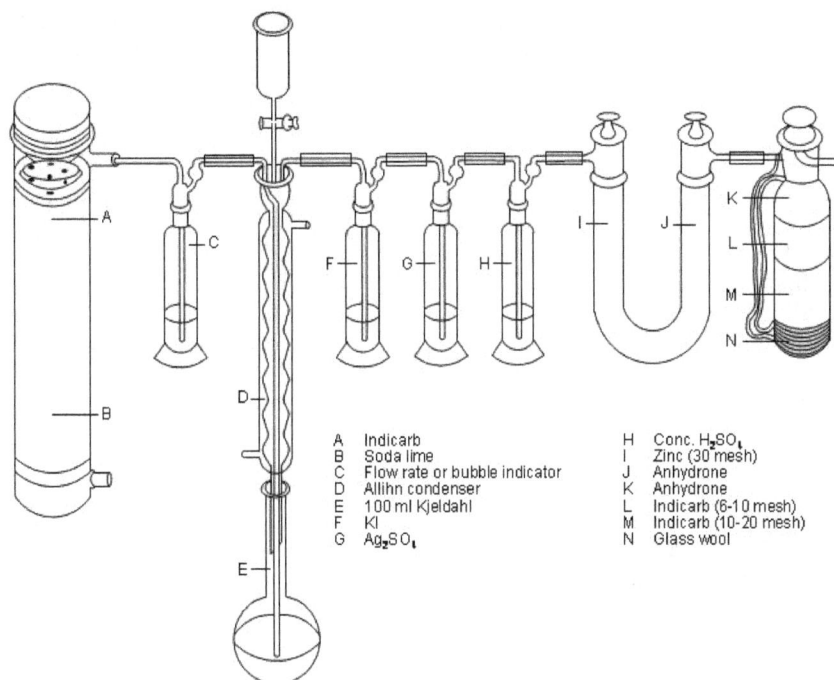

A	Indicarb	H	Conc. H_2SO_4
B	Soda lime	I	Zinc (30 mesh)
C	Flow rate or bubble indicator	J	Anhydrone
D	Allihn condenser	K	Anhydrone
E	100 ml Kjeldahl	L	Indicarb (6-10 mesh)
F	KI	M	Indicarb (10-20 mesh)
G	Ag_2SO_4	N	Glass wool

Figure 4 Apparatus for gravimetric organic carbon determination by wet combustion with potassium dichromate (6A1b).

Reagents
Digestion-acid mixture. Mix 600 ml concentrated H_2SO_4 and 400 ml 85-percent H_3PO_4.

Potassium dichromate ($K_2Cr_2O_7$), reagent grade.

Potassium iodide (KI). Dissolve 100 g KI in 100 ml water.

Silver sulfate (Ag_2SO_4), saturated aqueous solution.

Concentrated sulfuric acid (H_2SO_4).

Other reagents. Indicarb or Mikohbite, soda lime, 30-mesh zinc, and anhydrone (anhydrous magnesium perchlorate).

Procedure

Place a soil sample containing 20 to 40 mg carbon (usually 0.5 to 3 g ovendry soil) in digestion flask and add 1 to 2 g $K_2Cr_2O_7$. Wash the neck of the flask with 3 ml water and connect the flask to reflux condenser. Attach the weighed Nesbitt bulb to the system and open the valve at the top. Pour 25 ml digestion-acid mixture into funnel, let it enter the flask, and close the stopcock immediately to prevent loss of CO_2. Use digestion-acid mixture to lubricate the funnel stopcock. The tip of the air-delivery tube should extend about 0.5 cm below the surface of the acid during digestion. Adjust the "carrier stream" to a flow rate of one or two bubbles per second and maintain this rate during digestion. Heat with a gas flame of sufficient intensity to bring the sample to boiling in 3 to 4 min. Continue gentle boiling for a total heating period of 10 min (avoid excessive frothing). Heating is too rapid if white fumes of SO_3 are visible above the second bulb of the reflux condenser during boiling. At the end of the digestion period remove the flame and aerate for 10 min at the rate of six to eight bubbles per second. Then close the stopcock on the Nesbitt bulb, disconnect the bulb from the system, and weigh.

Calculations

Organic carbon (pct.) = $((A - B)/C) \times 27.3$

where
A = Final bulb weight (g)
B = Initial bulb weight (g)
C = Sample weight (g)

Report on ovendry basis.

References
Allison (1960).

Dry Combustion (6A2)
CO_2 Evolution, Gravimetric I (6A2a)

Apparatus
See Fig. 5.

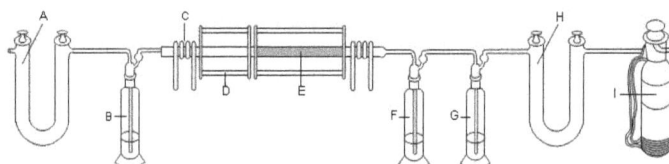

A	Schwartz absorption tube containing 8-20 mesh Caroxite	G	Gas washing bottle containing conc. H_2SO_4
B	Gas washing bottle containing conc. H_2SO_4	H	Schwartz absorption tube containing granular zinc in left arm and anhydrous magnesium perchlorate in right arm
C	Cooling coils		
D	Combustion furnace	I	Nesbit absorption bottle containing anhydrous magnesium perchlorate in top layers, 8-20 mesh Caroxite in middle layer, and glass wool in bottom layer
E	Platinum and asbestos catalyst		
F	Gas washing bottle containing saturated Ag_2SO_4		

Figure 5. Apparatus for organic carbon determination by dry combustion, carbon dioxide evolution I (6A2a)

628

Reagents
Powdered manganese oxide (MnO_2).

Procedure
Place 0.5 to 1.5 g soil that has been ground to 80 mesh in an Alundum boat containing 0.25 g powdered MnO_2. Insert the boat into the quartz tube of the multiple-unit combustion furnace shown. Before inserting the soil, preheat the long part of the quartz tube to 900°C or more (1000° or 1,100°C) and clear of CO_2 by passing CO_2 free oxygen through the combustion train until the weighing bottle shows a constant weight. While oxygen is passing slowly through the apparatus, heat to a temperature of 900° C or higher (15 to 30 min). Continue heating in a streaming oxygen atmosphere for 30 minutes more or until the Nesbitt absorption bulb has reached a constant weight.

Calculations
Report on ovendry basis as in 6A1b.

References
Robinson (1930)

CO_2 Evolution, Gravimetric II (6A2b)

Apparatus
See Fig. 6.

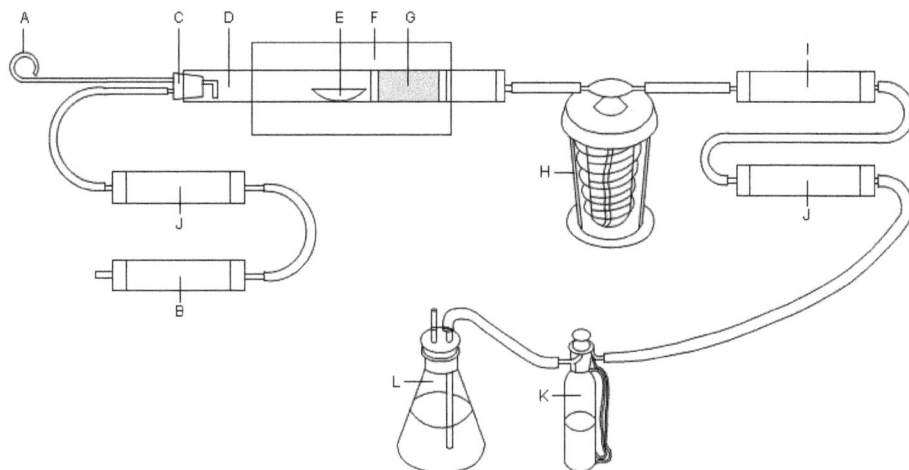

A   Push rod
B   Drying tube containing Indicarb
C   Robber stopper
D   Alundum combustion tube
E   Combustion boat
F   Combustion furnace
G   Cupric oxide wire
H   Milligan gas washing bottle containing conc. H_2SO_4
I   Tube containing ZnO_2
J   Tube containing Anhydrone
K   Nesbit absorption bulb containing Anhydrone and Indicarb
L   Bubble counter containing H_2SO_4

Figure 6. Apparatus used for organic carbon determination by dry combustion, carbon dioxide evolution II (6A2b).

Procedure

Heat tube to approximately 950°C. Sweep with oxygen until weight of Nesbitt bulb is constant. Remove rubber stopper in the oxygeninlet end of the tube and insert the boat containing 0.5 to 1.5 g soil. Reinsert the stopper and use the push rod to move the boat into the hot zone. Heat for 10 minutes, remove bulb, and record weight gain. Remove boat and repeat process with fresh sample, using the same Nesbitt bulb.

Calculations

Report on ovendry basis as in 6A1b.

References

Robinson (1930) and Post. (Post, G.J. A study of three methods for determination of organic carbon in Ohio soils of several great soil groups and the profile distribution of carbon-nitrogen ratios. M.Sc. thesis. The Ohio State University, 34 pp. 1956.)

CO_2 Evolution III (6A2c)

Apparatus

LECO 70-second carbon analyzer, model 750-100.

LECO induction furnace, model 521-000.

Reagents

Manganese dioxide.

Antimony.

1-g standard sample rings containing 0.870 percent carbon.

1-g standard sample rings containing 0.073 percent carbon.

Metal accelerator.

Iron chip accelerator.

Anhydrone.

Procedure

For noncalcareous soils, weigh approximately 1/2 g of <2-mm soil into crucibles in duplicate. Add to the soil in the crucibles one scoop of copper accelerator and one scoop of iron chip accelerator. Mix by stirring. Add an additional scoop of iron chips to the stirred mixture. Four standard soils containing 0.8, 2.1, 3.5, and 6.5 percent organic carbon are run with each group of soils. Follow LECO instruction manuals for instrument operation. Record readings from digital voltmeter as percent carbon.

References

Tabatabai and Bremner (1970).

Peroxide Digestion (6A3)
Gravimetric Weight Loss (6A3a)

Reagents
Hydrogen peroxide (H_2O_2), 6-percent.

Procedure
Digest soil for several hours in a covered beaker with 6-percent H_2O_2. Remove soluble material by washing three to five times with a Pasteur-Chamberlain clay filter, "F" fineness. Dry the beaker and soil, and weigh.

Calculations

Organic matter (pct.) = ((A + B)/C) x 100

where
A = Weight loss on heating (g)
B = Weight of dry matter in solution (g)
C = Sample weight (g)

Note that organic matter differs from organic carbon (see 6A1a).

References
North-Central Regional Research Committee on soils (1955).

Nitrogen (6B)
Kjeldahl Digestion I (6B1)

Reagents
Concentrated sulfuric acid (H_2SO_4).

Salt mixture:

Potassium sulfate (K_2SO_4), 1000 g.

Ferrous sulfate (anhydrous) $FeSO_4$, 55 g.

Copper sulfate (anhydrous) $CuSO_4$, 32 g.

Hengar granules (selenized).

Procedure
Weigh 5 g soil into 800-ml Kjeldahl flask, add 20 ml distilled water and let stand overnight. Add 10 g salt mixture, 2 or 3 Henger granules, and 30 ml H_2SO_4. Digest on Kjeldahl digestion heaters, rotating flasks frequently. Continue digestion 1 hr after mixture is clear.

References
Association of Official Agricultural Chemists (1945).

Ammonia Distillation (6B1a)

Reagents
Mixed indicator. Methyl red, 0.125-percent, and methylene blue, 0.0825-percent, in 95-percent ethanol.

Methyl red (optional), 0.25-percent.

Brom cresol green, 0.1-percent aqueous solution.

Boric acid (H_3BO_3), 4-percent.

HCl, standardized, 0.1 N or 0.05 N.

Procedure

Cool digestion flask (6B1) and dilute contents with about 400 ml water. Add 2 to 3 g mossy zinc, 5 drops antifoam mixture, and 70 ml concentrated NaOH solution. Connect flask to condenser and distill ammonia into 25 or 50 ml H_3BO_3 solution. Titrate with standard HCl to purple end point, using 10 drops mixed indicator and 2 drops brom cresol green or 3 drops brom cresol green and 1 drop methyl red.

Calculations

N (pct) = ((A - B)/C) x N x 1.4

where
A = Volume HCl sample (mL)
B = Volume HCl blank (mL)
C = Sample weight (g)
N = Normality of HCl

Report on ovendry basis.

Ammonia Distillation, Automatic Titrator (6B1b)

Reagents

Boric acid (H_3BO_3), 4 percent.

HCl, standardized, 0.1 N or 0.05 N.

Concentrated sodium hydroxide (NaOH) solution, 50 percent.

Antifoam mixture: Equal parts n-octyl alcohol and mineral oil.

Mossy zinc.

Procedure

Cool digestion flask (6B1) and dilute contents with about 400 ml water. Add 2 to 3 g mossy zinc, 5 drops antifoam mixture, and 70 ml concentrated NaOH solution. Connect flask to condenser and distill ammonia into 250-ml titrator beaker containing 50 ml H_3BO_3 solution. Titrate with standard HCl to end point pH setting of 4.60 on automatic titrator.

Calculations

N (pct.) = ((A - B)/C) x N x 1.4

where
A = Volume HCl sample (mL)

B = Volume HCl blank (mL)
C = Sample weight (g)
N = Normality of acid

Report on ovendry basis.

Semimicro Kjeldahl (6B2)

Apparatus
Aminco-Koegel semimicro rotary digestion rack and steam-distillation apparatus.

Reagents
Concentrated sulfuric acid (H_2SO_4).

H_2SO_4, 0.01 N, standardized.
Sodium hydroxide (NaOH), 50-percent.

Boric acid (H_3BO_3), 2-percent.

Mixed indicator. Mix 0.1 g methyl red and 0.1 g brom cresol green and dissolve in 250 ml ethanol.

Salt mixture. Mix 790 g potassium sulfate (K_2SO_4), 100 g ferrous sulfate ($FeSO_4$), 100 g copper sulfate ($CuSO_4$), and 10 g selenium metal.

Procedure
Using an analytical balance, weigh on a cigarette paper either 0.500 or 1.000 g ovendry soil that has been ground to about 0.2 mm. Roll soil in cigarette paper and drop into a 100-ml digestion-distillation flask. Add 2 g salt mixture 1 ml water, and 5 ml concentrated H_2SO_4. Swirl vigorously and digest, rotating the flask frequently until fumes are emitted. Continue digestion for at least 1 hour after mixture becomes white. Cool to room temperature and add 15 ml water. Shake until the contents of the flask are thoroughly mixed.

Ammonia Distillation (6B2a)

Procedure
Measure 10 ml 2-percent H_3BOa with an automatic pipet into a 125-ml flask and add 0.5 ml mixed indicator. Place this flask under delivery tube. Connect digestion-distillation flask containing soil digested according to method 6B2 to the distillation unit by the ground-glass connection. Start steam passing through the system and slowly add 15 ml 50-percent NaOH. Distill for 12 minutes, add 0.5 ml more mixed indicator, and titrate the absorbed ammonia with 0.01 N H_2SO_4.

Calculations

N (pct) = (A/B) x N x 1.4

where
A = Volume H_2SO_4 (mL)
B = Sample weight (g)
N = Normality of H_2SO_4

Report on ovendry basis.

Iron (6C)
Dithionite-Extraction (6C1)

Reagents

Sodium dithionite powder ($Na_2S_2O_4$).

Hydrochloric acid (HCl), 10-percent.

Apparatus

8-oz Pyrex nursing bottles or 250-ml flat-bottomed centrifuge bottles.

Procedure

      Place 4 g soil, ground to 80 mesh, in a nursing or centrifuge bottle. Add 4 g $Na_2S_2O_4$ and 75 ml water. Stopper and shake overnight or for 16 hours. Then adjust the pH to 3.5 to 4.0, if necessary, with 10-percent HCl. Let stand for no less than 1 hour, stirring four or five times. Transfer the suspension to a graduated cylinder, dilute to 200 ml with water, and mix. Centrifuge or filter a part of the suspension and transfer 50 ml of the clear solution to a 250-ml beaker.

References

Kilmer (1960).

Dichromate Titration (6C1a)

Reagents

Hydrogen peroxide (H_2O_2), 35-percent.

Ammonium hydroxide (NH_4OH), 1:1.

Hydrochloric acid (HCl), 1:1.
Stannous chloride ($SnCl_2$). Dissolve 1 g $SnCl_2$ in 2 to 4 ml concentrated HCl and dilute to 50 ml with water; prepare fresh each time.

Mercuric chloride ($HgCl_2$), saturated aqueous solution.

Phosphoric acid (H_3PO_4), 85-percent.

Potassium dichromate ($K_2Cr_2O_7$), 0.100 N, standard.

Barium diphenylaminesulfonate, 0.16-percent aqueous solution.

Procedure

      Add 10 to 15 ml H_2O_2 (6C1) to the solution to destroy any excess reducing agent. Cover the beaker with a watchglass and warm on a hot plate until the reaction starts. Set the solution aside until the reaction subsides and then boil for 10 to 15 minutes. Add a slight excess of 1:1 NH_4OH and boil the solution for 15 to 20 minutes to insure complete removal of H_2O_2. Dissolve $Fe(OH)_3$ by adding 1:1 HCl through the lip of the beaker. Usually 10 to 15 ml are enough. Heat the solution to 90° C and reduce by adding $SnCl_2$ by drops, stirring until the yellow color just disappears. Add three to four drops more. Cool the solution to room temperature and add 15 ml $HgCl_2$ solution all at once. A light silky precipitate of Hg_2Cl_2 forms if the proper amount of $SnCl_2$ has been added. Dilute the solution to 100 to 150 ml and add 5 ml H_3PO_4. Add 10 drops of barium diphenylaminesulfonate and titrate the solution with standard $K_2Cr_2O_7$ to a violet-blue end point.

Calculations

Fe (pct.) = (A/B) x N x (C/D) x 5.58

where
A = Volume $K_2Cr_2O_7$ (mL)
B = Sample weight (g)
N = Normality of $K_2Cr_2O_7$
C = Volume extract (mL)
D = Volume aliquot (mL)

Fe_2O_3 (pct.) = Fe (pct.) x 1.43

Report on ovendry basis.

EDTA Titration (6C1b)

Reagents
Hydrogen peroxide (H_2O_2), 35-percent.
Ammonium persulfate ((NH_4)$_2S_2O_8$).

Salicylic acid, 1-percent in 95-percent ethanol.

EDTA, standardized as g iron per ml EDTA. Prepare EDTA as described in 6N1a.

Iron standard, 0.500 g iron per liter.

Procedure
Pipet a 5- to 25-ml aliquot from the centrifuge tube of procedure 6C1 into a 250-ml beaker. Add 50 ml water to the beaker. Then add by drops 5 ml H_2O_2 and digest over low heat until bubbling from the decomposing H_2O_2 ceases. Remove immediately to avoid precipitation of Fe_2O_3 in samples high in iron. Caution: Add H_2O_2 slowly to prevent liberation of elemental sulfur from any remaining $Na_2S_2O_4$. Keep the volume in the beaker to about 50 ml during the digestion by adding water if necessary. Remove from heat and cool. Adjust the pH between 2.0 and 3.0 with a pH meter, using either concentrated acetic acid or a 20-percent NaOAc solution. Add a few milligrams (NH_4)$_2S_2O_8$ to the solution to insure total oxidation of iron. Then add 1 ml indicator (1-percent salicylic acid) and titrate with 0.02 N EDTA to a pale yellow or colorless end point.

Calculations

Fe (pct.) = (A/B) x V x (C/D) x 100

where
A = Volume EDTA (mL)
B = Sample weight (g)
V = Titer of EDTA in g Fe/ml EDTA
C = Volume extract (mL)
D = Volume aliquot (mL)

Fe_2O_2 (pct.) = Fe (pct.) x 1.43

Report on ovendry basis.

References
Cheng, Bray, and Kurtz (1953).

Dithionite-Citrate Extraction (6C2)

Reagents
Sodium dithionite ($Na_2S_2O_4$)

Sodium citrate

Superfloc flocculating agent, 0.2 percent in water

Procedure
Weigh 1 to 4 g of soil (approximately 0.2 g maximum extractable iron) into an 8-oz nursing bottle. Add 2 g sodium dithionite and 20 to 25 g sodium citrate. Make up to 4 oz with water, and shake overnight in a reciprocating shaker. Add 2 ml Superfloc solution to the suspension, make up to 8 oz with water, shake vigorously for 15 s and allow to settle for at least 1 hr. This extract is used for analysis of iron (6C2b), aluminum (6G7a), and manganese (6D2a).

References
Holmgren (1967).

Orthophenanthroline Colorimetry (6C2a)

Apparatus
Seligson pipet, 0.1-ml.

Reagents
Orthophenanthroline, 0.25-percent.

Iron solution, 1000 mg per liter, standard.

Sodium dithionite powder ($Na_2S_2O_4$).

Sodium citrate crystals.

Superfloc flocculating agent, 0.2-percent, in water.

Procedure
Add 5 drops Superfloc solution to the dithionite-treated soil suspension (6C2) and make to 8 oz. Shake vigorously for about 15 seconds and allow to settle. Pipet a 0.1-ml aliquot with a Seligson pipet into a 25-ml volumetric flask. Add water to about 10 ml. Using a small scoop, tap a pinch of dithionite and a pinch of sodium citrate into the flask. Add 0.5 ml 0.25-percent orthophenanthroline and make to volume. Shake and read in a colorimeter at 508 Mμ after 1 hour. To prepare the standards, pipet 5-, 10-, 25-, 50-, 100-, 150-, and 200-ml aliquots of standard iron solution (1000 mg/L) into 8-oz shaking bottles and make to 8 oz after adding reagents as in 6C2. Transfer 0.1-ml aliquots to 25-ml volumetrics and develop color by the above procedure.
Plot the standard curve as milligrams iron per 8-oz bottle against percentage transmission.

Calculations

Fe (pct.) = (A/B) x 10^{-1}

A = Fe in bottle (mg)
B = Sample weight (g)

Fe_2O_3 (pct.) = Fe (pct.) x 1.43

Report on ovendry basis.

References
Holmgren (1967).

Dithionite-Citrate-Bicarbonate Extraction (6C3)

Reagents
Sodium bicarbonate ($NaHCO_3$), 1 *M*.
Sodium citrate, 0.3 *M*.

Sodium chloride (NaCl), saturated solution.

Acetone.

Procedure
Weigh 4 g soil (1 g clay) into a 100-ml centrifuge tube. Add 40 ml 0.3 *M* Na-citrate and 5 ml 1 *M* $NaHCO_3$. Bring temperature to 80°C in water bath. Add 1 g solid $Na_2S_2O_4$, stir constantly for 1 minute and occasionally for 15 minutes. Add 10 ml NaCl solution and 10 ml acetone to promote flocculation. Mix, warm in water bath, and centrifuge 5 minutes at 1,600 to 2,200 rpm. Decant clear supernatant into 500-ml volumetric flask and make to volume.

References
Mehra and Jackson (1960).

Potassium Thiocyanate Colorimetry (6C3a)

Apparatus
Colorimeter.

Reagents
Hydrochloric acid (HCl), 6 *N*.

Potassium thiocyanate (KSCN), 20-percent.

Hydrogen peroxide (H_2O_2), 30-percent.

Procedure
Transfer suitable aliquot (0.5 to 3 ppm iron in final solution) to 50 ml-volumetric flask. Add water to 35 ml, 1 drop H_2O_2, 5 ml HCl, and 5 ml KSCN solution. Make to volume and read at 490 mμ in colorimeter.

Calculations

Fe (pct.) = (A/B) x (C/D) x 0.005

where

A = Fe from curve (mg)
B = Sample weight (g)
C = Volume extract (mL)
D = Volume aliquot (mL)

Fe_2O_3 (pct.) = Fe (pct.) x 1.43

Report on ovendry basis.

References
Jackson (1958).

Pyrophosphate-Dithionite Extraction (6C4)

Reagents
Pyrophosphate solution. Dissolve 89.2 g $Na_4P_2O_7 \cdot 10H_2O$ in 800 to 900 ml water. Adjust the pH of this solution to 8.0 by adding hydrogensaturated exchange resin. Decant or filter, wash the resin, and dilute the solution to 1000 ml to make 0.2 M $Na_4P_2O_7$.

Sodium dithionite ($Na_2S_2O_4$).

Digestion acid. 10 Parts concentrated HNO_3, 4 parts concentrated H_2SO_4, and 4 parts concentrated $HClO_4$.

Procedure
   Mix 80 ml pyrophosphate solution and 2.0 g solid sodium dithionite in a beaker and add this solution to 4 g soil in a centrifuge tube (pH 8.0 pyrophosphate solution and dithionite combined in this ratio result in a solution having a pH of about 7.3). Continue the extraction for 30 minutes at 50°C, shaking the suspension in the tube every 5 minutes. Centrifuge the suspension 5 to 10 minutes at 2000 rpm. Dilute the extract to 100 ml (solution A).
   Immediately transfer 5 ml solution A to a beaker. Add 1 to 2 ml digestion acid and heat on a hot plate until almost dry to destroy the organic and hydrolyze pyrophosphate to orthophosphate. Allow to cool, dissolve the salts in HCl, and dilute to 100 ml (solution B). Determine iron and aluminum in solution B by appropriate methods, such as 6C3a and 6G1a.

References
Franzmeier, Hajek, and Simonson (1965).

Sodium Pyrophosphate Extraction (6C5)

Reagents
Sodium pyrophosphate ($Na_4P_2O_7$), 0.1 M.

Superfloc solution, 0.4 percent.

Procedure
   Place 2 g soil into 250-ml centrifuge bottle (polypropylene). Add 200 ml 0.1 M $Na_4P_2O_7$, cap, and shake overnight. Add 5 to 10 drops 0.4-percent Superfloc, shake, and centrifuge at 2000 rpm (Int. No. II centrifuge). Transfer the supernatant liquid to a plastic or glass container and reserve for Fe and Al analyses.
   The supernatant liquid must be clear in reflected light. If fine colloids are visible, repeat the procedures. If fine colloids are still present, spin the suspension in a super centrifuge until the supernatant liquid is clear. Foam rubber can be used in the centrifuge cups as a cushion for the 250-ml flat-bottom plastic bottles.

References
Bascomb (1968).

Atomic absorption (6C5a)

Apparatus
Atomic absorption spectrophotometer.

Reagents
Standard Fe solution, 0 to 50 ppm.

Procedure
Establish standard curve and match readings from extract to curve readings. Dilute where necessary.

Calculations

Fe (pct.) = A x (B/C) x (1/10000) x dilution

where
A = Fe (ppm)
B = Volume extract (mL)
C = Sample weight (g)

Report on ovendry basis.

Ammonium Oxalate Extraction (6C6)

Reagents
Ammonium oxalate $(NH_4)_2C_2O_4$ 0.2 M, pH 3.0.

Adjust the pH of 0.2 M $(NH_4)_2C_2O_4$ to 3.0 with 0.2 M oxalic acid $(H_2C_2O_4)$.

Superfloc solution, 0.4 percent.

Procedure
Place 2 g soil into 250-ml centrifuge bottle (polypropylene). Add 200 ml 0.2 M $(NH_4)_2C_2O_4$, cap, and shake immediately in the dark for 4 hours. Add 5 to 10 drops 0.4-percent Superfloc, shake, and centrifuge at 2000 rpm (Int. No. II centrifuge). Transfer the supernatant liquid to a plastic or glass container. Store in the dark and reserve for Fe and Al analyses.

The supernatant liquid must be clear in reflected light. If fine colloids are visible repeat the procedure. If fine colloids are still present, spin the suspension in a supercentrifuge until the supernatant liquid is clear.

References
McKeague and Day (1965).

Atomic Absorption (6C6a)

Proceed as in 6C5a except use extract from 6C6.

Manganese (6D)
Dithionite Extraction (6D1)

Extract 4.00 g soil as described in 6C1.

Permanganate Colorimetry (6D1a)

Reagents
Concentrated nitric acid (HNO_3).

Hydrogen peroxide (H_2O_2), 30-percent.

Phosphoric acid (H_3PO_4), 85-percent.

Sodium para periodate ($Na_3H_2IO_6$) or sodium meta periodate ($NaIO_4$).

Purified water diluent. Add 100 ml 80-percent H_3PO_4 and 1 g $Na_3H_2IO_6$ to 1 liter water (Mn-free); heat to boiling and digest for 1 hour; stopper with foil-covered stopper. About 85 ml of this diluent is needed for each sample.

$KMnO_4$, standard.

Procedure
Take a 10- to 25-ml aliquot from the dithionite extract and place in a 150-ml beaker. Add 5 ml 30-percent H_2O_2, digest on hot plate, and evaporate until dry. Cool beaker and contents and add 3 ml concentrated HNO_3 and 2 ml 30-percent H_2O_2. Digest on hot plate for 30 minutes, using a close fitting cover glass, then raise cover glass, and evaporate until dry. Take up residue with 10 ml 85-percent H_3PO_4, heat to boiling, remove, and cool to about 50°C. Dilute with 10 ml water and add 0.2 g $Na_3H_2IO_6$. Cover beaker and heat to boiling. Cool to 50°C and add 62 ml purified water diluent and 0.1 g $Na_3H_2IO_6$. Digest at 90° C for 40 minutes or until no further color develops. Transfer the hot solution to a 100-ml volumetric flask, using purified water diluent to rinse the beaker. Cool, make up to volume with the diluent, stopper, and shake. Determine percentage transmittance with a photoelectric colorimeter at 540 mμ. Interpolate concentration from a standard absorbance concentration.

Calculations

Mn (pct.) = (A/B) x (C/D) x 54.9

where
A = MnO_4^- (meq/L)
B = Sample weight (g)
C = Volume extract (mL)
D = Volume aliquot (mL)

MnO (pct.) = Mn (pct.) x 1.291

Report on ovendry basis.

References
Jackson (1958).

Calcium Carbonate (6E)
HCl Treatment (6E1)
Gas Volumetric (semiquantitative) (6E1a)

This procedure uses a simple leveling device to measure the volume of gas released when the soil is treated with HCl. It has an inherent error caused by the solubility of CO_2 in the HCl solution. Data on file at the laboratory at Lincoln, Nebr., indicate that the results are about 10 percent (8 to 12 percent) low for $CaCO_3$ equivalents ranging from 40 percent to 6 percent (1-g basis). For 1-percent equivalents the values are about 20 percent low and for less than 1 percent, the values have doubtful significance.

References
Association of Official Agricultural Chemists (1945).

Manometric (6E1b)

Apparatus
Wide-mouth prescription bottles, 3-oz, with bakelite cap; drill 7/16-in hole in cap for serum bottle stopper. Rubber gasket, 1 3/8 in OD x 15/16 in ID.

Serum bottle stopper.

Mercury manometer and a 26-gauge hypodermic needle attached to manometer tube.

Gelatin capsule. 1/4 oz.

Reagents
Hydrochloric acid (HCl), 6 N.

Glycerin.

Procedure
Place 2 g of soil in prescription bottle and add 5 ml water. Moisten lip of bottle with a drop of glycerin to ensure a good seal with rubber gasket. Fill gelatin capsule with HCl, put cap in place and invert to seal cap on capsule. Place, capsule in bottle and immediately cap the bottle. In a minute or two the HCl will dissolve the capsule. After 1 hr insert hypodermic needle through serum stopper and read manometer. Compare reading with those for standards prepared by treating aliquots of standard Na_2CO_3 solution in same manner as samples.

Vary sample weight according to $CaCO_3$ content as follows: For <25 percent $CaCO_3$, use 2 g soil; for 25 to 50 percent $CaCO_3$, 1 g soil; and for >50 percent $CaCO_3$, 0.5 g soil. For trace amounts, add a few drops 6 N HCl to soil and observe under binocular microscope. Evolution of gas bubbles indicates the presence of $CaCO_3$.

References
Williams (1948).

Gravimetric (weight loss) (6E1c)

Apparatus
See Fig. 7.

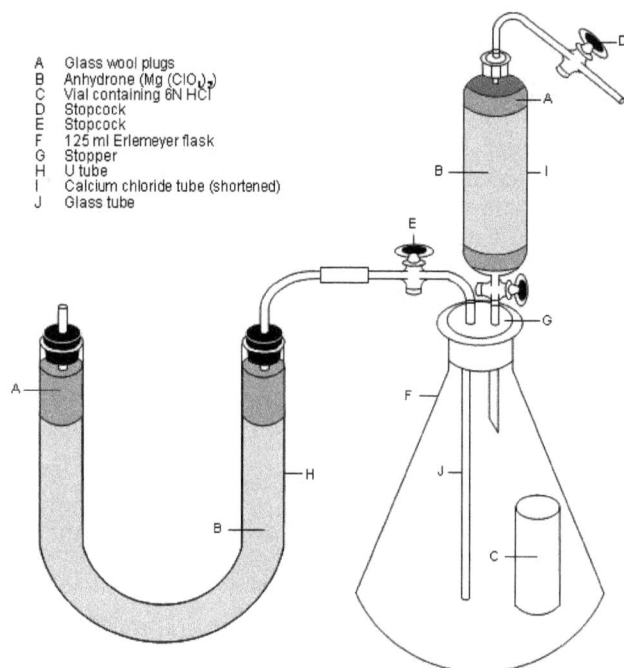

A   Glass wool plugs
B   Anhydrone (Mg (ClO$_4$)$_2$)
C   Vial containing 6N HCl
D   Stopcock
E   Stopcock
F   125 ml Erlemeyer flask
G   Stopper
H   U tube
I   Calcium chloride tube (shortened)
J   Glass tube

Figure 7. Apparatus used for calcium carbonate determination by weight loss (6E1c).

Reagents

Hydrochloric acid (HCl), 6 N.

Anhydrone (Mg(ClO$_4$)$_2$).

Procedure

Assemble apparatus as shown in Fig. 7. Place a sample of soil containing less than 1 g CaCO$_3$ equivalent in a 125-ml Erlenmeyer flask. Wash down the sides of the flask with 10 ml water. Place 7 ml 6N HCl into vial C and then place the vial upright in the flask without spilling any acid. Moisten stopper G with glycerin, sprinkle with a small amount of 180-mesh abrasive to overcome slipperiness, and place the apparatus with stopcocks D and E, tubes I and J, attached firmly in position in the flask. Close stopcocks D and E. Place the apparatus beside the balance. Wait 30 minutes before weighing to allow temperature of the apparatus to equilibrate with temperature of air in the balance. Do all weighing with stopcock D open since a change in temperature of the flask with the stopcock closed results in a change in weight of the apparatus. Use tongs to place apparatus on the weighing pan, open stopcock D, weigh to 0.1 mg, and then immediately close stopcock D. Check the weight 10 minutes later to be certain that the weight of the flask has stabilized. Open stopcock D and then shake apparatus to upset the vial, allowing the acid to react with the carbonates. After 10 minutes, attach the rubber tube from the airdrying vessel to stopcock E. Open stopcock E and apply suction at stopcock D to give 5 to 10 bubbles per second at the base of tube J to sweep out CO$_2$. Shake the flask after 10 minutes and again after 20 minutes. After 30-minutes sweeping time, stop the suction and close stopcocks D and E. Return apparatus to the balance. Delay weighing for 1 hour to allow the heat generated by absorption of water by the anhydrone to be dissipated. Weigh apparatus with stopcock D open. Check the weight after 10 minutes.

Calculations

Carbonate as CaCO$_3$ (pct.) = ((A - B)/C) x 228

where
A = Initial weight of flask (g)
B = Final weight of flask (g)
C = Sample weight (g)

Report on ovendry basis.

References
Erickson et al. (1947).

Gravimetric (weight gain) (6E1d)

Proceed as in 6E1c except add additional trap containing CO_2-absorbing Ascarite to end of gas train. Weigh Ascarite bulb before and after CO_2 evolution. Weight gain equals the CO_2 evolved from the sample. Better results are obtained if the Ascarite is size-graded so that CO_2 passes through the coarser material first. Indicarb can be used in place of Ascarite.

Titrimetric (6E1e)

Reagents
Hydrochloric acid (HCl), 0.5 N, standardized.

Sodium hydroxide (NaOH), 0.25 N, standardized.

Phenolphthalein, 1 percent in 60-percent ethanol.

Procedure
Place 5 to 25 g soil in a 150-ml beaker, add exactly 50 ml HCl, cover with a watchglass, and boil gently for 5 minutes. Cool, filter, and wash all the acid from the soil with water. Determine the amount of unused acid by adding 2 drops of phenolphthalein and back-titrating with NaOH.

Calculations

Carbonate as $CaCO_3$ (%) = ((50 x A - B x C)/D) x 5

where
A = Normality of HCl
B = Volume NaOH (mL)
C = Normality of NaOH
D = Sample weight (g)

Report on ovendry basis.

References
Richards (1954)

Warburg Method (6E1f)

Apparatus
Warburg manometer, mercury filled.

Warburg reaction vessel, 15-ml capacity, with vented stopper for sidearm.

Constant temperature bath.

Reagents
HCl, 1:1. Na_2CO_3 solution for standard curve. Dissolve 1.06 g Na_2CO_3 in water and make to 1 liter. Solution contains 1.06 mg Na_2CO_3 per ml or the equivalent of 1 mg $CaCO_3$ per ml. Obtain standard curve by measuring CO_2 pressure from 1, 2, 4, 6, 8, and 10 ml Na_2CO_3 solution.

Procedure
Weigh 100 mg sample of finely ground soil and transfer to Warburg reaction vessel. Be careful not to get any sample in center well. Pipet 1 ml water into vessel and mix well with sample. Pipet 1 ml 1:1 HCl into sidearm, insert greased stopper, and leave in vent-open position. Attach reaction vessel to manometer and fasten with rubber bands or spring supports. Place reaction vessel in constant temperature bath at 25°C for 5 to 10 minutes to bring flask contents to temperature of water bath. Remove flask from bath, close stopper vent, and fasten with rubber bands or springs. Tilt flask to allow acid to flow from sidearm into reaction vessel, mix contents, and return vessel to water bath. Let stand for at least 30 minutes before reading manometer. Use the standard curve to convert the difference between the two manometer arm readings (mm), to milligrams $CaCO_3$. Gently tap the manometer holder occasionally to prevent low readings caused by mercury adhering to manometer walls.

Sensitive Qualitative Method (6E2)
Visual, Gas Bubbles 6E2a)

Add few drops 6 N H_2SO_4 to soil and observe under binocular microscope. Evolution of gas bubbles indicates the presence of $CaCO_3$.

H_2SO_4 Treatment (6E3)
Gravimetric (weight gain) (6E3a)

Apparatus
See Fig. 4, procedure 6A1b.

Reagents
Sulfuric acid (H_2SO_4). Dissolve 57 ml concentrated H_2SO_4 and 92 g of $FeSO_4 \cdot 7H_2O$ in 600 ml water, cool, and dilute to 1000 ml. This solution is approximately 2 N in acidity and contains 5-percent $FeSO_4$ as antioxidant. Keep well stoppered.

Procedure
Place a 1- to 5-g sample of ovendry soil in the digestion flask E and connect condenser D. Weigh the Nesbitt bulb, attach to the system, and adjust the carrier stream to a flow rate of 1 or 2 bubbles per second. Pour 25 ml of the acid solution into the funnel and let it enter the digestion flask E. Close the stopcock immediately. Apply heat slowly and bring contents of flask to a boil in about 4 minutes. Continue gentle boiling for exactly 3 minutes more for a total heating period of 7 minutes. Remove the flame, adjust the carrier stream to 6 or 8 bubbles per second, and continue aerating for 10 minutes. Disconnect the Nesbitt bulb and weigh.

Calculations

Carbonate as $CaCO_3$ (%) = ((A - B)/C) x 227

where
A = Final weight of bulb (g)

B = Initial weight of bulb (g)
C = Sample weight (g)

Report on ovendry basis.

References
Allison (1960).

Gypsum (6F)
Water Extract (6F1)
Indirect Estimate (6F1b)

Add a weighed quantity of soil to enough water to dissolve all the gypsum by overnight shaking. The concentration of sulfate in this dilute soil:water extract should be <10 meq/L. Gypsum can be estimated by method 6F2. If crystals are observed or estimated gypsum content is >5 percent, the <2-mm sample should be ground to approximately 80 mesh. Determine total sulfate in this extract by any appropriate procedure. Also determine Ca and SO_4 in a saturation extract by any appropriate procedure.

Calculations

Gypsum = $(SO_4)_{DE}$ - $(SO_4)_{non\text{-}gypsum\ SE}$

but SO_4 $_{non\text{-}gypsum\ SE}$ = $(SO_4)_{SE}$ - $(SO_4)_{gypsum\ SE}$

∴ gypsum = $(SO_4)_{DE}$ + $(SO_4)_{gypsum\ SE}$ - $(SO_4)_{SE}$

$(SO_4)_{DE}$ = SO_4 in dilute water extract

$(SO_4)_{SE}$ = SO_4 in saturation extract

$(SO_4)_{gypsum\ SE}$ = 30 meq/L if SO_4 and Ca are ≥ 30 meq/L

= $(SO_4)_{SE}$ if $(Ca)_{SE}$ > $(SO_4)_{SE}$

= $(Ca)_{SE}$ if $(Ca)_{SE}$ < $(SO_4)_{SE}$

All quantities are reported in meq/100 g.

Gypsum (%)=Gypsum (meq/100g) x 0.0861 (g/meq)

References
Lagerwerff, Akin, and Moses (1965).

Ion Chromatograph (6F1c)

Apparatus
DIONEX Model 2110i ion chromatograph

Recorder (1 volt input).

Voltage stabilizer.

Reagents

0.1 M Na_2CO_3.

0.003 M $NaHCO_3$.

0.0024 M $Na2CO_3$.

1 N H_2SO_4.

Mixed standard solutions:
Fluoride 0.0125 to 5.0 meq/L.

Chloride 0.01 to 4.0 meq/L.

Nitrate 0.025 to 10.0 meq/L.

Sulfate 0.05 to 20.0 meq/L.

All solutions are filtered through a polycarbonate membrane having 0.4μm pore size. Soil extracts are filtered with a disposable filter unit (Millix^TM) having 0.22μm pore size.

Procedure

The soil extract is obtained as described in 6F1a. Fill a plastic syringe (3 to 10 cc) with a solution having a concentration within the range of the sulfate standard. Baseline is established using a full-scale μmhos setting of 3 before each determination. This setting is adjusted as needed, keeping in the range used for making the determinations on the mixed standard. Peak height readings are made on the mixed standard using eight concentrations. A curve fitting linear regression equation [y(meq/L) = a_1 (PKH) + a_0] is established for the sulfate standards. Sulfate concentration in the soil extracts is determined by this equation.

Calculations

See 6F1b.

Weight Loss (6F2)

Apparatus

Vacuum desiccator.

Aluminum dish.

Balance, 0.001-g sensitivity.

Reagents

Phosphorus pentoxide (P_2O_5).

Procedure

Place about 10 g of soil in a tared (Wt A) aluminum dish. Saturate sample with water and let stand overnight to air-dry. Place in a vacuum desiccator with P_2O_5 desiccant. Evacuate desiccator and allow to stand 48 hr. Remove dish from desiccator and weigh (Wt B), then place in oven at 105° C for 24 hr. Allow dish to cool in desiccator and weigh (Wt C).

Calculations

Gypsum (%) = ((WtB - WtC) x 100)/((WtB - WtA) x 0.1942)

The theoretical crystal water content of gypsum is 20.91 percent. However, Nelson et al. have determined that, in practice, this content averages 19.42 percent.

References
Nelson, Klameth, and Nettleton (1978).

Gypsum Requirement (6F5)

The amount of gypsum needed to replace all of the sodium on the exchange complex with calcium is the gypsum requirement.

Reagents
Saturated gypsum solution. Place about 25 g gypsum ($CaSO_4$-$2H_2O$) in 5 L water in a large flask, stopper, and shake by hand periodically for 1 hr or more. Let settle and decant through a filter into storage bottle. Determine calcium concentration by titration of an aliquot with standard EDTA solution using Eriochrome black T as indicator.

EDTA solution. Dissolve 1.25 g di-sodium ethylenediamine tetraacetate in water and dilute to 1 L. Standardize against solutions containing known concentrations of Ca and Mg.

Buffer solution. Dissolve 6.75 g ammonium chloride in about 400 ml water. Add 570 ml concentrated ammonium hydroxide and dilute to 1 L with distilled water.

Eriochrome black T indicator. Dissolve 1 g Eriochrome black T in 100 ml triethanolamine.

Procedure
Weigh 5 g soil into flask, add 100 ml saturated gypsum solution, stopper, and shake for 5 min in mechanical shaker. Filter through folded filter paper, discarding the first few milliliters of filtrate, which may be cloudy. Pipet a 5-ml aliquot of filtrate into a 125-ml Erlenmeyer flask and dilute to 25 or 30 ml with distilled water. Add 10 drops of buffer solution, 2 drops Eriochrome black T indicator, and titrate with standard EDTA solution to blue end point.

Calculations

Gypsum requirement (meq/100 g) = (A - B) x 2

where
A = Ca concentration of gypsum solution (meq/L)
B = Ca + Mg concentration of filtrate (meq/L)

References
Richards (1954).

Aluminum (6G)
KCl Extraction I (30 min) (6G1)

Reagents
Potassium chloride (KCl), 1 N.

Procedure

Weigh 10-g soil samples into 125-ml Erlenmeyer flasks. Add 50 ml 1 N KCl to each flask, mix several times, and let stand for 30 minutes. Filter through 5.5-cm Whatman No. 42 filter paper in Buchner funnel, using suction as necessary. Leach each sample as rapidly as possible with about five 9-ml portions of KCl, using the first to help transfer the remaining soil in the Erlenmeyer flasks to the Buchner funnels. Transfer the extract to 100-ml volumetric flasks and dilute to volume with the extracting solution. Or use Allihn leaching tubes and bring to standard weight in tared suction flasks.

References
Lin and Coleman (1960) and Pratt and Bair (1961).

Aluminon Colorimetry I, Hot Color Development (6G1a)

Reagents
Thioglycolic acid (HSCH$_2$COOH). Dilute 1 ml purified acid to 100 ml with water.

Aluminon reagent. Dissolve in separate containers 0.75 g

Aluminon (ammonium aurine tricarboxylate), 15 g gum acacia, and 200 g NH$_4$OAc crystals. To the NH$_4$OAc solution add 189 ml concentrated HCl, then the gum acacia, and finally the Aluminon. Mix, filter, and dilute to 1,500 ml with water. To get the gum acacia in suspension, add slowly to boiling water while stirring constantly.

Aluminum standard. Add 2.24 g AlCl$_3$·6H$_2$O per liter of water. This solution should be nearly 250 ppm aluminum. Check concentration of an aliquot containing 10 ppm aluminum by analyzing for chloride.

Procedure

If samples contain less than 5 meq per 100 g aluminum, pipet a 1-ml aliquot of each extract into numbered and calibrated test tubes. If more aluminum is present, dilute before the aliquot is taken. Dilute to approximately 20 ml with distilled water. Add 2 ml dilute thioglycolic acid to each tube, stopper, and shake all the tubes. Pipet 10 ml Aluminon into each tube and dilute to exactly 50 ml. The pH should be between 3.7 and 4.0. Stopper and shake all tubes. Place tubes in a rack and heat in a boiling-water bath for 4 minutes. Cool in running water to room temperature. Transfer samples to reading tubes and measure light transmittance at 535 mμ and compare with a standard curve.

Calculations

Al (meg/100g) = (A/B) x (C/D) x (9/5)

where
A = Al from curve (mg/L)
B = Sample weight (g)
C = Volume extract (mL)
D = Volume aliquot (mL)

Report on ovendry basis.

References
Chenery (1948) and Yoe and Hill (1927).

Aluminon Colorimetry II, HCl Predigestion (6G1b)

Procedure
Proceed as in 6G1a but first add 3 ml N HCl to the aliquot and heat for 30 minutes at 80 to 90° C.

References
Hsu (1963).

Aluminon Colorimetry III, Overnight Color Development (6G1c)

Proceed as in 6G1a except eliminate boiling-water bath, adjust pH to 4.0, and allow color to develop overnight before reading.

Fluoride Titration (6G1d)

Reagents
Potassium fluoride (KF), 1 N. Titrate with NaOH to a phenolphthalein end point. This eliminates the need for a blank correction in the Al titration.

Sodium hydroxide (NaOH), 0.1 N, standardized.

Sulfuric acid ($H_2SO_,$), 0.1 N, standardized.

Phenolphthalein, 0.1 percent.

Procedure
Add 6 to 8 drops phenophthalein to the leachate in the suction flask (6G1). Titrate with standard NaOH to a pink color that persists for 30 seconds or more. Correct for a KCl blank to obtain KCl extractable acidity. Then add 10 ml KF, and titrate with standard H_2SO_4 until the pink color disappears. Set aside while other samples are titrated and then complete to a lasting colorless end point. If there is a considerable amount of Al, add a few more drops of phenolphthalein.

Calculations

Acidity (meq/100 g) = (A/B) x N x 100

where
A = Volume NaOH (mL)
B = Sample weight (g)
N = Normality of NaOH

Al(meq/100) = (A/B) x N x 100

where
A = Volume H_2SO_4 (mL)
B = Sample weight (g)
N = Normality of H_2SO_4

References
Yuan (1959).

Atomic absorption (6G1e)

Apparatus
Perkin-Elmer Model 290 atomic absorption spectrophotometer with nitrous oxide burner attachment.

Reagents
Standard Al solution, 0 to 5 meq per liter.

Procedure
Dilute sample to within range of standard curve. Compare absorbance with standard curve.

Calculations

Al (meq/100 g) = (A/B) x dilution x (C/10)

where
A = Al from curve (meq/L)
B = Sample weight (g)
C = Volume extract (mL)

KCl Extraction II, Overnight (6G2)

Weigh 10 g soil into 125-ml Erlenmeyer flask. Add 50 ml 1 N KCl and let stand overnight. In the morning transfer to filter funnels and leach with an additional 50 ml KCl.

Aluminon Colorimetry I (6G2a)

Follow procedure for aluminum analysis described in 6G1a.

NH₄OAc Extraction (6G3)

Prepare soil as described in 5A1.

Aluminon Colorimetry III (6G3a)

Follow procedure of 6G1c.

NaOAc Extraction (6G4)

Prepare soil as described in 5A2.

Aluminon Colorimetry III (6G4a)

Follow procedure of 6G1c.

Sodium Pyrophosphate Extraction (6G5)

Prepare extract as described in 6C5.

Atomic Absorption (6G5a)

Apparatus
Atomic absorption spectrophotometer.

Reagents
Standard Al solution, 0 to 50 ppm or 0 to 160 ppm.

Procedure
Establish standard curve and match readings from extract to curve readings. Dilute where necessary.

Calculations

Al (pct.) = A x (B/C) x (1/10000) x dilution

where
A = Al (ppm)
B = Volume extract (mL)
C = Sample weight (g)

Report on ovendry basis.

Ammonium Oxalate Extraction (6G6)

Prepare extract as described in 6C6.

Atomic absorption (6G6a)

Analyze extract as described in 6G5a.

NH$_4$Cl, Automatic Extractor (6G8)

Prepare extract as described in 5A9.

Atomic Absorption (6G8a)

Apparatus
Atomic absorption spectrophotometer.

Reagents
Standard Al solutions, 0 to 6 meq/L

Procedure
Compare absorbance of samples from 5A9 with that of standards at 309.3 nm, diluting if necessary.

Calculations

Al (meq/100 g) = (A/B) x dilution x (C/10)

where
A = Al (meq/L)

B = Sample weight (g)
C = Volume extract (mL)

Report on ovendry basis.

Extractable Acidity (6H)
BaCl₂-Triethanolamine I (6H1)

Extractable acidity data are reported on some data sheets as exchange acidity and on others as extractable H^+.

Reagents

Buffer solution. Barium chloride, 0.5 N, and triethanolamine, 0.2 N. Add about 90 ml, 1 N HCl per liter to adjust pH to 8.2. Protect the buffer solution from CO_2 of the air by attaching a drying tube containing soda lime (sodium calcium hydrate) to the air opening at the top of the solution bottle.

Replacement solution. Barium chloride, 0.5 N. Add 5 ml buffer solution per liter. Protect the. replacement solution from CO_2 of the air by attaching a drying tube similar to that used for the buffer solution.

Procedure

Weigh 5 g soil into a 125-ml Erlenmeyer flask. Add 15 ml buffer solution and let stand for 30 minutes, swirling occasionally to mix. Use 35 ml buffer solution to transfer all the soil solution to a No. 4 Gooch crucible containing a moist Whatman No. 540 filter paper and filter into a 500-ml suction flask. The rate of filtration should be such that at least 30 minutes is needed to complete the filtering and leaching. Then leach the soil with 100 ml of the replacement solution, adding small amounts at a time. It may be necessary to use a larger amount of buffer solution to leach allophanic soils high in organic matter with extractable acidity of more than 35 meq per 100 g.

Back-Titration with HCl (6H1a)

Reagents

Hydrochloric acid (HCl), 0.2 N, standardized.

Brom cresol green, 0.1-percent aqueous solution.

Mixed indicator. Dissolve 1.250 g methyl red indicator and 0.825 g methylene blue in 1 liter 90-percent ethanol.

Procedure

Run a blank by adding 100 ml replacement solution, 2 drops brom cresol green, and 10 drops mixed indicator to 50 ml buffer solution. Titrate with HCl to a chosen end point in the range from green to purple. Add 2 drops brom cresol green and 10 drops mixed indicator to the leachate and titrate to the same end point chosen for the blank. Calculate exchange acidity *(EA)* as follows.

Calculations

EA (meq/100 g) = ((A - B)/C) x N x 100

where
A = Volume HCl blank (mL)

B = Volume HCl sample (mL)
C = Sample weight (g)
N = Normality of HCl

Report on ovendry basis.

References
Peech et al. (1947).

BaCl$_2$-Triethanolamine II (6H2)

Apparatus
Sulfur absorption tubes.

Whatman No. 41 filter paper or glass-fiber filter paper cut to fit sulfur absorption tubes.

Reagents
Buffer solution. BaCl$_2$, 0.5 N, and triethanolamine, 0.2 N as in 6H1.

Mixed indicator. Dissolve 1.250 g methyl red and0.825 g methylene blue in 1 liter 90 percent ethanol.

Celite.

Procedure
        Stopper bottom of sulfur absorption tubes with medicine-dropper bulbs and fit to a 300-ml suction flask with a rubber stopper. Place Whatman No. 41 filter paper in bottom of absorption tube, cover with 1/4 inch of acid-washed sand, and add exactly 25 ml buffer solution. Weigh 10 g soil and mix with teaspoonful of Celite. Add to the absorption tube by means of a funnel. After 30 minutes remove the medicine-dropper bulbs, wash bulbs out with a little water, and add washings to absorption tubes. Leach with 25 ml more buffer solution and then leach with 100 ml replacement solution in small increments. If necessary, use suction to facilitate leaching.

Back-Titration with HCl (6H2a)

Reagents
Same as in 6H1a.

Procedure
Titrate with standard HC1, using either 2 drops brom cresol green and 10 drops methyl red or 10 drops mixed indicator. Use same end point as that chosen for a blank run by leaching sand and Celite with 50 ml buffer solution and 100 ml replacement solution.

Calculations
Use same calculation as in 6H1a.

KCl-Triethanolamine (6H3)

Back-titration with NaOH (6H3a)

Procedure
        Leach 10 g soil with 50 ml KCl-triethanolamine solution and follow by washing with 50 ml unbuffered 1 N KCL. Add a known volume of standard acid to leachate and washings and

back-titrate with standard alkali (NaOH). Titrate an equal volume of acid to the same end point for a blank.

Calculations

EA (meq/100 g) = ((A - B)/C) x *N* x 100

where
EA = Extractable acidity
A = Volume NaOH sample (mL)
B = Volume NaOH blank (mL)
C = Sample weight (g)
N = Normality of NaOH

References
North-Central Regional Research Committee (1955).

BaCI₂-Triethanolamine III (6H4)

Apparatus
60-ml plastic syringe barrels.

Reagents
Buffer solution. Barium chloride, 0.5 *N*, and triethanolamine, 0.2 *N*. Add 1 *N* HCl (about 90 ml/L) to adjust pH to 8.2. Protect the buffer solution from CO_2 of the air by attaching a drying tube containing soda lime (sodium calcium hydrate) to the air opening at the top of the solution bottle.

Replacement solution. Barium chloride, 0.5 *N*. Add 5 ml of above buffer solution per liter. Protect the replacement solution from CO_2 of the air by attaching a drying tube similar to that used for the buffer solution.

"Celite" filter pulp.

Procedure
        Prepare syringe barrels as leaching tubes by forcing a 1-g ball of filter pulp into bottom of barrel with syringe plunger. Measure 1.5 g celite and 5 g soil sample into tube. Attach pinch clamp to delivery tube of syringe barrel and add approximately 25 ml buffer solution to sample. Let stand 30 min, stirring occasionally. Remove pinch clamp and filter with low suction into titrator beaker using a total of 50 ml buffer solution followed by 100 ml replacement solution.

References
Peech (1947).

Back-Titration with HCl, Automatic Titrator (6H4a)

Reagents
Hydrochloric acid (HCl), 0.33 *N*, standardized.

Procedure
        Titrate the leachate contained in the 250-ml beaker to an end-point pH setting of 4.60 with automatic titrator. Carry reagent blank through procedure.

Calculations

EA (meq/100 g) = ((A - B)/C) x N x 100

where
EA = Extractable acidity
A = Volume HCl blank (mL)
B = Volume HCl sample (mL)
C = Sample weight (g)
N = Normality of HCl

Report on ovendry basis.

Carbonate (6I)
Saturation Extract (6I1)
Acid titration (6I1a)

Reagents
Sulfuric acid (H_2SO_4), 0.05 N, standardized.

Phenolphthalein.

Procedure
Pipet an appropriate aliquot of saturation extract into a 250-ml Erlenmeyer flask or a porcelain crucible. The electrical conductivity (EC X 10_3) of the saturation extract (8A1a) can be used to determine the aliquot to be used for carbonate, bicarbonate, and chloride determinations. Where EC X 10_3 is 1.0 or less, use a 10-ml aliquot; if 1.0 to 10.0, use a 5-ml aliquot; if more than 10.0, use a 2-ml aliquot.
Make volume to 50 ml (10 ml for porcelain crucible) with water. To the 50 ml in the Erlenmeyer flask, add a drop or two of phenolphthalein. If a pink color is produced, titrate with 0.05 N H_2SO_4, adding a drop every 2 or 3 seconds until the pink color disappears. Use this solution to determine bicarbonate (6J1a).

Calculations

Carbonate (meq/L) = (A/B) x N x 2000

where
A = Volume H_2SO_4 (mL)
B = Volume aliquot (mL)
N = Normality of H_2SO_4

References
Association of Official Agricultural Chemists (1945) and Richards (1954).

Bicarbonate (6J)
Saturation Extract (6J1)
Acid Titration (6J1a)

Reagents
Sulfuric acid (H_2SO_4), 0.05 N, standardized.

Methyl orange, 0.01-percent aqueous solution.

Procedure

Use solution remaining from carbonate titration (6I1a). To the colorless solution from this titration or to the original solution if no color is produced with phenolphthalein, add 4 drops methyl orange and continue titration to the methyl orange end point without refilling the buret. Retain this solution for the chloride determination (6K1a). Make a blank correction for the methyl orange titration.

Calculations

Bicarbonate (meq/L) = $((A - (2 \times B))/C) \times N \times 1000$

where
A = Total volume H_2SO_4 (mL)
B = Volume H_2SO_4 from 6I1a (mL)
C = Volume aliquot (mL)
N = Normality of H_2SO_4

References
Association of Official Agricultural Chemists (1945) and Richards (1954).

Chloride (6K)
Saturation Extract (6K1)
Mohr Titration (6K1a)

Reagents
Potassium chromate (K_2CrO_4) indicator. Dissolve 5 g K_2CrO_4 in water and add a saturated solution of $AgNO_3$ until a permanent slight red precipitate is produced, filter, and dilute to 100 ml.

Silver nitrate ($AgNO_3$), 0.05 N, standardized.

Sodium bicarbonate ($NaHCO_3$), saturated solution (optional).

Nitric acid (HNO_3), 0.1 N (optional).

Procedure

To the solution from the bicarbonate titration (6J1a) add 6 drops K_2CrO_4 indicator and titrate with $AgNO_3$ to a reddish-orange end point. Make a correction with a blank of 50 ml water containing the indicators of both titrations. The laboratory at Riverside, Calif., modifies this procedure by adding saturated $NaHCO_3$ solution to a pink end point and neutralizing to a colorless end point with HNO_3 before adding the indicator.

Calculations

Chloride (meq/L) = $((A - B)/C) \times N \times 1000$

where
A = Volume $AgNO_3$ sample (mL)
B = Volume $AgNO_3$ blank (mL)
C = Volume aliquot (mL)
N = Normality of $AgNO_3$

References
Association of Official Agricultural Chemists (1945).

Potentiometric Titration (6K1b)

Apparatus
Silver Billet combination electrode, No. 39187.

Zeromatic pH meter (expanded scale).

Reagents
Standard silver nitrate ($AgNO_3$), 0.025 N.

Buffer solutions. Either potassium acid phthalate or trisodium citrate and citric acid. To prepare phthalate buffer, weigh 37.5 g potassium acid phthalate and bring to a volume of 500 ml with water; 4 ml of this buffer added to a 46-ml solution brings the pH to about 4. To prepare trisodium citrate buffer, weigh 43.8 g trisodium citrate and 43.3 g citric acid into 500-ml volumetric flask and bring to volume with water. Add a small amount of toluene to the solution for storage; 10 ml of this buffer added to a 40-ml solution brings pH to about 4.

Procedure
Standardize the pH meter by adjusting the needle to a convenient setting (about 0.8) on the expanded scale when the electrode is immersed in buffer solution (4 or 10 ml made to 50 ml) without chloride. To titrate the sample, pipet an aliquot containing as much as 2.0 meq chloride into a beaker and add 4 ml buffer. Make to 50 ml. Immerse the electrode and buret tip into the beaker and titrate with $AgNO_3$ to the end point previously established for the buffer without chloride.

Calculations

Chloride (meq/L) = (A/B) x N x 1000

where
A = Volume $AgNO_3$ (mL)
B = Volume aliquot (mL)
N = Normality of $AgNO_3$

Sulfate (6L)
Saturation Extract (6L1)
Gravimetric, BaSO$_4$ Precipitation (6L1a)

Reagents
Concentrated hydrochloric acid (HCl).

Barium chloride ($BaCl_2$), 10-percent.

Methyl orange, 0.01-percent.

Procedure
Pipet an aliquot of saturation extract into a 250-ml beaker. Dilute to approximately 100 ml with water. Add 2 drops methyl orange and 0.5 ml concentrated HCl to the beaker. Heat to boiling and add $BaCl_2$ solution by drops, stirring constantly until precipitation is complete. Let stand on hot plate for several hours. Remove from heat and let samples stand overnight. Filter through Gooch crucibles, which have been ignited and weighed. Dry in 105°C oven and ignite in muffle furnace at 1,200°F (650°C) for 30 minutes. Cool in desiccator and weigh.

Calculations

$$SO_4 \text{ (meq/L)} = (A/B) \times 8.568$$

where
A = $BaSO_4$ (mg)
B = Volume aliquot (mL)

References
Richards (1954).

EDTA Titration (6L1b)

Apparatus
Repipet, automatic dilutor, pipet range 0.1 to 1.0 ml.

Titration assembly including a 10-ml buret with magnetic stirrer.

Reagents
Thymol blue indicator, 0.04-percent.

Nitric acid (HNO_3), 0.4 N.

Calcium nitrate ($Ca(NO_3)_2$), 0.05 N. Dissolve 5.90 g $Ca(NO_3)_2 \cdot 4H_2O$ in 1 liter CO_2-free water. EC is 5.15 ± 0.15 mmhos per cm at 25°C.

Acetone, reagent grade, boiling range 55.5 to 57.5°C

Ethanol, 95-percent, reagent grade.

Hydrochloric acid (HCl), 0.01 N.

EDTA solution, 0.02 N. Standardize against $CaCl_2$.

Procedure
    Pipet an aliquot containing 0.01 to 0.05 meq SO_4 from soil-water extracts and transfer to a 100-ml beaker. Bring volume to 7.5 ± 0.5 ml with water. Add 2 drops 0.04-percent thymol blue and 0.4 N HNO_3 drop by drop until color changes from yellow to distinct red. Add 2 ml 0.05 N $Ca(NO_3)_2$, 20 ml acetone, and stir. Allow 30 minutes for the precipitate to flocculate. Place a 9.0-cm Whatman No. 42 filter paper in a 5.0-cm fluted funnel and fit snugly with water. Wash the sides of filter paper with 5 ml 95-percent ethanol from a wash bottle. Transfer the precipitate and supernatant to the filter paper with alcohol. Rinse the beaker twice and wash filter paper three times, using 3 to 5 ml ethanol per rinse. Allow the alcohol in the filter paper to evaporate. Wash the funnel stem thoroughly with water. Place the beaker that contained the $CaSO_4$ precipitate under the funnel and wash the filter paper with 3 to 5 ml portions of 0.01 N HCl until approximately 25 ml is leached. Proceed as in 6N1a, except eliminate carbamate and add an extra drop 4 N NaOH to neutralize the 25 ml 0.01 N HCl.
    The amount of sulfate is determined from the Ca^{++} content in the $CaSO_4$ precipitate.

Calculations

$$SO_4 \text{ (meq/L)} = (A/B) \times N \times 1000$$

where

A = Volume EDTA (mL)
B = Volume aliquot (mL)
N = Normality of EDTA

References
Bower and Wilcox (1965), Lagerwerff, Akin, and Moses (1965), and Nelson (1970).

NH₄OAc Extraction (6L2)

Obtain extract by procedure 5B1.

Gravimetric, BaSO₄ Precipitation (6L2a)

Proceed as in 6L1a. A greater quantity of acid will be needed to lower the pH. Otherwise the procedures are the same.

Nitrate (6M)
Saturation Extract (6M1)

Phenoldisulfonic Acid Colorimetry (6M1a)

Reagents
Phenoldisulfonic acid. Dissolve 25 g phenol in 150 ml concentrated H_2SO_4, add 75 ml fuming H_2SO_4 (13 to 15 percent SO_3), and heat at 100°C for 2 hours.

Standard potassium nitrate (KNO_3), 0.010 N.

Silver sulfate (Ag_2SO_4), 0.020 N.

Ammonium hydroxide solution (NH_4OH), 1:1, approximately 7 N.
Calcium oxide (CaO).

Procedure
First determine the chloride concentration in an aliquot of saturation extract as directed in 6K1a. Pipet another aliquot containing 0.004 to 0.04 meq of nitrate into a 25-ml volumetric flask. Add an amount of Ag_2SO_4 equivalent to the amount of chloride present, dilute to volume, and mix. Separate the precipitate by centrifuging the suspension in a 50-ml centrifuge tube. Transfer the solution to another centrifuge tube, flocculate any suspended organic matter by adding about 0.1 g CaO, and clear by centrifuging again. Pipet a 10-ml aliquot into an 8-cm evaporating dish. Evaporate the aliquot to dryness, cool, and dissolve the residue in 2 ml phenoldisulfonic acid. After 10 minutes, add 10 ml water and transfer to a 100-ml volumetric flask. Make alkaline by adding NH_4OH, dilute to volume, and mix. Measure light transmission through a 460 mμ filter of solution in an optical cell against that of water in a similar cell.
Prepare a calibration curve by pipetting 0-, 0.2, 0.4-, 0.8-, 1.2-, and 1.6-ml aliquots of standard KNO_3 into evaporating dishes and treating as for sample except for additions of Ag_2SO_4 and CaO and the clarifying procedure.

Calculations

NO_3 (meq/L) = (A/B) x 1000

where
A = NO_3 from curve (meq/L)
B = Volume aliquot (mL)

References
Richards (1954).

Diphenylamine (qualitative) (6M1b)

Use this procedure to test for nitrates if there is a significant excess of cations over anions in the extract. A quantitative measurement can be made if there is a positive indication of NO_3 (6M1a).

Reagents
Diphenylamine in H_2SO_4. Dissolve 0.05 g diphenylamine in 25 ml concentrated sulfuric acid. Store in polyethylene dropper bottle.

Procedure
Place a drop of extract in a spot plate and add 3 or 4 drops diphenylamine reagent. Nitrate is present if a blue color develops.

References
Treadwell and Hall (1943).

Calcium (6N)
Saturation Extract (6N1)
EDTA titration (6N1a)

Reagents
Sodium hydroxide (NaOH), approximately 4 N.

Calcium chloride ($CaCl_2$), 0.02 N. Dissolve calcite crystals in HCl and make to volume.

Murexide. Thoroughly mix 0.5 g ammonium purpurate with 100 g powdered potassium sulfate (K_2SO_4).

EDTA solution, 0.02 N. Standardize against $CaCl_2$

Sodium diethyldithiocarbamate, 1-percent.

Procedure
Pipet an aliquot containing 0.02 to 0.20 meq of calcium into a beaker. Add 5 drops carbamate, 1 drop NaOH for each 5-ml aliquot, and a suitable amount (15 to 20 mg for a 10-ml aliquot) of murexide, mixing after each addition. A magnetic stirrer is helpful. Titrate with EDTA to a lavender end point. A blank containing NaOH, murexide, carbamate, and a drop or two of EDTA helps to distinguish the end point. If the sample is overtitrated with EDTA, it can be back-titrated with standard $CaCl_2$. Retain solution for magnesium determination (6O1a).

Calculations

Ca (meq/L) = (A/B) x N x 1000

where
A = Volume EDTA (mL)
B = Volume aliquot (mL)
N = Normality of EDTA

References
Cheng and Bray (1951).

NH₄OAc Extraction (6N2)

Prepare NH₄OAc extract as described in 5A1. EDTA-alcohol extraction.

EDTA-Alcohol Separation (6N2a)

Reagents
Standard calcium chloride ($CaCl_2$), 5 mg per ml. Dissolve calcite crystals in HCl and make to volume.

Ethanol, 95-percent.

Standard EDTA. Dissolve 1.25 g disodium ethylenediaminetetraacetate in water and dilute to a volume of 1 liter. Standardize against solutions containing known amounts of calcium and magnesium. Run the standards through the separation procedure before titrating.

Sodium hydroxide (NaOH), 10-percent aqueous solution.

Calcon. Dissolve 1 g Calcon (Eriochrome Blue Black R) in 100 ml triethanolamine.

Procedure
Pipet 25-ml aliquots from the pH 7, NH₄OAc extracts obtained in the total exchange-capacity procedure (5A1) into 100-ml beakers and evaporate to dryness at moderate heat. Cool and add 3 ml N HNO₃ to dissolve the residue. Transfer the solution quantitatively to 50-ml conical centrifuge tubes with ethanol, using a wash bottle with a fine delivery tip. Add 1 ml 6 N H₂SO₄. While mixing the contents of the tube by swirling, add approximately 34 ml 95-percent ethanol. Cover the tubes and let stand overnight. The next morning remove the covers and centrifuge the tubes at about 2000 rpm (Int. No. II centrifuge) for 15 minutes. Decant the alcohol solution into 250-ml Erlenmeyer flasks and retain for the magnesium determination. Use the CaSO₄ precipitate for calcium determination.

Break up the CaSO₄ precipitate with a small steam of water from a wash bottle and transfer the precipitate and solution to 250-ml Erlenmeyer flasks. Dilute the solution to a total volume of about 100 ml. Place the sample on a magnetic stirrer, add 5 ml 10-percent NaOH, 2 drops Calcon indicator solution, and titrate with the standard EDTA solution to the blue color of a blank carried through the procedure. The pH of the solution should be about 12.5. The color change is from red to clear blue. Titrate until the color in the sample and in the blank are the same.

Calculations

Ca (meq/100 g) = (A/B) x N x (C/D) x 100

where
A = Volume EDTA (mL)
B = Sample weight (g)
N = Normality of EDTA
C = Volume extract (mL)
D = Volume aliquot (mL)

References

Barrows and Simpson (1962).

Oxalate Precipitation I, KMnO₄ Titration (6N2b)

Reagents

Oxalic acid ($C_2H_2O_4$), 5-percent aqueous solution.

Brom cresol green, 0.04-percent aqueous solution.

Ammonium hydroxide (NH_4OH), 1 N.

Sulfuric acid (H_2SO_4), 1 N.

Standard potassium permanganate ($KMnO_4$), 0.05 N

Wash solution, saturated calcium oxalate ($CaC_2 2O_4$)

Asbestos. Digest asbestos in 1 N HNO_3 solution containing just enough $KMnO_4$ to give a deep purple color. Add more permanganate if the color disappears; digest for 24 hours or until the permanganate color is permanent. Destroy the excess permanganate with oxalic acid and wash thoroughly on a Buchner funnel.

Procedure

Transfer an aliquot of the filtrate (5A1) to a 400-ml Pyrex beaker and evaporate to complete dryness. Cool, cover the beaker with a watchglass, and slowly add through the lip 10 ml concentrated HNO_3 and 2 ml concentrated HCl. Warm until the reaction has subsided and no more brown fumes are given off. Rinse the watchglass into the beaker. Evaporate to dryness at low heat to prevent spattering and continue to heat for about 10 minutes to dehydrate the salts. Then place the beaker in an electric muffle furnace at about 150°C, heat to 390° ± 10°, and hold at this temperature for about 20 minutes. Remove the beaker from the muffle furnace and cool. Treat the residue with 3 ml 6 N HCl, evaporate to dryness at low heat, and continue heating for about 30 minutes longer to dehydrate silica. Cool and dissolve the residue in 0.1 N HNO_3, using a rubber policeman to loosen the residue.

Add 5 ml oxalic acid, heat the contents of the beaker almost to boiling, and add 1 ml brom cresol green. Adjust the pH of the hot solution to approximately 4.6 by slowly adding 1 N NH_4OH, stirring constantly. Let digest at about 80° C for 1 hour or until the supernatant liquid is clear. Collect the CaC_2O_4 precipitate on a compact asbestos pad in a Gooch crucible or in a Whatman No. 42 filter paper in filter funnel. Rinse the beaker four times with water or water saturated with CaC_2O_4 and pour the washings into the crucible. Wash the precipitate five more times with water saturated with CaC_2O_4.

Remove the Gooch crucible from its holder, rinse the outside, and replace crucible in the beaker. If filter paper is used, pierce the paper and wash most of the precipitate into the beaker with 3.6 N H_2SO_4. Wash off excess H_2SO_4 with water and place filter paper on watchglass. Add 100 ml water and 7 ml concentrated H_2SO_4. Heat to 90°C and stir until CaC_2O_4 is dissolved. Titrate with standard $KMnO_4$ solution to a pink color. Add filter paper to solution and titrate to a permanent pink color.

Calculations

Ca (meq/100 g) = (A/B) x N x (C/D) x 100

where
A = $KMnO_4$ (mL)
B = Sample weight (g)

N = Normality of KMnO$_4$
C = Volume extract (mL)
D = Volume aliquot (mL)

Report on ovendry basis.

References
Peech et al. (1947).

Oxalate Precipitation II, KMnO$_4$ Titration (Fe, Al and Mn removed) (6N2c)

Proceed as in 6N2b but after muffle treatment and before oxalate precipitation, remove iron, aluminum, and manganese by the following procedure.

Reagents
Hydrochloric acid (HCl), 6 N.

Ammonium hydroxide (NH$_4$OH), 2 N.

Bromine water, saturated.

Ammonium chloride (NH$_4$Cl), 6 N.

Concentrated nitric acid (HNO$_3$).

Procedure
Dissolve salts and oxides by adding 5 ml 6 N HCl and heating on a hot plate until all salts and oxides are in solution. Add 75 to 100 ml water and heat the solution until it is nearly boiling. Immerse the pH electrodes into the hot solution and precipitate the hydroxides of iron, aluminum, and titanium by slowly adding 2 N NH$_4$OH until the meter indicates a pH of 6.2 to 6.4. Add 2 more drops of NH$_4$OH to neutralize the acidifying effect of the 15 ml saturated bromine water, which is slowly added next to precipitate manganese hydroxide. Since bromine water lowers the pH of the solution, readjust it to 6.2 to 6.4 with 2 N NH$_4$OH. Heat the solution with precipitate until it just begins to boil (1 or 2 min on a Bunsen burner) and remove from the heat.

Place on a hot plate at a temperature of 80° to 90°C for 1 hour. Filter when the breaker has cooled enough to handle easily. Use an 11-cm Whatman No. 42 filter paper or its equivalent. Collect the filtrate in a beaker of the same size as those used for precipitating calcium. Wash and police the beaker containing the precipitate with hot 2-percent NH$_4$Cl. Wash the precipitate on the filter with the same solution. Five washings are usually enough. To the filtrate add 10 ml concentrated HNO$_3$ and evaporate to dryness; add 5.0 ml 6 N HCl, take to dryness, and use high heat to dehydrate silica. Proceed with the calcium precipitation (6N2b).

References
Washington (1930) and Fieldes et al. (1951).

Oxalate Precipitation, Cerate Titration (6N2d)

Proceed as in 6N2b except substitute the following for the permanganate titration.

Reagents

Ammonium hexanitrate cerate ($(NH_4)_2Ce(NO_3)$) in molar perchloric acid ($HClO_4$), 0.1 N. Add 85 ml 70- to 72-percent perchloric acid to 500 ml water. Dissolve 56 g ammonium hexanitrate cerate in the acid solution and dilute to 1 liter.

Ammonium hexanitrate cerate in molar perchloric acid, 0.05 N. Follow the directions for the preparation of the 0.1 N solution but use only 28 g cerate.

Perchloric acid ($HClO_4$), 2 N. Add 170 ml 70 to 72-percent perchloric acid to 500 ml water and dilute to 1 liter.

Nitro-ferroin indicator solution. Dilute a solution of nitro-orthophenanthroline ferrous sulfate with water to a convenient working strength. Two to four drops of the solution should give a sharp color change at the end point.

Standardize the cerate solutions against accurately weighed quantities of primary standardgrade sodium oxalate. Convenient weights of sodium oxalate are 0.10 to 0.11 g for the 0.05 N solution and 0.10 to 0.18 g for the 0.1 N cerate solution. Dissolve the sodium oxalate in 100 to 150 ml 2 N perchloric acid and titrate as directed in the following procedure.

Procedure

Dissolve the filtered and washed (use water) calcium oxalate in 100 to 200 ml 2 N perchloric acid. If a paper filter has been used, macerate it before titration. Add 2 to 4 drops of nitro-ferroin indicator solution and titrate with 0.05 N or 0.1 N cerate solution, depending upon the amount of oxalate present. The solution changes from red to colorless at the end point.

Calculations

Ca (meq/100g) = (A/B) x N x (C/D) x 100

where
A = Volume Cerate (mL)
B = Sample weight (g)
N = Normality of Cerate
C = Volume extract (mL)
D = Volume aliquot (mL)

Report on ovendry basis.

NH₄Cl-Ethanol Extraction (calcareous soils) (6N3)

Apparatus
See Fig. 8.

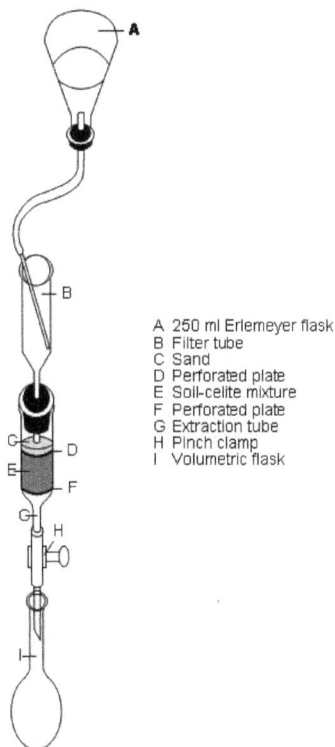

Figure 8. Apparatus for ammonium chloride-ethanol extraction for calcium (6N3)

A 250 ml Erlemeyer flask
B Filter tube
C Sand
D Perforated plate
E Soil-celite mixture
F Perforated plate
G Extraction tube
H Pinch clamp
I Volumetric flask

Reagents

Ammonium chloride (NH₄Cl), 1 *N*, in 60-percent ethanol. To make 9 liters of extraction solution, dissolve 482 g NH₄Cl in 2,835 ml water and add 5,985 ml 95-percent ethanol. Adjust pH to 8.5 with 140 to 145 ml NH₄OH.

Celite.

Procedure

Fill extraction tube with water, set tube upright in holder, and let most of the water drain out. Close screw clamp and place filter paper on plate with a stirring rod. Let remainder of the water drain out of tube. The filter paper provides enough tension to keep the bottom part of the tube filled with water. Place tube on the rack and add about 1 1/2 teaspoons washed sand. Place an extra perforated plate (inverted) on top of the sand and cover the plate with more sand. Place heaping teaspoon of Celite on the sand and pour about 20 ml extraction solution into the tube. Pour remainder of 400 ml extraction solution into a 500-ml Erlenmeyer flask. Add soil sample slowly and then stir with a rod to mix soil and Celite. Allow sample to settle and then place filter paper on top of the soil column. Put upper tube in place, stopper, and let stand overnight.

In the morning, place a 500-ml volumetric flask under the delivery tip and open screw clamp on lower extraction tube slowly. When level of liquid is a few milliliters above the soil, invert the 500-ml Erlenmeyer flask containing remainder of extraction solution (delivery tube in place), place glass tip in the upper tube, and open the pinch clamp. Use the screw clamp on lower tube to adjust flow rate through soil column. When all the extraction solution has passed through the soil column, remove volumetric flask, make to volume with water, and mix.

665

EDTA Titration (6N3a)

Pipet a 50-ml aliquot for determination of Ca and Mg into a 100-ml beaker and evaporate to dryness. Add 10-ml concentrated HNO_3 and 1 or 2 ml concentrated HCl. Cover with watchglass, place on hot plate, and heat until no more brown fumes are evolved. Remove cover glass, rinse into beaker, and evaporate solution to dryness. Take up residue with 3 ml N HNO_3. Quantitatively transfer solution with ethanol to a 50-ml conical centrifuge tube and proceed with determination of Ca according to 6N2a.

References
Tucker (1954).

KCl-Triethanolamine Extraction (6N4)

Prepare extract as in procedure 5B2.

Oxalate-Permanganate Titration (6N4a)

Proceed as in 6N2b.

EDTA Titration (6N4b)

Reagents
Sodium hydroxide (NaOH), 4 N

EDTA 0.02 N. Dissolve 3.723 g disodium dihydrogen ethylenediamine tetraacetate in water and dilute to 1 liter. Standardize the solution against standard $CaCl_2$ prepared in the TEA buffer solution.

Ammonium purpurate (murexide) indicator. Thoroughly mix 0.5 g ammonium purpurate with 100 g powdered potassium sulfate.

Eriochrome Black T (Erio T) indicator. Dissolve 0.5 g Erio T in 100 ml of triethanolamine.

Procedure
Pipet a 5-ml aliquot of extract from procedure 5B3 into a 100-ml beaker. Add 20 ml water, 5 drops 4 N NaOH, and 50 mg murexide. Titrate with standard EDTA using a 10-ml microburet. Approach the end point slowly (orange-red to lavender or purple). Save the solution for the Mg^{++} determination.

Calculations

Ca (meq/100 g) = (A/B) x N x (C/D) x 100

where
A = Volume EDTA (mL)
B = Sample weight (g)
N = Normality of EDTA
C = Volume extract (mL)
D = Volume aliquot (mL)

References
Bower and Wilcox (1965).

Atomic Absorption (6N4c)

Proceed as in 6N1b except use sample from KCl-TEA extraction.

HF Dissolution (6N5)

Obtain extract as in 7C3.

Atomic Absorption (6N5a)

Apparatus
Diluter.

Atomic absorption spectrophotometer.

Reagents
Standard Ca solutions, 0 to 30 meq/L.

Procedure
Dilute HF extracts from 7C3 and Ca standards fivefold to twentyfold with water. Compare absorbance of samples with that of standards at 442.7 nm.

Calculations

Ca (pct.) = (A x 10 x dilution x 20.04 mg/meq)/B

where
A = Ca (meq/L)
B = Sample weight (mg)

CaO (pct.) = Ca (pct.) x 1.40

Report on ovendry basis.

Magnesium (6O)
Saturation Extract (6O1)
EDTA Titration (6O1a)

Reagents
Buffer solution. Mix 33.75 g NH_4Cl with 285 ml concentrated (15 N) NH_4OH, add 5 g disodium Mg-versenate and dilute to 500 ml.

Eriochrome Black T indicator. Dissolve 0.5 g Eriochrome Black T (F241) and 4.5 g hydroxylamine hydrochloride ($NH_2OH \cdot HCl$) in 100 ml 95-percent ethanol or dissolve 1.0 g

Eriochrome Black T in 100 ml triethanolamine.

EDTA 0.02 N. Standardize with magnesium solution.

Procedure
To the sample just titrated for calcium (6N1a) add 3 or 4 drops concentrated HCl, stir until the murexide is destroyed, add 1 ml $NH_4Cl \cdot NH_4OH$ buffer solution, 1 or 2 drops Eriochrome Black T indicator, and complete the titration for magnesium, using EDTA. The end point should be a clear blue with no tinge of red.

Calculations

Mg (meq/100 g) = (A/B) x N x (C/D) x 100

where
A = Volume EDTA (mL)
B = Sample weight (g)
N = Normality of EDTA
C = Volume extract (mL)
D = Volume aliquot (mL)

Report on ovendry basis.

References
Cheng and Bray (1951).

NH$_4$OAc Extraction (6O2)

Prepare NH$_4$OAc extraction as described

EDTA Titration, Alcohol Separation (6O2a)

Reagents
Buffer solution. Dissolve 67.5 g NH$_4$Cl in about 400 ml water. Add 570 ml concentrated NH$_4$OH and dilute to 1 liter with water.

Hydroxylamine hydrochloride (NH$_2$OH·HCl), 5-percent aqueous solution. Prepare fresh solution every 10 days.

Potassium ferrocyanide (K$_4$Fe(CN)$_6$·3H$_2$O), 4-percent aqueous solution.

Triethanolamine, U.S.P.

Eriochrome Black T. Dissolve 1 g Eriochrome Black T

(Superchrome Black TS) in 100 ml triethanolamine.

Standard magnesium solution, 5.0 mg per milliliter. Transfer 2.500 g unoxidized reagent-grade magnesium metal to a 500-ml volumetric flask. Add 150 ml water and 20 ml concentrated HCl. When in solution, make to volume with water and mix. Dilute an aliquot of this solution to get a solution containing 0.5 mg magnesium per milliliter.

EDTA, 0.02 N. Standardize with magnesium standard solution.

Procedure
Place the Erlenmeyer flasks containing the alcohol solution retained from the CaSO$_4$ separation (6N2a) on a hot plate and evaporate the alcohol at moderate heat. Do not evaporate to complete dryness. Cool and dilute to 100 ml with water and add 5 ml buffer solution and 10 drops each of hydroxylamine hydrochloride, potassium ferrocyanide, and triethanolamine. Stir and let stand 5 to 10 minutes. Place the sample on the stirrer, add 2 drops Eriochrome Black T, and titrate with standard EDTA to the ice-blue end point. The color change is from red through wine to ice blue. A blank carried through this procedure usually requires 0.3 to 0.8 ml EDTA to

get the proper ice-blue color. Correct for a blank carried through this procedure and use the corrected titration to calculate the magnesium in the sample.

Calculations

Mg (meq/100 g) = (A/B) x N x (C/D) x 100

where
A = Volume EDTA (mL)
B = Sample weight (g)
N = Normality of EDTA
C = Volume extract (mL)
D = Volume aliquot (mL)

Report on ovendry basis.

References
Barrows and Simpson (1962).

Phosphate Titration (6O2b)

Reagents
Sodium hydroxide (NaOH), 0.1 N, standardized. Protect from CO_2 of the air with a sodalime trap.

Sulfuric acid (H_2SO_4), 0.1 N.

Ammonium hydroxide (NH_4OH), concentrated.

Diammonium hydrogen phosphate (($NH_4)_2 HPO_4$), 10-percent solution.

Brom cresol green, 0.1-percent aqueous solution.

Hydrochloric acid (HCl), 1:1.

Carbon-dioxide-free water. Boil water in a 5-liter round-bottom boiling flask for about 15 minutes. Cool and protect from CO_2 of the air with a sodalime trap.

Procedure
Transfer the filtrate from the calcium determination (6N2b, 6N2c, or 6N2d) to a 400-ml beaker, add 10 ml concentrated HN_3, cover with a 3.5-inch Speedyvap watchglass and evaporate to dryness. Dissolve the residue in 5 ml 1:1 HCl and transfer to a 250-ml Erlenmeyer flask, policing twice and rinsing the beaker twice after final policing. The volume of solution should be about 75 ml or more. Using 3 to 4 drops brom cresol green indicator, neutralize the solution with concentrated NH_4OH added by drops. Add 5 ml 10-percent $(NH_4)_2HPO_4$ and 10 ml concentrated NH_4OH. Heat the solution just to boiling, cool, stopper, and let stand overnight.
Filter through a 9-cm Whatman No. 40 filter paper, pouring the solution down a stirring rod. Rinse the flask five times with 1 N NH_4OH and pour the rinsings onto the filter. Wash the precipitate on the filter five more times with 1 N NH_4OH. Place the wet filter paper with precipitate on a watchglass and let dry at no more than 40°C until free of ammonia. Place the dry filter in the original flask, add 5 drops brom cresol green and 10 ml 0.1 N H_2SO_4 or more if necessary to dissolve the precipitate. The solution should be yellow. After most of the precipitate has dissolved, add 50 ml CO_2-free water, stopper the flask, and shake vigorously

until the filter paper is macerated. Remove the stopper and rinse it and the flask walls with CO_2-free water. Back-titrate with standard 0.1 N NaOH to pH 4.5. To determine the correct end point, prepare a color standard by pipetting 5 ml potassium dihydrogen phosphate (2-percent solution) into a 250-ml Erlenmeyer flask, adding 65 ml water, 5 drops brom cresol green, and a macerated filter paper.

Calculations

Mg (meq/100 g) = ((A - B)/C) x N x (D/E) x 100

where
A = Volume NaOH blank (mL)
B = Volume NaOH sample (mL)
C = Sample weight (g)
N = Normalitiy of NaOH
D = Volume extract (mL)
E = Volume aliquot (mL)

Report on ovendry basis.

References
Peech et al. (1947).

Gravimetric, Magnesium Pyrophosphate (6O2c)

Reagents
Diammonium hydrogen phosphate ((NH_4)$_2$$HPO_4$), 10-percent solution.

Nitric acid (HNO_3), concentrated.

Ammonium hydroxide (NH_4OH), concentrated.

Ammonium hydroxide (NH_4OH), 1:1.

Hydrochloric acid (HCl), 6 N.

Procedure
Continue analysis on filtrate from oxalate precipitation (6N2b). This filtrate will probably fill a 150-ml beaker. Place cover glass on filtrate and heat at a low temperature. When volume has been reduced, add 20 ml concentrated HNO_3. Evaporate to complete dryness and wash cover glass and sides of beaker with water. Dissolve residue in 5 ml 6 N HCl and then dilute to about 75 ml. Add 2 or 3 drops brom cresol green and bring pH to 4.6 with 1:1 NH_4OH. Add 5 ml 10-percent diammonium hydrogen phosphate (make up fresh each time). Add 10 ml concentrated NH_4OH, stir solution vigorously until a precipitate forms, and let stand overnight.

On the next day filter on a 11.0-cm Whatman No. 42 filter paper, rinse beaker five times with 1 N NH_4OH, and pour washings into the filter. Wash the precipitate in the filter five more times with 1 N NH_4OH. Place filter in oven to dry (2 to 3 hours) and evolve NH_4OH to prevent any explosion in the muffle furnace. Place crucibles (Coors 000) with filters containing magnesium precipitate in muffle furnace. Raise temperature gradually to 1000°C and hold at 1000°C for 1 hour. Allow muffle furnace to cool down and remove crucibles. Place in desiccator and dry over phosphorus pentoxide (P_2O_5). Weigh $Mg_2P_2O_7$ and record.

Calculations

Mg (meq/100 g) = (A/B) x (C/D) x 1.797

where
A = $Mg_2P_2O_7$ (mg)
B = Sample weight (g)
C = Volume extract (mL)
D = Volume aliquot (mL)

NH_4Cl-Ethanol Extraction (calcareous soils) (6O3)

Proceed as in 6N3.

EDTA Titration (6O3a)

Proceed as in 6N3a except determine magnesium in alcohol extract by method 6O2a.

KCl-Triethanolamine Extraction (6O4)

Prepare extract as described in 5B2 or 5B3.

Phosphate Titration (6O4a)

Proceed as in 6O2b except use extract from 5B2 or 5B3.

EDTA Titration (6O4b)

Reagents
Concentrated hydrochloric acid (HCl).

Concentrated ammonium hydroxide (NH_4OH).

EDTA 0.02 *N*. Dissolve 3.723 g disodium dihydrogen ethylenediaminetetraacetate in water and dilute to a volume of 1000 ml. Standardize the solution against standard $MgCl_2$.

Eriochrome Black T (Erio T) indicator. Dissolve 0.5 g Erio T in 100 ml triethanolamine.

Procedure
Add 4 or 5 drops concentrated HCl to the solution used for the calcium determination (6N4b). Set aside until the murexide turns colorless. Add 15 to 20 drops concentrated NH_4OH. This should bring the pH between 10.0 and 10.3. Add 1 drop Erio T and titrate with EDTA to a clear blue end point.

Calculations

Mg(meq/100g) = (A/B) x *N* x (C/D) x 100

where
A = Volume EDTA (mL)
B = Sample weight (g)
N = Normality of EDTA
C = Volume extract (mL)
D = Volume aliquot (mL)

Report on ovendry basis.

Atomic Absorption (6O4c)

Proceed as in 6O1b except use samples from the KCl-TEA extract.

HF Dissolution (6O5)

Obtain extract as in 7C3.

Atomic Absorption (6O5a)

Apparatus
Diluter.
Atomic absorption spectrophotometer.

Reagents
Standard Mg solutions, 0 to 10 meq/L.

Procedure
Dilute HF extracts from 7C3 and Mg standards fivefold to twentyfold with water. Compare absorbance of samples with that of standards at 285.2 nm.

Calculations

Mg (pct.) = (A x 10 x dilution x 12.16 mg/meq)/B

where
A = Mg (meq/L)
B = Sample weight (mg)

MgO (pct.) = Mg (pct.) x 1.66

Report on ovendry basis.

Sodium (6P)
Saturation Extract (6P1)
Flame Photometry (6P1a)

Apparatus
Beckman Model DU spectrophotometer with flame attachment.

Reagents
Standard sodium solutions, 0.0 to 2.0 meq per liter.

Concentrated hydrochloric acid (HCl).

Hydrochloric acid (HCl), 6 N.

Hydrochloric acid (HCl), 0.4 N.

Concentrated nitric acid (HNO_3).

Procedure

Pipet an aliquot of appropriate size (5 to 25 ml) of the saturation extract into a 100-ml beaker and evaporate to dryness on a hot plate. Treat the residue with 1 ml concentrated HCl and 3 ml concentrated HNO_3 and again evaporate to dryness on the hot plate. Repeat the acid treatment on the residue. Add 5 ml 6 N HCl to the residue and bring to dryness. Then raise the temperature to high for 20 minutes to render the silica insoluble. Wash and filter the residue into 50-ml volumetric flasks, using 0.4 N HCl. Determine flame luminosity of samples appropriately diluted and compare with luminosity of standard solutions made up with 0.4 N HCl. The evaporation and dehydration steps are used only where there is enough silica to clog the burner. If they are not used, merely dilute the sample.

Calculations

Na (meq/L) = A x dilution

where
A = Na from curve (meq/L)

NH$_4$OAc Extraction (6P2)
Flame Photometry (6P2a)

Proceed as in 6P1a except make standard solutions in NH_4OAc. The evaporation and dehydration steps can be eliminated.

Calculations

Na (meq/100 g) = (A/B) x dilution x (C/10)

where
A = Na from curve (meq/L)
B = Sample weight (g)
C = Volume extract (mL)

Report on ovendry basis.

References
Fieldes et al. (1951).

HF Dissolution (6P3)

Obtain extract as in 7C3.

Atomic Absorption (6P3a)

Apparatus
Diluter.

Atomic absorption spectrophotometer.

Reagents
Standard Na solutions, 0 to 20 meq/L in HF and boric acid.

Procedure

Dilute HF extracts from 7C3 and Na standards fivefold to twentyfold with water. Compare absorbance of samples with that of standards at 589 nm.

Calculations

Na (pct.) = (A x dilution x 23.00 mg/meq)/B

where
A = Na (meq/L)
B = Sample weight (mg)

Na_2O (pct.) = Na (pct.) x 1.35

Report on ovendry basis.

Potassium (6Q)
Saturation Extract (6Q1)
Flame photometry (6Q1a)

Apparatus
Beckman Model DU spectrophotometer with flame attachment.

Reagents
Standard potassium chloride (KCl) solutions ranging from 0.0 to 1.0 meq per liter.

Procedure
Proceed as in 6P1a. Determine flame luminosity of potassium at 768 mμ and compare with that of the standard solutions.

Calculations

K (meq/L) = A x dilution

where
A = K from curve (meq/L)

References
Fieldes et al. (1951).

NH_4OAc Extraction (6Q2)
Flame Photometry (6Q2a)

Proceed as in 6Q1a except make up standards in NH_4OAc solution.

Calculations

K (meq/100 g) = (A/B) x dilution x (C/10)

where
A = K from curve (meq/L)
B = Sample weight (g)
C = Volume extract (mL)

Report on ovendry basis.

Sulfur (6R)
NaHCO₃ Extract, pH 8.5 (6R1)
Methylene Blue Colorimetry (6R1a)

References
Kilmer and Nearpass (1960).

HCl Release (sulfide) (6R2)

Apparatus
Nitrogen tank (water pumped).

Scrubber. 250-ml Erlenmeyer flask equipped with three-hole rubber stopper to accommodate entry and exit for sweep gas, and a 4-foot glass tube to serve as a manometer.

Reaction flask. 250-ml Erlenmeyer equipped with three-hole rubber stopper to accommodate entry and exit for sweep gas, and a buret for adding acid. Reaction flask sits on a magnetic stirrer.

Collection bottles. Two 500-ml bottles (No. 8 rubber stopper), each fitted with a two-hole rubber stopper to accommodate entry and exit tubes for sweep gas. The entry tubes should be detachable below the rubber stopper (Fig. 9). Attach a pinch clamp to the exit tube to help control the flow rate of gases.

Procedure
Place 10 ml zinc acetate solution in collection flask. Add water to 150 ml volume and place flask in train. Add moist sample (collected as in 1A2b) in tared reaction flask (250-ml Erlenmeyer), introduce N_2 gas (unless sample is run immediately), stopper, and weigh. Determine moisture content on a separate sample. Place flask in collection train. Sweep with N_2 gas for about 5 minutes. Reduce flow until pressure drops enough so that 50 to 60 ml of 6 N HCl can be added to reaction flask. Adjust flow of N_2 to about 4 bubbles per second in collection flask and turn on stirrer. Collect sample for 45 to 60 minutes. Second collection bottle should be a blank. Cut flow, disconnect entry tube but leave in collection bottle, remove collection bottles, and stopper until ready to titrate.

Iodine Titration (6R2a)

Apparatus
Iodine applicator, approximately a 50-ml reservoir with stopcock delivery in a two-hole rubber stopper (No. 8). Fit a glass tube for air exit through the stopper.

Buret for thiosulfate.

Magnetic stirrer.

Reagents
Iodine 0.1 N, standardized.

Sodium thiosulfate 0.1 N, standardized.

Starch indicator.

Hydrochloric acid (HCl), 6 N.

Procedure

Mix an aliquot of standardized iodine solution and 5 ml of 6 N HCl in iodine applicator. Place applicator on bottle and add acidified iodine. Wash contents of applicator, quantitatively, into bottle. Remove applicator, stopper bottle, and swirl so that iodine enters the top of the entry tube from collection train. Any white precipitate of ZnS should dissolve off entry tube. Remove stopper and titrate with standardized thiosulfate until iodine color becomes faint. Add 1 or 2 ml starch indicator and titrate until blue color changes to clear. The end point is abrupt.

Stopper bottle and again swirl so that solution passes through the entry tube. Blue color should reappear. Again titrate to the end point. Magnetic stirrer can be used to mix the sample.

Calculations

S (meq/100 g) = ((A - B)/C) x N x 100

where
A = Volume thio for blank (mL)
B = Volume thio for sample (mL)
C = Sample weight (g)
N = Normality of thio

References
Pierce and Haenisch (1955), Johnson and Ulrich (1959), and Chapman and Pratt (1961).

SO2 Evolution (6R3)
KIO$_3$ titration (6R3a)

Apparatus
LECO induction furnace model 521.

LECO automatic sulfur titrator model 532.

LECO crucibles and lids.

Oxygen tank and regulator.

LECO starch dispenser and 0.2-ml scoop.

Reagents
Potassium iodate (KIO$_3$).

Potassium iodide (KI).

Arrowroot starch.

Hydrochloric acid (HCl) 7.7 N.

Hydrochloric acid (HCl) 0.18 N.

Magnesium oxide, (MgO).

Iron-chip accelerator.

Copper metal accelerator.

Procedure

Into a tared crucible, weigh approximately 1/2 g of 60-mesh soil, recording gross weight. Where high sulfur content might be present, either 1/4 or 1/10 g sample should be run. Add 2 scoops of MgO and a scoop of iron chips. Mix thoroughly. Add a half scoop of copper accelerator and a scoop of iron chips. Magnesium oxide scoops are heaping; all others are level. A cover is placed on the crucible, which is placed on the pedestal and raised into the combustion tube for ignition. The LECO instruction manual is followed in setting up the furnace and titrator. The timer is set to 8 min and grid tap switch to midposition. These settings should be adjusted as needed to get complete fusion of the mixture in the crucible; however, plate current should not exceed 350 mA. When the burette reading does not change for 2 min and plate current has achieved 300 to 350 mA, the titration is complete and the titer is recorded. A blank is run using all ingredients except soil. Sulfate removal before analysis may be desirable in some instances. Sample is leached with 50 ml of 7.7 N HCl followed by 500 ml of distilled water.

Calculations

The KIO_3 burette is direct reading in percent for a 1-g sample containing up to 0.2 percent sulfur, provided the KIO_3 concentration is 0.444 g/L. With 1.110 g KIO_3/L, multiply burette readings by 5 (1/2-g sample, 0.005 to 1.00-percent sulfur range).

References
Smith (1974).

Phosphorus (6S)
Perchloric Acid Digestion (6S1)

Perchloric acid is extremely hazardous and subject to explosion if improperly handled. Do not attempt this procedure unless the hazards are well understood and the laboratory is specially equipped to handle perchloric acid digestion.

Reagents

Perchloric acid ($HClO_4$), 60-percent.

Concentrated nitric acid (HNO_3).

Concentrated hydrochloric acid (HCl).

Procedure

Weigh 2.000 g ovendry soil, ground to approximately 100 mesh, into a 300-ml Erlenmeyer flask, add 30 ml 60-percent $HClO_4$, and boil until the soil is white. Continue boiling 20 minutes longer to insure complete extraction. Soils high in organic matter should be pretreated with HNO_3 and HCl to destroy the readily oxidize organic matter.

Molybdovanadophosphoric Acid Colorimetry (6S1a)

Apparatus
Spectrophotometer.

Reagents
Solution I. Dissolve 20 g ammonium molybdate (($NH_4)_6Mo_7O_{24}\cdot4H_2O$) in 250 ml water.

Solution II. Dissolve 1.25 g ammonium metavanadate (NH_4VO_3) in 300 ml boiling water, cool, and add 425 ml 60-percent $HClO_4$. Mix solution I and II and dilute to 1 liter in a volumetric flask. Store in a brown bottle.

Standard phosphorus solution. Weigh out 0.2194 g ovendry KH_2PO_4 and dilute to 1 liter. This solution contains 50 ppm phosphorus.

Concentrated nitric acid (HNO_3) for samples high in organic matter only.

Concentrated hydrochloric acid (HCl) for samples high in organic matter only.

Procedure

Transfer the extract into 250-ml volumetric flasks, bring to volume, and let residue settle out. Pipet a 25-ml aliquot into a 50-ml volumetric flask, add 10 ml molybdovanadate reagent, bring to volume, and mix.

After 10 minutes, the color is fully developed on most samples and can be read at 460 mμ. Prepare a standard curve covering the range O to 5 ppm phosphorus in 50 ml solution. Plot on semilog paper.

Calculations

Total P (pct.) = (A/400) x (250/B)

where
A = P from curve (ppm)
B = Volume aliquot (mL)

Comments

The color developed is molybdovanadophosphoric acid and is very stable, lasting 2 weeks or more.

To destroy organic matter in samples high in organic matter, add 15 ml HNO_3 and 5 ml HCl. When brown fumes stop coming off, add $HClO_4$, and follow the usual procedure.

Sediment disturbance during aliquot removal makes it impossible to take more than one aliquot a day. If more aliquots are necessary, remove the sediment by filtering the suspension into a 250-ml volumetric flack, using Whatman No. 50 filter paper.

Comparison of results by Na_2CO_3 fusion and by perchloric acid on lava samples indicates that extraction may not be complete for some silicate minerals. Extraction by $HClO_4$ should be complete on common phosphate minerals.

The volume of molybdo-vanadate reagent added is not critical but must be constant. The presence of chlorides slows down color development but does not interfere otherwise.

References
Sherman (1942), and Kitson and Mellons (1944), and Jackson (1956).

Adsorption Coefficient (6S2)

Apparatus
Automatic extractor, 24 place.

Syringes, 60 cc polypropylene; use one sample tube and one extraction syringe per sample.

Reagents

Extractant. Dissolve 4.5 g ammonium fluoride (NH_4F) and 85.6 g ammonium chloride (NH_4Cl) in about 4 L of distilled water, add 92 ml glacial acetic acid and 10 ml concentrated HCl, make to 8 L and mix.

Sulfuric-molybdate-tartrate solution. Dissolve 100 g ammonium molybdate [$(NH_4)_6Mo_7O_{24}·4H_2O$] and 2.425 g antimony potassium tartrate [$K(SbO)C_4H_4O_6·1/2H_2O$] in 500 ml distilled water, heating if necessary but not to exceed 60°C. Slowly add 1,400 ml concentrated H_2SO_4 and mix well. Cool, dilute to 2 L with water and store in refrigerator in polyethelene or Pyrex bottle.

Ascorbic acid solution. Dissolve 88.0 g ascorbic acid in distilled water, dilute to 1 L, mix, and store in glass bottle in refrigerator.

Phosphorus stock standard, 100 ppm. Weigh 0.4394 g dried monobasic potassium phosphate (KH_2PO_4) into a 1-L volumetric flask, dissolve, and make to volume with extractant solution.

Phosphorus working standards, 2 to 10 ppm. Pipette 2, 4, 6, 8, and 10-ml aliquots of phosphorus stock standard into a series of 100-ml volumetric flasks and make to volume with extractant solution. The standards contain 2, 4, 6, 8, and 10 ppm P.

Saturate stock solution. Dissolve 4.394 g dried monobasic potassium phosphate (KH_2PO_4) in distilled water and make to 1 L.

Saturate working solution. Pipette 20 and 80-ml aliquots of Saturate stock solution into two 1-L volumetric flasks. The resulting solutions contain 20 and 80 ppm P.

Color solution. Measure 40 ml ascorbic solution and 80 ml sulfuric-molybdate-tartrate solution into 2 L of distilled water. Bring to 4 L, mix, and store in refrigerator.

Apparatus

Colorimeter.

Automatic extractor.

Shaker.

Procedure

A. Saturation

Weigh three 2-g subsamples of ovendried soil into 50-ml Erlenmeyer flasks. To the first add 2 ml distilled water. To the second add 2 ml 20 ppm P solution. To the third add 2 ml 80 ppm P solution. Let stand for 1 hr then place in oven at 60° C and dry overnight.

B. Extraction

To each of the dried samples in the 50-ml Erlenmeyer flasks, add 20 ml extractant reagent, and shake for 20 min (Burrell shaker). Extract samples using the automatic extractor.

C. Developing the color

Standard curve. Using 50-ml Erlenmeyer flasks, pipette aliquots from the phosphorus working standards as follows:

Flask 1--2 ml extractant
Flask 2--2 ml 2 ppm P

Flask 3--2 ml 4 ppm P
Flask 4--2 ml 6 ppm P
Flask 5--2 ml 8 ppm P
Flask 6--2 ml 10 ppm P

Samples. For each sample extracted in part B, pipette 2 ml of extract into clean 50-ml Erlenmeyer flasks corresponding with sample numbers. To all flasks, standards, and samples, add 25 ml of color solution, swirl to mix, and let stand for 15 min to allow color to develop. After color has developed fully, transfer to colorimeter tubes.

D. Reading the color
Using a wavelength setting of 880 μm, set colorimeter to 100 percent transmittance (T) with No. 1 standard containing 2 ml extractant. Read percent transmittance of remaining standards and samples. Generally, the standard curve is around the following values:

(ppm)	%T
0	100
2	77
4	59
6	45
8	33
10	24

E. Calculations

1. Develop standard curve by the least squares analysis using concentration of standards as a f(ln%t). This results in the equation:

$$concentration = m(ln\%t) + b$$

2. Use this equation to determine solution concentrations of unknowns (leachate). Concentration of leachate x 10 is desorbed P in ppm of dry soil.

3. P retained of that added = P added (desorbed P at that conc. minus desorbed P at zero P addition).

4. Pa (adsorption coefficient) is the slope of the least square regression of P retained as a function of phosphorus added, f(P added).

Example

P (ppm)	T (%)	Added P (ppm)	T (%)
0	100		
2	77		
4	59		
6	45	0	84
8	33	20	73
10	24	80	45

1. Conc.= -7.023(ln%t) + 32.5158

2. Concentration
= -7.023 (ln84) + 32.5158 = 1.40
= -7.023 (ln73) + 32.5158 = 2.38
= -7.023 (ln45) + 32.5158 = 5.78

Desorbed P (ppm) of dry soil
= 1.40 x 10 = 14.0
= 2.38 x 10 = 23.8
= 5.78 x 10 = 57.8

3. P (ppm) retained of that added
= 0 - (14.0 - 14.0) = 0
= 20 - (23.8 - 14.0) = 10.2
= 80 - (57.8 - 14.0) = 36.2

4. y = 0.4478(P added) + 0.5083
Pα= 0.4478

References
Mehlich (1978).

Boron (6T)
Saturation Extract (6T1)
Carmine Colorimetry (6T1a)

Refer to USDA Handbook 60, method 17 (p. 100) and method 73b (p. 142).

Silicon (6V)
HF Dissolution (6V1)

Obtain extract as in 7C3.

Atomic Absorption (6V1a)

Apparatus
Diluter.

Atomic absorption spectrophotometer.

Reagents
Standard Si solutions, 400 and 800 mg/L in HF.

Procedure
Dilute HF extracts from 7C3 and Si standards fivefold to twentyfold with water. Compare absorbance of samples with that of standards at 252 nm.

Calculations

Si (pct.) = (A x 10 x dilution)/B

where
A = Si (mg/L)
B = Sample weight (g)

SiO_2 (pct) = (pct) x 2.14

Report on ovendry basis.

MINERALOGY (7)

Instrumental Analyses (7A)
Preparation (7A1)

The treatment to be used in preparing samples depends on analysis objective and sample composition.

Carbonate Removal (7A1a)

Reagents
Sodium acetate (NaOAc), N, pH 5.0. Dissolve 82 g NaOAc, 27 mL glacial acetic acid L^{-1}, adjust to pH 5.0.

Procedure
Place 5 g soil in a 90-ml centrifuge tube, add 40 ml N NaOAc, pH 5.0, and heat at 95°C for 30 minutes, stirring occasionally. Centrifuge at 1,600 rpm for 5 minutes and decant supernatant liquid. Repeat if necessary until carbonates are removed. Wash twice with N NaOAc, pH 5.0. Save decantates for calcium and magnesium analysis. One washing is enough to prepare neutral or basic noncalcareous soils for optimum hydrogen peroxide treatment.

References
Jackson (1956).

Organic-Matter Removal (7A1b)

Reagents
Hydrogen peroxide (H_2O_2), 30- to 35-percent.

Procedure
Transfer sample to a 250-ml tall breaker, using a minimum of water, and add 5 ml H_2O_2. When frothing subsides, heat at 90° C. Continue to watch for frothing. Add 5- to 10-ml aliquots of H_2O_2 each hour until 25 to 30 ml H_2O_2 has been used. Wash five times, removing water by filter candles. Transfer to a 90-ml centrifuge tube.

References
Kilmer and Alexander (1949).

Iron Removal (7A1c)

Reagents
Sodium bicarbonate ($NaHCO_3$), N.

Sodium citrate ($Na_3C_6H_5O_7$), 0.3 M.

Sodium dithionite ($Na_2S_2O_4$) powder.

Procedure

If iron reduction is intended, add 5 ml N NaHCO$_3$ and as much as 40 ml 0.3 M sodium citrate. Heat to 70°C (not above 80°C) in a water bath and add 1 to 2 g Na$_2$S$_2$O$_4$, stir for 1 minute, and then stir intermittently for 15 minutes. Decant supernatant liquid and save for iron analysis. Repeat treatment as needed if soil is high in iron. Wash twice with 0.3 M sodium citrate.

References

Aguilera and Jackson (1953) and Mehra and Jackson (1960).

Disaggregation and Particle-Size Fractionation (7A1d)

Reagents

Sodium bicarbonate (NaHCO$_3$), pH 9.5 to 10.0.

Sodium metaphosphate (NaPO$_3$). Prepare as in 3A1.

Procedure

Use NaHCO$_3$ solution or sodium hexametaphosphate-sodium bicarbonate as a dispersing agent. Use hexametaphosphate carefully with amorphous materials since phosphates may be precipitated. A dilute HCl treatment may be useful for some highly allophanic soils. Do not use mechanical blenders for disaggregation if silt and sand are to be studied because they fracture large quartz grains. Separate sand from silt with a 300-mesh sieve and further separate the sands, using a nest of sieves (3A1). Separate clay (<2μ) from silt by centrifuging at 750 rpm for 3 minutes (International No. 2 centrifuge with No. 240 head and solution depth of 10 cm). If silt and sand are to be studied, save these fractions in small vials after drying and weighing. If interested in separating fine clay (<0.2μ) from the coarse clay (0.2μ to 2μ), centrifuge at 2,400 rpm for 30 minutes with a solution depth of 10 cm. Adjust time according to temperature. Add 50-ml aliquots of clay suspension after each centrifugation until the required amount of clay is obtained. Make each suspension of coarse and fine clay to known volume and determine its concentration and the concentration of the whole clay.

References

Kilmer and Alexander (1949) and Jackson (1956).

Particle-Size Distribution Analysis (PSDA) Pretreatment (7A1e)

Mineralogical analysis can be performed on samples from the particle-size distribution analysis (3A1). These samples have undergone peroxide digestion and sodium metaphosphate dispersion.

X-ray Diffraction (7A2)

The minerals in soil clays of greatest interest are mostly flaky or platy, e.g., kaolinite, illite (mica), vermiculite, chlorite, and montmorillonite. They are most readily identified and distinguished from one another by observing the effect of different treatments on the interplanar spacings along the axis perpendicular to the platy surfaces. X-ray diffraction produces peaks on a chart corresponding to the various angles (2R) of goniometer from which the crystallographic spacing of the mineral or minerals can be calculated by Bragg's law. Tables of spacings corresponding to angles have been published in U.S. Geological Survey Circular 29 (Switzer et al. 1948).

The pretreatment used to distinguish montmorillonite from vermiculite and chlorite and to identify illite is saturation of the exchange complex of the clay with magnesium and treatment with ethylene glycol or glycerol. With this treatment, montmorillonite has a distinctive interplanar spacing of 17 Angstroms (17 A) to 19 A. Chlorite and vermiculite keep a 14 A spacing and mica a spacing of 10 A. To distinguish vermiculite from chlorite and to identify kaolinite, which has a 7 A spacing, the pretreatment consists of saturating the clay with potassium and heating on a glass slide at 500°C. Intermediate heat treatments 110 and 250°C, can be used to study interlayering in the collapsing minerals or other special problems. After the 500°C treatment, vermiculite and montmorillonite collapse completely to 10 A kaolinite becomes amorphous and chlorite still shows 14 A and sometimes 7 A peaks. Interstratified forms of these minerals are indicated by spacings intermediate between those of the individual components.

Clay suspensions are dried as thin films so that the plates are parallel to one another (preferred orientation). This results in greater X-ray diffraction peak intensities. For identification and semiquantitative estimation of nonplaty minerals such as quartz, feldspars, and crystalline iron and aluminum oxides, randomly oriented dry-powder samples can be used. This dry-powder method was used for nearly all analysis, including clay fractions, before 1951.

Various techniques are used to prepare the ion-saturated clays and to improve their parallel orientation. Details can be obtained from the soil survey laboratories.

References
Brindley (1951), Brown (1961), Brunton (1955), Grim (1953), Jackson (1956), and Switzer et al. (1948).

Thin film on Glass, Solution Pretreatment (7A2a)

Reagents
Potassium chloride (KCl), N.

Magnesium acetate (Mg(OAC)$_2$), N.

Magnesium chloride (MgCl$_2$), $1\ N$.

Glycerol, 10-percent in ethanol by volume.

Procedure
Place an aliquot containing 50 mg clay in a 50-ml centrifuge tube. All a few ml 1 N KCl, centrifuge, and discard the clear supernatant. Combine sediments if necessary to get 50 mg in the tube. Wash four times by suspending and centrifuging in 20-ml portions N KCl. After the last washing with N KCl, wash with water until some of the clay remains suspended after centrifuging. Add a few drops of acetone or centrifuge at higher speed, or both, to flocculate the clay. Discard the supernatant. Clays are now free of chloride. Suspend the sediment in water and adjust the volume of the suspension to yield the desired weight of clay per slide. For most clays 50 mg per slide (27 by 46 mm) gives maximum intensity of reflection with minimum peeling of clay films. For amorphous clays, 25 mg per slide is adequate if glass slides are dried in a low-humidity atmosphere.

For magnesium saturation and glycerol solvation place an aliquot containing 100 mg clay in a 50-ml centrifuge tube. Wash twice with N Mg(OAC)$_2$ acetate and then three times with N MgCl$_2$. Wash the suspension free of chloride or until clay disperses. Place 2.5 ml clay suspension containing 50 mg clay on a glass slide (25 mg clay if the clay is amorphous). Solvate the remaining clay in the test tube with glycerol (about 1/2 ml of 10-percent glycerol in ethanol per 50 mg clay). Mix well and pipet 50 mg clay onto the glass slide. The slide should be moist but not wet. Or prepare the glycerol slide by adding 10-percent glycerol, a drop at a time, to the slides until the clay film is moist.

Thin Film on Glass, Resin Pretreatment (7A2b)

Reagents
Potassium-charged resin (Dowex 50W-X8).

Magnesium-charged resin (Dowex 50W-X8).

Glycerol, 10-percent in ethanol by volume.

Procedure
Add 1/4 teaspoon K-charged resin to 50 mg clay in a 1 ml volume in a 50-ml centrifuge tube. Mix and transfer a 1-ml aliquot to a glass slide (27 by 46 mm). Take the aliquot from the top of the suspension to avoid removing the resin.

Magnesium-clay and Mg-glycerol-clay slides can be prepared using a Mg-charged cation exchange resin. Add 1/2 teaspoon Mg-charged resin to the clay suspension (100 mg clay in a 4-ml volume) in a 50-ml centrifuge tube. Mix with the suspension, remove 1-ml aliquot, and place it on a glass slide. Add approximately 1/2 ml 10-percent glycerol in ethanol to the tube. Mix and transfer a 1-ml aliquot to a glass slide or use a Mg-clay slide for both Mg and Mg-glycerol solvated slides. Record a diffraction pattern for the Mg-saturated clay film. After solvating the clay film with 10-percent glycerol solution, record a second X-ray pattern.

References
Rex (1967).

Thin film on Glass, Sodium Metaphosphate Pretreatment (7A2c)

Shake soil overnight in sodium metaphosphate solution (3A2). Centrifuge to separate the clay or siphon off the clay. Pipet about 50 mg clay to a glass slide (47 by 26 mm). Concentrate the clay suspension if necessary. Scan the clay film at room temperature, again after heating to 500°C. The clay film is Na^+ saturated. The sodium metaphosphate peaks do not interfere with peaks of the more common clay minerals in this quick check method.

Thin Film on Tile, Solution Pretreatment (7A2d)

Apparatus
Ceramic tile (porous precipitate drying plate, sawed into 27- by 46- by 7-mm blocks).

Procedure
Prepare clay suspensions as in 7A2a except dry the suspensions on ceramic tile blocks. Clay suspensions dry in a few seconds on tile, preventing particle-size segregation. Partly immerse the Mg-saturated clay films in a 10-percent glycerol solution. The porous tile rapidly transfers the glycerol to the clay film. Blot off excess glycerol before recording the X-ray pattern.

Thin Film on Tile, Resin Pretreatment (7A2e)

Prepare clay suspensions as in 7A2b. Dry on ceramic tile blocks as in 7A2d. Solvate with glycerol as in 7A2d.

Thin Film on Tile, Sodium Metaphosphate Pretreatment (7A2f)

Prepare the sample as in 7A2c. Pipet the clay onto ceramic tile blocks as in 7A2d. Follow method 7A2c for the other treatments. Or solvate with glycerol as in 7A2d.

Powder Mount, Diffractometer Recording (7A2g)

Distinguishing dioctahedral and trioctahedral minerals requires random orientation of the sample. There is no completely satisfactory method for preparing a random mount, but several techniques are used.

Pack the sample in a box mount against a glass slide. When the box is full, tape the back of the box. Invert the box and remove the slide to expose the sample to X-rays. For more random packing, sprinkle the dry sample (ground to <100 mesh) on double stick tape fixed on a glass slide or on a thin film of Vaseline on a glass slide. Scan the sample by X-ray and measure the reflections with a geiger, proportional, or other counter.

Quick checks for whole samples, particularly for nonlayered minerals, can be made with a modified powder mount. Form the sample into a thick slurry, apply to a glass slide, and let dry. This for convenience rather than random orientation.

Powder Mount, Camera Recording (7A2h)

Photographic plates are still the best means of identifying minerals. Mount the sample in the center of a circular X-ray camera. Record the X-ray reflections on photographic film placed in a cylindrical film holder inside the camera. All diffraction peaks are recorded simultaneously.

Thin Film on Glass, NaPO₃ Pretreatment II (7A2j)

Apparatus
Hypodermic syringe (1.0 cc).

Glass slides 24 x 46 mm or 14 x 19 mm.
International No. 2 centrifuge with a No. 240 head.

100-ml centrifuge tubes (plastic).

Reagents
Glycerol-water mixture (1:8 glycerol-water).

Sodium hexametaphosphate solutions.

Procedure
Shake approximately 5 g ovendried soil (<2 mm) overnight with 5 ml sodium hexametaphosphate solution (3A1) and 35 ml of water in a 100-ml centrifuge tube, centrifuge at 750 rpm for 3 min for a 10-cm suspension depth, and decant clays. Draw about 0.5 cc of clay suspension into the syringe. Expel approximately 0.2 cc of the clay suspension onto an area approximately 20 x 27 mm in a band across the middle of a 46- x 27-mm slide or expel approximately 0.1 cc of clay suspension, containing approximately 6 mg of clay, onto and covering the 14- x 19-mm slide. Prior to the deposition of the clay suspension, one small drop of glycerol-water mixture is placed on the slide which is to be solvated. Prepare four slides for X-ray diffraction: 1) Na^+--room temperature, 2) Na^+--solvated, 3) Na^+--heated 2 hr at 300°C, and 4) Na^+--heated for 2 hr at 500°C.

Powder Mounts (7A2k)

Two procedures are used for random orientation of mineral separates. In the first procedure, double-stick tape is affixed to a glass slide, a surplus of the sample is sprinkled onto the tape, the excess material is removed, and the slide is scanned by X-ray analysis. In the second procedure, a <2-mm soil sample is ground finer than 100 mesh prior to slide preparation. A thin film of Vaseline is applied to a glass slide, the 100-mesh sample is added, the excess removed, and the slide is scanned by X-ray analysis.

For quick check of a <2-mm sample, particularly for nonlayered minerals, a small portion is ground to less than 100 mesh and placed on a glass slide. Water is applied a little at a time until a thick slurry is formed. The slurry is allowed to dry and the slide is scanned by X-ray analysis. This method is also applicable for specific mineral separates, very fine sands or silts.

References
Brown (1961), Jackson (1956).

Differential Thermal Analysis (7A3)

Differential thermal analysis (DTA) is a measurement of the difference in heat absorbed by or evolved from a sample of soil material and a thermally inert material as the two are heated simultaneously at a constant rate. Thermocouples are in contact with two platinum pans; one pan contains an unknown and the other pan contains an inert material of similar composition. If a reaction occurs, a difference in temperature is registered on a stripchart recorder or photographically. The magnitude of the difference depends on the nature of the reaction and amount of reacting substance in the unknown. The temperature at which the reaction occurs identifies the substance if enough is known about the sample to predict the possibilities.

Apparatus
Columbia scientific instrument (CSI) system 200.

Mortar and pestle.

Analytical balance.

Desiccator.

Reagents
Reference sample, calcined kaolinite, 2 to 20 μ.

Ethyl alcohol, 95 percent.

Magnesium nitrate $(Mg(NO_3)_2 \cdot 6H_2O$

Procedure
The decanted clay from 7A2i or 7A2j is air-dried, ground in alcohol to approximately 100 mesh, and stored in a desiccator with $Mg(NO_3)_2 \cdot 6H_2O$. A 3- to 7-mg sample is placed on a small platinum pan in the sample holder. The temperature of the kaolinite reference sample and clay sample is increased at a rate of 20°C per minute to a maximum of 900°C. The sample can be heated in air or nitrogen.

The common endothermic reactions studied or recorded are loss of structural water in gibbsite, goethite, and kaolin and loss of carbon dioxide in carbonates. Change of state or rearrangement of crystal lattices can be either exothermic or endothermic. Oxidation reactions such as burning of carbon and oxidation of ferrous iron are exothermic.

Loss of structural hydroxyls can be measured quantitatively by calibrating areas of peaks of known mixtures of standard minerals, as is done commonly to determine the percentage of kaolin and gibbsite in soils. The standard curves are prepared by running the known mixtures under the same conditions as the unknowns. Kaolin has an endotherm at 500° to 600°C and gibbsite, at 310°C. Each worker should prepare a set of standard curves.

Endotherms at about 120°C indicate surface-adsorbed water. Montmorillonite produces a double peak at a low temperature if saturated with a divalent cation. The proportion of this mineral can be estimated if samples are kept in an atmosphere with a high (70 to 80 percent) relative humidity for 24 hr or more before analysis. Allophane has a broad endotherm at about 160°C.

Samples can be any well-powdered material, whole-soil, or separated fractions. Organic matter is objectionable because it produces irregular exothermic reactions that obscure the important peaks. If a clay separate is used, it must be washed free of hygroscopic salts or salts containing water of crystallization.

References
Grim (1968), McKenzie (1957), and Tan and Hajek in Dixon and Weed (1977).

Thermal Gravimetric Analysis (7A4)
CSI Stone Model 10002B (7A4a)

Thermal gravimetric analysis is the detection and measurement of weight changes in a sample of soil material as the sample is being heated or cooled over a specific temperature range.

Apparatus
CSI Stone Model lOOOB used in conjunction with an RC-202 recorder-controller. Furnace is water cooled, with a rapid cooling Kanthal element. Furnace is capable of operation at temperatures of up to 1,200°C.

Procedure
Prepare sample as described in 7A3 and place in balance pan suspended above thermocouple assembly. Heat sample at rate of 20°C/min to desired temperature. If a weight loss occurs, it is registered on a stripchart recorder. The magnitude of the weight loss depends on the reaction and the amount of reacting substance in the unknown. The temperature at which the reaction occurs usually identifies the substance.

Infrared Analyses (7A5)

Soil or clay samples (7A2j) are incorporated into a potassium bromide (KBr) pellet for infrared analyses. Sample concentration in the pellet ranges from 0.1 to 1 percent.

Reagents
Potassium bromide, spectroscopic grade.

Apparatus
Infrared spectrometer. Perkin Elmer Model 283.

Pellet die.

Hydraulic press.

Analytical balance.

Procedure
Mix 0.30 g KBr and 1 mg of sample in mortar and pestle. Transfer the mixture to the pellet die, and place die in hydraulic press. Apply 8 tons of pressure for 1 min. Place pellet in instrument holder and scan for 12 min. Peaks produced on chart recorder are used to identify the substance.

References
J. L. White in Dixon and Weed (1977).

Optical Analyses (7B)
Grain Studies (7B1)
Grain mounts, Canada balsam (7B1b)

For Canada balsam, heat slide plus balsam for 15 min at 125°C. Add mineral grains, stir, heat for an additional 5 min, place cover glass in position and press firmly, remove slide from hot plate, and cool.

The refractive index of Canadian balsam is close to that of quartz, which helps to distinguish quartz from other colorless minerals, particularly the feldspars. Other available commercial media cover the refractive index range of 1.53 to 1.55. Piperine with a refractive index of 1.68, which is close to that of many of the common heavy minerals, is best for mounting them.

Electron Microscopy (7B2)

Electron microscopy gives information on particle size and morphology of clay-size particles. Evidence of clay formation or weathering can also be seen. Positive identification of halloysite often depends on observation of rolled structures under the electron microscope.

Procedure
Place a drop of dilute clay suspension on a 200-mesh copper grid. After drying, insert this grid in the microscope.

Total Analysis (7C)
Chemical (7C1)

The procedures follow the standard procedures for rock analysis set forth by Hillebrand and Lundell (1929) and modified by Robinson (1920) and by Shapiro and Brannock (1956).

X-ray Emission Spectrography (7C2)

X-ray emission spectrography is elemental analysis by measuring the X-ray fluorescence produced by bombarding a sample with high-energy X-rays. Each element yields fluorescent radiation of unique wave lengths, one of which is selected for measurement by using an analyzing crystal that diffracts according to Bragg's law. The intensity of the fluorescent radiation is generally proportional to the amount of the element present, but this is affected by sample homogeneity, particle size, and the absorption and enhancement of radiation by any other elements present in the sample (matrix effects). These effects can be overcome or compensated for by (1) comparing the intensities with those of standards of similar composition prepared in a similar manner, (2) fusing both samples and standards in borax or lithium borate to eliminate particle-size effects and to reduce matrix effects, and (3) making matrix corrections by calculation the absorption-enhancement coefficient of the sample for the particular radiation being measured.

References
Vanden Heuvel (1965).

Surface Area (7D)
Glycerol retention (7D1)

Apparatus
Weighing cans.

Reagents
Glycerol, 2-percent.

Procedure
Ovendry a clay sample (about 0.2 g) at 110°C for 2 hours. Cool and weigh. Add 5 ml 2-percent glycerol solution and mix. Heat in oven containing free glycerol at 110°C to constant weight. Record weight.

Calculations
To calculate the percent of glycerol retained subtract weight of ovendry sample from weight of glycerol and ovendry sample, divide by weight of ovendry sample, and multiply by 100. For the surface area of noncollapsible minerals (m_2/g), multiply glycerol retained by 19.1.

References
Kinter and Diamond (1958).

MISCELLANEOUS (8)

Saturation Extract, Mixed (8A)
Saturation Extract (8A1)

Apparatus
Richards or Buchner funnels.

Filter rack or flask.

Filter paper.

Vacuum pump.

Extract containers such as test tubes or 1-oz bottles.

Procedure
Transfer the saturated soil paste to a filter funnel with a filter paper in place and apply vacuum until air begins to pass through the filter. Collect the extract in a bottle or test tube. If carbonate and bicarbonate are to be determined on the extract, add 1 drop of 1000 ppm sodium hexametaphosphate solution for each 25 ml of extract to prevent precipitation of calcium carbonate on standing.

References
Richards (1954).

Conductivity of Saturation Extract (8A1a)

Apparatus
Conductivity bridge.

Conductivity cell.

Procedure
　　　　Determine temperature of the saturation extract obtained by methods 8A1 or 8B1. Draw the extract into the cell and read the meter. Correct for temperature and cell constant using Table 1 (Table 15, Richards 1954) and report as electrical conductivity, mmhos per centimeter at 25°C. If the instrument fails to balance, dilute the extract 1:9 with distilled water and redetermine. The conductivity of the diluted extract is approximately one-tenth the conductivity of the saturation extract.

References
Richards(1954).

Conductivity of Saturation Extract (quick test) (8A1b)

Apparatus
Extractor, miniature Richards-type (Fig. 10).

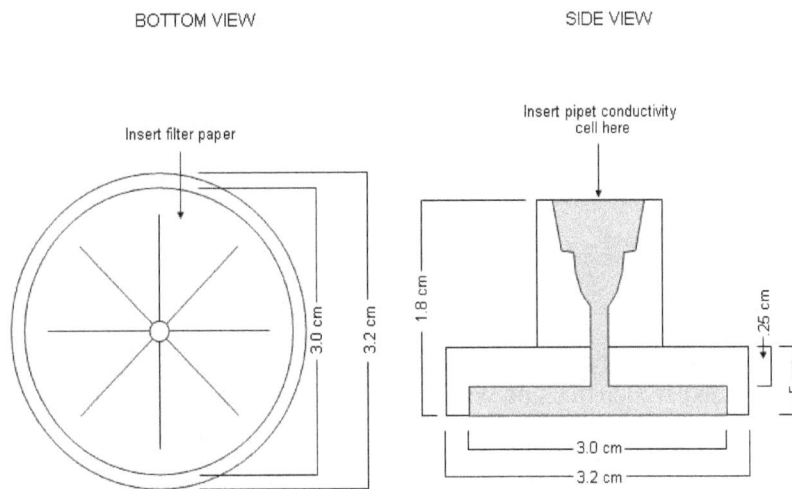

BOTTOM VIEW　　　　　　　　　　SIDE VIEW

Insert filter paper

Insert pipet conductivity cell here

1.8 cm

3.0 cm

3.2 cm

.25 cm

.5 cm

3.0 cm

3.2 cm

Figure 10. Miniature Richards-type extractor made of polymethyl methacrylate (Lucite).

Conductivity cell, micropipet.

Filter paper, glass fiber, 3.0 cm.

Vacuum pump.

Procedure
　　　　Add 1 tablespoon soil to a 100-ml beaker or container. Make a saturated paste as in 8A. Place filter paper in recess of extractor and moisten with water. Insert the tip of the pipet

conductivity cell through the other end of the extractor and into the paste. Apply suction to the cell until full. Proceed as in 8A1a.

Bureau of Soils Cup, Resistance (8A2)

Apparatus
Wheatstone bridge.

Bureau of Soils electrode cup.

Procedure
Rinse the soil cup with water, dry, and fill with soil paste (8A). Jar cup to remove air bubbles strike off excess paste so the cup is level full, and connect cup to the bridge. Record resistance (ohms) and temperature of soil paste (°F).

Calculations
Convert resistance of the soil paste in ohms to percentage of soluble salt by using the tables and formulas on pages 346-349, Soil Survey Manual.

References
Richards (1954) and Soil Survey Staff (1951).

Saturated, Capillary Rise (8B)
Apparatus

Sand table. Mariotte bottle.

Filter paper.

Polyethylene dish with lid.

Procedure
Weigh 250 g airdry soil into cups made from Whatman No. 52 (15-cm) filter paper and place them on a sand table wetted at 5-cm tension with water. The sand table used consists of two nested plastic dishpans. The outer pan holds distilled water, which is kept at a constant level by a Mariotte bottle. The inner pan, containing medium to fine (35 to 80 mesh) pure quartz sand, rests on rubber stoppers and is suspended in the distilled water. Its perforated bottom is covered with a fine cloth-mesh screen that permits water to move upward by capillarity through the sand to the table surface. The sand on the table surface is then smoothed and covered with an absorbent paper towel. Lightweight porous firebricks can be used in place of the sand table.

Keep the samples on the sand table 16 to 18 hours, remove them, and weigh. Water adsorption drops rapidly after an initial wetting of 2 hours and the rate becomes very slow after 6 to 9 hours. Moisture moves toward the top and center of the sample, which is wetted last, insuring retention of soluble salts in the soil. Calculate moisture at saturation from the wet- and dry-soil weights, correcting for the wet and dry filter paper weights. Add airdry moisture percentage to moisture at saturation and report on ovendry basis. After the wet weighing, transfer the sample to a pint polyethylene refrigerator dish, mix briefly with a spatula, and determine the pH. Keep a lid on the dish whenever possible to reduce evaporation.

References
Longenecker and Lyerly (1964).

Saturation Extract (8B1)

Proceed as in 8A1, using the saturated paste obtained by method 8B.

Conductivity of Saturation Extract (8B1a)

Proceed as in method 8A1a except use saturation extract obtained by method 8B1.

Reaction pH (8C)
Soil Suspensions (8C1)
Water dilution (8C1a)

Procedure

For 1:1 dilution add an equal weight of water to 20 or 30 g soil in a 50-ml beaker or paper cup. Stir at regular intervals for about an hour. Measure pH of the soil suspension with a glass electrode, stirring well just before immersing the electrodes in the suspension. For other dilutions vary the amount of soil, keeping the volume of water constant.

KCl (8C1c)

Procedure

Proceed as in method 8C1a except use N KCl instead of water.

CaCl$_2$ (8C1e)

Proceed as in 8C1a except use 0.01 M CaCl$_2$. This procedure can be combined with 8C1a by adding an equal volume of 0.02 M CaCl$_2$ to the soil suspension prepared for the water pH. Stir twice at 15-minute intervals before reading. The soil-solution ratio will be 1:2, but the pH difference between 1:1 and 1:2 suspensions is negligible.

References
Schofield and Taylor (1955) and Peech (1965).

Literature Cited
Aguilera, N.H., and M.L. Jackson. 1953. Iron oxide removal from soils and clays. Soil Sci. Soc. Amer. Proc. 17: 359-364.
Allison, L.E. 1960. Wet combustion apparatus and procedure for organic and inorganic carbon in soil. Soil Sci. Soc. Amer. Proc. 24: 36-40.
American Society for Testing and Materials. 1970. Book of ASTM standards, pp. 212-219.
Association of Offlcial Agricultural Chemists. 1945. Official and tentative methods of analysis, 6th ed. 932 pp. Washington, D.C.
Barrows, H.L., and E.C. Simpson. 1962. An EDTA method for the direct routine determination of calcium and magnesium in soils and plant tissue. Soil Sci. Soc. Amer. Proc. 26: 443-445.
Bascomb, C.L. 1968. Distribution of pyrophosphateextractable iron and organic carbon in soils of various groups. J. Soil Sci. 19: 251-268.
Bower, C.A., and L.V. Wilcox. 1965. Soluble salts. *In* Methods of soil analysis, Monog. 9. v. 2, pp. 933-951. Amer. Soc. Agron., Madison, Wis.
Brasher, B.R., D.P. Franzmeier, V. T. Valassis, and S. E. Davidson. 1966. Use of Saran resin to coat natural soil clods for bulk-density and water-retention measurements. Soil Sci. 101 108.
Brindley, G.W., ed. 1951. X-ray identification and crystal structures of clay minerals. 345 pp. Mineralogical Society, London.
Brown, G., ed. 1961. The X-ray identification and crystal structures of clay minerals. 544 pp. Mineralogical Society (Clay Minerals Group), London.
Bruton, G. 1955. Vapor pressure glycolation of oriented clay minerals. Amer. Min. 40: 124-126.

Cady, J.G. 1965. Petrographic microscope techniques. *In* Methods of soil analysis, Monog. 9. v. 2, pp. 604-631. Amer. Soc. Agron., Madison, Wis.

Chapman, H.D., and P.F. Pratt. 1961. Methods of analysis for soils, plants, and waters. Univ. of Calif., Div. of Agr. Sci. 309 pp.

Chenery, E.M. 1948. Thioglycollic acid as an inhibitor for iron in the colorimetric determination of aluminum by means of aluminon. Analyst 73: 501-502.

Cheng, K.L., and R.H. Bray. 1951. Determination of calcium and magnesium in soil and plant material. Soil Sci. 72: 449-458.

Cheng, K.L., R.H. Bray, and T. Kurtz. 1953. Determination of total iron in soils by disodium dihydrogen ethylenediaminetetraacetate titration. Anal. Chem. 25:347-348.

Erickson, A.E., L.C. Li, and J.E. Gieseking. 1947. A convenient method for estimating carbonates in soils and related materials. Soil Sci. 63:451-454.

Fieldes, M., P.J.T. King, J.P. Richardson, and L. Swindale. 1951. Estimation of exchangeable cations in soils with the Beckman flame spectrophotometer. Soil Sci. 72: 219-232.

Fieldes, M., and K.W. Parrott. 1966. The nature of allophane in soils. III. Rapid field and laboratory test for allophane. New Zealand J. Sci. 9: 623-629.

Franzmeier, D.P., B.F. Hajek, and C.H. Simonson. 1965. Use of amorphous material to identify spodic horizons, Soil Sci. Soc. Amer. Proc. 29: 737-743.

Franzmeier, D.P., and S.J. Ross, Jr. 1968. Soil swelling: Laboratory measurement and relation to other soil properties. Soil Sci. Soc. Amer. Proc. 32: 573-577.

Fry, W.H. 1933. Petrographic methods for soil laboratories. U.S. Dept. Agr. Tech. Bull. 344. 96 pp.

Grim R.E. 1953. Clay mineralogy. 384 pp. McGraw-Hill, New York.

Grossman, R.B. 1964. Composite thin sections for estimation of clay-film volume. Soil Sci. Soc. Amer. Proc 28:132-133

Grossman, R.B., B.R. Brasher, D.P. Franzmeier, and J.L. Walker. 1968. Linear extensibility as calculated from natural-clod bulk density measurements. Soil Sci. Soc. Amer. Proc. 32: 570-573.

Grossman, R.B. and J.L. Millet. 1961. Carbonate removal from soils by a modification of the acetate buffer method. Soil Sci. Soc. Amer. Proc. 25: 325-326.

Hillebrand, W.F., and G.E.F. Lundell. 1929. Applied inorganic analysis, with special reference to the analysis of metals, minerals and rocks. 929 pp. John Wiley and Sons, New York.

Holmgren, G.G.S. 1967. A rapid citrate-dithionite extractable iron procedure. Soil Sci. Soc. Amer. Proc. 31: 210 21 1.

Holmgren, G.G.S. 1968. Nomographic calculation of linear extensibility in soils containing coarse fragments. Soil Sci. Soc. Amer. Proc. 32: 568-570.

Hsu, P.H. 1963. Effect of initial pH, phosphate, and silicate on the determination of aluminum with aluminon. Soil Sci. 96: 230-238.

Jackson, M.L. 1956. Soil chemical analysis, advanced course. 991 pp. Privately published, Madison, Wis. 1958. Soil chemical analysis. 498 pp. Prentice-Hall, Inc., Englewood Cliffs, N.J.

Johnson, C.M., and A. Ulrich. 1959. Calif. Agr. Exp. Sta. Bull. No. 766. 78 pp.

Kilmer, V.J. 1960. The estimation of free iron oxides in soils. Soil Sci. Soc. Amer. Proc. 24: 420-421.

Kilmer, V.J., and L.T. Alexander. 1949. Methods of making mechanical analyses of soils. Soil Sci. 68: 15-24.

Kilmer, V.J., and J.F. Mullins. 1954. Improved stirring and pipetting apparatus for mechanical analysis for soils. Soil Sci. 77: 437-441.

Kilmer, V.J., and D.C. Nearpass. 1960. The determination of available sulfur in soils. Soil Sci. Soc. Amer. Proc. 24: 337-339.

Kinter, E.B., and S. Diamond. 1958. Gravimetric determination of monolayer-glycerol complexes of clay minerals. Proc. 5th Natl. Conf. on Clays and Clay Minerals, Natl. Acad. Sci. Natl. Res. Council Publ. 566, pp. 318-333

Kitson, R.E., and M.E. Mellons. 1944. Colorimetric determination of phosphorus as molybdivanadophosphoric acids. Indus. and Engr. Chem. 16: 379.

Krumbein, W.C., and F.J. Pettijohn. 1938. Manual of sedimentary petrography. 549 pp. D. Appleton-Century Co., New York.

Lagerwerff, J.V., G.W. Akin, and S.W. Moses. 1965. Detection and determination of gypsum in soils. Soil Sci. Soc. Amer. Proc. 29: 535-540.

Lin, C., and N.T. Coleman. 1960. The measurement of exchangeable aluminum in soils and clays. Soil Sci. Soc. Amer. Proc. 24: 444-446.

Lockwood, W.N. 1950. Impregnating sandstone specimens with thermosetting plastics for studies of oilbearing formations. Bull. Amer. Assoc. Petrol. Geol. 34: 2001-2067.

Longenecker, D.E., and P.J. Lyerly. 1964. Making soil pastes for salinity analysis: a reproducible capillary procedure. Soil Sci. 97: 268-275.

McKenzie, R.C. 1957. The differential thermal investigation of clays. Mineralogical Society, London.

McKeague, J.A., and J.H. Day. 1966. Dithionite and oxalate extractable Fe and Al as aids in differentiating various classes of soils. Can. J. Soil Sci. 46: 13-22.

Mehra, 0.P., and M.L. Jackson. 1960. Iron oxide removal from soils and clays by a dithionite-citrate system buffered with sodium bicarbonate. Proc. 7th Natl. Conf. on Clays and Clay Minerals, pp. 317-327. Permagon Pres, New York.

Milner, H.B. 1962. Sedimentary petrography, 4th ed. v. 2, pp 643 and 715. Macmillan Co., New York.

Nelson, R.E. 1970. Semimicro determination of sulfate in water extracts of soils. Soil Sci. Soc. Amer. Proc. 34:343-345

Peech, M. 1965. Lime requirement. *In* Methods of soil analysis. Monog. 9. v. 2, pp. 927-932. Amer. Soc. Agron., Madison, Wis.

Peech, M., L.T. Alexander, L.A. Dean, and J.F. Reed. 1947. Methods of soil analysis for soil fertility investigations. U.S. Dept. Agr. C. 757, 25 pp.

Pierce, W.C., and E.L. Haenisch. 1955. Quantitative analysis, pp. 141-143. John Wiley and Sons, New York.

Pratt, P.F., and F.L. Bair. 1961. A comparison of three reagents for the extraction of aluminum from soils. Soil Sci. 91: 357-359.

Reed, F.S., and J.L. Mergner. l953. The preparation of rock thin sections. Amer. Miner. 38: 1184-1203.

North-Central Regional Research Committee on Solis. 1955. Loess-derived gray-brown podzolic soils in the Upper Mississippi River Valley. Univ. of Ill. Bull. 587, 80 pp.

Rennie, P.J. 1957. Routine determination of the solids, water, and air volumes within soil clods of natural structure. Soil Sci. 84: 351-365.

Richards, L.A., ed. 1954. Diagnosis and improvement of saline and alkali soils. U.S. Salinity Laboratory, U.S. Dept. Agr. Hbk. 60, 160 pp.

Robinson, W.0. 1930. Method and procedure of soil analysis used in the division of soil chemistry and physics. U.S. Dept. Agr. C. 139, 19 pp.

Rogers, A.F., and P.F. Cherry. 1933. Thin-section mineralogy. MSGR-Hill Book Co., New York.

Schofield, R.K., and A.W. Taylor. 1955. The measurement of soil pH. Soil Sci. Soc. Amer. Proc. 19: 164-167.

Shapiro, L., and W.W. Brannock. 1956. Rapid analysis of silicate rocks. U.S. Geol. Survey Bull. 1036-C: 19-56.

Sherman, M.S. 1942. Calorimetric determination of phosphorus in soils. Indus. and Engr. Chem. 14: 182.

Shields, L.G., and M.W. Meyer. 1964. Carbonate clay: measurement and relationship to clay distribution and cation exchange capacity. Soil Sci. Soc. Amer. Proc 28: 416-419.

Soil Survey Staff. 1951. Soil Survey Manual. U.S. Dept Agr. Hbk. 18, 503 pp.

Switzer, G., J.M. Axelrod, M.L. Lindberg, and E.S. Larson. 1948. Tables of d spacings for angle 2R. U.S. Geol., Survey C. 29.

Treadwell, F.P., and W.B. Hall. 1943. Analytical Chemistry, Vol. 1: Qualitative analysis, 9th ed. John Wiley and Sons, New York.

Tucker, B.M. 1954. The determination of exchangeable calcium and magnesium in carbonate soils. Austral. J. Agr. Res. 5: 706-715.

Tyner, E.H. 1939. The use of sodium metaphosphate for dispersion of soils for mechanical analysis. Soil Sci. Soc. Amer. Proc. 4: 106-113.

Vanden Heuvel, R.C. 1965. Elemental analysis by X-ray emission spectrography. *In* Methods of soil analysis. Monog. 9. v. 2, pp. 771-821. Amer. Soc. Agron. Madison, Wis.

Walkley, A. 1935. An examination of methods for determining organic carbon and nitrogen in soils. J. Agr. Sci. 25: 598-609.

Washington, H.S. 1930. The chemical analysis of rocks, 4th ed. pp. 168-172. John Wiley and Sons, New York.

Williams, D.E. 1948. A rapid manometric method for the determination of carbonate in soils. Soil Sci. Soc. Amer. Proc. 13: 127-129.

Yoe, J.H., and W.L. Hill. 1927. An investigation of the reaction of aluminum with the ammonium salt of aurinetri-carboxylic acid under different experimental conditions, and its application to the colorimetric determination of aluminum in water. J. Amer. Chem. Soc. 49:2395-2407.

Young, K.K. 1962. A method of making moisture desorption measurements on undisturbed soil samples. Soil Sci. Soc. Amer. Proc. 26: 301.

Young, K.K. and J.D. Dixon. 1966. Overestimation of water content at field capacity from sieved-sample data. Soil Sci. 101: 104-107.

Yuan, T.L. 1959. Determination of exchangeable hydrogen in soils by a titration method. Soil Sci. 88: 164-167.

Yuen, S.H., and A.G. Pollard. 1952. The determination of nitrogen in agricultural materials by the Nessler reagent. I, preparation of the reagent. J. Sci. Food and Agr. 3: 441-447.

SOIL **Hayter silt loam** SOIL Nos. **S63Ky-74-6** LOCATION **McCreary County, Kentucky**

SOIL SURVEY LABORATORY **Beltsville, Maryland** LAB Nos. **63776-63781**

Depth (in)	Hori-Zon	Total Sand (2-0.05)	Total Silt (0.05-0.002)	Total Clay (<0.002)	Very Course (2-1)	Course (1-0.5)	Med. (0.5-0.25)	Fine (0.25-0.1)	Very Fine (0.1-0.05)	Silt (0.05-0.02)	Int. III (0.02-0.002)	Int. II (0.2-0.02)	(2-0.1)	3B2 >2, Vol pct. of whole soils	>2, Wt pct. of whole soils	2-20 pct. of <75 mm	20-75 pct. of <75 mm
1/2-5	Ap2	15.0	59.0	26.0	3.8	1.9	0.6	1.6	7.1	15.7	43.3	23.9	7.9	23	36	21	--
5-10	B1	15.0	57.4	27.6	3.0	1.9	0.6	1.1	8.4	14.6	42.8	23.7	6.6	36	52	22	4
10-19	B21	15.6	57.0	27.4	3.4	1.9	0.6	1.5	8.2	14.7	42.3	24.0	7.4	34	47	17	2
19-34	B22t	14.3	57.0	28.7	3.4	2.1	0.8	1.2	6.8	13.4	43.6	20.9	7.5	26	37	12	--
34-48	B23t	15.9	53.7	30.4	4.8	2.9	1.0	1.6	5.6	9.2	44.5	15.7	10.3	47	59	16	18
48-60	B3t	24.1	50.8	25.1	8.1	4.9	1.7	2.8	6.6	9.5	41.3	17.8	17.5	51	62	27	7

Depth (in.)	Organic Carbon Pct.	Nitrogen Pct.	C/N	Carbonate as CaCO3 Pct.	Ext. iron as Fe Pct.	1/3 bar g/cc	Oven drt g/cc	LE Pct.		1/3 bar Pct.	15 bar Pct.	WRD in/in		8C1c (1:1) KCl IN.	8C1a (1:1) H2O
1/2-5	1.86	0.180	10		2.9	1.36	1.39	0.7		26.6	12.0	0.15		4.4	4.9
5-10	.87	.100	9		2.9	1.35	1.39	0.6		26.8	11.5	0.13		3.9	4.6
10-19	.35	.058	6		2.3	1.54	1.61	0.9		21.6	11.1	0.11		3.9	5.4
19-34	.21				2.4	1.57	1.63	0.9		21.5	12.0	0.11		3.9	5.4
34-48	.22				2.5	1.60	1.63	0.3		21.7	12.4	0.08		3.8	5.4
48-60	.10				2.9	1.68	1.75	0.6		18.4	10.1	0.07		3.8	5.3

Depth (in.)	Ca	Mg	Na	K	Ext. acidity	CEC 5A3a cations	6G1d A1						Sum cations Pct.	Pct.
1/2-5	4.3	1.5	0.1	0.5	12.8	19.2	0.4						33	
5-10	1.6	1.1	0.1	02	10.8	13.8	1.5						22	
10-19	1.8	1.6	0.1	02	7.8	11.5	0.7						32	
19-34	1.4	2.4	0.1	.2	7.8	11.9	1.0						34	
34-48	1.5	3.2	Tr.	.2	7.6	12.5	1.2						38	
48-60	1.1	3.0	Tr.	.2	6.9	13.0	1.2						33	

Figure 1. Laboratory data sheet for Hayter silt loam, McCreary County, KY.

Soil Type:	Hayter silt loam
Soil No.:	S63Ky-74-6
Location:	McCreary County, Kentucky, North off Hwy. 759 about 2 miles east of U.S. Hwy. 27.

| Vegetation and land use: | Hickory, persimmon, yellow poplar. |

| Slope and land form: | 50 percent. |

| Drainage: | Well drained. |

| Parent Material: | Colluvium from sandstone and shale. |

| Sampled by and date: | D.P. Franzmeier, E.J. Pedersen, C.R. Gass, L. Manhart, G. Chapman; October 15, 1963. |

| Described by: | J.H. Winsor, C.K. Losche. |

Horizon and
Beltsville
Lab. No.

| 01 | 1-1/2 to 0 inches. Hardwood leaf litter. |

| Ap1 | 0 to 1/2 inch. Very dark grayish brown (10YR 3/2) silt loam; moderate fine granular structure; very friable; 12 percent sandstone fragments (> 3 in. diameter); many roots; pH 7.0. |

| Ap2 63776 | 1/2 to 5 inches. Brown (10YR 4/3) silt loam; weak medium granular structure; very friable; 12 percent sandstone fragments; many roots; pH 5.0. |

| B1 63777 | 5 to 10 inches. Brown (7.5YR 4/4) silt loam; weak to moderate fine subangular blocky structure; friable; 25 percent sandstone fragments; many roots; pH 5.0. |

| B21 63778 | 10 to 19 inches. Brown (7.5YR 4/4) silt clay loam / silt loam; moderate medium blocky structure; friable; 25 percent sandstone fragments; common roots; pH 5.0. |

| B22t 63779 | 19 to 34 inches. Brown to dark brown (7.5YR 4/4 - 3/2) silty clay loam; moderate medium blocky structure; friable; common clay films; 20 percent sandstone fragments; few roots; pH 5.0. |

| B23t 63780 | 34 to 48 inches. Brown (7.5YR 4/4) silty clay loam; moderate medium subangular blocky structure; friable to firm; 30 percent sandstone fragments; common clay films; few roots; pH 5.0. |

| B3t 63781 | 48 to 60 inches. Brown (7.5YR 5/4) silty clay loam; weak to moderate medium subangular blocky structure; friable to firm; common clay films; 35 percent sandstone fragments; few roots; pH 5.0. |

Notes: Colors are given for moist soil. The B21 and B23t layers were sampled for the Bureau of Public Roads. Reaction was determined by Soiltex.

Figure 2. Profile description of Hayter silt loam, McCreary County, KY.

SUBJECT INDEX: CURRENT METHODS ONLY

SUBJECT INDEX: CURRENT METHODS ONLY

www.ingramcontent.com/pod-product-compliance
Lightning Source LLC
Chambersburg PA
CBHW061327190326
41458CB00011B/3925